W9-AJQ-261

STATE UNIVERSITY OF NEW YORK
COLLEGE OF ARTS AND SCIENCE AT GENESEO
THE MILNE LIBRARY
WITHDRAWN
MILNE LIBRARY
S.U.N.Y. GENESEO

WINDPOWER

A Handbook on Wind Energy Conversion Systems

Other Books by V. Daniel Hunt

- ENERGY DICTIONARY, 1979, Van Nostrand Reinhold, New York, N.Y.
- ENERGY ISSUES IN HEALTH CARE, 1979, The Energy Institute, Washington, D.C.
- HANDBOOK OF ENERGY TECHNOLOGY - TRENDS AND PERSPECTIVES, 1981, Van Nostrand Reinhold, New York, N.Y.
- HANDBOOK OF CONSERVATION AND SOLAR ENERGY - TRENDS AND PERSPECTIVES, 1981, Van Nostrand Reinhold, New York, N.Y.

WINDPOWER

A Handbook on Wind Energy Conversion Systems

V. Daniel Hunt

DIRECTOR

THE ENERGY INSTITUTE

6143556

VNR VAN NOSTRAND REINHOLD COMPANY

NEW YORK CINCINNATI ATLANTA DALLAS SAN FRANCISCO

LONDON TORONTO MELBOURNE

TK
1541
H86

Van Nostrand Reinhold Company Regional Offices:
New York Cincinnati Atlanta Dallas San Francisco

Van Nostrand Reinhold Company International Offices:
London Toronto Melbourne

Copyright © 1981 by Litton Educational Publishing, Inc.

Library of Congress Catalog Card Number: 80-12581
ISBN: 0-442-27389-4

All rights reserved. No part of this work covered by the copyright hereon may be reproduced or used in any form or by any means - graphic, electronic, or mechanical, including photocopying, recording, taping, or information storage and retrieval systems - without permission of the publisher.

Manufactured in the United States of America

Published by Van Nostrand Reinhold Company
135 West 50th Street, New York, N. Y. 10020

Published simultaneously in Canada by Van Nostrand Reinhold, Ltd.

15 14 13 12 11 10 9 8 7 6 5 4 3 2 1

Library of Congress Cataloging in Publication Data

Hunt, V Daniel.
Windpower: A handbook on wind energy conversion systems.

Includes bibliographical references and index.
1. Windpower. I. Title.
TK1541.H86 621.4'5 80-12581
ISBN 0-442-27389-4

DEDICATION

This book is dedicated to my wife Janet Claire Hunt without whose support, assistance and love this book could not have been completed.

Notice

This book was prepared as an account of work sponsored by Van Nostrand Reinhold Company. Neither Van Nostrand Reinhold nor The Energy Institute, nor any of their employees, nor any of their contractors, subcontractors, or their employees, make any warranty, expressed or implied, or assume any legal liability or responsibility for the accuracy, completeness or usefulness of any information, apparatus, product or process disclosed, or represent that its use would not infringe on privately owned rights.

The views, opinions and conclusions in this book are those of the author and do not necessarily represent those of the United States Government or the United States Department of Energy.

Public domain information and those documents abstracted, edited or otherwise utilized are noted in the acknowledgments or on specific illustrations.

Preface

In September of 1977, I attended the Third Wind Energy Workshop sponsored by the Department of Energy (DOE). One of the clarion calls was for the collection, simplification and dissemination of technical and programmatic information regarding wind power, and the discussion of current developments by the private sector and the Federal Government. My interest in wind power was aroused; however, I found little data and less organization of the material than I had expected. With the hope of correcting this situation, *Windpower–A Handbook on Wind Energy Conversion Systems* was begun.

This handbook represents several years of data collection, research, conversations with wind energy advocates and conferences, plus assimilation and utilization of the best available public information, to present in one comprehensive professional reference book an overview of wind energy.

In a rapidly growing field such as wind power, there is seldom agreement on the content of any new effort. It is hoped that this book, with its wide scope and its emphasis on the application of technology, will prove to be a valuable tool.

V. Daniel Hunt
Fairfax Station, VA.

Acknowledgments

The information in *Windpower–A Handbook on Wind Energy Conversion Systems* has been obtained from a wide variety of authorities who are specialists in their respective fields of wind energy.

I wish to thank Mr. Louis Divone and his staff at the Department of Energy for providing data, reports, photographs and illustrations, which are used extensively in this book. I am also grateful to the many contributors to the book, whose cooperation was indispensable. In particular, this book is based on *Wind Power for Farms, Homes and Small Industry* and *A Sitting Handbook for Small Wind Energy Conversion Systems*, financed and prepared under the direction of the Department of Energy.

The preparation of a book of this magnitude is dependent upon an excellent staff, and I have been fortunate in this regard. Judith A. Anderson has effectively served as coordinator and technical editor of this effort. B. A. "Dusty" Rhoades and Allen Higgs and their staff at Guild, Inc. have done a very fine job in preparing the graphics. Special thanks are extended to Janet C. Hunt, Joan I. Drinnen, Michelle M. Donahue, Anne Potter, Betty A. Mundt, and Grace C. Fisher for the composition and typing of the manuscript.

Credits for the photographs and illustrations are indicated where appropriate throughout. Specific acknowledgments by chapter are noted below:

A Siting Handbook for Small Wind Energy Conversion Systems, prepared by Harry L. Wegley, Montie M. Orgill and Ron L. Drake, of DOE's Battelle Pacific Northwest Laboratory, was used as the basis for the siting portion of the wind characteristics material in Chapter 2. In July of 1979, Mr. Wegley provided detailed revisions to the subject text, as well as illustrations.

Chapter 4 utilized material from *Wind Power for Farms, Homes, and Small Industry* prepared by Jack Park and Dick Schwind for the Department of Energy.

Robert E. Wilson of Oregon State University in Corvallis, Oregon, and Peter B. S. Lissaman of Aerovironment, Inc. of Pasadena, California, produced a report entitled "Applied Aerodynamics of Wind Power Machines." This effort was originally supported by the National Science Foundation, Research Applied to National Needs (RANN) under Grant No. GI-41840. Over the past several years, this report has served as the focal point for clearly describing the applied aerodynamics of wind power machines. With permission, we have included this report as a baseline description of applied aerodynamics for wind energy conversion systems (Chapter 5).

I am indebted to each of the organizations which provided data, verification of information and photographs of their wind energy conversion systems for

Chapter 8. I also acknowledge the excellent effort by the *Wind Power Digest* in creating and presenting, in their *Wind Access Catalog*, timely technical information on a variety of wind energy conversion systems.

The Federal Wind Energy Program described in Chapter 10 is based on material received from the Department of Energy. I acknowledge use of the Wind Energy Systems Program Summary Report, DOE/ET-0093, and the Wind Energy Mutli-Year Program Plan.

Material from DOE's Commercialization Strategy Reports for small-scale and large-scale wind systems was used for Chapter 11. The Task Force Chairman responsible for these commercialization reports was Louis V. Divone of DOE.

For Chapter 12, material from the *Environmental Development Plan for Wind Energy Conversion*, published by DOE in March 1977, was used, as were *Legal-Institutional Implications of WECS*, published by the NSF in their report NSF/RA-770204, Chapter 7 of DOE's *Wind Power for Farms, Homes and Small Industry*, published in DOE report RFP-2841/1270/1270/78/4 dated September 1978, and the NSF report entitled *Public Reactions to Wind Energy Devices*.

International energy organizations graciously contributed to Chapter 13, on international developments. Credit is extended to Federal Minister Volker Hauff of the Federal Republic of Germany for providing information on Germany's wind program. Also appreciated is the material from R. J. Templin of Canada's National Research Council. DEFU provided current descriptions of the wind program in Denmark, and I thank Ir J. Pelser, the Technical Managing Director of ECN, the Netherlands Energy Research Foundation, for his personal attention. The International Energy Agency and DOE also provided input for this chapter.

For Chapter 14, I selected individuals and organizations which have been involved with the research, design, development and commercialization of wind energy conversion systems to provide their views on the future of wind power. We thank all of these contributors whose views are expressed in this Chapter.

The following individuals have been quoted in this chapter: R. Nolan Clark, Frank R. Eldridge, Kenneth M. Foreman, Herman Kahn, Marcellus Jacobs, Amory B. Lovins, Richard A. Oman, Joseph Savino, Volta Torrey and Dr. Vas.

Contents

WINDPOWER

A Handbook on Wind Energy Conversion Systems

1
Introduction

THE SCOPE OF THIS HANDBOOK

The major topics for this handbook are history; wind characteristics; fundamental operation of wind energy conversion systems (WECS); system design characteristics; towers; electric power; conversion and storage systems; wind energy conversion systems applications; the Federal Wind Energy Program; commercialization; environmental, institutional and legal barriers; international development; and the future of wind power. A glossary and a list of abbreviations and acronyms, as well as other reference data, are also included for the convenience of the reader.

This introduction provides a brief overview of the wind energy field and describes the contents of the handbook. In Chapter 2, an historical view of the development of wind power is given, from the first known systems installed in Persia as early as 644 A.D., to the latest WECS being manufactured in the U.S. and around the world. Since a fundamental problem with wind power is the variability of the winds, Chapter 3 provides basic meteorology and climatology and discusses the impacts of the wind on wind energy conversion systems.

Chapter 4 describes, in basic terms, the fundamental operation of water-pumping and electricity-producing wind energy conversion systems. The applied aerodynamic aspects of the major WECS configurations, which include the horizontal-axis rotor, the Darrieus vertical-axis rotor, the Savonius rotor and the new, unique configurations, are discussed in Chapter 5. Chapter 6 gives a brief summary of the characteristics of towers and the installation of your WECS.

Chapter 7 covers the conversion and storage of wind energy. The technique to determine electric power requirements is described in a step-by-step procedure. Conversion systems such as the inverter are discussed from a theoretical as well as a practical basis. Information on the care and use of batteries for wind energy storage is covered in depth.

Chapter 8 provides the latest information available on the variety of water-pumping and electricity-producing wind energy conversion systems produced today. Each manufacturer and model machine is presented in a common data format and supplemented by a photograph of the system, along with a plot of the power versus the wind speed.

The various applications of wind energy conversion systems are examined in Chapter 9. This chapter presents the information necessary to select an appropriate system for specific application requirements. In Chapter 10, the Federal Wind Energy Program is described.

The commercialization of small-scale and large-scale wind energy conversion systems depends on cost-effective systems which are economic in relation to existing power. Chapter 11 describes the commercialization strategy and economics of wind energy conversion systems.

The environmental, institutional and legal barriers to installation of wind energy conversion systems are described in Chapter 12. The section on legal barriers provides detailed information on zoning and other restrictions which could impede the commercialization or installation of a wind system.

A summary of international efforts to promote wind power is given in Chapter 13. This chapter includes current U.S. agreements with international partners and specific descriptions of countries which are aggressively developing wind energy.

The last chapter presents the views of a wide variety of advocates of wind power, including the American Wind Energy Association, Frank Eldridge, Marcellus L. Jacobs, C. G. Justus, Herman Kahn, Amory B. Lovins, Donald J. Mayer, Richard Oman, Pasquale M. Sforza and DOE, expressing their views on the future of wind power.

BACKGROUND

Historical Perspective

During the Middle Ages in Europe, manor rights usually included the right to prohibit construction of windmills. This compelled the tenants to have their grain ground by the lord of the manor. Other legal requirements banned planting of trees near windmills to ensure "free access to the wind." Similar laws are still enforced in Holland. By the fourteenth century, the Dutch had improved windmill designs and made extensive use of windmills to drain marshes of the Rhine delta. Around 1600, the first papermill powered by wind was built in Holland to meet the great demand for paper that had been created by the invention of the printing press. Over 20 windmills of about 40 kilowatts (kW) each were used to drain Beemster Polder. At the end of the sixteenth century, sawmills in Holland were powered by the wind and used to process imported timber. In Denmark, toward the end of the nineteenth century, there were 3000 industrial windmills and 30,000 other types in use for homes and farms. These had a total power output of about 200 megawatts (MW).*

More than six million small windmills (less than 1 kW) have been used in the U.S. since the 1850's to pump water and to generate electricity. Roughly 150,000 are still in operation.** Such machines produced about one billion kilowatt-hours (kWh) of energy annually as early as 1860.

Sales of the 1879 windmill industry were about $1 million, increasing to about $10 million by 1919. In 1889, there were 77 windmill factories in the U.S., but by 1919 this number had decreased to only 31.† By 1900, windmills had become a significant factor in exports, and State Department consuls reported substantial demands for American windmills nearly everywhere except in Europe.

Experiments with large wind power machines were also conducted prior to 1950. The largest wind machine every built to generate electricity was the Smith-Putnam 1.25-MW unit installed in Vermont. This machine had a rotor diameter of 53 meters (m). It delivered utility power intermittently from 1941 to 1945, when a damaged blade broke and could not be repaired because of wartime material shortages.

Large wind machines were used in other countries around 1955. Nearly 30,000 wind power plants were in operation in the U.S.S.R. The Gedser wind turbine was operated in Denmark until the 1960's.†† It could produce 200 kW of electrical power, had a rotor 27 m in diameter, and produced 400,000 kWh of electricity annually. However, the U.S. interest in wind power generally declined in the 25 years following 1950, and only recently has serious attention again been given to large-scale collection of energy from the wind.

Basic Technical Concepts

The wind derives its energy from the solar radiation that reaches the earth's surface. Uneven heating of the earth from the equator to the poles and over the oceans and the continents drives the motions of the atmosphere. The wind resource is difficult to determine

*Solar Energy Research Institute's *Annual Review of Solar Energy*, November 1978.

**Frank R. Eldridge, *Wind Machines*. The MITRE Corporation prepared for the National Science Foundation (Grant AER-75-12937), October 1975. Van Nostrand Reinhold, 1979.

†Volta Torrey, *Wind Catchers* (Brattleboro, Vermont: The Stephen Greene Press, 1976).

††Eldridge, *Wind Machines*.

precisely, and some published estimates of available wind energy differ by a factor of 10,000. A reasonable estimate between these two extremes would place the maximum U.S. resource at roughly 2 trillion watts—approximately equal to our 1972 mean rate of energy usage from all sources.* However, this would require using wind turbines thinly scattered over 3 percent of the U.S. land area with the best winds—a total area equal in size to the state of Colorado. Still, only a very small fraction of this total land area would be dedicated to the machines; also, the machines are compatible with other land uses, e.g., agriculture. The ultimate amount of wind energy that will be used is difficult to predict; but the resource is large, and the technology is available, if not yet optimum in terms of cost, durability, esthetics, safety and convenience.

The primary method proposed for using wind energy is to convert the kinetic energy of the wind into mechanical energy and then into electrical energy (minor efforts on other concepts are also in progress). Wind turbines are used to transform the airflow into rotary power. The major designs use:

- Propellers with two, three or many blades.
- Vertical-axis Darrieus ("egg-beater") turbines.
- Various types of concentrators to shape the airflow and increase turbine efficiency.

The energy output from a turbine is available in several forms. Mechanical energy can be used directly for several purposes, including heating and pumping of fluids. Electrical energy can be produced as direct (dc) or alternating (ac) current, and dc energy can be stored in batteries or used directly for heating and lighting, or to operate dc motors and appliances (if these were more commonly available to the consumer or industry); ac energy can be supplied to the utility grid and/or can be used directly at the point of collection. Because the wind is not always available, the use of wind energy by utility grids presents special problems.

Modern designs to use wind power cover a wide range of sizes and technologies. The smallest machines being developed and sold today are smaller and considerably lighter than a subcompact car. The largest machines to be built in the Federal program are larger than a jumbo jet. Components of fiberglass, steel, aluminum, concrete, wood, plastics and other materials are in use. Deployment plans range from units for a single home, to multiunit farms approaching the capacity of a nuclear power facility with a billion dollar cost. The wind environment from which the power is to be taken varies just as much. Continental scale weather patterns and climatic trends will influence the energy collected. Wind fluctuations that occur as frequently as several times a second, and wind variations over distances of only a few tens of meters, will also affect machine performance and endurance.

Wind Energy Conversion Systems Development

Wind energy conversion system (WECS) manufacturing firms have existed in the U.S. for a relatively long time. The second half of the 1800's and the early 1900's saw an estimated six million small wind machines—under 1 horsepower (hp) or 0.75 kW—built and used. Some 150,000 are still in use, mostly for water-pumping. The industry virtually disappeared until the 1970's, when renewed public interest in wind power encouraged research, development and production, resulting in a new period of growth.

The new WECS are used primarily for pumping water and generating electricity. Other uses, such as heat generation, are being studied on a very small scale. According to one report, manufacture and sales of wind machines for water-pumping during the years 1975 and 1976 were stable at 2500 units per year. Another source estimates that 25,000 (up to 2 kW or 2.6 hp) water-pumping units were sold in 1976, a larger number of which were exported. The capital cost (1977) of small water-pumping units in 1977 dollars was

*M. R. Gustavson, *Wind Energy Resource Parameters,* M77-29, The MITRE Corporation, METREK Division, February 1977.

reported to be between $1300/kW and $1500/kW.

It has been reported that the manufacture and sales of electric wind generators increased over 50 percent in 1976, from a base of 750 in 1975. These figures include imports, which have been comprising a high percentage of sales, especially in the larger (2–10 kW) electric generators. Small electric generating systems, up to 2 kW, are being produced in the U.S., and research and development of the larger machines (2–10 kW) are in progress. Another report estimates that production of machines around 5 kW in 1976 was between 100 and 500 units. Two companies have their wind turbine generator (WTG) machines on the market for about $1533/kW and $2000/kW, respectively.

One source reports 23 manufacturers and distributors for 1976, of which 11 manufacture and distribute, 7 distribute, 1 produces prototypes and 4 are system designers. Another source reports 21 manufacturers of wind turbine generators alone. Still another publication lists 22 companies as of the fall of 1977. Four of these companies are foreign, and five of the U.S. companies manufacture water-pumpers.

A very rough estimate of the WECS production breakdown is as follows: 40 percent foreign, 40 percent U.S. (includes rebuilt machines), and 20 percent miscellaneous (prototypes, home production, etc.).

It is clear that reports on manufacturing activity and the number of manufacturers are conflicting and, in many cases, incomplete. The amount of production varies from report to report, and information is not always readily available from the proper sources. However, rapid growth of the industry does seem to be one trend that is generally accepted.

Goals and Current Federal Program

The goals* of the Federal program are to:

- Assess the national wind energy potential.
- Determine expected regional needs, wind resources and wind energy costs.
- Study social and environmental issues.
- Improve turbine siting methods and develop equipment design requirements.
- Improve equipment performance and lower capital costs.
- Explore innovative wind energy conversion methods.
- Develop small machines (less than 100 kW) for agricultural or other uses.
- Develop intermediate machines (100–1000 kW) for community, industrial and utility uses.
- Develop large machines (1 MW or more) for utility-grid applications.

The program is divided into five major program elements:

- Program development and technology.
- Farm and rural use (small systems).
- 100-kW-scale systems.
- MW-scale systems.
- Large-scale multi-unit systems.

The program development and technology element is further subdivided into six areas:

- Mission analysis (definition of the national resource, possible uses of wind energy and R&D requirements).
- Applications of wind energy (economic and technical information for producing electric utility power).
- Legal, social and environmental issues (local and Federal laws, public acceptance, environmental problems).
- Wind characteristics (local and regional wind resources, equipment siting, wind measurements and data).
- Technology development (fabrication, components and mechanical and electrical subsystems).
- Advanced and innovative concepts (non-propeller designs).

Organization of the program involves the Department of Energy (DOE), other Federal agencies and several national laboratories. The laboratories are used to provide basic and applied research and to perform program management functions. Most of the R&D tasks

*Wind Energy Systems Program Summary, DOE/ET-0093, December 1978.

are performed by various academic and industrial contractors. Large-scale propeller-type systems are assigned to the NASA-Lewis Research Center. The Darrieus vertical-axis system is assigned to the Rockwell International/Rocky Flats Plant. Wind characteristics studies are assigned to the Battelle Pacific Northwest Laboratory. Mission analysis and applications studies are assigned to the Charles Stark Draper Laboratory. Specific tasks in other areas and overall program management are handled at DOE.

Recent Principal Developments

The Fourth Biennial Conference and Workshop on Wind Energy Conversion Systems was held in October 1978 to review program accomplishments during the previous several years. The recent growth in U.S. wind energy research was reflected in the conference's results. Papers covering approximately 90 studies were presented.* The subject areas included large and small wind turbines, economics, environmental and institutional concerns, meteorology and siting, large arrays of turbines and innovative design concepts.

The final reports of two mission analysis studies** were also completed during fiscal year (FY) 1977, giving estimates of the national wind resource, the future economic impact of wind energy, its possible applications and markets and the initial cost projections for equipment.

Wind characteristics research in the last couple of years† focused on improved methods for siting equipment, evaluation of data requirements for machine design and localized wind resource assessment. Handbooks of information for use in machine design and performance evaluation were begun, and site selection handbooks were prepared for siting of both small and large turbines. Additional activities included the following:

- An eigenvector technique for analysis of vector wind fields was developed. This may result in significant reductions in the number of computer simulations required in siting methodologies.
- A meteorological field experiment was conducted to obtain wind information on a physical scale directly associated with large propeller-type wind turbines.
- A synthesis of three previous National Wind Energy Assessments produced maps estimating the annual and seasonal distributions of wind energy in the contiguous U.S.

Development of 100-kW-scale and MW-scale propeller-type wind turbines progressed through work on improved rotor designs, testing of machines in utility grids, design contracts for large turbine components and data collection at candidate sites for future machine field tests across the U.S.††

Contracts for the 200-kW test turbine were awarded to supply rotor blades and to install machines at Clayton, New Mexico, and at Culebra, Puerto Rico. The Clayton installation and checkout were completed, and the first operation was accomplished in November 1977.

General Electric completed machine design of a 2-MW turbine, and procurement of components was initiated. Operation commenced in November 1978, on a mountaintop near Boone, North Carolina. The Boeing Corporation was selected as the contractor to design and install one or more "second-generation" 2.5-MW turbines.

Installation of meteorological towers at 17 candidate sites was completed, and data are

Proceedings: The Fourth Biennial Conference and Workshop on Wind Energy Conversion Systems, October 1979, DOE sponsored, published by JBF Scientific Corporation.

**A. Coty, *Wind Energy Mission Analysis,* SAN/1075-76/1, Lockheed California Company for the Energy Research and Development Administration, Contract No. AT (04-3)-1075, September 1976. Also John A. Garate, *Wind Energy Mission Analysis,* COO/2578-1/1, General Electric Space Division for the Energy Research and Development Administration; Contract No. EY-76-C-02-2578, February 1977.

†C. E. Elderkin and J. V. Ramsdell, *Annual Report of Wind Characteristics Program Element for the Period April 1976 Through June 1977,* BNWL-2220 WIND-10, Battelle Pacific Northwest Laboratories for the Energy Research and Development Administration, Contract No. EY-76-C-06-1830, July 1977.

††Personal communication, W. Robbins, NASA-Lewis Program Office, to D. Hardy, Solar Energy Research Institute, January 1978.

being obtained for all sites. A mobile data system for monitoring startup of wind turbines at all the different sites was procured.

Kaman Aerospace was selected to design and fabricate a 45-m composite rotor blade. Urethane and prestressed concrete blade studies were also completed, and a wooden rotor blade study contract was awarded.

The 17-m vertical-axis Darrieus wind turbine was installed in March 1977 at the Sandia Albuquerque Laboratory and tested in a two-bladed configuration. Its performance* (both structurally and from a power output standpoint) was excellent and agreed with prior analyses. A low-cost Darrieus turbine fabrication program was initiated. This program called for redesign of the 17-m system by a commercial organization, to introduce as many cost-saving processes as possible. An RFP was issued and four responses received. A parametric optimization of the Darrieus turbine study was initiated to identify the most cost-effective configurations and sizes. During FY 1978, the 17-m turbine was tested in a three-bladed configuration, and a contract for low-cost commercial fabrication of the Darrieus turbine will be awarded.

The goals of the Small Wind Energy Conversion Systems (SWECS) Program are to stimulate manufacture and sales, increase public use and reduce the cost of energy from wind turbine generators (WEC's).** Energy costs can be reduced by decreasing WEC capital cost, improving performance, increasing reliability or extending equipment lifetime.

A Test Center was established and a total of eight different WEC's were mounted on towers for testing. At the end of FY 1979, five were undergoing tests; two had been returned to the manufacturers for retrofit of design improvements; and one had been destroyed in a windstorm. Specific design improvements were identified and implemented on two WEC's as a direct result of testing.

Technology development subtasks for small wind machines were begun for development of 1-kW high-reliability, 8-kW and 40-kW WEC's. Two contracts were announced for 8-kW WEC development (with Windworks, Inc. of Mukwonago, Wisconsin, and United Technologies Research Center of East Hartford, Connecticut). Two contracts were also awarded for the 1-kW and 40-kW size turbines. The 40-kW machines have a goal of $500/kW (including tower but excluding installation); the 8-kW WEC's have a goal of $750/kW. The high-reliability WEC's have as a primary goal the capability of operating unattended for one year in a severe environment, and a secondary cost goal of $1500/kW.

Efforts in standards development resulted in an informal survey of the wind energy community's opinions. The American Wind Energy Association (AWEA), with assistance from Rockwell International and DOE, will pursue establishment of standards for small WEC performance evaluation.

Problems, Uncertainties, Dissenting Views

The Federal Wind Energy Program has grown rapidly in funding since 1973. These funds have been used to initiate a wide variety of studies. The rapid growth in funding does not mean, however, that additional funds are not necessary or could not be productively used. Also, the great diversity of the present R&D effort does not mean that the different areas of study are now treated as equally important or that the optimum method of coordinating the many separate tasks has been found.

The greatest importance in the present program is given to providing utility-grid power from propeller-type machines of the largest possible physical size. The sizes of these proposed machines approach real limits determined by manufacturing and economic factors.† For example, the largest machine planned will have blades with a radius of 45 m. Rail transport from a manufacturing center to a field installation is limited to 53-m sec-

*Personal communication, R. Braasch, Sandia Albuquerque Program Office, to D. Hardy, Solar Energy Research Institute, January 1978.

**Personal communication, T. Healy, Rocky Flats Program Office, to D. Hardy, Solar Energy Research Institute, January 1978.

†*Proceedings: The Fourth Biennial Conference and Workshop on Wind Energy Conversion Systems.*

tions, and the largest standard heavy construction cranes are 61 m tall. Thus, transportation and field assembly would be much more costly if even larger sizes were attempted.

Extensive plans exist for field tests of horizontal-axis prototypes at different sites (and hence climatic conditions) across the U.S.

Production of utility power is considerably more difficult to evaluate than are non-utility applications. Utility grids are required to provide power on demand even if an intermittent power source is used. The grids must also have very exact control of electrical voltage and frequency. Providing this control restricts options available to the designer and makes optimum conversion of the wind to other energy forms more difficult. Synchronization with the grid frequency can be easily accomplished if wind turbines do not contribute more than 10 percent of the local system's power. Above this level, the stability (stable frequency) of the system may be reduced, and more complex utility engineering problems occur.

The utility companies also face a complex economic situation. The cost of generating electricity varies greatly during the day because different fuels and equipment are used to meet the changing hour-by-hour demands for power. The wind is intermittent and cannot guarantee power at a given hour when the most expensive "peak load" power is generated. Thus, previous economic studies have largely concentrated on using wind energy for long-term fossil fuel savings or compared wind energy to inexpensive base-load power costs. The cost of conventional fuels are now a "pass-through" cost to consumers in many cases. Saving fossil fuels, although important nationally, has a reduced economic importance to utilities under these conditions.

The value of wind energy for electric utility purposes, therefore, is a complex issue, involving daily and seasonal power demands, fuel costs, rate structures and the "mix" of existing (or future) equipment available for power generation. The value of wind energy locally used as a direct substitute for other energy forms can be more easily determined. Applications exist where the energy "storage" is the "product"; e.g., irrigation or process heat. Wind energy also offers a wide range of power levels, from a few kilowatts to several megawatts, and can provide heat, mechanical power or electricity. It is a versatile energy source not inherently limited to any one type or scale of application.

The optimum size for individual wind turbines is unclear. Greater effort on relatively small turbines began in FY 1977 and will be accelerated in FY 1978, as a result of the Rockwell International/Rocky Flats program. Early studies predicted that the lowest costs would occur for very large machines with power ratings of a few megawatts. The potential importance of small wind machines, according to *Science* magazine,* has not been thoroughly assessed. Intermediate size machines with power ratings of approximately 100–1000 kW would be useful in distributed energy systems or in clusters for heating and cooling, pumping fluids and community generation of electricity. It has been suggested that the market for such applications, measured in total power delivered or in fossil fuel saved, might be comparable to large machines in centralized electric power generation.

All recent cost projections assume lower future costs due to mass production. To achieve the same power delivery, more small machines are required, and the potential for lower manufacturing costs may be greater. The small and intermediate turbines are close to the scale of other industrial goods that have shown economies in mass production. The largest proposed turbines are larger than jumbo jets, which are produced in limited quantities. Thus, the optimum turbine sizes and the applications for the economic use of wind energy are yet to be determined.

Standards regarding wind turbine safety and performance are inadequate at present, and building codes do not cover these products. Product liability insurance will be difficult or impossible to purchase until "zones of safety" surrounding a machine can be clearly defined. Support has been given for a Federal initiative to define standards that would be modified to

---

*William D. Metz, "Wind Energy: Large and Small Systems Competing," *Science* 197: 971–973, September 2, 1977.

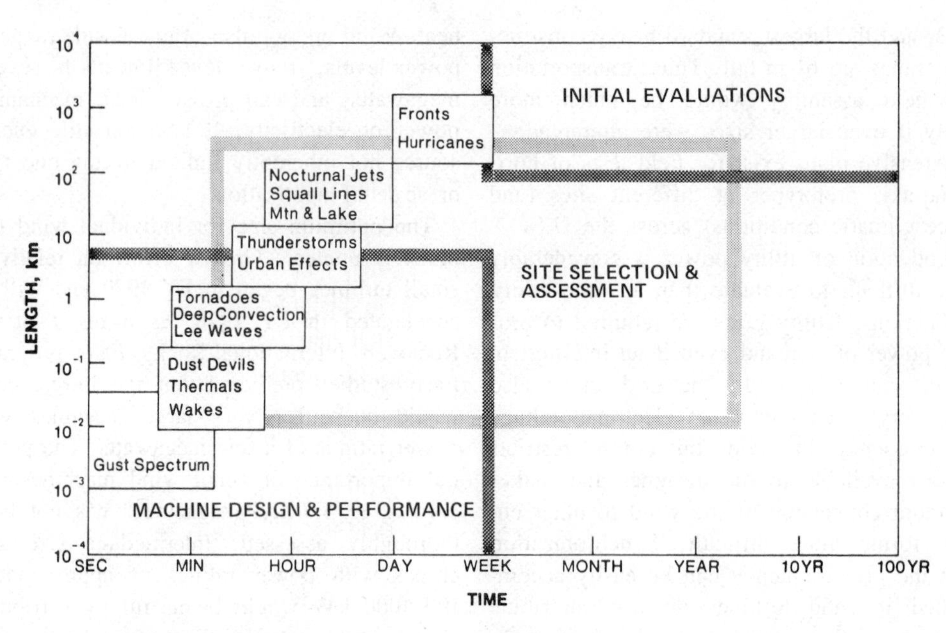

Figure 1-1. The time and space scales important to wind energy development.

Source: Hardy and Walton, "Wind Energy Assessment," presented at Miami International Conference on Alternative Energy Sources, Dec. 1977.

suit each state. Equipment warranties, specifications and safety standards are also important considerations in consumers' tax credits or other incentives.

Regardless of the turbine design or size, the energy delivered by a wind system will depend on where the system is sited. Predicting the output of existing (or future) machines across the U.S. cannot be done with certainty now. The output of a machine depends on the mean wind speed and on variations about the mean. Thus, identical machines placed in two different parts of the nation with equal mean wind speeds may produce different outputs. At any location, the energy output will also vary with time—from one year to the next, by season of the year, by day and by hour within the day. This is true because many weather phenomena occur over distances or time periods important to wind energy collection (Figure 1-1).

Assessment of the wind resource is vital and complex. In any region of the U.S. the available wind energy will change with time. Energy needs also are time dependent. The match between the two will be important. Other solar resources will also be locally available. The wind must therefore be compared with other solar options, and an optimum choice or combination of solar technologies selected. The best choice will not be the same for each region of the nation.

2
Historical Development

Wind and water were among the earliest energy sources to be harnessed for simple, useful tasks. Man's ability to make use of water power, as well as the extent of his use, has steadily increased from the time of the industrial revolution. By contrast, his ability to harness the wind has been less successful. Even the coming of the age of steel and the development of an extensive electronic technology have not been sufficient to effect the large-scale use of wind power.

Many similarities exist between these two energy resources. Both are continuously in the process of being renewed at nearly constant rates by energy supplied from the sun. The energy capture produces no pollution. But although the fuels themselves are free, the energy capture devices involve high capital costs. As a form of solar energy, wind energy has the liabilities of low density and intermittency.

Because water power is an energy resource of much greater density than the wind, and because many favorable situations exist for storing the energy potential in water, there is more extensive use of water power. One is immediately confronted by three problems in any program to make use of wind power on any significant scale: 1) its low energy density necessitates a relatively large capture unit; 2) the energy must be captured from moment to moment and its availability varies; and 3) energy storage involves considerable additional costs. These factors have limited the historical use of wind power almost exclusively to very small-scale systems.

EARLY WIND ENERGY CONVERSION SYSTEMS (WECS)

Historically, wind energy conversion systems can be considered as one of man's truly basic machines. Early documents refer to use of windmills, as depicted in Figure 2-1 in Persia in 644 A.D. These early WECS were really windmills, because they were used to grind grain.

These primitive wind machines stayed the same until the twelfth century, when almost simultaneously in France and England the horizontal-axis or Dutch-type windmill made its appearance (Figure 2-2). Dutch settlers brought the windmill to America in the mid-1700's (see Figure 2-3). These windmills typically ground grain and pumped water. Through the years, the design of these windmills changed only superficially.

The Turn of the Century

In 1890, the first of the modern windmills for producing electricity was designed and built, and put into service in Denmark. By 1908, several hundred wind power stations producing 5–25 kW dotted the Danish landscape.

Wind machines (Figure 2-4) played a significant role in rural America prior to the 1930's, when the Rural Electrification Act (REA) provided cheap and very convenient electricity to the farmers.

More than six million small-scale WECS (less than 1 kW) had been used in the U.S. since the 1850's to pump water and generate

Figure 2-1, Persian vertical-axis windmill circa 640 A.D.

Figure 2-2. DeZwann Dutch windmill.

electricity. These machines produced about one billion kWh of energy annually as early as 1860.

Sales of the 1879 windmill industry were about $1 million, increasing to about $10 million by 1919.

WIND ENERGY CONVERSION SYSTEMS IN THE TWENTIETH CENTURY

The Jacobs Design

Early engineering started on the Jacobs wind-operated electric generating plant in 1925. After several years of testing different types of windmills, the three-blade airplane type of propeller was found to be far superior in power output. By means of a flyball-governor-operated, variable pitch speed control, the maximum speed of the propeller was accurately and easily controlled, to prevent excessive speeds in high winds and storms. The three-blade propeller was found to be necessary (as compared to the two-blade type) to prevent excessive vibration whenever the shift in the wind direction required the plant to change its facing direction on the tower.

A propeller diameter of 15 ft produced ample power for electric generator operation to develop 400–500 kWh/month, based on the available winds in most areas of the states in the western half of the U.S. This required 10- to 20-mph winds for two or three days each week. A specially designed six-pole battery charging shunt generator was developed to operate at a speed range of 125–225 rpm for direct connection to the governor hub of the propeller. It was designed so that its load factor would exactly parallel the power output curve of the wind-driven propeller when operating in the 7- to 20-mph range that it was felt produced the most hours of wind per month. Wind plants that require higher than 20-mph winds to deliver their rated output will find many areas where there are too many days with winds below that speed each month; thus, their effective average monthly output in many areas is below expectations.

Jacobs' experience with plants installed in many parts of Alaska, Canada, Finland, the northwestern U.S. and a number of special installations (such as the plant for the jointly operated U.S. and U.K. weather station at Eureka in the Arctic Circle and with the Byrd expedition at Little America) has shown that aluminum-painted (copper-edged), sprucewood propellers have considerably less trouble with frost and ice formation than when they are varnished or when other types of coatings are used.

Generators located on high steel towers are subject to considerable static discharge from the armature through the ball or roller bearings, and excessive charges from nearby lightning will often arc through a bearing and weld spots on the balls and race, causing it to break up soon. Jacobs found the revolving propellers collected discharges into the direct connected armature and the lightning pick-up effect of the propellers was frequent and of

Figure 2-3. Robertson windmill reconstructed in Williamsburg, Virginia. (*Courtesy of Colonial Williamsburg.*)

Figure 2-4. Rural American wind machine. (*Courtesy of DOE.*)

considerable intensity. To correct this, dual sets of heavy grounding brushes were installed on the armature shaft. With the additional use of a large-capacity oil-filled condenser connected across the generator brushes and frame, Jacobs practically eliminated any damage to the generators from lightning—so much so that with high-grade ample insulation used throughout the generator and the grounding brushes and condensers, they gave an unconditional five-year guarantee with every generator against burn-out from any cause. Many thousands of generators have been built during the past 20 years using this construction, without any replacements ever being required because of lightning damage or burn-out.

The price received at the factory for the 2500-watt, 32-volt plant was $490, less the cost of a suitable tower and batteries, which could often be secured in the country or area to which the plant was shipped. The 21,000-watt hourglass cell Lead/acid storage battery with a ten-year guarantee cost $365. A 50-ft self-supporting steel tower was supplied for $175, making a total cost for the plant of $1025. (This is about $400/kW for the manufacturing cost of the plant.) Shipping and installation costs were additional. The installation cost required only the labor of two men for two days and a small amount of cement to put into the anchor holes when the tower was built. No special equipment or training were necessary. Regular installation and operating instructions were prepared and sent with each plant.

Operating and maintenance costs of this plant were largely limited to the replacement of the storage battery which, on a ten-year basis, was about $36/year; from records kept of more than 1000 plants over a ten-year period, the maintenance cost of repairs was less than $5/year. Some of the owners of the plants bought the Edison-type battery and, after 20 years, were still using the same battery.

The 100-kW Russian WECS

One of the first of the large experimental machines was the 100-kWh wind turbine that was built in 1931 by the Russians (Figure 2-5). It was located at Balaclava near Yalta on the Black Sea. The rotor was 100 ft in diameter, and the tower was 100 ft high. Maximum rated power, 100 kW, was obtained at wind speeds in excess of 24.6 mph. The average wind speed at the site was 15 mph. The rotor drove a 100-kW, 200-V induction generator, which was connected by a 6300-V line to a 20-MW steam power station located in Savastopol, some 20 miles away. Although this wind machine was very primitive—that is, the blade surface was roofing metal and the main gears were made of wood—one year the plant did achieve an output of 279,000 kWh. This gave a power utilization yield (actual power output divided by total possible power output) of 32 percent. The generator and controls were located in the housing on top of the tower. Regulation was accomplished by pitch control of the blade. The wind thrust was absorbed by the inclined strut. The ground portion of this strut rested on a carriage which sat on a circular track. The carriage was automatically driven to keep the rotor facing into the wind. In addition to this machine, many smaller machines have been installed in Russia to supply power to agricultural communities.

Power from the Smith-Putnam WECS

In the twentieth century, the search for power led several countries to attempt to tap this widespread source of "free" energy. For the first time, comparatively large wind machines were also constructed and tested.

Figure 2-5. 100-kW Russian WECS. (*Courtesy of DOE/NASA LRC.*)

One of the largest wind machines (1250 kW) was started in 1934, when an engineer, Palmer C. Putnam, began to consider wind-driven generators for reducing the cost of electricity at his Cape Cod home. In 1939, Putnam presented his ideas and the results of his preliminary work to the S. Morgan Smith Company of York, Pennsylvania. The company agreed to fund a wind energy project, and the Smith-Putnam wind turbine experiment was born. The wind machine was to be connected to the existing system of the Central Vermont Public Service Corporation. Out of some 50 Vermont sites considered, a 2000 ft hill, Grandpa's Knob (located in Rutland), was selected. Engineers from several universities participated in the project. On August 29, 1941, less than two years after the original meeting, the blades were rotated for the first time.

The Smith-Putnam machine (shown in Figure 2-6) was physically the largest wind machine ever built and tested. The tower was 110 ft high, while the rotor was 175 ft in diameter and had an 11-ft, 4-in. chord. Each blade weight 8 tons and consisted of stainless steel ribs covered by a stainless steel skin. The blade pitch was adjustable, so that a constant rotor speed of 28.7 rpm could be maintained. This rotational speed was maintained in wind speeds as high as 70–75 mph. At higher wind speeds, the blades were feathered, and the machine was brought to a stop. The rotor turned an ac synchronous generator that produced 1250 kW of power at wind speeds greater than 30 mph. This power was fed into the power company network.

Shortly after the system had undergone its initial checkout and was brought on line, a main bearing failed. Since World War II was in progress and this project had low priority, it took several years to obtain a new main bearing. The new bearing was installed early in 1945. Following the installation, the machine was in operation only a few months when an overstressed blade failed. Total intermittent running time achieved was 1100 hours. The project was reviewed, and although considered to have been a technical success, was not considered to have been an economic suc-

Figure 2-6. Smith-Putnam 1250 kW WECS in 1941.

cess. The original installation cost data indicated that additional machines in small quantities would have cost approximately \$190/kW. The target price in 1945 was \$125/kW. The project was stopped, and the wind machine was dismantled.

The Percy H. Thomas Concept

The technical results of the Smith-Putnam wind turbine caused Percy H. Thomas, an engineer with the Federal Power Commission, to spend approximately 10 years in a detailed analysis of wind power generation of electricity. Mr. Thomas, primarily using the economic data from the Grandpa's Knob operation, initially concluded that a 5000–10,000-kW wind-driven machine was necessary for economic feasibility. He designed two large machines, one for 6500 kW and the other for 7500 kW.

The 6500-kW machine is shown in Figure 2-7. In 1951, the Federal Power Commission tried to interest Congress in funding a prototype of this machine. Because the Korean War was on, the project was not funded and was subsequently cancelled. Some details of the proposed system were as follows: The tower height was to be 475 ft, and each of the rotors 200 ft in diameter; the rotors were to drive dc generators which would produce 6500 kW at wind speeds greater than 28 mph, and the dc power was to feed to a dc-to-ac synchronous converter, which would supply the electrical network. All generating equipment was to be housed at the top of the tower. Mr. Thomas estimated the capital costs for this machine as \$75/kW.

The Enfield-Andreau WECS

The English also had a fairly extensive wind energy program from 1945 through 1960. One machine, shown in Figure 2-8, was the Enfield-Andreau wind turbine. This machine was built in England and set up at St. Albans in the early 1950's. It was designed to put out 100 kW of ac power in a 30-mph wind. The tower was 100 ft high, and the rotor measured 79 ft from tip to tip. This machine was particularly interesting in that, unlike conventional wind turbines, it used air rather than gears to transmit the propeller power to the generator. The propeller blades were hollow, and when they rotated, they acted as centrifugal air pumps. The air entered ports in the lower part of the tower; passed through an air turbine, which turned the electric generator; went up through the tower; and went out the hollow tips of the blades. Unfortunately, friction losses in the internal air duct of the machine were large enough to minimize any advantages achieved by elimination of mechanical coupling.

The Danish Gedser System

The Danish also had a wind energy effort during the 1950's. The result of some of this work, the Danish Gedser wind turbine, is shown in Figure 2-9. This machine, built in 1957, produced 200 kW in a 33.6-mph wind. It was connected to the Danish public power system and produced approximately 400,000 kWh/year. The tower was 85 ft high and the rotor 79 ft in diameter. The generator was located in the housing on the top of the tower. The installation cost of this system was approximately $205/kW. This wind turbine ran until 1968, when it was stopped because the power produced was not cost-competitive. However, under a joint program sponsored by the DOE and the Danish government, the Gedser WECS is running again to provide comparative test data on the current capabilities of such large-scale WECS.

The French Nogent Le Roi WECS

The French also did some wind energy work during the 1950's. They built at least two large machines, one with an output of 130 kW and a blade diameter of approximately 70 ft and the other, located at Nogent Le Roi, had an output of 300 kW, with a blade diameter of approximately 100 ft.

Hütters Wind Turbine

The Germans, under the direction of Dr. Ulrich Hütter, did some fundamental large-scale system work in the 1950's and 1960's. The first machine produced 100 kW of power in an 18-mph wind. Previous machines required

Figure 2-7. Proposed 6500-kW Percy Thomas twin-wheel WECS. (*Courtesy of DOE/NASA LRC.*)

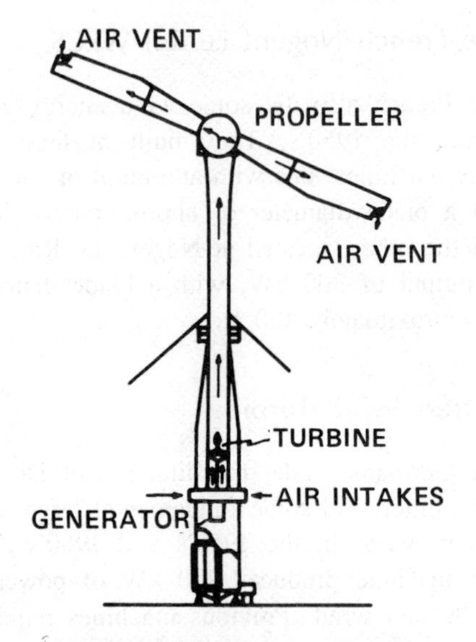

Figure 2-8. Enfield-Andreau WECS.

Figure 2-9. Denmark's three-bladed Gedser wind electric system. (*Courtesy of DOE.*)

much higher wind speeds. This machine used lightweight 115-ft-diameter fiberglass blades and a simple hollow pipe tower supported by guy wires. The blade pitch could be changed at higher wind speeds to keep the propeller rotation constant. Dr. Hütter's machines ran from September 1957 to August 1968. During this period, he obtained more than 4000 hours of full rated power operation. He also made substantial contributions to the design of high-speed wind turbine rotors. The German effort constituted the most advanced work on large scale wind machines.

Wind Power in the 1980's

Refer to Chapter 8, which describes currently available WECS. Chapter 10 provides an in-depth overview of the Federal Wind Energy Program (see Figure 2-10) and summarizes the development of small-scale and large-scale WECS.

Although WECS have been buillt and tested in a number of countries around the world, these systems have been dismantled after running for a while. The problem has been that the installation cost per kilowatt has been too high when compared with the costs of other methods of producing electric power. In addition, because of wind variability, it is usually not sufficient to have a WECS alone; some form of energy storage must also be considered.

Today's increasing cost of fuel, coupled with potential fuel scarcities, has provoked the re-examination of wind energy as a future source of power.

In summary, the wind contains a large amount of available energy. Recognizing this fact, several countries, including the U.S., have begun aggressive R&D programs to build and test small-scale and large-scale WECS since 600 A.D., as shown in Figure 2-11. These machines have shown the technical feasibility of power from the wind, but the following problems have yet to be solved:

- Installation (and hence, operating costs) are high when compared with those of conventional fuel systems.

- A major sustained effort is needed to make WECS competitive with other energy-producing systems.
- Energy storage systems are required for application where an electric power grid or standby generators are insufficient.

For an in-depth narrative of the historical development of wind power, read *Wind-Catchers, American Windmills of Yesterday and Tomorrow,* by Volta Torrey.

THE ENERGY PROBLEM TODAY

Energy has been relatively inexpensive in the U.S. The nation has been blessed with abundant reserves of coal, oil and natural gas. Americans used these resources liberally to raise their living standards; the supply or price of energy was not a major concern. Consequently, the country has produced homes, cars, appliances and factory equipment that use energy wastefully by today's standards.

Figure 2-10. DOE MOD-O LWECS at NASA Plumbrook Site. (*Courtesy of DOE/NASA LRC.*)

In the past, economic growth and energy consumption have been very closely related. Between 1960 and 1977, the U.S. economy grew 81 percent, from $737 billion to $1.3 trillion in constant 1972 dollars. During the same period, U.S. energy consumption grew 70 percent, from 45 quads to 76 quads. Before the energy crisis in late 1973, the relationship was even closer. Between 1960 and 1972, GNP grew at an average annual rate of 3.9 percent. Energy consumption grew at an average annual rate of 4.0 percent during those years.

The complacency about energy ended rather abruptly in 1973, with the oil embargo. The embargo, and the quadrupling of imported oil prices, showed Americans the implications of depending on imported oil.

Rising energy costs have been one cause of inflation. A higher price for this basic resource is quickly felt throughout the economy. These higher costs have made it more difficult to reach full employment by rendering uneconomic a portion of the country's stock of capital equipment. The high cost of imported energy has undermined the value of the dollar abroad, creating more inflationary pressure, and possibly weakening American foreign policy.

While the link between energy consumption and economic growth (as shown in Figure 2-12) will persist, the tie between them now should not be as close. Energy consumption is likely to grow far more slowly than the economy. Many European countries have provided their people a high living standard, with much lower energy consumption than that of the U.S. Energy prices may induce the U.S. economy to grow more energy efficient in regard to energy use.

Even though U.S. energy consumption should grow less than the gross national product and well below historical growth rates,

600 Earliest vertical wind machine developed in Persia for water-pumping.
900 Gardens irrigated by horizontal wind wheels in Persia.
1100 Dutch windmills used to grind grain and pump water.
1200 Europeans use vertical sails on towers and posts to grind grain.
1400 Mediterranean sail windmill developed.
1500 Holland utilized windmills to reclaim lowland.
1600 Colonists build windmills on eastern seaboard.
1700 Wind-driven wheels replaced by steam engines.
1850 American farm windmill designed to pump water.
1870 Chicago becomes center of windmill industry.
1880 Windmills used to irrigate American desert.
1900 Small windmills used to power home radios.
1926 Flettner rotor concept demonstrated.
1929 Savonious vertical-axis rotor designed.
1930 Optimum wind machine designed by the Jacobs brothers.
1930 Windmills replaced by central power plants and REA transmission lines.
1931 The National Research Council evaluates Darrieus rotor design.
1941 The Smith-Putnam wind-driven turbine built at Grandpa's Knob, Vermont.
1950 Venturi-shrouded wing generator evaluated by Electrical Research Association in England.
1950 Atomic energy reduces interest in higher cost techniques such as wind power.
1954 Enfield-Andreau 100-kW wind turbine developed in France.
1961 Advanced aerodynamic wind turbine designed by Hütter.
1970 Energy crisis revives interest in the wind as a clean source of power.
1974 Princeton sailwing airfoils used on wind machine.
1974 The Clyde bicycle wheel wind turbine developed for water-pumping and electric power generation.
1974 Noah wind rotor developed as international project to provide underdeveloped areas power.
1975 NASA 100-kW wind turbine started in Sandusky, Ohio.
1976 U.S. funds $8 million for research and development of wind power.
1977 DOE/NASA Federal Wind Program demonstrates large-scale WECS MOD-O.
1977 17-m Darrieus vertical-axis WECS operational at Sandia Laboratories.
1978 DOE awards contract for 1-kW, 8-kW and 40-kW SWECS.

Figure 2-11. Wind energy chronology

energy demand may still surpass 100 quads by 1990. The mix of energy sources will probably change in meeting this demand. Oil and gas may provide a smaller share of the total supply, and coal and nuclear power may provide an increased share. The contributions of other energy sources could be relatively constant, although the actual amount of energy they provide may rise. By the year 2000, however, alternate energy sources could constitute a large and rapidly rising share of U.S. energy.

Since 1945, petroleum and natural gas have provided more energy in the U.S. than has any other source. Abundant supplies from easily accessible domestic fields induced the nation to rely less on coal and more on convenient fuels. During the late 1940's, the U.S. began to augment its domestic supplies of oil with imports to satisfy rising demand. By 1960, imported oil amounted to 18 percent of U.S. petroleum consumption, and imported gas amounted to 1 percent of demand for that product. By 1977, these figures had risen to 45 percent and 5 percent, respectively.

Despite new oil and gas supplies from Alaska, offshore drilling and synthetics, the country's dependence on imports could still grow, as shown in Figure 2-13. Some projections indicate that imports could account for 51 percent of U.S. oil consumption and 17 percent of gas consumption by 1990. Other studies predict that world oil demand may begin to outrun supply as early as 1985. Recent discoveries of oil and gas, however, could push those dates into the more distant future.

The absolute worldwide exhaustion of oil

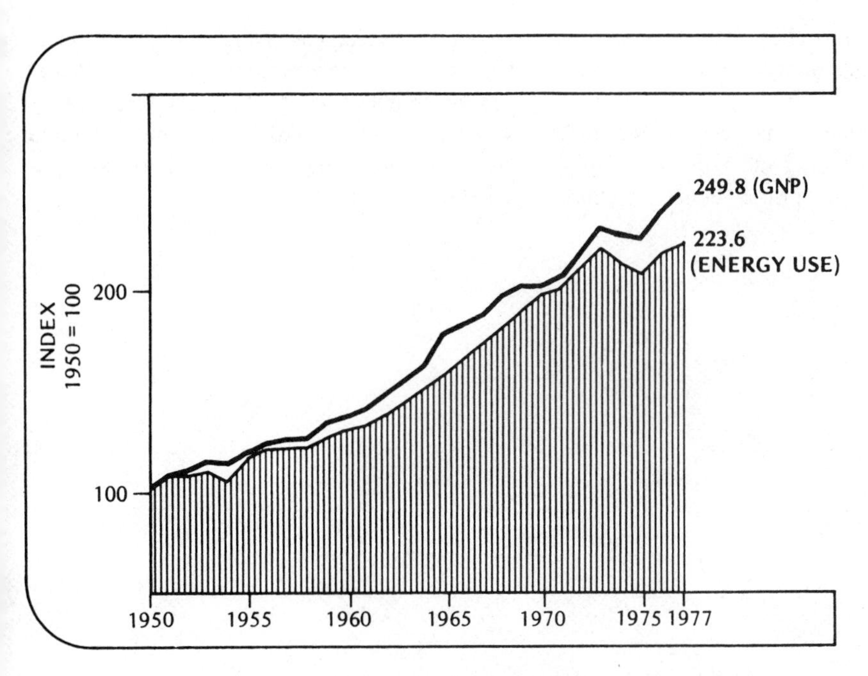

Figure 2-12. Growth of GNP and energy use.

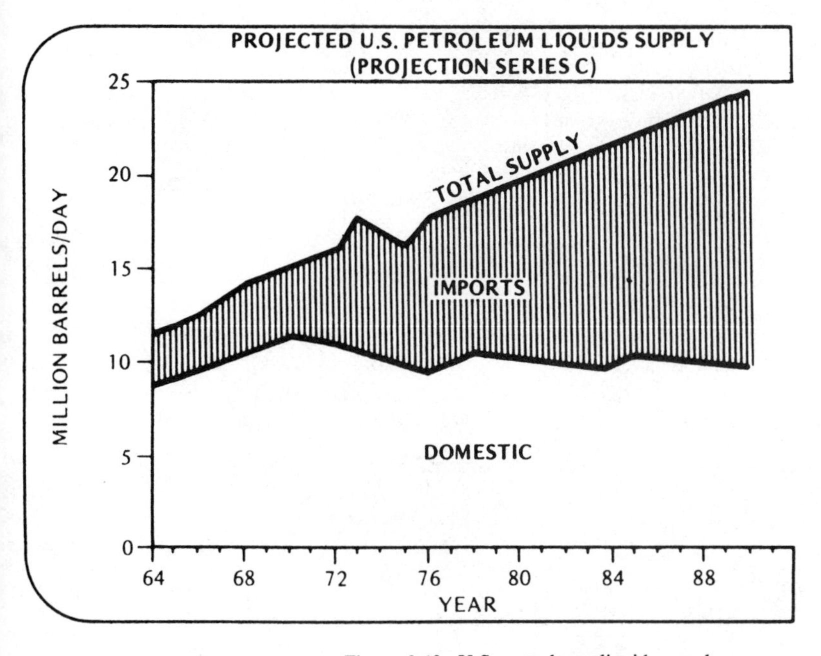

Figure 2-13. U.S. petroleum liquids supply.

and gas resources will probably never occur. As the demand for petroleum rises and the supply fails to keep pace, its price will rise. Energy users will switch away from petroleum and producers will find it economic to use more expensive methods of recovering oil. As the process continues, the use of petroleum will become more and more restricted.

Because of this, the U.S. is searching for other energy sources, such as wind energy. The nation is also seeking ways to conserve energy. This would allow the economy to expand without the danger of having its energy lifeline cut.

The energy situation is one of the most difficult problems facing the U.S. The supply of energy, and its price, concern almost everyone. Society is being asked to adapt very quickly to new relationships and to adopt a more frugal way of life. The transition is not easy. It involves giving up habits acquired over a lifetime, and may lead to lowered expectations for the future.

3

Wind Characteristics and their Impact

THE WIND

Generation of the Wind

Wind may be traced ultimately to the effect of the sun on the earth, including the lower portions of the atmosphere. Most changes in weather involve large-scale, approximately horizontal, movement of air. Air in such motion is called wind. This motion is produced by differences of atmospheric pressure, which are mostly attributable to differences of temperature.

The sun provides the heat required to warm the air above the earth. Therefore, wind is actually solar energy. Solar energy, in its broadest sense, includes almost every form of energy alternative. As this solar radiant energy from the sun arrives at the earth, about 43 percent is reflected back into space by the atmosphere; about 17 percent is absorbed in the lower portions of the atmosphere; and the remaining 40 percent reaches the surface of the earth, where much of it is re-radiated into space.

This radiation from the earth is in long waves relative to the short-wave radiation from the sun, since it emanates from the earth, which is cooler. Long-wave radiation, being readily absorbed by the water vapor in the air, is primarily responsible for the warmth of the atmosphere near the earth's surface. Thus, the atmosphere acts much like the glass on the roof of a greenhouse. This phenomenon is usually called the "greenhouse effect." The atmosphere allows part of the incoming solar radiation to reach the surface of the earth, but it is heated by the terrestrial radiation passing outward. Over the entire earth and for long periods of time, the total outgoing energy must be equivalent to the incoming energy, or the temperature of the earth—including its atmosphere—would steadily increase or decrease.

The more nearly perpendicularly the sun's rays strike the earth, the more heat energy per unit area is received at that place on the earth. The tropics receive the most heat per unit area and the polar regions receive the least heat per unit area. The process which brings about the required transfer of heat is the general wind circulation in the atmosphere.

Two percent of all solar energy reaching the earth is converted to wind energy. Surface winds over the U.S. available for conversion are sufficient to supply about 30 times the total energy consumption of the U.S. To reduce the scale of our thinking, let us first look at a few further generalities about wind.

The U.S. and other parts of the north and south temperate zones experience a general westerly wind (Figure 3-1). Changes occur in the weather with the alternate passage of high and low pressure systems. These cause barometer readings to fluctuate. The various pressure systems tend to migrate from west to east and bring about wind shifts, temperature changes, rain and other weather features.

Along with this general trend are the regional and local weather effects that are often strongly influenced by temperature differences

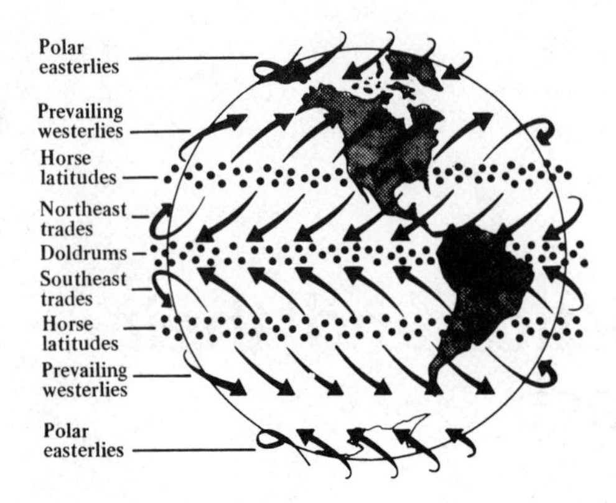

Figure 3-1. Worldwide wind circulation. (*Courtesy of NSF*.)

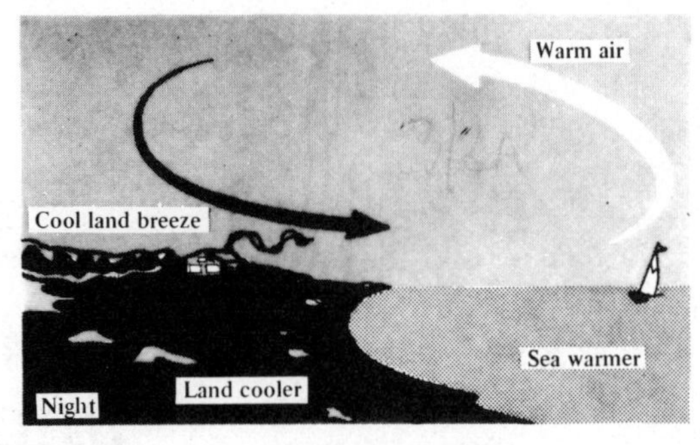

a. Land breeze (night).

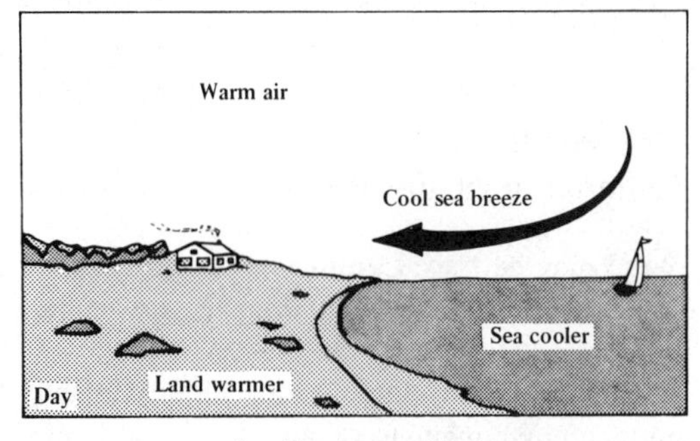

b. Sea breeze (day).

Figure 3-2. Local wind circulation example. (*Courtesy of NSF*.)

between the air, the land and the water. For example, in an area of mountains and valleys, daytime sunshine heats the mountain air, which rises and is then replaced by cooler air from the valley. This creates valley winds moving uphill. At night, the air cools by radiation to the night sky and moves downhill, creating a mountain breeze.

Another example of regional or local wind occurs along coastal areas. Daily temperature differences between the surfaces of the sea and the land cause alternate sea and land breezes (Figures 3-2*a* and *b*). (Sea-land and valley-mountain winds are described in more detail later in this chapter.) We can see that wind available for conversion is a result of both the motion and the atmosphere over a huge area and the local effects of terrain and temperature.

Wind Power Variations. A 16-ft-diameter wind turbine might produce 1 kW of power in a steady 15-mph wind, depending on the design. This 1 kW of power produced continuously for 30 days amounts to 720 kWh. This amount of energy is about that used in the typical American home. The wind, however, is not constant. In fact, it is even more erratic than one would expect. The actual power available from the wind is proportional to the cube of the wind speed. If the wind speed is doubled, there is eight times the power available to the rotor. This magnifies the effect of the fluctuating wind. For instance, a 20-mph wind has 2.37 times more power available than a 15-mph wind:

$$\frac{20 \times 20 \times 20}{15 \times 15 \times 15} = 2.37.$$

Regular daily fluctuations in wind speed can be large or small. Figure 3-3 shows the typical daily fluctuations of hourly readings of wind speeds at three locations. Oak Ridge, Tennessee (five-year average record) is in a southeastern U.S. interior location. Most of that area has a low average annual wind speed. The winds at the Muskegon Coast Guard Station, Michigan (one-year record) are considerably enhanced by a clear sweep from Lake Michigan and by strong lake breezes. The Livermore, California (one-month record) location is in a mountain pass into the great Sacramento-San Joaquin Valley, so a daily mountain-valley wind cycle occurs.

Everyone is aware of good and bad years for rain. Wind power also varies from year to year. Dodge City, Kansas has an average wind speed of 15.5 mph and an average wind power of 336 W/m²—about the highest for any city in the U.S. In the ten-year period of 1955–1964, the yearly deviations from the average power were +5, −9, +5, −22, +26, +3, −13, −15, +18.5 and +33.5 percent.

Monthly and annual wind power is determined from the use of computers to take the cube of regularly sampled wind speeds, usually hourly or every third hour. The thousands of values are then averaged to obtain these monthly or annual averages. Many of the hourly wind readings are available from the National Weather Service in a form ready to be processed by a computer.

There are several approximate methods that can be used to determine average wind power. First, the average speed can be developed from the power equation. However, Figure 3-4 shows the large error that can occur if this is done. The shaded band in this figure represents the actual wind power divided by the power calculated from taking the cube of the average wind speed for 90 percent of the weather stations in the U.S. The average of the cube of each of many fluctuating readings is considerably greater than the cube of the simple average wind speed. The effect is largest at low annual wind speeds. The band in Figure 3-4 contains 90 percent of the values, so you have a 90 percent chance of being within this band. The average values shown in Figure 3-4 are as follows:

Average annual wind speed, mph	8	10	12
Correction factor*	3.2	2.7	2.4

Average annual wind speed, mph	14	16	18
Correction factor*	2.1	2.0	2.0

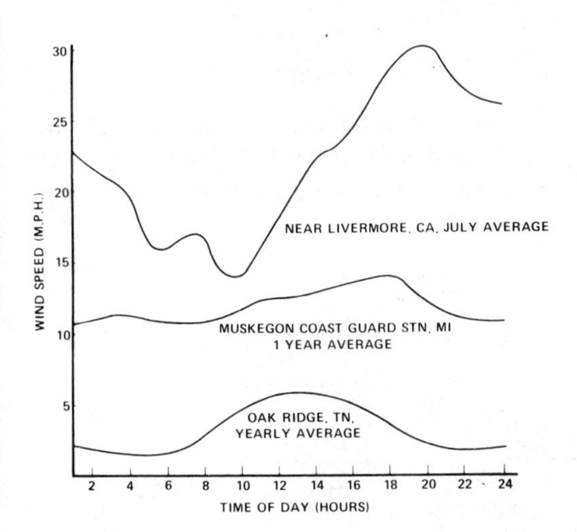

Figure 3-3. Sample daily wind variations. (*Courtesy of DOE.*)

Another way to visualize the variations in the wind is to use a wind duration plot. Some typical wind duration curves are shown in Figure 3-5. Each point on these curves shows the number of hours in the year which the speed equals or exceeds the hours indicated directly below. For instance, a 10-mph wind speed is exceeded 3500 hours a year at Plumbrook, Ohio, and is exceeded 6950 hours a year at Amarillo.

The average winds vary across the U.S. as summarized in Table 3.1. (Refer to the reference section of this handbook for detailed average wind speeds for the U.S. and Southern Canada.

Influences on Airflow. Pressure systems (frequently 500–1000 miles or more in diameter), which are associated with large-scale wind patterns, migrate from west to east across North America. As the air in the large-scale wind pattern moves through local areas, its speed and direction may be changed by the local topography and by local heating or cooling. At a particular WECS site, trees, buildings or other small-scale influences may further disturb the wind flow. The combined effects of these three scales of influence produce highly variable winds.

The surface over which the wind flows affects wind speed near that surface. A rough surface (such as trees and buildings) will produce more friction than a smooth surface (such as a lake). The greater the friction, the

$$\text{*Correction factor} = \frac{\text{Average annual wind power}}{\text{Fictitious power calculated from average annual wind speed}}$$

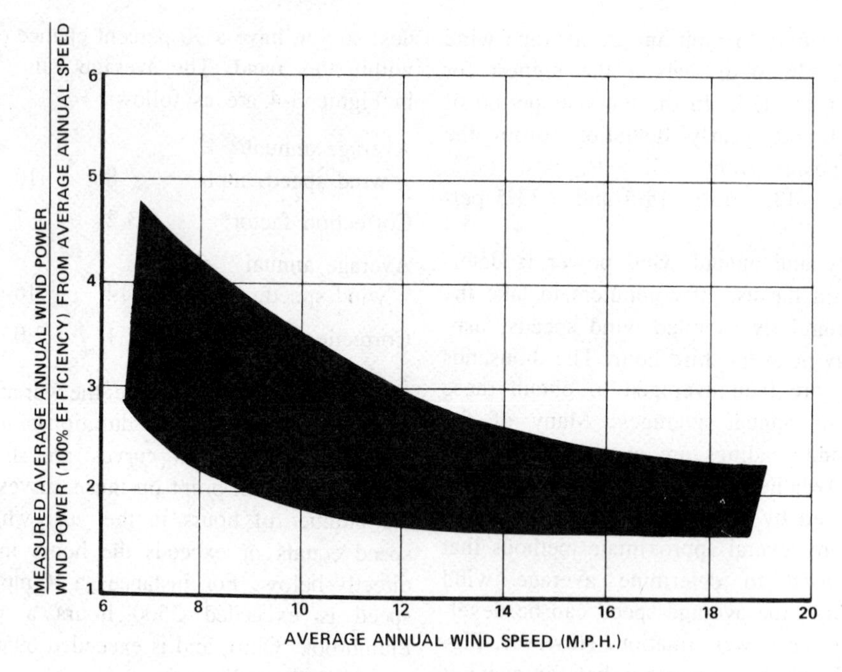

Figure 3-4. Average annual wind speed, mph. (*Courtesy of DOE.*)

more the wind speed is reduced near the surface.

Figure 3-6 illustrates how surface roughness affects wind speed by means of a vertical wind speed profile—simply a picture of the change in wind speed with height. Within 10 ft of the surface, wind speed is greatly reduced by friction. Wind speed increases, however, between the surface and 1000 ft as the effects of surface roughness are overcome. Knowing how the surface roughness affects the vertical wind speed profile is extremely valuable when determining the most beneficial WECS tower height.

Sources of Wind Climatology Data

At over 1000 locations in the U.S., a daily log sheet is filled out with hourly weather observations of the one-minute average wind speed and direction. These records are sent to the National Climatic Center (NCC) in Asheville, North Carolina, where these one-minute averages for every third hour are entered onto computer magnetic tape. Various monthly and yearly summaries are prepared, and all the original data are stored in archives. Each station receives summaries of its data and these are available for inspection. Therefore, the NCC at Asheville is currently the best source of wind data. NCC will, for the cost of reproduction (usually a few cents per copy), provide available summaries for sites in or near a locality. These data may be obtained by writing to:

Director
National Climatic Center
Federal Building
Asheville, North Carolina 28801.

The wind summaries are generally similar to that given in Table 3-2. Frequently, wind roses have been constructed for stations. Fig-

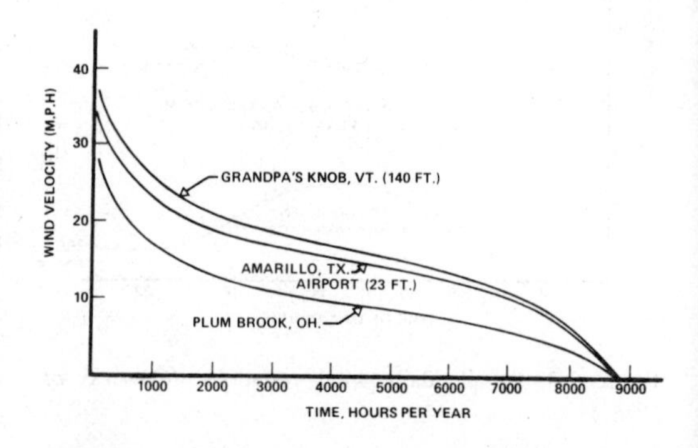

Figure 3-5. Wind duration curves. (*Courtesy of DOE.*)

Table 3-1. Average wind speeds for selected cities.

CITY	AVERAGE WIND SPEED
Fargo, North Dakota	14.4
Wichita, Kansas	13.7
Boston, Massachusetts	13.3
New York, New York	12.9
Fort Worth, Texas	12.5
Des Moines, Iowa	12.1
Honolulu, Hawaii	12.1
Milwaukee, Wisconsin	12.1
Chicago, Illinois	11.2
Minneapolis, Minnesota	11.2
Indianapolis, Indiana	10.8
Providence, Rhode Island	10.7
Seattle-Tacoma, Washington	10.7
San Francisco, California	10.6
Baltimore, Maryland	10.4
Detroit, Michigan	10.3
Denver, Colorado	10.0
Kansas City, Missouri	9.8
Atlanta, Georgia	9.7
Washington, D.C.	9.7
Philadelphia, Pennsylvania	9.6
Portland, Maine	9.6
New Orleans, Louisiana	9.0
Miami, Florida	8.8
Little Rock, Arkansas	8.7
Salt Lake City, Utah	8.7
Albuquerque, New Mexico	8.6
Tucson, Arizona	8.1
Birmingham, Alabama	7.9
Anchorage, Alaska	6.8
Los Angeles, California	6.8

ure 3-7 illustrates a typical wind rose. Each arrow shaft is proportional in length to the percentage of time that the wind blows along the arrow. Numbers at the head of each arrow indicate the average wind speed for that direction.

An index has been developed which lists all sites for which wind summaries are available. These sites include past and present National Weather Service Stations, Federal Aviation Administration and Civil Aeronautics Administration sites and military installations. The index, entitled *Index—Summarized Wind Data,* by M. J. Changery, W. T. Hodge and J. V. Ramsdell, BNWL-2220 WIND-11 (September 1977), can be obtained from the National Climatic Center.

Wind climatology may also be obtained from utilities operating nuclear power plants. The power utilities' wind summaries include wind speed frequencies by direction, graphs of wind speed versus duration of speed, height and location of the wind sensor, the average wind speed, the available wind power and descriptions of the site and the surrounding terrain.

The *Selective Guide to Climatic Data Sources,* which is available from the U.S. Government Printing Office, is also a useful guide for more wind data.

Other possible sources of wind data include the United States Soil Conservation Service; the Agricultural Extension Service; United States and State Forest Services; some public utilities; airlines; industrial plants; and agricultural and meterological departments at local colleges and universities.

A Use of Wind Summaries. Wind summaries for a potential WECS site are ex-

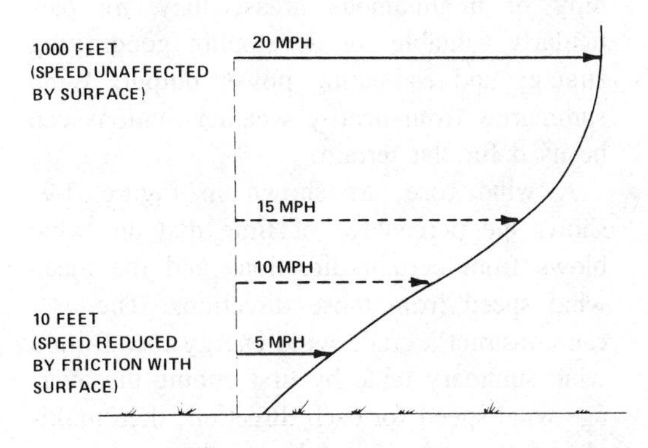

Figure 3-6. Effect of surface friction on low-level wind. (*Courtesy of DOE.*)

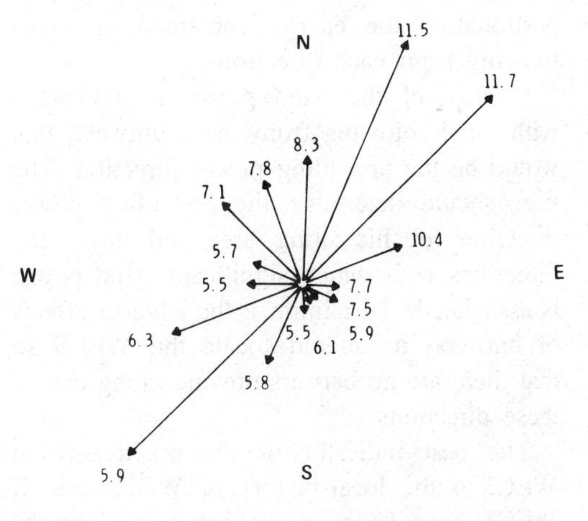

Figure 3-7. Typical wind rose. (*Courtesy of DOE.*)

Table 3-2. Sample wind summary – percentage frequencies of wind direction and speed: windspeed intervals (mph).

DIRECTION	0-3	4-7	8-12	13-18	19-24	24-31	32-38	TOTAL
N	1	1						2
NNE	1	2	1					4
NE	3	8	3					14
ENE	1	5	2					8
E	1	2						3
ESE	1	2						3
SE	1	3	2					6
SSE		3	2	1				6
S	1	3	3	1				8
SSW	1	3	5	5	1			15
SW	1	4	5	5	2			17
WSW	2	2	1					5
WNW	1	1						2
NW		1						1
NNW	1							1
Calm	3							
Total	20	41	24	12	3	0	0	100

tremely useful. In complex terrain, such as hilly or mountainous areas, they are particularly valuable for developing good siting strategy and estimating power output. Wind summaries from nearby weather stations can be used for flat terrain.

A wind rose, as shown in Figure 3-7, shows the percentage of time that the wind blows from certain directions and the mean wind speed from those directions. The user can construct a crude wind energy rose from a wind summary table by first cubing the average wind speed for each direction, then multiplying the cubed speeds by the percentage frequency of occurence for each wind direction. The derived numbers are roughly proportional to the energy contained in winds blowing from each direction.

If most of the wind power is associated with winds blowing from the southwest, that would be the prevailing power direction. The user should determine the prevailing power direction for his siting area and any other directions with which significant wind power is associated. To minimize the adverse effects of barriers, he should locate the WECS so that there are no barriers upwind along any of these directions.

The best indicator of the practicality of WECS is the local history of WECS use. If WECS have been or are being used in the vicinity, users can supply useful information about the type, size and application of their WECS; adequacy of the power output; siting procedures used; and accuracy of the estimated power output.

If there is no local history of WECS use, areas where available wind power is above 100 W/m^2 merit further investigation. Good WECS sites do exist in regions where available power is less than 100 W/m^2, but are generally limited to small areas of locally enhanced winds, such as hills, mountains, ridges or seacoasts. Figure 3-8 illustrates this by presenting available power for only the higher elevations. The figure indicates that considerable wind energy is available even in the southeast and southwest regions which are often considered as low power regions.

Before deciding against using wind energy, one should examine the local landforms. If the annual average wind speeds at nearby weather stations are less than 8 mph, and if there are no local terrain features to enhance the wind, small WECS are probably not practical.

Power from the Wind

You are fortunate if you live in flat, open terrain and have a neighbor with a WECS. You are more fortunate if he happens to be performing the same tasks with his machine that you wish to perform, such as pumping water or producing electricity. In effect, he has been measuring his wind power for a long time. The questions that you should consider are these: How adequate is the wind for him? How does your demand compare with his? How likely are you to have an appreciably lower (or higher) wind power per square foot (or square meter) than he has?

More likely, there is no wind turbine close enough to provide any useful experience for you. You will have to make a decision about how much time, effort and money to invest to increase your knowledge of your wind. There are three parts to this process:

Step 1. Making a preliminary estimate of your wind power.

Step 2. Measuring your winds or wind power.

Step 3. Comparing your measurements with nearby meteorological stations to determine your long-term average from your short-term data.

Step 1 is the very least you must do. That rough estimate of your wind power can be combined with results from a preliminary load survey (as described in Chapter 9); equipment selection can be aided by reviewing the manufacturer's WECS descriptions in Chapter 8; and a preliminary cost analysis (as described in Chapter 9) can be performed. You may find that you have only a fraction of the necessary wind available to make your investment a sound one, and that only a minimal effort should be made to determine whether your wind power is much greater than your initial estimate.

Preliminary Estimate of Wind Power. As a first step in making an estimate of the wind power available, you should check for tall, unavoidable obstructions, particularly trees. Typical tower heights for wind generators in the 1–10 kW class are up to 100 ft. Two simple methods for measuring tree height, if the top is visible, are illustrated in Figure 3-9. For the first method, find a time in the morning or afternoon when the tree shadow falls across flat ground. Set up a vertical stick of known height and compare the length of the shadow from the stick to that from the tree.

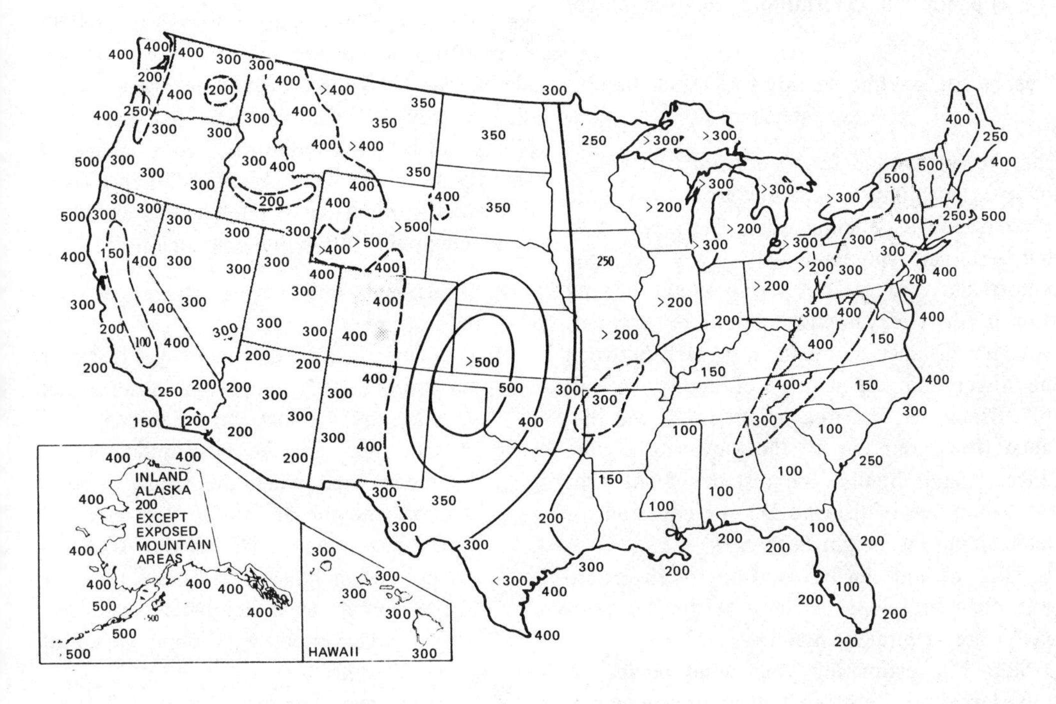

Figure 3-8. Annual average wind power at 50 m above higher elevations, in W/m². (*Courtesy of DOE.*)

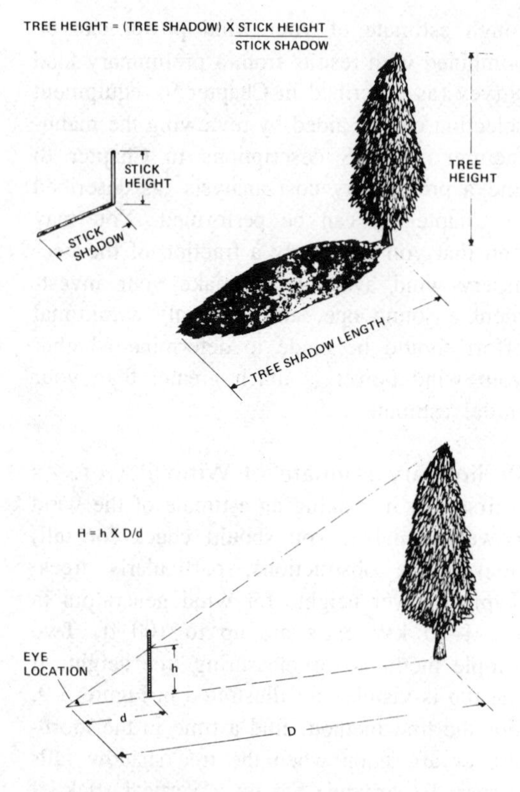

Figure 3.9. Two methods for estimating the height of a tree. (*Courtesy of DOE.*)

The equation for determining the tree height is:

Tree height = (tree shadow) × (stick height) ÷ (stick shadow).

The second method for measuring tree height involves attaching a yardstick to a small pole or to the edge of some other fixed object so that you can simultaneously sight both the bottom and top of the tree without moving your head. The yardstick should be vertical. Note the distance along the yardstick between the lower and upper lines of sight; we call this distance h; the tree height is H; the distance from your eye to the yardstick is distance d; and, finally, the distance from your eye to the tree is distance D. The equation for determining the height is then $H - h \times D \div d$. This method can be used on rough terrain with changing slope, as long as the measurements are accurately made.

Step 2 in estimating your wind power involves finding the closest meteorological stations (listed in this chapter and in the reference section). Compare their annual wind power averages. If they are at all close and there are no reasons why your value would not be about the same, you can take an average or place more weight on some of the readings than on others, according to your circumstances, to get a first estimate of your wind power. For open plains areas, this will work well, while in hilly or mountainous terrain, the results will most likely be poor.

Step 3 in obtaining a preliminary estimate of wind power is to simply start developing a better awareness of the wind and to make comparisons of the wind at your location with that where there are wind-measuring stations. Table 3-3 shows the Beaufort Scale, which relates wind speeds to easily recognizable phenomena.

Hand-held wind anemometers, such as that shown in Figure 3-10, are available at many boating, outdoors and aircraft supply stores. These are the least expensive of all wind-measuring devices and can be used to help calibrate your sense of the wind.

Palmer Cosslett Putnam and his team that designed and built the Smith-Putnam wind turbine studied many promising wind sites in New England. They were mostly frustrated in their efforts to develop general rules for predicting the best wind sites. One powerful indicator they *did* discover was that trees and plants can be greatly deformed in a consistent way by the wind. This is called flagging (it is described in more detail later in this chapter). They made the following conclusions:

- Occasional very severe storms do not deform trees.
- Tree deformation is a poor yardstick of maximum icing, although absence of breakage by ice may be significant.
- Balsam trees are the best indicators of the mean wind when the mean velocity at a tree height of 30–50 ft reaches 17 mph. The other end of the scale is reached when balsam is forced to grow like a carpet, at a height of one foot, which indicates a mean wind speed of about 27 mph.
- In this range, between 17 and 27 mph,

Table 3-3. Beaufort Scale.

BEAUFORT NUMBER*	KNOTS	MPH	M/SECOND	KM/HOUR	SEAMAN'S TERM	WORLD METEOROLOGICAL ORGANIZATION (1964)	ESTIMATING WIND SPEED EFFECTS OBSERVED ON LAND
0	Under 1	Under 1	0.0-0.2	Under 1	Calm	Calm	Calm; smoke rises vertically.
1	1-3	1-3	0.3-1.5	1-5	Light air	Light air	Smoke drift indicates wind direction; vanes do not move.
2	4-6	4-7	1.6-3.3	6-11	Light breeze	Light breeze	Wind felt on face; leaves rustle; vanes begin to move.
3	7-10	8-12	3.4-5.4	12-19	Gentle breeze	Gentle breeze	Leaves, small twigs in constant motion; light flags extended.
4	11-16	13-18	5.5-7.9	20-28	Moderate breeze	Moderate breeze	Dust, leaves and loose paper raised up; small branches move.
5	17-21	19-24	8.0-10.7	29-38	Fresh breeze	Fresh breeze	Small trees in leaf begin to sway.
6	22-27	25-31	10.8-13.8	39-49	Strong breeze	Strong breeze	Larger branches of trees in motion; whistling heard in wires.
7	28-33	32-38	13.9-17.1	50-61	Moderate gale	Near gale	Whole trees in motion; resistance felt in walking against wind.
8	34-40	39-46	17.2-20.7	62-74	Fresh gale	Gale	Twigs and small branches broken off trees; progress generally impeded.
9	41-47	47-54	20.8-24.4	75-88	Strong gale	Strong gale	Slight structural damage occurs; slate blown from roofs.
10	48-55	55-63	24.5-28.4	89-102	Whole gale	Storm	Seldom experienced on land; trees broken or uprooted; considerable structural damage occurs.
11	56-63	64-72	28.5-32.6	103-117	Storm	Violent Storm	Very rarely experienced on land; usually accompanied by widespread damage.
12 13 14 15 16 17	64-71 72-80 81-89 90-99 100-108 109-118	73-82 83-92 93-102 104-114 115-125 126-136	32.7-36.9 37.0-41.4 41-5-46.1 46.2-50.9 51.0-56.0 56.1-61.2	118-133 134-149 150-166 167-183 184-201 202-220	Hurricane	Hurricane	

*NOTE: Since January 1966, weather map symbols have been based upon wind speed in knots, at five-knot intervals, rather than upon Beaufort number.

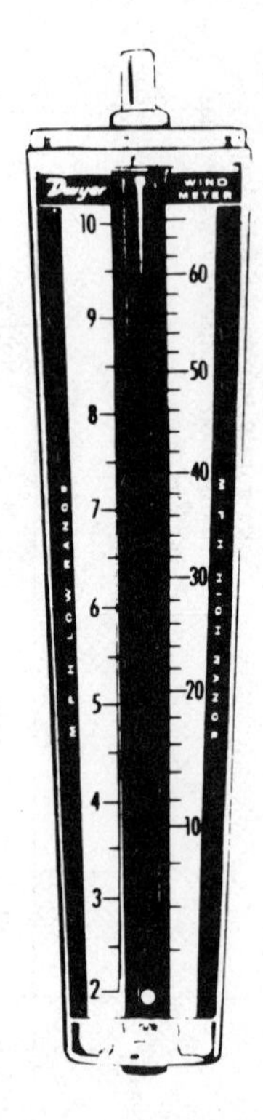

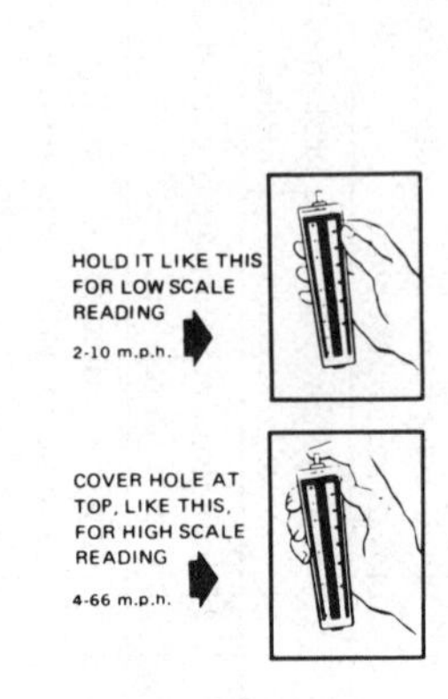

To operate, simply, remove the Dwyer portable wind meter from its clear plastic case and hold at eye level with back of the unit to the wind. The position of the white ball in the tube indicates wind speed in miles per hour. For high range use, cover the hole at the top with your finger and read.

Figure 3-10. Dwyer Instruments, Inc.'s hand-held wind anemometer. (*Courtesy of Dwyer Instruments, Inc.*)

there are five easily recognized types of progressive deformation: brushing, flagging, throwing, clipping and carpet.

- Tree deformation is a sensitive indicator of the unpredictable wind flows through and over mountains. Local transitions from prevailing very high winds (which hold balsams to a carpet), to prevailing winds so moderate that the balsams reach normal growth without deformation, occur within a matter of yards of one another! Many gardeners know that even moderate winds can have a strong effect on most vegetable plants. A government-sponsored study is under way for determining wind effects on vegetation in the northeastern section of the U.S.

Finally, combine all your information to make the best estimate of your wind power. Use this value for the first cut at sizing your wind system and estimating its costs.

Wind and Wind Power Measurement Equipment. To measure your wind potential, mount a sensor on a pole high enough, and far enough away from buildings and trees, to have a clear sweep of the wind. Figure 3-11 shows the DOE/NASA wind-measurement tower. For SWECS applications, a smaller makeshift tower can be used. Ten feet is an absolute minimum height. Considering the time and money that will be invested in the survey, 25–50 ft is probably easily justifiable. Television antenna towers provide a good method to obtain these heights and can be as inexpensive as $1.50–$2.00/ft.

Most standard meteorological equipment for commercial use will appear to be quite expensive to the individual homeowner. However, new, relatively inexpensive items are rapidly being developed by manufacturers involved in the WECS market. Many of these items can also be rented from the manufacturer or distributor. For discussion purposes, this equipment can be divided into: 1) sensors for actually measuring the wind velocity; 2) meters for indicating the speed and/or direction; and 3) recording equipment that processes the data in one of several possible ways and records the results.

Several kinds of sensors of particular interest are shown in Figures 3-12 through 3-15. The wind-cup anemometer shown in Figure 3-12 is the most popular type for surveys. It measures wind velocity but not direction. Another type of wind speed indicator is shown in Figure 3-13; this is often used for SWEC wind-measurements. To measure the direction, a wind vane needs to be added, as shown in Figure 3-14. Both wind speed and direction can be obtained with the propeller-type device attached to a tail vane (shown in Figure 3-15). If you expect that the direction of the wind will be important to the placement of your wind turbine, you will need more than the basic wind-cup anemometer.

A note of caution must be made here. It is not unusual for an anemometer to give readings that are considerably in error, particularly after extended use. A 10 percent wind speed error will produce about a 30 percent error in expected wind power, so some care is appropriate in selecting and maintaining the anemometer.

The simplest way of displaying the wind

Figure 3-11. Wind-measurement tower for DOE/NASA test site. (*Courtesy of DOE/NASA LRC.*)

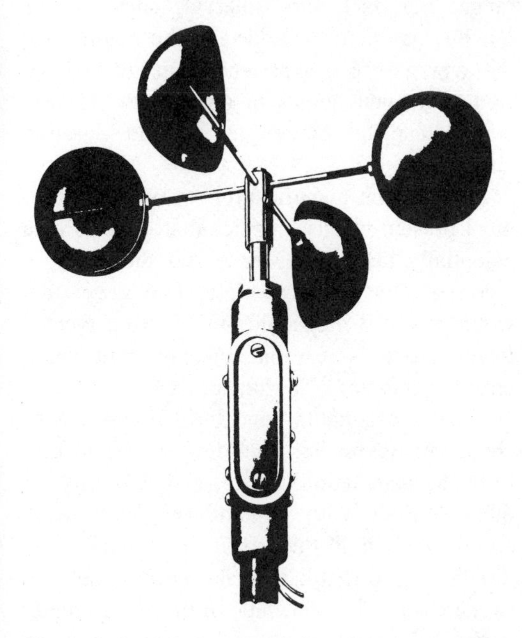

Figure 3-12. M.C. Stewart wind-cup anemometer. (*Courtesy M.C. Stewart, Ashburnham, Mass*).

Figure 3-13. Dwyer Mark II wind speed indicator. (*Courtesy of Dwyer Instruments.*)

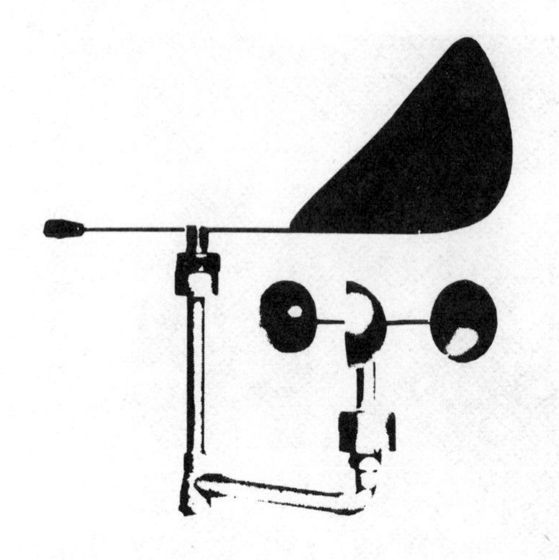

Figure 3-14. Wind-cup anemometer with wind vane. (*Courtesy of DOE.*)

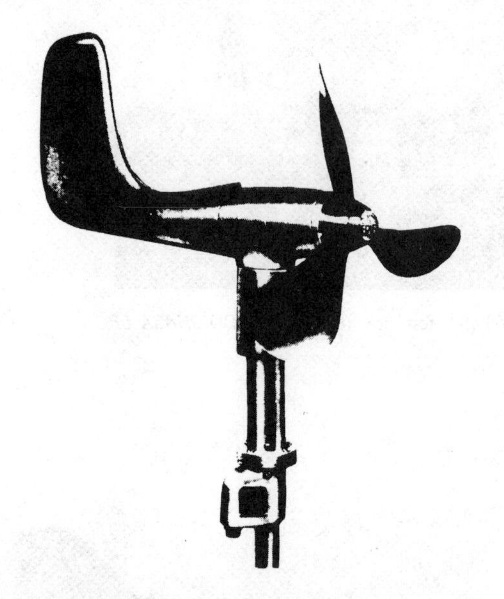

Figure 3-15. Propeller-type wind anemometer. (*Courtesy of DOE.*)

speed and direction is on meters. You must, however, read and log the values on a regular basis, preferably hourly during your hours awake. These sensors can generate either a dc signal, pulses or an ac signal. If a dc signal is used by the sensor, the wire length will probably be important, while the pulse type is unaffected by wire length as long as the signal received has adequate strength.

The simplest recorder is a counter that displays the run-of-the-wind. It totals up the miles of wind that blow past the anemometer. Thus, if it records 240 miles in a 24-hour period, the average wind velocity is 10 mph. You only need to read it once a day, but it would be better to read it more frequently to establish daily wind patterns. A homemade wind vane can be used to advantage for estimating the wind direction.

Long-term recording devices need no attention on a daily basis. One type prints the wind speed and/or direction on a paper chart. Such continuous recordings as obtained at meteorological stations require expensive apparatus and are not usually necessary. At a fraction of the cost is a small recorder that prints a dot onto a slowly advancing paper chart recorder, indicating velocity every few seconds. A roll of paper will usually last a month or more. To use the data, one can see an average for each time period, such as quarter-, half- or one-hour intervals, and then record and process the data by hand. Owner's manuals and equipment manufacturer's publications detail simple methods for data reduction and use.

Several recorders are available that divide wind speeds into ranges (e.g., 0–3, 4–7, 8–11) and have counter displays for each range. For each time interval, such as one minute, one count is added to the counter for the average speed experienced during that period. The total counts in one month, for example, can be plotted as a wind duration curve.

At least one manufacturer of WECS equipment presently sells a device that collects data essentially the same way as did the previous recorders, but then calculates the power that would result if a specific WECS were located there. There is just one display: total wind energy generated. The manufacturer will help you select a suitable wind turbine and use its characteristics in this recording device to perform the calculation. It can predict results for other WECS with a similarly rated wind speed. Although this device is simple and direct, it is difficult to make correlations between its results and data from nearby meteorological stations.

Determining Wind Power. If you asked, "How long do I need to take wind data?" a meteorologist might tell you, "If you really want to know your wind, about five years is required!" He would be right *and* wrong. His answer could be right for some utility company which is considering the investment of millions of dollars in a group of wind turbines when there is no long-term wind data for that region. His answer would not be appropriate if you are considering installing a $6000 WECS.

A better question is, "How much will I gain by taking data—an extra week, an extra month or an extra six months—and how much will I gain by using wind survey equipment with more capability?" Measuring your wind for one month and claiming you have determined your average wind power is as absurd as measuring your rainfall for one month, multiplying by 12, and claiming you now know your average annual rainfall. However, by comparison with local weather station records, a few months worth of rainfall records can often provide you with a good indication of your annual average rainfall. Likewise, your average annual wind power is best estimated by comparing your data with weather station records.

You might now ask, "Where is the tradeoff between the time and money invested in a wind survey and the time and money invested in the windmill? What is the value of a wind survey to me? How much should I be willing to pay?" Let us look at three examples.

For the first example, assume you live in flat country, about 20 or 30 miles from the nearest meteorological station. You have done a preliminary wind survey and you think you have a good site. The wind power at the nearest meteorological station is within ±50 percent of that at this local station. We will assume that you have evaluated your energy needs. You have tentatively selected several wind systems, the smallest one of your wind power is 50 percent greater than your estimate, and the largest windmill if your wind power is 50 percent less than the estimate. The largest unit has (100 + 50) ÷ (100 − 50) = 3 times the capacity of the smallest unit. They range in price from $3000–$7000.

From your study of the economics involved, you have decided that $7000 is your break-even point compared to extending the power lines and using public utility power. You feel that spending $700 for a survey would be worthwhile. Say that the survey shows you have 25 percent more power available than your original estimate showed, with a ±15 percent uncertainty. You would simply buy a $4000 system. The $700 survey cost has, in effect, bought you assurance (in a sense, insurance) that you need not spend $7000.

As another example, suppose you have a summer cabin in the woods and you use it approximately one month a year. You have estimated the available power to be ±40 percent. Since your entire need for electrical energy is only for recreation, during an unusually calm spell of weather, you could simply not go to the cabin. Using the high (140 percent) and the low (60 percent) wind power and your estimated need, you come up with two systems, mostly composed of used equipment, with prices $1200 and $1600. The summer season is coming up and you are not there to take a survey and would have to contract it out. You decide to go ahead with a larger system since you feel you cannot get enough useful wind survey information for the few hundred dollars difference and the time allowed.

As a final example, consider that you are a farmer or rancher who needs a lot of power. You have made a preliminary survey and selected two systems, based on your estimate of maximum and minimum wind power, that will cost $10,000 and $25,000. Obviously, spending several thousands of dollars for a good wind survey will be a worthwhile investment.

The important features of all three of these examples are: 1) a preliminary estimate of wind power with an estimate of your accuracy; 2) an estimate of energy requirements (at least a preliminary estimate); 3) an estimate of your wind energy system low and high wind power costs; and 4) the maximum sum to place on the site survey. These four common features of the above examples are the first four steps shown in Figure 3-16, a logic diagram showing the steps to be taken in accomplishing a wind power survey.

Concerning the decision on the value of the wind survey, we return to the question posed at the beginning of this section: "How much will I gain by taking data—an extra week, an extra month, or an extra six months—and how much will I gain by using wind survey equipment with more capability?" There is probably no definite answer, but if you are interested in a small system and you live in flat country near a weather station with wind records, and your preliminary survey shows your estimated average annual wind speed is quite high, two or three months of data with one wind-cup anemometer will probably produce adequate results. For the potentially larger and more expensive system, more sophisticated wind survey equipment and longer data collection times are appropriate. Also where the estimated average annual wind speed is less, there is a greater need for more than just run-of-the-wind readings. This is due to the larger scatter in wind power at the lower annual wind speeds shown in Figure 3-4.

Equipment alternatives include renting or purchasing of the wind sensor equipment. Some wind turbine manufacturers/distributors will rent anemometer equipment. If you buy equipment, you can expect to recover some of your costs by selling it when you have finished your site survey. Also, you may wish to have a professional meteorologist, meteorology firm or knowledgable distributor install the wind-measuring equipment and perform the wind survey for you.

The next step in the process of determining your wind power (see Figure 3-16) is to compare your wind recordings with the readings that have been obtained simultaneously at nearby meteorological stations. You will then

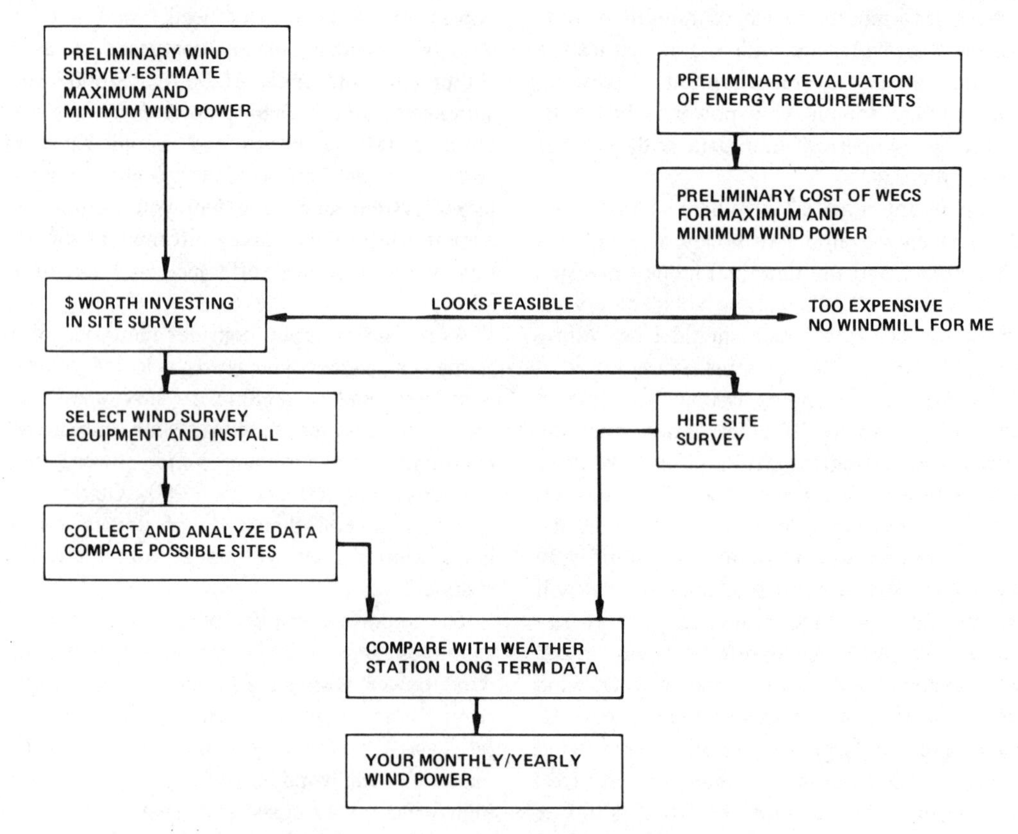

Figure 3-16. Logical steps to determine your wind power. (*Courtesy of DOE.*)

be able to estimate your long-term average wind power by using their summaries of years of data.

At each of the stations listed in the reference section, an hourly record of weather conditions is written out on form WBAN 10A, one sheet for each day. This sheet includes hourly observations of wind direction and speed. These are made by estimating, on the hour, the average wind speed for a one-minute period, by observing a dial or strip chart recording. Visit or write your nearby station(s) to obtain copies of the records for the days of interest to you, for comparison with your data. The WBAN 10A sheets (and the strip chart records) are sent to the National Climatic Center, and copies can be ordered from them.

Much of the data since the mid-1940's has been prepared for, and processed by, computers to obtain various long-term wind averages. Each station has a set of average monthly and yearly wind speeds for its location. These sheets are titled "Percentage Frequency of Wind Direction and Speed." They show the percentage of the hourly readings in each of 11 different speed ranges (in knots) and 16 directions. In the right column is the mean wind speed (same as average wind speed) for each wind direction (remember that is the direction from which the wind blows). Along the bottom are summations for each column and the average wind speed for the entire period. You will want a copy of the annual summary and the sheets for your critical months, when your expected demand is greatest compared to the available wind power. All the information you need for making up a wind rose or a wind duration curve is on this form.

The method for using the data you collect from a wind-cup anemometer registering just the run-of-the-wind is quite simple. Read your meter at the beginning of each month. Follow the manufacturer's procedure for determining the miles of wind passing the meter during the previous month. Divide this number by the hours that have elapsed (for a 30-day month, 24 x 30 = 720) and you have your average wind speed for that month. Now, add up all the hourly wind speeds recorded at your nearest meteorological station for the month and divide that by the same number of hours. You have the average wind speed for the meterological station for the same period. For instance, if your reading at the end of the first month is 5760 miles, dividing by 720 hours gives 8.0 mph. If the sum of the weather station values for the same month were 6261, then the average wind speed would be

$$6261 \times 1.15 \div 720 = 10.0 \text{ mph.}$$

The 1.15 factor converts knots to mph (all weather station readings are in knots). Your wind speed was 80 percent of that measured by the weather station, and your wind power is $(0.8 \times 0.8 \times 0.8 \times 100) = 51$ percent of that at the station. Now, look in the reference section for the average annual wind power for that station. For example, this value is 200 W/m^2. Your expected average annual wind power based on your single reading for one month would be $0.51 \times 200 - 102$ W/m^2. This might be a satisfactory estimate, or it might be a very poor value, depending on the terrain factors. If subsequent monthly comparisons show large differences in your wind power compared to the weather station wind power, the single month value was, of course, not good. However, if the values for the next couple of months are close to the first value, the average wind power based on these three months could be quite good.

To find a WECS site with the most available wind power, it is essential to understand the variation of power with wind speed. The following equation defines this relationship:

$$\text{Available power} = 0.5 \times D \times A \times S^3$$

where

D = air density
A = area of the rotor disc
S = the wind speed ($S^3 = S \times S \times S$, cube of wind speed).

Rotor discs are illustrated in Figure 3-17 for three different types of WECS. Since air density (D) at a site normally varies only 10 percent or less during the year, the amount of power available depends primarily on the area (A) of the rotor disc and the wind speed (S). Increasing the diameter of the rotor disc (by increasing the blade length) will allow the WECS to intercept more of the wind, and thereby harness more power.* Since the available power varies with the cube of the wind speed, choosing a site where wind speed is greatest is desirable. Table 3-4 demonstrates how even a small change in wind speed results in a large change in available power. Suppose that one computation of available power at a site had been based on a wind speed estimate of 10 mph when the actual speed was 9 mph. The actual available power would be almost 30 percent less than the estimated power, due solely to a 1-mph error in the estimated wind speed.

WIND HAZARDS

Weather hazards may influence the economic feasibility of a WECS or the selection of a particular machine. For example, if salt spray at a coastal site reduces the expected lifetime of a WECS by one-half, the cost of wind energy to the user sharply increases. Good siting strategy, therefore, will not only maximize the wind speed, but also reduce hazards.

Many WECS weather hazards cannot be avoided. In such cases, the user must either purchase a WECS designed to survive in the local environment or in some way protect the WECS from the hazard. The potential economic impact of either approach must be evaluated.

Turbulence

Air turbulence consists of rapid changes in speed and/or direction of the wind. The turbulence most harmful to WECS is the small-scale, rapid fluctuation often caused by the wind flowing over a rough surface or a barrier. Turbulence has two adverse effects: 1) a decrease in harnessable power, and 2) vibrations and unequal loading on the WECS that may eventually weaken and damage it.

*The choice of WECS size should not be made solely on this basis, but in conjunction with the WECS dealer and other considerations discussed in this handbook.

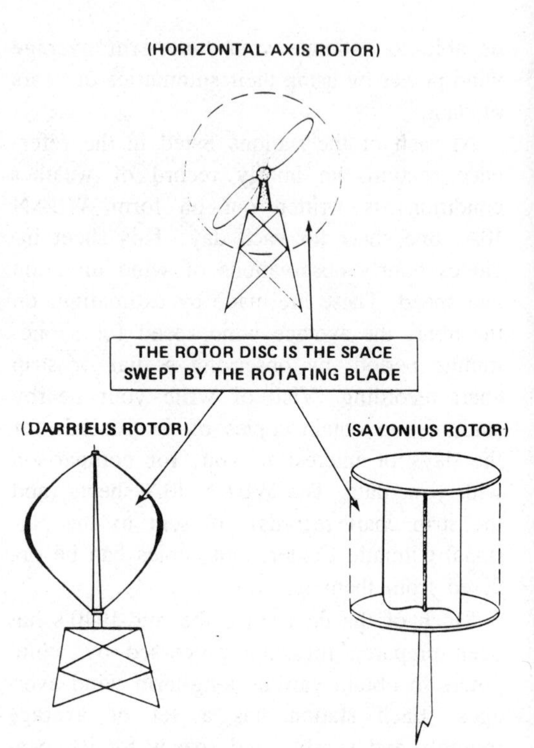

Figure 3-17. Definition of the rotor disc. (*Courtesy of DOE.*)

Table 3-4. Percentage change in available power with changes in wind speeds.

SPEED (MPH)	PERCENT POWER CHANGE FROM POWER AT BASE SPEED OF 10 MPH
5	−88
6	−78
7	−66
8	−41
9	−27
10	0
11	+33
12	+73
13	+120
14	+174
15	+238

To characterize the turbulence at a site, the user should determine the prevailing wind power directions.** When the prevailing wind is blowing, the predominant areas of tur-

**If more than one wind direction frequently occurs, the user should investigate each to fully understand the turbulence hazard to the WECS.

bulence at a proposed WECS site can be detected by one or more 4-ft lengths of ribbon tied to a mast or a kite string, as shown in Figure 3-18. How much the ribbons flap indicates the amount of turbulence. The expected location and intensity of turbulence produced by barriers and landforms are described (at least qualitatively) in this chapter.

Strong Wind Shear

Strong wind shear may pose a hazard to small WECS in some locations. Wind shear is simply a large change in speed or direction over a small distance. If a large change occurs over a distance less than or equal to the diameter of the rotor disc, then unequal forces will be acting on the blades (see Figure 3-17 for definition of rotor disc). Over a period of time these forces could damage the WECS.

Generally, the longer the blades, the more susceptible the WECS is to shear hazards. However, shear can be a hazard to any WECS whose rotor disc is too near the ground, a canyon wall, a steep mountainside or the top of a flat-topped ridge (refer to Figure 3-43).

Extreme Winds

WECS blades and the supporting towers are both susceptible to damage from high winds. The blades become vulnerable if the protection systems designed into many WECS fail in extreme winds. Towers must be capable of supporting the WECS in all wind speeds which normally occur in the local area.

Users in or near mountains should obtain extreme wind speeds from nearby weather stations when planning a WECS.

The WECS dealer may assist in selecting the best tower but, before it is purchased, the user should contact local building inspectors to ensure compliance with existing codes.

Thunderstorms

Thunderstorms produce several weather hazards, such as severe winds, heavy rains, lightning, hail and possibly tornadoes. Figure 3-19 shows that thunderstorms occur more than 40 days per year in most parts of the U.S. The largest number of most intense thunderstorms occur in Florida and the Great Plains states of Kansas and Oklahoma.

Although the frequency of lightning is not known, it can be partially inferred from the thunderstorm occurrences shown in Figure 3-19. Considering its cost, a WECS should be protected from lightning strikes wherever it is located.

Hail often causes heavy damage to buildings; it may also cause damage to a wind machine and its support structure. Large hail is most frequently observed in Texas, Oklahoma, Kansas and Nebraska, as shown in Figure 3-20.

Tornadoes occur most often in the central part of the U.S. in an area called "tornado alley," extending from southwestern Texas to northern Illinois. Figure 3-21 shows the approximate risk of a tornado strike for different areas of the continental U.S. Since WECS, like houses, are not designed to withstand tornadoes, the prospective buyer must assess the risk of tornado damage.

Icing

Ice accumulated on blades, towers and transmission lines can cause hazards or reduce the efficiency of wind machines. There are two types of icing: rime ice and glaze ice.

Rime ice differs from glaze principally because of its source. It forms from frost or freezing fog, rather than from rain. Rime icing occurs mainly at high elevations. It is drier, less dense and, therefore, less haz-

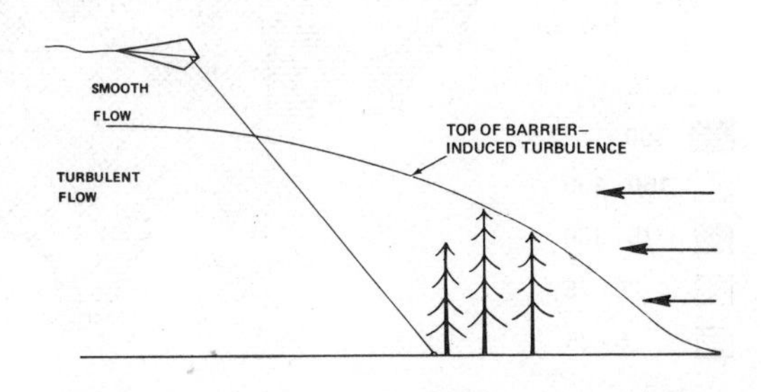

Figure 3-18. Simple method of detecting turbulence. (*Courtesy of DOE.*)

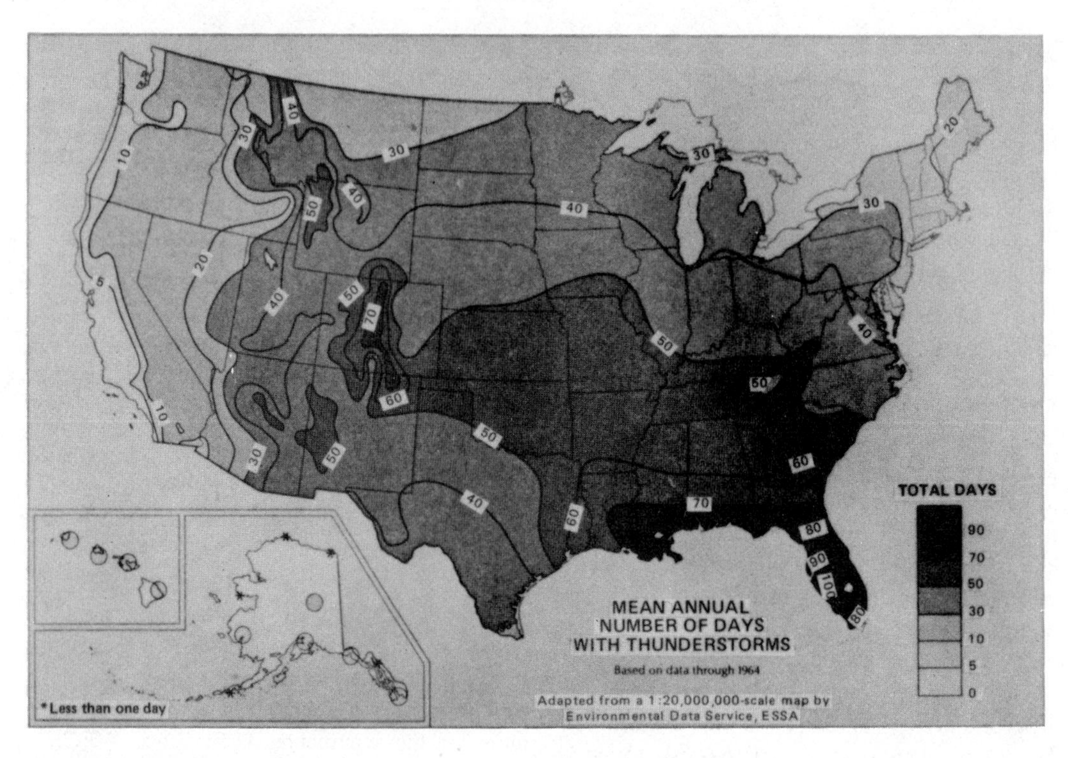

Figure 3-19. The mean annual number of days with thunderstorms. (*Courtesy of DOE.*)

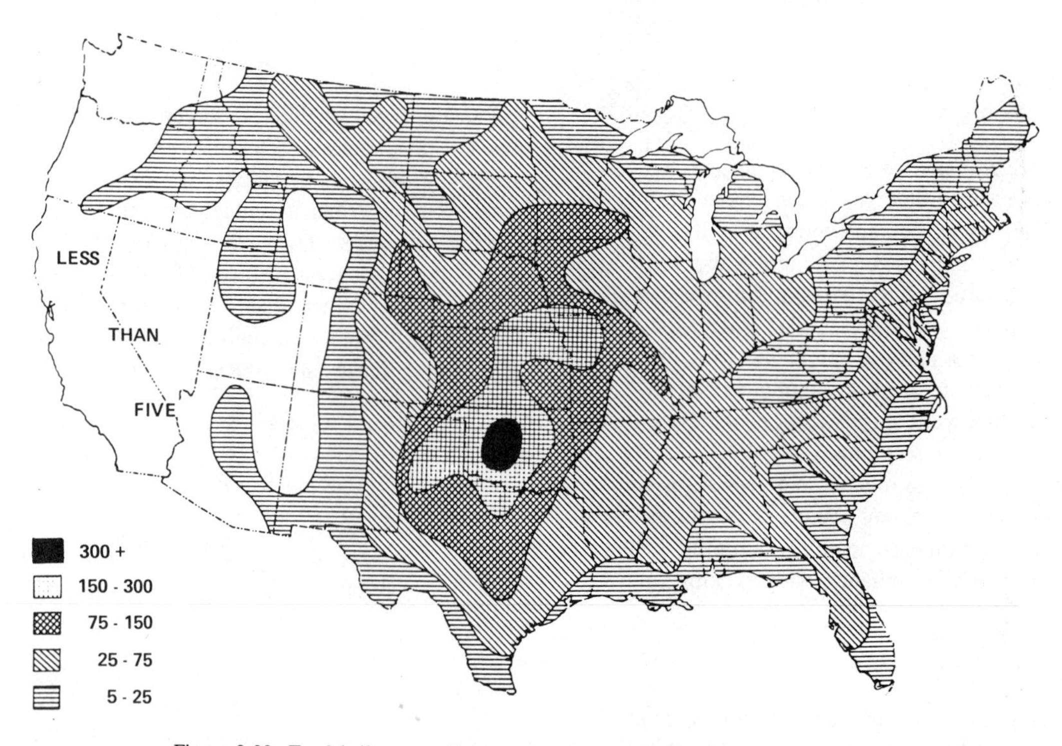

Figure 3-20. Total hail reports ¾ in. and greater, 1955–1967. (*Courtesy of DOE.*)

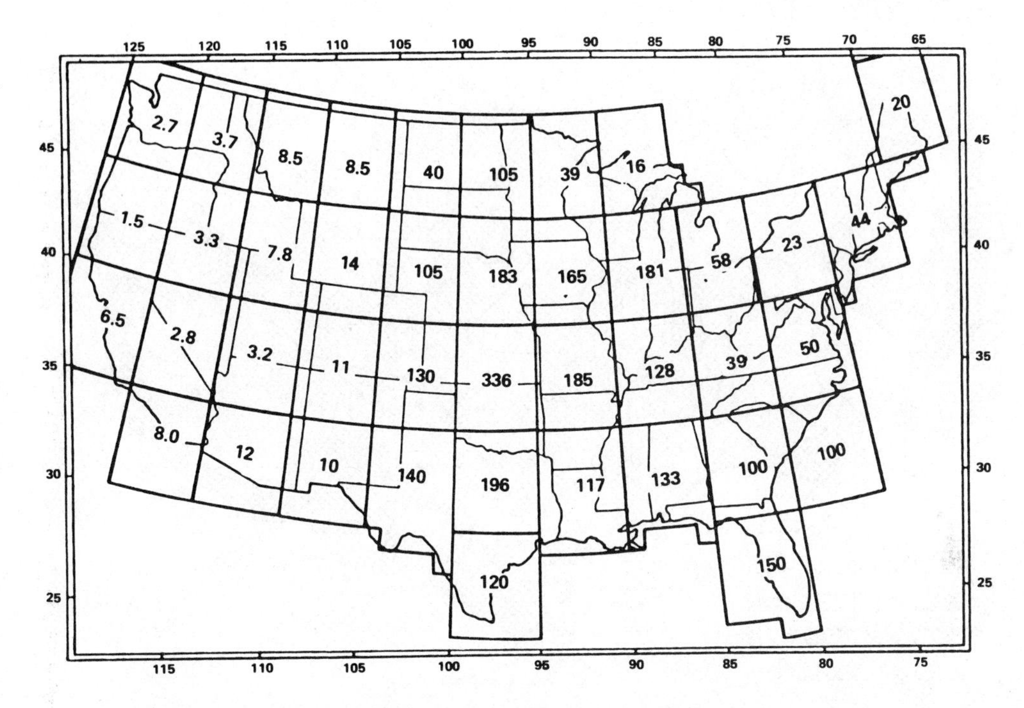

Figure 3-21. Tornado strike probability within five-degree squares in the Continental U.S. (The numbers represent chances in 100,000 of a tornado strike during a one-year period.) (*Courtesy of DOE.*)

ardous than glaze; however, it can, over a period of time, build up large accumulations.

Glaze icing, formed from freezing rain, occurs most frequently in valleys, basins and other low elevations. When rain falls through a subfreezing layer of air at the ground, the drops freeze on contact with the surface. Under favorable conditions, freezing precipitation can rapidly accumulate on a cold surface to thicknesses of more than 2 in. Data gathered by the Association of American Railroads, Edison Electric Institute, American Telephone and Telegraph and other organizations on ice accumulation on transmission lines in the U.S. have been analyzed for the period 1911–1938; the number of times that icing greater than 0.25 in. occurred is shown in Figure 3-22.

Heavy Snow

Snow causes three principal hazards to a WECS: 1) service and maintenance can be made difficult by excessive snow depths; 2) excessively heavy snowfall may damage parts of the turbine; and 3) blowing snow may infiltrate the machine parts and cause breakage from freezing and thawing.

Figure 3-23, which shows the maximum snow depth for a storm period, is provided as a guide for estimating snowfall. However, in some mountain regions, much more snowfall has been recorded than is shown on the map. How long a typical storm lasts and how long snow remains on the ground are also important considerations.

As the figure illustrates, the high wind areas on the eastern sides of Lake Superior and Lake Michigan receive more snow (as much as 60 in. or more per year) then the area beyond these snowbelts. A potential user considering a site on the eastern sides of the Great Lakes should therefore consider the damaging effects of heavy snowfalls and blowing snow.

Floods and Landslides

WECS users should be aware of floods and slides. In general, all structures should be kept out of floodplains. If an ideal wind site is located in a river valley, the user should

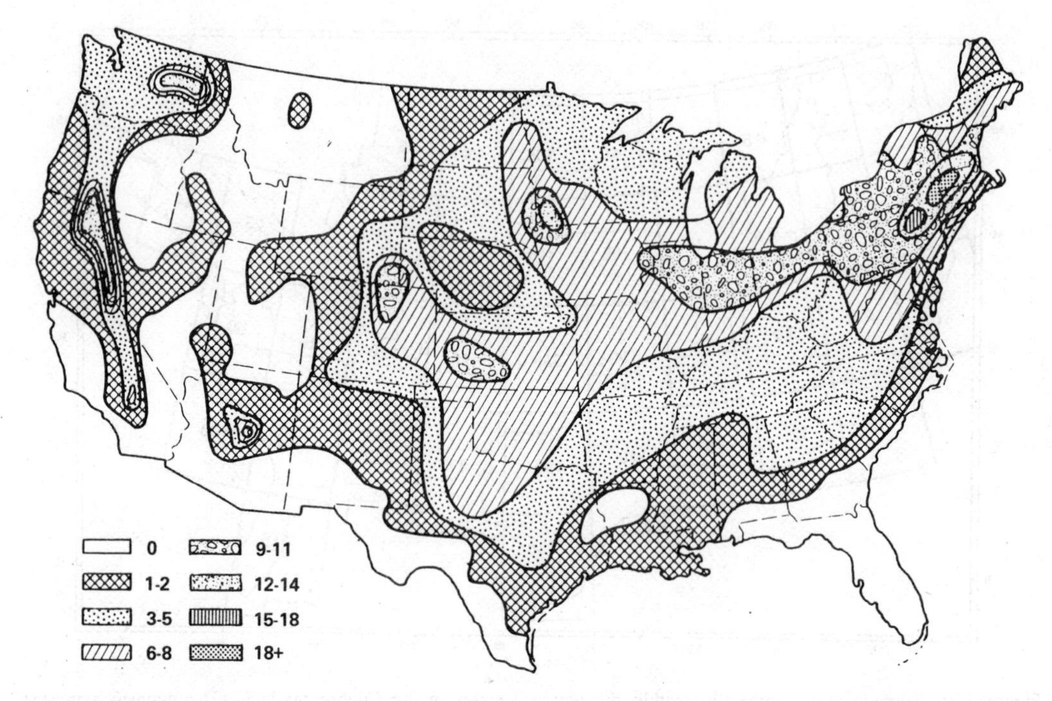

Figure 3-22. Number of times ice 0.25 in. or more thick was observed during the nine-year period of the Association of American Railroads Study. (*Courtesy of DOE.*)

build a structure to withstand flood conditions. He should also investigate the potential for earth slides and the stability of the soil foundation at any potential wind site.

Extreme Temperatures

Extremely high or low temperatures will adversely affect most WECS. Lubricants frequently freeze in very cold temperatures, causing rapid wear on moving parts. Many paints, lubricants and other protective materials deteriorate in high temperatures. The user should review the local climatology and then consider the possible added expense of protecting the WECS against extreme temperatures.

Salt Spray and Blowing Dust

Salt spray and dust may damage a WECS unless the machines are properly constructed and maintained. The corrosive properties of salt spray should be taken into account for any site within 10 miles of the sea.

Blowing dust may damage the system if it penetrates the moving parts, such as the gears and turning shafts. Many diverse regions of the country (urban, agricultural, desert, valley and plain areas) are subject to suspended dust. However, mountainous, forested and coastal regions have few major dust storms. The highest frequency of dust occurs in the southern Great Plains, but blowing dust also occurs often in portions of the western states, northern Great Plains, southern Pacific Coast and the southeast (see Figure 3-24).

SITING OF WIND ENERGY CONVERSION SYSTEMS

Siting in Flat Terrain

Choosing a site in flat terrain is not as complicated as choosing a site in hilly or mountainous areas. Only two primary questions need to be considered:

- What surface roughnesses affect the wind profile in the area?
- What barriers might affect the free flow of the wind?

Terrain can be considered flat if it meets

the following conditions, as shown in Figure 3-25: the elevation difference between the site and the surrounding terrain is less than 200 ft for 3–4 miles in any direction, and the ratio of h ÷ l in Figure 3-25 is less than 0.03. The potential user can determine if his site meets these conditions either by inspecting it or by consulting topographical maps.

The conditions given for determining flat terrain are very conservative. If there are no large hills, mountains, cliffs, etc. within a mile or so of the proposed WECS site, the data in this section can be used for siting. However, if nearby terrain features might influence his choice of a site, the user should refer to the portion(s) of the section of this chapter dealing with these features to better understand the local airflow.

Wind rose information can also guide the user in determining the influence of nearby terrain. For example, suppose a 40-ft-high hill lies 1/2 mile northeast of the proposed site (this classifies the terrain as non-flat); also assumes the wind rose indicates that winds blow from the northeast quadrant only 5 percent of the time with an average speed of 5 mph. Obviously, so little power is associated with winds blowing from the hill to the site that the hill can be disregarded. If there are no terrain features upwind of the site along the principal wind power direction(s), the terrain can be considered flat.

Uniform Roughness. Surface roughness describes the texture of the terrain. The rougher the surface, the more the wind flowing over it is impeded. Flat terrain with uniform surface roughness is the simplest type of terrain for a WECS site. A large area of flat, open grassland is a good example of uniform terrain. Providing there are no obstacles (i.e., buildings, trees or hills), the wind speed at a given height is nearly the same over the entire area.

The only way to increase the available power in uniform terrain is to raise the machine higher above the ground. A measurement or estimate of the average wind speed at one level can be used to estimate wind speed (thus the available power) at other levels. Table 3-5 provides estimates of wind speed changes for several surface roughnesses at various tower heights. The numbers in the

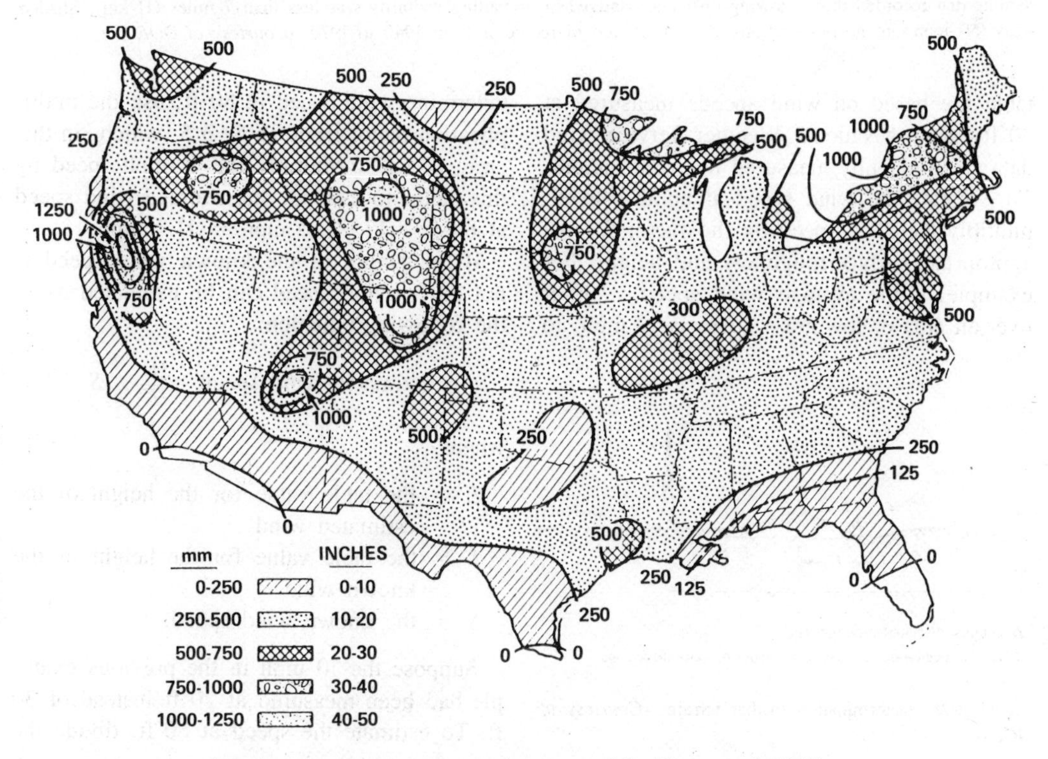

Figure 3-23. Extreme storm maximum snowfall. (*Courtesy of DOE.*)

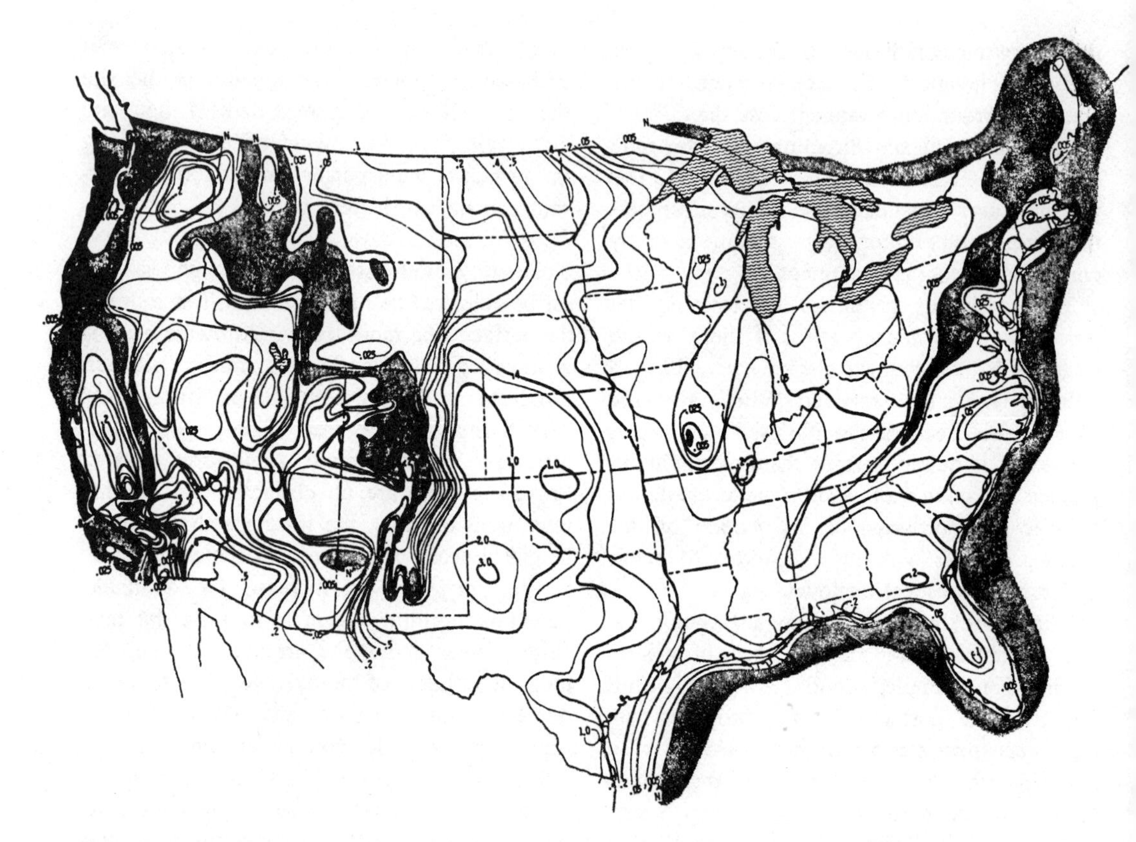

Figure 3-24. Annual percent frequency of dusty hours, based on hourly observations from 343 weather observation stations that recorded dust, blowing dust and sand when prevailing visibility was less than 7 miles (11 km). Shaded areas (N) represent no observations of dust. Period of record is from 1940 to 1970. (*Courtesy of DOE.*)

table are based on wind speeds measured at 30 ft because National Weather Service wind data are frequently measured near that height. To estimate the wind speed at another level, multiply the 30-ft speed by the factor for the appropriate surface roughness and height. For example, if the average wind speed at 30 ft over an area of low grass cover is 10 mph, to determine wind speed at 80 ft, use the multiplication factor from Table 3-5 (which, in this case, is 1.17). Multiply the 10-mph speed by this factor to estimate the average wind speed at 80 ft: 1.17 × 10 mph = 11.7 mph.

If the height of the known wind speed is not 30 ft, wind speed can be estimated using the following equation:

$$\text{Estimated wind speed} = \frac{E}{K} \times S$$

where

E = the table value for the height of the estimated wind

K = the table value for the height of the known wind

S = the known wind speed.

Suppose the 10 mph in the previous example had been measured at 20 ft instead of 30 ft. To estimate the speed at 80 ft, divide the

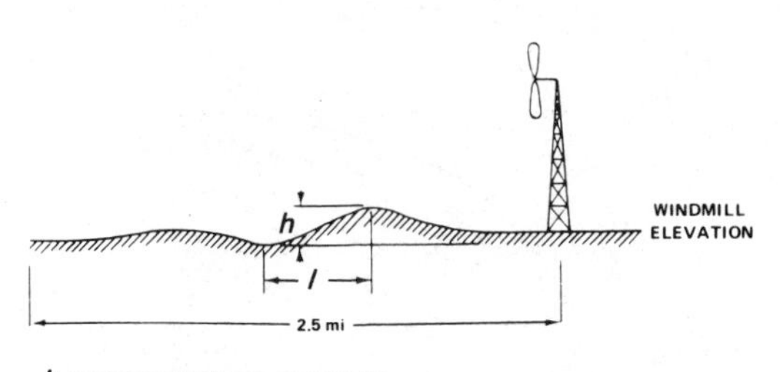

Figure 3-25. Determination of flat terrain. (*Courtesy of DOE.*)

factor for 80 ft (1.17) by the factor for 20 ft (0.94) to obtain the corrected factor (1.24); then multiply this corrected factor by the known wind speed (10 mph) to estimate the 80-ft wind speed (12.4 mph). This calculation is shown in equation form below (using the equation above):

$$\frac{E}{K} \times S = \frac{1.17}{0.94} \times 10 \text{ mph} =$$

$$1.24 \times 10 \text{ mph} = 12.4 \text{ mph}.$$

Table 3-6 provides available wind power changes between levels.* If the height of the known wind is 30 ft, the percentage change of available power between this level and another can be read directly from the table. If the known height is other than 30 ft, this equation can be used to compute the available power change:

$$\text{Fractional power change} = \frac{E - K}{100 + K}$$

where

E = the table value for the estimated wind height

K = the table value for the known wind height.

Computing the available power change for the previous example (i.e., extrapolating from 20 ft up to 80 ft over low grass), K is -17 and E is 60. The fractional power change is:

$$\frac{E - K}{100 + K} = \frac{60 - (-17)}{100 + (-17)} =$$

$$\frac{60 + 17}{100 - 17} = \frac{77}{83} = 0.93.$$

To express the available power change as a percent, simply multiply by 100 (0.93 × 100 = 93 percent increase in available power by raising the WECS from 20 ft to 80 ft above a low grass surface).

The heights in Tables 3-5 and 3-6 should not always be considered as heights above ground. Over areas of dense vegetation (such as an orchard or a forest), a new "effective ground level" is established at approximately the height where branches of adjacent trees touch. Below this level, there is little wind; consequently, it is called the *level of zero wind*. In a dense corn field, the level of zero wind would be the average corn height; in a wheat field, the average height of the wheat, etc. The height at which this level occurs is called the "zero displacement height," and is labeled "d" in Figure 3-26. If "d" is less than 10 ft, it can usually be disregarded in estimating speed and power changes. However, if ground level is used when "d" is actually 10 ft or more, changes in speed and power from one level to another will be underestimated. Tables 3-5 and 3-6 express all heights above the "d" height, rather than above ground.

Changes in Roughness. Often roughness varies upwind of the WECS. Figure 3-27 shows how a sharp change in roughness affects the wind profile. If a WECS were sited at the first level in Part A of this figure, the user would be greatly under-utilizing wind energy, since roughness changes cause a sharp increase in wind speed slightly above the first level. Part B of the figure shows that in smooth terrain, little, if anything, would be gained by increasing tower height from the first level to even as high as the third. One principle stands out: the user will gain more in terms of available power by increasing the height of a WECS tower located in rough terrain than he will by increasing the height in smoother terrain.

When siting in areas of varying roughness, determining the winds at one height from those measured at another presents a new problem. Which upwind surface roughness is influencing the wind profile at the height of the WECS? As the figure demonstrates, the answer to this question can tell the user if he can significantly increase available power by increasing the tower height. In addressing this question, it is crucial to know which wind directions are associated with the most power. Roughness changes along the most powerful wind directions will have the greatest effect on power availability at the site.

To estimate the level at which a dramatic change in wind speed might be expected, the

*Available wind power should be used only to compare sites, not to estimate output power because no WECS can harness all available power.

Table 3-5. Extrapolation of the wind speed from 30 ft to other heights over flat terrain of uniform roughness.*

ROUGHNESS CHARACTERISTIC	20	40	60	80	100	120	140	160**	180**	200**
Smooth surface ocean, sand	0.94	1.04	1.10	1.15	1.18	1.21	1.24	1.26	1.29	1.30
Low grass or fallow ground	0.94	1.05	1.12	1.17	1.21	1.25	1.28	1.31	1.33	1.35
High grass or low row crops	0.93	1.05	1.13	1.19	1.24	1.28	1.32	1.35	1.38	1.41
Tall row crops or low woods	0.92	1.06	1.16	1.23	1.29	1.34	1.38	1.42	1.46	1.49
High woods with many trees	0.89	1.08	1.21	1.32	1.40	1.47	1.54	1.60	1.65	1.70
Suburbs, small towns	0.82	1.15	1.39	1.60	1.78	1.95	2.09	2.23	2.36	2.49

*The table was developed using power law indices obtained from C. Huang, Pacific Northwest Laboratory, Richland, Washington.
**These three columns should be used with caution because extrapolation to levels more than 100 ft above or below the base height may not be completely reliable.

Table 3-6. Power change due to extrapolation to a new height* (Base height = 30 ft).

CHARACTERISTIC ROUGHNESS	20	40	60	80	100	120	140	160**	180**	200**
Smooth surface	-17	12	33	52	64	77	91	100	115	120
Low grass	-17	16	40	60	77	95	110	125	135	146
High grass	-20	16	44	69	91	110	130	146	163	180
Tall row crops	-22	19	56	86	115	141	163	186	211	231
High woods	-30	26	77	130	174	218	265	310	349	391
Suburbs	-45	52	169	310	464	641	813	1009	1214	1444

*The user is likely to be using National Weather Service (NWS) wind data. Since most NWS wind data is measured at about 30 ft, that level was chosen as the base height for this table. The table was developed using power law indices obtained from C. Huang, Pacific Northwest Laboratory, Richland, Washington.
**These three columns should be used with caution because extrapolation to levels more than 100 ft above or below the base height may not be completely reliable.

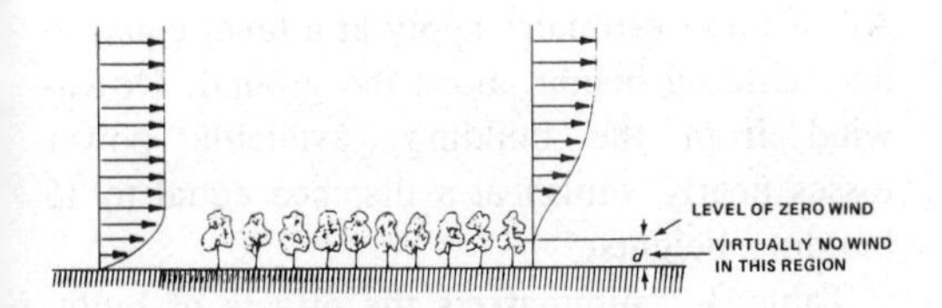

Figure 3-26. Formation of a new wind profile above ground level. (*Courtesy of DOE.*)

user must estimate the height to which upwind surface roughnesses affect the wind profile. Figure 3-28 provides this estimate, called the transition height.

The diagram in Figure 3-29 shows how data from Figure 3-28 can be used to take advantage of transition height. Since the terrain changes from an upwind "a" (water) to a downwind "b" (low grass) roughness, the upper portion of Figure 3-28 shows that Curve 1 should be used. Curve 1 in the graph indicates that the transition height at 500 ft downwind of the shoreline is about 40 ft above the ground. Since the smoother surface (water) is upwind, wind speed should increase sharply around the 40 ft level, 500-ft downwind from the roughness change. In this example, the WECS should be located above the transition height, since that location has more available power. Had the rougher surface been upwind, there would be less to gain by locating the WECS above the transition height.

The transition height curves in Figure 3-28 are simplified approximations of a very complex phenomenon. Gradual, rather than sharp, roughness changes may cause the transition to occur in a layer of 10–20 ft or more, instead of at a distinct level. Consequently, the information in this section should be used only to make estimates of the wind profile, which then can be used to select possible WECS sites and tower heights. The best way to verify the wind profile near a change in terrain roughness is to make a few wind measurements at various heights during prevailing wind conditions. The information in this chapter will help determine where to take these measurements to gain the most useful information about the wind.

Barriers in Flat Terrain. Barriers produce disturbed areas of airflow downwind, called wakes, in which wind speed is reduced and turbulence increased. Because most wind generators have relatively thin blades which ro-

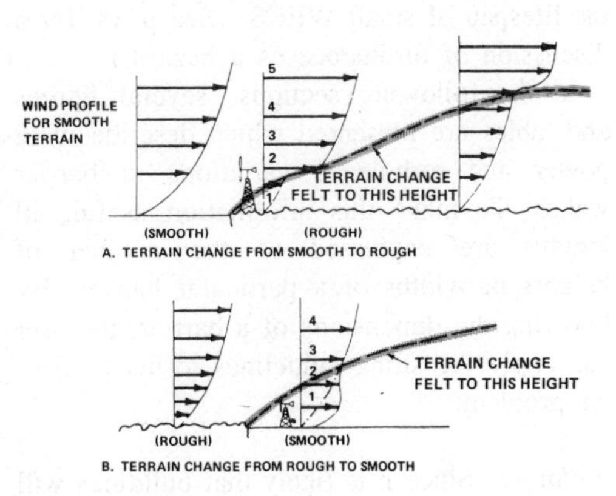

Figure 3-27. Wind speed profiles near a change in terrain. (*Courtesy of DOE.*)

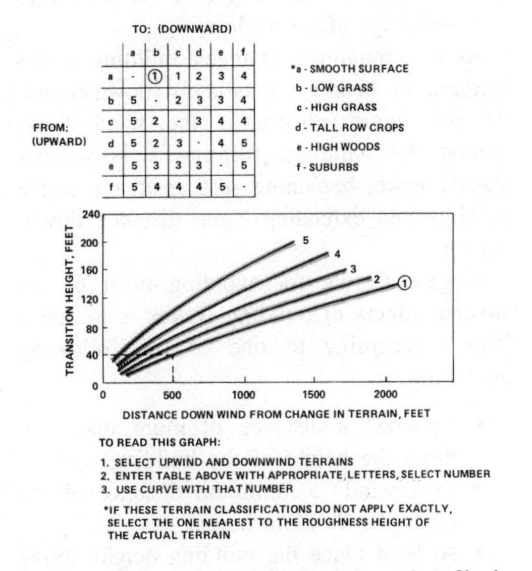

	a	b	c	d	e	f
a	·	①	1	2	3	4
b	5	·	2	3	3	4
c	5	2	·	3	4	4
d	5	2	3	·	4	5
e	5	3	3	3	·	5
f	5	4	4	4	5	·

Figure 3-28. Transition height in wind speed profile due to a change in roughness. (*Courtesy of DOE.*)

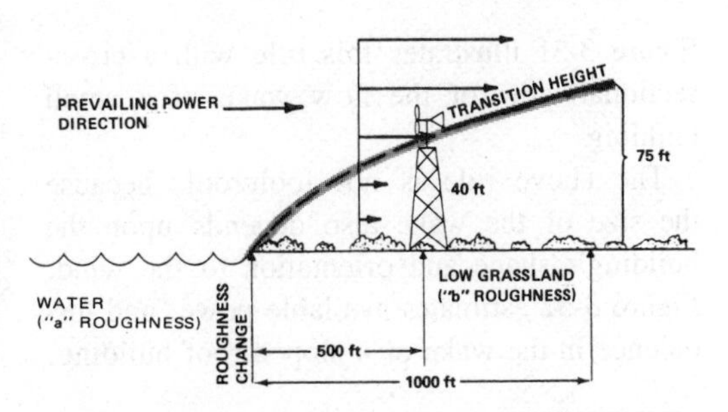

Figure 3-29. Example of a transition height diagram depicting one change in roughness. (*Courtesy of DOE.*)

tate at high speeds, barrier wakes should be avoided whenever possible, not only to maximize power, but also to minimize turbulence. Exposure to turbulence may greatly shorten the lifespan of small WECS. (See p. 00 for a discussion of turbulence as a hazard.)

In the following sections, several figures and tables are presented which describe wind power and turbulence variations in barrier wakes. To make this information useful, all lengths are expressed as the number of heights or widths of a particular barrier. By knowing the dimensions of a barrier, the user can apply the siting guidelines to his particular problem.

Buildings. Since it is likely that buildings will be located near a WECS candidate site, it is important to know how they affect airflow and available power. Figure 3-30 illustrates how buildings affect airflow.

As with roughness changes, building wakes increase in height immediately downstream. As the figure illustrates, the wind flows around the building, forming a horseshoe-shaped wake, beginning just upstream of the building and extending some distance downstream.

A general rule for avoiding most of the adverse effects of building wakes is to site a WECS according to one of the following guidelines.

- Upwind* a distance of more than two times the height of the building.
- Downwind* a minimum distance of ten times the building height.
- At least twice the building height above ground if the WECS is to be mounted on the building.

Figure 3-31 illustrates this rule with a cross-sectional view of the flow wake of a small building.

The above rule is not foolproof, because the size of the wake also depends upon the building's shape and orientation to the wind. Figure 3-32 estimates available power and turbulence in the wake of a sloped-roof building.

*Upwind and downwind indicate directions along the principal power direction.

All of these estimates apply at a level equal to one building height above the ground. Downwind from the building, available power losses nearly vanish at a distance equal to 15 building heights.

Table 3-7 summarizes the effects of building shape on wind speed, available power and turbulence, for buildings oriented perpendicular to the wind flow. Building shape is given by the ratio of "width divided by height." As might be expected, power reduction is felt farther downstream for wider buildings. At 20 times the height downwind, only very wide buildings (those in which width ÷ height = 3 or more) produce more than a 10 percent power reduction. The speed, power and turbulence changes reflected in Table 3-7 occur only when the WECS lies in the building

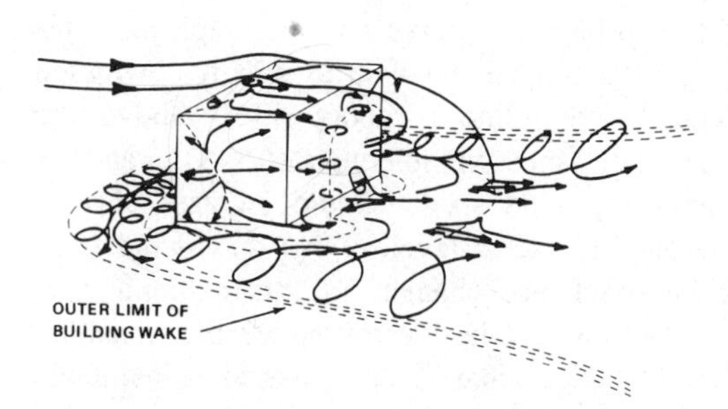

Figure 3-30. Airflow around a block building. (*Courtesy of DOE*.)

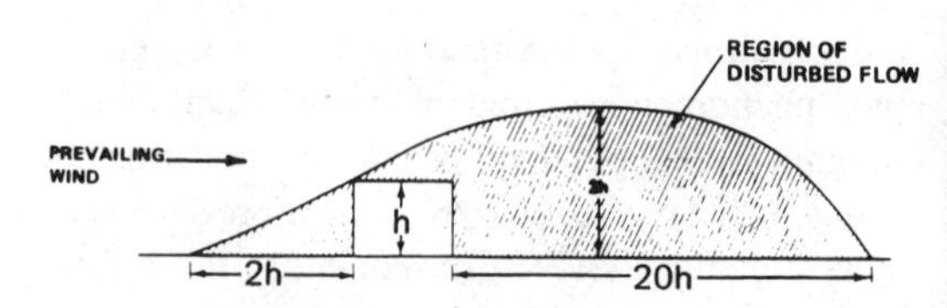

Figure 3-31. Zone of disturbed flow over a small building. (*Courtesy of DOE*.)

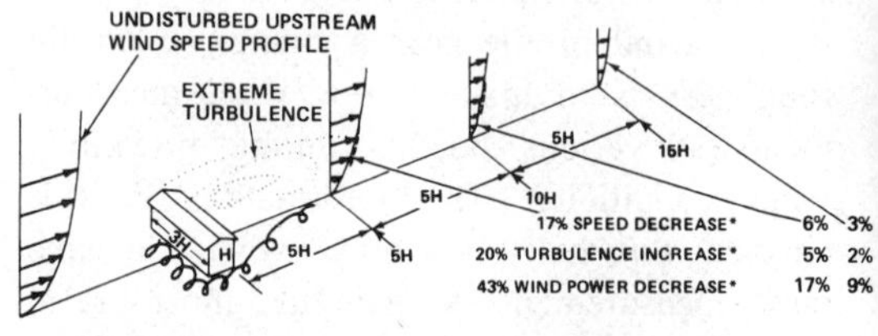

Figure 3-32. The effects of an undisturbed airflow encountering an obstruction. (*Courtesy of DOE*.)

Table 3-7. Wake behavior of variously shaped buildings.

	DOWNWIND DISTANCES (IN TERMS OF BUILDING HEIGHTS)								
	5H			10H			20H		
BUILDING SHAPE (WIDTH ÷ HEIGHT)	PERCENT SPEED DECREASE	PERCENT POWER DECREASE	PERCENT TURBULENCE INCREASE	PERCENT SPEED DECREASE	PERCENT POWER DECREASE	PERCENT TURBULENCE INCREASE	PERCENT SPEED DECREASE	PERCENT POWER DECREASE	PERCENT TURBULENCE INCREASE
4	36	74	25	14	36	7	5	14	1
3	24	56	15	11	29	5	4	12	0.5
1	11	29	4	5	14	1	2	6	
0.33	2.5	7.3	2.5	1.3	4	0.75			
0.25	2	6	2.5	1	3	0.50			
Height of the wake flow region (in building heights)		1.5			2.0			3.0	

wake. Wind rose information will indicate how often this actually occurs. Annual percentage time of occurrence multiplied by the percentage power decrease in the table will give the net power loss.

If a tower is located on the roof of a building, the turbulence near the roof should be considered. A slanted roof produces less turbulence than does a flat roof and may actually increase the wind speed over the building. The zone of speed increase may extend up to twice the building height if the building is wider than it is tall and is perpendicular to the prevailing wind. However, since wide buildings are generally not very high, the roof is only exposed to the lower wind speeds near the ground. Rather than attempting to use the power in the wind accelerated over such a building, it is generally wiser to raise the WECS as high as is economically practical, taking advantage of the fact that winds usually increase and turbulence decreases with height.

Shelterbelts. Shelterbelts are windbreaks usually consisting of a row of trees. When selecting a site near a shelterbelt, the user should:

- Choose a site far enough upwind/downwind to avoid the disturbed flow.
- Use a tower of sufficient height to avoid the disturbed flow.
- Minimize power loss and turbulence by examining the nature of the windflow near the shelterbelt and choose a site accordingly, if the disturbed flow at the shelterbelt cannot be entirely avoided.

The degree to which the wind flow is disturbed depends on the height, length and porosity of the shelterbelt. Porosity is the ratio of the open area in a windbreak to the total area (expressed here as the percentage of open area).

Figure 3-33 locates the region of greatest turbulence and wind speed reduction near a thick windbreak. How far upwind and downwind this area of disturbed flow extends varies with the height of the windbreak. Generally, the taller the windbreak, the farther the region upwind and downwind that will experience a disturbed airflow.

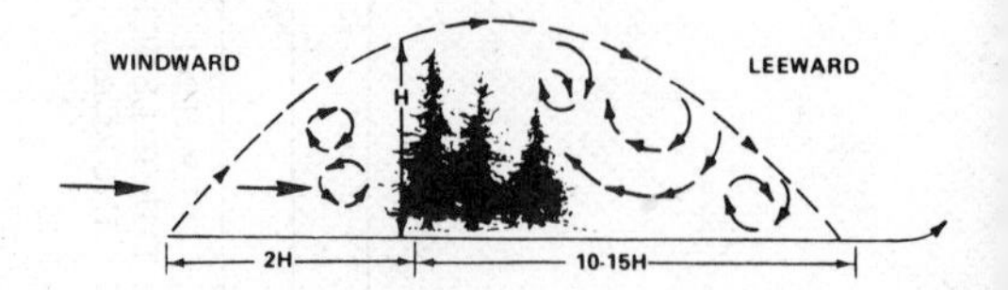

Figure 3-33. Airflow near a shelterbelt. (*Courtesy of DOE.*)

Figure 3-34 illustrates the effect of a row of trees on the wind speed at various heights and distances from the windbreak. The wind speeds are expressed as percentages of undisturbed upwind flow at several selected heights. All heights and distances are expressed in terms of the height of the shelterbelt to make application to a particular siting problem easier.

When examining this figure, the reader should note that loose foliage actually reduces winds behind the windbreak more than dense foliage. Furthermore, medium-density foliage reduces wind speeds farther downwind than either loose or dense foliage.

For levels 1½ H or less, the wind speed

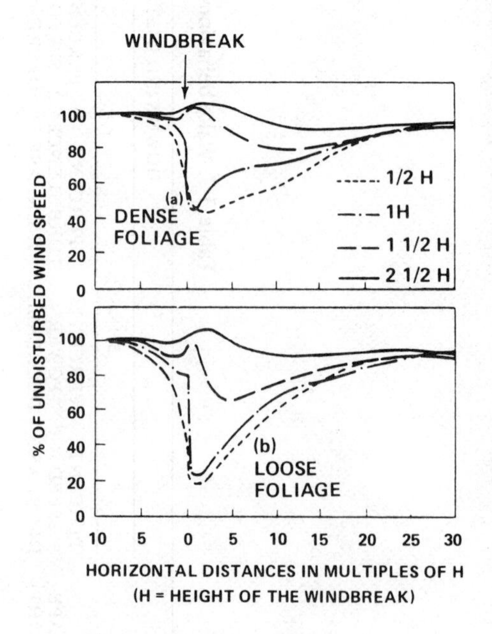

Figure 3-34. Percent wind speed at different levels above the surface behind a row of trees of height, H. (*Courtesy of DOE.*)

begins to decrease at 5 or 6 H upstream of the shelterbelt. Therefore, if the shelterbelt is 30 ft high and the WECS tower is 45 ft high, the site should be at least 150 ft (5 H) upstream of the windbreak to entirely avoid the speed decrease and turbulence on the windward side.

At a distance of 2½ H downwind, the wind speed at the 2½ H level (for both dense and loose foliage) increases approximately 5 percent. At first glance, this appears to be a good WECS site. However, there is a turbulent zone downwind from the shelterbelt that may make this site undesirable, particularly if the tower is too short.

To capitalize on the acceleration of the wind over a shelterbelt, the entire rotor disc must be located above the turbulent zone. To determine where this turbulent zone is located, the user should study turbulence patterns during prevailing wind conditions. One should also study other frequently occurring wind directions. If significant turbulence or power loss is possible when the wind blows from the most powerful directions, another site should be selected.

Table 3-8 provides information on the wind speed/available power reductions and turbulence increases for sites in the lee of the shelterbelt. Speed, power and turbulence changes are expressed as upwind percentages. The porosity of the windbreak can be estimated visually, then Table 3-8 can be used to determine how far downwind the site should be located to minimize power loss and turbulence. Speed, power and turbulence changes expressed in the table occur only when the WECS lies in the shelterbelt wake. Wind rose information will indicate how often this actually occurs. Annual percentage time of occurence multiplied by the table percentage will give the net change.

Individual Trees. The trees near a prospective WECS site may not be organized into a shelterbelt. In such cases, the effect of an individual tree or of several trees scattered over the surrounding area may be a problem.

The wake of disturbed airflow behind individual trees grows larger (but weaker) with distance, much like a building wake. However, the highly disturbed portion of a tree wake extends farther downstream than does that of a solid object. Table 3-9 may be used to estimate available power loss downstream. For example, consider a 30-ft-wide tree having fairly dense foliage. At 30 tree widths (or 900 ft) downstream, the table indicates a 9 percent loss of available power whenever the WECS is in the tree wake. The numbers in the bottom two rows of the table provide estimates of the width and height of the tree wake. The velocity and power losses expressed in the table occur only when the WECS lies in the tree wake.

If available, wind rose information can be used to estimate the percentage of time a site will be in the tree wake, and thereby the total power loss due to the tree. For instance, suppose that 50 percent of the time, the wind direction places the site in the tree wake. In the example above, the tree produced a 9 percent loss of available power. If the loss occurred 50 percent of the time, 4.5 percent (50 percent × 9 percent) of the available power would be lost annually.

Scattered Barriers. The advantages of increasing tower height are evident from this example, especially if scattered trees or buildings are in the vicinity. Since choosing a site not located in any barrier wake will probably be impossible in these areas, the WECS should be raised above the most highly disturbed airflow. To avoid most of the undesirable effects of trees and other barriers, the rotor disc should be situated on the tower at a minimum height of three times that of the tallest barrier in the vicinity. However, this is usally impractical (for economic or other reasons). There are two alternatives: find the minimum height required to clear the region of highest turbulence by using the turbulence detection techniques, or choose the site so that the WECS will clear the highest obstruction within a 500-ft radius by a least 25 ft.

Siting in Non-flat Terrain

Any terrain that does not meet the criteria listed in Figure 3-25 is considered to be non-flat or complex. To select candidate sites in

Table 3-8. Available power loss and turbulence increase downwind from shelterbelts of various porosities.

POROSITY* (OPEN AREA ÷ TOTAL AREA)	DOWNWIND DISTANCES IN TERMS OF SHELTERBELT HEIGHTS								
	5 H			10 H			20 H		
	PERCENT SPEED DECREASE	PERCENT POWER DECREASE	PERCENT TURBULENCE INCREASE	PERCENT SPEED DECREASE	PERCENT POWER DECREASE	PERCENT TURBULENCE INCREASE	PERCENT SPEED DECREASE	PERCENT POWER DECREASE	PERCENT TURBULENCE INCREASE
0% (no space between trees)	40	78	18	15	39	18	3	9	15
20% (with loose foliage such as pine or broadleaf trees)	80	99	9	40	78		12	32	
40% (with dense foliage such as Colorado Spruce)	70	97	34	55	90		20	49	
Top of turbulent zone (in terms of shelterbelt height)	2.5			3.0			3.5		

*Determine the porosity category of the shelterbelt by estimating the percentage of open area and by associating the foliage with the example tree type.

such terrain, the potential user should identify the terrain features (i.e., hills, ridges, cliffs, valleys) located in or near the siting area.

In complex terrain, landforms affect the airflow to some height above the ground in many of the same ways as surface roughness does. However, topographical features affect airflow on a much larger scale, overshadowing the effects of roughness. When weighing various siting factors by their effects on wind power, topographical features should be considered first; barriers second; and roughness third. For example, if a particular section of a ridge is selected as a good candidate site, the location of barriers and surface roughness should only be considered to pinpoint the best site on that section of the ridge.

Ridges. Ridges are defined as elongated hills rising about 500–2000 ft above surrounding terrain and having little or no flat area on the summit (see Figure 3-35). There are three advantages to locating a WECS on a ridge: 1) the ridge acts as a huge tower; 2) the undesirable effects of cooling near the ground are avoided; and 3) the ridge may accelerate the airflow over it, thereby increasing the available power.

The first two advantages are not unique to ridges, but apply to all topographical features having high relief (hills, mountains, etc.).

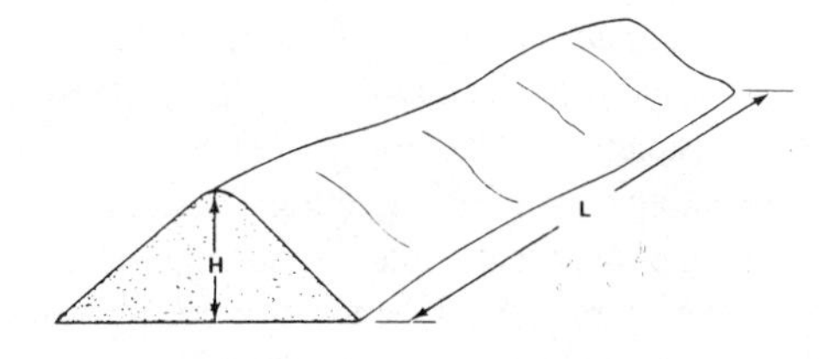

1. H = 500 TO 2000 ft
2. L = AT LEAST 10 x H[10]
3. ROUNDED OR PEAKED TOP (NOT FLAT)

Figure 3-35. Definition of a ridge. (*Courtesy of DOE.*)

Winds generally increase with height. A ridge, then, like a tower, raises a WECS into a region of higher winds. In addition, daily temperature changes affect the wind profile. At night, as the earth's surface cools, the air near the surface cools. This cool, heavy air drains from the hillsides into the valleys and may accumulate into a layer several hundred feet deep by early morning. This cool dome of air disengages from the general wind flow above it to produce the cool, calm mornings that lowlands often experience. Because of this phenomenon, a WECS located on a hill or a ridge may produce power all night, but one located at a lower elevation may not.

A similar, but more persistent, situation may occur in the winter, when cold air moves into an area. Much like flowing water, cold air tends to fill all the low spots. This may

Table 3-9. Speed and power loss in tree wakes.

		DISTANCE DOWNWIND (IN TREE WIDTHS)				
		5	10	15	20	30
Dense foliage tree (such as a Colorado spruce)	Maximum percent loss of velocity	20	9	6	4	3
	Maximum percent loss of power	49	25	17	13	9
Thin foliage tree (such as a pine)	Maximum percent loss of velocity	16	7	4	3	2
	Maximum percent loss of power	41	18	12	8	6
Height of the turbulent flow region (in tree heights)		1.5	2.0	2.5	3.0	3.5
Width of turbulent flow region (in tree widths)		1.5	2.0	2.5	3.0	3.5

cause extended periods of calm in the lowlands while the surrounding hills experience winds capable of driving a WECS.

By siting at higher elevations, such as on a ridge, the user can take advantage of more persistent winds. And, since a WECS located on a ridge produces more energy, it can reduce the amount of energy storage capacity needed (such as batteries) and provide a more dependable and economical source of power.

The third advantage is that the acceleration of the wind flowing over the ridge can greatly increase available power. Figure 3-36 shows how air approaching the ridge is squeezed into a thinner layer, which causes it to speed up as it crosses the summit.

The orientation of a ridge relative to the prevailing wind direction is an important factor in determining the amount of wind acceleration over the ridge. Figure 3-37 depicts various ridge orientations and ranks their suitability as WECS sites. However, when comparing ridges, it is important to remember that a ridge several hundred feet or more higher than another should have significantly stronger winds simply because the wind increases with height. This is true even if the higher ridge is slightly less perpendicular to the prevailing wind than the lower ridge.

Part A of Figure 3-37 shows the ideal orientation of a ridge to the prevailing wind. The maximum acceleration at the ridge summit occurs when the prevailing wind blows perpendicular to the ridge line. The acceleration lessens if the ridge line is not perpendicular, as in Part B of the figure. When the ridge line is parallel to the prevailing wind, as in Part C, there is little acceleration over the ridge top; however, the ridge may still be a fair to

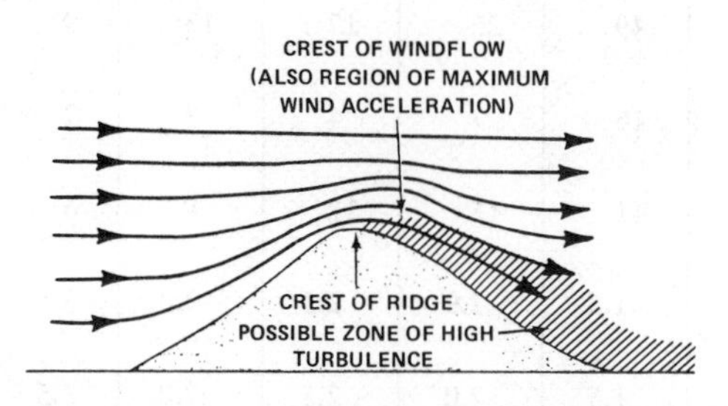

Figure 3-36. Acceleration of wind over a ridge. (*Courtesy of DOE*.)

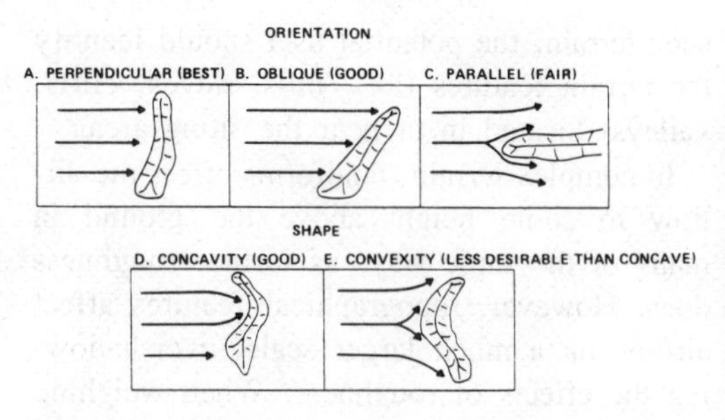

Figure 3-37. The effects of ridge orientation and shape upon WECS site suitability. (*Courtesy of DOE*.)

good wind site because it acts as an isolated hill or peak.

The orientation of concave or convex ridges (or such portions of a ridge) can further modify the wind flow. Part D of Figure 3-37 shows how concavity on the windward side may enhance acceleration over the ridge by funneling the wind. On the other hand, convexity on the windward side (Part E) reduces acceleration by deflecting the wind flow around the ridge.

Figure 3-38 shows the cross-sectional shapes of several ridges and ranks them by the amount of acceleration they produce. Notice that a triangular-shaped ridge causes the greatest acceleration, and that the rounded ridge is a close second. The data used in ranking these shapes were collected in laboratory experiments using wind tunnels to simulate real ridges. Although few wind experiments have been conducted over actual ridges, the results are similar to tunnel simulations. Both indicate that certain slopes, primarily in the nearest few hundred yards to the summit,* increase the wind more effectively than do others. Table 3-10 classified smooth, regular ridge slopes according to their values as wind power sites.

Figure 3-39 gives percentage variations in wind speed for an ideally-shaped ridge. Since these numbers are taken from wind tunnel experiments, they should not be taken too literally; nevertheless, the user should expect similar wind speed patterns along the path of flow. Generally, wind speed decreases significantly at the foot of the ridge, then acceler-

*This portion of the ridge has the greatest influence on the wind profile immediately above the summit.

Table 3-10. WECS site suitability based upon slope of the ridge.

WECS SITE SUITABILITY	SLOPE OF THE HILL NEAR THE SUMMIT: PERCENT GRADE*	SLOPE OF THE HILL NEAR THE SUMMIT: SLOPE ANGLE
Ideal	29	16°
Very good	17	10°
Good	10	6°
Fair	5	3°
Avoid	Less than 5 Greater than 50	Less than 3° Greater than 27°

*Percent grade as used above is the number of feet of rise per 100 feet of horizontal distance.

ates to a maximum at the ridge crest. It only exceeds the upwind speed on the upper half of the ridge.

Another consideration in choosing a site on a ridge is the turbulent zone which often forms in the lee of ridges (refer to Figure 3-36). The steeper the ridge slope and the stronger the wind flow, the more likely it is that turbulence will form in the lee of the ridge. Thus, it is safest to site at the summit of the ridge, both to maximize power and to avoid lee turbulence.

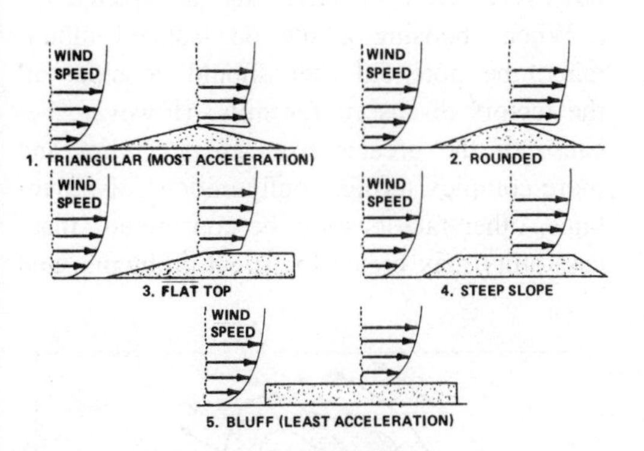

Figure 3-38. Ranking of ridge shape by amount of wind acceleration. (*Courtesy of DOE.*)

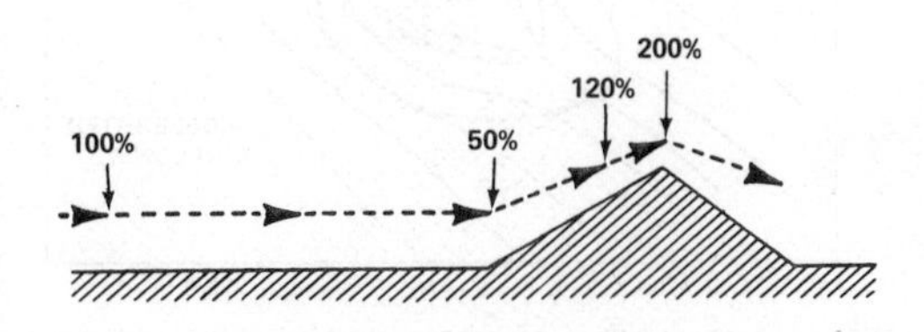

Figure 3-39. Percentage variation in wind speed over an idealized ridge. (*Courtesy of DOE.*)

Shoulders (ends) of ridges are often good WECS sites. Even for a very long ridge, as much as one-third of the air approaching at low levels may flow around, rather than over, the ridge. To move such a volume of air around the ridge, the wind must accelerate as it flows around the ends. No quantitative estimates of this acceleration are available at this time, but it appears that, from the standpoint of available wind power, the ends of ridges may rank second behind the ridge crest as the best potential WECS sites.

Flat-topped ridges present special problems because they can actually create hazardous wind shear at low levels, as Figure 3-40 illustrates. Consequently, the slope classifications used in Table 3-10 do not apply to these ridges. The hatched area at the top of the flat ridge indicates a region of reduced wind speed due to the "separation" of the flow from the surface. Immediately above the separation zone is a zone of high wind shear. This shear zone is located just at the top of the shaded area in the figure. Siting a WECS in this region will cause unequal loads on the blade as it rotates through areas of different wind speeds and could decrease performance and the life of the blade. The wind shear problem can be avoided by increasing tower height to allow the blade to clear the shear zone or by moving the WECS toward the windward slope.

As in the case of flat terrain, the effects of barriers and roughness should not be overlooked. Figure 3-41 shows how a rough surface upwind of a ridge can greatly decrease

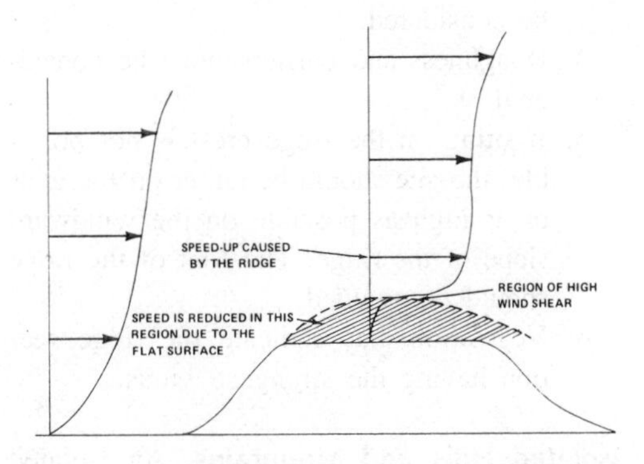

Figure 3-40. Hazardous wind shear over a flat-topped ridge. (*Courtesy of DOE.*)

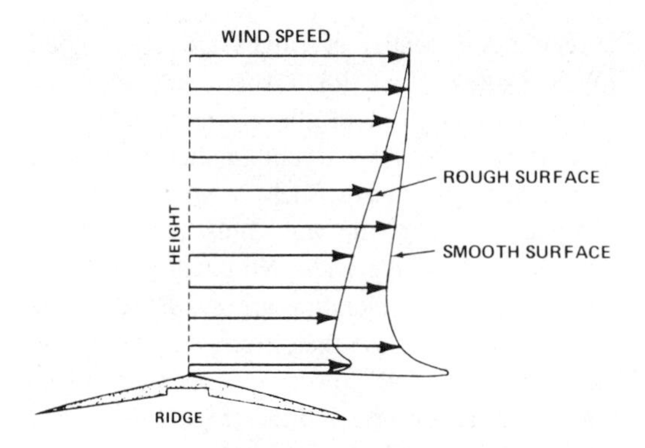

Figure 3-41. Effect of surface roughness on wind flow over a low sharp-crested ridge. (*Courtesy of DOE.*)

the wind speed. After selecting the best section of a ridge, based upon its geometry, the potential user should consider the barriers, then the upwind surface roughness.

The most important considerations in siting WECS on or near ridges are summarized below.

1. The best ridges or sections of a single ridge are the most nearly perpendicular to the prevailing wind. (However, a ridge several hundred feet higher than another and only slightly less perpendicular to the wind is preferable.)
2. Ridges or sections of a single ridge having the most ideal slopes within several hundred yards of the crest should be selected (use Table 3-10). Ridge sites meriting special consideration are those with features such as gaps, passes or saddles.
3. Sites where turbulence or excessive wind shear cannot be avoided should not be considered.
4. Roughness and barriers must be considered.
5. If siting on the ridge crest is not possible, the site should be either on the ends or as high as possible on the windward slope of the ridge. The foot of the ridge should be avoided.
6. Vegetation may indicate the ridge section having the strongest winds.

Isolated Hills and Mountains. An isolated hill is 500–2000 ft high, is detached from any ridges and has a length of less than 10 times its height. Hills greater than 2000 ft high will be referred to as mountains.

Hills, like ridges, may accelerate the wind flowing over them (but not as much as ridges will), since air tends to flow around the hill (Figure 3-42). Not enough information is currently available to make quantitative estimates of wind accelerations either over or around isolated hills. However, Table 3-10 can be used to rank hills according to their slope.

Two benefits are gained by siting on hills: 1) airflow can be accelerated, and 2) the hill acts as a huge tower, raising the WECS into a stronger airflow aloft and above part of the nocturnal cooling and resulting calm periods.

The best WECS sites on an isolated hill may be along the sides of the hill tangent to the prevailing wind (shown as hatched areas in Figure 3-42). However, further research is required to verify this supposition. Currently, simultaneous wind recordings are the surest method of comparing hillside and hilltop sites.

Table 3-11 ranks the suitability of WECS sites on hills. (The effects of surface roughness and barriers should also be weighed, however, before a WECS site is selected.)

When choosing a site on isolated mountains, the potential user should consider all the factors discussed for hills. However, because of the greater size, greater relief and more complex terrain configurations of mountains, other factors must be considered. Inaccessibility may create logistical problems, and

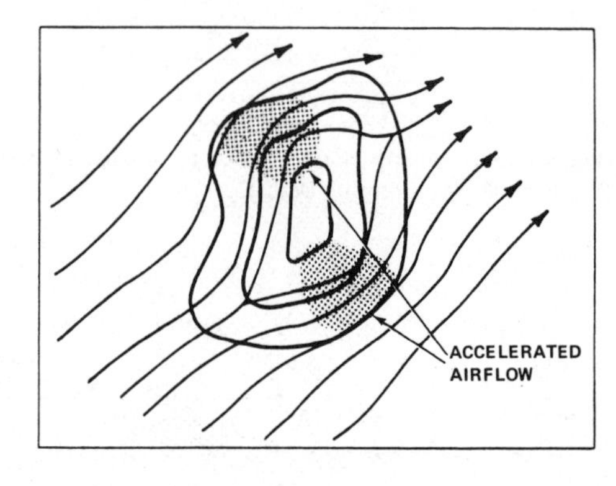

Figure 3-42. Airflow around an isolated hill (top view). (*Courtesy of DOE.*)

Table 3-11. WECS site suitability on isolated hills.

SUITABILITY	LOCATION	FLOW CHARACTERISTICS
Better	Upper half of hills where prevailing wind is tangent.	The point of maximum acceleration around the hill.
Good	Top of hills.	The point of maximum acceleration over the hill.
Fair	Upper half of the windward face of the hill.	A slight acceleration of flow up the hill.
Avoid	Entire leeward half of hills.*	Reduced wind speeds and high turbulence.
	The foot and lower portions of hills.	Reduced wind speeds.

*Under certain conditions, the strongest winds may occur on the leeward slopes of larger hills and mountains (such as on the east slopes of the Rocky Mountains). However, these winds are usually gusty and localized, and generally represent more of a hazard than a wind resource.

thunderstorms, hail, snow and icing hazards will occur more frequently than at lower elevations.

In spite of the drawbacks, an isolated mountain may still be the most promising WECS site in an area. To select the best site(s) in the favorable areas of the mountain, use the criteria for hills in Table 3-11. For mountains, these favorable areas may be very large, containing many different terrain features, barriers and surface roughnesses. To pinpoint the best site(s), consider the largest terrain features first; then evaluate the barriers and surface roughness.

Passes and Saddles. Passes and saddles are low spots or notches in mountain barriers. Such sites offer three advantages to WECS operations. First, since they are often the lowest spots in a mountain chain, they are more accessible than other mountain locations. Second, because they are flanked by much higher terrain, the air is funneled as it is forced through the passes. Third, depending upon the steepness of the slope near the summit, wind may accelerate over the crest as it does over a ridge.

Factors affecting airflow through passes are orientation to the prevailing wind; width and length of the pass; elevation differences between the pass and adjacent mountains; the slope of the pass near the crest; and the surface roughness. At this time, there has not been sufficient research to allow classification of WECS site suitability in terms of these factors. However, some desirable characteristics of passes are listed below.

1. The pass should be open to the prevailing wind (preferably parallel to the prevailing wind).
2. The pass should have high hills or mountains on both sides (the higher, the better).
3. The slope (grade) of the pass near the summit should be sufficient to further accelerate the wind (see Table 3-10 for slope suitability).
4. The surface should be smooth (the smoother, the better). (If the pass is very narrow, the user should consider the roughness of the sides of the pass.)

Figure 3-43 shows two views of the wind profiles in a pass. The top half of the figure is a view through the pass. A core of maximum wind (denoted by the innermost circle) is located in the center of the pass, well above the surface. In the bottom part of the figure is a view looking across the pass. A strong increase in wind from the ground up to the wind maximum is clearly shown. The WECS should be sited near the center of the pass at a

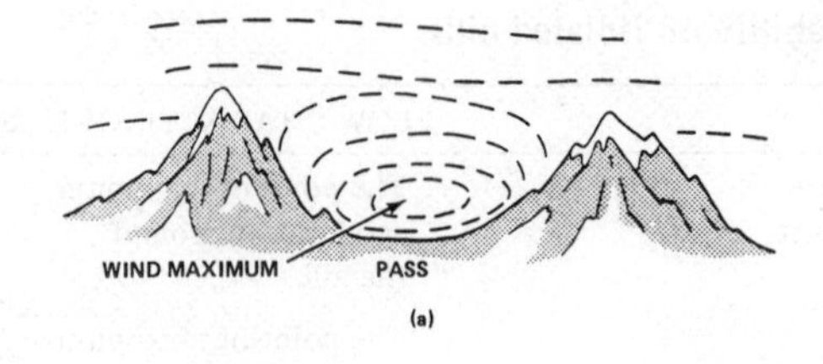

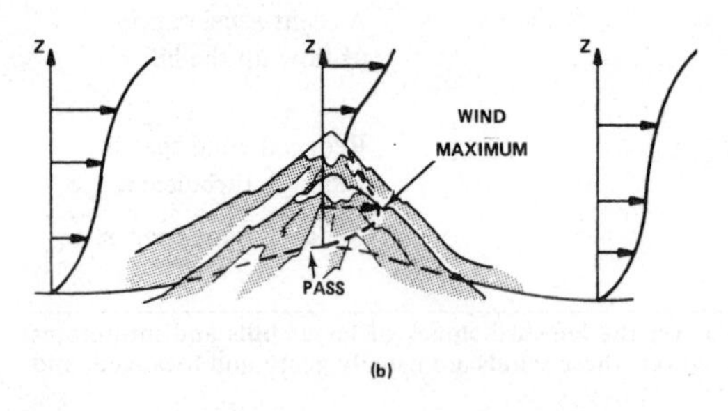

Figure 3-43. A schematic of the wind pattern and velocity profile through a mountain pass. (*Courtesy of DOE.*)

level as near the core of maximum winds as possible. Below this level there may be very strong vertical wind shear and much turbulence. Since the location of the core will vary from pass to pass, wind measurements are recommended before a final decision on WECS placement is made.

Passes to avoid are those not open to the prevailing wind (because there will be much less flow through them) and passes, or portions of passes, which are extremely narrow and canyon-like (because these may have turbulence and strong horizontal wind shear).

Gaps and Gorges. In some areas, rivers and streams have eroded deep gaps or gorges through mountain chains and ridges. The Columbia River Gorge in Oregon and Washington is an example. Since these gaps are frequently the only low-level paths through mountain barriers, much air is forced through them (Figure 3-44).

The problem of siting WECS in gaps and gorges is much like that of siting in passes and saddles. However, there are a few important differences. On the positive side, gaps and gorges are generally deeper than passes and can significantly enhance even relatively light winds. A river gorge can augment mountain-valley or land-sea breezes, providing a reliable source of power. Gaps and gorges are also usually more accessible than mountain passes. The chief drawback to sites in gaps and gorges is that, because they are narrow, there is often much turbulence and wind shear. In addition, since streams usually flow through them, there may be no land near the center on which to locate a WECS.

Valleys and Canyons. The airflow pattern in a particular valley or canyon depends on such factors as the orientation of the valley to the prevailing wind; the slope of the valley floor, the height, length and width of the surrounding ridges; irregularities in the width; and the surface roughness of the valley.

Valleys and canyons which do not slope downward from mountains are usually not good sites. Perhaps the only benefit to siting in non-sloping valleys is the possible funneling effect when the large-scale prevailing wind blows parallel to the valley. Funneling occurs only if the valley or canyon is constricted at some point. Unless the valley is constricted, the surrounding ridges will provide better WECS sites than will the valley floor.

Three types of flow patterns occur in val-

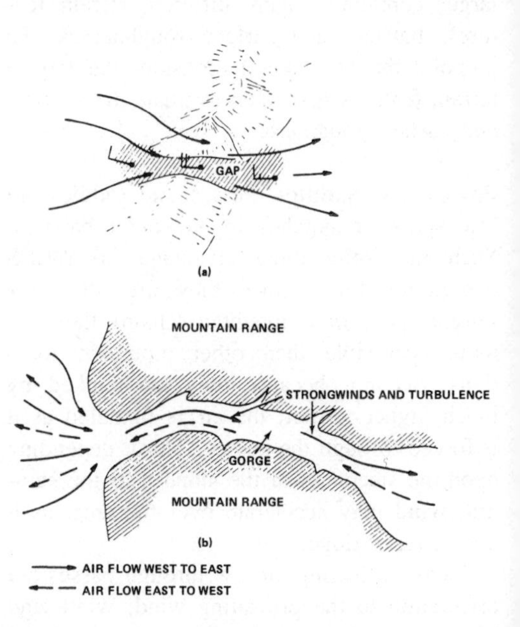

Figure 3-44. A schematic illustration of flow patterns that may be observed through gaps and gorges. (*Courtesy of DOE.*)

ley-mountain systems. The first, known as valley (mountain)-slope winds, occurs when the large-scale wind over the area is weak, and the daily heating and cooling cycle dominates. This happens most often during the warmer months (May to September).

The daily sequence of valley (mountain)-slope winds is shown in Figure 3-45. Shortly after sunrise, when the valley is cold and the plains are warm, upslope winds (white arrows) and the continuation of the mountain winds (black arrows) combine (Part A). At forenoon, when the plains and the valley floor are the same temperature, the slope winds are strong and there is a transition from mountain to valley winds (Part B). At noon and during early afternoon, the slope winds diminish. The valley wind is fully developed and the valley is warmer than the plains (Part C). In late afternoon, the slope winds cease and the valley winds continue. The valley is still warmer than the plains (Part D). Shortly after sunset, when the valley is only slightly warmer than the plains, downslope winds begin and the valley winds weaken (Part E). In early night, downslope winds are well developed. The valleys and plains are at the same temperature. This overall condition is characteristic of the transition period between valley and mountain winds (Part F). In the middle of the night, the valley is colder than the plains. Hence, the downslope winds continue and the mountain wind is fully developed (Part G). From late night to morning, when the valley is colder than the plains, downslope winds cease and the mountain wind fills the valley (Part H).

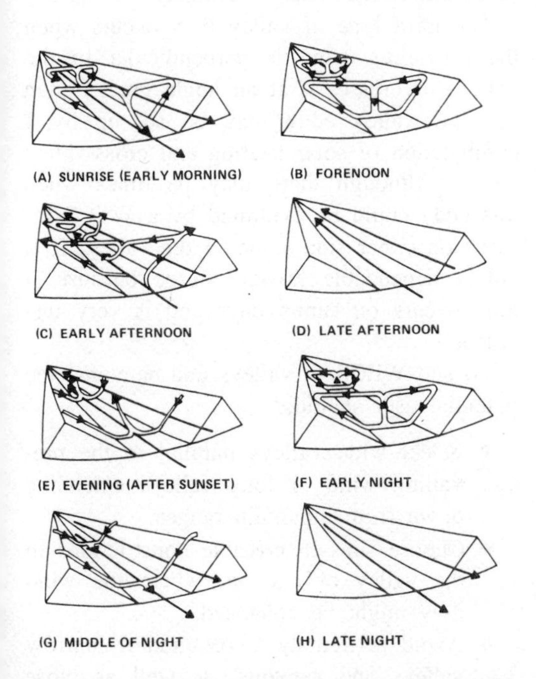

Figure 3-45. The daily sequence of mountain and valley winds. (*Reprinted by permission of the American Meteorological Society.*)

The winds of greatest interest for small WECS users are the mountain wind at night (Parts A, G and H of Figure 3-45) and the valley wind during the afternoon (Parts C, D and E). Figure 3-46 illustrates a wind profile observed for mountain winds in Vermont. The wind accelerates down the valley, with the strongest mountain winds occurring at the mouth (lower end) of the valley, and the lightest winds at the head (upper end). In the vertical direction, the wind speed increases upward from the valley floor and has reached a maximum in the center of the valley at about two-thirds the height of the surrounding ridges. At the point of maximum wind, the speed may reach as high as 25 mph. The mountain wind is generally well developed for valleys between high ridges and/or rather steeply sloping valley floors. The upper half of the wind profile is very smooth, while the lower half occasionally becomes gusty and turbulent.

The daytime wind blowing up the valley tends to be more sensitive to factors such as

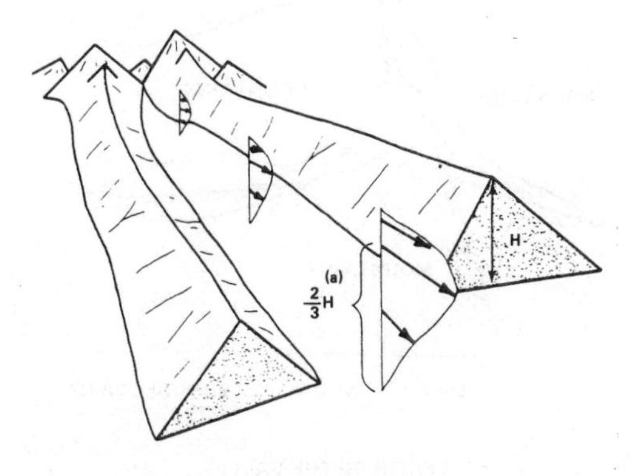

Figure 3-46. Vertical profile of the mountain wind. (*Courtesy of DOE.*)

heating of the sun (the driving force for this wind) and the winds blowing high overhead. As a result, the valley winds are more variable, and often weaker, than the mountain winds. Unlike the mountain wind, which is strongest near the center of the valley, valley winds are normally greatest along the side slope most directly facing the sun. Figure 3-47 shows how to take advantage of mountain and valley winds.

The second type of flow pattern in mountain-valley systems occurs when moderate to strong prevailing winds are parallel to (or within about 35° of) the valley. In this case, broad valleys surrounded by mountains can effectively channel and accelerate the large-scale wind.

Figure 3-48 shows possible wind sites, where valley channeling enhances the wind flow. In (a), a funnel-shaped valley on the windward side of a mountain range is shown. The constriction (or narrowing) near the mouth produces a zone of accelerated flow. In this example, the valley is large (approximately 60 miles wide) and open to the prevailing wind. In (b), a narrow valley in the lee of a mountain range is illustrated. It is parallel to the prevailing wind and constricted slightly near its mouth.

A valley which is parallel to the prevailing

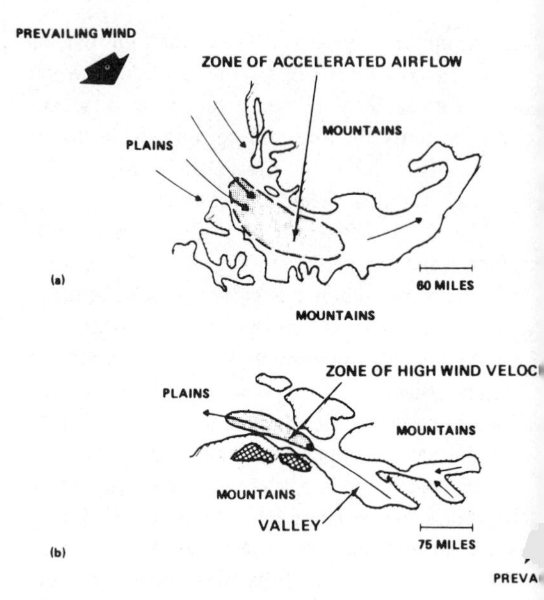

Figure 3-48. Possible WECS sites where prevailing winds are channeled by valleys. (*Courtesy of DOE*.)

wind and experiences mountain-valley winds will provide sites which are dependable sources of power. Moderate to strong prevailing winds in winter and spring will drive the WECS. During the warmer months, mountain-valley winds can be utilized.

The third type of valley flow occurs when the prevailing wind is perpendicular to the valley (or crosses it at an angle greater than 35°). A valley eddy may be set up by a combination of solar heating and cross-valley winds. Although there may be times when this eddy could be exploited by a WECS located on either side slope of the valley, it is not a dependable power source because it only occurs on sunny days and is very turbulent.

To site WECS in valleys and canyons, the potential user should:

- Select wide valleys parallel to the prevailing wind or long valleys extending down from mountain ranges.
- Choose sites in possible constrictions in the valley or canyon, where the wind flow might be enhanced.
- Avoid extremely short and/or narrow valleys and canyons, as well as those perpendicular to the prevailing winds.
- Choose sites near the mouth of the val-

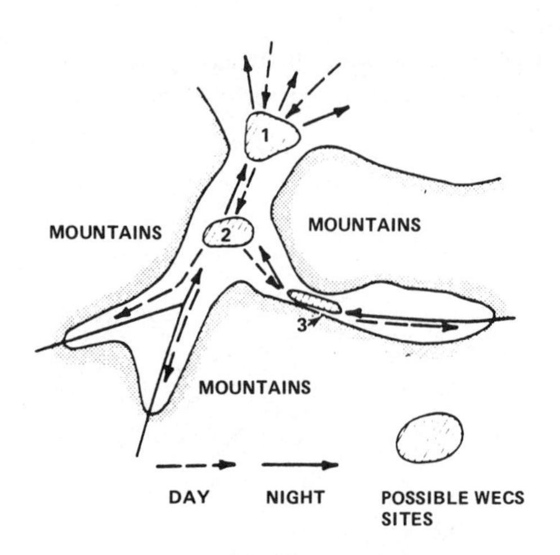

Figure 3-47. Possible WECS sites in sloping valleys and canyons. (*Courtesy of DOE*.)

ley, where mountain-valley winds occur.

- Ensure that the tower is high enough to place the WECS as near to the level of maximum wind as is practical.
- Use vegetation to indicate high wind areas.
- Consider nearby topographical features, barriers and surface roughness (after favorable areas in the valley or canyon are located).

Basins. Basins are depressions surrounded by higher terrain. Large, shallow inland basins (such as the Columbia Basin in southeast Washington) may have daily wind cycles during the warmer months of the year which can be used to drive small WECS. The flow into and out of a basin is similar to the mountain-valley cycle in Figure 3-45. In fact, valleys sloping down into basins may provide sufficient channeling to warrant consideration as WECS sites.

The flow of cool air from surrounding mountains and hills into the basin during the night is usually stronger than the flow out of the basin caused by daytime heating. Well developed night-time flow into a basin may average 10–20 mph for several hours during the night, and occasionally more than 25 mph for periods of one or two hours. Afternoon flow out of the basin is generally lighter, averaging 5–15 mph.

Winter and spring storms, combined with the summer wind cycles, may provide sufficient wind power in basins for most of the year. However, in the fall and portions of the winter, basins frequently fill with cold air. During these periods, the air in the basin may be stagnant for days or even weeks. Consequently, WECS in basins may require larger energy storage systems or possibly backup power for the calm period.

The following guidelines are helpful when siting WECS in basins.

1. Consider only large, shallow inland basins.
2. Use vegetation indicators of wind to locate areas of enhanced winds in basins.
3. Consider all topographical features, barriers and surface roughness effects.

Cliffs. A cliff, as discussed in this report, is a topographical feature of sufficient length (10 or more times the height) to force the airflow over rather than around its face. For such long cliffs, the factors affecting the airflow are the slope (both on the windward and lee sides), the height of the cliff, the curvature along the face and the surface roughness upwind.

Figure 3-49 shows how the air flows over cliffs or different slopes. The swirls in the flow near the base and downwind from the cliff edge are turbulent regions, which must be avoided. Turbulent swirls (which we will call areas of flow separation) become larger as the face of the cliff leans more into the wind. When the cliff slopes downward on the lee side, as in (c), the zone of turbulence moves more downwind from the face. Part of the turbulence can be avoided by siting a WECS very close to the face of such hill-shaped cliffs. Selecting a section of the cliff having a more gradual slope (a) is sometimes advantageous because the tower height required to clear the turbulent zone is reduced.

Any curvature along the face of a cliff should also be considered. Figure 3-50 illustrates a top view of a curved cliff section. The curvature of the face channels the winds into the concave portions. Although no estimates are available of how much wind speed is enhanced in these concave areas, they are

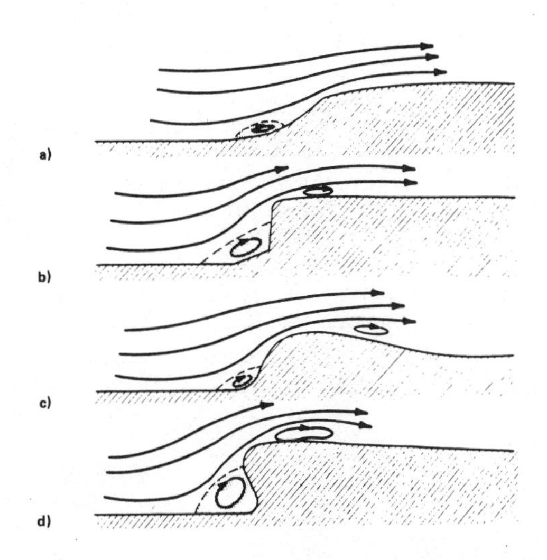

Figure 3-49. Airflow over cliffs having differently sloped faces. (*Courtesy of DOE.*)

probably better WECS sites than convex areas, since, more air may be forced through them.

Laboratory and field experiments both indicate that cliffs do enhance the wind speed (much as ridges do). Figure 3-51 shows the vertical wind profile of air flowing over a cliff. The longer arrows in Profile 3, compared to those in Profile 1, illustrate how wind speed is enhanced. The dotted regions show turbulent areas of flow separation. Wind speed rapidly increases near the top of the flow separation. This region of shear should be avoided, either by choosing a new site or by raising the WECS so that the rotor disc is above the shear zone.

Since this turbulent zone continually changes size and shape, it is wise to choose as high a tower as is practical (this will also increase available power). To estimate the size of the zone, follow the procedures for turbulence detection (p. 36). Measurements should be made on several different days when the prevailing wind is blowing. In general, sunny days will produce larger turbulent zones. If the turbulence extends too high, consider sites very near the cliff edge.

Other factors to consider when siting on cliffs are the surface roughness upstream and the prevailing wind direction. For maximum

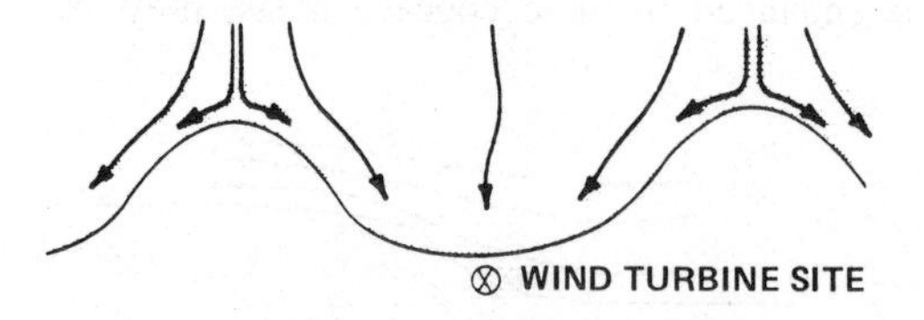

Figure 3-50. Top view of airflow over concave and convex portions of a cliff face. (*Courtesy of DOE*.)

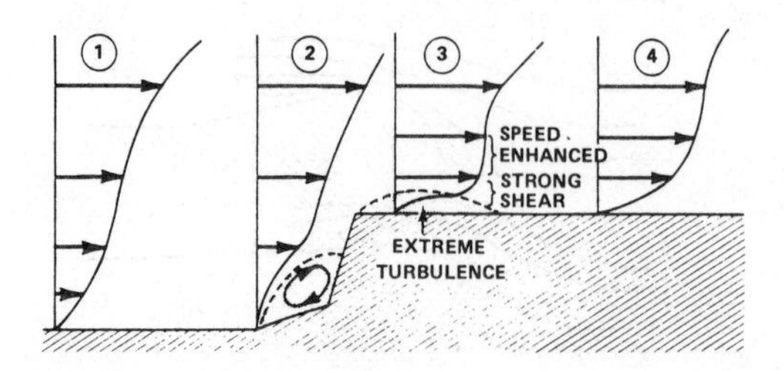

Figure 3-51. Vertical profiles of air flowing over a cliff. (*Courtesy of DOE*.)

enhancement of the wind speed, the prevailing wind direction should be perpendicular to the cliff section on which the WECS will be located.

Studies of airflow over cliffs made in wind tunnels and with theoretical models show that the location of the zone of strongest winds depends on the height of the cliff. Provided the user can site above the separation zone, the best location on a cliff appears to lie between 0.25 and 2.5 times the cliff height downwind.

Since the size of the separation zone varies greatly, because of a complex combination of local influences, the best strategy is to make short-term wind measurements to locate the least turbulent site. Some later data indicate that even this "conservative" measure may not be sufficient to avoid the wind shear and turbulence for high cliffs when strong winds occur frequently.

Cliffs may be too turbulent for siting small WECS, especially high cliffs or those with strong winds. We do not recommend siting on a cliff unless the user is sure that, even during strong wind storms, his site will not experience rapid changes of wind speed. It appears that the conservative measure may be to site as far downwind of the cliff as possible—and even then it is wise to make some wind measurements.

In summary, when choosing a site on a cliff, the following major points should be considered:

1. The best cliffs (or portions of a single cliff) are well exposed to the wind (i.e., they are not sheltered by tall trees.
2. The best cliffs (or portions of a single cliff) are oriented perpendicular to the prevailing winds.
3. If the face of the cliff is curved, a concave portion is the best location (Figure 3-50).
4. The shape and slope of the cliff (or section of a cliff) which cause the least turbulence should be selected (Figure 3-49).
5. General wind patterns near cliffs may be revealed by the deformation of trees and vegetation.

6. The entire rotor disc should clear the zone of separation.

Mesas and Buttes. Mesas and buttes should be evaluated in the same way as cliffs. In many cases, they are too turbulent for siting of SWECS. In the United States, they are found almost exclusively in the western half of the country, primarily in the Southwest. Although they are generally high enough to intercept the stronger winds aloft, they are often found in regions of relatively light winds and frequently are inaccessible due to their steep sides.

Smaller buttes (those less than 2000 ft in height, and less than about five times as long as they are high) can be considered flat-topped hills. Consequently, they may have considerable turbulence and wind shear at low levels (Figure 3-40). The best WECS sites on such buttes appear to be along the windward edge.

Figure 3-52 shows some flow patterns over and around mesas and buttes. In (a), the wind accelerates over the top, although not as much

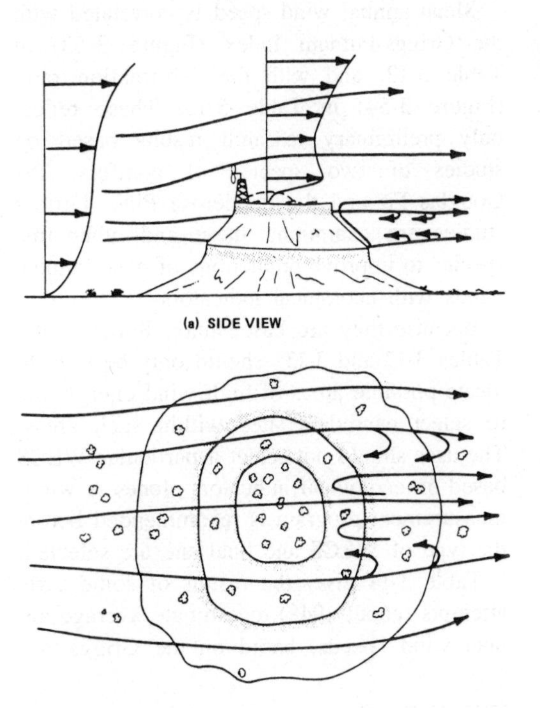

Figure 3-52. Flow around and over buttes and mesas. (*Courtesy of DOE.*)

as over triangular or rounded hills. When a mesa or butte is located in an area where the winds are already enhanced by valley funneling or other effects, additional power benefits may be gained.

If the mesa or butte is more than 10 times longer than it is high, there should be enough flow over it (rather than around it) to be treated as a cliff. Very large mesas (those more than 2000 ft high and more than 6 or 7 miles long) may also produce mountain-type effects.

If there is no prominent prevailing wind direction, a very tall tower will provide some protection against turbulence and wind shear while the WECS is in the lee.

Ecological Indicators of Site Suitability

Vegetation deformed by average winds can be used both to estimate the average speed (thus power) and to compare candidate sites. This technique works best in three regions: 1) along coasts; 2) in river valleys and gorges exhibiting strong channeling of the wind; and 3) in mountainous terrain. Ecological indicators are especially useful in remote mountainous terrain, not only because there are little wind data, but because the winds are often highly variable over small areas and difficult to characterize. The most easily observed deformities of trees (illustrated in Figure 3-53) are listed and defined below.

- *Brushing*—Branches and twigs bend downwind like the hair of a pelt which has been brushed in one direction only. This deformity can be observed in deciduous trees after their leaves have fallen. It is the most sensitive indicator of light winds.
- *Flagging*—Branches stream downwind, and the upwind branches are short or have been stripped away.
- *Throwing*—A tree is windthrown when the main trunk and the branches lean away from the prevailing wind.
- *Clipping*—Because strong winds prevent the leader branches from extending up to their normal height, the treetops are held to an abnormally low level.

- *Carpeting*—This deformity occurs because the winds are so strong that every twig reaching more than several inches above the ground is killed, allowing the carpet to extend far downwind.

Figure 3-53 is one of the best guides to rank tree deformities by wind speed. Both a top view and a side view of the tree are shown to demonstrate the brushing of individual twigs and branches and the shape of the tree turnk and crown. The figure uses the Griggs-Putnam classification of tree deformities described by indices from 0 to VII. When WECS sites are ranked by this scheme, only like species of trees should be compared, because different types of trees may not be deformed to the same degree.

Another good indicator of relative wind speeds is the deformation ratio. It also measures how much the tree crown has been flagged. Figure 3-54 shows the two angles, *a* and *b,* that must be measured to compute the deformation ratio *D*. To measure these angles, the trees can either be photographed or sketched to scale. (The user might sketch the tree on clear acetate while he looks at it through the acetate.) He should draw or take the tree pictures while viewing the tree perpendicularly to the prevailing wind direction, so he can see the full effects of flagging.

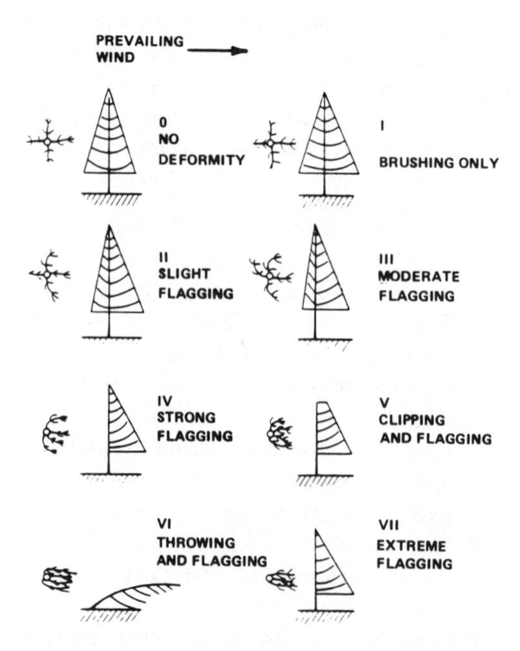

Figure 3-53. Wind speed rating scale based on the shape of the crown and degree twigs, branches and trunk are bent (Griggs-Putnam Index). (*Courtesy of DOE.*)

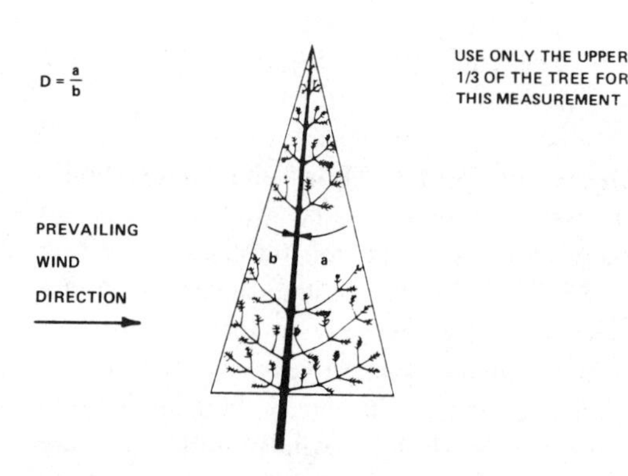

Figure 3-54. Deformation ratio computed as a measure of the degree of flagging. (*Courtesy of DOE.*)

To compute *D,* the two angles shown in the figure (*a* on the downwind side and *b* on the upwind side) should be measured in degrees, using a protractor, and then divided ($D = a \div b$). The larger the value of *D,* the stronger the average wind speed.

Mean annual wind speed is correlated with the Griggs-Putnam Index (Figure 3-53) in Table 3-12, and with the deformation ratio (Figure 3-54) in Table 3-13. These reflect only preliminary research results based on studies of two species of conifers, the Douglas Fir and the Ponderosa Pine. Further studies are examining these and other tree species to improve predictions of mean annual winds with ecological indicators.

Because they are based upon limited data, Tables 3-12 and 3-13* should only be used to locate possible areas of high wind energy and to select candidate sites within such areas. The user should not select a particular WECS based on ecological indicators alone. A wind-measurement program is recommended before the type of WECS and final site are selected.

Table 3-14 gives the results of some early attempts (about 1948) to estimate average annual wind speeds, based on the Griggs-Put-

*The data in these tables were prepared by E. W. Hewson, J. E. Wade and R. W. Baker, of Oregon State University.

Table 3-12. Mean annual wind speed versus the Griggs-Putnam Index.

GRIGGS-PUTNAM INDEX (as in Figure 3-42)	I	II	III	IV	V
Probable mean annual wind speed range (mph)	6-10	8-12	11-15	12-19	13-22

Table 3-13. Mean annual wind speed versus the deformation ratio.

DEFORMATION RATIO (as in Figure 3-43)	I	II	III	IV	V	VI	IV
Probable mean annual wind speed range (mph)	4-8	7-10	10-13	12-15	14-18	15-21	16-24

nam Index, for certain evergreens in the Northeast. Although different species are often deformed to different extents by the same winds, very strong winds (those capable of causing strong flagging, clipping or carpeting) affect different species the same way. Comparison of the data presented in Table 3-12 (based on Douglas Fir and Ponderosa Pine) and in Table 3-14 (based primarily on Balsam) bears this out.

Although the presence of one type of deformity (or a combination of deformities) may indicate an area of high average winds, and the degree of deformity may give estimates of the relative strengths of the winds, there are still pitfalls to rating sites according to tree deformity. Because past or present growing conditions can greatly affect the size and shape of trees, only isolated trees appearing to have grown under similar conditions should be compared. For example, a tree in or near a dense stand of timber should not be compared to an isolated tree. Another fact to be aware of is that limbs are stripped from trees not only by strong flagging. They can be damaged by man, disease, other trees that once grew nearby, or, possibly, ice storms. Misinterpreting such signs could lead to the wrong assumptions about the prevailing wind direction and the average speed. Common sense, however, should reveal whether or not all the deformities observed in an area fit together into a consistent pattern.

The following guidelines summarize this section and suggest how to use ecological indicators effectively.

1. Detect ecological indicators of strong wind.
2. Compare isolated trees within the strong wind areas to select candidate sites.
3. Consider flow patterns over barriers, terrain features and surface roughness in the final selection.
4. Measure the wind to ensure that the best site in complex terrain is selected.
5. Base selection of a particular WECS and estimation of its power output on wind measurements, not on ecological indicators alone.

Site Analysis Methodology

If only the annual, average power output is important, the site evaluation process for WECS applications is completed once the feasibility is established and the best site is chosen. If WECS applications have occurred in the immediate vicinity, little analysis is required, since experience will be the best guide. However, if more precise economic or performance information is required, additional analysis of wind data should be undertaken.

Table 3-15 presents three general approaches to site analysis and the respective advantages and disadvantages of each. These methods range from expending a few dollars

Table 3-14. Griggs-Putnam Index versus average annual wind speed for conifers in the Northeastern U.S.*

TYPES OF DEFORMATION (GRIGGS-PUTNAM INDEX)	DESCRIPTION	TREE HEIGHT (FT)	VELOCITY AT TREE HEIGHT (MPH)
Carpeting (VIII)	Balsam, spruce and fir held to 1 ft.	1	27.0
Clipping (V)	Balsam, spruce and fir held to 4 ft.	4	21.5
Throwing (VI)	Balsam thrown.	25	19.2
Flagging (IV)	Balsam strongly flagged.	30	18.6
Flagging (III)	Balsam moderately flagged.	30	17.9
Flagging (II)	Balsam minimally flagged.	40	17.3
Brushing (I)	Balsam not flagged	40	15.5
Flagging (II)	Hemlock and white pine show minimal flagging.	40	10.6

*Adapted from *Power from the Wind,* by Palmer Cosslett Putnam, ©1975 by Allis Chalmers Corporation.

and a few hours analyzing existing data, to collecting and analyzing on-site data for an entire year. Each approach has different levels of analysis that can be performed, depending upon the user's needs for information, his budget and the format of the wind data.

Use of Available Data. Method 1 uses only wind data collected at a representative weather station, which is a station that can be expected to have wind characteristics similar to the WECS site because of similar exposure to prevailing winds. Determining whether a nearby weather station is representative is not simple; even in areas such as the Great Plains, wind conditions can vary significantly over short distances. The relationship of the site and the weather station to local terrain is very important when using data from a nearby weather station. For example, a shallow river valley will usually have lower average wind speeds than will the surrounding higher elevations. Lower winds are particularly prevalent in depressions during the night and early morning, because cold, heavy air drains into the depressions and isolates them from the regional winds. Therefore, a weather station located in such an area could have lower average wind speeds than would a site located at a higher elevation.

As a very general rule, sites within 10–20 miles of one another in large regions of relatively flat terrain should have similar wind characteristics, provided they have similar exposures to the prevailing wind directions. In very flat areas, this distance may be extended to 30 miles or more. In rugged, hilly or mountainous terrain, however, the winds from a nearby station are usually not applicable for a site analysis.

The amount of information that can be gleaned from available wind data depends upon the form in which the data are summarized. Summaries that give wind speed versus direction can be used to estimate annual power output and to identify potential wind barriers. If monthly average wind speeds or averages by time of day are listed, other valuable statistics can easily be computed. For example, if monthly average wind speeds are given, each monthly average can be used to obtain an estimate of the average monthly power output of a WECS. If power needs to be available during certain seasons, such as for crop irrigation, then the summarized monthly average wind speeds can be used to estimate how well the WECS power output will match the seasonal demand.

For some WECS applications, the hourly variations in WECS power output must match the hourly load, such as when a WECS is used to reduce the amount of electrical energy purchased from a utility. In this situation, WECS economics may be greatly affected by utility rate structures. A utility might charge WECS-owning customers time-of-day rates; that is, the cost of electricity will be higher during the utility's peak demand time(s) than during other times. Likewise, the price of excess power produced by the WECS and sold back to the utility could vary with time of day. Under these conditions, the economic viability of WECS might depend upon how much of the WECS power is produced during the "high cost" hours of the day rather than on annual average power output. If hour-of-day (diurnal) wind speed averages are summarized, as in Table 3-16, this type of analysis can be performed.

If wind data are only available in an unsummarized form, the needed diurnal or monthly wind summaries can usually be produced, but more time will be required to organize the data into the proper format. The user should weigh the time and money needed to properly summarize such data against his need for answering monthly or hourly load-matching questions.

Limited On-site Data Collection. The second method of site analysis might be considered whenever nearby weather stations may not adequately represent the WECS site. Weather stations may not be representative for the following reasons:

1. If they have slightly different exposures to the prevailing winds than the WECS site; or
2. If they have the same exposure but may be too far away to adequately represent the WECS site.

In this approach, the wind characteristics of the weather station are corrected by the ratio method. First, the site is instrumented and data collected over a specified time interval. Three months is the suggested minimum. Then, the wind speed at the site is averaged and that average is divided by the average wind speed for the same period at a nearby weather station. The result of dividing the site's average wind speed by that of the weather station is the correction factor, which is then applied to the weather station's long-term wind speed averages to make those averages more representative of the WECS site.

Since the siting method requires only an average wind speed for the period of data

Table 3-15. Various approaches to site analysis.

METHOD	APPROACH	ADVANTAGES	DISADVANTAGES
1	Use wind data from a nearby station; determine power output characteristics.	Little time or expense required for collecting and analyzing data. If used properly, can be acceptably accurate.	Only works well in large areas of flat terrain where average annual wind speeds are 10 mph or greater.
2	Make limited on-site measurements; establish rough correlations with nearby station; compute power output using adjusted wind data.	If there is a high correlation between the site and the station, this method should be more accurate than the first method.	Of questionable accuracy, particularly where there is seasonal modulation variation in correlation between the WECS site and the nearby station.
3	Collect wind data for the site and analyze it to obtain power output characteristics.	Most reliable method. Works in all types of terrain.	Requires at least a year of data collection. Added costs of wind recorders. Data period should represent typical wind conditions.

Table 3-16. Example of local climatological data summarized by hour of day.

	AVERAGES*							RESULTANT WIND	
HOUR LOCAL TIME*	SKY COVER (IN TENTHS)	STATION PRESSURE (IN.)	DRY BULB (°F)	WET BULB (°F)	RELATIVE HUMIDITY (%)	DEW POINT (°F)	WIND SPEED (MPH)	DIRECTION	SPEED (MPH)
01	8	29.59	42	40	84	38	9.1	17	8.1
04	8	29.59	41	40	88	38	9.1	18	7.2
07	8	29.61	42	40	89	38	9.0	18	7.5
10	8	29.62	46	43	80	39	11.0	21	7.4
13	7	29.61	50	45	68	39	11.3	22	7.9
16	8	29.59	51	45	66	38	10.4	24	3.5
19	7	29.59	48	43	71	38	9.4	24	3.5
22	7	29.60	45	42	79	38	8.8	18	4.6

*Averages are given for every third hour of the day.

collection, a wind-run anemometer is adequate. However, the anemometer should be sited using the same guidelines as those used in siting an actual WECS. Ideally, the anemometer would be placed at the same location and at the same height as the planned WECS.

Although the ratio method has been used widely in short-term wind data collection programs, research indicates that the accuracy of this method is very erratic. Furthermore, predicting how well the method will work for any given site is not currently possible, because ratio-type correction factors at many locations have been observed to vary from year to year and from season to season. Consequently, the ratio obtained may represent only the season of the year in which measurements were taken, or it may not represent the true long-term ratio for any season. These possibilities can be reduced slightly be either measuring for the three months of the year that are believed to have the most wind energy, or by collecting data for the months of peak power demand, if seasonal load matching is important.

Extended On-site Data Collection. The third method of site analysis involves extended on-site wind measurements, usually for a full year or more. Although this method is more reliable, it is also more expensive and time consuming. However, costs may vary depending on the type of instrument required, and on the costs of installation, maintenance and data analysis.

When planning an on-site measurement program, a WECS dealer, a manufacturer or a meteorologist should be consulted. These individuals can help determine the actual cost of an extensive wind-measurement program, the type of data analysis that can be performed and the information that can be obtained from the study. However, this type of analysis is not economically feasible for most small-WECS users.

A suggested procedure for establishing a wind-measurement program includes:

- A listing of the information needed to evaluate WECS economics and performance.
- An estimate of the time and money available for data analysis.
- The actual siting of the instrument.

Once these items have been considered, the wind instrument that meets all data needs at an affordable cost should be selected.

In determining the information that is needed, the user might want to consider the importance of having wind energy available in certain seasons or at certain times of the day. If it is considered important, he should select a wind instrument that will permit averaging the power output of a WECS by season and/or by the hour of the day. Furthermore, if the user is considering an energy storage system, such as batteries, he may need to estimate the maximum expected return time (MERT); i.e., the maximum time the wind might remain below the cut-in speed of the WECS. Since no power would be produced during this time, either the storage system must be sufficient to meet energy needs for this period, or there must be some form of backup power.

Estimating the time and money available for data analysis may result in trade-offs. For example, a "smart" data logger may actually perform the analysis automatically as the data is collected, but the cost would be more than for a simpler instrument. On the other hand, a simple wind-run anemometer can provide useful information providing the user is willing to read it frequently and regularly, such as every six hours, and to perform a great deal of arithmetic.

If the MERT for the cut-in speed of the WECS is needed, some wind stations have statistics available that might help. However, return times can be estimated if the user decides to collect on-site data. If a wind-run anemometer is used, it must be read frequently during periods of low wind in order to define the time that the WECS would not have been generating power. Some sophisticated data loggers can be programmed to measure return times automatically.

An instrument should be sited carefully and placed at the same height as the WECS. The instrument should be durable enough to withstand the environmental conditions to which it will be exposed.

Although Method 3 is the most accurate approach to site analysis, some uncertainty exists as to how well the year (or more) of collected data represents the true long-term winds at the WECS site. The entire year of data collected, or one of the seasons during the year, may have been abnormally windy or calm. Consulting a meteorologist who is familiar with the area, or a long-term resident, may give some qualitative insight into whether the site analysis will show more or less power output from the WECS than might be expected in an average year.

To date, statistical comparisons of a site with a nearby station have not proven sufficiently reliable to correct the winds collected on-site (for a year or more) before doing a detailed economic analysis. Therefore, the year of data collected should be used for the economic analysis. If the economic value of WECS appears marginal after the site data have been analyzed, the user may want to weigh his final decision against representativeness of the year of data.

Site Analysis Considerations. Except in situations where it is the only obvious solution to a power generating problem, the decision to purchase a WECS must depend upon some level of economic and performance analysis. The purchaser must be convinced that the cost of the power generated by the WECS will be cheaper over the life of the machine than the power generated by other alternatives, or that any greater cost would be outweighed by other considerations, such as the desirability of achieving energy independence. In some situations, the cost of WECS power may have to be considerably cheaper than the alternatives, because the purchaser may prefer to apply a premium for the convenience and historical reliability of central grid power. Obviously, the behavior of the wind at the machine site has an important bearing on the ultimate cost of the power generated. The accuracy to which these wind characteristics must be known, and the resulting accuracy of the economic and performance analysis, will depend upon the application of the machine and the size of the investment in wind systems.

4 Fundamental Operation of Wind Energy Conversion Systems

WIND POWER—HOW IT WORKS

Basically, wind turbines extract power from the wind when their rotors are pushed around by moving air. There are two primary ways in which the wind can accomplish this. One way is illustrated in Figure 4-1—a diagram of a parachute that tugs on a rope that in turn lifts a bucket of water from a well; i.e., a wind machine. Important here, though, is the parachute tugging. It is caused by drag, which is the same force you experience while holding your hand in the breeze outside your car while driving along the highway. Wind is actually pushing the parachute along.

The drag effect was used by early windmill builders to great advantage. A diagram of a simple panemone, like the machines they built, looks something like Figure 4-2. Notice that one vane is broadside to the wind. On this side of the machine, wind force (drag) will be strong. On the other side of the center shaft, the vane swings around edgewise to the wind and the drag is much less. Thus the machine turns, pivoting about the center shaft. Most drag machines work in this way, although not all of them feature the pivot-mounted vanes.

The other way in which wind can exert its force on a wind machine is by the aerodynamic action called lift. Lift is a force produced on airplane wings in flight (Figure 4-3). Notice that airflow around the airfoil-shaped blade tends to change direction slightly. A low-pressure area (like suction) forms over the curved side (topside) of the airfoil, and a high-pressure area (pushing upward) forms on the bottom. The result is a force upward and perpendicular to the wind direction; this makes the lift arrow point slightly forward in

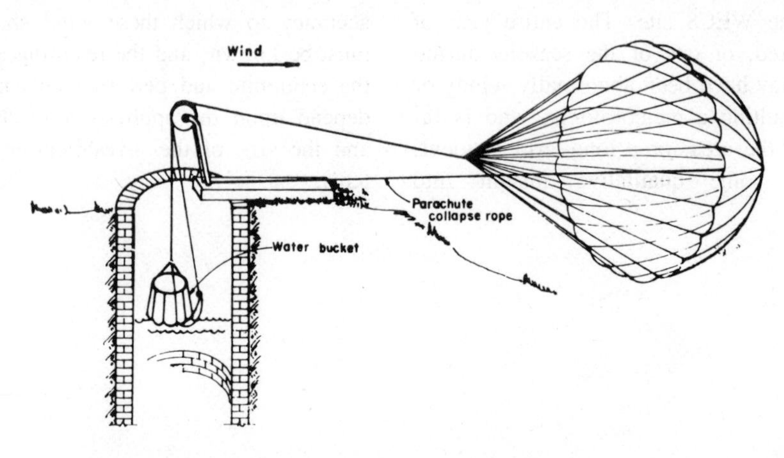

Figure 4-1. Simple wind-powered water pump. (*Courtesy of the United Nations.*)

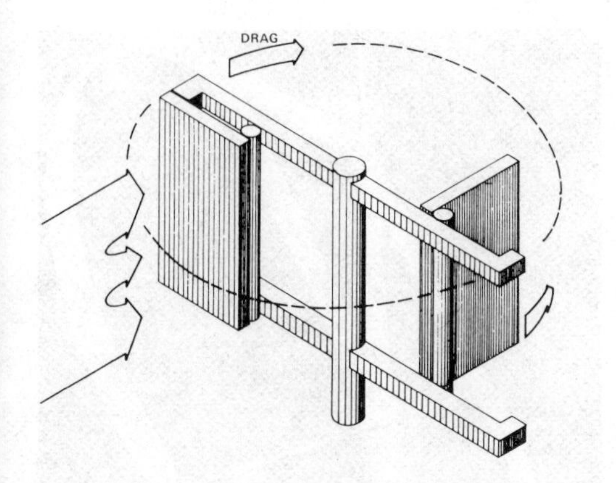

Figure 4-2. Simple "Panemone" type wind turbine. (*Courtesy of DOE.*)

Figure 4-3. Drag is also produced because the wind is being slowed slightly in the process of creating lift.

You can perform simple experiments to verify lift. Take a sheet of stiff cardboard or a wood slat about the size of a 3- by 5-inch filing card. During a trip in an automobile, at about 30 mph hold the sheet or slat out in the wind.

You are now ready to experience lift and drag. Gripping one end only, hold the board with its long dimension pointing outward from the car. Hold it edgewise to the wind. Let us call the edge at the front end the leading edge. Point the leading edge slightly upward; you should experience a slight upward lift. Point it slightly downward; you should now experience a slight downward force (also called lift by engineers). Somewhere between upward and downward lift is an angle that produces no lift at all. See if you can find this angle of zero lift. It may not be parallel to the ground because your car bends the airflow around the windshield and fenders and over the hood.

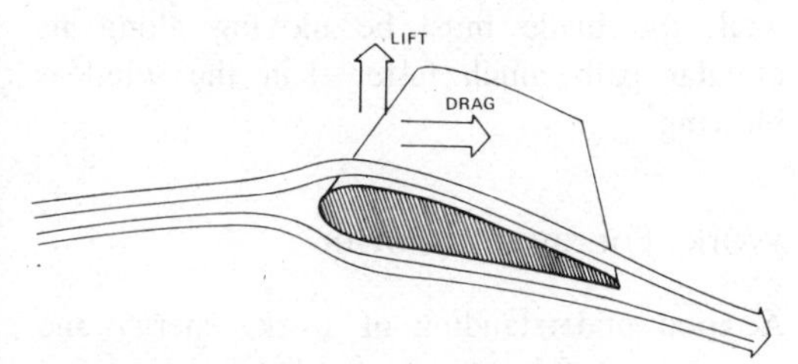

Figure 4-3. Forces acting on a wind turbine blade. (*Courtesy of DOE.*)

At the angle of zero lift, notice that a slight amount of drag is produced. Drag will tug the board aft. Now, tilt the board about 90°, leading edge up. Notice that the drag has greatly increased. You might drop the board at this point if you are moving too fast.

Now, to discover lift and drag working together, return the leading edge to the zero-lift position. Slowly rotate the board leading edge upward. Notice lift increasing. Notice drag also increasing. Lift will increase a little faster than drag, then suddenly drop substantially while drag continues to rise. This occurs with the leading edge somewhere near 20° above the zero-lift angle.

During this experiment, you will realize that at some particular angle, lift is much greater than drag. Lift is the force used to power wind machines designed for high efficiency, and this region of highest lift with low drag is very important to WECS designers.

How does a wind machine use lift as the power-producing force? Notice in Figure 4-4 that the blade performs its slight bending action on the windstream, with low-pressure and high-pressure sides similar to those shown in Figure 4-3. Lift is produced, as illustrated, generally in a direction that pushes the blade along its path.

Drag is also produced, as you might expect, and this force tries to bend the blades and slow them down in their travel around the center power shaft. In addition, the drag force

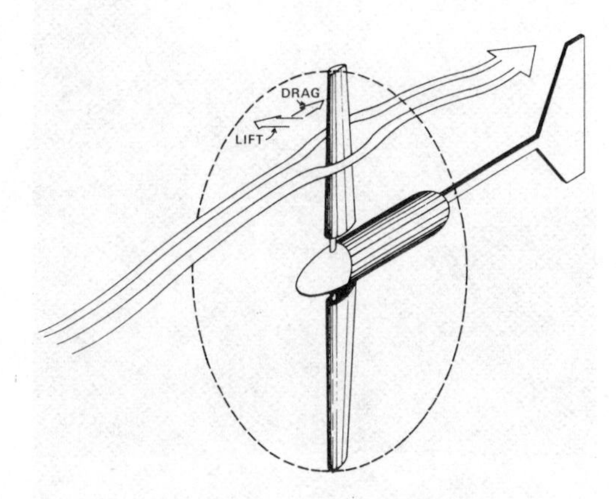

Figure 4-4. From lift to power—horizontal-axis wind turbine. (*Courtesy of DOE.*)

tries to topple the tower at low drag for this type of WECS. Figure 4-5 illustrates a typical wind generator designed for high-lift, low-drag operation.

Figure 4-6 shows a Darrieus wind turbine, sometimes called an "egg-beater" type of wind machine. Unlike the propeller type in Figure 4-5, with its power shaft pointing horizontally into the wind, the power shaft of the Darrieus is pointed across the wind. It could be horizontal, as long as it pointed across the wind, but designers have found it to be more practical for the shaft to be vertical.

Chances are you found it easy to see how the propeller blade of Figure 4-4 was pulled along, but the functioning of the Darrieus blade of Figure 4-6 is less obvious. If you visualize how a sailboat has wind first on one side of the sails, then on the other, as it travels around, you can begin to understand how the Darrieus works.

In the propeller case (Figure 4-4), the lift force always pushes the blade along at about the same force, pivoting it about the shaft. With the Darrieus (Figure 4-6), the lift force almost always tugs the blade along its path but never with a constant force. At two areas along the path, lift is very weak. You can see that this occurs when the blade is pointed directly into the wind and directly downwind. At all other points along the blade path, lift tends to be much stronger and generally pulls the blade along its path. For this to work well, the blade must be moving along its circular path much faster than the wind is blowing.

Figure 4-5. Dunlite high-lift, low-drag WECS.

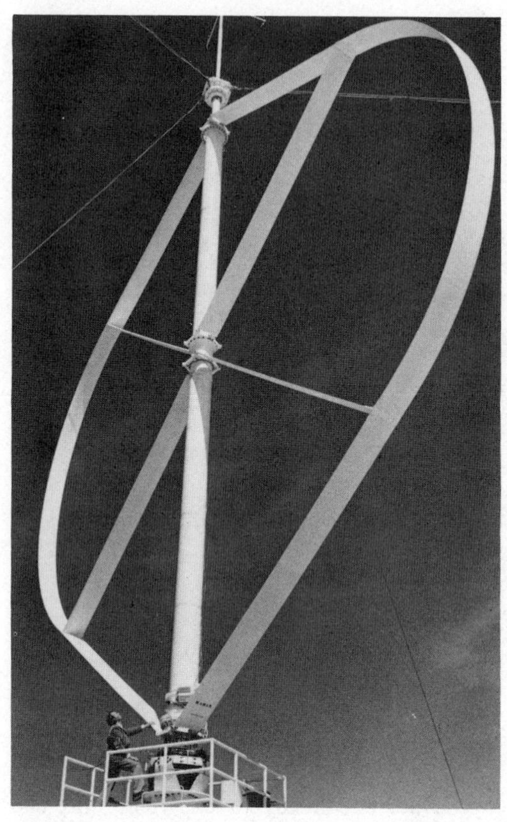

Figure 4-6. DOE Darrieus vertical-axis WECS. *(Courtesy of DOE/Sandia.)*

Work, Energy and Power

A good understanding of work, energy and power is useful to understand the basic wind power terms. For instance, does a 12-V, 100-A-hour battery store power or energy?

"Power" and "energy" are often used interchangeably; however, they have different and important meanings.

A length and force are needed to describe amounts of work. Work and velocity are needed to describe power, and power multiplied by time equals energy.

Work is performed when a force is used to lift, push or pull some object through a distance. The amount of work done is determined by multiplying the force applied by the distance traveled (assuming the direction of the force is the same as the direction of travel). For instance, raising a 550 lb rock (250) kg 1 ft (0.305 m) requires 550 ft-lb of work.

Mechanical power is the rate at which that work is performed. That is, the force applied to an object times the velocity of the object (in the direction of the force), gives the power applied to it. For instance, if a windmill raises the 550 lb of water at the rate of 1 ft/second, it is doing 550 ft-lb of work per second, which is 1 hp (1 horsepower).

Electric power, as distinguished from the mechanical power just discussed, is measured in watts (W), kilowatts (kW= 1000 W) and (by a power company) in megawatts (MW= 1000 kW). One hp equals 746 W of electric power.

If we operate a 1-hp motor for 10 hours at full capacity, 10 hp-hours of energy will have been consumed, assuming that the motor is 100 percent efficient. Similarly, since 1 hp equals 746 W, then 746 × 10 = 7460 W-hours, or 7.46 kWh (kilowatt-hours) of electric energy will have been used, again assuming 100 percent efficiency.

More realistically, let us assume the electric energy is consumed at 50 percent efficiency. This means that half the power going into the motor is wasted. Then, to get 1 hp-hour out (7.46 kWh), we need to put twice this amount in (2 hp-hour, or 14.92 kWh).

Power is usually measured in hp (mechanical) and W (electrical), while energy is usually measured in hp-hour (mechanical) and kWh (electrical). Electric energy consumption at constant power is simply power (kW) multiplied by the length of time involved (hours).

For our purposes, we can conveniently categorize energy and power as electrical, mechanical or thermal (heat). Mechanical power is converted to electricity by a generator. A motor, for instance, converts electricity back to mechanical power. Because inefficiencies in both devices produce some heating, they usually need ventilation to remove that heat. An electric heater, of course, converts the electricity (more accurately, electric power) that passes through it to heat. Heat energy is converted to mechanical energy by an engine, such as an internal combustion engine.

Mechanical power is most often generated and transferred through a rotating shaft (such as an auto drive shaft or a motor shaft). Mechanical power is described as a force times a velocity. For rotating machinery, mechanical power is calculated from shaft torque times rpm.*

Mechanical energy can be either potential or kinetic energy. A weight held in your hand above the ground has potential energy. It can potentially do some damage if you drop it, because the potential energy due to its height above the ground is continually transformed to kinetic energy as the weight speeds up. This kinetic energy is the result of speed and weight. In fact, kinetic energy increases with the square of the speed (speed times speed) while the weight is falling. That means that if the speed doubles while the weight is falling, the weight then has four times the kinetic energy.

Energy from the Wind

The kinetic energy in the wind (energy contained in the speeding air) is proportional to the square of its velocity (just as for a falling weight). Kinetic energy in the wind is partially transformed to pressure against an object when that object is approached and air slows down. This pressure, added up over the entire object, is the total force on that object.

Power, we noted earlier, is force times velocity. This also applies to wind power. Since

*A useful expression relating torque and rpm to hp is as follows: hp = 0.190 × torque × rpm ÷ 1000, where the torque is in ft=lb.

wind forces are proportional to the square of the velocity, wind power is proportional to wind speed cubed (multiplied by itself three times). If the wind speed doubles, wind power goes up by a factor of eight. This is an extremely important concept in wind power generation.

Wind turbine blades take energy from the air rather than put energy in like a propeller, so when wind speed doubles, the power that can be extracted is eight times as great. When the wind speed triples, the power that can be extracted is 27 times greater. This tremendous effect of cubing the velocity can place great importance on the process of determining the best wind turbine location and emphasize the selection of the correct machine for your wind speeds.

The power that wind turbine blades can extract from the wind is given by the following expression:

$$\text{Power} = \frac{1}{2} e \times k \times A \times \rho \times V^3$$

where:

- e = *efficiency of the blades*
- k = conversion factor for units (e.g., if units on the right side are ft, lb and seconds, and results are desired in kW
- A = area swept out by the blades ($\pi \times$ radius^2 for conventional wind turbines)
- V = wind velocity, far enough upstream so as not to be affected by the wind turbine
- ρ = Greek letter rho; equals the density of air.

The efficiency of the blades in converting the kinetic energy in the wind to rotational power in the shaft needs some careful consideration. If all the kinetic energy in the air approaching a wind turbine were extracted by the spinning blades, the air would stop, like a car losing all its kinetic energy when it crashes into a wall. However, the air cannot stop, otherwise all the rest of the air behind it would have to spill around the rotor. Nature does not work that way. The air senses any solid object that it is approaching and moves around it, like the airflow around your automobile. When air approaches a partially solid object, such as the disc created by a spinning rotor, some of the air moves around it. The rest slows down as power is extracted by the windwheel. Figure 4-7 illustrates how air, starting far upstream of a conventional windmill rotor, travels past the rotor. This airstream starts out being somewhat smaller than the windwheel but gradually expands to the windmill rotor size as it passes through. At this point, some of the power is taken from the wind. The power extracted by the windwheel, divided by the power in the undisturbed wind passing through a hoop the same size as the rotor, is called the rotor power coefficient, or, more commonly, the rotor efficiency. Because some of the wind passes around, rather than through, the windwheel, the efficiency must be less than 100 percent.

Using the laws of physics, engineers have shown that the maximum efficiency of a conventional wind system cannot exceed 59.3 percent. Wind system efficiencies that are claimed to be greater than this are suspect.* We have been discussing, however, horizontal-axis machines without a tip vane or surrounded with sheet metal to direct the flow. More power can be extracted from the blades if a duct is placed around the rotor, but then, if the maximum duct cross-sectional area is used in the equation, rather than the blade rotor area, the maximum efficiency possible is still about 59.3 percent.

Well designed blades operating at ideal conditions can extract most, but not all, of the 59.3 percent maximum power available. About 70 percent of this 59.3 percent is typical. Thus, a wind turbine rotor might have an advertised power coefficient, or efficiency, of $0.7 \times 0.593 = 41.5$ percent. Also, gear box, chain drive or pulley losses, plus generator or

*There are currently no manufacturer's standards established for rating wind turbines. Usually, wind turbines are described in terms of power, not efficiency. Occasionally, an efficiency may be stated in terms of percentage of this 59.3 percent of theoretical maximum power available. Thus, a 70 percent efficiency would mean 41.5 percent true efficiency.

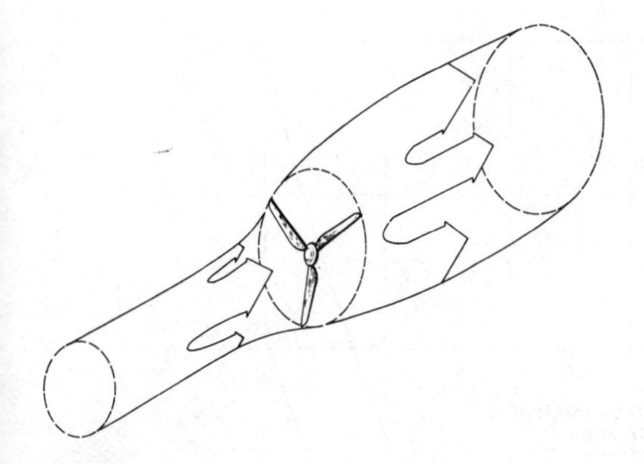

Figure 4-7. Airstream expansion through a rotor. (*Courtesy of DOE.*)

pump losses, could decrease overall wind turbine efficiency to about 30 percent. This is about the maximum coefficient possible from a conventional, well designed wind turbine, operating at its best condition.

The density of air at 60°F at sea level is 0.0763 lb/ft³ (1.22 kg/m³). The densities at various altitudes divided by the sea level density (we will use the symbol DRA for Density Ratio at Altitude) are as shown below.

Altitude (ft)	0	2500	5000	7500	10,000
DRA (at 60°F)	1	0.912	0.832	0.756	0.687

The densities at various temperatures divided by the density at 60° (we use the symbol DRT for Density Ratio at Temperature) are as follows.

Temperature (°F)	0	20	40	60	80	100	120
DRT	1.130	1.083	1.040	1	0.963	0.929	0.897

To determine the true density at some particular altitude and temperature, we multiply together the appropriate DRA, DRT and standard density of 0.0763. For example, at 100°F and 5000 ft elevation, the density $\rho = 0.832 \times 0.929 \times 0.0763 = 0.0590$ lb/ft³.

The symbol k is simply a number, depending on the units used for density, velocity and area. To simplify the above equation, 1/2 k × standard density are grouped together and labeled K, so:

$$\text{Power} = K \times e \times \text{DRA} \times \text{DRT} \times A \times V^3.$$

The most common units for K are given below.

Power	*Area*	*Velocity*	*Value for K*
W	ft²	mph	0.00508
W	ft²	m/second	0.00569
W	m²	mph	0.0547
W	m²	m/second	0.6125
W	ft²	knots	0.00776
hp	ft²	mph	0.00000681
hp	ft²	m/second	0.0000763

The above equation has been used to calculate the curves in Figure 4-8 for DRA and DRT equal 1.0, and e = 30 percent.

As an example, consider a 15-ft diameter rotor operating in a 20 mph wind at an altitude of 500 ft and at 80°F. The windmill efficiency (including generator and transmission losses) is 30 percent. What is the power?

$$\begin{aligned} \text{Area} &= 3.14 \times 15^2 \div 4 = 176.6 \text{ ft}^2 \\ \text{DRA} &= 0.0832,\ \text{DRT} = 0.963 \\ \text{Power} &= 0.00508 \times 0.30 \times 0.832 \times 0.963 \\ &\quad \times 176.6 \times 20^3 \\ &= 1725 \text{ W} \\ &= 1.725 \text{ kW.} \end{aligned}$$

Notice that Figure 4-9 shows a power of 2.2 kW for this rotor at sea level and 60°F.

Finally, you want to buy power, not efficiency. If two wind turbines have the same power output in the same wind conditions and cost and reliability are the same, it is relatively unimportant that one may be more efficient than the other.* Placing the wind turbine in the best wind location available to you and matching the power-producing velocity range of the wind turbine to your wind conditions and load is most important.

As a first estimate, Figure 4-8 or 4-9 can be used to estimate the power output of any wind turbine in any wind. These curves use an overall efficiency for the rotor, transmission and generator of 30 percent, which is a typical value for wind turbine generators. For

*Everything else being the same, the more efficient wind turbine will have a smaller rotor diameter. This could reduce its weight and cost.

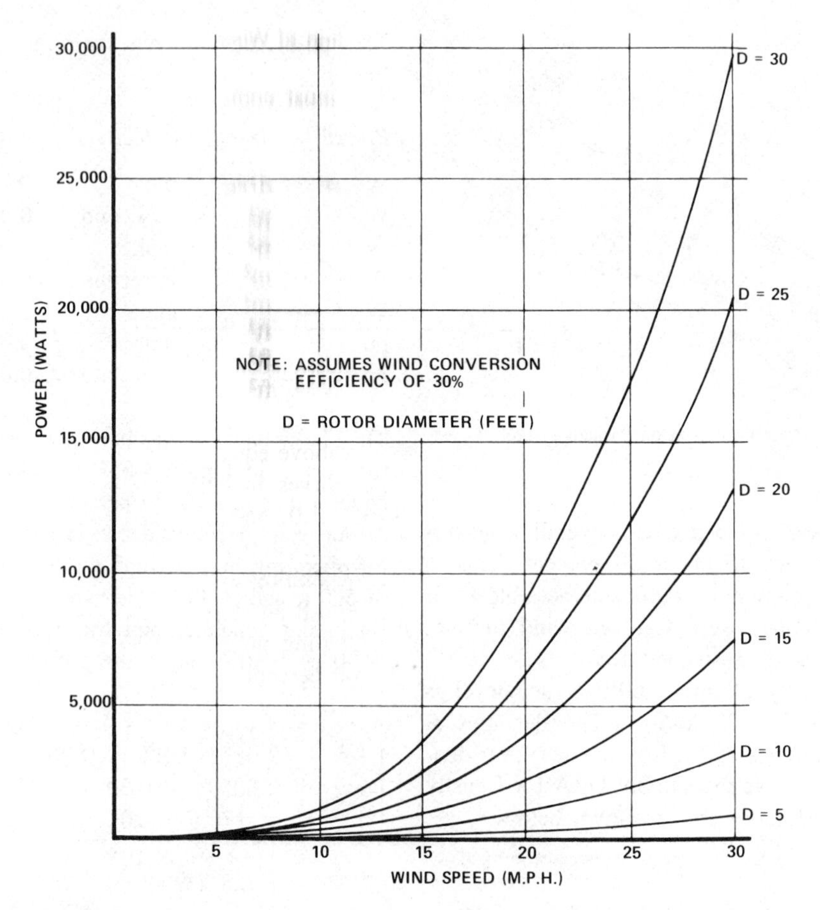

Figure 4-8. Typical wind power curves with wind conversion efficiency of 30 percent (calculated using the formula: Power = $K \times e \times$ DRA $\times$ DRT $\times A \times V^3$). (*Courtesy of DOE.*)

example, assume that the blade diameter is 20 ft (6m) and the wind speed is 10 mph (4.5 m/second). From Figure 4-9, this results in approximately 500 W for a typical wind turbine. To convert W to hp, divide by 746; so hp = 0.67.

Manufacturers of wind turbine generators will supply sales literature containing power curves similar to those in Figure 4-10, with which you can make a more accurate determination of power and energy. Chapter 8 contains power curves for all available WECS. To evaluate any wind turbine for its power and energy yields, it is important to consider the rated wind speed of the machine. This is the wind speed at which rated power is achieved. Also, you should know the cut-in speed, which is the wind speed at which the generator begins to produce power.

Figure 4-10 illustrates the characteristics and power curves of two hypothetical wind turbine generators of the 1000-W size. Notice that WECS *A* is rated at 32 mph, while WECS *B* is rated at 20 mph. You can expect that WECS *B* is considerably larger in diameter than WECS *A*.

As an example of the comparison of energy yields, Figure 4-11 is a hypothetical wind duration curve for one month. By dividing the 720 hours of that month into 20-hour segments and finding (on the graph) the average wind speed for each 20-hour segment, we can estimate the energy yield of each of the two wind turbines, using Table 4-1. Note that we are showing this calculation as an example of energy estimation.

All that is required to make such a chart is to write down, for each 20-hour section of the curve, the average wind speed and the power (W) at that speed for each wind turbine (from

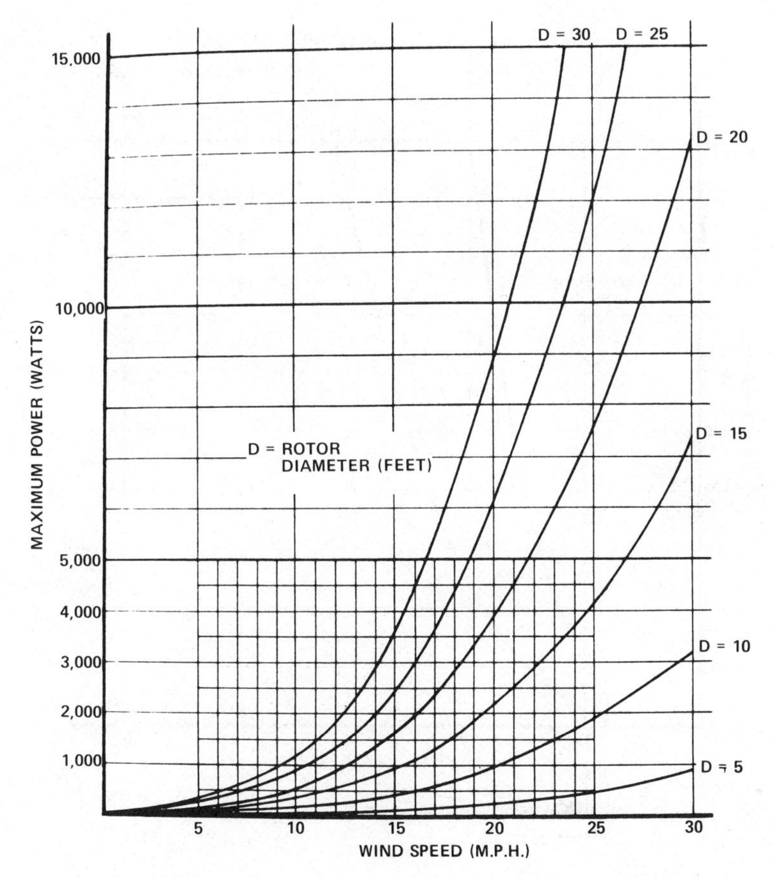

Figure 4-9. Typical wind power curves. (*Courtesy of DOE.*)

Figures 4-11 and 4-10; you can use time intervals other than 20 hours if you wish). Then multiply each power value times the 20-hour duration. This yields W-hours. Add up all the W-hours produced by each machine, and convert to kWh by dividing by 1000. In this example, WECS *B* yields roughly 230 kWh; WECS *A* produced 95 kWh.

As indicated, a lower-rated speed implies a higher energy yield. If, for example, in the case just illustrated, all characteristics were the same, and WECS *C* were added to the comparison with a rated wind speed equal to the average wind velocity (which, in this case, is about 13 mph—value at 360 hours), the yield would be considerably greater. WECS *A* is about 6 ft in diameter; WECS *B* is about 12 ft in diameter; and WECS *C* is about 20 ft in diameter.

You can expect the initial cost per kW of rated power to increase with decreasing rated wind speed; at the same time, yield (kWh) will increase, unless your wind distribution shows a considerable number of hours with wind speeds greater than 20 mph. For this reason, you need to know more than price and power rating. As you can see, rated wind speed is a valuable tool in wind turbine comparison.

BASIC ELEMENTS OF WIND ENERGY CONVERSION SYSTEMS

Different Types of Rotors

Figure 4-12 shows a farm windmill, and Figure 4-13, a Savonious rotor. While these two types of wind systems are very different, both rotors present a large surface area to the wind in relation to the width and height of the machine. Notice that almost the entire disc area of the farm windmill is covered by blade

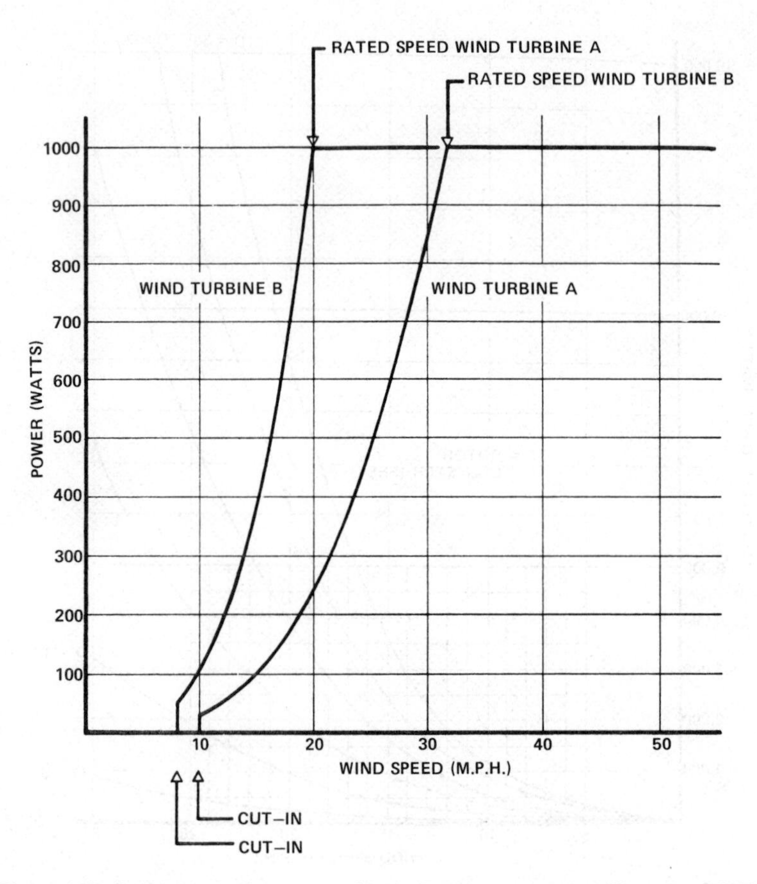

Figure 4-10. Power curves for two sample wind turbine generators. (*Courtesy of DOE*.)

surface; this presents a solid appearance to the wind. The appropriate term for this is solidity, which is the ratio of blade or rotor surface area to rotor swept area, the area inside the perimeter of the spinning blades. Thus, solidity for the two machines illustrated in Figures 4-12 and 4-13 is nearly 1.0.

Solidity = blade surface area ÷ rotor swept area.

To calculate rotor swept area, look at Figures 4-14 and 4-15. Swept area for a vertical-axis machine like the Savonious rotor is simply height times width. Swept area for horizontal axis disk-shaped rotors is calculated from the following formula:*

$$A = \pi \times D^2 \div 4$$

where:

A = swept area, in ft² or M²
D = diameter in ft or m.

For example, the swept area of a 16-ft-diameter horizontal axis rotor is calculated as follows:

$$A = 3.14 \times 16 \times 16 \div 4 = 200.9 \text{ ft}^2$$

Mechanical drive applications, such as water-pumping, demand a very high starting torque from the rotor. The pump may have a load of water it is trying to lift from a deep well at the same time the rotor is starting to turn. Further, high rpm operation (high revolution rates from the rotor) is not required, because it is generally better to pump a large quantity of water slowly than it is to pump a small quantity rapidly. This reduces resistance to water flow in the pipes. Larger-diameter, slower-moving pumps require slow-turning, high-torque rotors, such as shown in Figure 4-12.

Electric generators operate by moving magnets past coils of wire. Two methods are

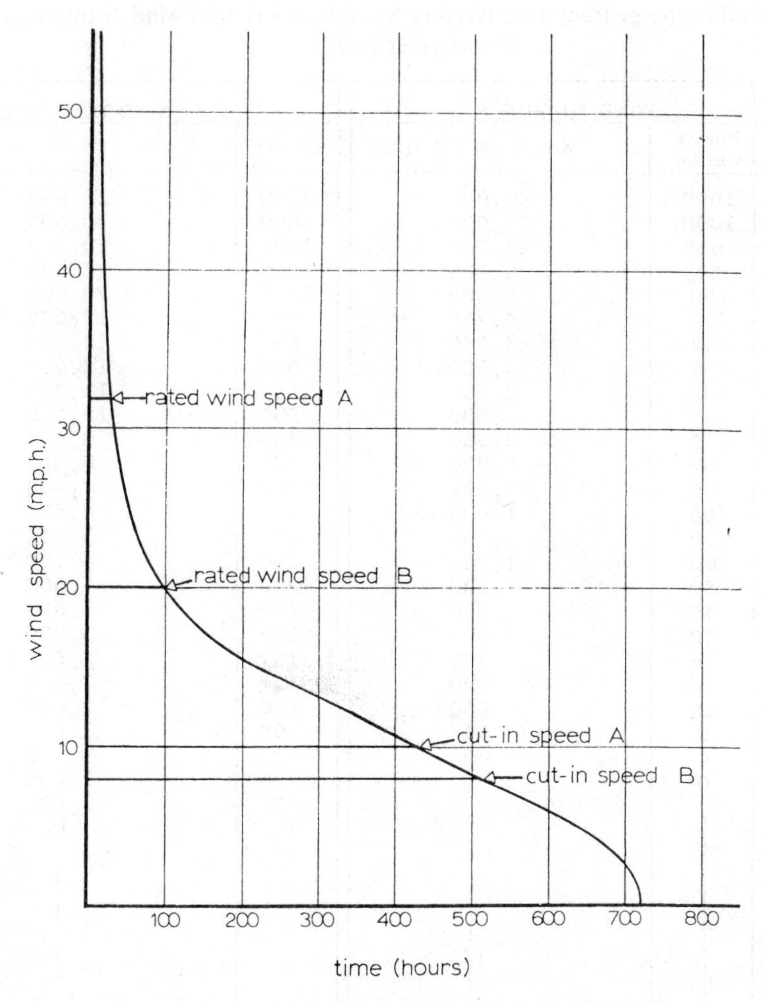

Figure 4-11. Sample wind duration curve. (*Courtesy of DOE*.)

available to get the required power from a generator:

1. Large coils and strong magnets
2. High-speed motion of magnet past coil.

Most generators are actually a balance of these two design methods. However, to get 2 kW out of a generator that turns at 200 rpm, the large magnets and coils might weigh as much as 300 lb (135 kg). The same 2 kW can be generated by a smaller generator, which weighs about 50 lb (22.5 kg), by spinning that generator at about 2000 rpm.

From this we can see that a lightweight, low-cost wind turbine requires a fast-turning rotor with much lower solidity, as shown in Figure 4-16.

One further relationship is needed to complete the discussion of solidity: tip speed ratio, the speed at which the rotor perimeter is moving divided by the wind speed. If the wind is blowing at 20 mph (9m/second), and a rotor is turning so that the outer tip of the blade is also moving at 20 mph around its circular path, tip speed ratio equals 1. Rotors such as that in Figure 4-12 operate at tip speed ratios of about 1.

Suppose the tip were moving at 200 mph. With a wind speed of 20 mph, the tip speed ratio would equal 10. Low-solidity rotors operate at tip speed ratios much greater than 1, usually between 5 and 10. We can now see a relationship between tip speed ratio, which is a measure of rpm, and solidity. High-solidity

Table 4-1. Monthly energy from wind turbines A and B for previous wind distribution curve. *(Courtesy of DOE.)*

		WIND TURBINE A		WIND TURBINE B	
No.	V mph	Power watts	Watts × 20 Hrs.	Power	Watts × 20 Hrs.
1	>40	1000	20,000	1000	20,000
2	35	1000	20,000	1000	20,000
3	26	650	13,000	1000	20,000
4	22	320	6,400	1000	20,000
5	20	250	5,000	1000	20,000
6	19	238	4,760	950	19,000
7	17.5	162	3,240	650	13,000
8	17	150	3,000	600	12,000
9	16.5	138	2,760	550	11,000
10	16	125	2,500	500	10,000
11	15.5	113	2,260	450	9,000
12	15	100	2,000	400	8,000
13	14.5	90	1,800	360	7,200
14	14	80	1,600	320	6,400
15	13.5	70	1,400	280	5,600
16	13	60	1,200	240	4,800
17	12.5	50	1,000	200	4,000
18	12	45	900	180	3,600
19	11.5	40	800	160	3,200
20	11	35	700	140	2,800
21	10.5	30	600	120	2,400
22	10	25	500	100	2,000
23	9.5	0	0	95	1,800
24	9	0	0	87	1,740
25	8.5	0	↓	70	1,400
26	8	↓		50	1,000
27	7.5			0	0
28	7			↓	↓
29	6.5				
30	6				
31	5.5				
32	5				
33	4.5				
34	4				
35	3.5				
36	2				
Total watt hours			95,420 = 95.4 kwh		229,940 = 229.9 kwh

Figure 4-12. Water-pumping farm windmill.

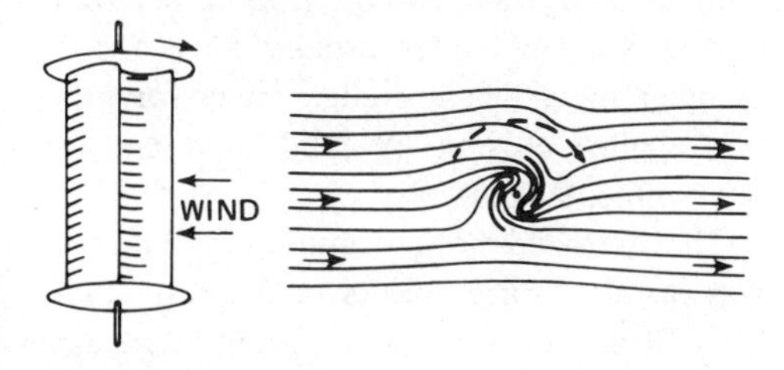

Figure 4-13. Savonious rotor. (*Courtesy of NSF*.)

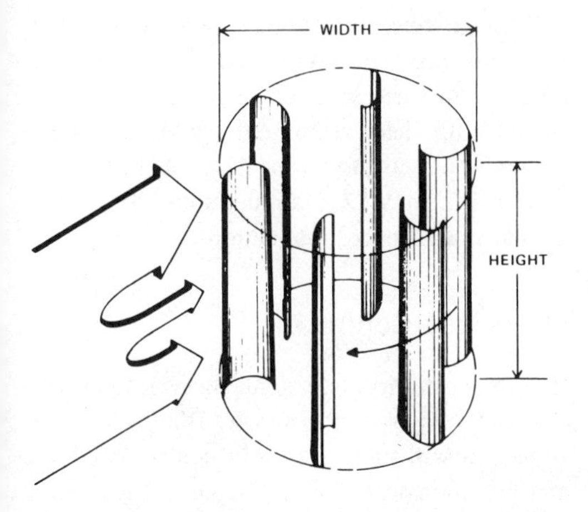

Figure 4-14. Vertical-axis wind turbine. (*Courtesy of DOE*.)

rotors spin slowly compared to low-solidity rotors.

Figure 4-16 shows how the relative torque of various WECS decreases with increasing tip speed ratio. As we noted previously, high torque requires a high solidity, and that type of WECS works best at low tip speed ratios. Figure 4-17 shows how the best operating tip speed ratio changes with solidity.

A wide variety of WECS are sketched in Figures 4-18 and 4-19. Many of the types shown are presently the subject of advanced concept studies.

The relative efficiencies of the types of

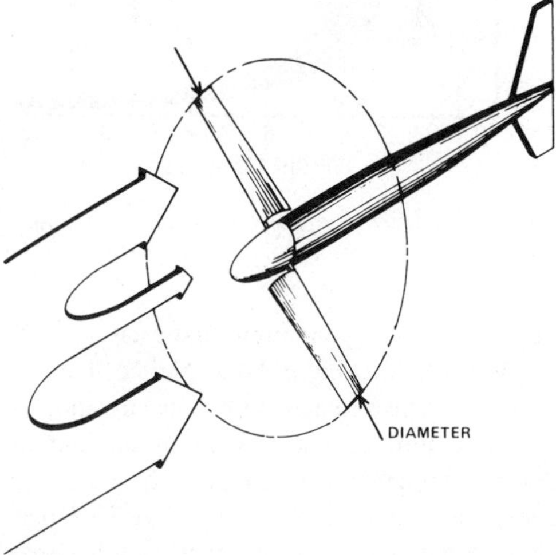

Figure 4-15. Horizontal-axis wind turbine. (*Courtesy of DOE*.)

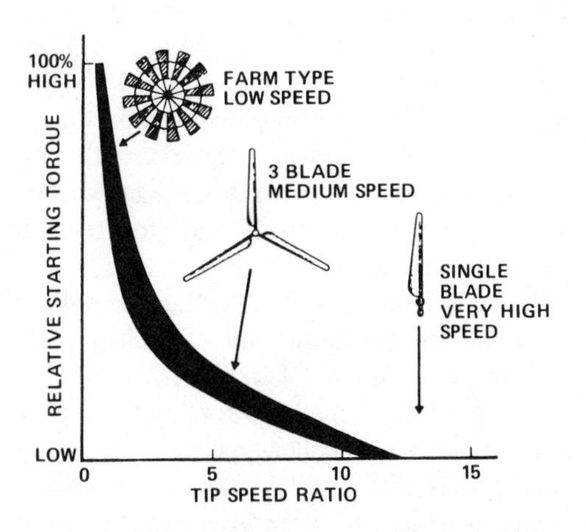

Figure 4-16. Relative starting torque. (*Courtesy of DOE*.)

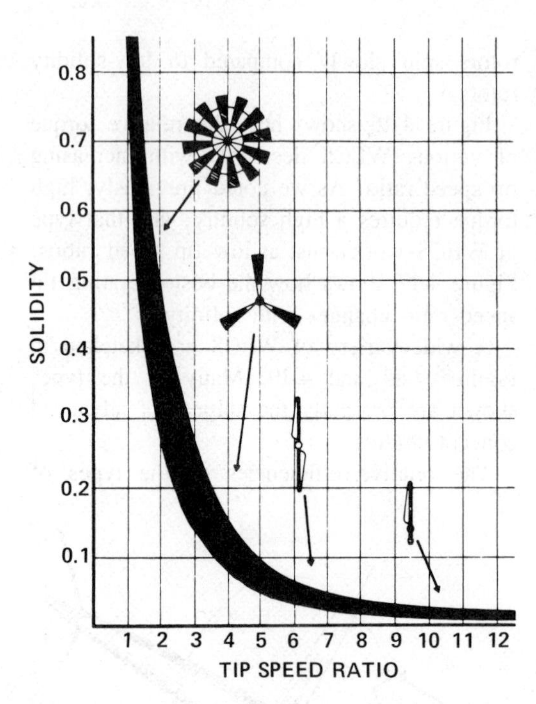

Figure 4-17. Solidity of several wind machines. (*Courtesy of DOE*.)

WECS in which you might have an interest are illustrated in Figure 4-20. Notice that the efficiency is also related to tip speed ratio, as is starting torque. The maximum amount of power a simple rotor (without a shroud or tip vanes) can extract from the wind is 59.3 percent of the wind power that would pass through that windwheel. From Figure 4-20, you can see that no windwheel actually extracts 59.3 percent.

Solidity affects design appearance in its relation to the number of blades a WECS has. High-solidity wind turbines have many blades; low-solidity machines have few, usually four or less. Figure 4-21 illustrates a wind turbine of intermediate solidity used for electric power generation.

There is more to a WECS than the solidity of the blades, the torque, the efficiency or the load the rotor drives. Figures 4-21 and 4-22 show two distinct methods for controlling the direction of the propeller-type machine: 1) upwind blades with a tail fin, and 2) downwind blades that use the drag effect of the windwheel to keep the machine aimed directly into the wind.

Many of the wind turbines are designed with three or more blades. Two blades are occasionally used, but small two-bladed wind turbines usually need a larger tail fin than an equivalent three-bladed machine, or special weights to make the windwheel behave as a four-bladed unit. Small two-bladed machines exhibit a choppy motion in yaw (aiming into the wind). This is due to the natural resistance of spinning blades to changes in direction—something like a wobbling gyroscope.

Small two-bladed WECS have governor-control mechanism weights in a position where another set of blades would otherwise be installed. For small machines, this approach is practical. For larger two-bladed machines, yaw (aiming) controls are more appropriate.

The major construction variations you find when selecting a WECS generally will involve the blades. The diagrams and discussion that follow are suited to propeller as well as Darrieus machines. One popular blade material is wood, either laminated or solid, with or without fiberglass coatings (Figure 4-23). Uncoated wooden blades usually have a copper or other metal leading edge cover for protection against erosion by sand, rain and other environmental factors.

The extruded hollow aluminum blade first was installed on a WECS in the early 1950's. This blade construction is being used again, especially for the "egg-beater" Darrieus machines.

Built-up fiberglass blades with honeycomb or foam cores, or hollow cores except for a tubular structural spar, are also being used, as are built-up sheet aluminum blades. All these methods of construction have a history of service life in WECS applications, as well as in many aerospace applications.

Overspeed Control

It is important to understand the various methods of rotor speed control. Blades are designed to withstand a certain centrifugal force and a certain wind load. The centrifugal force tends to exert a pull on the blades, whereas wind loads tend to bend the blades (Figure 4-24). A control is needed to prevent overstressing the WECS in high winds. Ob-

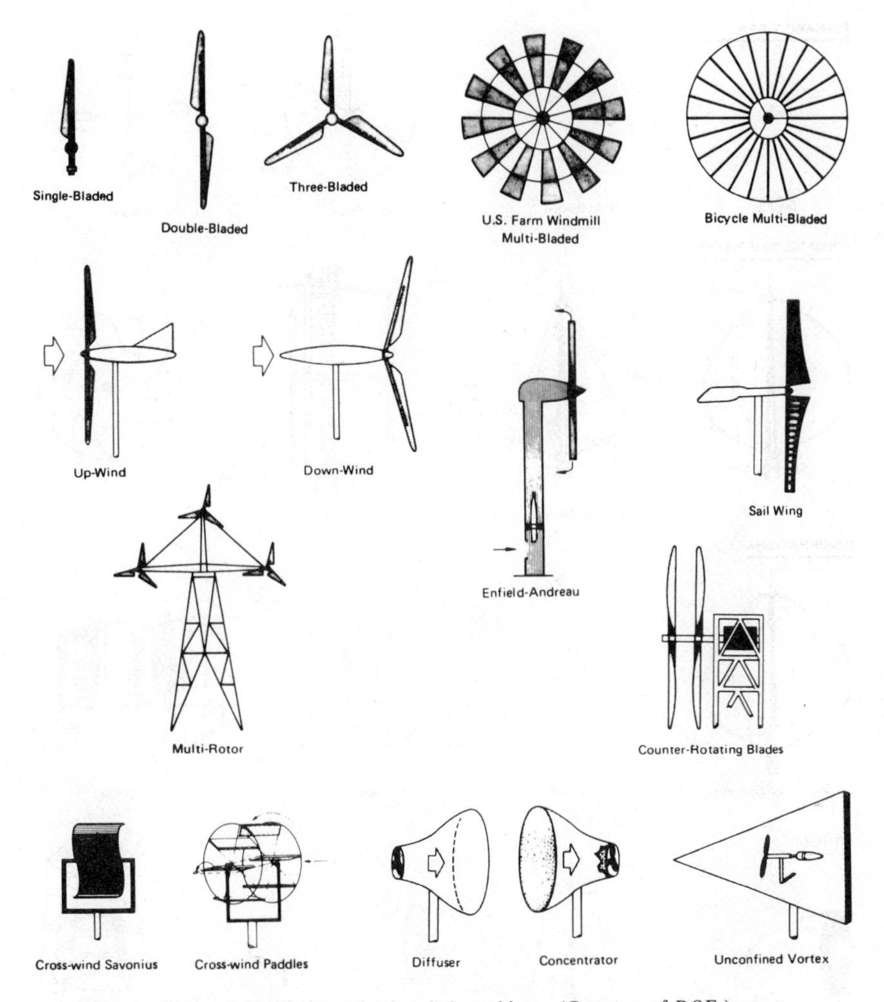

Figure 4-18. Horizontal-axis wind machines. (*Courtesy of DOE*.)

viously, one could design a wind turbine strong enough to withstand the highest possible wind, but this would be an expensive installation compared to a more fragile unit having a good control system.

Two primary methods exist for controlling a wind turbine: 1) tilting the windwheel out of excessive winds, and 2) changing the blade angles (feathering) to lower their loads.

Figure 4-25 illustrates these two methods commonly used for shut-off control.

An early method of control that was used extensively in Nebraska* is illustrated in Figure 4-26. Here, a wind fence is raised to block wind from flowing through the rotor. Appropriate for the type of windmill illustrated, this method has fallen into disuse with the invention of more sophisticated mechanisms. The Greeks, long before folks moved to Nebraska, controlled their sail windmills (as shown in Figure 4-27) by taking in or letting out sail cloth. They knew when to take in sail because a whistle, mounted at the tip of one blade, would emit a loud noise whenever the machine was turning too fast.

Figure 4-28 illustrates a simple mechanical mechanism used to control blade angle (sometimes called blade pitch angle) for feathering the blades. Notice that the leading edge of the blade in its normal position is at an angle suitable to cause blade motion in the direction indicated. As blade rpm increases, centrifugal

*Barbour, Erwin Hinckley, *The Homemade Windmills of Nebraska,* 1899. Reprinted by Farallones Institute, Occidental, California.

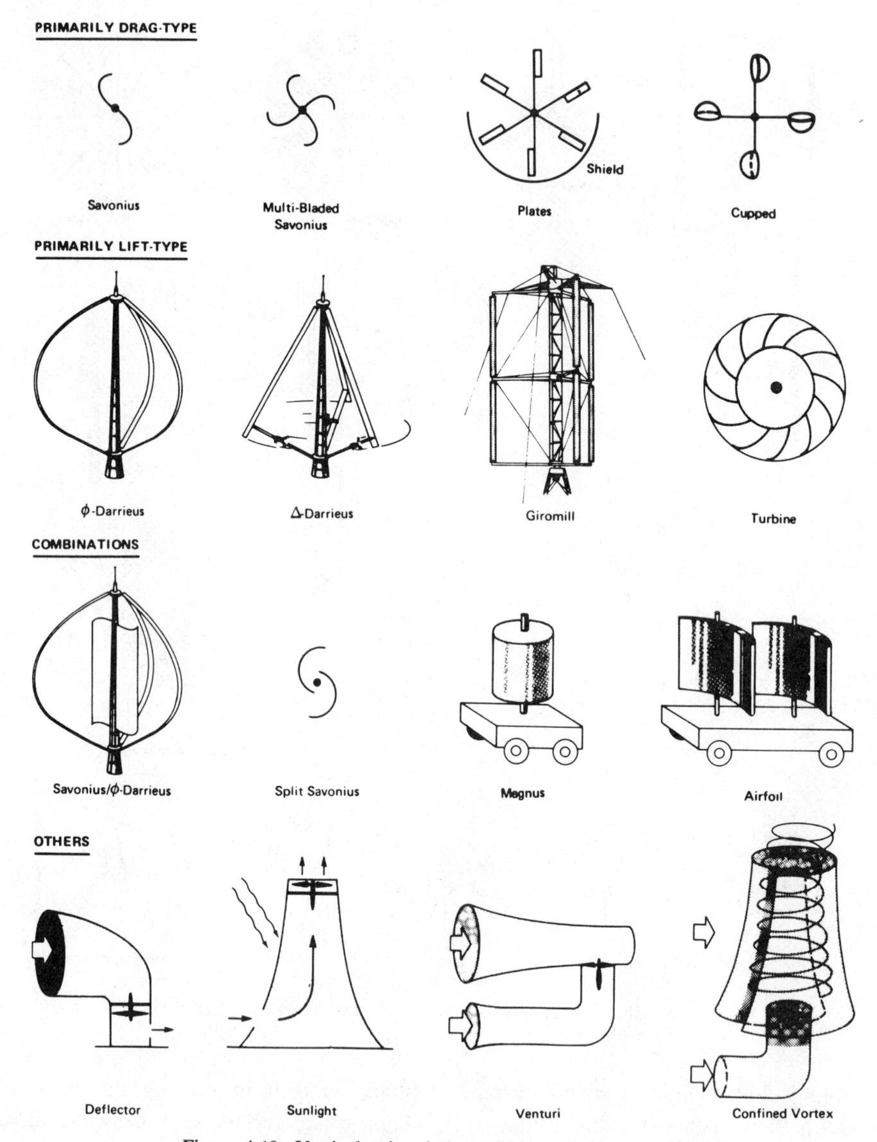

Figure 4-19. Vertical-axis wind machines. (*Courtesy of DOE*.)

forces on the flyweight cause the weight, which is connected to the blade, to move around the blade center pivot shaft, and cause the blade to pitch toward the feathered position. The feathered position pulls the leading edge of the blade into the wind to reduce or eliminate its driving force.

Overspeed ground control or shut-off or reset function can be combined with any of the design types to provide manual shutdown of the WECS if required, such as during icing conditions or high winds. Other methods of blade control, such as automatic drag spoilers and hydraulic brakes, can also be used.

Overspeed ground control for the vertical-axis machines (such as the Savonious type) can be accomplished by methods such as blocking the machine from the wind (as was done on old machines), by venting the S-shaped vanes or by changing the S-shape to reduce torque.

The Darrieus rotor, which is commercially

available, can be controlled using any one of two main methods:

1. A unique aerodynamic characteristic of the Darrieus rotor permits the blades to stall—that is, to quit lifting or pulling—which slows it down. Stall is caused by overloading the blades with the generator, which requires an electronic control system.
2. Drag spoilers mounted either on the blades or on the support structure can be mechanically or electronically actuated to slow the speed.

Darrieus rotors designed with straight blades (Figure 4-29) are able to use the blade pitch control in addition to the above methods. The particular Darrieus machine illustrated uses variable blade pitch to control the rotational speed.

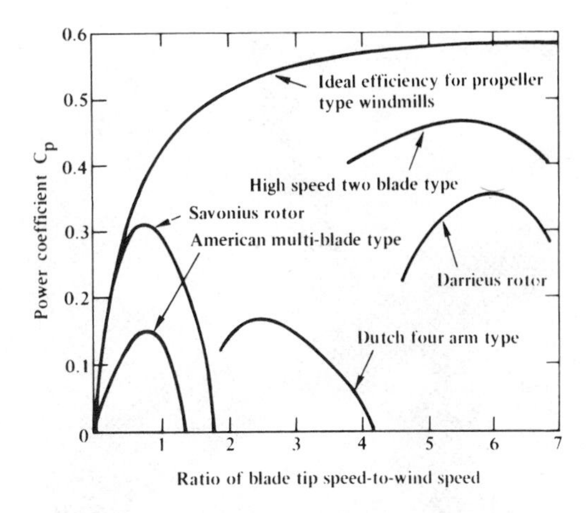

Figure 4-20. Typical performance of several wind machines. (*Courtesy of DOE.*)

Another WECS characteristic you should consider is its furling speed—the wind speed at which the WECS is automatically or man-

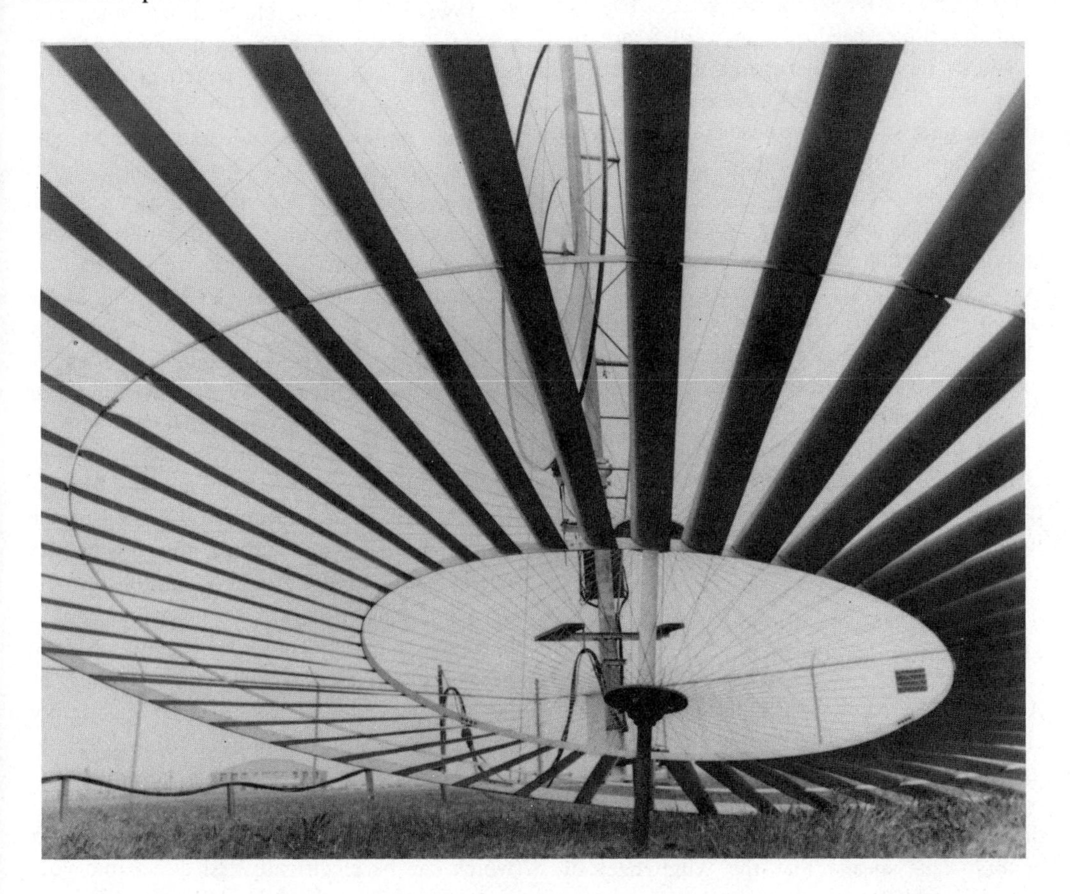

Figure 4-21. Upwind rotor of intermediate solidity with tail fin control. (*Courtesy of DOE.*)

Figure 4-22. Downwind Rotor.

ually shut off. First, not all WECS have or need a furling speed. These WECS are designed to survive while operating in even the most severe winds. For these WECS, the manufacturer's specifications should state the highest speed for which their product is designed or has been tested.

WECS that must be furled will do so by automatic mechanical or electronic control, or you will be required to operate a ground shutoff control when high winds are anticipated. On modern machines, the tail vane may be locked sideways to turn the machine out of the wind; a brake is locked; or blades are locked in the feathered position.

Thus, the type of control system used in particular WECS will play an important part in the furling operation.

Electric Power Generation

As stated before, WECS used to produce electricity tend toward low-solidity designs that have high tip speed ratios for high rpm operation. Before rural electrification, some of the early wind turbines used direct-drive generators. This means that the windwheel directly turned the generator, which was a heavy, low-speed unit. Many of the newer machines have transmissions mounted between the rotor and the generator to increase the rotor rpm by a factor called gear ratio—typically 4, 5 or more. Thus, for example, a rotor speed of 100 rpm is increased by the transmission to 400 rpm or more. This allows for a lighter, lower-cost generator but requires the additional weight, cost and maintenance of a transmission.

Lower cost is not the only factor enhanced by low weight. Heavier units can be more difficult to hoist onto the tower than equivalent lighter units. However, a factor in favor of direct drive is the lower maintenance required by elimination of the transmission. You can see that the choice between direct drive or a transmission drive involves many trade-offs.

WECS have been built most often with the generator and blades mounted at the top of the tower. It is possible that machines will eventually have drive shafts at the bottom of the tower, where the generator will be mounted. Another possibility is to mount an hydraulic pump at the rotor and use hydraulic pressure to cause fluid flow through tubes down the tower to a hydraulic motor. With this concept, it becomes possible to have several rotors powering one central generator.

Generators installed in WECS can be of the dc or ac type.

Alternating current is generated in an ac generator (or alternator) by passing coils near alternate poles of magnets. The ac current generated is fed directly to the wires outside the unit. Here, you should understand that the frequency of the ac current is governed by the rpm of the generator [utility power is 60 Hz (cycles per second) throughout the U.S.] The faster the coils of wire pass the magnet poles, the higher the frequency. To establish a fixed or constant frequency, the rotor rpm must be held constant, regardless of wind speed. For small WECS, the blade control device required to hold a constant rpm can be an expensive mechanism.

Generators used to produce ac power (which can be electrically fed direct into your load at the same time the utility ac power is wired in) are synchronous generators. A

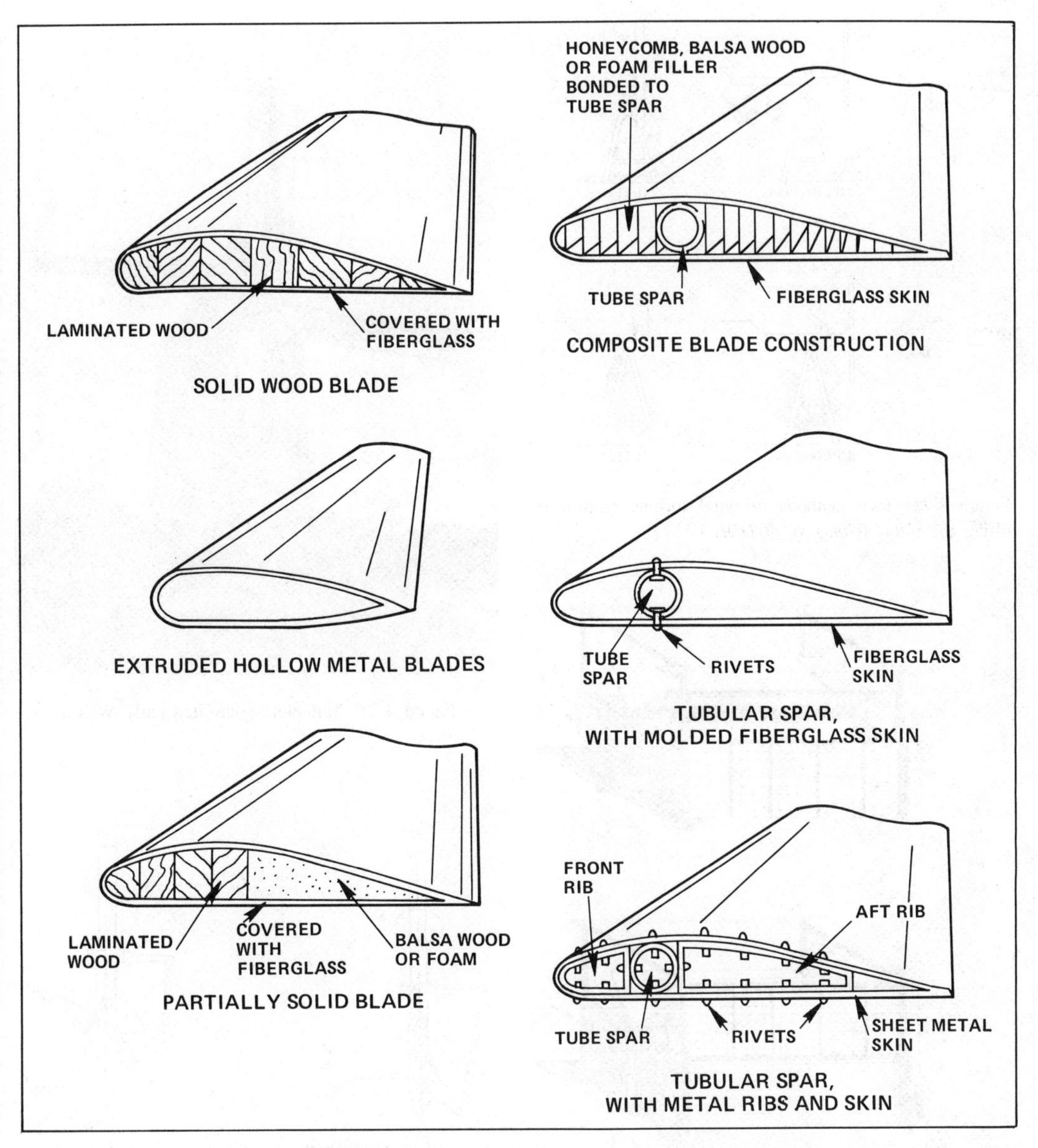

Figure 4-23. Different blade construction methods. (*Courtesy of DOE.*)

WECS generator operating in a synchronous mode would be generating ac power at exactly the same frequency and voltage as the utility mains, or other source of power. This type of operation, as mentioned, increases the complexity of blade control systems and, therefore, increases the cost of the wind machine. On large wind turbines, synchronous generation is usually practical.

Special generators are being developed that produce a constant frequency but allow for

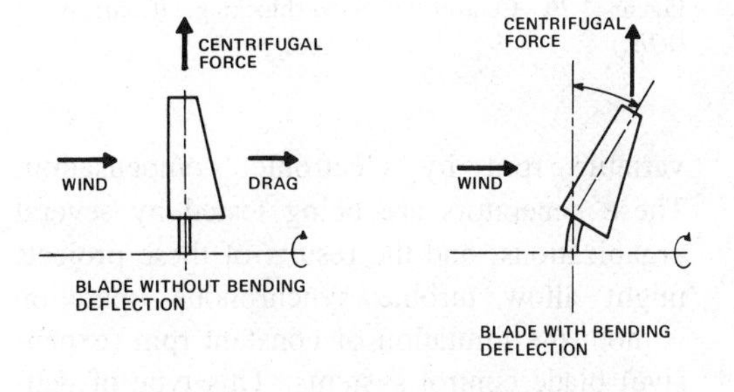

Figure 4-24. Loads on a wind turbine blade. (*Courtesy of DOE.*)

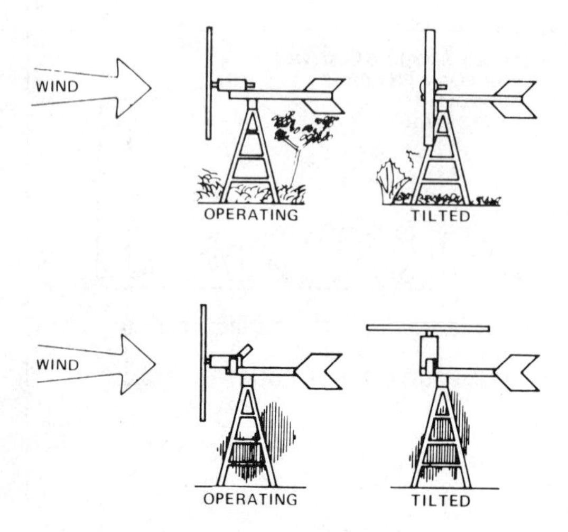

Figure 4-25. Two methods of wind turbine control by tilting the rotor. (*Courtesy of DOE.*)

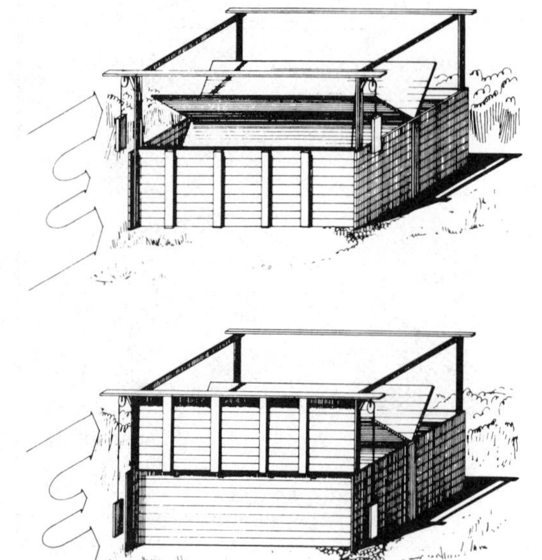

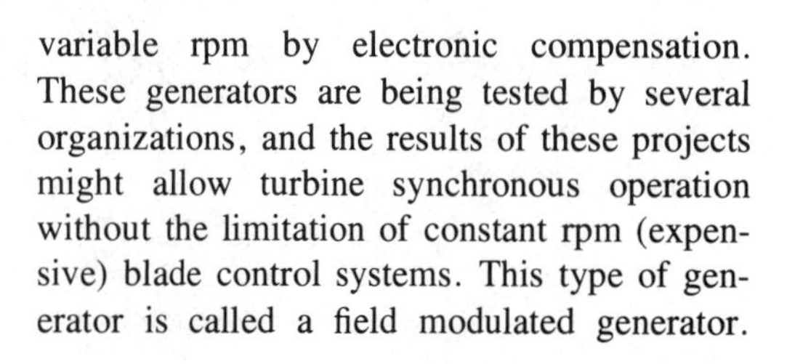

Figure 4-26. Control by wind blocking. (*Courtesy of DOE.*)

Figure 4-27. Sail cloth-controlled early WECS.

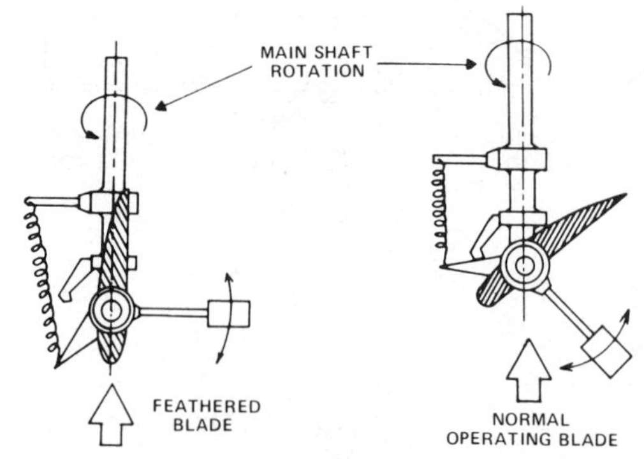

Figure 4-28. Blade feathering control. (*Courtesy of DOE.*)

variable rpm by electronic compensation. These generators are being tested by several organizations, and the results of these projects might allow turbine synchronous operation without the limitation of constant rpm (expensive) blade control systems. This type of generator is called a field modulated generator. Another method for generating fixed-frequency ac is to generate dc, and change the dc into ac by means of an inverter. Some inverters are designed to permit synchronous operation; others are not.

Generation of dc usually involves generation of ac inside the generator, then conver-

Figure 4-29. Cycloturbine vertical-axis Darrieus configuration. *(Courtesy of Pinson Cycloturbine.)*

sion of the ac to dc by means of brushes and a commutator. This has been the common design for dc generators until recently. A method more commonly used now, because of improvements in diode technology, is to rectify the ac output of the alternator to dc. This technique eliminates the need for brushes and a commutator and takes advantage of the superior low-speed characteristics of the alternator over those of the dc generator. Virtually all new automobiles use the diode-alternator combination in their electrical systems. Thus, the ac power generated internally is fed to the battery and other loads as dc. Direct current is the only type of current that can be stored in a battery. Variable-frequency ac, as generated with a small wind generator, can be used without diode rectification to dc for many applications (such as electric resistance heating). The current flow for each of these three generators is diagrammed in Figure 4-30.

Some experimenters have used the current before it is rectified to dc. Alternating current can be fed to a transformer, which steps up the voltage while lowering the current (amps). This reduces line loss that can occur on long wire runs from a wind turbine to a load. Thus, a 12-V alternator is stepped up to, say, 100 V. At the other end of the long run, another transformer (Figure 4-31) steps down the high voltage to an appropriate value, where the ac is rectified to dc. This method of transmission is subject to losses (about 5 percent) from the transformers. Transformers are designed for best operation at one frequency, while small WECS alternators generate variable frequency according to the rpm of the rotor. Some transformers are designed for one frequency (60 Hz or 400 Hz), while others are designed to operate over a 50–400-Hz range.

Direct current generators have brushes made with carbon, graphite or other materials, to transfer electric power from rotating windings to the stationary case of the unit. These brushes transfer the full electric power of the generator. Some alternators have brushes as well. In contrast to dc generators, however, these brushes transfer only the field current, a

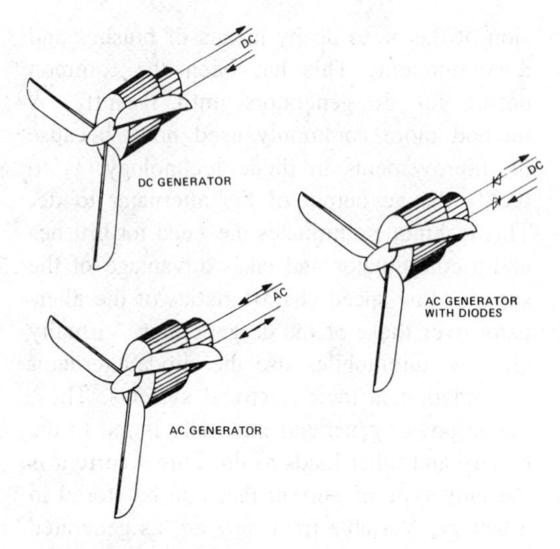

Figure 4-30. Three types of generators. (*Courtesy of DOE.*)

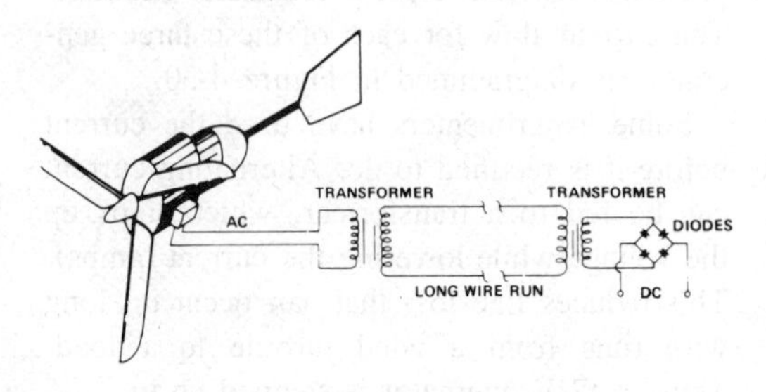

Figure 4-31. Using transformers and wind generated ac for long wire runs. (*Courtesy of DOE.*)

small percentage of the total alternator output. Alternators are also available without brushes (brushless units) at a somewhat higher cost than equivalent brush-type units.

Some generators (or alternators) are available with permanent magnets. These magnets cause the electric current to flow as they spin past the coil windings. Other generators are available field-wound, which means that electromagnetic coils requiring energizing current are installed in place of permanent magnets.

Three methods are used for regulating or controlling the electric output of the generator:

1. Voltage regulators are used on field-wound units to control the strength of the field coils, which in turn control the output voltage.
2. Voltage controllers may be used on permanent magnet units to adjust voltage levels according to the output of the generator and the needs of the system.
3. No regulation at all. The output of the permanent magnet generator is used as is, while that of the wound field is fed back to the field, either directly or through a resistor to give a variable-strength field according to the strength of the generator output.

Generators and alternators are selected or designed by WECS manufacturers according to their own criteria, which include cost, weight, performance and availability. Units with specially made generators, as well as units with truck, automotive and industrial alternators, are available. Availability of spare parts and industrial equipment varies. In some cases, the WECS manufacturer designs his own generator as a means of improving the overall system performance.

Figure 4-32 is a wiring diagram for a simple electric WECS. The WECS will charge the batteries, which, in turn, will supply power to the two loads illustrated. Battery storage would be sized to store as much energy as is needed to make up for periods when the wind is lower than required or the power demand exceeds the wind generator capacity. The WECS would be sized to supply at least enough kWh of energy as needed for the loads.

Suppose that the WECS supplies more kWh

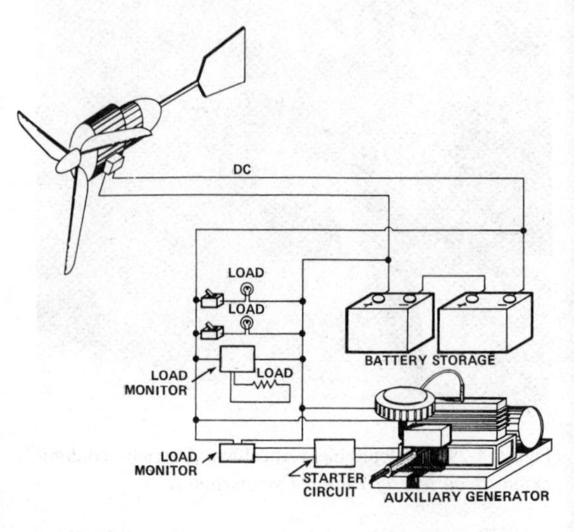

Figure 4-32. Simple wind electric wiring diagram. (*Courtesy of DOE.*)

than are needed. The batteries would be overcharged, and energy would be wasted. To preclude this situation, a load monitor is used (see Figure 4-33). The load monitor senses situations when the WECS creates more power than the electric system needs, and reacts by switching on Load C. Load C might be a resistance electric heater immersed in a water heater tank. It may be another battery bank, or any other load that will use the excess power. The load monitor thus prevents energy waste, and in so doing improves the energy utilization of the simple battery system.

Load monitors can be used another way. Suppose that the WECS generator does not supply the required energy. Perhaps a week of no wind occurs, and the batteries are nearly discharged. A load monitor can be used to sense this condition and activate a backup system.

The backup system could be a gasoline-powered generator, or another set of batteries. The load monitor can control the source. In the case of the gasoline-powered generator, the load monitor can flash a light, ring a bell or otherwise warn you of the situation, or it can energize the starter circuit on the auxiliary generator to bring it on-line. Schematically, this could be done as shown in Figure 4-34.

Energy Storage

The key to a viable electric WECS using battery storage is a low-cost, high-efficiency storage battery. Table 4-2 presents many of the important characteristics of three prominent types of batteries.

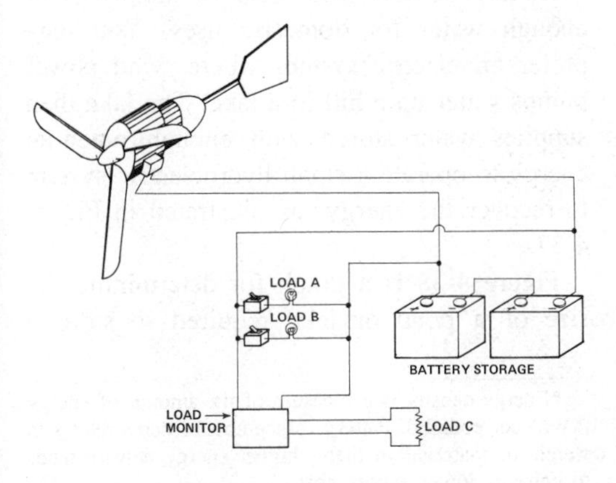

Figure 4-33. Wind electrical system with load monitor. (*Courtesy of DOE.*)

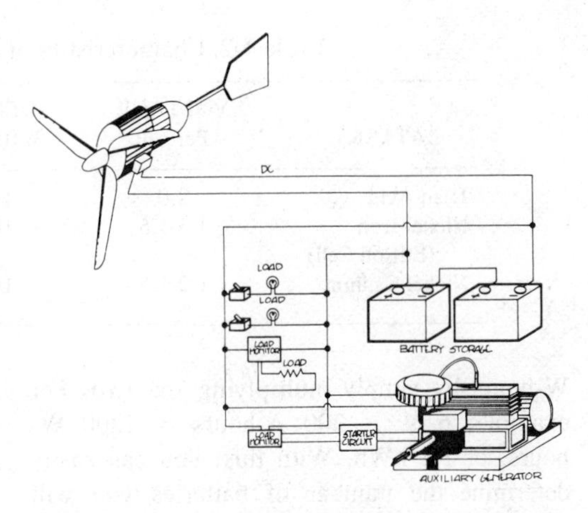

Figure 4-34. Complete wind-electrical system with backup generator. (*Courtesy of DOE.*)

Lead-acid batteries of the automotive type are among the lowest-efficiency storage batteries available. Auto batteries usually retain about 40–50 percent of the energy your wind turbine generator will charge them with. A low-efficiency storage should be assumed for such batteries when evaluating their usefulness.

Golf cart batteries are one of the most suitable type available today, as are industrial batteries used for electric forklifts and units for standby power for computers, telephones and electronic instrumentation. Standby batteries cost much more than do golf cart batteries, but are designed for higher reliability and longer life. Golf cart and standby batteries are designed to be deep-cycled (discharged to very low charge levels) while auto batteries are not.

Batteries are rated by their voltage and by their storage capacity (A-hours). For example, a typical golf cart battery might be a 6-V unit, rated at 200 A-hours. It is important to know that the A-hour rating is based on a certain discharge rate. The typical rating is 20 hours for golf cart batteries. If 200 A-hours were discharged over a period of 20 hours, the discharge rate would be 10 A/hour ($200 \div 20 = 10$ A). Greater discharge rates will result in a slightly reduced A-hour capacity. You can get performance curves from battery manufacturers that illustrate this fact. You can convert the A-hour and voltage ratings into

Table 4-2. Characteristics of different types of batteries.

BATTERY	VOLTAGE Per cell	DENSITY Watt-hrs/lb	CYCLE LIFE	STORAGE EFFICIENCY
Lead-Acid	2.0	10-20	200-2000	50-80%
Nickel-Iron (Edison Cell)	1.3-1.5	10-25	2000	60-80
Nickel-Cadium	1.2-1.5	10-20	2000	80

W-hours by simply multiplying the two. For example, 6 V × 200 A-hours = 1200 W-hours, or 1.2 kWh. With this, you can easily determine the number of batteries you will need.

Example 1: The storage capacity needed is 30 kWh. If we use 6-V batteries rated at 200 A-hours, 30,000 W-hours ÷ 6 V = 30,000 ÷ 6 = 5000 A-hours. With these 200-A-hour batteries, we need 5000 ÷ 200 = 25 batteries connected in parallel. These batteries would be wired as shown in Figure 4-35.

Notice that connecting batteries in parallel increases the A-hour rating of the entire battery bank (simply add up the total A-hours available from each battery), while the output voltage remains the same as that of an individual battery. All batteries in a system must have the same voltage rating. An advantage to this arrangement is that any number of batteries can be taken away or added at any time to adjust your storage capacity to your needs.

Example 2: Storage capacity needed is 10 kWh. For a 100-V system which uses 2-V batteries, it would require 100 ÷ 2 = 50 batteries wired in series to make 100 V out of 2-V cells. Each battery must have an A-hour rating of 10,000 W-hours ÷ 100 V = 100 A-hours. Thus, 50 2-V, 100-A-hour batteries satisfy the 10-kWh storage capacity requirement. These batteries would be wired in series as shown in Figure 4-36.

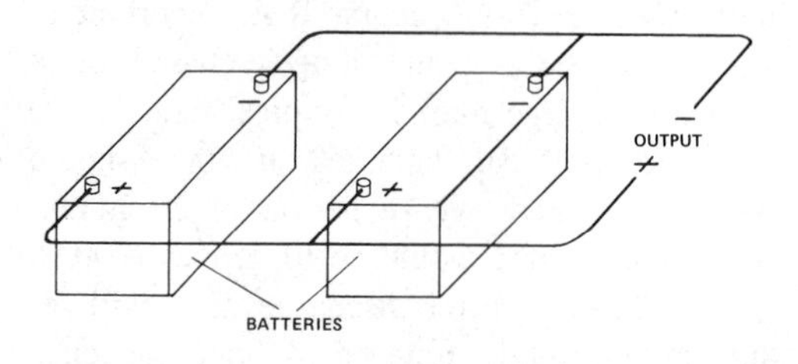

Figure 4-35. Battery storage bank—parallel wiring. (*Courtesy of DOE.*)

Notice that connecting batteries in series increase the voltage of the battery bank while the A-hour rating of the bank remains the same as that of the smallest A-hour rated battery in the bank. To increase the storage capacity of, say, a 100-V battery bank, either increase the size of individual batteries or wire another 100-V bank in parallel.

The cost of a battery storage system is relatively high. As a result of an increasing demand for electric personal transportation vehicles, a large amount of battery research has begun to reduce these costs. Research supported both by private and government funds is rapidly closing the gap between the batteries available on the market and potentially lower-cost, higher-efficiency units. Figures such as double and triple the energy density* are often quoted. Batteries made with such materials as nickel-zinc and exotic metals are being tested and developed to increase electric car performance. Wind electric system energy storage per dollar invested is also expected to improve.

It may be that you wish to simply store enough water for domestic uses. You may prefer an electric system where wind power pumps water up a hill to a lake. The lake then supplies water stored with enough potential energy to operate a small hydroelectric system to recover the energy, as illustrated in Figure 4-37.

Figure 4-38 is a graph for determining the size of a pond or lake required to store a

*Energy density is a measure of the amount of energy (kWh) per pound of battery. Since cost of items relates to weight of materials in them, higher energy density tends to enhance lower energy cost.

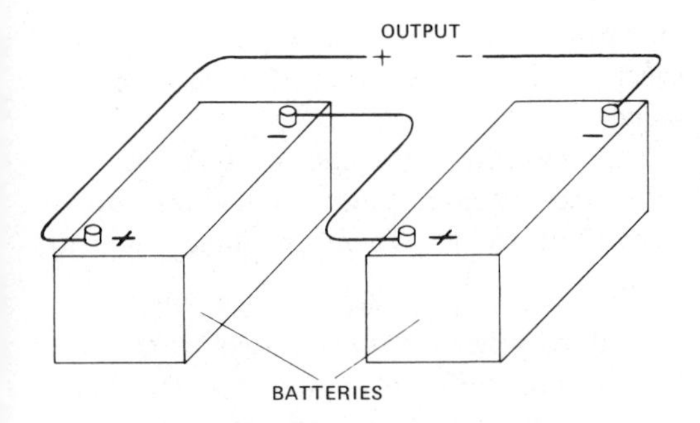

Figure 4-36. Battery storage bank—series wiring. (*Courtesy of DOE.*)

given amount of energy. Notice the large values. We sometimes hear of the idea that a few 55-gallon drums up in the attic ought to hold enough water to keep the lights burning all night, while the water trickles out of the drums, turning a water turbine along its path. Unfortunately, this would not produce much power.

The decision to use pumped water storage must be based on availability of land and such factors as these: whether the required amount of land would be better used for something else; cost (bulldozing a lake can be expensive); and the end use of the pumped water. Electricity generation is just one use; irrigation, fire prevention, stock watering, fish farm and recreation are others.

In the case of electric energy storage, a small hydroelectric system will probably exceed the cost of an equivalent battery system.

Wind power can be used to heat water for energy storage by splashing. Figure 4-39 schematically illustrates this method. Friction of water being pumped and splashed introduces all the energy as heat without the losses of a generator. As with any other heating system, heat loss can be kept to a minimum by insulating the tank. Splashing paddles can substitute for pumps. The cost-effectiveness of this method has yet to be determined.

Water may easily be heated by wind electricity as diagrammed in Figures 4-40 and 4-41. In fact, this is perhaps the most efficient means of storing energy once the electricity has been generated. Where batteries are 60–80 percent efficient, electric heaters approach 100 percent. The only losses are those through poor insulation of the storage tank, and electrical line losses.

A wind-powered heating system is called a wind furnace. The heat supplied is best used for domestic or agricultural heating, as well as lower-temperature industrial applications. As illustrated in Figure 4-40, the wind-powered generator can be either ac or dc; regulated or unregulated. This provides some latitude in selection of WECS for a wind furnace, although manufacturers of the WECS, for many reasons, may not supply all

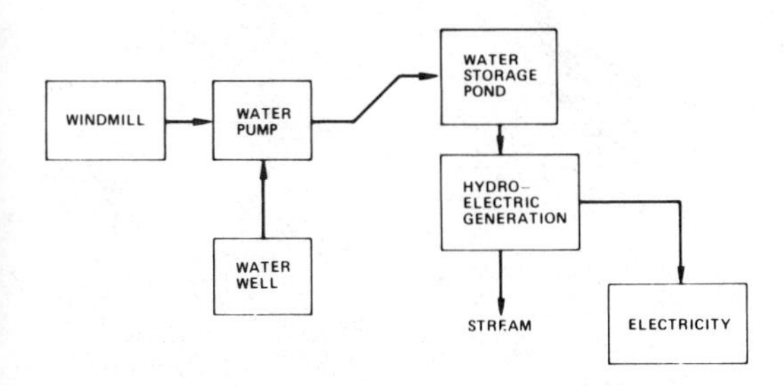

Figure 4-37. Water storage for electricity. (*Courtesy of DOE.*)

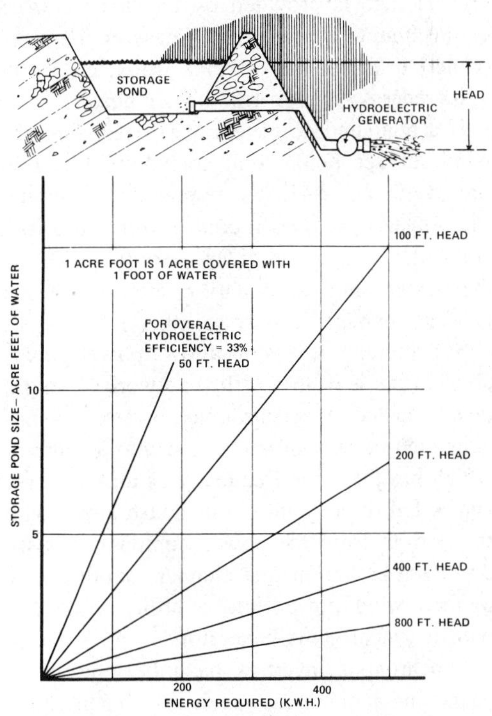

Figure 4-38. Pond size for energy storage. (*Courtesy of DOE.*)

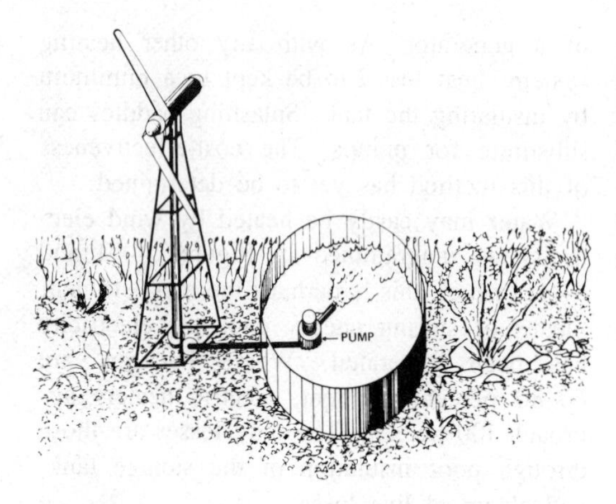

Figure 4-39. Water splash heating system. (*Courtesy of DOE.*)

of the options. For them, it is more efficient to manufacture just one or a few types of WECS that will serve the greatest number of applications.

The resistance electrical heater unit can be the air type, as with baseboard electric home heaters, or the water (or other liquid) immersion type. By using the immersion type, energy storage is provided by the thermal mass of the liquid, while thermal mass of the room (concrete floors, tile or brick walls, etc.) provides energy storage for the air heater.

If a regulated dc generator is used, then the wind furnace could look something like Figure 4-40 or 4-41. A temperature monitor (thermostat) provides a control input to a load controller, which provides priority power to the heater, and, secondarily, after the heater is warm enough, power to other loads.

Synchronous inversion allows power generation in phase with a utility network. Using a device called a synchronous inverter, wind-generated dc current is charged into ac current which has the same frequency as utility power and is fed directly into your house along with the current from the power company. As we discussed earlier in this chapter, another way to feed wind power into a utility system is with a synchronous generator.

Synchronous inverters have been used for years in applications such as regenerative drives for elevators. Here, the elevator is powered by the utility lines during its upward travel. During the downward travel, the motor becomes a generator that returns a portion of the upward power to the utility grid through a synchronous inverter while the elevator descends.

The immediate reaction to this idea is, "But the electric meter would run backwards!" It would. There are three cases to note:

1. Wind generator not operating or not producing sufficient power for the load. (Utility meter runs forward—its normal direction.)
2. Wind generator supplying just enough power for the load. (Utility meter stops.)
3. Wind generator supplying surplus of power. (Extra power will be fed to the grid. This will cause the meter to run backwards.)

The next response is usually, "But will my utility company allow this?" The answer to this question varies from state to state, and

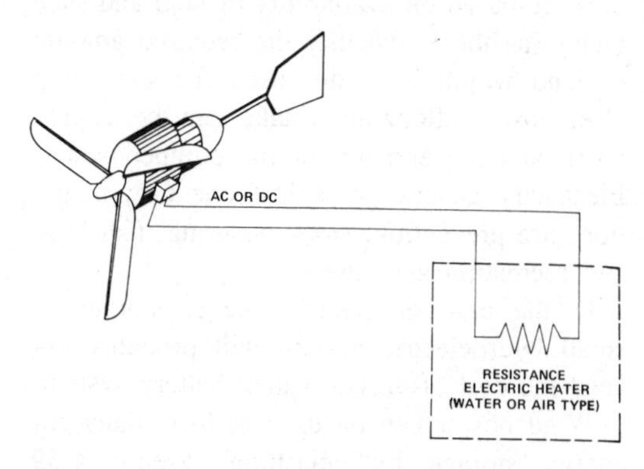

Figure 4-40. Simple electrical heating system. (*Courtesy of DOE.*)

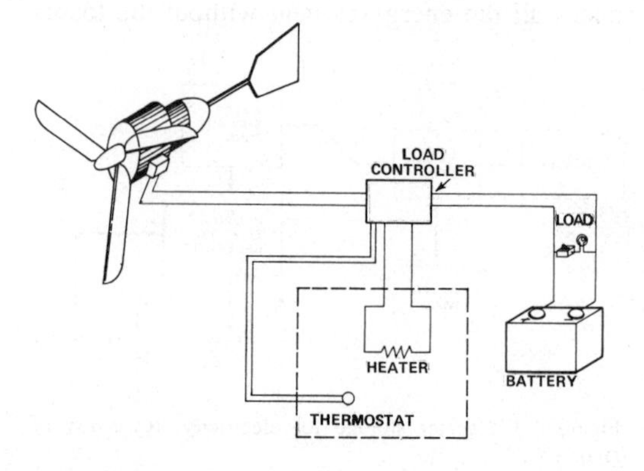

Figure 4-41. Complete electrical system with wind furnace. (*Courtesy of DOE.*)

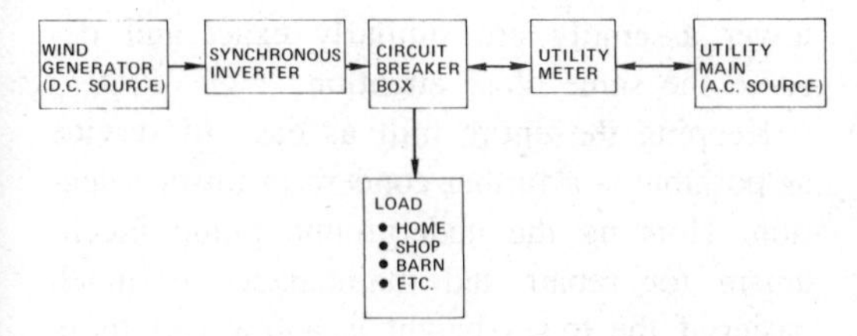

Figure 4-42. Synchronous inverter electrical system. (*Courtesy of DOE.*)

from utility company to utility company, and is rapidly changing. The key question is development of a fair billing procedure for a complex problem. Other questions have been or are being resolved, such as safety, power quality and power factors. Regenerative drives (electric elevators, for example) have been used for years, and the addition of WECS applications should not pose insurmountable technical problems.

Of great importance to some WECS installations is the fact that utility mains can be replaced in the diagram (Figure 4-42) by an ac generator powered by gasoline or diesel fuel. Certain technical details must be observed (the generator must have a higher power capacity than the WECS), but the synchronous inverter will run well with an ac generator as the source of frequency and voltage. This is because the synchronous inverter uses the mains, or ac generator, as a reference for conversion of dc to ac.

Synchronous inverters are discussed here under energy storage because, in effect, you are using the utility grid as a storage cell for your excess energy. With the gasoline or diesel ac generator instead of the grid, energy "storage" results from less fuel burned by the generator.

OPERATION OF WATER-PUMPING WIND ENERGY CONVERSION SYSTEMS

In 1854, Daniel Halladay invented a device for pumping well water with wind power. This simple machine, called a windmill, underwent a few improvements over the years, but the basic design remained unchanged. For many decades, windmills were among the most familiar sights on the American prairie. Then, with the coming of rural electrification in the 1930's, they began falling into disuse.

Today, windmills are staging a comeback in the face of rising conventional energy prices and the threat of power shortages. A piece of equipment which requires only wind to operate and which, properly installed and maintained, can give over 40 years of reliable service, is a very attractive investment indeed.

Figure 4-43 shows Daniel Halladay's wind-driven water pump.

Major Components

The basic components of the water-pumping wind energy conversion system are:

- Wheel, gear box and tail (top assembly)
- Tower
- Drive shaft
- Swivel
- Red rod, polished rod and sucker rod (pump rod assembly)
- Packerhead
- Discharge pipe
- Well seal
- Drop pipe
- Cylinder and screen

Figure 4-43. Daniel Halladay style water-pumping windmill.

Figure 4-44 diagrams the basic components of the water-pumping windmill rig.

Wheel and Top Assembly. The diameter of the wheel is a major factor in determining windmill size. The diameters of commercially produced wheels range from 6–16 ft in 2-ft increments. Wheels 20 ft across are available on special order.

The wheel connects to a gear box, as shown in Figure 4-44, which converts its rotary motion into vertical motion for pumping. The gear box contains a spring mechanism that allows the tail to fold parallel to the wheel in a high wind. This effectively shuts down the mill and prevents its self-destruction. When the gale subsides, the tail is released so it may turn the wheel back into the wind.

Towers. The most important consideration in choosing a tower is the need to get the wheel above wind obstructions. The tower should be at least 10 ft higher than any tree, hill or building within a hundred yards of the mill site. Commercially produced towers come in 6-ft increments from 21–47 ft high.

It is also crucial for a tower to be strong enough to support the wheel mounted on top of it. Manufacturers' recommendations on matching towers with wheels are invariably detailed and precise, and they should be followed to the letter. The instruction sheets for tower assembly are similarly exact and deserve the same close attention.

Keeping the entire mill as easy to service as possible is a further concern in tower selection. Hoisting the underground pump mechanism for repair and maintenance is much easier if the tower height is a few feet more than the longest section of drop pipe or sucker rod.

Well Seal and Pump Rod Assembly. Figure 4-45 details the pump rod assembly and a well seal with a packerhead. The well seal serves a double purpose. It keeps dirt, insects, thirsty rodents, lizards and other small animals from falling into the well and causing contamination. Also, it acts as a base for holding the drop pipe firmly in place.

The uppermost part of the pump rod assembly is the red rod. It is designed to be the weakest link in the pump train and, as such, should consist of a wooden spar no thicker than 2 in. If anything goes wrong above or below it, the red rod is supposed to break in order to minimize danger to other, more expensive or hard to reach parts.

Extending downwards from the red rod through the packerhead and well seal is the polished rod. It is usually made of brass to reduce friction. Also, brass resists corrosion, and this is important for a part that works in both air and water.

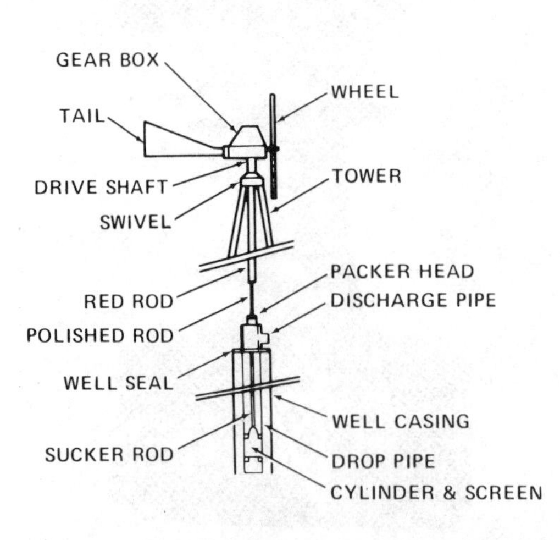

Figure 4-44. Basic components of water-pumping windmill. (*Courtesy of New Mexico Energy Institute.*)

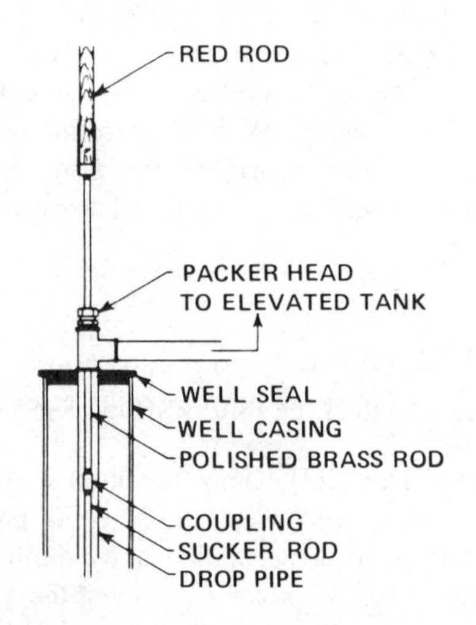

Figure 4-45. Well seal and pump rod assembly (*Courtesy of New Mexico Energy Institute.*)

At its bottom end, a coupling fastens the polished rod to the sucker rod. In general, mills which pump water no more than 100 ft above well level have solid steel sucker rods. Those which raise water 100–250 ft use hollow steel. Deep well rigs take advantage of the lightness and buoyancy of wooden sucker rods to provide lift capacities of more than 250 ft.

The Packerhead. A packerhead is a fitting, usually made of brass, which goes over the top of the drop pipe to prevent overflow. Figure 4-45 locates the part in a diagram, and Figure 4-46 shows it photographically. A packerhead is necessary to pump water any higher than the inlet to the discharge pipe. A check valve at the same point may also be useful to keep water flowing to an elevated tank instead of spilling back into the well.

If there is any danger of freezing, however, it is best either to eliminate this check or install it along with the packerhead in a protected underground housing. A key factor in designing any windmill system is making sure that water will not remain too long in exposed uninsulated lines, where it might freeze.

To pump water to a nearby tank below the level of the discharge pipe, no packerhead is necessary. Still, it is often worthwhile to install one. The fitting is not expensive, and it helps avoid well contamination.

Figure 4-46. Packerhead pipe interface.

The Drop Pipe. The well driller usually cases the well, and the windmill owner may then lower the drop pipe into place as part of the set-up procedure. A good grade of galvanized pipe in any standard size may serve as a drop pipe. Though other diameters of pipe may be preferable for some purposes, 2 in. is common for supplying domestic water.

It is important in choosing a drop pipe to make sure that it is smooth on the inside. Otherwise, replacement leathers are likely to be damaged when the plunger is lowered back into the cylinder. In extremely deep wells, paying an extra 10–15 percent for the added smoothness of reamed and drifted pipe can be a wise investment. For lesser depths, however, good quality galvanized pipe usually proves satisfactory.

Cylinder and Screen. Figure 4-47 shows a cylinder. Its diameter and the length of the plunger stroke inside it are major factors in determining the windmill's pumping capacity. Standard cylinders range from 1 7/8–4 in. in diameter in increments of 1/4 in. Other sizes are available on special order.

The stroke of a windmill is the distance which the plunger moves up and down. A short stroke enables the mill to begin pumping in a light breeze, but in stronger breezes a long stroke causes more water to be pumped. Many gearboxes are designed to permit stroke adjustment, thus assuring optimum performance for a given windmill model under a variety of conditions.

In essence, the cylinder is the bottom of the

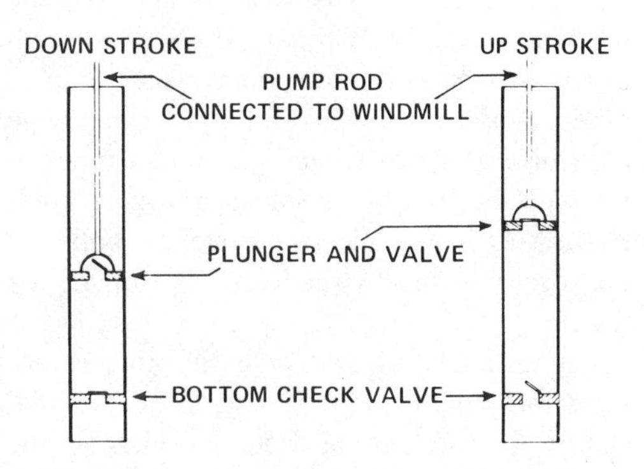

Figure 4-47. The function of the cylinder and screen. (*Courtesy of New Mexico Energy Institute*.)

drop pipe. It usually screws into the latter and is about 1/8 in. smaller in diameter. The size difference is a maintenance feature designed to assure the plunger enough leeway to move freely through the larger drop pipe. This facilitates changing the leathers that fit around the plunger and serve to assure a tight fit between it and the cylinder.

The smaller cylinder also provides a way to pull the bottom check valve if necessary. A set of threads in the plunger and the check valve makes it possible to connect the two and raise them as a single unit.

It usually is wise to put a screen just below the cylinder to prevent sand and other sediment from getting into the pump system and shortening its useful life.

Lift, Transport and Storage

The elements of water lift, transportation and storage associated with the water-pumping wind energy conversion system are discussed below.

The Elevated Discharge Pipe. The best way to build up water pressure is to raise the storage tank. If there is a hill nearby, this may be done by placing the water storage tank high on the hill. Otherwise, a tower is necessary. Each foot of elevation produces an additional 0.433 lb of pressure per square inch (psi) in the line from tank bottom to outlet. Since a domestic faucet must have a pressure of at least 10 psi to work properly, it requires a minimum water elevation, or "head," of 23.1 ft. Dishwashers and washing machines demand 18 psi or 41.48 ft of head.

The easiest way to pump water into a raised tank with a windmill is to extend the drop pipe as far as it can go without interfering with the swivel. As shown in Figure 4-48, this makes it possible to elevate the outlet to the discharge pipe virtually to the top of the tower.

An assembly of this type has the advantages of minimizing the length of the discharge line and lifting water as high as the windmill can pump it without a packerhead. The main disadvantages are elimination of the

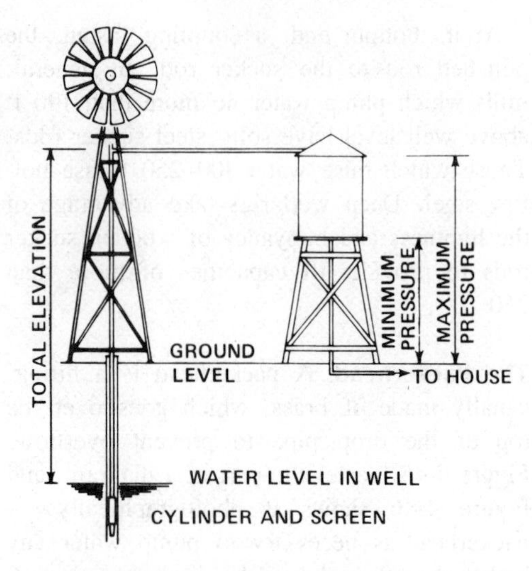

Figure 4-48. Elevated water outlet. (*Courtesy of New Mexico Energy Institute*.)

red rod as a safety feature and the system's inability to pump water any higher than the tower.

The Booster Mill. Booster mills can provide added capacity for transporting water some distance uphill or overland. While the main mill draws water from a well, the booster straddles a holding tank. It is thus in a position to reinforce the pumping process and maintain flow in a long line.

Figure 4-49 schematically shows the relation between two windmills operating in tandem. Ordinarily, the booster mill has a lower tower in order to avoid wind interference between the two rigs. Though the distance between the two mills may vary considerably, they are unlikely to work well if placed closer than 50 ft apart.

Storage Tanks. Tanks come in a variety of shapes, sizes and types, and windmills may be used to supply them all. The ordinary stock tank is low and open across the top so that animals can drink from it. The key consideration in choosing one is that the holding capacity be sufficient to water the herd which uses it.

Tanks storing water for human consumption involve a number of factors. Not only must they be well covered to prevent contamination

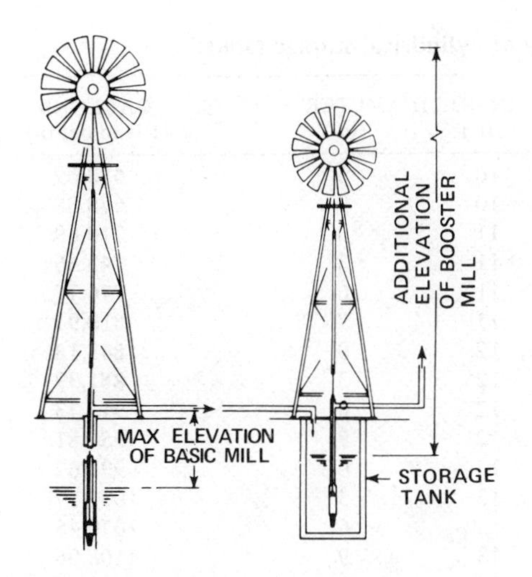

Figure 4-49. Tandem connected booster mill. (*Courtesy of New Mexico Energy Institute.*)

from bird droppings and airborne debris, but they must, as noted in the discussion of lift, be high enough to assure good water pressure.

To determine the size tank needed for a domestic water supply, it is first necessary to estimate the number of gallons a household uses on an average day. (The reference information in the back of this book provides detailed water consumption data.) Windmill salesmen customarily assume a daily per capita consumption of 50 gallons, and studies by academic experts on rural life also indicate that this figure is reasonably accurate. It remains, then, to decide how much reserve water a family ought to keep on hand.

A tank able to store up to a ten-day water supply offers some important advantages. The main one, of course, is that it greatly reduces the likelihood that the system will ever run dry. At the same time, it provides a substantial reserve for fire fighting, should the need ever arise, and offers a surplus for such purposes as keeping a small garden.

On the other extreme, minimum tank capacity to ensure against periods of windlessness varies greatly from place to place. In the Southwest, studies indicate that, in low wind areas, a reserve water supply equal to as many as seven days of average household use may be most desirable. In high wind areas, it may be possible to get by with as little as a three-day supply.

As in all matters concerning windmills, the best source of advice is usually a nearby owner or someone who remembers a time when windmills were in use in a particular locale. County agents constitute another possible information source. As a further aid in tank selection, Table 4-3 shows the content of cylindrical tanks for each foot in depth.

Siting and Sizing

This section provides information on the traditional siting recommendations for water-pumping rigs; where to place them; and proper choice of the water-pumping windmill.

Traditional Approaches. In the late nineteenth and early twentieth centuries, windmill sales boomed, and windmill lore was widespread. Many well drillers in rural areas also sold and installed windmills. The purchase prices they quoted normally included set-up costs. Telling a buyer what size mill to buy and where to put it was a matter of drawing on rule-of-thumb knowledge plus a considerable store of vernacular wisdom about local water availability and wind conditions.

Unfortunately, little was ever written about traditional siting and sizing methods. This presents modern buyers, who often prefer to set up their own rigs, with something of an information gap. Engineers seeking to fill it combine what knowledge they can recover with the empirical findings of their own observations and research.

Placement of the Windmill. It obviously makes no sense to put a windmill where there is no water, but where there is a choice of sites for drilling wells, factors concerning the windmill may be worth thinking about.

Deciding whether there is enough wind in an area to meet a specific water need with a windmill is largely a matter of particulars. Statements about site selection are usually generalizations, and they is no substitute for familiarity with local conditions.

The rule of thumb in New Mexico is that most windmills will pump the equivalent of 8 hours/day at the rate produced by a 15-mph

Table 4-3. Water storage capacity of cylindrical storage tanks.

INSIDE DIAMETER (ft.)	(in.)	GALLONS (1 ft in depth)	INSIDE DIAMETER (ft.)	(in.)	GALLONS (1 ft in depth)
1	0	5.87	10	6	653.69
1	3	9.17	10	9	678.88
1	6	13.21	11	0	710.69
1	9	17.98	11	3	743.36
2	0	23.49	11	6	776.77
2	3	29.73	11	9	810.91
2	6	36.70	12	0	848.18
2	9	44.41	12	3	881.39
3	0	52.86	12	6	917.73
3	3	62.03	12	9	954.81
3	6	73.15	13	0	992.62
3	9	82.59	13	3	1031.17
4	0	93.97	13	6	1070.45
4	3	103.03	13	9	1108.06
4	6	118.93	14	0	1151.21
4	9	132.52	14	3	1192.69
5	0	146.83	14	6	1234.91
5	3	161.88	14	9	1277.86
5	6	177.67	15	0	1321.54
5	9	194.19	15	3	1365.96
6	0	211.44	15	6	1407.51
6	3	229.43	15	9	1457.00
6	6	248.15	16	0	1503.62
6	9	267.61	16	3	1550.97
7	0	287.80	16	6	1599.06
7	3	308.72	16	9	1647.89
7	6	330.38	17	0	1697.45
7	9	352.76	17	3	1747.74
8	0	375.90	17	6	1798.76
8	3	399.76	17	9	1850.53
8	6	424.36	18	0	1903.02
8	9	449.21	18	3	1956.25
9	0	475.80	18	6	2010.21
9	3	502.65	18	9	2064.91
9	6	530.18	19	0	2121.58
9	9	558.45	19	3	2176.68
10	0	587.47	19	6	2233.52
10	3	617.17	20	0	2349.46

wind. However, April tends to have the strongest winds and August the lightest. Maximum storage capacity serves under these circumstances as an important safeguard against periods of light breezes.

In open areas, the choice of a specific windmill site is usually fairly simple. The main consideration is, as already noted, to get the wheel at least 10 ft above and 300 ft away from all wind obstructions. Even where there are no obstructions, though, tower height is worth thinking about, since the wheel is likely to catch more wind at greater elevations. Wind availability at 50 ft, for example, may be 7 or 8 percent greater than at 30 ft.

Uneven terrain alters the process of site selection. Narrow valleys produce a funnelling effect, which results in higher winds at lower elevations. Wide valleys tend to have light winds. In any case, long-time residents of an area are often the best source of wind information, and this is doubly true if they have experience with, or memories of, windmills in a given locale. Additional advice may be obtainable through the nearest weather station, a county agent, an airport or a univer-

Table 4-4. Total elevation in feet and capacities in gallons/per hour for windmills.

CYLINDER	SIZE OF WINDMILL													
	6 FT.		8 FT.		10 FT.		12 FT.		14 FT.		16 FT.		20 FT.	
SIZE	ELEV.	G.P.H.	ELEV.	G.P.H.	ELEV.	G.P.H.	ELEV.	G.P.H.	ELEV.	G.P.H.	ELEV.	G.P.H.	ELEV.	G.P.H.
1⅞	120	115	175	119	260	103	390	121	570	103	920	138	1200	162
2	95	122	140	135	215	117	320	137	456	118	750	157	1026	184
2¼	77	165	112	170	170	148	250	174	360	149	590	199	903	232
2½	65	204	94	210	140	182	210	214	300	184	490	245	896	287
2¾	56	247	80	255	120	221	175	259	260	222	425	296	692	347
3	47	294	68	303	100	263	149	308	220	264	360	353	603	413
3¼	39	345	55	356	87	308	128	362	186	311	305	414	496	485
3½	34	400	49	412	75	357	111	420	161	360	265	480	390	562
3¾	29	459	42	474	65	411	96	482	141	413	230	551	310	646
4	27	522	38	539	57	467	85	548	124	470	200	627	252	734

Note: Table is based on a 15 MPH wind with minimum cylinder stroke settings. For 10 MPH wind reduce capacities by 38%. 18 ft. windmill is not available. Cylinder size is diameter in inches. Size of windmill is diameter of wheel in feet.

sity. (Refer to Chapter 3 for more specific information on wind characteristics.)

Choosing the Correct Size. Given adequate wind, the key to windmill selection is determining the amount of water which must be pumped and the height to which it must be lifted. The distance water must travel from the bottom of the pump cylinder to the top of the tank is the total elevation. Thus, picking a windmill is a matter of deciding what size system will pump the desired average number of gallons/hour to a particular total elevation.

Table 4-4 shows the pumping capacities in a 15-mph wind of several sizes of windmills at minimum stroke settings. Lengthening the stroke on mills designed for such adjustment results in more gallons of pumping per hour, but it decreases the total elevation to which water may be raised. Increasing the diameter of the wheel results in pumping to greater total elevation.

It is usually best to choose the largest wheel and the smallest cylinder consistent with need. This not only starts pumping in lighter winds, but it also minimizes mechanical strain on the system as a whole.

5
Applied Aerodynamics

INTRODUCTION

Recent interest in wind machines has resulted in the re-invention and analysis of many of the wind power machines developed in the past. Because of the considerable time period since the last large-scale interest in wind power in this country, which occurred over 25 years ago, much of the information that was published is out of print or not generally available. It is the purpose of this chapter to review the aerodynamics of various types of wind power machines and to indicate advantages and disadvantages of various schemes for obtaining power from the wind.

The advent of the digital computer makes the task of preparing general performance plots for wind machines quite easy. Simple one-dimensional models for various power-producing machines are given, along with their performance characteristics, and are presented as a function of their elementary aerodynamic and kinematic characteristics. Propeller-type wind turbine theory is reviewed to the level of strip theory, including both induced axial and tangential velocities. It is intended that this material be of use in rapid evaluation and comparative analysis of the aerodynamic performance of wind power machines.

Role of Aerodynamics in Wind Power

The success of wind power as an alternate energy source is obviously a direct function of the economics of production of wind power machines. In this regard, the role of improved power output through the development of better aerodynamic performance offers some potential return; however, the focus is on the cost of the entire system, of which the air-to-mechanical-energy transducer is but one part. The technology and methodology used to develop present day fixed and rotating wind aircraft appears to be adequate to develop wind power.

One of the key areas associated with future development of wind power is rotor dynamics. The interaction of inertial, elastic and aerodynamic forces will have a direct bearing on the manufacture, life and operation of wind power systems, while at the same time having a minor effect on the power output. Thus, the aerodynamics of performance prediction, quasi-static in nature, is deemed adequately developed, while the subject of aeroelasticity remains to be transferred from aircraft applications to wind power applications.

Since 1920, there have been numerous attempts in designing feasible WECS for large-scale power generation in accordance with modern theories. This section describes representative types of these designs.

It is convenient to classify wind-driven machines by the direction of their axis of rotation relative to wind direction, as follows:

- *Wind-axis machines*—machines whose axis rotation is parallel to the direction of the wind.
- *Cross-wind-axis machines*—machines whose axis of rotation is perpendicular to the direction of the wind.

Cross-Wind-Axis Machines

Savonius Rotor. The Savonius Rotor, in its most simplified form, appears as a vertical cylinder sliced in half from top to bottom, the two halves being displaced as shown in Figure 5-1. It appears to work on the same principle as a cup anemometer, with the addition that wind can pass between the bent sheets. In this manner, torque is produced by the pressure difference between the concave and convex surfaces of the half facing the wind, and also by recirculation effects on the convex surface that comes backwards upwind. The Savonius design was fairly efficient, reaching a maximum of around 31 percent, but it was very inefficient with respect to the weight per unit power output, since its construction results in all the area that is swept out being occupied by metal. A Savonius rotor requires 30 times more surface for the same power as a conventional rotor blade wind turbine. Therefore, it is only useful and economical for small power requirements.

Madaras Rotor. The Madaras Rotor works on the principle of the Magnus effect. In essence, it involves a boundary layer control technique which attempts to suppress boundary layer formation by reduction of the relative velocity between the fluid and the solid boundary. The simplest way to achieve the Magnus effect involves the rotating of a cylinder. Figure 5-2 shows the flow pattern which exists about a rotating cylinder placed in a

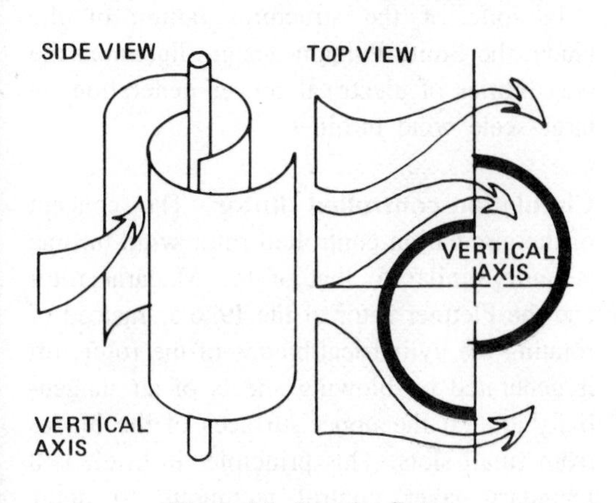

Figure 5-1. Savonius Rotor.

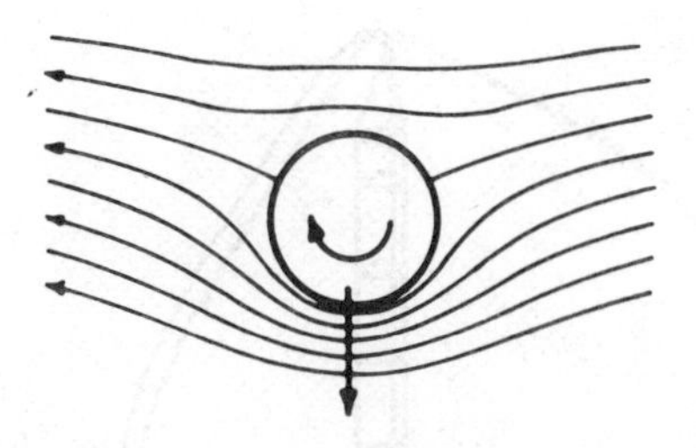

Figure 5-2. Magnus effect.

stream at a right angle to the flow. On the upper half of the cylinder surface, when the flow and the cylinder are moving in the same direction, separation is completely eliminated. On the lower side, separation is only partly developed. Thus, circulation is induced, causing a lift force perpendicular to the flow and the axis of the cylinder.

Madaras proposed to construct a circular track around which rotating cylinders, mounted vertically on flat cars, would move. Each cylinder was to have been 90 ft high and 18 ft in diameter. It would be driven by an electric motor. The Magnus effect would propel the cars around the track and drive generators connected to the car axles. However the system's poor aerodynamic design, mechanical losses and electrical losses, coupled with its unsuitability for use on mountaintop locations, resulted in very little being done with this design. A single full-sized cylinder was built in Burlington, New Jersey. DOE is performing additional analysis of the Madaras concept.

Darrieus Rotor. Georges Darrieus of Paris filed a U.S. patent in 1926 for a vertical-axis rotor (sketched in Figure 5-3).

The Darrieus rotor has been investigated by South and Rangi of the National Research Council of Canada in Ottawa. This rotor has performance near that of a propeller-type rotor and requires power input for starting. The simplicity of design and associated potential for low-cost production make it a promising candidate for economical power production. The ability to scale the Darrieus-type rotor to higher levels of power production, 100 kW is under evolution. The Sandia Laboratory is

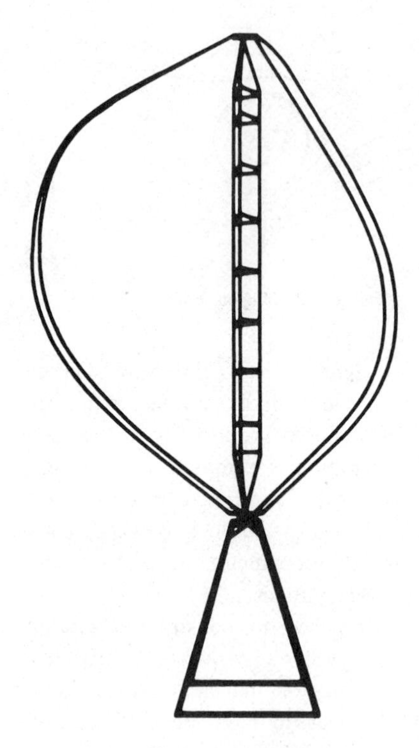

Figure 5-3. Darrieus rotor.

currently responsible for a progressive evolution of Darrieus rotor designs.

Wind-Axis Machines

Ducted Rotor. In 1954, the British built an experimental windmill with two hollow, airplane-type blades, as shown in Figure 5-4. Unlike conventional machines, it had no coupling between the propeller and the generator. As the blades were turned by the wind, centrifugal force pulled air from the hollow tower through the blade tips. At the same time, the pressure difference between the tip of the rotor and the blade pedestal also drew up air through the semi-vacuum created in the 100-ft-high tower. As air flowed through the tower, it passed through a turbine that drove a generator. The blade was 80 ft in diameter and was capable of producing 100 kW in a 35-mph wind at 95 rpm.

In order to maintain constant rotor speed, hydraulic motors were used to vary the blade pitch and were effective at wind speeds of 30–60 mph. The blades were designed so that they could flap under wind pressure of heavy gusts. The motion of the rotor to face into the wind was aided and controlled by a power-operated system. The main advantage of this system was that the power generating equipment was not supported aloft.

Smith-Putnam Design. The Smith-Putnam windmill built at Grandpa's Knob in Vermont was the largest ever constructed. The rotor diameter was 175 ft and consisted of two stainless steel blades using NACA 4418 airfoil sections. The rotor and generator weighed about 250 tons and were supported by a 100-ft tower.

The pitch control was automatic, keeping the blades at a constant speed of 28.7 rpm at wind velocities of 18 mph and above. As the wind velocity increased, the blades began to feather by turning edgewise. The blades were designed with an ability to come up to 20° to guard against sudden gusts and still maintain a reasonably constant speed. The coning was itself damped by oil-filled cylinders. The power plant was designed to withstand wind up to 120 mph and 100 mph with 6 in. of ice on the leading edge. The wind turbine was intended to generate 1,000 kW.

The turbine, shown in Figure 5-5, was erected in 1941 and operated as a test unit until February 1943, when the 24-in. main bearing failed and a replacement could not be secured for two years. In 1945, one of the blades flew off and ended experimentation with this design.

In spite of the structural failure of the blade, the Smith-Putnam design illustrated the possibilities of electrical power generation by large-scale wind turbines.

Circulation-controlled Rotor. The concept of the circulation-controlled rotor wind turbine is quite similar to that of the Madaras rotor and the Flettner rotor of the 1920's. Instead of rotating the cylindrical blades of the rotor, lift is generated by blowing sheets of air tangentially around the upper surfaces of the blades from small slots. This principle, in brief, is a boundary layer control technique to delay flow separation. Blowing re-energizes the low

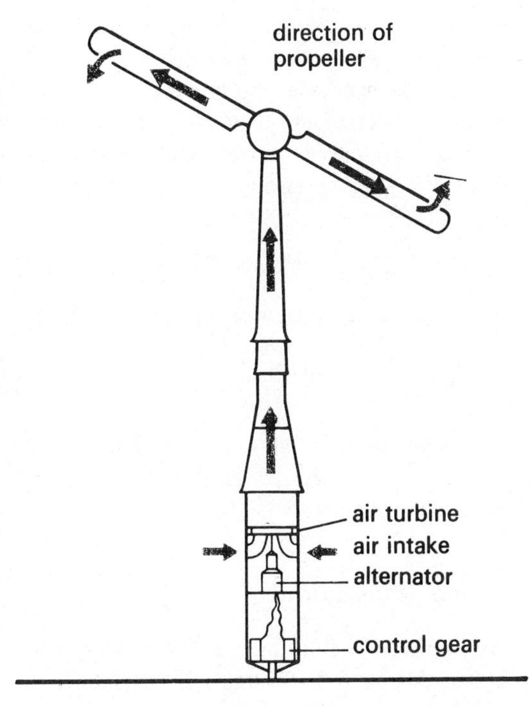

Figure 5-4. Enfield-Andreau ducted rotor.

energy boundary layer of the upper surface of the cylinder, thereby moving the point of separation farther back on the cylinder. Consequently, the pressure drag is reduced, but there is an accompanying increase in viscous drag. At the same time, circulation is induced by blowing and there is an increase in suction on the upper surface and a decrease in suction on the lower surface, all of which generate lift.

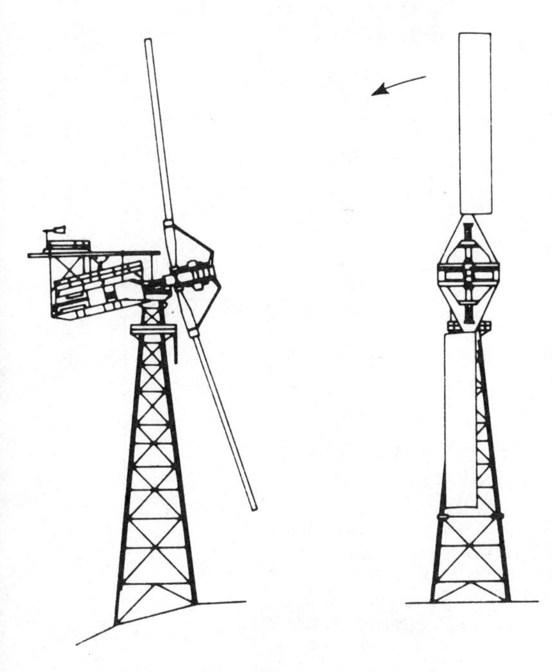

Figure 5-5. Smith-Putnam wind turbine.

This design possesses a number of advantages. First, at zero lift, the cylinder is insensitive to gusts; therefore, the rotor would not tend to speed up with sudden gusts. Second, no flapping or coning is needed, because the blade can be mounted rigidly to the hub without the difficulties of a conventional propeller blade that was solidly fastened. The large moment of inertia of a cylindrical cross-section of this type of blade causes it to be very stiff. The spanwise constant lift coefficient is achieved by adjusting the location of the slot, thereby foregoing complicated pitch controls. This design provides for easy construction and control, and supplies a very rigid structure to cope with its operating environment.

An analytical investigation of this design was made at Oregon State University, where it was found that at high-tip speed ratios, the compressor power to drive the jet was greater than power output from the rotor, while at low-tip speed ratios, the required rotor solidity (rotor projected area divided by the disc area) was large enough to offset the structural simplicity of a circular rotor.

Figure 5-6 gives a performance comparison of the various types of rotors that have been constructed.

TRANSLATING WIND POWER MACHINES

Drag Translators

Perhaps the simplest type of wind power machine is the device that moves in a straight line under action of the wind.

Historically, wind-driven translating devices have been used for propulsion rather than for power extraction. Analysis of translating lift-driven and drag-driven devices can be useful in examining various rotary machines, since the translation can be considered as an instantaneous blade element of a rotating machine. First, consider the machine to be driven by drag. Figure 5-7 illustrates the action of the elementary drag device.

For such a device, the power extracted, P, is the product of the drag and the translation velocity. The drag device sees a relative velocity $V_\infty - \nu$, so that the power is expressed by

$$P = D\nu = (1/2)\rho(V_\infty - \nu)^2 C_D S\nu, \quad (5\text{-}1)$$

which yields a maximum power coefficient,

$$C_{P_{\max}} \equiv \frac{P_{\max}}{(1/2)\rho S V_\infty^3} = (4/27)C_D, \quad (5\text{-}2)$$

at a velocity ratio $\nu/V_\infty = 1/3$. Further, the velocity of the device must always be less than the wind velocity, V_∞.

Lifting Translators

By contrast, the lifting translator does much better. Figure 5-8 illustrates a translating airfoil subject to lift and drag. The lifting surface sees a free-stream velocity, W, that makes an angle α with the wind, defined by

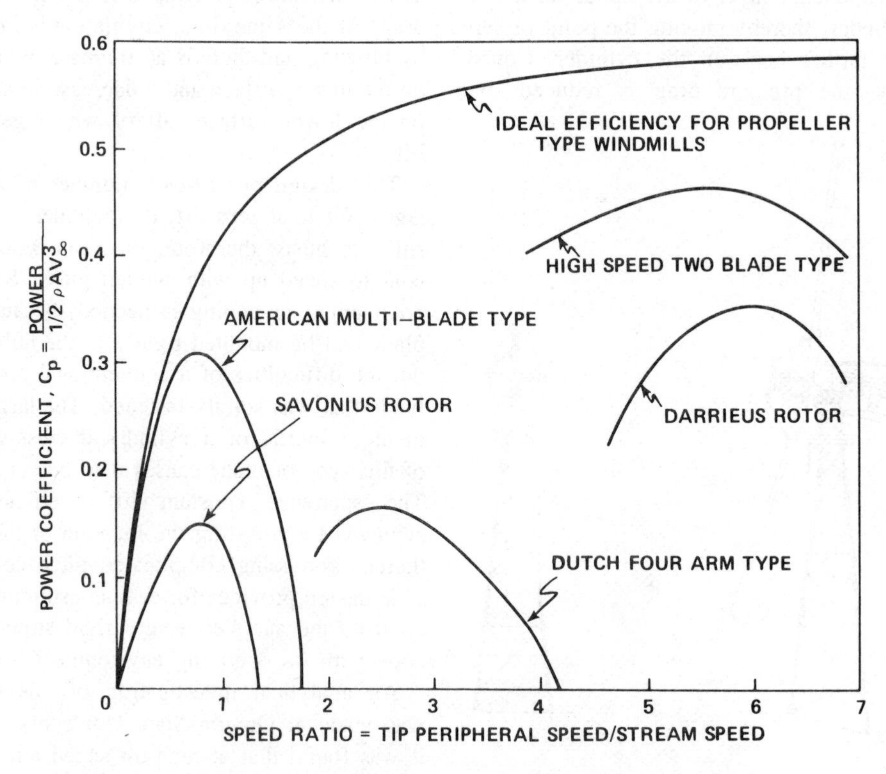

Figure 5-6. Typical performance of wind power machines.

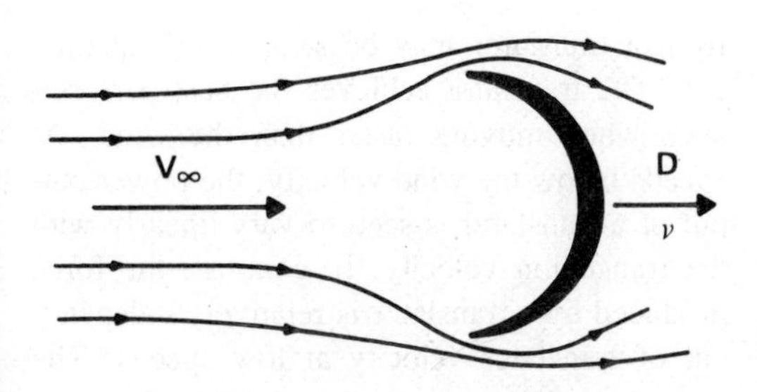

Figure 5-7. Translating drag device.

the relation tan $\alpha = \nu/V_\infty$. As the translator velocity increases, it may be noted that the lift and drag forces will rotate with the relative wind. Analysis of the free body yields the power extracted as

$$P = C_L(1/2)\rho V_\infty^3 S \left\{1 - \frac{\nu/V_\infty}{E}\right\} \frac{\nu}{V_\infty} \sqrt{1 + \nu^2/V_\infty^2}, E \equiv L/D, \tag{5-3}$$

so that the maximum power coefficient based on the airfoil projected area is given approximately by the expression

$$C_{P_{max}} \equiv \frac{P_{max}}{(1/2)\rho V_\infty^3 S} \doteq C_L \frac{2E}{9} \sqrt{1 + \frac{4E^2}{9}}. \tag{5-4}$$

The velocity ratio for maximum power coefficient is approximately

$$\left(\frac{\nu}{V_\infty}\right)_{C_{P_{max}}} \doteq \frac{2E}{3}. \tag{5-5}$$

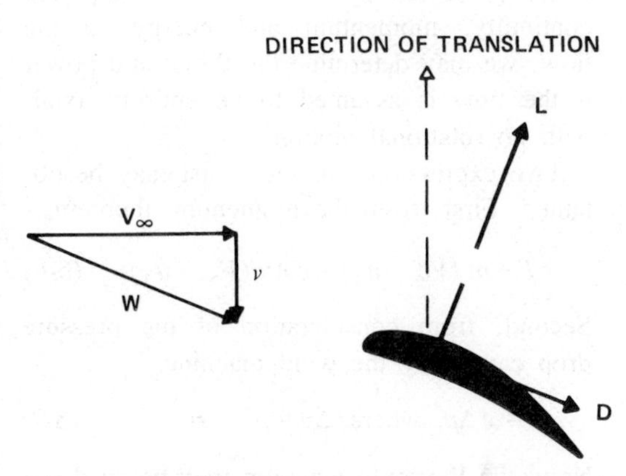

Figure 5-8. Translating airfoil.

Figure 5-9 illustrates the power extracted from a translating airfoil as a function of the lift to drag ratio and the velocity ratio ν/V_∞. A similar analysis can also be easily made for the case where the wind velocity is not perpendicular to the translation velocity. The power extracted for a lift to drag ratio of 10 is shown as a function of velocity ratio and wind angle β in Figure 5-10. It may be noted that the angle β that maximizes the power extracted is near zero, being closely approximated by $\beta \cong -1/(2E)$. The speed ratio for maximum power remains adequately approximated by Equation 5-5. Deviation of the angle β from zero produces a rapid reduction in the power extracted.

Comparison of lift and drag as mechanisms to remove power from the wind readily shows the advantages of using lifting surfaces. Lifting devices will operate at velocities in excess of the wind velocity, an advantage that is greatest for rotating machines. It is further noted that any rotary machine using drag must have a maximum power coefficient, based on maximum projected surface area, of less than $(4/27)C_{D_{max}}$.

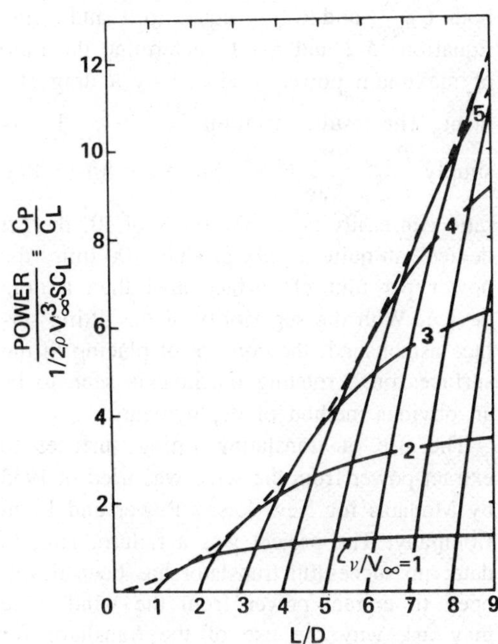

Figure 5-9. Power from a translating airfoil versus lift to drag ratio.

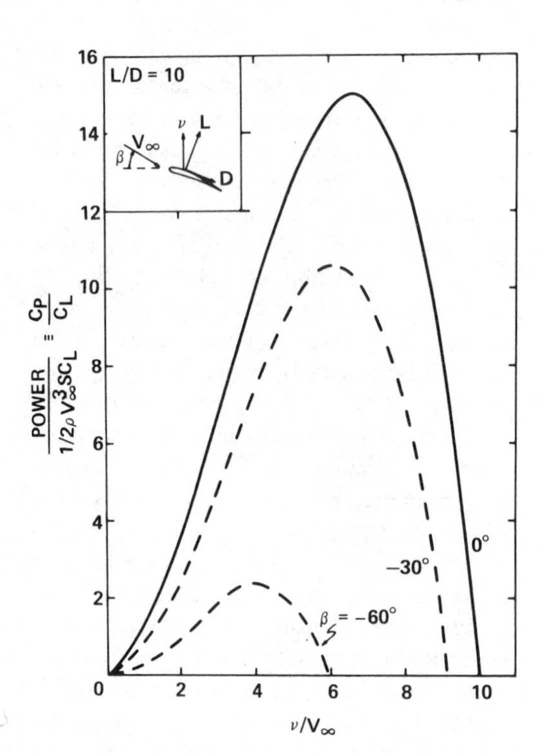

Figure 5-10. Translating airfoil with relative wind.

The relative efficiency of lifting elements to drag elements may be estimated by setting both $C_{D_{max}}$ and $C_{L_{max}}$ equal to 1 and using Equations 5-2 and 5-4 to determine the ratio of maximum power produced by a drag element. The result, assuming $\frac{4E^2}{9} >> 1$, is simply $\frac{P_{max}\text{ lift}}{P_{max}\text{ drag}} = E^2$. Since the lift-to-drag ratio can easily be on the order of 10, the lift device can quite readily produce 100 times the power per unit of surface area than a drag device. With the superiority of the lifting surface established, the concept of placing lifting surfaces on a rotating machine is seen to be an obvious method of deployment.

The use of translating lifting surfaces to extract power from the wind was tried in 1933 by Madaras for New Jersey Power and Light Company. The project was a failure, and, to date, no successful translator has been developed to extract power from the wind. One may ask why the use of the translator for propulsion has been so successful while the efforts in power have been few and unsuccessful. Part of the difficulty obtaining power from a translator may be seen from Equation 5-5. The translator achieves the best performance when moving faster than the wind. At speeds below the wind velocity, the power output of a translator is seen to vary linearly with the translation velocity. In contrast, the force produced by a translator is relatively independent of translator velocity at low speeds. The large speeds required for the translator to achieve high power extraction rates are the chief disadvantage, as large speeds mean extensive capital investment in machines and land. Other disadvantages of translators are proximity to the ground and sensitivity to changes in wind direction.

WIND-AXIS ROTORS: GENERAL MOMENTUM THEORY

Now let us turn our attention to wind turbines. The propeller-type windmill or wind turbine remains today, as in 1940, the most efficient machine and the leading candidate for large-scale wind power production. As a first step, we will consider a one-dimensional analysis of the output of a wind turbine and then proceed to a more detailed approach, linking blade geometry to power output.

Rankine-Froude Theory

Starting with the axial momentum theory originated by Rankine and W. and R. E. Froude, consider flow past a wind turbine as shown in Figure 5-11. The free stream wind is V_∞, which is slowed by a wind device. Applying continuity, momentum and energy to the flow, we may determine the thrust and power if the flow is assumed to be entirely axial, with no rotational motion.

Two expressions for the thrust may be obtained. First, from the momentum theorem:

$$T = m\,(V_\infty - u_1) = \rho A u\,(V_\infty - u_1). \qquad (5\text{-}6)$$

Second, from consideration of the pressure drop caused by the wind machine:

$$T = A\Delta p, \text{ where } \Delta p = p^+ - p^-. \qquad (5\text{-}7)$$

Now, the Bernoulli equation may be used between free stream and the upwind side of the turbine, and again between the downwind side of the turbine and the wake, so that

$$T = \rho \frac{A}{2} (V_\infty^2 - u_1^2). \tag{5-8}$$

Together with the momentum expression, we obtain

$$u = \frac{V_\infty + u_1}{2}. \tag{5-9}$$

In other words, the velocity at the disc is the average of the initial and final velocities. If we denote $V_\infty - u = aV_\infty$, note that $V_\infty - u_1 = 2aV_\infty$, and the final wake velocity change, $V_\infty - u_1$, is twice the velocity change at the disc. The thrust is not immediately of great importance; however, the power is. From the first law of thermodynamics, assuming isothermal flow, with $P_1 = P_\infty$:

$$P = \rho A u \left\{ \frac{V_\infty^2}{2} - \frac{u_1^2}{2} \right\} = \frac{\rho A u}{2} (V_\infty + u_1)(V_\infty - u_1)$$

or

$$\frac{P}{1/2 \rho A V_\infty^3} = 4a(1-a)^2, \tag{5-10}$$

which has a maximum when $a = 1/3$:

$$\frac{P_{\max}}{1/2 \rho A V_\infty^3} = \frac{16}{27} = 0.593. \tag{5-11}$$

Thus, a maximum power is defined. The term a is known as the axial interference factor and is a measure of the influence of the turbine on the air. The minimum final wake velocity is zero, so as $u_1 = V_\infty(1 - 2a)$, we obtain $a_{max} = 1/2$.

When examining Equation 5-11, it may be noted that the denominator is the kinetic energy of the wind contained in an area equivalent to that swept out by the rotor. Equation 5-11, however, does not represent the maximum efficiency, since the mass flow rate through the disc is not AV_∞ but Au. Hence, the efficiency, power output divided by power available, is given by

$$\frac{P}{\rho A u \frac{V_\infty^2}{2}} = 4a(1-a). \tag{5-12}$$

The maximum efficiency is 100 percent at $a = 1/2$, which yields a power coefficient of 0.5. The efficiency at maximum power coefficient is 88.8 percent.

Further one-dimensional modeling can be accomplished with the additional consideration of wake rotation. As the initial stream is not rotational, interaction with a rotating wind machine will cause the wake to rotate. In the case of a propeller, the wake rotates in the direction of the propeller; in the case of an energy-extracting device (windmill), the wake rotates in the opposite sense. If there is rotational kinetic energy in the wake, in addition to translational kinetic energy, then from thermodynamic considerations, we may expect lower power extraction than in the case of the wake having only translation.

The following simple example will relate wake rotational kinetic energy to rotor angular velocity.

Initial kinetic energy $= E_{T_1}$
Power Extracted $= P$
Final kinetic energy $= E_{T_2} + E_{R_2}$
(Translation) (Rotation)

From thermodynamics:

$$P = E_{T_1} - E_{T_2} - E_{R_2}.$$

As $P =$ (torque) $\times$ (torque) $\times$ (angular velocity), note that increased torque produces greater wake angular momentum and thereby greater wake rotational kinetic energy, so, for a given amount of initial energy E_{T_1}, the greatest power extraction will occur when E_{R_2} is low (which means high angular velocity and low torque).

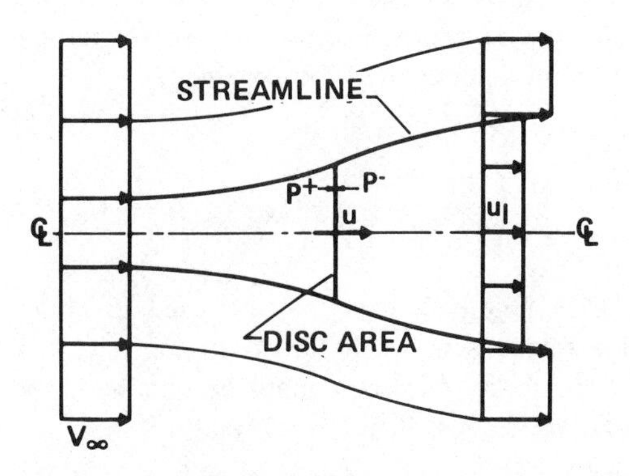

Figure 5-11. One-dimensional flow past a wind turbine.

Effect of Wake Rotation

Joukowski considered the effect of wake rotation in the analysis of propellers. Adopting his notation to the analysis of wind turbines, the effect of wake rotation on power removal may be estimated. The wake flow model, if assumed to be irrotational, produces unrealistic rotational velocities near the rotation axis. However, the contribution of the regions of high angular velocities may be subtracted out and a rotational core inserted, yielding a simple model which affords utility to the results in establishing bounds.

Using a streamtube analysis, equations can be written that express the relation between the wake velocities, both axial and rotational, and the corresponding velocities at the rotor. In addition, for certain special cases, an expression for the power coefficient can be obtained. The main outcome of this approach is a measure of the effects of rotation on the relative values of the induced velocities at the rotor and in the wake.

Figure 5-12 illustrates the streamtube. The resulting equations are given below.

Continuity: $urdr = u_1 r_1 dr$. (5-13)

Moment of momentum: $r^2\omega = r^2{}_1\omega_1$ (5-14)

(where ω and ω_1 are the rotor and wake angular velocities of the fluid). In addition, we may obtain an energy equation.

$$1/2(u_1 - V_\infty)^2 = \left(\frac{\Omega + \frac{\omega_1}{2}}{u_1} - \frac{\Omega + \frac{\omega}{2}}{u} \right) u_1 \omega_1 r_1^2 \tag{5-15}$$

(where Ω is the angular velocity of the rotor). Finally, an expression for the radial gradient in axial velocity may be obtained.

$$\frac{d}{dr_1}\left(\frac{V_\infty^2 - u_1^2}{2} \right) = (\Omega + \omega_1)\,\frac{d}{dr_1}\,(\omega_1 r_1^2). \tag{5-16}$$

These four equations may be used to obtain the relations between thrust, torque and flow in the wake. Closure cannot be obtained, and one needs specification of one of the variables, say ω, in order to obtain a solution.

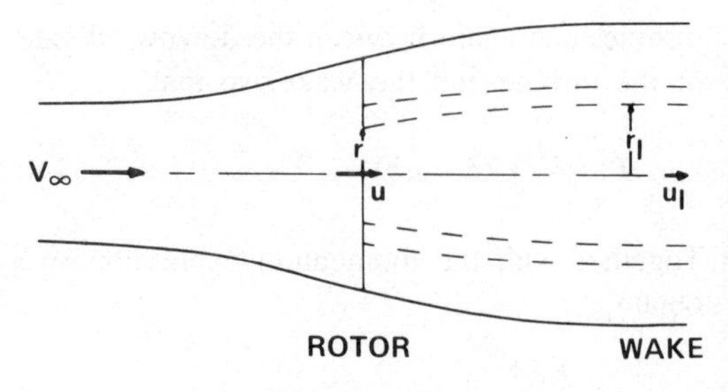

Figure 5-12. Streamtube geometry.

The particular forms of the momentum equation used are Bernoulli's equation and Euler's equation.

Several features of the flow may be noted.

1. The pressure varies across the wake due to the rotational velocity.
2. The rotor and wake axial velocities vary radially.
3. The angular velocity of the fluid, which is opposite the direction of rotation of the rotor, changes discontinuously at the rotor.
4. Fluid drag has been assumed to be zero.

Expressions for the torque and thrust for an annular element may also be obtained.

Torque: $dQ = pur^2\omega dA$.

Thrust: $dT = p\left(\Omega + \frac{\omega}{2}\right) r^2\omega dA$.

From the expression for the wake radial velocity gradient, it may be seen that when $r^2\omega$ is constant, the wake axial velocity is constant. Defining

$$u_1 \equiv V_\infty(1 - b),$$

$$u \equiv V_\infty(1 - a),$$

we may obtain

$$a = \frac{b}{2}\left(1 - \frac{(1-a)b^2}{4X^2(b-a)}\right) \tag{5-17}$$

and

$$C_P \equiv \frac{\text{power}}{1/2\rho V_\infty^3 A} = \frac{b^2(1-a)^2}{b-a}.$$

Figure 5-13 illustrates the variation of the ratio a/b as a function of a and X. It may be

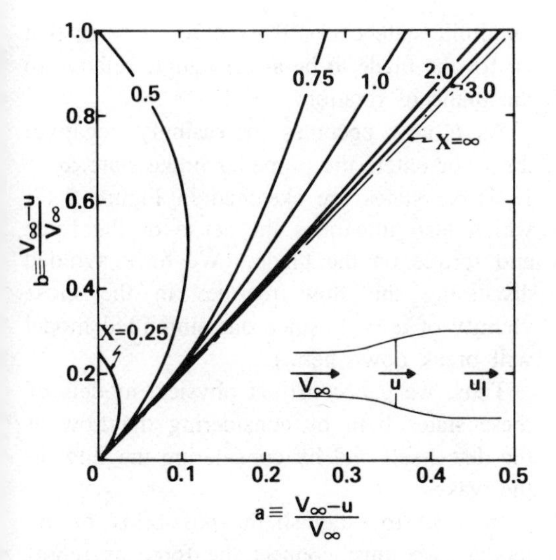

Figure 5-13. Effect of tip speed ratio on the induced velocities for flow with an irrotational wake.

observed that the axial velocity change at the disc is always approximately 1/2 the value in wake for tip speed ratios above 2.

The power coefficient requires some modification since $r^2\omega$ = constant produced infinite velocities near the axis. In lieu of an irrotational vortex wake, we may substitute a Rankine vortex wake. Letting $N \equiv \Omega/\omega_{max}$, we obtain

$$C_P = \frac{b(1-a)^2}{b-a}\left[2Na + (1-N)b\right]. \quad (5\text{-}18)$$

The maximum power coefficient for a rotor with a Rankine vortex wake is shown in Figure 5-14. As would be expected, the highest values of power coefficient occur at high tip speed ratios, where the torque, and, consequently, the wake rotation, are the least.

The flow model used to arrive at these results requires the flow to occur in annular, non-interacting steam tubes. Goorjian has recently criticized this flow model. In spite of the difficulties associated with this model, it affords some insight into the effect of neglecting wake rotation in blade element theories of wind turbines.

Simple Model of Multiple Flow States

In the previous analysis, it has been tacitly assumed that the device is operating as a draglike power extraction device; that is, $0 < a < 1$. For $a < 0$, it is quite simple to continue the analysis to show that the device will act as a propulsor, producing thrust and adding energy to the wake flow. This flow regime is typical of that type of a propeller.

A particularly interesting case occurs for $a > 1$. This may be physically modeled by considering a powered propeller with its pitch adjusted so that it induces a forward flow; that is, a propeller in the reverse thrust (or brake) state. An idealized streamlike pattern is shown in Figure 5-15.

Continuing the analysis using the same approach as in the Section on the Rankine-Froude theory, we find in this case that

$$C_P = -4a(1-a)^2 \quad (5\text{-}19)$$

$$C_T \equiv \frac{\text{force}}{1/2\rho A V_\infty^2} = -4a(1-a). \quad (5\text{-}20)$$

Thus, all three cases can be written in the form

$$C_T = 4a\,|1-a| \quad (5\text{-}21)$$

$$C_P = 4a\,|1-a|(1-a). \quad (5\text{-}22)$$

We note that positive C_T represents a draglike force and positive C_P energy taken from the wind stream.

Thus, by using simple momentum theory alone, it is possible to construct three induced flow patterns or values of a given thrust coefficient. These are illustrated in Figure 5-16. While this is a highly idealized model of the flow, it can still be exploited to provide an insight into the flow states. In practice, such states will generally occur when a rotor operates at tip speed ratios appreciably different than the design tip speed ratio, when the tips may be driven into the propeller brake state.

Figure 5-16 was developed considering the wake flow only. A further insight is obtained by considering the flow at the actuator itself, and observing how the different states might occur by varying the loading system at the actuator; for example, by changing the angle of attack by force-producing elements.

While these modes can occur on any actuator device, it is helpful to discuss them in

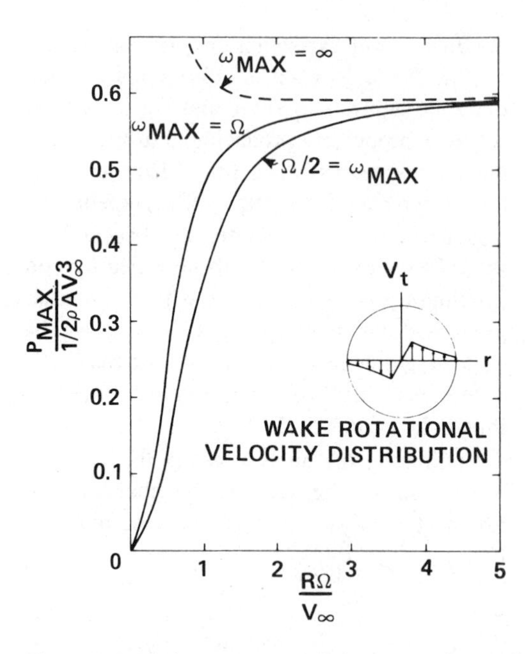

Figure 5-14. Maximum power coefficient versus tip speed ratio for a rotor with a Rankine vortex wake.

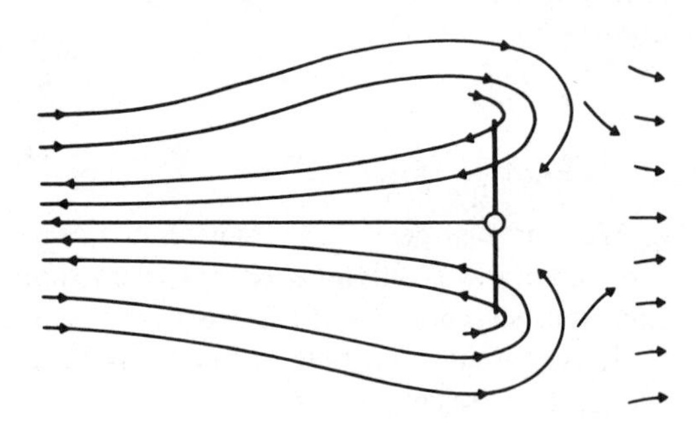

Figure 5-15. Propeller brake state.

terms of the familiar wind-axis-type propeller. Consider a rotor of variable pitch, driven at fixed angular velocity in an airstream of fixed speed, as in Figure 5-17. Assuming the blade angle θ is initially set so as to produce thrust, we will now find that $a < 0$ and that power must be provided to the system to maintain angular velocity. Now, as the blade angle θ is reduced, the thrust and power will reduce until the zero slip condition is reached. Here no torque is required and the rotor is producing no thrust. For further reductions in θ, the rotor enters the windmill state and power must be extracted from the shaft to keep angular speed constant, the maximum power occurring at $a = 1/3$. It should be noted that the windmill state can still exist for $\theta = 0$; that is, for the blade to be at zero angle relative to the plane of rotation.

As θ now becomes increasingly negative, the rotor enters the propeller brake state, $a > 1$. These states are sketched in Figure 5-17, which also illustrates the sense of the force and torque on the blade. (We have avoided discussing the flow regimes in the close vicinity of $a = 1$, since our simplified model will break down here.)

Thus, we can construct physical models of these states both by considering the flow at the disc itself and by considering the flow in the wake.

In order to establish the possibility of the modes, we must connect the force as represented by wake momentum to that as represented by lifting forces on the blade elements themselves. For our simple model, we will consider, as an example, the wind-axis rotor (propeller) and use conventional blade element theory, ignoring swirl terms and assuming the wake induced flow is twice that at the disk itself. This model is sketched in Figure 5-18.

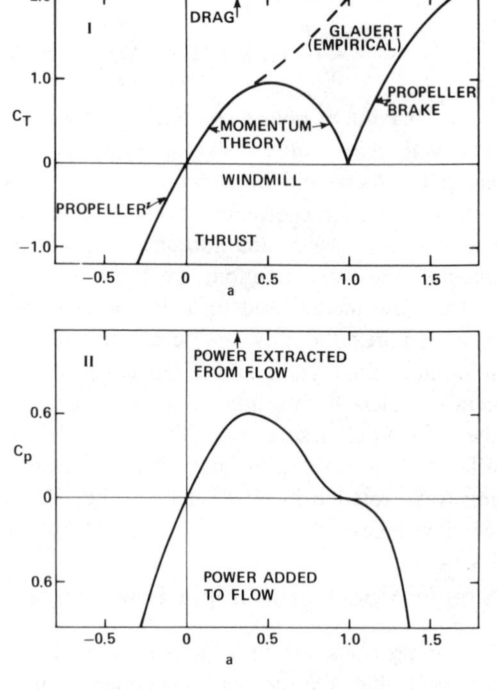

Figure 5-16. Rotor operation modes.

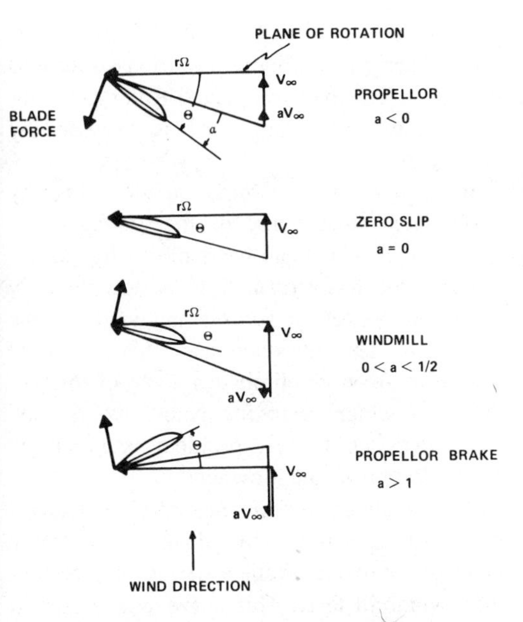

Figure 5-17. Blade force for various modes.

Assuming we are in the propeller or windmill state, $a < 1$. By momentum theory, the force on the annulus is given by

$$dT = \rho V_\infty^2(1 - a)2a \cdot 2\pi r dr \quad (5\text{-}23)$$

and the local thrust coefficient is given by

$$C_T = 4a(1 - a). \quad (5\text{-}24)$$

Now, considering flow at the blade element itself, we get the circulation, from

$$\Gamma = \rho \pi c\,[V_\infty(1 - a)\cos\theta - \Omega r \sin\theta\,]. \quad (5\text{-}25)$$

Thus, the force on the annulus is given by

$$dT = \Omega r \Gamma dr, \quad (5\text{-}26)$$

$$C_T = \frac{xc}{r}\,[(1 - a)\cos\theta - x\sin\theta\,] \quad (5\text{-}27)$$

where x is the local tip speed ratio.

For the propeller brake state, $a > 1$, we get by momentum theory $C_T = -4a(1 - a)$, while the blade force is given by the same result as previously. We can define a local solidity σ as $\sigma = cdr/\pi$rdr. Thus, we can write, for all a:

$$4\pi x\sigma\,[(1 - a)\cos\theta - x\sin\theta\,] = 4a\,|1 - a|. \quad (5\text{-}28)$$

The nature of solutions to this equation can most easily be seen from Figure 5-19. Note that for $\theta < \theta_1$, the simple powered thrusting propeller occurs, while for $\theta < \theta_2$, the propeller brake mode occurs. The angles in the intermediate range $\theta_2 > \theta > 0$ exhibit three possible equilibrium states, two windmill modes and one propeller brake mode.

It is of interest to note that the slope of the blade force line is a function of solidity and tip speed ratio.

For the triple mode case, it appears that the point shown as "B" is unstable, and that "A" and "C" are both stable and occur depending upon how the state is approached. A simplified explanation of why state "B" is unstable is as follows: Assume that at "B", the induced flow a is slightly increased; now the drag force on the disk (following θ = constant) becomes much larger than that represented by the wake momentum, and thus this wake momentum is further reduced and the system moves towards $a = 1.0$. On the other hand, "A" and "C" are stable according to these arguments. Thus, a working assumption in blade element theory is that no solutions with $1/2 < a < 1$ can occur.

It should be stressed that the above is an idealized model and that it inevitably involves flow inconsistencies. For example, it can be seen that a model giving $a < 1$ on an annulus which has inner and outer annuli with the value of $a < 0.5$ will somehow violate flow continuity.

We note that states of $a > 0.5$ should not occur in the major design range of a windmill. However, in cases where it is necessary

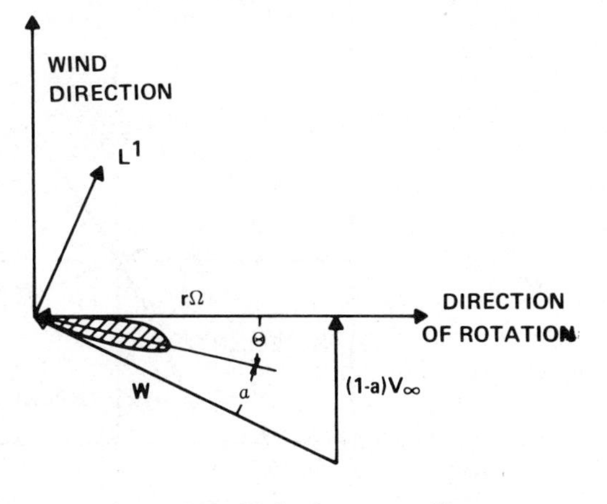

Figure 5-18. Blade element coordinates.

to prevent rotor overspeed due to high incoming winds or reduced shaft torque loads, it may be possible to use the confused flow of the propeller brake state to dump energy. This is a method of speed control which is quite distinct from the normal blade feather technique in which a is reduced. Note that this behavior is not the same as blade stall, which occurs at the low X region of the characteristic.

We note here that the present analysis should be considered a small perturbation model and thus should not be considered valid for $a > 0.5$. For example, for $0.5 < a < 1$, the simple one-dimensional model developed here implies flow reversal in the far wake and zero wake velocity somewhere between the actuator disc and downstream infinity, a streamline configuration which is physically unacceptable. Again, the simple propeller brake analysis must be considered quite inadequate in the vicinity of $a = 1$.

No satisfactory theories exist for flow in this region, although quite extensive research has been done on this problem in connection with helicopter rotor theory. In helicopter analysis, this region is that associated with a lifting descending rotor, where the anomalous states of the parachute brake, the turbulent wake and the vortex ring states occur (Shapiro).

For most normal windmill operating modes, a is less than 0.5; thus, it is seldom necessary to analyze conditions for $a > 0.5$. However, for off-design conditions, spurious solutions with $a > 0.5$ may well ocur. Rotor performance for the entire range of a is discussed by Wolkovitch.

It is of interest to note, as described by Wolkovitch, that many of the anomalies of flow near $a = 1.0$ can be removed by assuming that the free-stream flow is not precisely axial, but yawed at some small angle to the rotor axis. The introduction of this additional degree of freedom eliminates some of the singularities which occur for axial flow. A classical approach to this problem is given by Lock, Bateman and Townend.

A generalized performance curve of C_T versus a was constructed by Glauert, using these concepts and data from a series of free-running windmill tests. This curve was shown in Figures 5-16 and 5-19. Since these were free-running, these tests correspond to $C_P = 0$, or in helicopter terminology, the autorotative state. It should be noted that in the autorotative state, one portion of the rotor is driving the remainder, and thus, in fact, the rotor is subjected to non-uniform a, and the value of axial perturbation given is the *mean* a for the disc.

Ducted Actuators

Shrouds or ducts are frequently used to increase the static thrust of powered propellers.

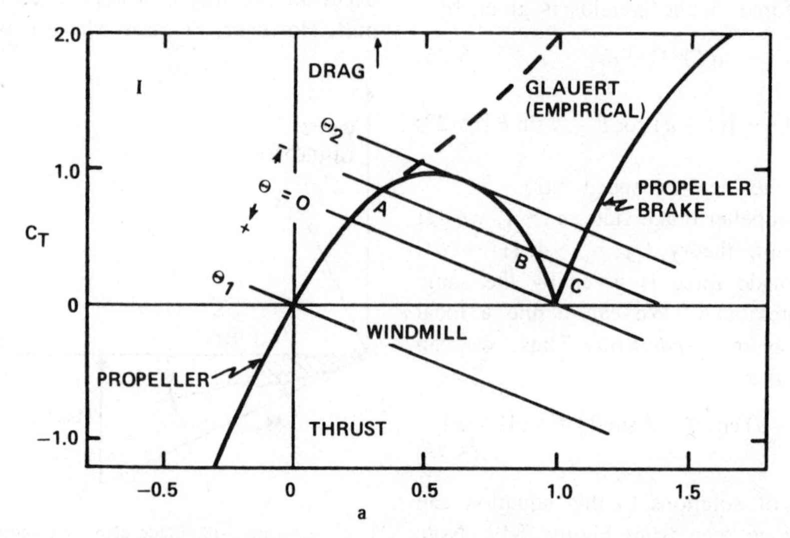

Figure 5-19. Blade element states for various blade pitch angles.

It has been well established that a duct can quite effectively reduce the slipstream contraction of a thrusting propeller and can thus increase its thrust to power ratio, at least at zero forward speed. It can be shown that, even ignoring skin friction, the effectiveness of the shroud reduces as the forward speed is increased, and when duct drag and other duct pitching moments are taken into account, the shrouded propeller is not effective technically. In calculations of ducted propeller performance, it is usual to assume that the flow leaves the duct exit at freestream static pressure; consequently, there is no further change in slipstream velocity, and for purposes of calculation, the duct exit area may be taken as the ultimate wake cross-section. In the case of a static free propeller, the ultimate slipstream is one-half the propeller area. Thus, any duct which causes the final slipstream contraction to be less than this will increase the thrust to power ratio of the system. It is of interest to observe that even a cylindrical duct of the same cross-section as the propeller will increase the thrust, and there will have been no slipstream contraction. It should be noted that the increased force is represented by a forward thrust on the duct, and a major part of this contribution is the force on the leading edge and entry area of the duct due to the low pressures there.

Because of the improvement in thrusting propeller performance due to a duct, it has frequently been suggested that a ducted windmill might have superior performance. A comprehensive analysis of ducted windmills is given by Lilley and Rainbird.

From a physical viewpoint, the effect of a duct will be to increase the wake expansion. We have showed that for a free windmill the optimal wake cross-section should be twice that of the windmill disc. Thus, if it is possible to cause the optimal wake cross-section to be larger than this, while still keeping the wake axial induced flow at the optimal level of two-thirds the freestream velocity, then, based on rotor area, the power coefficient will exceed the free rotor limit of 0.593. In effect, the duct has caused more flow to be drawn through the rotor and increased its power extraction capacity. A simple analysis of this follows, in which it is shown that unlike the free rotor, a momentum-type analysis cannot be made on this device without assumptions which are quite hard to justify.

In Figure 5-20, we show a typical ducted windmill system. Assuming the mass flow through the system is m, we can immediately write the power extracted as

$$P = m\Delta H = mV_\infty^2\, 2a(1 - a). \quad (5\text{-}29)$$

This force on the entire system (rotor and duct) may be written as

$$T = mV_\infty 2a. \quad (5\text{-}30)$$

We note that these equations are not closed in that we do not have an expression for m. For free actuator theory, the remaining equation is readily obtained by stating that the force on the system is the force on the actuator, which is given by $T = A\Delta p - A\Delta\Pi$, where A is the actuator area. This immediately gives the result $m = pAV_\infty(1 - a)$ for the free propeller case.

For our case, with the duct, it is still true that the propeller force is given by $A\Delta p$, but the duct force cannot be determined by simple momentum theory, since the pressure field on the outside of the duct is not known. By one-dimensional theory, the pressure on the duct interior can be calculated, except for the region very close to the leading edge. Thus, using momentum theory, one additional assumption is required.

We can consider this to be satisfied by assuming the velocity at the duct exit to be $V_\infty(1 - b)$. If we assume, as is done in powered ducted propeller theory, that the pressure at the duct exit is freestream static, then we get $b = 2a$ and the mass flow can be determined as $m = A_e pV_\infty(1 - 2a)$. Then,

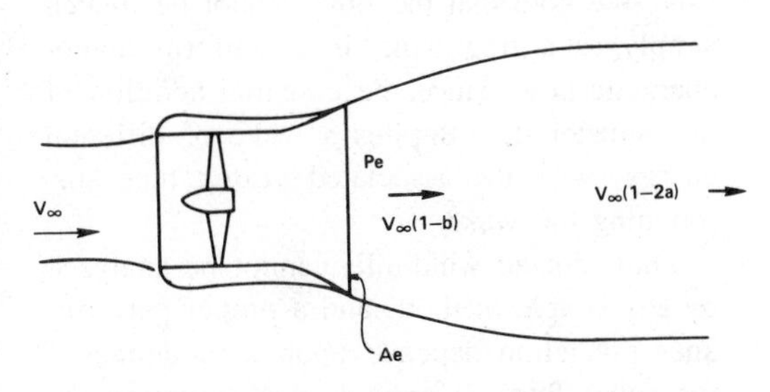

Figure 5-20. Ducted windmill geometry.

basing the power coefficient on the duct exit area A_e, we obtain

$$C_P = 4a(1 - a)(1 - 2a). \quad (5\text{-}31)$$

This expression can be maximized to give $C_{p_{max}} = 0.385$ at $a = 0.211$. If we write power coefficient in terms of rotor area, then we get

$$C_P = 4a(1 - a)(1 - 2a)A_e/A. \quad (5\text{-}32)$$

Observe that if the duct to rotor area ratio exceeds 1.54, then the power coefficient of the ducted system, based on rotor area, will exceed that of the rotor alone. At this level of analysis, all performance characteristics are determined by the assumption of the exit flow condition for the duct. This can be expressed by writing the power coefficient (based on exit area) and the duct exit pressure coefficient C_P*, which gives us

$$C_P = 4a(1 - a)(1 - b) \quad (5\text{-}33)$$

and

$$C_P^* = -(2a - b)(2 - 2a - b). \quad (5\text{-}34)$$

It will be seen that, assuming the exit pressure is lower than freestream static, which must be the case, wake expansion downstream of the duct has a higher mass flow and a higher power coefficient. In studying Lilley and Rainbird's paper, it must be noted that the performance is plotted in terms of the assumed duct exit pressure.

As described in the previous sections, it is possible in principle to compute the wake shape by potential flow techniques, assuming a contour, computing internal and external flows and ensuring pressure continuity on the wake bounding surface. Evidently, the details of the duct geometry must enter into this analysis. We note that the duct cannot be treated simply as a ring wing in a uniform homoenergetic flow, since the essential addition of the actuator disc implies a wake of different energy, with the associated vortex tube surrounding the wake.

Thus, ducted windmills cannot be analyzed by any simple method, and a proper performance prediction depends upon a modeling of the entire flow. It appears that assuming the exit pressure coefficient is a poor approximation, since the result is directly dependent on this quantity, which will vary notably for every duct rotor system, and even for a given system at different rotor loadings.

WIND-AXIS ROTORS: VORTEX/STRIP THEORY

Vortex Representation of the Wake

The wake of a windmill system consists of a flow of different total head from the mainstream. For an inviscid flow, the discontinuity in head may be represented by a sheet of vorticity. The mode of generation of this vorticity, and its geometry, can be of great assistance in developing models of the flow. In more advanced wake models, we usually stipulate the wake vorticity distribution and then use the Biot-Savart Law to calculate the induced flow of this wake. It is then possible to compute the pressure and flow fields on the wake to determine whether it is in equilibrium. Thus, a proper solution of the inviscid wake must involve both the *kinematics* and the *dynamics* of the flow. In other words, the wake shape and strength must generally be determined by an iterative process, where the initial geometry and strength are assumed and the induced flow is checked to assure that the wake streamline and pressure fields are consistent. A similar situation occurs in ordinary wing theory, although the interactive nature of the problem is usually removed by *assuming* the vortex wake leaves the wing parallel to the freestream flow. This implies that there will be downwash flow *through* the wake, a kinematically inconsistent situation. However, it is only in cases of very highly loaded wings that it is necessary to account for wake deformation.

Analogous assumptions are used for the actuator disc, where it is assumed that the wake vortex tube is parallel to the freestream flow.

If we consider the prototype actuator, a disc which may arbitrarily be switched from zero to infinite porosity, we can create a model of the vortex ring shedding process. Assume that the disc is oscillated forwards and backwards and is solid during the forward motion (against the mainstream) and fully porous during the rearward motion. The disc will now shed a series of ring vortices which will be convected downstream with the free-stream, as shown in Figure 5-21.

In the limit, if we assume a vortex tube of constant strength is developed, and using an hypothesis of light loads, the vortex tube will have the same diameter as the actuator disc.

Standard methods are available to compute the induced flow of a semi-infinite vortex tube at its end. We will not go into these here, except to state that the solution gives a uniform induced axial flow over the cross-section, although the radial flows are infinite at the tube edge. If another vortex tube of appropriate strength were added to this system, then the singular radial flows would be removed and the axial flow would become twice that at the end of a semi-infinite tube. This is another way of demonstrating the result already obtained from the momentum analysis, that the induced flow in the downstream wake is twice that at the disc.

It will be observed that this system has no tangential velocity in the wake and hence there is no torque. For this to be an approximate model of a propeller-type windmill, the tip speed ratio must be large so that for a given power the torque is in fact low.

The next refinement to add to this simple model is one which introduces torque. Consistent with actuator disc theory, we can model this with a large number of radial vorticity lines in the plane of the disc, representing a many-bladed system of constant blade circulation. In order to satisfy Helmholtz's Laws on the kinematics of vortex lines, we see that this implies a central vortex of finite strength with distributed streamwise vorticity along the wake cylinder (Figure 5-22).

A variant of this model is to assume that the actuator disc is an annulus. Then the surface of the inner vortex tube consists of ring and spanwise vortex lines of similar geometry, connected by radial vorticity at the disc itself.

For a lightly loaded system in which the wake boundaries may be considered right circular cylinders parallel to the mainstream, this is a fully self-consistent model with axial and tangential perturbations entirely confined to this annular cylinder. In other words, the induced flow of such a system does not affect other annuli, as can be seen by superimposing two circular vortex tube systems. Another annulus of completely different induction could be located inside or outside this one without affecting the induced flows in the first. Thus, the induced flow system of each annulus is a function only of the blade geometry in that annulus, and the angle of attack or chord of the blade in neighboring annuli can be changed without affecting adjacent induced flows. This interesting result of annular independence is the basis of blade element theory, which assumes that annuli do not interact. We note that this is different from the situation in wing theory, where changes in geometry at one spanwise station will affect induced flows at all other stations. It is apparent that it is the idealization of continuous streamwise vorticity on the vortex wake tube which effectively isolates the induction of an annulus. Thus, for

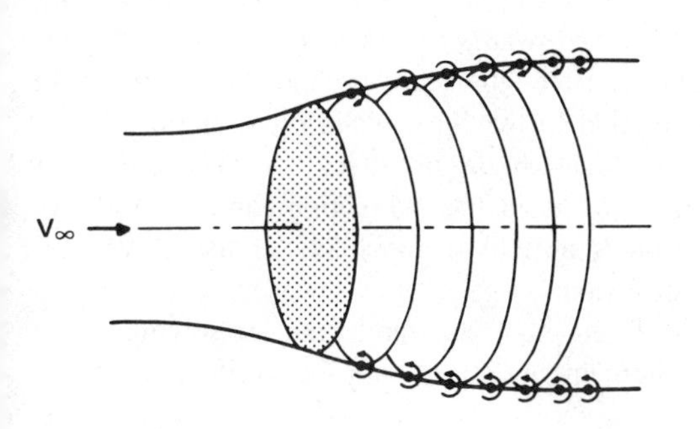

Figure 5-21. Actuator disc.

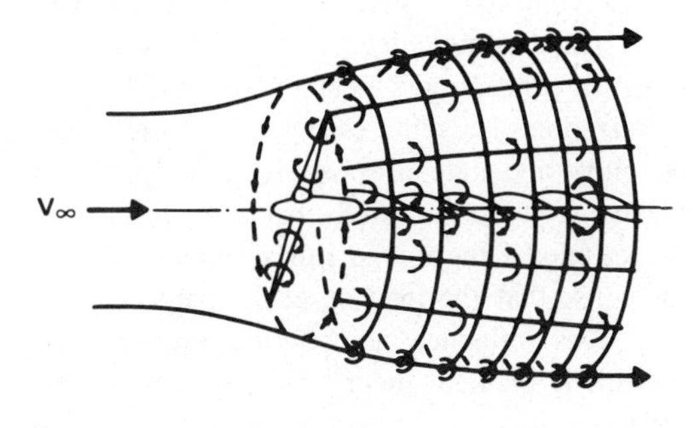

Figure 5-22. Vortex lattice system for a multi-bladed rotor (only two blades are shown).

non-interactive blade element theory to be valid, the product of the number of blades and the tip speed ratio should be large.

We note also that the concept of a continuous bounding vortex sheet composed of vortex rings permits differences in total head between the flows separated by the sheet. It permits the wake flow to be of reduced total head, as assumed in simple models.

If we now consider a more realistic rotor system having a finite number of blades and rotating at a finite velocity, a somewhat different situation occurs. Assuming that vortex shedding occurs only at the root and tips and that the vortex lies parallel to the local flow, then the wake vortex geometry becomes as shown in Figure 5-23. The helix angle of the vortices is directly related to the tip speed ratio.

We note that this finite bladed model contains somewhat similar structure to that of the wake of Figure 5-22, where the ring and streamwise vortex systems could be considered as *components* of the helix system of Figure 5-23. We note also that a large tip speed ratio, or a large number of blades, will cause the finite helix system to be more densely packed, so that the idealization of a continuous bounding vortex sheet becomes more realistic.

However, examination of Figure 5-23 will illustrate that, near each blade, the flow and vortex system is similar to that of a high-aspect ratio wing. Consequently, a vortex of finite strength cannot be shed from the tips, since this would imply infinite induced washes there. Thus, the local situation becomes quite similar to that of a yawing wing, and a continuous sheet of vorticity is shed from the trailing edge. Generally, this vorticity is concentrated near the tip so that the idealization of a finite strength tip vortex may be quite adequate a short distance from the blade. It will be noted that for the finite-bladed model there is spanwise interaction, in the sense that the load on each spanwise section does influence neighboring sections, so that blade element theory must be considered an approximation for a rotor with few blades at low advance ratios.

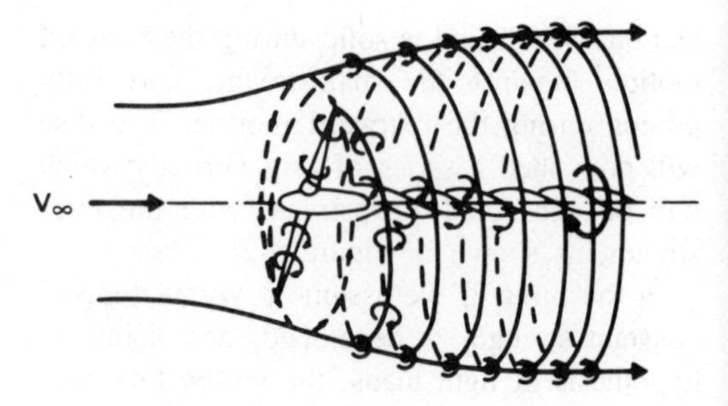

Figure 5-23. Idealization of vortex system of a two-bladed rotor.

An analysis for a two-bladed rotor system at very low advance ratio is given by Kuchemann, where the rotor is modeled as a rolling high-aspect ratio wing.

The model in which the blades sheds a system of helical vortex sheets is generally termed the Goldstein model. This elegant model is more complicated than most and we will not discuss it here.

Annulus Flow Equations

A frequently used and accurate method for performance calculations for propellers and helicopter rotors is to assume that the flow through the rotor occurs in non-interacting circular streamtubes. This method, when used in conjunction with the induced velocities, has been called by a variety of names, including modified blade element theory, blade element theory, vortex theory and strip theory. The method, which can be seen to assume locally 2-D flow at each radial station, proceeds as described below.

The element of a wind turbine rotor illustrated in Figure 5-24 is viewed from the tip looking towards the axis of rotation in Figure 5-25. Here, the relative wind, W, is shown in relation to the local blade pitch angle, θ, and the local angle of attack, α. The plane of rotation is in the x-direction, and the y-direction is normal to the blade in the downwind direction.

From the diagram, the following trigonometric relations may be verified:

$$\alpha = \phi - \theta \tag{5-35}$$

$$\tan \phi = \frac{1-a}{1+a'} \frac{V_\infty}{r\Omega} \tag{5-36}$$

$$C_y = C_L \cos \phi + C_D \sin \phi \tag{5-37}$$

$$C_x = C_L \sin \phi - C_D \cos \phi \tag{5-38}$$

(where C_L and C_D are the sectional lift and drag coefficients based upon the local relative velocity, W, and the local angle of attach, α).

A relation between the axial interference factor, a, and the forces developed on the blade may be obtained by equating the axial force, dT, generated in an annular element of thickness, dr, by momentum considerations to the axial force predicted from blade element aerodynamic considerations. From momentum:

$$dT_M = \rho(2\pi r dr)u(V_\infty - u_1), \tag{5-39}$$

while for B blades each having local chord c,

$$dT_B = Bc\tfrac{1}{2}\rho W^2 C_y dr. \tag{5-40}$$

Equating these two expressions and assuming that the local wake axial interference factor $b = 2a$, one obtains

$$\frac{a}{1-a} = \frac{BcC_y}{8\pi r \sin^2 \phi}. \tag{5-41}$$

In a similar manner, the torque determined from angular momentum considerations is equated to the torque developed from the blade element in an annular differential streamtube.

From the moment of momentum theorem, one obtains

$$dQ = \rho(2\pi r dr)ur(2a'\Omega), \tag{5-42}$$

where the angular velocity imparted to the slipstream has been assumed to be twice the angular velocity at the rotor disc. The blade-produced torque is

$$dQ = \rho Bc \frac{W^2}{2} C_x dr. \tag{5-43}$$

Combining these relations,

$$\frac{a'}{1+a'} = \frac{BcC_x}{4\pi r \sin^2 \phi}. \tag{5-44}$$

If suitable airfoil sectional performance data

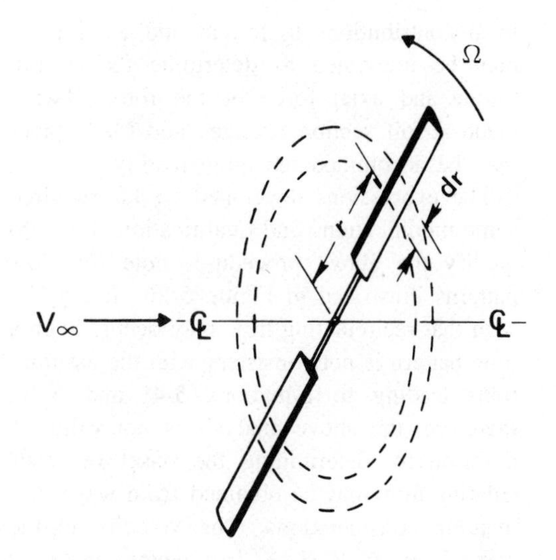

Figure 5-24. Rotor blade element.

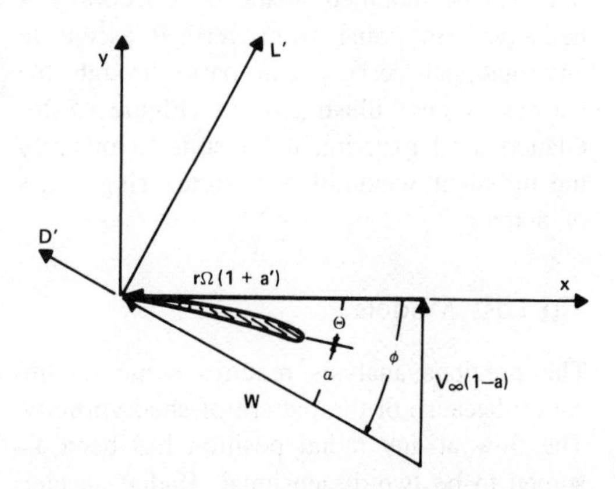

Figure 5-25. Velocity diagram for a rotor blade element.

are available, then the local flow conditions at a given radial station r may be determined by the following procedure:

Given r, c, $C_L(\alpha)$, $C_D(\alpha)$, θ, V_∞ and Ω:

1. Guess a and a' ($a = a' = 0$ is acceptable to start).
2. Calculate ϕ.
3. Calculate α.
4. Calculate C_L and C_D.
5. Calculate C_x and C_y.
6. Calculate a.
7. Calculate a'.
8. Go back to Step 2 and repeat.

Once the above interaction converges, the sectional flow properties are known and the local contributions to torque and axial force may be integrated to determine the overall torque and axial force of the rotor. Twist, blade airfoil section changes and blade taper may be accommodated quite readily.

The expressions developed so far required some modifications and qualification. First, to qualify the above procedure, note the flow patterns illustrated in Figure 5-26. It may be seen that recirculating flow may occur. Such a flow pattern is not consistent with the assumptions leading to Equations 5-41 and 5-44; therefore, the above analysis is not valid. A criterion for determining the onset of recirculating flow may be obtained from wake momentum considerations. The velocity in the wake is $\mu_1 = V_\infty(1 - 2a)$; hence, for $a > 1/2$, recirculation can occur. (This consideration will be modified in the next section.) A helicopter, in going from vertical ascent to autorotational descent can pass through the various states illustrated in Figure 5-26. Glauert used experimental results to quantify the turbulent windmill and vortex ring states of a rotor.

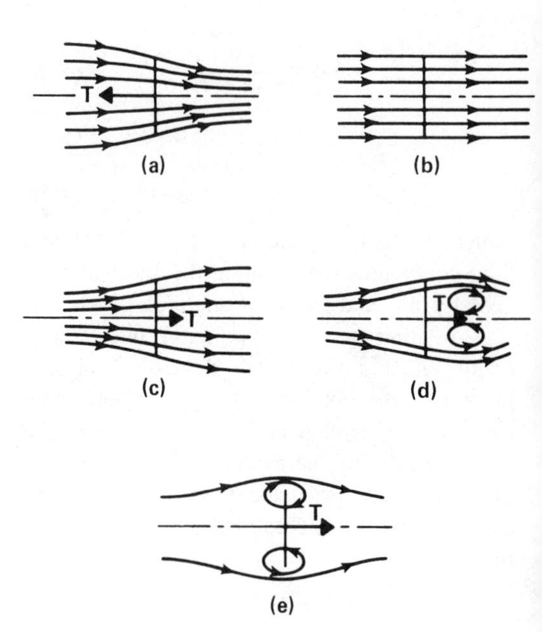

Figure 5-26. Working states of a rotor: (a) propeller; (b) zero-thrust; (c) windmill; (d) turbulent windmill; (e) vortex ring.

Tip Loss Models

The previous analysis requires some modification because of the pattern of shed vorticity. The flow at any radial position has been assumed to be two-dimensional. Radial acceleration and wake-induced flow at the tip can alter the assumed flow pattern. The effects of radial acceleration can be neglected for most wind power machines; however, the wake effects cannot be neglected. So-called tip losses have been treated with a variety of approaches, the simplest of these being to reduce the maximum rotor radius to some fraction of the actual radius, characteristically on the order of 97 percent of the actual radius. Prandtl and Goldstein have analyzed flow about lightly loaded propellers (negligible wake contraction) and developed models for the reduction of circulation due to wake interaction at the tips. The result of Prandtl and Goldstein's approach is a circulation-reduction factor F, such that

$$F \equiv \frac{B\Gamma}{\Gamma_\infty}, \tag{5-45}$$

where B is the number of blades; Γ is the circulation at a radial station r; and Γ_∞ is the corresponding circulation for a rotor with an infinite number of blades. The factor F is a function of tip speed ratio, number of blades and radial position. Of the two models, the Goldstein model is the more accurate. However, since Goldstein's flow model involves an infinite series of modified Bessel functions, it is more difficult to use. Since there is little difference in results for situations involving three or more blades, the Prandtl model, which yields a simple solution, can be used.

The incorporation of the tip loss factor into the equation for induced velocities proceeds as outlined below.

The physical meaning of the tip correction is virtually that the maximum change of axial velocity, $(V_\infty - u_1)$ or $2aV_\infty$, in the slipstream, occurs only on the vortex sheets, and the average velocity change is only a fraction (F) of this velocity. Thus, the velocity change $2aV_\infty$ becomes $2aFV_\infty$, and, in similar manner, the angular velocity change is written $2a'F\Omega$. Equations 5-41 and 5-44 then become

$$\frac{a}{1-a} = \frac{\sigma C_y}{8F\sin^2\phi} \tag{5-46}$$

$$\frac{a'}{1+a'} = \frac{\sigma C_x}{8F \sin \phi \cos \phi} \qquad 5\text{-}47$$

(where F is the Goldstein or Prandtl tip correction factor and the quantity σ is a local solidity given by $\sigma \equiv \frac{Bc}{\pi r}$). A further refinement of the analysis can be made in the axial flow velocity, u, through the rotor disc in Equation 5-39, which is assumed to vary in the same manner as the wake velocity. The average flow velocity through the rotor is then given by $u = V_\infty(1 - aF)$. Equation 5-47 remains the same. However, Equation 5-46 becomes a quadratic:

$$(1 - aF)aF = \frac{\sigma C_y}{8 \sin^2 \phi}. \qquad (5\text{-}48)$$

The use of Equation 5-48 in lieu of 5-46 yields slightly higher performance and significant reduction in the number of iterations required for convergence. It may be noted that the criterion for recirculating flow becomes $aF > 1/2$.

The Goldstein tip correction for a heavily loaded rotor may be determined following the method of Lock. Lock's approach bases the calculation of F on the local value of ϕ, so that $F = F(\phi, r/R)$. The angle ϕ defines a local speed ratio via the relation $\mu = \cot^{-1}\phi$. The corresponding tip speed ratio is $\mu_\infty = R\mu/r$, and thus the Goldstein tip correction factor, $F = F(\mu, \mu_\infty)$, can be determined. As a practical consideration, it may be noted that at low tip speed ratios, the tip loss is appreciable over the entire blade. In such cases, this approach ceases to be a tip loss correction, instead being a dominant factor in the calculations. Prandtl's F factor is given by

$$F = \frac{2}{\pi} \cos^{-1} [\exp(-f)] \qquad (5\text{-}49)$$

where

$$f = \frac{B}{2} \frac{R - r}{R \sin \phi}. \qquad (5\text{-}50)$$

As the factor F has been derived for a frictionless rotor with optimum distribution of circulation along the blade, the approximate nature of the previous analysis should be noted.

Figure 5-27 gives the calculated power coefficient of the Smith-Putnam wind turbine as a function of tip speed ratio. The Smith-Putnam turbine employed an NACA 4418 airfoil which had discontinuously twisted 5° along a 65-ft length. The turbine diameter was 175 ft with an 11-ft, 4-in. chord. The Goldstein tip correction was used to develop the curve.

The effect of pitch angle can be seen in Figure 5-27. Increased pitch reduces the maximum power but can increase the power available at low tip speed ratios. Figure 5-27 also can be used to illustrate some generalizations concerning wind machines. At low tip speed ratios, the power coefficient is strongly influenced by the maximum lift coefficient. The angle ϕ is large at low tip speed ratios, and much of the rotor, particularly the inboard stations, can be stalled when operating below the design speed. At tip speed ratios above the peak power coefficient, the effect of drag becomes dominant. A high drag coefficient will result in a rapid decrease in power with increasing angular velocity. Finally, at some large tip speed ratio, the net power output will become zero. If the slope of the power curve at $C_P = 0$ is negative, the rotor operation zero power output (feathered) will be stable, since for constant wind velocity, decreased rotational speed will result in positive power output, which, in turn, will return the rotor to its original speed. The steeper the curve, the greater the stability.

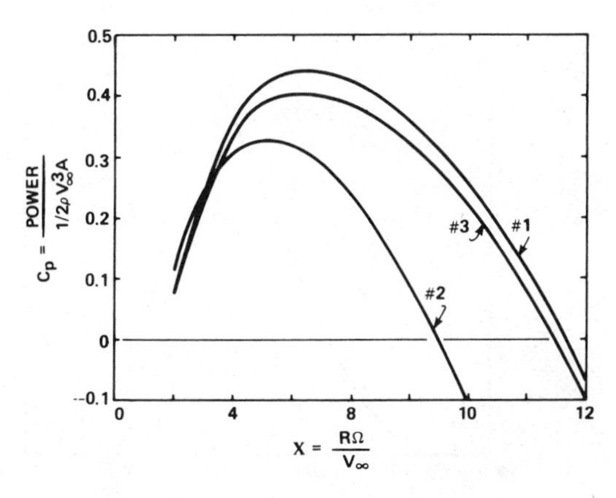

Figure 5-27. Calculated performance of the Smith-Putnam wind turbine.

1. $\theta_{pitch} = 0°$, Equations 5-47 and 5-48.
2. $\theta_{pitch} = 5°$, Equations 5-46 and 5-47.
3. $\theta_{pitch} = 0°$, Equations 5-46 and 5-47.

A plot of power coefficient versus tip speed ratio yields information concerning power output, efficiency and rotation speed for a given wind velocity. Another type of display that illustrates rotor performance is a plot of power versus wind velocity. Retaining C_P and X as our variables, the power is directly proportional to C_P/X^3, while velocity is given by $1/X$. Figure 5-28 illustrates the Smith-Putnam wind turbine calculations shown in Figure 5-27.

Note that at constant rpm and pitch angle, the stall controls the maximum power output and the drag controls the starting velocity. Another point of considerable importance is the location of the maximum power coefficient. Operation near the point of the maximum power coefficient will give the greatest increase in power for a given increase in wind velocity, and hence the greatest sensitivity to wind speed fluctuations.

The Optimum Rotor: Glauert

Glauert has developed a simple model for the optimum windmill. The approach used is to treat the rotor as a rotating actuator disc (i.e., it corresponds to a rotor with an infinite number of blades) and to set up an integral for the power. The power integral is made stationary subject to an energy constraint, the results yielding the maximum power output for a given tip speed ratio.

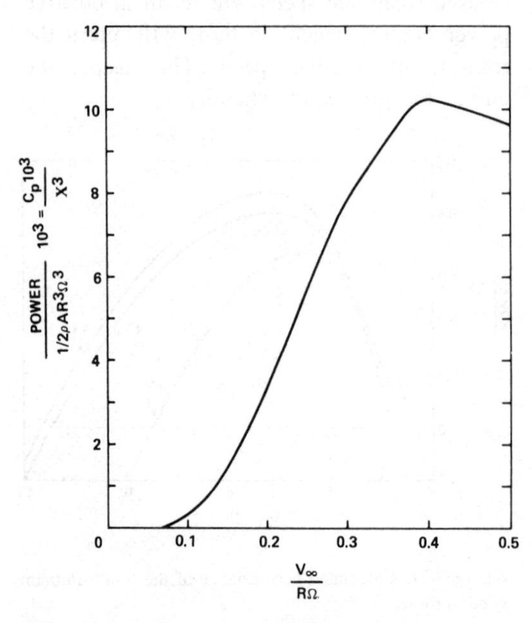

Figure 5-28. Power output versus wind speed for the Smith-Putnam wind turbine $\theta_p = 0°$.

The relation for the power coefficient is

$$C_P \equiv \frac{P}{1/2\rho V_\infty^3 \pi R^2} = \frac{8}{X^2}\int_0^X (1-a)a'x^3\,dx \tag{5-51}$$

where

$$x = \frac{r\Omega}{V_\infty},\ X = \frac{R\Omega}{V_\infty},\ a = \frac{V_\infty - u}{V_\infty} \text{ and } a' = \frac{\omega}{2\Omega}.$$

Since the integral for the power involves two dependent variables, another relation is required. This is the energy equation:

$$a'(1-a')x^2 = a(1-a). \tag{5-52}$$

Perhaps the most unusual way of illustrating this relation is to consider the velocities at the rotor plane (Figure 5-29). The flow is assumed to be uniform in annular streamtubes with no circumferential variations. Under these conditions, two-dimensional flow may be assumed.

In the absence of drag, the velocity induced at the rotor must be due to lift and hence be perpendicular to the relative velocity. Two expressions for tan ϕ may be developed under the condition that the total induced velocity is normal to the relative velocity (Equation 5-53):

$$\tan\phi = \frac{(1-a)V_\infty}{(1+a')r\Omega} = \frac{a'r\Omega}{aV_\infty} \tag{5-53}$$

so that

$$a'(1+a')x^2 = a(1-a). \tag{5-54}$$

The variational problem is now posed:

$$C_P = \int_0^X F(a, a', x)dx \tag{5-55}$$

with

$$G(a, a', x) = 0 = a'(1+a')x^2 - a(1-a).$$

The solution yields

$$a' = \frac{1 - 3a}{4a - 1} \qquad (5\text{-}56)$$

so that

$$a'x^2 = (1 - a)(4a - 1). \qquad (5\text{-}57)$$

Hence, $1/3 \geq a \geq 1/4$. The variations in a, a', $a'x^2$ and x are given in Table 5-1. Since high-speed rotors easily reach tip speed ratios of 7 or more, it can be seen that most of an ideal rotor will operate with $a = 1/3$ and the rotational velocity distributed in the form of an irrotational vortex; i.e., as

$$x \to \infty, a'x^2 \to (2/9) = \frac{\omega r^2}{2\Omega} \frac{\Omega^2}{V_\infty^2} = \frac{rV_t}{2} \frac{\Omega}{V_\infty^2}.$$

The power coefficient for various tip speed ratios is given in Table 5-2. At low tip speed ratios, the power coefficient is low because of the large rotational kinetic energy in the wake. At large tip speed ratios, the power coefficient approaches 0.593 and the wake rotation approaches zero. The variation of C_P with tip speed ratio is illustrated in Figure 5-6.

Further information may be obtained from this model using the blade element theory. As the quantities a and a' are known for each radial position, the relative velocity and the angle ϕ may be determined. Figure 5-5 may be used to illustrate the velocities and forces in relation to the blade configuration. Of course, since we have assumed that the drag is zero, the only force that acts on the blade is lift.

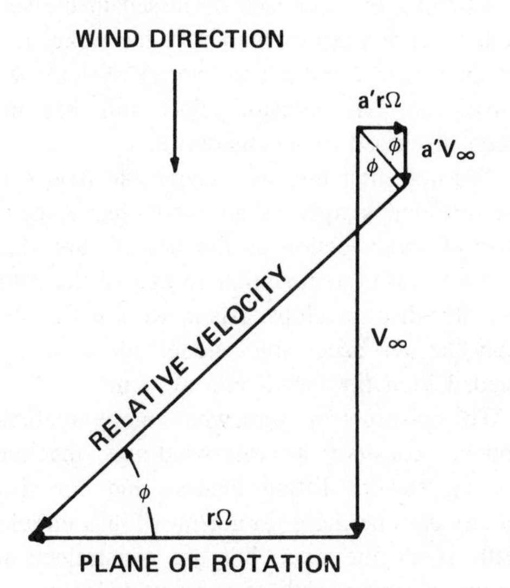

Figure 5-29. Velocity diagram.

Table 5-1. Flow conditions for the optimum actuator disc.

a	a'	$a'x^2$	x
0.25	∞	0	0
0.27	2.375	0.0584	0.157
0.29	0.812	0.1136	0.374
0.31	0.292	0.1656	0.753
0.33	0.031	0.2144	2.630
1/3	0	0.2222	∞

The incremental thrust and torque acting on an annulus containing B blades each having chord c are given by

$$dT = \frac{Bc}{2} \rho W^2 C_L \cos\phi \, dr \qquad (5\text{-}58)$$

and

$$dQ = \frac{Bc}{2} r\rho W^2 C_L \sin\phi \, dr. \qquad (5\text{-}59)$$

The momentum expressions yield (assuming $b = 2a$)

$$dT = 4\pi\rho V_\infty^2 (1 - a) a dr \qquad (5\text{-}60)$$

$$dQ = 4\pi r^3 \rho V_\infty \Omega (1 - a) a' dr \qquad (5\text{-}61)$$

so that

$$\frac{a}{1 - a} = \frac{BcC_L \cos\phi}{8\pi r \sin^2\phi} \qquad (5\text{-}62)$$

$$\frac{a'}{1 + a'} = \frac{BcC_L \sin\phi}{8\pi r \sin\phi \cos\phi} \qquad (5\text{-}63)$$

Now a and a' are known as a function of x so that the shape of the blades may be determined. Table 5-3 gives the results. It may be noted that an optimum blade for a given X and constant C_L will have a chord that approaches a maximum at $x \cong 0.7$.

Vortex Theory

The flow over real rotors differs in many respects from the flow model used to describe the optimum actuator disc. A frequently used model involves the use of bound vortices to

Table 5-2. C_P versus X for the optimum actuator disc.

$\frac{R\Omega}{V_\infty}$	C_P
0.5	0.288
1.0	0.416
1.5	0.480
2.0	0.512
2.5	0.532
5.0	0.570
7.5	0.582
10.0	0.593

represent lift. Following the concepts of vortex theory as applied to wings, each blade of the rotor is modeled as a bound vortex line. This simple scheme enables the induced flow at each section to be determined via the Biot-Savart Law. However, one may note that the induced flow will vary chordwise over the blade section. In order to fully represent the flow, the blade should be replaced by a bound vortex sheet in lieu of a vortex line. Since most windmill rotors have very low solidity, the chordwise variation in flow may be neglected without loss of accuracy.

In this scheme, the bound vorticity serves to produce the local life on the blade while the trailing vortex filaments induce velocities at each element of the blade. Several solutions for the induced velocity at a blade element have been obtained by solving partial differential equations, but the most straightforward method is a direct integration of the Biot-Savart Law. Now, as straightforward as this method may appear, it requires as an input the knowledge of the trajectory of the vortex filaments in the wake. Since the wake will consist of the superposition of a large array of vortex filaments, each acting on one another, the vortex trajectory (or configuration) cannot be established unless all the vortices are coupled. In Prandtl Lifting-Line Theory, the wake is assumed to lie in the plane of the wing, and although one can calculate physically impossible velocities which flow *through* the vortex sheet wake, the results of Prandtl's theory gives very acceptable answers. Just as the wake from a wing is a vortex sheet (which happens to roll up a short distance downstream of the wing), the wake shed by a propeller may also be considered as a vortex sheet (which also rolls up in the wake). This approach may be likened to that used in elementary strength of materials, where one assumes a deformation geometry and calculates forces—here we assume wake geometry and calculate induced velocities.

Table 5-3. Blade parameters for the optimum actuator disc.

ϕ	x	$\frac{Bc\Omega C_L}{2\pi V_\infty}$
50	0.35	0.497
30	1.00	0.536
20	1.73	0.418
15	2.43	0.329
10	3.73	0.228
7	5.39	0.161
5	7.60	0.116

For an optimum rotor using vortex theory, the Betz criterion may be used. This criterion requires the wake to move back as a rigid screw surface. The writings of Betz, Theodoresen, Lerbs and Weinig cover analytical techniques required to define the optimum propeller.

CROSS-WIND-AXIS MACHINES

Vortex Modeling of the Wake

Continuing the approach discussed in the section on vortex representation of the wake, it is of interest to construct the vortex system of a cross-wind-axis actuator, since this has not been discussed in the literature.

We note first that we assume the device to be modeled simply as an oscillating actuator disc of cross-section as the devices are shed and a wake system similar to that of the rotor actuator disc develops. Again we see that this may be an acceptable model for a many-bladed high tip speed ratio system.

To construct a somewhat more realistic model, consider a cross-wind-axis machine having slender lifting blades, and for simplicity assume these do not move in a circular path about the axis, but are constrained to follow a square path at constant velocity and at zero angle of attack relative to the path.

This model is shown in Figure 5-30. As a blade moves up the leeward sector, it sheds a starting vortex and a trailing pair as shown, and finally sheds its bound vortex as it assumes zero lift over the upper portion of the path. On passage across the forward portion, a similar situation occurs. Thus, the final wake system appears as shown in Figure 5-31. Note that the criss-cross system on the sides will converge to a simple ring-type system; that is, the streamwise vorticity component will cancel as the tip speed ratio and blade number are increased.

It can be shown that the solution for the induction of an infinite vortex tube of arbitrary cross-section is the same as that of one of circular cross-section, a uniform internal axial flow and zero external flow.

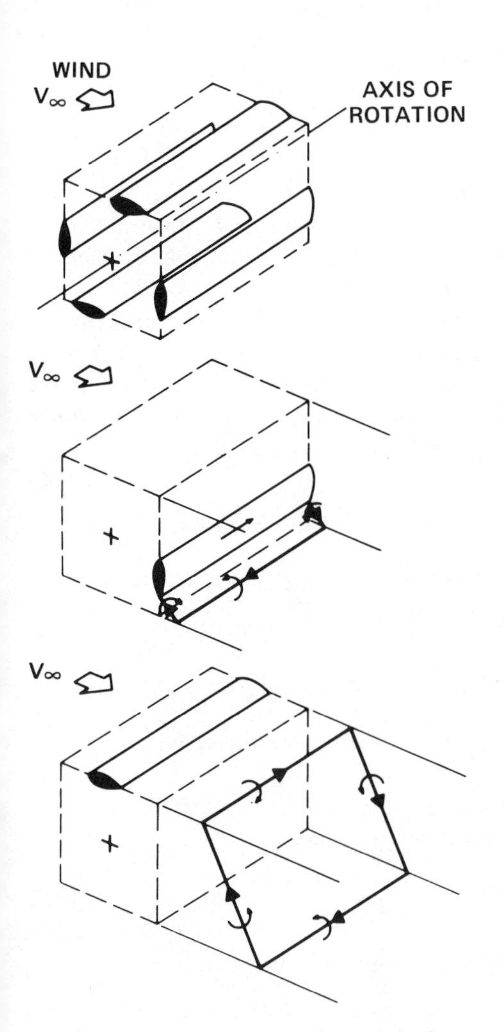

Figure 5-30. Vortex shedding of cross-wind-axis actuator.

If we now consider the more realistic case of a cross-wind-axis system where the blades rotate about a fixed axis, then the lift—and, consequently, the shed trailing and starting vorticity—is continually changing. Adopting arguments similar to those used for the square path system and assuming high advance ratios, we now obtain a wake vortex system as sketched in Figure 5-31. This is importantly different from the previous case, since there is internal spanwise vorticity within the tube. It can easily be shown that this spanwise vorticity is linearly distributed across the tube and that the induced internal axial flow is not uniform. Thus, it appears that even an ideal cross-axis machine cannot achieve the ideal power coefficient of a wind-axis system, since the induced axial flow is not uniform.

Darrieus Rotor

To analyze a Darrieus-type cross-wind-axis device, we adopt the standard approach of wing theory, which is to express the forces on the system by a momentum analysis of the wake, as well as by an airfoil theory at the lifting surface itself. The expression for these forces contains unknown induced flows. By equating the wake and wing forces, one obtains sufficient equations to determine the induced flows.

For the device considered, we assume that each spanwise (parallel to the axis) station behaves quasi-independently in the sense that the forces on the device at each station may be equated to the wake forces. In general,

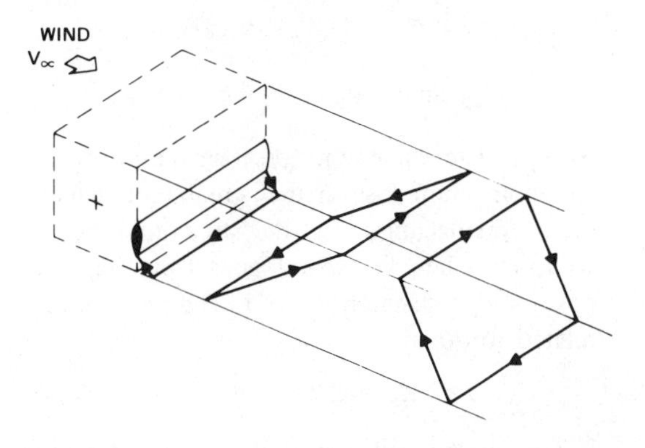

Figure 5-31. Vortex system of single-bladed, cross-wind-axis actuator.

these devices can experience a windwise as well as a cross-wind force, so that the wake can be deflected to the side.

Consistent with vortex theory, we will assume the induced flows at the device are one half their value in the wake. Thus we obtain that if the wake windwise perturbation is $2aV_\infty$, then at the device itself the incoming flow has velocity $V_\infty(1 - a)$, giving the flow system illustrated in Figure 5-32.

In order to simplify the analysis, we shall adopt the following assumptions:

$$\beta = 0$$
$$C_D = 0$$
$$C_L = 2\pi \sin \alpha$$
$$c << R.$$

Our results will than be limited to an inviscid analysis at high tip speed ratios where the maximum angle of attach, α, is small. The low tip speed ratio performance requires numerical analysis to model the non-linear aerodynamics near stall. Using the above assumptions and starting with the Kutta-Joukowski Law, we can write

$$L = \rho W \Gamma = 1/2 \rho W^2 c C_L \quad (5\text{-}64)$$

so that

$$\Gamma = \frac{c}{2} W C_L = \pi c W \sin \alpha. \quad (5\text{-}65)$$

Since the force on the airfoil can be expressed as

$$\vec{\mathbf{F}} = \rho \vec{\mathbf{W}} \times \vec{\Gamma}, \quad (5\text{-}66)$$

we obtain

$$\vec{\mathbf{F}} = \rho \pi c \left[-V_a V_t \sin^2 \theta \hat{j} - (V_a^2 \sin \theta + V_a V_t \sin \theta \cos \theta) \hat{i} \right]. \quad (5\text{-}67)$$

Now we can equate the force on the airfoil to the momentum lost in the streamtube which the airfoil occupies. Let the streamtube be of width dx when the airfoil goes from angular position θ to position $\theta + d\theta$. The width dx is related to $d\theta$ by

$$dx = R d\theta \,|\sin \theta|. \quad (5\text{-}68)$$

The process will repeat itself every revolution so the time interval of our analysis shall be one period which is $2/\Omega$. Of this time period, the airfoil will spend a time increment of $d\theta/\Omega$ in the front portion of the streamtube and another time increment of $d\theta/\Omega$ in the rear portion of the wake. Since the streamwise force contribution from Equation 5-67 is seen to be symmetrical with respect to the angles $\pm\ \theta$, we may write the blade force equation for the time period $2/\Omega$ as

$$dF_{\text{blade}} = 2 \rho \pi c V_t V_a \sin^2 \theta \frac{d\theta}{\Omega}. \quad (5\text{-}68)$$

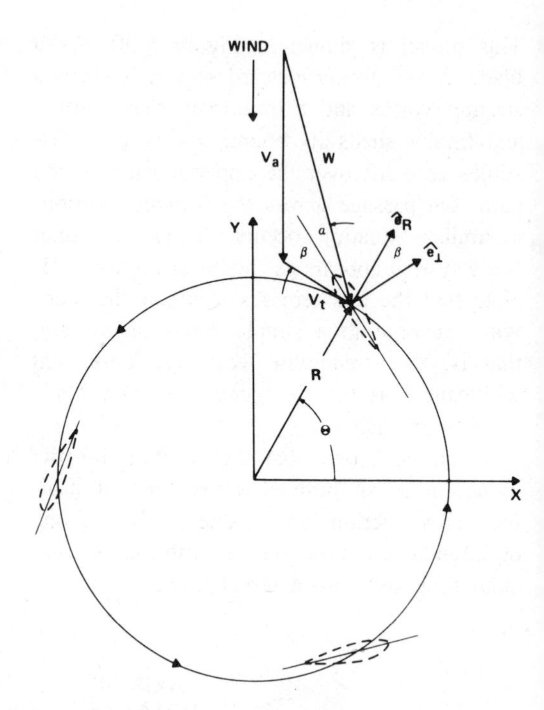

Figure 5-32. Flow system for a cross-wind-axis turbine.

Now the momentum equation yields the force in the streamtube as

$$dF_{\text{momentum}} = \rho R d\theta \,|\sin \theta|\, (1 - a) V_\infty 2 V_\infty a \frac{2\pi}{\Omega} \quad (5\text{-}69)$$

Equating these two forces under the assumption that $V_a = V_\infty(1 - a)$ and $V_t = R\Omega$ yields an expression for the axial interference factor a for one blade,

$$a = \frac{c}{2R} \frac{R\Omega}{V_\infty} |\sin \theta|, \quad (5\text{-}70)$$

or for B blades,

$$a = \frac{Bc}{2R} \frac{R\Omega}{V_\infty} |\sin \theta|. \quad (5\text{-}71)$$

Now that a is defined, the blade force may be resolved into torque and radial components. The torque is given by

$$Q = \rho\pi cR\, V_\infty^2 (1 - a)^2 \sin^2 \theta. \quad (5\text{-}72)$$

The average torque for a rotor with B blades is

(5-73)

$$\overline{Q} = [\rho\pi BcRV_\infty^2]\left[1/2 - \frac{4}{3\pi}\frac{BcX}{R} + \frac{3}{32}\left(\frac{BcX}{R}\right)^2\right]$$

and the corresponding sectional power coefficient is given by

(5-74)

$$C_P = \pi X\frac{Bc}{R}\left[1/2 - \frac{4}{3\pi}\frac{BcX}{R} + \frac{3}{32}\frac{B^2c^2X^2}{R^2}\right].$$

This expression yields a maximum power coefficient of 0.554 when the quantity $BcX/2R = a_{max} = 0.401$. Further refinements can be made with consideration of drag and maximum angle of attack. The maximum angle of attack occurs approximately at the point $\theta = \pi/2$, where

$$\tan\alpha = \frac{1 - a_M}{X} = \frac{1}{X}\frac{Bc}{2R}. \quad (5\text{-}75)$$

When α is set equal to α_{max}, we may rearrange Equation 5-75 to express the starting tip speed ratio. Using $\alpha_{max} = 14°$, we obtain

$$X_{\text{start}} \doteq \frac{4}{1 + 2\dfrac{Bc}{R}} \quad (5\text{-}76)$$

so that a three-bladed rotor with a 1-ft chord and a 20-ft radius would have a starting tip speed ratio of about 3.

Since this type of rotor will not operate at low tip speeds, the drag losses may be approximated by assuming that the local velocity is $W \cong R\Omega$. The drag torque is then

(5-77)

$$\overline{Q}_D = -\frac{C_D}{2}\rho R^2\Omega^2 BcR = -\frac{C_D}{2}\rho V_\infty^3\frac{BcX^3}{\Omega}$$

and the contribution to the power coefficient is

$$\Delta C_P = -C_D\frac{Bc}{2R}X^3. \quad (5\text{-}78)$$

At this point, a solidity may be defined as $\sigma\ Bc/2R$, the ratio of blade circumference to disc diameter. The power coefficient becomes

(5-79)

$$C_P = \pi\sigma X - \frac{16}{3}\sigma^2X^2 + \sigma^3X^3\left(\frac{3\pi}{4} - \frac{C_D}{\sigma^2}\right)$$

and it may be seen that

$$C_P = C_P\left(\sigma X, \frac{C_D}{\sigma^2}\right).$$

The Circular Rotor

At the high rotational speeds required for the Darrieus-type rotor, the inertial loads are large and result in substantial bending loads in the blades. These bending loads may be removed by deploying the blade in a shape similar to the caternary so that the loads are entirely tensile. The required shape has been investigated by Blackwell and given the name *troposkien*. The curve is described by elliptic integrals and is approximated by a sine curve or parabola. The effect on performance caused by bringing the blades closer to the axis of rotation is substantial, since both the rotational speed and the usable component of the lift are reduced. Figure 5-33 illustrates the troposkien curve and the local angle γ between the blade tangent and the axis of rotation.

If we analyze a unit height of the rotor, the expression for a becomes

$$a = \sigma X \cos\gamma\,|\sin\theta\,|, \quad (5\text{-}80)$$

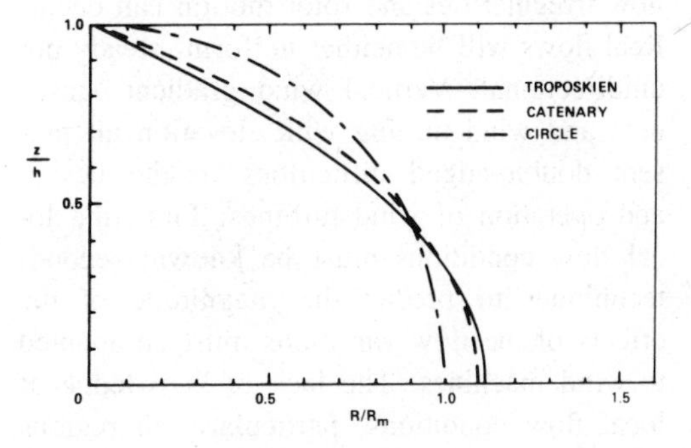

Figure 5-33. Troposkien, circle and caternary of equal length.

where the product σX may be taken as the solidity and tip speed ratio at the point $R - R_{max}$, since this product is independent of R. The torque generated by a slice dz along the rotor axis is

$$\frac{d\overline{Q}}{dz} = \rho \pi B c V_\infty^2 \cos \gamma \left[1/2 - \frac{8}{3\pi} \sigma X \cos \gamma + \frac{3}{8} \sigma^2 X^2 \cos^2 \gamma \right] \tag{5-81}$$

and the incremental power coefficient is

$$\frac{dC_P}{dz} = \frac{d\overline{Q}}{dz} \frac{\Omega}{1/2 \rho V_\infty^2 A} = \frac{4\pi\sigma X}{A} R \cos \gamma \left[1/2 - \frac{8}{3\pi} \sigma X \cos \gamma + \frac{3}{8} \sigma^2 X^2 \cos^2 \gamma \right]. \tag{5-82}$$

The integration of Equation 5-82 for an arbitrary geometry may be accomplished; one simple case is the circular blade for which a maximum power coefficient of 0.536 occurs at $\sigma X = a_{max} = 0.461$. The effects of drag and stall can be included in the above model by development of a blade element theory similar to that developed for the wind-axis rotor.

FORCES AND MOMENTS DUE TO VERTICAL WIND GRADIENT

Introduction

In the analysis of rotors covered previously, the relative wind was assumed to be uniform and parallel and to be perpendicular to the plane of rotation of the rotor. In reality, both flow irregularities and rotor motion can occur. Real flows will be neither uniform, steady nor unidirectional. Vertical wind gradient, gustiness and wind turning with elevation all present double-edged difficulties to the design and operation of wind turbines. First, the local flow conditions must be known; second, techniques to predict the magnitude of the effects of the flow variations must be adapted to wind machines. The lack of knowledge of local flow conditions, particularly in regions of rough terrain, represents a considerable barrier. While some wind gradient data exists for flow over rough terrain, there is little or no data on turbulence spectra and wind turning. Slade has reported the presence of considerable wind turning in the atmospheric surface layer over rough terrain. By contast, the knowledge of flow over flat terrain is much more complete.

Extensive studies have been made of wind structure in the atmospheric surface layer over flat terrain. Monin and Obukhov have developed a relation for the mean flow that encompasses stable, neutral and unstable stratification. Their relation involves three parameters: the surface friction velocity; the surface roughness; and a stability parameter, the Monin-Obukhov length. By contrast, the mean flow over rough terrain is frequently approximated by a power law relation with height

$$\frac{V_\infty}{V_R} = \left(\frac{Z}{h}\right)^n \tag{5-83}$$

where the coefficient n is less than 1. Because of the simplicity of Equation 5-83, since it requires fewer parameters, and the fact that the wind variation over a limited range ($\sim$ 100–200 ft) is required for wind turbines, we shall use the above relation.

The departures from the flow studied previously have no first-order effects on the turbine mean output. However, periodic variations in torque, time-dependent side forces and pitching moments can occur. These forces and moments will affect the overall system dynamics and hence both the design and operation of a wind turbine. In addition to the aerodynamic forces and moments, certain mechanical forces and moments are present. This section will deal only with the aerodynamic loads due to wind gradient.

Both rotor yaw and flapping can induce large forces and moments. Rotor yaw can be treated using the analysis of Ribner.

The Effects of Vertical Wind Gradient

A vertical wind gradient will induce forces and moments as illustrated in Figure 5-34.

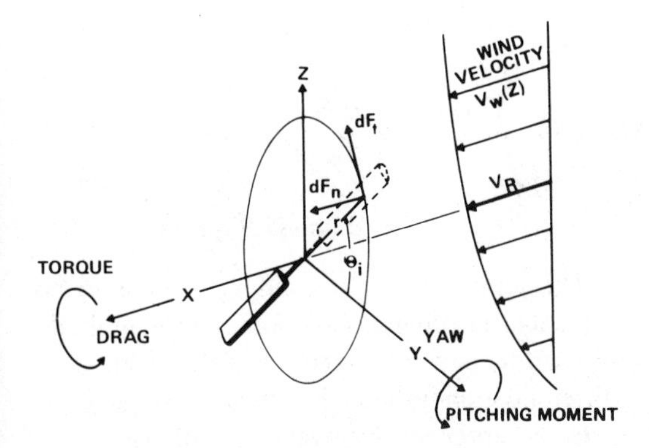

Figure 5-34. Rotor in a wind gradient.

The largest of these are the torque variation and the pitching moment. As would be expected, the magnitude of these moments is dependent upon scale, since it is the velocity difference between the top and bottom of the rotor that is significant. Before proceeding further, it should be noted the incremental forces on a blade element have been designated normal (n) and tangential (t) in order to avoid confusion with the coordinates xyz. Thus the (x,y) coordinates identified earlier are now the (t,n) coordinates.

Figure 5-35 illustrates the blade velocity diagram. This velocity diagram differs from previous illustrations in that the freestream velocity, V_∞, has been replaced with the local wind velocity, V_w.

For a blade in the upper half of the rotor disc, the axial velocity will be higher than for a blade in the lower half of the disc. The increase in V_w increases both the resultant velocity, W, and the angle of attack, α. At

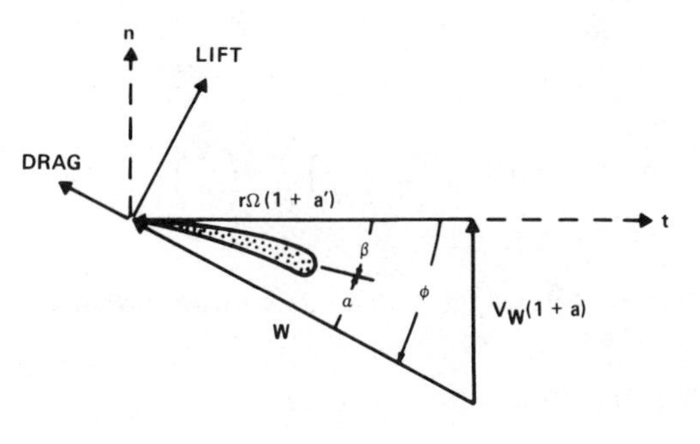

Figure 5-35. Blade velocity diagram.

high tip speed ratios, it may be seen that the principal effect of increased velocity (due to gradient or gust) will be an increased angle of attack. The variation in angle of attack in turn will cause variations in the force dF_t and dF_n with the angle of rotation. Expressions for the first- and second-order forces and moments can be generated from the steady-state performance aerodynamics in the manner described below. The differential force on a rotor element may be expressed in terms of a Taylor Series about the rotor hub, so that

(5-84)

$$dF_t|_z = dF_t|_0 + dF_{t_u} u_z|_0 \Delta Z + [dF_{t_{uu}} u_z^2 + dF_{t_u} u_{zz}] \frac{\Delta Z^2}{2} \ldots$$

where

$$dF_{t_u} \equiv \frac{\partial F_t}{\partial u}, \quad dF_{t_{uu}} \equiv \frac{\partial^2 F_t}{\partial u^2}\ ; \ u \equiv \frac{V_W}{V_{\text{ref}}}$$

A similar expression may be obtained for dF_n. These forces change from their hub values ($Z = 0$) due to the variation in wind velocity with elevation. The distance $\Delta Z = r \sin \theta_i$, where θ_i refers to the angle of rotation of the ith blade. Rewriting Equation 5-84 in the form

$$dF_t = T_0 + T_1 \Delta Z + T_2 \frac{\Delta Z^2}{2} \ldots \quad (5\text{-}85)$$

and expressing dF_n in the same manner,

$$dF_n = N_0 + N_1 \Delta Z + N_2 \frac{\Delta Z^2}{2} \ldots, \quad (5\text{-}86)$$

we obtain the following forces and moments by integrating over the blade and summing over B blades, where $B \geq 2$.

(5-87)

$$\textit{Torque: } Q = \int r T_0 dr + \int r^2 T_1 dr \sum_{i=1}^{B} \sin \theta_i \;(= 0)$$

$$+ \int r^3 T_2 dr \sum_{i=1}^{B} \sin^2 \theta_i.$$

Pitching moment:

$$M_y = \int rN_0 dr \underbrace{\sum_{i=1}^{B} \sin\theta_i}_{0} + \int r^2 N_1 dr \sum_{i=1}^{B} \sin^2\theta_i + \int \frac{r^3 N_2}{2} dr \underbrace{\sum_{i=1}^{B} \sin^3\theta_i}_{0 \text{ unless } B \text{ is odd}}. \qquad (5\text{-}88)$$

The summation over B equally spaced blades has been evaluated and it may be noted that wind gradient induces no first- or second-order yawing moment.

Table 5-4 gives the values of the summations for various numbers of blades.

The terms N_1, N_2, T_1 and T_2 remain to be evaluated in order to determine the magnitude of the forces and moments. As the method of differentiation is straightforward, let us indicate the approach by evaluation of N_1:

$$N_1 = \frac{\partial(dF_n)}{\partial u} \frac{\partial u}{\partial Z}\bigg|_{Z=0} \Delta Z = 1/2\rho V_R^2 c \left[\frac{\partial}{\partial u}\left\{(u^2+v^2)(C_L \cos\theta + C_D \sin\theta\right\} \frac{\partial u}{\partial Z}\right]_{Z=0} \Delta Z, \quad (5\text{-}89)$$

$$\text{where} \quad \frac{\partial C_L}{\partial u} = \frac{\partial C_L}{\partial \alpha}\frac{\partial \alpha}{\partial \phi}\frac{\partial \phi}{\partial u} = C_{L_\alpha}\frac{\partial \phi}{\partial u}. \qquad (5\text{-}90)$$

The final result is

$$N_1 = 1/2\rho V_R^2 cv \left(C_{L_\alpha} \cos\phi + C_L \sin\phi + \frac{C_D(1+\sin^2\phi)}{\cos\phi}\right)\bigg|_{Z=0} \frac{\partial u}{\partial Z}\bigg|_0 \Delta Z. \qquad (5\text{-}91)$$

The variation in torque coefficient may be evaluated from the preceding analysis,

$$\frac{\Delta Q}{1/2\rho V_R^2 \pi R^3} \equiv \Delta C_Q = \left\{I_1\left(\frac{R}{V_R}\frac{dV_W}{dz}\bigg|_{Z=0}\right)^2 + I_2\left(\frac{R^2}{V_R}\frac{d^2V_W}{dz^2}\bigg|_{Z=0}\right)\right\}\sum_{i=1}^{B}\sin^2\theta_i, \qquad (5\text{-}92)$$

where the integrals I_1 and I_2 are evaluated as

$$I_1 = \frac{1}{\pi}\int_{R_{hub}}^{R} \left[(2-\cos^2\phi)C_L \sin\phi + 2\cos\phi\, C_{L_\alpha} - 2\cos^3\phi\, C_D\right](1-a)^2\left(\frac{r}{R}\right)^3 \frac{c}{R} d\left(\frac{r}{R}\right) \qquad (5\text{-}93)$$

and

$$I_2 = \frac{X}{\pi}\int_{R_{hub}}^{R} \left[(1+\sin^2\phi)C_L + C_{L_\alpha}\sin\phi - C_D \sin\phi\right](1+a')(1-a)\left(\frac{r}{R}\right)^4 \frac{c}{R} d\left(\frac{r}{R}\right). \qquad (5\text{-}94)$$

Similar integrals are obtained by the other forces and moments. As can be seen from Equations 5-93 and 5-94, considerable simplification can be made if the angle ϕ is small and the values of C_L and c are constant over the outer portions of the blades. In this case, closed form approximations can be obtained.

Approximate Relations

At high top speed ratios, $\sin\phi \cong \phi$, and by neglecting drag, it may be shown that the following expressions can be obtained:

Pitching moment:
$$\frac{M_y}{1/2\rho V_R^2 \pi R^2} = \frac{\sigma X C_{L\alpha}}{4}(1-a)\frac{R\,dV_W}{V_R\,dZ}\bigg|_{Z=0} \sum_{i=1}^{B} \frac{\sin^2\theta_i}{B}. \tag{5-95}$$

Torque change:
$$\frac{\Delta Q}{1/2\rho V_R^2 \pi R^3} = \left\{ I_1 \left(\frac{R\,dV_W}{V_R\,dZ}\right)^2\bigg|_{Z=0} + I_2\left(\frac{R^2 d^2 V_W}{V_R\,dZ^2}\bigg|\right)\right\}_{Z=0} \sum_{i=1}^{B} \frac{\sin^2\theta_i}{B} \tag{5-96}$$

where $I_1 \cong \dfrac{\sigma C_{L\alpha}(1-a)^2}{2B}$

and $I_2 \cong \left\{\dfrac{\sigma X C_L}{5B} + \dfrac{(1-a)\sigma C_{L\alpha}}{4B}\right\}(1-a).$

Some representative values may be obtained by using the data from the Smith-Putnam wind turbine. The long-term wind data yield is

$$\frac{1\,dV_W}{V_R\,dZ}\bigg|_{Z=0} = \frac{0.104}{h_h}, \frac{1\,d^2V_W}{V_R\,dZ^2}\bigg|_{Z=0}$$

$$= -\frac{0.14}{h_h{}^2}, h_h = 120,$$

and, using the values listed below,

$\sigma = 0.083$ $R = 87.5$ ft

$X = 6$ $B = 2$

$C_{L\alpha} = 5.5$ $C_L = 0.6$

$a = 0.33$

we obtain (neglecting the flapping motion of the Smith-Putnam machine):

Pitching moment:
$$\frac{M_y}{1/2\rho V_R^2 \pi R^3} = 0.0173\,(1-\cos 2\Omega t). \tag{5-97}$$

Torque variation:
$$\frac{\Delta Q}{1/2\rho V_R^2 \pi R^3} = -0.00153\,(1-\cos 2\Omega t). \tag{5-98}$$

The numbers are difficult to judge. Accordingly, let us refer the forces to the drag of the wind turbine and the moments to the torque:

$$D = C_T 1/2\rho V_R^2 \pi R^2, C_T \cong 2/3$$

$$Q = C_Q 1/2\rho V_R^2 \pi R^3, C_Q = \frac{C_P}{X}, C_P = 0.4.$$

Table 5-4. Trigonometric sums.

	$B=2$	$B=3$	$B=4$
$\sum_{i=1}^{B} \sin^2\theta_i$	$1-\cos 2\Omega t$	$3/2$	2
$\sum_{i=1}^{B} \sin^2\theta_i \cos\theta_i$	0	$-\frac{3}{4}\cos 3\Omega t$	0
$\sum_{i=1}^{B} \sin^3\theta_i$	0	$-\frac{3}{4}\sin 3\Omega t$	0

Pitching moment:

$$\frac{P}{Q} = 0.259\,(1 - \cos 2\Omega t) \quad (5\text{-}99)$$

Torque variation:

$$\frac{\Delta Q}{Q} = -\,0.023\,(1 - \cos 2\Omega t). \quad (5\text{-}100)$$

The yaw and drag forces are quite insignificant, while the pitching moment is seen to be quite appreciable, amounting to a variation of 52 percent of the value of the torque. The torque variation is seen to be up to a 4.6 percent decrease in torque and hence also power. The torque variation itself requires more discussion. Equation 5-96 gives an expression for the torque change, which may be modified to express the percentage torque change.

Adopting the power law profile given by Equation 5-83, we obtain

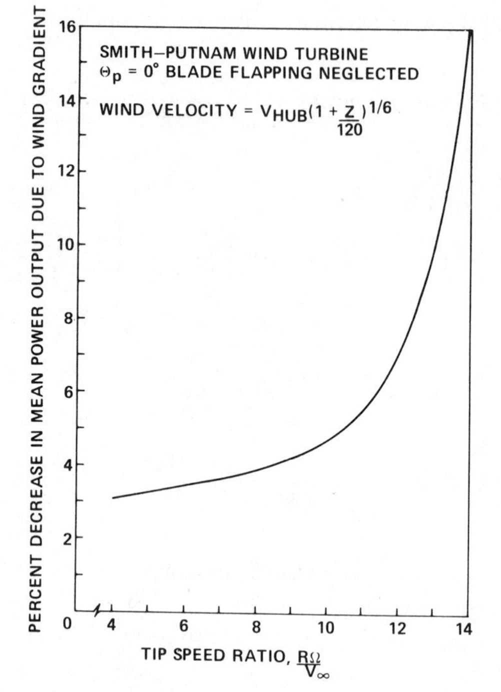

Figure 5-36. Percent reduction in power output due to wind gradient.

$$\frac{\Delta C_Q}{C_Q} = -\frac{X\sigma(1-a)^2 C_{L_\alpha}}{4C_P}\left(\frac{R}{h}\right)^2 \left\{-2\eta^2 + \left\{\frac{4XC_L}{5(1-a)C_{L_\alpha}} + 1\right\}\eta(1-\eta)\right\}\sum_{i=1}^{B}\frac{\sin^2\theta_i}{B} \quad (5\text{-}101)$$

(where h is the height of the rotor hub). The effects of scale, tip speed ratio, solidity and load may be estimated from Equation 5-101. For example, a rotor operating at constant rpm at wind speeds below the design point will have large X and low C_P. The torque variation for large scale will be appreciable.

It may be noted that the bracketed expression in Equation 5-101 is approximately equal to $\eta - 3\eta^3$. This expression has a maximum value when $\eta = 1/6$. Experimental evidence for flow over smooth terrain yields $\eta \cong 0.17$.

The expressions developed for the torque variation may also be used to evaluate the change in power output due to wind gradient. The variation in torque due to the wind gradient is approximately constant over a wide range of tip speed ratios. The net output of a wind turbine, however, changes appreciably with tip speed so that the percentage variation in turbine output due to wind gradient (or gust) increases greatly as the net turbine output approaches zero. Figure 5-36 illustrates the percentage decrease in mean turbine output due to wind gradient for the Smith-Putnam wind turbine. Flapping motion of the blades was not included in this example. The absolute magnitude of the power variation due to gradient may be obtained by using the results of Figure 5-36 along with Figure 5-28.

The material in this chapter based on the work of Robert E. Wilson and Peter B. S. Lissaman is being updated. Refer to *Aerodynamic Performance of Wind Turbines*, prepared at Oregon State University, Corvallis, Oregon. NTIS report PB-259-089.

6

Towers and Systems Installation

This chapter provides a brief introduction to WECS towers and installation. The purpose of the supporting tower is to hold the WECS securely in place at a height where it is in a clear windstream, free of turbulence or wind shadows created by nearby buildings, trees, sharply rising hills and other obstacles, and reasonably free of the surface drag of the earth. Normally, the proper height is at least 50 ft higher above the nearest obstacles within a 100 yd radius in all directions.

Towers for the larger wind plants are made of galvanized steel to limit the risks of corrosion during the average 20-year life of the wind plant. They must be designed to withstand the considerable lateral thrusts produced by high winds acting on the large blades, and they must be equipped with special top adaptors designed to fit each particular model wind plant. They should be equipped with lightning protection kits and, in many instances, with anti-climb sections.

Standard wind plant towers are available in heights up to 90 ft, and higher towers can and have been built for special situations. The choice of proper height for a specific installation will depend upon wind characteristics at the proposed site, as discussed in Chapter 3.

A note of caution: Erecting a 60-ft tower and mounting on it a 500 to 700-lb Dunlite or Enertech WECS requires an experienced supervisor and the right kind of equipment—a mobile crane or a specially designed ginpole or davit, with winch and long cable. Most wind plant distributors refuse to guarantee a wind plant or tower unless one of their field supervisors has been present during the installation.

WIND SYSTEM TOWERS

Supporting a wind turbine that weighs several hundred pounds, for a small WECS, and many thousands of pounds, for large WECS, is no simple task and requires a rigid support structure. Towers are subjected to two types of loads, as illustrated in Figure 6-1: WECS weight, which compresses the tower downward, and drag, which tries to bend the tower downwind.

Towers are made in two basic configurations: guy-wire supported, as shown in Figure 6-2, and cantilever (sometimes called

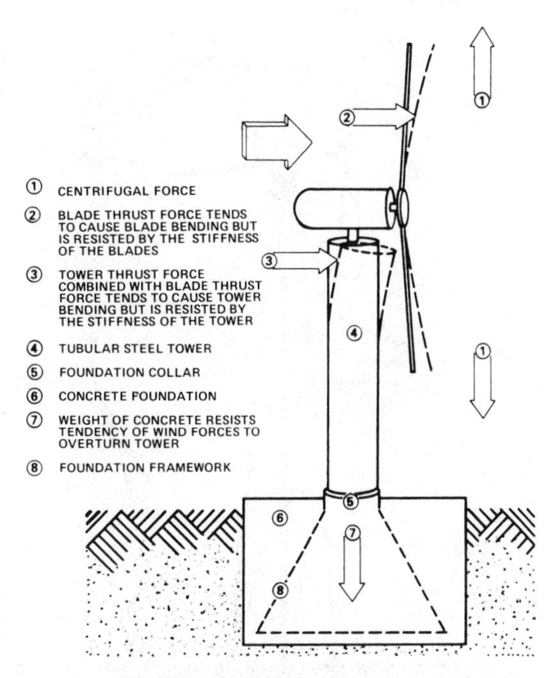
① CENTRIFUGAL FORCE
② BLADE THRUST FORCE TENDS TO CAUSE BLADE BENDING BUT IS RESISTED BY THE STIFFNESS OF THE BLADES
③ TOWER THRUST FORCE COMBINED WITH BLADE THRUST FORCE TENDS TO CAUSE TOWER BENDING BUT IS RESISTED BY THE STIFFNESS OF THE TOWER
④ TUBULAR STEEL TOWER
⑤ FOUNDATION COLLAR
⑥ CONCRETE FOUNDATION
⑦ WEIGHT OF CONCRETE RESISTS TENDENCY OF WIND FORCES TO OVERTURN TOWER
⑧ FOUNDATION FRAMEWORK

Figure 6-1. Structural load imposed by wind forces. (*Courtesy of Energy Development Co.*)

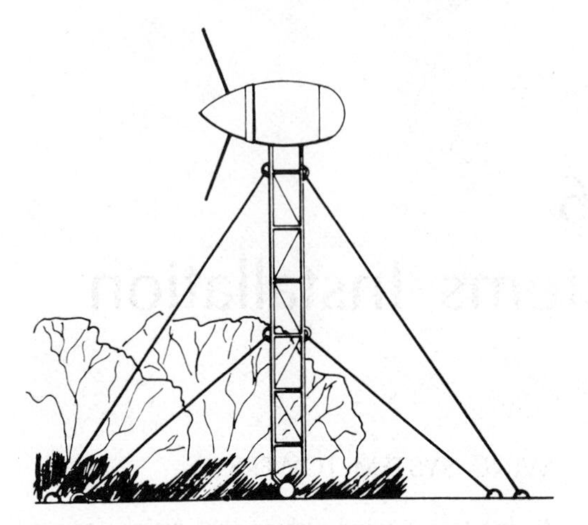

Figure 6-2. Guy-wire supported tower. (*Courtesy of DOE.*)

freestanding towers), as shown in Figure 6-3. Also, given these two basic structural support configurations, towers are made with telephone poles; concrete structures, as shown in Figure 6-4; steel bridge trusses, as shown in Figure 6-5; or other single column structures,

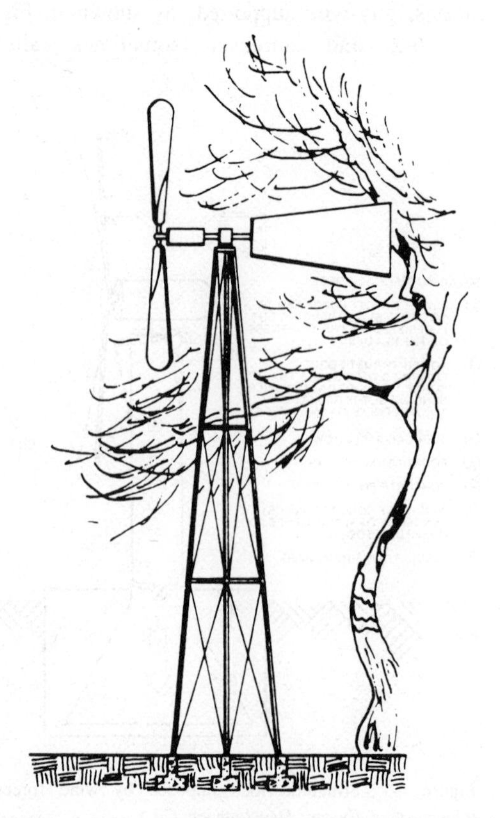

Figure 6-3. Freestanding WECS tower. (*Courtesy of DOE.*)

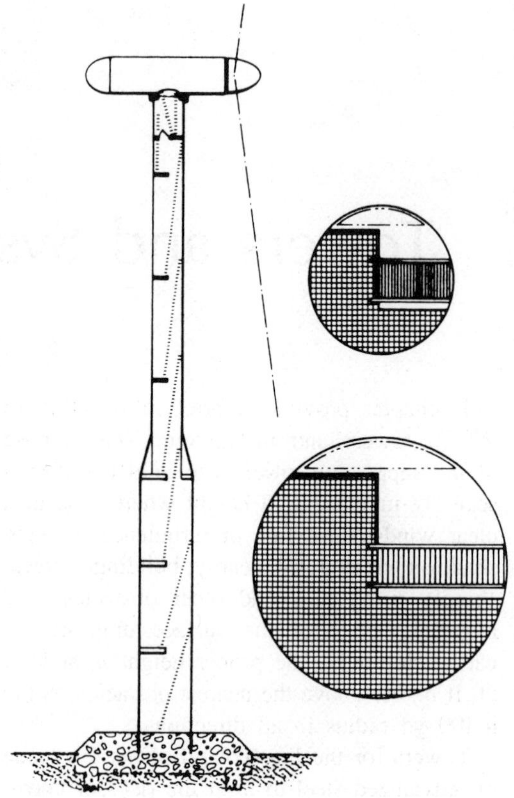

Figure 6-4. Concrete shell tower. (*Courtesy of NASA.*)

as well as lattice frameworks of pipe (as shown in the octahedron tower in Figure 6-6, or wooden boards.

Regardless of which tower design you select, the overriding consideration is the selection of a tower that can support the WECS you select in the highest wind possible at your site. Of course, cost must also be considered. The load that causes many towers to fail is a combination of WECS and tower drag. A pair of skinny windmill blades may not look like they could cause much drag, but when extracting full power at rated speed, they create nearly the drag of a solid disc the diameter of the rotor.

The ways in which towers typically fail are these:

- Freestanding towers buckle due to higher-than-design wind drag load from the rotor.
- A footing that anchors the tower to the ground becomes uprooted.
- A bolt somewhere along the tower fails

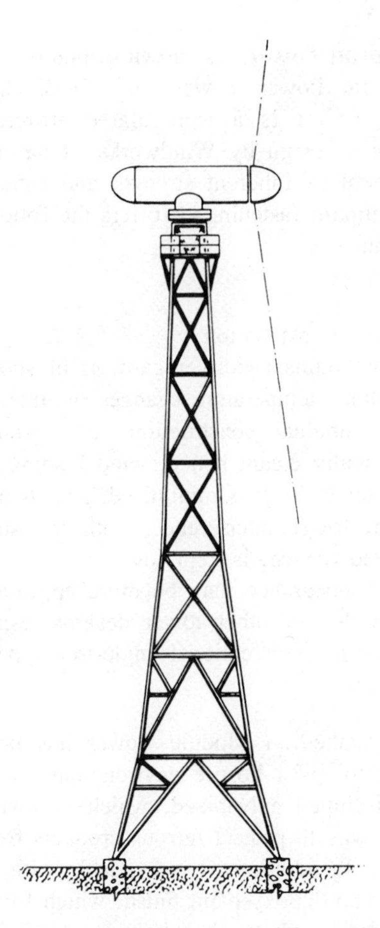

Figure 6-5. Self-supporting steel truss tower. (*Courtesy of NASA.*)

due to improper tightening (or falls out becase it is not tightened at all), resulting in a tower weak point that eventually fails.

- Guy-wire braced towers buckle from improper spacing of the wires up the tower. A tower that requires three sets of cables spaced evenly along the length of the tower gets only two sets, resulting in intercable spans greater than design specifications.
- Guy wires fail from improper wire size, improper tension fasteners or damage.
- Guy-wire anchors uproot from the ground or come away from the structure to which they are attached.

When selected a tower, consider the difference between guy-wire braced and freestanding. Here, you must know something about the structure and about the soil supporting the tower. The county agent should be consulted about the type of soil you have and its ability to act as a foundation.

Building codes for the area will detail the basis for foundation design, and the wind turbine dealer or manufacturer should have drag data for the product selected. A registered professional structural engineer can perform any calculations necessary to ensure that a particular tower installation will support the windmill. The cost of professional services in the area of tower selection and design should be considered cheap insurance for a sound installation. Vibratory loads induced by the wind or wind machine also should be considered and professional advice may be required.

One other point to consider is the potential hazard of guy wires, particularly to children playing.

Figure 6-6. Octahedron tower configuration. (*Photo courtesy of Natural Power Tower, Inc.*)

Specific Types of Towers

This section provides more detailed information on specific manufacturer's towers.

Unarco-Rohn. Unarco-Rohn has been manufacturing both self-supporting and guyed towers for use with WECS for several years.* Towers have been fabricated for use with the following WECS: Dunlite, Elektro, Enertech, Jacobs, Winco, Sencenbaugh, Kedco, Aeropower, Grumman, Dakota Wind Electric, Aerowatt, Zephyr and phototype WECS.

The standard WECS tower is designed for a 30-psf (87-mph) wind load condition. Figure 6-7 shows wind load condition zones across the U.S. Special towers for higher wind loads or ice loading conditions (Figure 6-8) are available on special order. The Models 25G and 45G are the model numbers for the guyed towers, and the SSV is the model of WECS self-supporting towers. Figure 6-9 shows a Unarco-Rohn 100-ft self-supporting SSV tower with a Jacobs 2,400-W WECS installed at the manufacturers plant. Tower prices are quoted from Unarco-Rohn based on height, wind loading and the type of mechanical interface between tower and WECS.

General Unarco-Rohn tower prices are given below.

SSV Self-supporting Tower Series

20 ft	$885.00
40 ft	1335.00
60 ft	1825.00
80 ft	2455.00
100 ft	3295.00

Model 45G Guyed Tower Series

26 ft	$645.00
36 ft	735.00
46 ft	980.00
56 ft	1090.00
66 ft	1165.00
76 ft	1270.00
86 ft	1380.00
96 ft	1655.00

*Personal communication between Philip W. Metcalfe, Unarco-Rohn and author.

Octahedron Tower. As shown in Figure 6-6, the Natural Power Tower, Inc. Octahedron Module Tower is a triangulated structure based on a design by Windworks. It has the dual benefit of inherent strength and rigidity with minimum fastening. It offers the following advantages:

- Ease of installation.
- It maintains rigidity regardless of shock loading, temperature changes or time.
- Its tubular construction is aerodynamically clean; hence, wind loading of the tower itself is minimized, and, therefore, tower interference, with the supported device, is kept low.
- The appearance may be more appealing than that of other tower designs, especially the conventional angle-iron type.

The Octahedron Module Tower has been designed to resist hostile environments. It is fully hot-dipped galvanized, which is a very effective way to protect ferrous products from corrosion. In addition, it is available with an optional vinyl-epoxy paint finish, which forms a tight bond with the hot-dipped galvanizing to provide virtually total protection. This finish is particularly suitable for use in marine or other corrosive atmospheres and, in addition, it can be obtained in one's choice of several colors.

The tower has been designed to operate in wind loads of 40 lb/ft^2 with simultaneous lateral thrust of 1,000 lb at the top. This is roughly equivalent to a wind loading of 125 mph against the tower when supporting a WECS with a 20-ft diameter feathered blades. A general safety factor of 1.7, in accordance with standards of the American Institute of Steel Manufacturers, has been employed throughout the design. Natural Power Tower, Inc. feels that this tower design is uniquely suited for supporting complex structures under extreme wind conditions and under all normal environmental conditions. It is especially suited for supporting WECS. The tower is available in heights of 25, 34, 43, 52, 70 and 89 ft. It is designed for unguyed installation

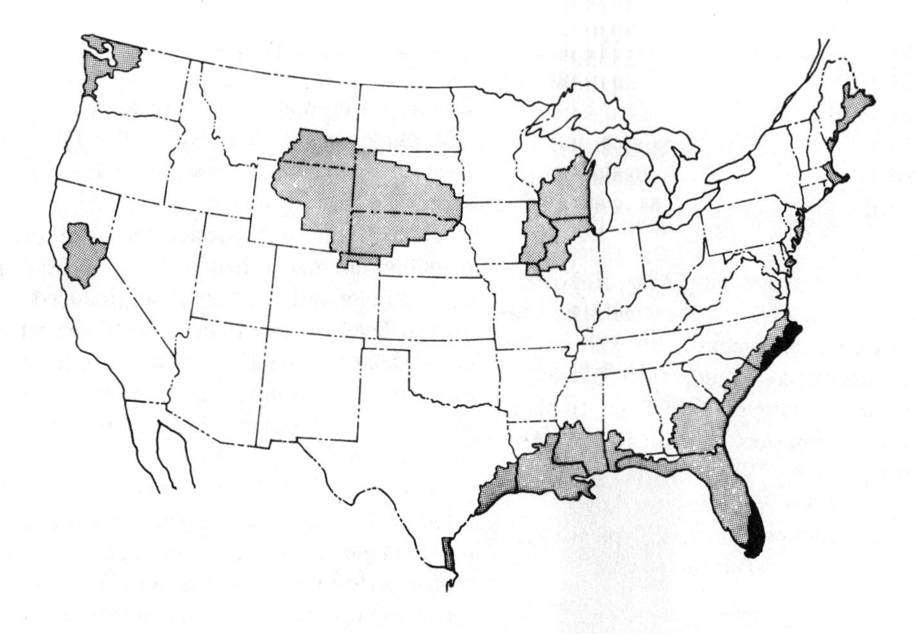

WIND LOADS FOR VARIOUS AREAS IN U. S.			
Tower Height	ZONE		
	A	B	C
Under 300 Ft.	30 Lbs/Sq. Ft.	40 Lbs/Sq. Ft.	60 Lbs/Sq. Ft.
301 Ft. to 650 Ft.	35 Lbs/Sq. Ft.	48 Lbs/Sq. Ft.	50 Lbs/Sq. Ft.
Over 650 Ft.	40 Lbs/Sq. Ft.	55 Lbs/Sq. Ft.	70 Lbs/Sq. Ft.
MAP CODE			

Figure 6-7. Wind loading zones for WECS towers.

on footings, or it can be used with the same basic advantages as a guyed tower for temporary installation.

The Octahedron Module Tower is easy to assemble and erect; its modular design enables it to be constructed, if necessary, from the ground up, with each section constructed upon the preceding one, eliminating the necessity for cranes, gin poles or other heavy erection equipment.

The Natural Power Tower's basic features are simplicity, lightness, high inherent rigidity and strength, ease of assembly, environmental resistance and aerodynamic purity. Engineering data, drawings and test results are available upon request from Natural Power Tower, Inc. Prices for the Octahedron Module Tower from Natural Power Tower, Inc. are as shown below.

Octahedron Module Tower, Standard Weight

NPTI 25 ft	$885.00
NPTI 34 ft	1110.00
NPTI 42 ft	1370.00
NPTI 51 ft	1670.00
NPTI 61 ft	2020.00
NPTI 70 ft	2340.00
NPTI 80 ft	2755.00
NPTI 89 ft	3290.00

Octahedron Module Tower, Heavy Weight

NPTI 25 ft	$1110.00
NPTI 34 ft	1405.00

NPTI 42 ft	$1720.00
NPTI 51 ft	2095.00
NPTI 61 ft	2525.00
NPTI 70 ft	2920.00
NPTI 80 ft	3335.00
NPTI 89 ft	3810.00

Towers for Water-pumping WECS. Towers for water-pumping WECS are made by the primary WECS manufacturer, as well as by tower manufacturers. Figure 6-10 illustrates the typical water-pumping WECS tower manufactured by Dempster Industries, Inc. The other water-pumping WECS towers are similar. Figure 6-10 also provides the key design features of the Dempster tower. Typical physical size is noted in Table 6-1.

Figure 6-8. The Enertech 1500 being tested for endurance and ruggedness on top of Mt. Washington, New Hampshire. (*Photo courtesy of ENERTECH.*)

Correct Tower Height

Enertech Corporation has prepared an excellent guide called *Planning a Wind-powered Generating System*. Below are quoted the paragraphs regarding towers.

"The two most important considerations in planning the tower height for a WECS are avoidance of turbulent air flow produced near ground level by the 'roughness' of the terrain over which the wind flows, and avoidance of excessive ground drag which lowers wind velocity near the ground and severely restricts the performance of a WECS."

"*Turbulence:* A WECS must never be located such that it is subject to excessively turbulent air flow, Light turbulence will decrease performance since a WECS cannot react to rapid changes in wind direction, while heavy turbulence may reduce expected equipment life or result in WECS failure."

"Turbulence may be avoided by following a few basic rules:

1. If possible the WECS should be mounted on a cleared site free from minor obstructions such as trees and buildings for at least 100 yds. in all directions and free from major obstructions such as abrupt land forms for at least 200 yds. Even over clear ground, however, the minimum recommended tower height is 50 feet.
2. If it is not possible to avoid obstructions as above, tower height should be increased to a value of approximately 30 feet greater than the height of obstructions within 100 yds."

"*Ground Drag:* The avoidance of ground drag will increase performance dramatically. Generally speaking, the least expensive way in which to increase your power output from a WECS is to increase tower height."

"A generally recognized 'rule-of-thumb' is that wind speed increases as the one-seventh power of the height above ground."

Disturbances in the air caused by objects in the immediate area (barns, houses, trees) can cause severe stresses on the blades and bearings which could decrease generator efficiency by as much as 50 percent. "It is apparent

Table 6-1. Water-pumping Tower Size.

STYLE A

HEIGHT, FEET	SPREAD AT BASE OUTSIDE TO OUTSIDE STYLE A	APPROXIMATE WEIGHT, LBS.	
		FO 12-FOOT MILLS	FOR 14-FOOT MILLS
30	7′9¾″	935	1000
40	10′6¾″	1230	1295

STYLE B

DIMENSIONS AT BASE,		OUTSIDE TO OUTSIDE SPREAD OF CORNER POSTS	APPROXIMATE SHIPPING WEIGHT, POUNDS	
HEIGHT, FEET	AT TOP END OF ANCHOR POSTS		2-INCH	2½-INCH
22	4′3¾″	4′4¾″	340	415
28	5′5″	5′6″	430	515
33	6′6¼″	6′7¼″	516	616
39	7′7¼″	7′8³/₈″	616	726

from the foregoing that the first principle in choosing a tower height is the avoidance of excessive turbulence." Assuming for the moment that at your site turbulence considerations can be met by the use of a relatively short 50 or 60 foot tower, what will be the most economical choice of tower height?

"Using the 1/7 power of height rule which relates tower height to wind velocity or energy shows that the addition of successive 10 foot increments to a tower will result in substantial increases until eventually there is practically no change in successive increments. It is clear that at some point the gain in performance is so slight that it is not worth adding further tower height."

"The most economical tower height will be the height at which the percentage gain in power output per tower increment equals the percentage increase in system cost per tower increment."

"Before you can make this calculation for yourself you must have a relatively accurate estimate for the total installed cost of all other components of your system (all components except tower), plus an estimate of the installed cost of each increment of tower height. From these numbers you will be able to calculate the percentage of the total system cost that each new increment of tower height will represent."

"For each successive 10 foot tower increment the increase in power is greater than the increase in price and is therefore justified up to 170 feet."

"Although by economic considerations alone we have established an optimum height of 170 feet, there may be overriding considerations such as local zoning restrictions, FAA limitations or restrictions, etc. These possibilities will have to be checked for each individual site by the owner. In many cases, since towers have become so inexpensive by comparison to other wind-system components, tower height will be limited more by zoning and other restrictions rather than by simple economics."

It is imperative that the plant be installed as high as economically possible to reach undisturbed air. Placing WECS up a minimum of 50 feet above ground will greatly increase the amount of power available to the propeller. In all cases, the propeller tips must be 15–20 ft above all obstacles within a 300-ft radius. Objects within a 300-ft radius have a very disturbing effect on the air and cause whirling eddy currents that greatly affect plant performance. It should also be noted that tall trees behind a plant interfere as well as trees in front. Another important fact to remember is that the WECS should be installed as close to the battery system as possible. On 12-V systems, this should not be over 360 ft. With 115-V systems, runs up to 1000–1500 ft are

Figure 6-9. Unarco-Rohn Model SSV 100-ft self-supporting tower. (*Photo courtesy of Unarco-Rohn.*)

reasonable. However, costs should always be estimated to increase in vertical height gain, as the most economical approach over a distance. For example, it may be cheaper to go an additional 20–30 ft in vertical tower height than to run costly wire a long horizontal distance.

INSTALLATION OF TOWER AND WECS

A WECS dealer should probably install the entire WECS, or at least the tower and wind turbine, because raising this equipment can be dangerous. If you plan to have anyone else assist, check your insurance policy. Most homeowners' policies do not cover this kind of activity.

The most important aspects of WECS are the design of the WECS and the tower, the two items subjected to wind loads. Good design, however, is not enough. These units must be properly installed, including appropriate grounding for lightning strike protection. After installation, maintenance must be performed as required to ensure continued reliable service. Each item, if performed properly, will contribute to the ultimate safety and efficiency of the wind system.

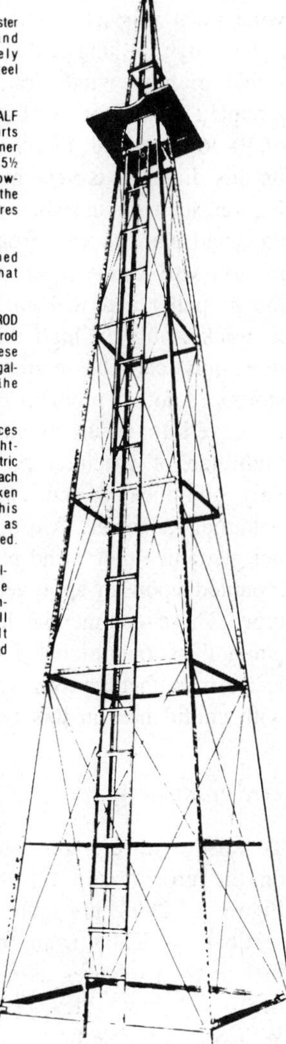

WOOD PLATFORM—of all Dempster towers insures safety and accessibility. It is securely fastened to the tower by steel angle supports.
GIRTS EVERY FIVE AND ONE-HALF FEET—Heavy, angle steel girts extend horizontally from corner post to corner post, every 5½ feet, from the top of the tower to the bottom. Placing the girts so close together assures great strength.
LADDER—Towers are furnished with a complete ladder that is absolutely safe to climb.
ADJUSTABLE SWINGING PUMP ROD GUIDES—keep the wood pump rod in line with the pump. These guides are made from round galvanized steel, bolted to the wood pump rod and to the girts.
ECCENTRIC WASHER—Wire braces are easily and quickly tightened by means of the eccentric washer on the upper end of each brace, and they never slacken back. By simply turning this washer, the brace is made as tight or as loose as desired.
CONVENIENT PULLOUT—The pullout is arranged to provide the proper leverage which enables you to pull the mill out of the wind easily. It is a long wood lever attached to the corner post.
WOODEN GIRT—The bottom girt is made of wood because a steel girt becomes bent from the countless times it is stepped on and thus pulls the tower in at the bottom. The wood girt prevents such buckling. Steadies tower when erected on ground and raised to position. 2" x 2" or 2½" angle steel.
All girts are below the splice.
ANCHOR POSTS—Substantial anchor posts and plates, made from angle steel, provide a strong anchorage. Each post has two angle plates which give ample support for anchoring the tower. Galvanized, the plates and posts cannot rust.

Figure 6-10. Dempster Industries water-pumping WECS tower.

Figure 6-11 diagrams the step-by-step sequence of WECS ownership. If any step in this diagram is omitted, a potential ownership problem is created. Consider the neighborhood resident who hoists aloft a wind turbine without considering his neighbors' feelings; something like the problems which arose early in the history of television antennas. The reaction is likely to be: "I don't have one, so why should you?" or "That's an ugly machine, can't you hide it over behind that tree?" or "That's a very noisy propeller, isn't it?" We have heard these comments before; some are legitimate, others are not.

In at least one U.S. protectorate, it would be illegal to have any form of auxiliary power source, wind included, if the utility mains exist at the edge of your property. This is not the case in the U.S., but local building ordinances and codes may prohibit installation of towers tall enough to make wind power practical, or they may require a tower designed to withstand loads so high that the tower cost makes the entire system economically impractical.

Another possibility is that the entire proposed system meets all requirements but cannot be installed due to lack of space. For example, perhaps the tower cannot be raised within the confines of the area, or there are too many tall trees.

The possible dangers involved in raising a tower and a WECS cannot be emphasized enough, so be very careful if you do it yourself. Make sure your plans are approved by a knowledgable person.

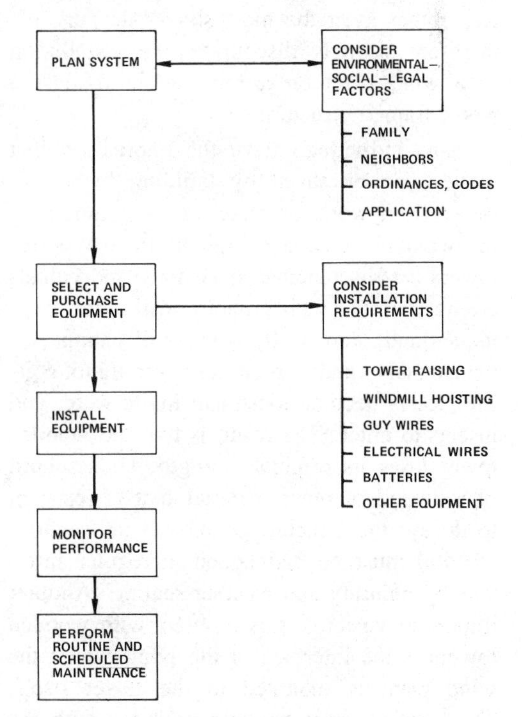

Figure 6-11. Planning a WECS installation. (*Courtesy of DOE.*)

Tower design and installation are as important as the site selection, and a great deal of attention is given to building a support strong enough to handle both anticipated static vertical and horizontal loads from the WECS body itself, and the dynamic thrust loads developed from the disc area of the propeller at its highest anticipated wind speed. A large safety factor must also be included in the tower design.

Manufacturers discourage the installation of these units on home roofs. The loads upon a 12- or 14-ft-diameter propeller are so great in high winds that they could possibly cause serious structural damage to the roof. Those with wood frame structures should take note that even wind plants with 6–8-ft-diameter propellers can cause noise to be transmitted into the structure, even though the plant is balanced and smooth-running. Wood frame structures often have natural frequencies in the 2–20 CPS range. Windmill speeds of 120–1200 rpm will generally produce driving frequencies which may cause noise to be transmitted throughout the structure. This noise, which resembles a low howl or groan, can bother even the most sound sleeper. We therefore strongly discourage the installation of a wind plant on a roof and/or against a wood framed structure.

Many individuals have the impression that money can be saved by building their own tower from wood or metal or even mounting the wind plant upon a high fir tree. Wooden towers are discouraged since most individuals are not familiar with proper construction techniques and, above all, wood will change dimensionally and structurally throughout its life. Joints become loose and invite water and insects to enter. The result is that the wooden tower loses its original strength. Unless built of redwood or other material that is resistant to decay, the structure is subject to weathering and must be maintained at regular intervals by painting and weather-sealing. Another important aspect of this problem with wooden towers is the interface or the point where the wind plant is mounted to the tower itself. This section must be perpendicular with the axis of rotation of the wind plant and must be secured mechanically. Many wood tower failures occur here, with the result that the entire wind plant is swept away in a high wind.

Although a single tree may appear very stable in high winds, once a wind plant is mounted at the top, and a propeller diameter of 10–16 ft is used, the cross-sectional loading on this diameter is very high. This high loading on such an unstable platform will cause the wind plant to yaw from side to side. The motion will cause power to be spilled from the propeller and will not allow the machine to track stably in high winds. This unstable operation can be extremely dangerous in storms. Another problem is that the tree must be guyed to ensure a stable platform. This multitude of potential problems results in a very complicated and expensive method of obtaining a tower. Most manufacturers will not guarantee their wind plants unless they are mounted upon an approved tower design and type. These commercial towers have a large amount of operational history behind them and a manufacturer can warranty his machine with confidence on this type of structure.

Tower Raising

To raise a tower, you can either assemble it on the ground and tilt it up (as shown in Figure 6-12) or assemble it by stacking the sections vertically from the ground up. The first method requires assembly of all components and guy wires, and as much of the WECS as possible on the ground. The base of the tower is then fixed to a pivot to prevent the tower from sliding along the ground, and a rope is tied from the tower, over a gin pole, to a car or a winch. Moving the car pulls the tower up. The gin pole serves in the initial stages to improve the angle at which the rope pulls on the tower.

In the case of the tower being pulled up by a rope tied to a car bumper, it might be well to pull with the car backing up so the driver maintains a clear view of the action. Also, an effective, foolproof communications link must be established and maintained between the driver and the person who is directing the operation. If not, towers pulled over center, bent and broken cables and a host of other crises are likely to beset the tower crew.

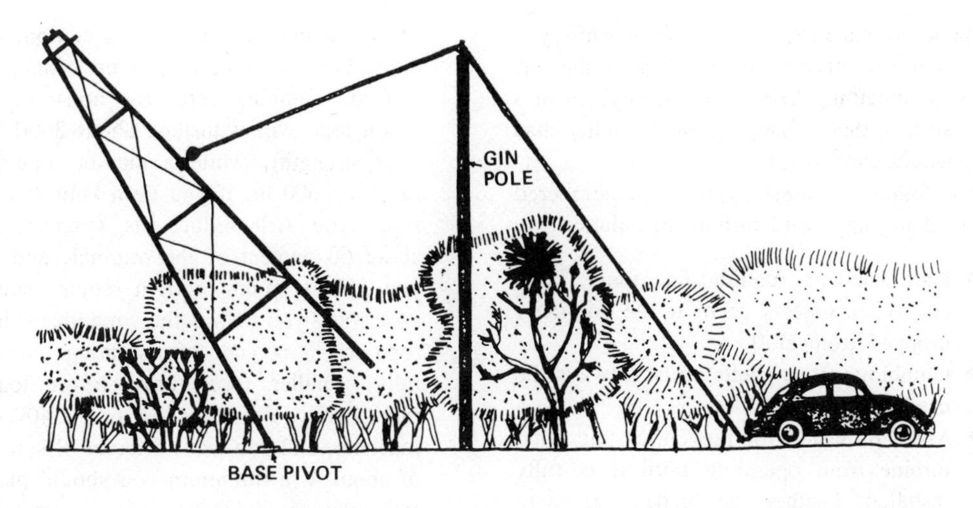

Figure 6-12. Assembling a freestanding tower. (*Courtesy of DOE.*)

Tower-raising techniques such as this are usually described in the owner's manual or the installation instructions that come with the tower.

Many towers, such as the freestanding octahedron tower, can be erected in place. This is usually done by assembling the first few bays on the ground, standing them up, then assembling the remaining bays while standing on each successive lower bay.

WECS Raising

To raise a wind turbine, you can hoist a completely assembled machine up an already erected tower; hoist a partially assembled machine up an already erected tower, completing the assembly aloft; or tilt the tower up with the wind turbine already installed. The first two methods are traditional; the last, in many cases, is not safe.

Personal experience will tell, but generally the amount of enthusiasm one has for doing anything atop a tower decreases rapidly with increasing tower height. This serves as a token justification for ground-level assembly of the tower and wind turbine, but you should consider the hazards.

Tilting up a tower with the wind machine already installed imposes additional loads on the tower. The compressive load at the base pivot and the bending loads where the rope is attached will be much greater. Most likely, you will have to provide extra bracing, but in any case you should consult the manufacturer about these loads or rely on a competent installation crew.

Consider also that if anything goes wrong, you stand to lose the entire tower and wind machine. Risks go up rapidly as tower height and wind machine weight increase. This is not to say, however, that this method will not work; it does, but the individual installation will dictate the method. For sturdy water-pumping WECS towers of 20–40 ft, experience has shown that tilting the whole works aloft can work.

For taller towers, you can expect to hoist the wind turbine up an already erected tower. This will require a block-and-tackle supported aloft and an extra rope to the WECS. The block-and-tackle is used to lift, while the rope is tugged at from the ground to keep the wind machine from banging into the tower as it journeys upward.

Consider that with a hand-operated block-and-tackle, you can feel what is happening. Tail vanes snagged on a guy wire may not be detected until damage has occurred if you use a winch or auto-pulled hoist. The support structure that holds the block-and-tackle to the tower top must not bend or otherwise yield to the loads of the WECS. You can test it by hoisting up the WECS with a volunteer adding extra weight. Remember that you or another person must depend on this hoist to suspend the machine over its mount while you bolt it down, maybe 60 or 80 ft in the air. This is something like changing engines in a Volkswagen that is hanging much higher than you would care to fall.

The following points should be remembered while doing any wind turbine installation:

- Hard hats are required, as tools, bolts and other objects seem to be routinely dropped from aloft.
- Climbing safety belts must always be used.
- Make provision for preventing the wind turbine from operating until it is fully installed. Feather the blades, tie them with a rope or otherwise lock them. One of Murphy's laws says that whatever can go wrong, will. While you may not detect a breeze on the ground, there may be enough wind aloft to create an unpleasant surprise about halfway into the installation process.
- Perform installations when no wind is expected, and start early. Installations always take longer than expected; bolting a wind turbine aloft after dark is to be avoided.
- Plan and practice the entire installation process very carefully. The process should cover such details as who has which bolt in which pocket, when the bolt is to be installed with what tool, and by whom.
- Create an alternative plan. This plan is designed to be used as a contingency if something goes wrong (e.g., the main bolt gets dropped and the wind is coming up).
- Tools and parts can best be carried aloft in a carpenter's tool belt, available at most hardware stores.
- Gloves and a warm jacket with lots of pockets will be useful in keeping you warm and able to work and finish the installation, even if a wind starts to come up; the pockets will save tiring trips up and down the tower.
- Pay close attention to the strength of ropes, pulleys or other auxiliary equipment you may use in hoisting equipment aloft. For example, if you use standard 7/16-in. climbing rope for hoisting, a nylon rope will withstand about 3900 lb (wet strength), while a manila rope is rated at 2600 lb. If you tie a knot in the rope, you will reduce its strength to about 60 percent of the original, and if you pull it around a tight radius—such as a bolt—you reduce the strength of the rope to 80 percent of its rating. Naturally, smaller ropes have lower load ratings. If you are hoisting aloft a 400-lb machine, and you want a factor of safety of about 4 (a minimum you should plan for), you are going to need a rope and other equipment capable of hoisting $4 \times 400 = 1600$ lb. If you have a manila rope rated at 2600 lb and you tie a knot at its attachment, the rope is really good for $0.6 \times 2600 = 1560$ lb. This rope is the minimum strength to consider for the job.

Wiring

Wire size, wire routing and lightning protection are important considerations. Wire size is determined by the current (A) that will flow and the length of wire. In general, you select wire sizes to limit the line voltage loss to a small percentage (Figure 6-13). Using Figure 6-13 and the following simple equation, you can calculate the wire size you need.

For aluminum wire: Circular area size = $35 \times A \times$ ft of length/V line loss.

For copper wire: Circular area size = $22 \times A \times$ ft of length/V line loss.

Circular area size is converted to wire gauge size from Table 6-2.

For example, calculate the wire size required for a copper wire run of 25 ft, carrying 25 A at 24 V, with a 1-percent line voltage drop.

Solution: From Figure 6-13, 1% = 0.24 V. Then, circular area size = $22 \times 25 \times 25 / 0.24 = 57{,}291$. From Table 6-2, 57,291 is

Table 6-2. Wire gauge size.

WIRE GAUGE (AWG)	CIRCULAR AREA SIZE (CIRCULAR MILLS)
14	4,017
12	6,530
10	10,380
8	16,510
6	26,250
4	41,740
3	52,640
2	66,370
1	83,690
1/0	105,500
2/0	133,100
3/0	167,800

between wire gauges 2 and 3. Select wire gauge 2 for conservative selection.

Wire routing, except for any deviations necessary for lightning protection, is a matter of direct routing, adequate support to prevent wind or mechanical damage to the wind turbine or wire insulator and adequate separation for prevention of electrical short circuits. Wires that are routed down the tower should be tied to the tower every few feet or fed through a conduit to preclude wind damage.

Lightning Protection

The secret of protecting wind power equipment is to install a good ground wire. This means that you must electrically tie the tower if it is metal, or tie the WECS itself (if it is mounted on a wooden tower) to the earth. All tower guy wires must be grounded by the methods discussed here, and electrically connected to each other as well as to the tower.

The National Electrical Code specifies ways in which various towers and antennas are to be grounded, and the information is useful for WECS. In general, an underground metal water pipe is desirable to use as the ground. In its absence, one or several (use several in dry ground) heavily galvanized pipes (1 1/2-in. diameter is adequate, 1 1/4-in. minimum), or a 1/2-in. copper rod, are driven into the ground to a minimum depth of 8–10 ft. Instead of rods, sheets of copper-clad steel or galvanized iron about 3 by 3 ft in dimension can be buried about 8 ft deep horizontally and connected to one another as well as to the tower. This forms an electrical "ground plane." When more than one rod is used, all should be electrically connected to one another as well as to the tower. Wire size for grounding should be number 6 or larger, and it is usually a bare wire.

Electrical wires can be protected by a spark arrester (Figure 6-14). These are available at electrical supply houses. In the case of a ground wire on a wooden tower, this wire will protect both the WECS and the electrical

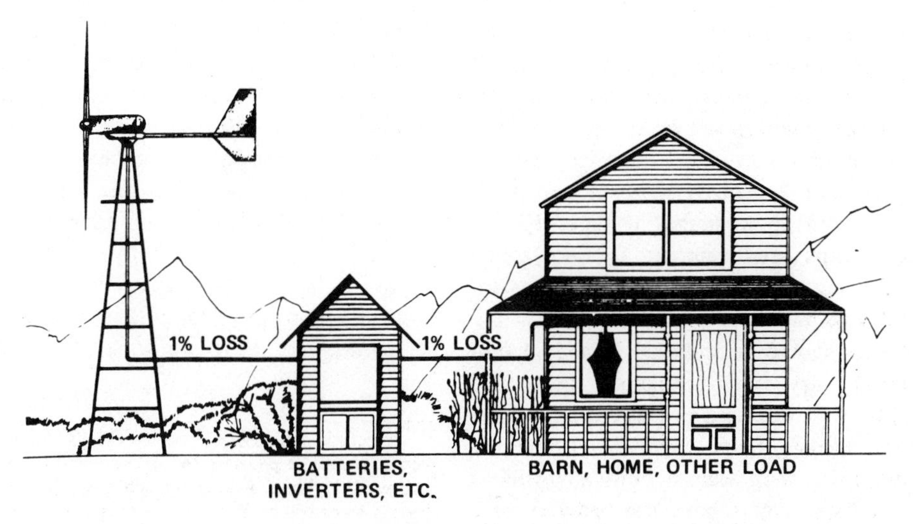

Figure 6-13. Typical line voltage loss allowance. (*Courtesy of DOE.*)

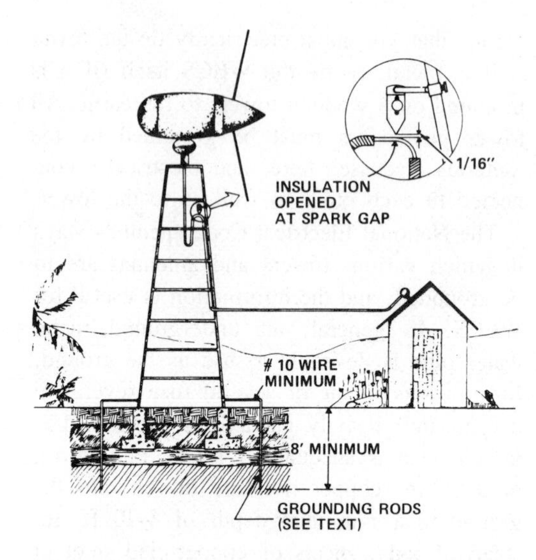

Figure 6-14. Lightning protection. (*Courtesy of DOE.*)

wires. You should consider adding a ground wire up the entire height of a metal tower, as corrosion eventually weakens the ground connection of these towers. This ground wire should have a cross-sectional area at least as great as the total of the two wires it is protecting.

Installing Other Equipment

A wind system that generates electricity to be stored in batteries is a good deal more complex in its installation requirements than a farm-type water-pumper. Provisions must be made to install batteries, inverters, controls, wires and perhaps other equipment. In all cases, follow the manufacturer's recommendations. A few items should be considered: It is generally desirable to install batteries near the wind generator, especially if lower voltage is to be inverted up to higher voltage. Higher voltage means lower current for any given load. This means wires can be smaller in size, and line loss is reduced.

Batteries should be installed in a cool (but not cold), dry, well-ventilated space and should be well insulated to prevent large temperature changes. Some installations have the batteries in a small lean-to built alongside a home or barn; others have the batteries in a basement.* Basement installation is reasonable, but gas formed by the batteries produces an explosion hazard if there are open flames or sparks in the same area. Special caps, as shown in Figure 6-15, can be purchased for lead-acid batteries that reduce gassing by catalytic conversion of the hydrogen gas back into the water. Some batteries are even offered without vents. You must know the gassing characteristics of your battery before selecting an installation site. Inverters of the motor-generator type should be bolted to a bench or mounting pad. These units do not vibrate much, but their bearings last longer with a solid installation. Static (electronic) inverters generate heat, which means ventilation is the prime installation requirement, as is a dry space free from heavy dust exposure. Most electrical controls and load monitor equipment may be installed in the same space as inverters and batteries.

Maintenance

From an historical standpoint, you might purchase a good automobile and drive it an average of 50 mph for 100,000 miles. This translates to 2000 hours of operation. It is likely that you would have the car serviced every 5000 miles, or 100 hours. Sales brochures for small WECS, on the other hand, sometimes speak of 20 years of trouble-free operation. Factory representatives talk of customers asking how long their wind turbine will last before it needs fixing.

There are 8760 hours in a year. If your WECS operates just one-fourth of the hours in one year—a reasonable number—it will have as many hours on it in one year as your automobile does when you trade it in. It is also reasonable to expect to change the oil, grease a bearing or change the brushes just once a year to get a long-life performance out of a machine. Here are some of the factors

*Electrical equipment, and especially large battery banks, should be protected by a locked door from vandals and small children. A large wrench dropped across large battery terminals could result in an explosion or a fire.

you should consider before making the final selection of your system:

- Maintenance history of WECS components.
- Can routine maintenance be easily performed up on a tower, or must the machine be lowered?
- Frequency of expected routine maintenance.
- Nature of monthly or yearly expected routine maintenance: lubrication, component replacement and inspection.
- Number of different tools required to perform maintenance tasks.
- Availability and cost of spare parts.
- Completeness of owner's manual/maintenance documentation.
- Relative safety. Can machine be shut off? Is there sufficient blade clearance from maintenance personnel? Are there exposed shafts, wires and potential hazards?
- Is a factory-trained, experienced installation/maintenance organization available?

Answers to many of these questions are available directly from dealers and other users. Some questions will never have answers but are subject to your best estimate during product evaluation.

The air mass flowing past your wind machine is full of dust and grit, which, over a long period of time, gets into the various components, including bearings, transmissions and generators. Changing the oil, greasing the bearings and inspecting generators or pumps are the ways in which you or your mechanic can monitor the system and prevent rapid wear from such environmental conditions. Evidence of rust, loose wires, worn bushings and so on should be on an inspection list, which is used each time the WECS is inspected.

Blades or vanes which are exposed to the wind are subject to impact from hail, rain, ice and rocks. Any inspection should include examination of these blades. Wooden blades might need fresh paint; fiberglass and metal blades might also need similar service.

Vibrations in the wind turbine can cause bolts and nuts to loosen and parts to fatigue and fail, wires to break, etc. This is another area for thorough examination. Properly bolted joints will not fatigue. These are items that should be inspected as a routine, preventative procedure. This type of maintenance should be done yearly and should be done on a non-windy day. Most of the items listed above rarely, if ever, require any maintenance action, but they should be inspected anyway. Someowners climb their tower once a month, just to see that everything is in order. In any event, manufacturers usually have a recommended inspection routine.

Maintenance should be scheduled following any extreme wind or hailstorm. These conditions warrant a brief inspection.

Electrical equipment, such as batteries and inverters, require cleaning, water checks and terminal inspection. Water pumps need to be checked for leaks. The list seems endless, but each check is necessary. Time spent in inspection, cleaning and lubricating will be returned in extended service life.

Figure 6-15. Tyco Laboratories "Cell Ceal" prevents excessive battery gassing.

Environmental Considerations

Planning a WECS installation involves consideration of its impact on the environment.

Small systems are not suspected of producing harmful environmental effects, based on almost a century of experience with hundreds of thousands of wind machines. Studies are being conducted to test the impact of wind turbine rotors on TV reception. Large-scale systems, such as in Block Island, Rhode Island, have nominal interference because the TV source is at an extended range. Again, small systems are not really suspect here, unless one installs a dozen or so of them, in which case all of these units collectively might affect electromagnetic waves. These effects would be highly local in nature.

Typically, wind turbines are installed far enough from dwellings that ambient wind noise is higher than machine or rotor noise. You should keep this characteristic in mind and determine for yourself the noise characteristics of the wind machine you plan to use. Noise comes from blade tips, transmissions, bearings and generators. Some machines are noisy; others are not.

Towers with guy wires usually require care to preclude guy wires from encroaching upon existing or planned easements. Tower footings may extend deep into the ground. It is usually unacceptable to install a tower directly over a septic tank or water main, but it has been done.

The visual impact of a wind system is an area for personal taste—not just your taste but that of your neighbors. No words of caution written here will substitute for your own investigation into the potential reaction to your planned wind system.

All of these notes have been derived from the author's experience and from interviews with respected wind energy technicians and consultants. While many of the points raised here were written as warnings, the frequency of occurrence is low for any problem area mentioned. We feel that such occurrences will remain low and owners of small-scale WECS will enjoy years of satisfactory service from their machines if careful consideration is given to these and other factors related to WECS ownership.

7

Energy Conversion and Storage

OVERVIEW

This chapter expands the technical descriptions of energy conversion systems such as the synchronous inverter and storage systems (with emphasis placed on the lead-acid battery) and their utilization with WECS.

Because of the intermittent nature of the wind, the growth and acceptance of wind power and the rapid commercialization of WECS will be dependent upon the use of compatible energy storage and conversion systems.

Batteries directly convert electrical to stored chemical energy through reversible chemical reactions (see Table 7-1). The cells in a battery provide direct current and consist of positive and negative electrodes in electrical contact through an electrolyte medium. Cells may be self-contained, as with commercial lead-acid and nickel-cadmium systems, or they may be comprised of separate storage and electrode (power) capability with a pumped electrolyte, as with zinc-chlorine and redox batteries.

Because of the modular nature of batteries, potential applications exist for a wide variety of WECS applications, as noted in Figure 7-1.

Mechanical storage refers to storing energy in the form of potential or kinetic energy (see Table 7-2).

Flywheels and inertial systems store kinetic energy in objects (conventionally discs) fabricated from high-strength metals or composites, and spun to high angular velocities. The stored energy density is limited by the rotational speed and thus by the cohesive strength and defect density of the flywheel material. Efficiency is determined primarily by the coupling to the primary power source and the bearing losses of the flywheel.

Aboveground pumped hydrostorage is the only economical mode of large-scale energy storage now available to utilities. It operates like conventional hydroelectric power generation, except to operate the turbines, the water must first be pumped uphill by using electricity generated by off-peak system capacity. This application is primarily suitable for utility load-leveling applications. A system to integrate a large-scale wind system with the pumped hydroelectric potential of Western rivers has been started by the Bureau of Reclamation.

Underground pumped hydrostorage is similar to aboveground pumped storage, with one or both reservoirs located below ground surface level to permit a greater number of siting options and reduce the area requirements for aboveground installations. Although primarily a large-scale storage system, underground systems could be adapted to commercial and neighborhood use.

Underground compressed air storage is primarily for utility applications. This approach stores off-peak energy in the form of compressed air in large, underground, airtight reservoirs. The compressed air can be stored in a constant volume or (using hydrostatic techniques) in a constant pressure mode. Although turbines can be run directly on the expanding air during discharge, the preferred design mode is to inject a compressed air/oil mixture into a gas turbine. The design of "no-oil" second generation systems has started. One option is to store the heat of compression;

Table 7-1. Battery storage summary.

	System Efficiency (%)	Useful Life (Years)	Operating Temperatures (°C)	Theoretical Cell Energy Density (Wh/kg)	Design/Current Cell Energy Density (Wh/kg)	Design Volumetric Energy Density (kWh/m³)	Depth of Discharge (%)	Current Density 10 hr rate** (mA/cm²)	Capital Costs ($/kW)	Capital Costs ($kWh)	Demonstrated Cell Size (kWh)	Demonstrated Cell Life (cycles)	Critical Material
Lead Acid (Pb/PbO_2)	70-75	10	20-30	240	25	45	25	10-15	60-100	25-110	>20	>2000	lead
Sodium Sulfur (Na/S)	70-80	10-25	300-350	790	115/80-100	150	85	75	60-100	15-60	0.5	400	—
Sodium-Antimony trichloride ($Na/SbCl_3$)	70-80	10-25	200	770	110	120	80-90	25	70	15-25	0.02	175	antimony
Lithium-metal sulfide (LiS/FeS_2)	70-80	10-25	400-450	950	190/70-90	210	80	30	60-100	15-60	1.0	1000	lithium
Zinc-chlorine (Zn/Cl_2)	70-80	10-25	50	460	100	60	100	40-50 (5 hr rate)**	70	12-30	1.7	100	ruthenium (catalyst)
Zinc Bromine (Zn/Br_2)	70-80	10-25	30-60	430	60-70	90	90	30	—	—	0.01	2000	
Hydrogen Chlorine (H_2/Cl_2)	70-80	10-25	30-60	990	110	20	95	300	70	25-30	.001	50	platinum ruthenium (catalysts)
Iron-redox* (Fe/Fe^{+3})	—	20	20-50	155	45-75	60	100	40-60	100-200	5-10	10	>1000	—

*GEL, Inc., 1511 Peace Street, Durham, N.C. 27701. Timothy Gooley, Private Communication.

**Five and 10 hr rate refer to the rate at which the battery is discharged. At a 10 hr rate the cycled capacity of the battery (total capacity x depth of discharge) is increased by a low discharge rate.

Sources: OTA Report, "Applications of Solar Technology to Today's Energy Needs," Vol. II, Chapter XIV.
EPRI (EM-264) and ERDA E(11-1)-2501 joint project report prepared by PSE and GCO, Newark, N.J., "An Assessment of Energy Storage Systems Suitable for Use by Utilities."

Table 7-2. Mechanical storage summary.

Energy Storage Concept	System Efficiency (%)	Key Characteristics Useful Life (Years)	Nominal Range of Plant Size Power Rating (MW)	Energy Rating (MWh)	Capital Costs ($/kW)	($/kWh)	Energy Density (kWh/m^3)	Special Hazard	Major Limitations Potential	Availability Development Status	Development Requirements
Above-Ground Pumped Hydro	70-75	50	100-2K	1K-20K	90-180	2-12	1.4	None	Siting, environmental	Current State-of-the-Art	Incidental improvements only
Underground	70-75	50	200-2K	1K-20K	90-180	2-12	1.4	Flooding	Siting, environment	Basic Technology available	Higher head equipment
Underground Compressed Air	45-75	20-30	200-2K	2K-20K	100-210	4-30	0.7-35	Methane accumulation in cavern	Siting, cavern character	Initial implementation underway	Application design
Pneumatic Storage	55-65	20-30	up to 25kW	up to 100kWh	225	800	3.5-17	Rupture of high pressure tanks, high temp. discharge	Small scale	Proof of concept stage	System & component development
Inertial Storage Flywheel	70-85	20-30	<10kW to 10MW	<50kWh to 50MWh	65-120	50-300	17-70	Wheel disintegration	System complexity	Conceptual designs & experimental prototypes	Composite flywheel & system development
Super-Conducting Magnetic Storage	70-90	20-30	>10K	>1K	50-60	30-140	—	Possible magnetic field effects	Suitable siting	Conceptual components under development	Further concept development

Sources: OTA Report, "Applications of Solar Technology to Today's Energy Needs," Vol. II, Chapter XIV.
NSF Report #77SD4245 (1977) prepared by General Electric, "Applied Research on Energy Storage and Conversion for Photovoltaic and Wind Systems. EPRI (EM-264) & ERDA E(11-1)-2501 joint project report prepared by PSE & GCO, Newark, N.J., "An Assessment of Energy Storage Systems Suitable for Use by Utilities."

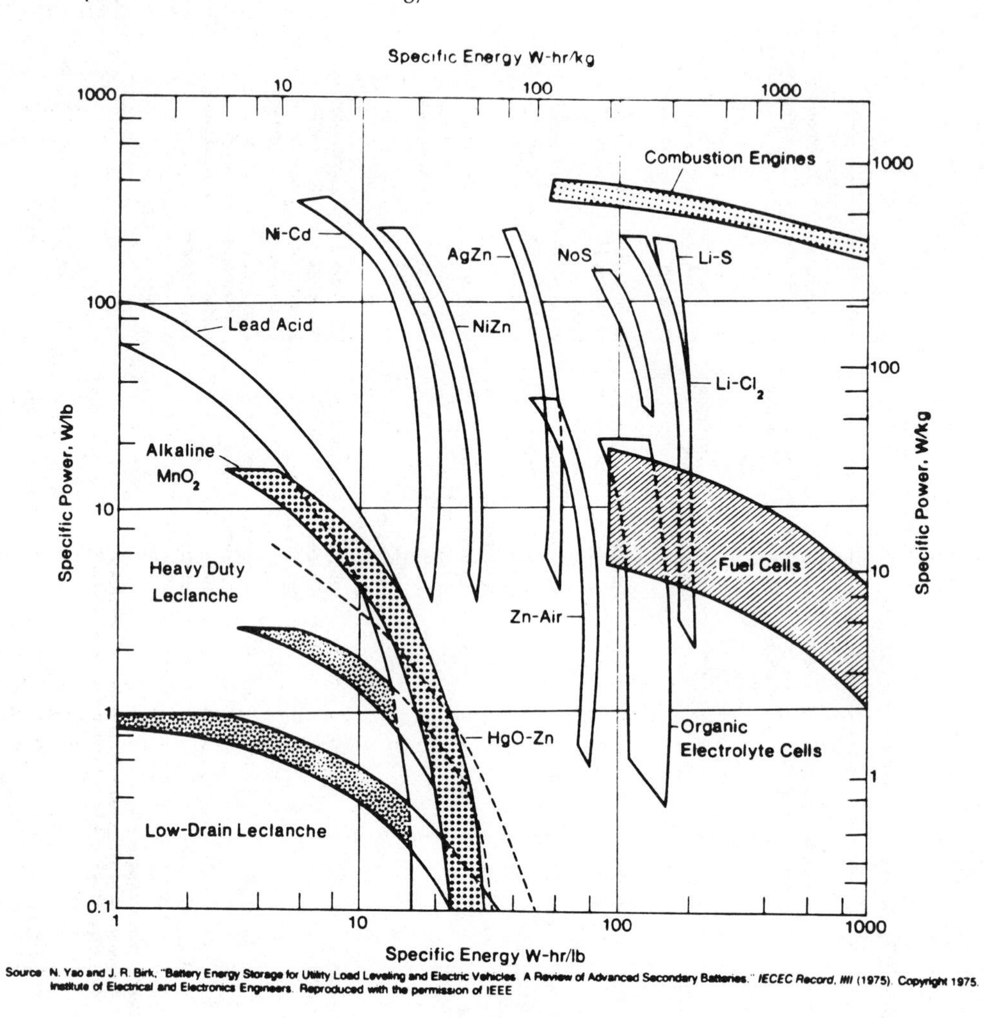

Source N. Yao and J. R. Birk, "Battery Energy Storage for Utility Load Leveling and Electric Vehicles A Review of Advanced Secondary Batteries," *IECEC Record*, *IIII* (1975). Copyright 1975. Institute of Electrical and Electronics Engineers. Reproduced with the permission of IEEE

Figure 7-1. Performance capability of various battery systems.

another is to use a coal combustor to eliminate the need for oil in the power generation step.

Pneumatic storage describes residential and commercial scale application of compressed air storage using a constructed tank and air-driven turbine. Efficiencies are low unless there is recovery of the heat of compression in a total energy system.

ENERGY CONVERSION

The variability of the wind can be resolved by providing electrical or mechanical storage or conversion systems, as noted in the preceding discussion. This section describes in detail the theory and application of the dominant energy conversion system—the synchronous inverter.

Synchronous Inverters

Mr. Hans Meyer, of Windworks, Inc., has authorized the incorporation of portions of the *GEMINI Synchronous Inverter Brochure* in this chapter.

The Gemini Synchronous Inverter is a line-commutated, line-feeding inverter which, when interposed between a variable voltage dc power source and an ac power source, converts the dc power to ac at standard line voltages and frequencies.

In operation, all the available dc power is

converted to ac. If more power is available from the dc source than is required by the load, the excess flows into the ac lines, where it can be used by other consumers connected to the same ac system. If less power is available than required by the load, the difference is provided by the ac lines in the normal fashion.

For variable or intermittent power sources such as WECS, the use of a Gemini Synchronous Inverter allows simple, safe and inexpensive interfacing with conventional ac electrical power systems such as utility grids and engine-generator sets.

By interfacing the alternate source with conventional power lines, the need for storage can be eliminated, as well as the need for separate circuits and special loads capable of functioning with the unregulated or dc power. The energy produced by the alternate source allows a corresponding reduction in the fuel requirements at the point of ac generation.

The nature of the Gemini control circuitry permits each unit to be individually programmed for optimum extraction of energy from the alternate source. Voltage, current and current slope controls allow selection of 1) the point at which power conversion begins; 2) the rate of conversion; and 3) the maximum quantity to be converted.

Thermal, chemical or electrical energy storage may be added to the system, if desired. When storage is used, or when the elimination of power feedback to the ac lines can be economically justified, controls are available that permit monitoring the net consumption of power from the ac lines, so that any surplus energy which would otherwise be fed back to the ac lines can be directed to either a storage system or to some other "non-critical" load, such as an auxiliary water heater or a preheater for a standard hot water heating system.

The technological base for the Gemini is derived from solid state industrial drive circuitry which has been developed and put to widespread use over the past 20 years. Conservatively estimated, some 1500 MW of regenerative and 6000 MW of non-regenerative solid state drives have been put on-line in the past 10 years. A regenerative drive takes energy from the ac lines to energize a piece of equipment or machinery. Alternately, for control or braking purposes, it can extract energy from the machinery and return it to the ac lines. Even though the drive may at times supply power to the ac lines, it is still a net energy consumer. This characteristic distinguishes solid state drive equipment from synchronous inverter systems which are net energy producers. Additionally, drive equipment is almost always associated with mechanical motion, while a synchronous inverter operates equally well with rotating and non-rotating sources.

Issues that must be considered when interfacing two power sources include safety, power quality and rate structures.

Safety is of concern in that there is a possibility for power to flow from one source to the other even if the other source has been de-energized. Energy fed into a downed power line could jeopardize the safety of a serviceman who has failed to take the usual precautions to ensure a dead line. The Gemini has features built in to minimize risks of this type. The very nature of line-commutated inverters makes it impossible to produce ac in the absence of ac voltage from the lines, and a line voltage-activated contactor (relay) disconnects the dc source from the Gemini and the ac lines whenever the line voltage drops 20 percent or more. Fusing is provided at both the dc and ac sides of the Gemini to protect against high currents and potential fire damage under fault conditions.

Gemini power quality falls well within the range of values found for other typical household and industrial loads and is identical to that of solid state drive equipment. The Gemini represents an inductive load to the ac lines, similar to that of an induction motor with a comparable power rating. Field tests indicate that electromagnetic interference is well within standards established by the Federal Communications Commission. Where justified, power quality can be improved using the same techniques which have been developed to improve power quality in industrial applications.

Rate structure becomes important when the ac and dc power sources are not owned by the same entity, such as the case when tying an alternate energy system to a utility grid. A further distinction can be made between installations where the prime interest is to decrease the consumption of ac power as opposed to those installations where the desire is to feed back significant quantities of energy and to be reimbursed for it. The first type of installation is referred to as "supplemental" and generally requires very little modification of the rate structure, since its net effect of being "on-line" is no different than a simple reduction in loading of the ac power source. The second type of installation, where significant energy is fed back to the ac lines, is often referred to as "co-generation," and generally requires a specific contract with the utility company involved.

The simple circuitry employed in the Gemini provides high reliability and high efficiency at the lowest cost per kW for dc to ac power conditioning equipment.

Wind energy is the nearest term, most cost-effective solar electric energy source presently under development in this country. Department of Energy estimates suggest that 5 percent of the nation's electrical energy needs can be supplied by wind systems by the year 2000. It is conceivable that as much as half of this capacity will be supplied by small residential size wind systems interfaced with local utility grids, while the other half will come from large, utility owned and operated systems. The Gemini Synchronous Inverter provides an economical and fail-safe means of interconnecting small wind systems with existing ac circuitry, thereby eliminating the need for redundant circuitry, specialized appliances or battery storage.

Theory of Operation. Many industrial or commercial electric motor drive systems are connected to mechanical loads that store large quantities of kinetic energy in the form of moving or rotating mass. When such a load is accelerated to operating speed, a large amount of temporary power is required. Similarly, when the load speed is reduced to zero, the kinetic energy must be extracted from the moving mass. Again, the power involved in deceleration can be very large.

Other types of loads store energy in the form of potential energy resulting from a change in vertical location of large masses. Cranes and hoists are typical examples of this situation. In both cases, the exchange of energy with the load takes place in both directions, and very early in the history of electric drive systems it was discovered that while electrical or mechanical braking could be used to absorb the energy when necessary, undesirable heat was produced; efficiency suffered; and the process was difficult to control with any degree of precision. Regenerative drive systems were developed to minimize these problems. "Regeneration" is the process of feeding excess energy from load into conventional ac power lines.

Electronic switching devices, such as mercury tubes, thyratons or thyristors, can control power flow in either direction by instantaneously connecting the ac line to a dc motor during a selected portion of each cycle or half cycle of the ac line voltage. The variation of conduction periods is referred to as "phase control" and is also used to control power to welders, heaters, industrial arc furnaces, battery chargers and numerous other loads. All three types of switches have one common characteristic. The turn "on" signal can be removed and the switch will continue to stay "on" until external conditions reduce the current through the switch to zero. In some cases, this occurs naturally at some time in the cycle; in other cases, additional circuitry is needed to force the current to zero to turn the switch "off."

A greatly simplified explanation of how power can flow in either direction can be seen in Figure 7-2, showing the ac voltage and current waveforms of a single phase line supplying an inductive load. Although a purely inductive load consumes no power, current flows through the load and power is instantaneously traded back and forth with the ac lines.

During intervals marked "D", the voltage and current are in the same direction, and power flows from the line to the load. During

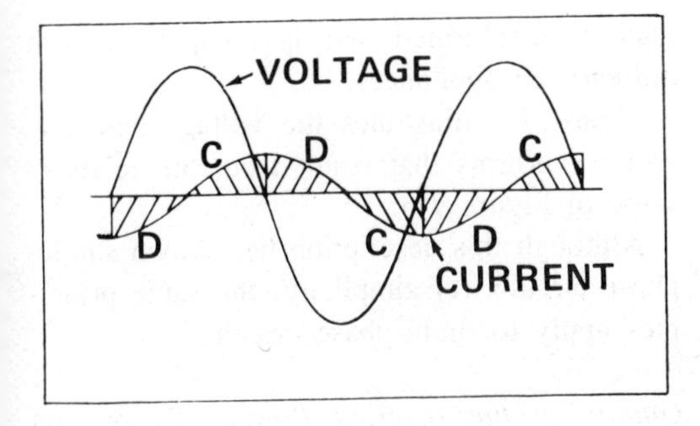

Figure 7-2. Waveform (ac).

intervals marked *D*, voltage and current oppose each other, and power flows from the load to the line. If the load is a pure inductance, areas *C* and *D* are equal, and the average power is zero.

A modern dc drive system uses this relationship in combination with high-speed switches to allow current to flow during one of these times, and to block the flow of current during the other time. If power is required to flow from the line to the dc motor, the switch is closed during the times marked *C* and opened during the times marked *D*. Similarly, if it is desired to extract power from the dc motor, it is connected to the line during *D* intervals and disconnected during *C* intervals.

Synchronous Inversion Concept. If the basic concept of regenerative power feedback used for retrieval of load energy in a dc drive is adapted for general conversion of dc power to ac, such a system can be used to efficiently take advantage of alternate energy sources. All that is required is to first convert the alternate energy to dc. If the alternate energy is mechanical, it can drive a dc generator or an alternator whose output is then rectified. Solar cells can convert sunlight directly to dc power. Other forms of energy are already dc and can be connected directly with such a system to an ac source.

In its simplest form, a single phase ac line is connected to a source of dc power through what looks like an ordinary bridge of thyristors. Notice, however, that the polarity of the dc side of the bridge is reversed when compared to a thyristor bridge intended for converting ac to controlled dc. Figure 7-3 shows this basic connection.

Figure 7-4 shows two alternate paths for current flow from the dc source to the ac lines, arranged so that each path is suitable for power flow to the line in one of the half-cycles during the interval that corresponds to the *D* area in Figure 7-2.

Figure 7-5 shows waveforms of ac line voltages and the dc source for the path that is shown schematically in Figure 7-4A. While an arbitrary value of dc voltage is shown, the actual magnitude can be any value from zero to the peak of the ac wave. During the positive half-cycle, there are two intervals, *A* and *B*, where the dc voltage is instantaneously more positive than the ac line voltage, and current flow from the dc source to the line will oppose the line voltage, so that the flow of power is to the line. Current flow in the same direction is also possible during the negative half-cycle, but since the ac voltage has reversed, current no longer opposes the line voltage, and power flow is in the wrong direction—from the line to the dc source.

Time intervals *A* and *B* have one significant difference. During interval *A*, the difference between the ac and dc voltages is initially high and decreases to zero. This condition is useful when thyristors are used as the switches, since it automatically reduces the current in the thyristors to zero and the thyristor turns off or "commutates" naturally. In interval *B*, however, the voltage differential is zero initially and increases with time until it reaches a large value at the end of the interval. To operate in this time interval, a switching device must therefore have an independent

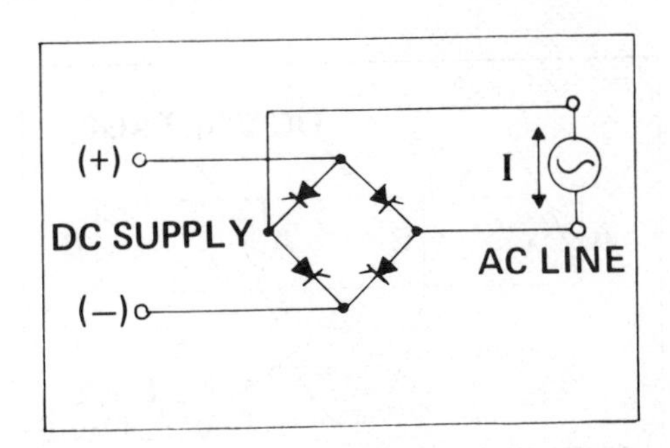

Figure 7-3. Bridge circuit.

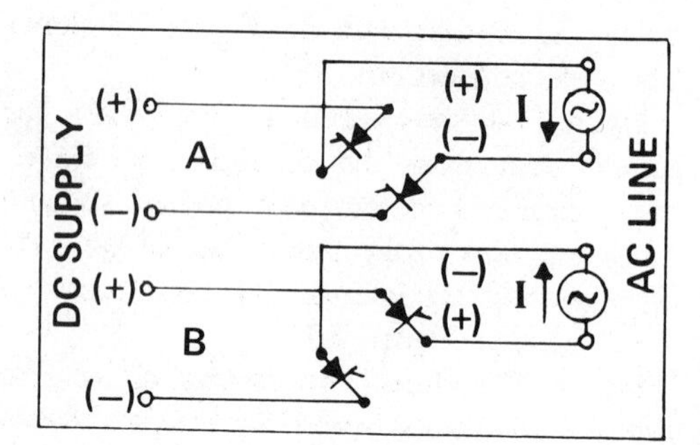

Figure 7-4 A and B. Alternate paths for current flow.

means of commutating to an off state. With transistors, this is relatively simple; however, thyristors require complex commutating circuitry. For this reason, when thyristors are used as the switching element, the conversion period is generally limited to time interval *A*, and the inverter is known as a "line-commutated inverter", since the line voltage itself provides the reduction in current that turns the device off.

While the circuit of Figure 7-4A and the waveforms of Figure 7-5 demonstrate the transfer of power from the dc source to the ac line, the current is of a single polarity, and the dc power it represents would not generally be useful, would tend to saturate transformers and could not be transformed to distribution voltage levels. The alternative conducting path shown in Figure 7-4B provides a reversed polarity of the dc source for conduction in a similar period of the other half-cycle.

The waveforms of Figure 7-5 now become those of Figure 7-6, showing the effects of the reversed dc voltage in the negative half-cycle. The current is now truly ac, can be readily transformed and is compatible with ordinary ac appliances.

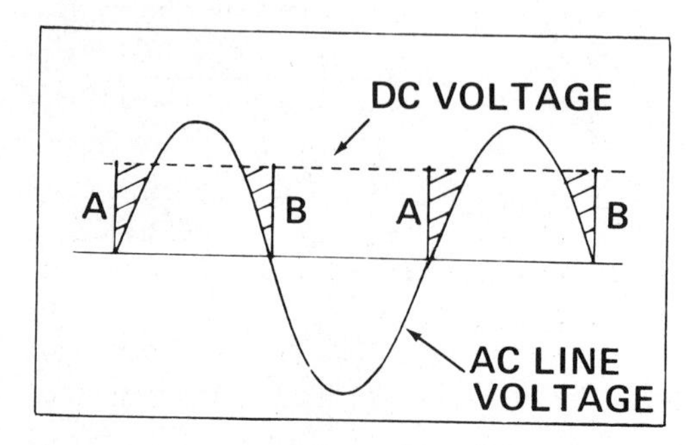

Figure 7-5. Line voltage (ac).

Figure 7-7 illustrates the voltage and current waveforms that result from the relationships in Figure 7-6.

Although this description has shown single-phase circuits for simplicity, the same principles apply to multi-phase circuits.

Quality of the Inverted Power. To prevent disturbing the waveform of the ac line voltage with this type of inverter, the impedance of the dc source must be many times larger than that of the ac line, since connecting any two voltage sources together results in a terminal voltage that is divided between them in a manner proportional to their individual impedances. A relatively large dc source impedance allows the connection to the ac line to be made without changing the line waveform, and the voltage at the output terminals of the synchronous inverter is therefore the normal line voltage, both in magnitude and frequency, without distortion.

The normally inductive impedance of a dc generator fits this requirement. An inductive source, in addition, tends to spread out the

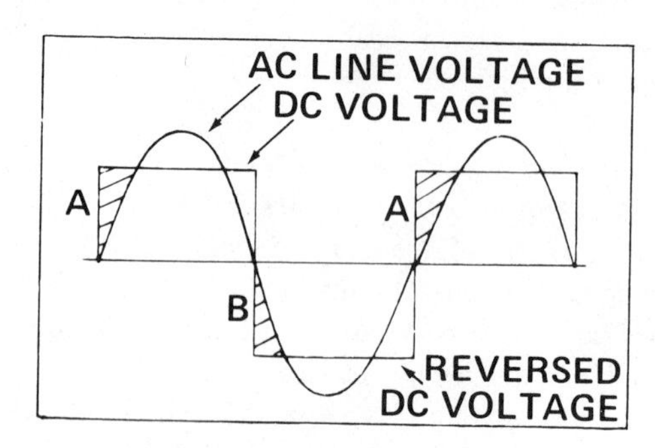

Figure 7-6. Reversed dc voltage.

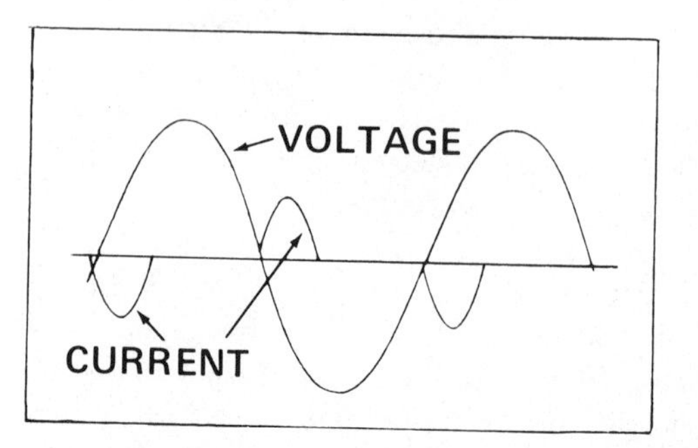

Figure 7-7. Voltage and current waveforms.

conduction periods, resulting in lower resistive losses due to improved form factor. Any installation of a synchronous inverter of this type must be engineered for proper impedance ratios, and inductance added to the dc source, if required, to maintain line voltage integrity.

While the voltage waveform is very likely a nearly perfect sine wave, the output current of the inverter is an entirely different matter. Even though ac in nature, it is delivered in pulses, once each half-cycle, and the shape of these pulses is dependent on both the magnitude of the dc voltage and the total impedance in the dc and ac sources. Prediction of the shape of these waveforms is a very complex problem.

The current of a synchronous inverter occurs with the same polarity and during the same part of the cycle as would an inductive load current. For this reason, the introduction of power in this manner is seen by the power line as an additional lagging load, and the effects are generally the same.

Tests indicate that standard power factor correction techniques are as effective for inverted power as they are for normal inductive loads.

Metering accuracy of standard kWh meters has been compared to a digital meter having an accuracy of 0.1 percent. Even with the pulsed currents, no significant inaccuracy was found in the standard kWh meters. It is assumed that metering accuracy will be maintained as long as voltage waveforms are not distorted, since the voltage coil of the meter has a slow response which may not be able to react to the harmonic voltages as easily as the faster current-measuring coils can respond to the harmonic current.

Other Considerations. A synchronous inverter that takes advantage of the line voltage for commutation of its switching devices cannot function without the line voltage present and must be automatically disconnected when the line fails. If not, the thyristors may cause a fault at the dc source and the power lines could possibly have dc voltages that would be dangerous to personnel working on the lines.

Gemini Synchronous Inverters prevent both of these problems by using a contactor to isolate the dc source from the inverter and the line in the event of power failure. The coil to this contactor is energized by the ac lines, and therefore the contactor will open immediately on loss of ac line power.

The switching action of the thyristors is quite capable of causing radio or TV interference, and filters to prevent this are part of the design. An interference filter consisting of a line reactor and a capacitor is included in the line voltage terminals of a Gemini inverter to minimize these types of interference.

Comparison with Other Methods. Most other methods of adding power to the utility-grid system use rotating machinery to generate the ac by direct connection to the power lines. The generators may be synchronous alternators, squirrel cage induction motors or wound rotor induction motors.

The advantage of any of these methods over the synchronous inverter lies in the low harmonic content of the generated current. With a synchronous alternator, there are the additional advantages of controlled power factor and the ability to function as an independent source of ac in the event of a power line failure.

With either a synchronous alternator or a squirrel cage induction generator, the operating speed is either fixed or must remain within a narrow range. This constraint is not consistent with the variable nature of waste energy sources. In the case of the synchronous alternator, additional complications arise when momentary loss of source energy requires a disconnect, followed by a subsequent resynchronization.

The wound rotor induction motor can operate above synchronous speed over a fairly wide range of speeds, but control of secondary power can be complex, possibly best accommodated by the use of the versatile synchronous inverter.

Advantages of the synchronous inverter include handling wide voltage variations of both rotating and non-rotating sources, active and passive control for maximizing extraction of energy from the source, highly efficient operation and low cost.

Installation. The installation of a Gemini Synchronous Inverter is accomplished by wiring the unit in parallel with the ac source. For single phase units of 40 A capacity or less, this can be accomplished by simply plugging the unit into a standard electrical range outlet (see Figure 7-8). This is a particularly convenient approach in that it provides for a positive disconnect whenever servicing is required.

Three-phase units can be wired in permanently to the circuit breaker box serving the ac loads or facility in question, as shown in Figure 7-9. For both single-phase and three-phase applications, the Gemini should be isolated on its own circuit and circuit breaker. Power from the dc source is then fed through the Gemini to the circuit breaker box, where it is automatically distributed to any ac loads or fed back to the ac source.

Before operating the system, the Gemini must be programmed to load the dc source for maximum extraction of energy. Three variables must be specified to accomplish the programming: the voltage cut-in, that point at which the unit begins converting dc to ac; the current slope, the rate at which the unit converts power as a function of some variable; and the current limit, the maximum current to be converted in normal operation.

Standard Gemini control circuitry requires feedback of a key system parameter to accomplish the variable loading. Typically, for small wind systems, this variable is dc terminal voltage which is generally proportional to

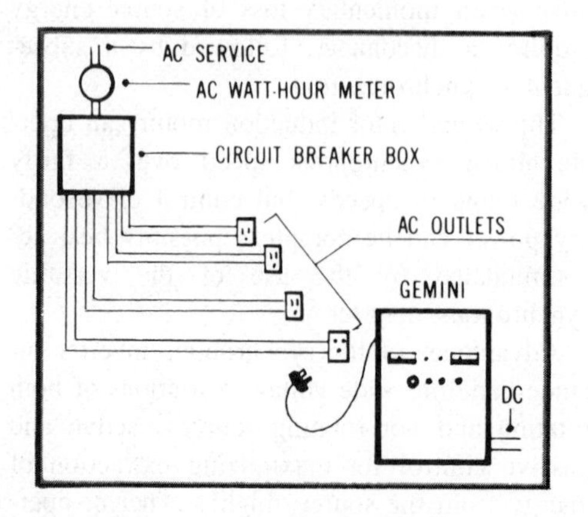

Figure 7-8. Simple installation of synchronous inverter.

rotor speed. Some other signal, such as rpm or wind speed, can be used where it enhances the overall system performance.

Safety. A synchronous inverter that takes advantage of the line voltage for commutation of its switching devices cannot function without the line voltage present and must automatically disconnect when the line fails. If it does not, the thyristors may cause a fault at the dc source and the power lines on the secondary side of the service transformer could possibly have dc voltages that would be dangerous to personnel working on the lines.

Care has been taken in the design of the Gemini to avoid these problems by using a contactor to isolate the dc source from the inverter and, hence, from the line in the event of power failure. The contactor coil is energized by the ac lines so that loss of power will cause the contactor to open.

Further care has been taken to fuse both the ac and dc sides of the Gemini. In the extremely unlikely event that the contactor should fail in the closed position, the dc resistance of the secondary of the distribution transformer or a pair of thyristors would act as a virtual short circuit to the dc source, causing the ac or dc fusing in the Gemini to open and again provide protection.

The fusing of the input and output also serves to protect the dc source, the internal wiring and the ac lines from overload conditions.

The nature of a line-commutated inverter results in the dc source being connected to first one side and then the other side of the ac service. In that the utility service is tied to ground, it is very important that neither side of the dc source be tied to ground to avoid a fault condition.

Under normal conditions, this approach is acceptable and is the most economical. When special testing is taking place or when the particular nature of the dc source requires grounding of one leg of the dc circuit, isolation transformers may be used.

An ac power switch and ac and dc power indicator lights are located on the exterior of Gemini Synchronous Inverters to inform the user of the operating condition of the unit. A

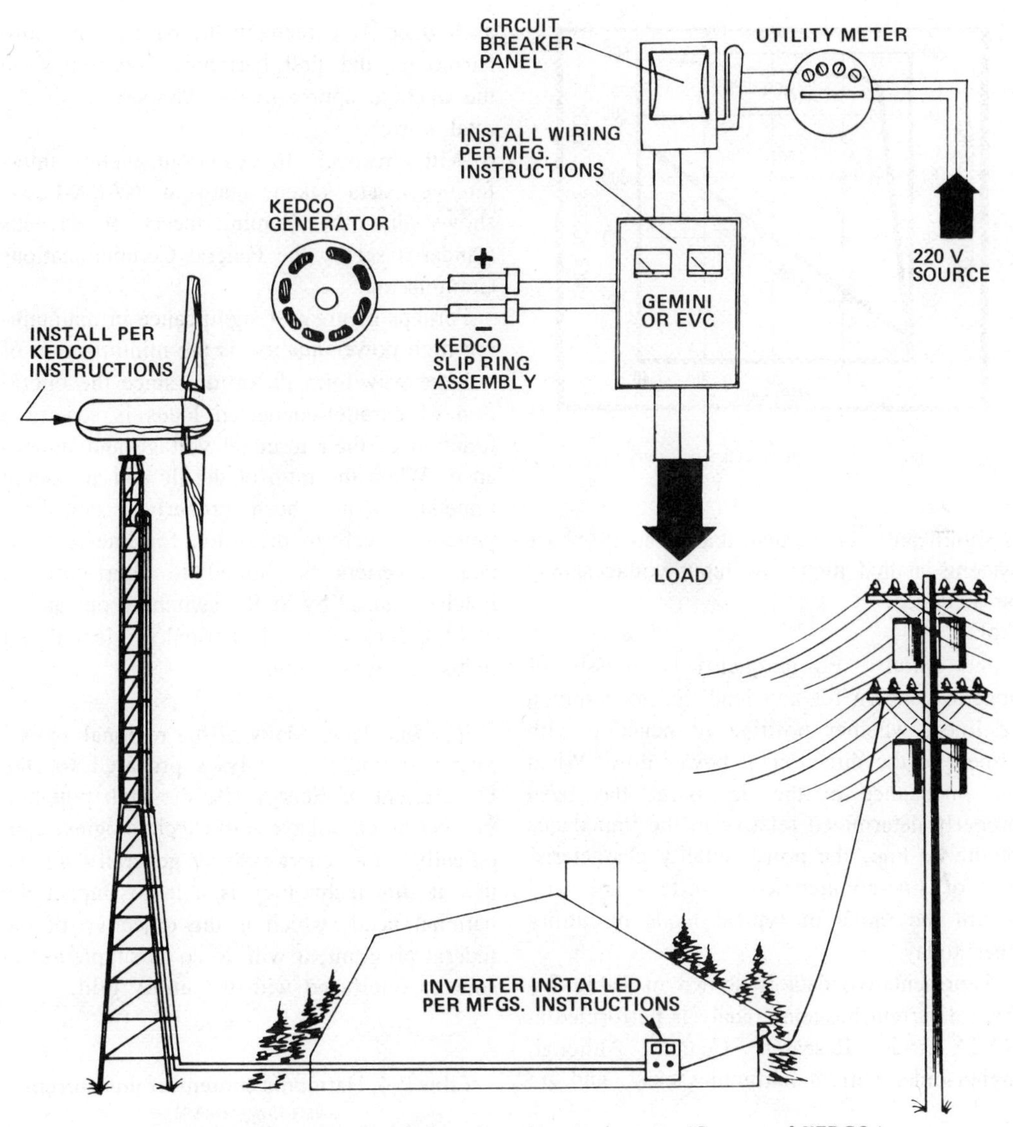

Figure 7-9. Three-phase installation of synchronous inverter. *(Courtesy of KEDCO.)*

dc ammeter and a voltmeter are also externally mounted to enable monitoring of the system performance and to permit easy programming of the control circuitry.

Performance. The losses in a line-commutated inverter are essentially proportional to the current being converted. The power output, on the other hand, is a function of both current and voltage. The losses incurred during operation can therefore be minimized by running at the highest possible voltage for a given power level.

Shown in Figure 7-10 is a typical efficiency curve for an 8-kW Gemini tied to a wind system employing a shunt-wound dc generator.

In addition to minimizing system losses associated with current level, operating at higher voltages results in a minimum VAR load imposed on the ac lines.

Data collected in tests conducted by Wisconsin Electric Power Company on a single-phase Gemini at various power loadings and for various ac loads are shown in Table 7-3. The performance of three-phase systems

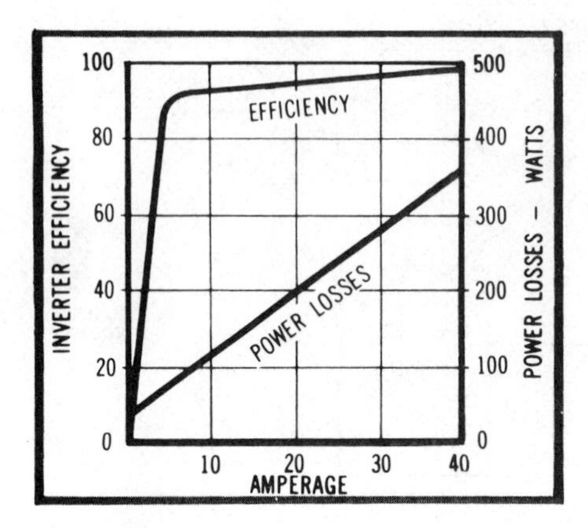

Figure 7-10. Typical efficiency curve.

is significantly better than that of single-phase systems in that higher dc input voltages may be achieved.

Power Quality. Power quality is an issue of general concern for any load tied to common ac lines, whether positive or negative with respect to the direction of power flow. When the impedance of the dc source has been properly determined relative to the impedance of the ac line, the power quality characteristics of line-commutated inverters are well within the range of typical loads on utility lines today.

Representative data is shown in Table 7-4 from a current harmonic analysis performed at NASA-Lewis Research Center. Although higher-order current harmonics exist, and are as high as 11 percent in the case of the third harmonic, the first harmonic contributes on the average approximately 98 percent of the total power.

With respect to electromagnetic interference, data taken again at NASA-Lewis shows that the Gemini meets or exceeds standards set by the Federal Communications Commission.

Perhaps of greatest significance in maintaining high power quality, is the minimization of voltage waveform distortion, since the operation of parallel-connected loads is strictly a function of their terminal voltage and impedance. When the ratio of the dc and ac source impedances has been properly established, voltage waveform distortion for line-commutated inverters is limited to short-duration notches caused by SCR's switching on, and is of little consequence in normal residential and industrial applications.

Utility Interface. Many of the regional assessments and market analyses prepared for the Department of Energy cite the high potential of distributed solar electric technologies, specifically wind energy. It is generally agreed that if this technology is able to impact the national need, which is the objective of the federal program, it will do so as supplemental sources interfaced with the utility grid.

Table 7-3. Gemini Synchronous Inverter supplying ac line with additional loads.

	Gemini Output: 2 KW			Gemini Output: 6 KW		
	No Load	Res Load	Res Cap Load	No Load	Res Load	Res Cap Load
Line						
AC Volts	235	232	233	238	235	235
AC Amps	14.3	14.0	7.5	37.5	28.3	20.0
KVA	3.36	3.25	1.75	8.93	6.65	4.70
KW	− 2.035	1.925	1.935	− 5.93	− 1.90	− 1.90
KVAR	2.84	2.709	0.125	6.559	6.231	3.612
Load						
KW	0	3.927	4.00	0	4.015	4.015
Gemini						
DC Volts						
Input	190.8	190.5	190.6	180.0	180.6	180.4
DC Amps						
Input	12.2	12.2	12.2	35.5	35.5	35.5
Output KW	2.03	2.047	2.057	6.0	6.0	6.0

Table 7-4. Harmonic content of line current.

RMS TOTAL 99.94

A RMS Line Current						
A	33.5	29.2	35.8	32.6	37.8	
	Percent Harmonic (RMS)					Ave.
1st	98.94	98.76	99.09	98.57	99.51	98.97
2nd	2.26	2.11	2.15	2.23	2.47	2.24
3rd	12.31	12.85	10.69	14.33	7.52	11.54
4th	– –	– –	1.32	– –	1.25	.51
5th	5.05	6.06	5.57	7.02	4.11	5.56
6th	– –	– –	1.34	– –	– –	.27
7th	3.72	3.9	2.98	3.81	2.55	3.45
8th	– –	1.28	1.46	– –	– –	.55
9th	2.58	2.74	2.53	2.49	1.81	2.43
10th	– –	1.09	1.26	– –	– –	.47
11th	1.96	2.69	1.98	– –	1.57	1.64
12th	– –	1.10	1.62	– –	1.03	.75
13th	1.69	1.87	1.60	– –	1.05	1.24

Rate structure becomes an important issue when the ac and dc sources are not owned by the same entity. When this is the case, a distinction can be made between supplemental installation, where the objective is to decrease consumption of energy from the ac source, and cogeneration installations, where payback for energy delivered to the ac source plays a significant role in the economics of the system.

For supplemental systems, proper sizing of the source in light of the available resource and in consideration of the load demand generally results in a very low percentage of the generated energy being fed back. The net effect of the system is a reduction in load seen by the ac source and generally requires little modification of the rate structure. For cogeneration installations, the ramifications of trading significant amounts of energy generally require a specific contract with the utility involved.

Numerous Gemini installations have been put on-line in the past several years in cooperation with various utilities. In some cases, rate structures have been developed that provide for reimbursement of the customer for energy feedback, while in other cases, the customer is serviced under the standard rate but with a racheted meter and no payback. Data are being collected on a majority of these installations to provide a basis for establishing equitable rate structures.

The ramifications of interfacing are complex. Technical and economic considerations vary with the generation mix and load demand of the utility, the availability of the alternate resource and the percent penetration of the alternate source in terms of generating capacity.

Load Management. Alternating current dumping circuitry, as shown in Figure 7-11, is presently under development. It provides for the consumption of any surplus generated power that would otherwise be fed back to the ac source. A wattmeter circuit senses the magnitude and direction of power flow between the ac loads and the ac source and provides a signal to the dumping circuitry which controls

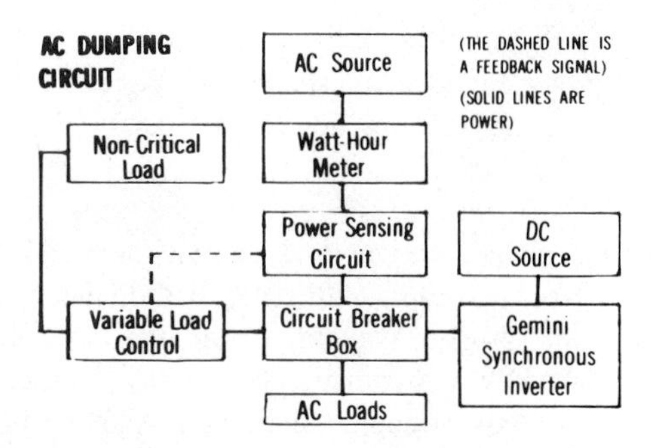

Figure 7-11. Dumping circuits (ac).

a variable load. In that the amount of surplus generated power and the time of its occurrence cannot be predicted, the dumping circuit must work with non-critical loads such as secondary hot water heaters.

Direct current dumping circuitry is also under development, as shown in Figure 7-12. This permits adding dc electrical storage to a supplemental energy system. Again, the wattmeter circuit is used to sense power flow between the ac loads and the source, only in this case a signal is fed to the Gemini to convert only that amount of energy necessary to null out consumption from the ac source. Any surplus energy coming from the dc source in excess of that required to null out consumption goes to the battery which is wired in parallel with the Gemini. When the battery reaches a minimum charge state and sufficient power is not available from the dc source to meet the load, the system automatically shuts down until such time as it can again supply the load. If desired, charging of the batteries from the ac line can be accomplished with little additional circuitry.

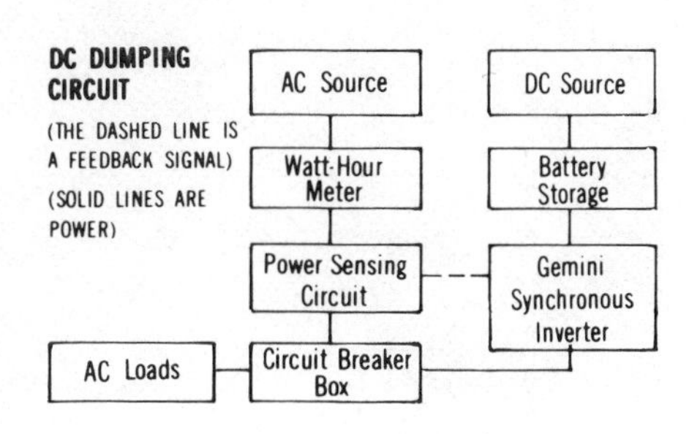

Figure 7-12. Dumping circuits (dc).

Typical Specifications. The Gemini Synchronous Inverter as shown in Figure 7-13 includes the features described below.

Basic features:

- Voltage, current and current slope controls permit matching the load demand to the power available from the source, thereby maximizing the energy extracted. For example, in a wind system, the power available from the wind varies with the cube of the wind speed. The Gemini controls permit loading the wind system to extract the available energy to within a few percent of what could be extracted under ideal conditions.
- Inverter efficiency at rated output is 96 percent for single-phase systems and 98 percent for three-phase systems (exclusive of optional filters and transformers).
- The no-load power draw for the inverters is typically less than one-half of 1 percent of rated capacity, thereby maximizing the net energy production of the power source.
- A dc contactor energized by the ac lines automatically disconnects the dc source from the ac lines during utility outages and automatically reconnects it when the ac power is restored. This feature ensures the safety of a utility lineman while servicing a downed line to which the Gemini is interfaced.
- Input and output fuses are installed in all Gemini Synchronous Inverters to protect internal wiring, the dc source and the ac lines from severe overload conditions.
- Inverters rated at 40 A or less may be installed by plugging into a standard electric range outlet of suitable rating.

Figure 7-13. Gemini Synchronous Inverter.

- All units have a dc ammeter and voltmeter, ac and dc power indicator lights and an ac power switch externally mounted for convenience and safety.
- Test circuitry is built into the Gemini controls to facilitate programming of the unit and for use in troubleshooting.
- All units are housed in wall or floor mounted steel cabinets.

Optional features:

- Interface filters—When the dc source is of low impedance, the use of an inductive or an inductive/capacitive filter may be required. Optimally sized air core reactors and electrolytic capacitors are used to fabricate interface filters for photovoltaic arrays, wind systems which employ alternators or low-impedance generators, battery storage applications and other low impedance sources. For experimental installations, multiple-tap air core reactors and multi-valued capacitors can be supplied to permit determination of the filter configuration, which results in optimum system performance.
- Maximum power trackers—Automatic power tracking circuitry is available for use with concentrating and non-concentrating photovoltaic arrays. The tracker continuously seeks that operating level which results in the maximum power conversion for a given climatic condition. A wattmeter circuit monitors power output to provide information to a power tracker circuit, which varies the loading conditions seen by the array. This nearly instantaneous variation of load permits optimization of the energy output at very little additional cost and with essentially no increase in circuitry losses.
- Isolation transformers—For application in which the dc source voltage is too high or too low with respect to the ac line voltage for optimum performance and for applications requiring grounding of the dc source for testing or precautionary reasons, isolation transformers of appropriate rating can be provided.
- dc field supplies—For systems requiring the use of shunt-wound generators, the field supply can be built into and controlled by the Gemini. When the dc source is not in operation, the field is de-energized to minimize losses. Field excitation can be either fixed or variable for optimum system performance.
- Starting circuit—For Darrieus wind systems, a starting circuit has been developed which permits utilizing ac power to accelerate the turbine. When generating speed is reached, a reversing contactor permits the Gemini to extract power in the normal fashion. The starting sequence may be initiated by a timer, a wind signal input or a manual control.

Table 7-5 summarizes the basic specifications for the single-phase system, and Table 7-6 provides the specifications for three-phase Gemini Synchronous Inverters.

dc/ac Inverters

Inverters are devices which convert the dc voltage (power) to ac power. It is an age-old question whether dc or ac electricity is better. We do not propose an answer but suggest that many appliances are not designed to run on dc. Most motors, some stereos, TV sets and certain other devices usually require 110 V ac. Many WECS are rated at 12, 24, 32 or maybe 110 V dc. An inverter can also be used to change dc to ac.

Some inverters use a dc electric motor to drive an ac generator. By driving the generator at constant rpm, a constant frequency, usually 60 Hz (Hertz or cycles per second) is generated. These units are called rotary inverters. Other inverters, as shown in Figure 7-14, are called static inverters, and these are solid state, using transistors to switch dc into ac. Some units, the lower-cost variety, generate a square wave output, while more expensive units and rotary inverters generate sine wave outputs. The most desirable output is sine wave, especially if a stereo or TV set is to be operated. Square waves powering a stereo sometimes distort the sound.

Inverters are rated by their maximum continuous capacity in W. A small surge ca-

Table 7-5. Single-phase synchronous inverter.

MODEL	POWER CAPACITY (KW)	INPUT VOLTAGE (VDC)	MAXIMUM AMPERAGE (AMPS)	OUTPUT VOLTAGE (VAC)
PCU-40-1	2 4	0-100 0-200	20 20	120 240
PCU-80-1	4 8	0-100 0-200	40 40	120 240
PCU-150-1	15	0-200	75	240

Table 7-6. Three-phase synchronous inverter.

MODEL	POWER CAPACITY (KW)	INPUT VOLTAGE (VDC)	MAXIMUM AMPERAGE (AMPS)	OUTPUT VOLTAGE (VAC)
PCU-200-3	20 20	250 500	80 40	240 480
PCU-400-3	40 40	250 500	160 80	240 480
PCU-500-3	50 50	250 500	200 100	240 480
PCU-1000-3	100 100	250 500	400 200	240 480

pability is possible for most models. Thus, a typical 500-W continuous transistor inverter might be rated to 700 W for 10 seconds or even a minute, depending on the unit. Surge capability is needed, especially for inverters operating motorized appliances such as refrigerators, because electric motors require considerable extra power for starting.

Selection of a suitable inverter involves another important factor—the efficiency of the inverter, and, in more expensive models, automatic power adjustment. With a low-cost inverter, as would be available from most recreation vehicle supply stores, the inverter will draw (from the battery) almost the maximum rated power, regardless of the load the inverter is driving. Thus, a typical 500-W inverter may draw 400 or more W from the

Figure 7-14. NOVA Inverter.

storage system, while only powering one 100-W light bulb. Higher-cost inverters offer the important option of a load monitor, which automatically adjusts the current draw by the inverter, according to the load.

A typical efficiency curve for a static inverter looks like Figure 7-15. From this, you can see that, wherever possible, it is best to select inverters that will operate near their maximum rated capacity. This could mean using several small inverters for various loads or one large automatic inverter for the entire system. In any case, cost of such inverters may dictate which inverter is selected. Ultimately, the cost of the energy supplied is the primary consideration.

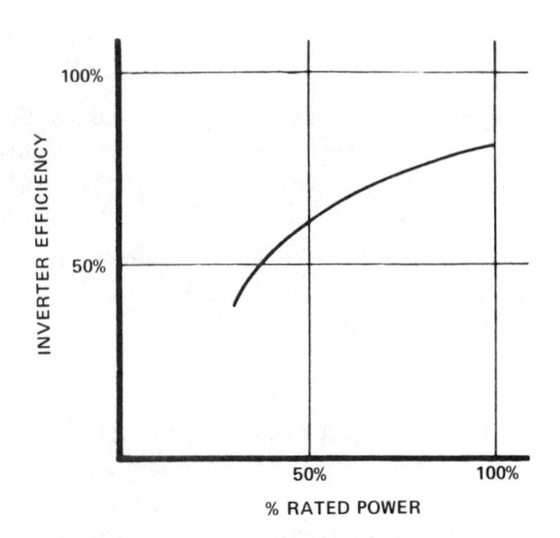

Figure 7-15. Typical static inverter efficiency.

ELECTRIC STORAGE SYSTEMS

There are many ways to store energy converted from the wind—compressed air storage, thermal storage, pumped storage, hydrogen storage, flywheel storage and battery storage. Of these, the most reliable, versatile and proven technique for small wind systems is battery storage.

Battery Storage

Battery storage systems offer a number of attractive features: they are capable of storing large quantities of power; they can deliver this power at slow or extremely high rates without damage; they are rugged and reliable; and they are a source of clean power in the sense that the output from a battery is extremely free from electronic fluctuations or "noise."

While there are numerous types of storage batteries, the most widely known is the lead-acid rechargeable type. Since lead-acid batteries are the most well suited to stationary wind system applications, we have confined the discussion to this type.*

Lead-acid Battery. A battery is a group of electrochemical cells connected together to supply a nominal voltage of dc power, as shown in Figure 7-16. The cell in a lead-acid battery consists of two sets of lead-alloy plates, suspended in a container or jar of electrolyte (battery acid-dilute sulphuric acid). These plates are arranged in parallel, alternately "positives" and "negatives." All negatives are joined together and all positives are joined together by special alloy straps which lead to external posts, one positive and one negative. The size (capacity) of a cell is determined by the thickness, width, length and number of plates. The positive and negative plates are prevented from touching each other

Figure 7-16. Mule battery for WECS application.

*This material on batteries has been provided through the courtesy of ENERTECH and ESB Incorporated, Exide Power Systems Division.

by "separators" made from various non-conductive micro-porous materials.

When a dc voltage is applied across the positive and negative posts, the plates "charge" by changing chemically. The surface lead of the positive plates changes to lead oxide, while the negative plates remain unoxidized. When the battery is discharged, this process reverses.

The electrolyte is the chemical media required to allow the above reaction to occur. During charging, the acid of the electrolyte is all "re-formed," while during discharge, the acid is broken down. Thus, a charged battery has a high concentration of acid (measured by high specific gravity) and a discharged battery has a low concentration of acid (measured by low specific gravity).

Battery Characteristics

Voltage. Almost any combination of dissimilar metals in a conducting electrolyte will produce some electrical voltage. The lead-acid battery has an inherent cell voltage of approximately 2 V. This voltage, however, will vary with changes in specific gravity resulting from temperature variations, physical conditions of the battery and charge or discharge rate. Nominally, a 3-cell battery is a 6-V battery, a 6-cell battery is a 12-V battery and so on. For wind power applications, we generally use multiples of 6-V batteries for 114-V systems; multiples of 12-V batteries for lower voltage, high amperage systems' and, occasionally, multiples of single-cell 2-V batteries for systems designed for special applications.

Specific Gravity. The electrolyte in a lead-acid cell is a dilute solution of sulfuric acid and water (preferably distilled). The ratio of acid to water is measured as specific gravity. Pure water has a specific gravity of 1.000. The specific gravity of a battery is a matter of engineering design and may vary with many factors. The minimum requirement that must be met is that there be a sufficient amount of sulfuric acid to meet the operative chemical requirements of theh cell, but not so much that the acid will have adverse effects on the plates themselves.

Table 7-7 is a generalized table relating specific gravity to other battery characteristics.

Capacity. A battery's ability to deliver energy is called capacity and is measured in amp-hours. Simply, put, capacity equals discharge in amps over a duration of time (usually expressed in hours).

Capacity is affected by various factors—the discharge rate, temperature, specific gravity and final voltage.

The discharge rate will vary from application to application. In a wind system, it is generally moderate, because the capacity of the system is usually relatively large and the use of a sustained heavy load is not usual. A sustained high discharge rate will result in excessive voltage drop in the cells and possible damage to the battery.

Generally speaking, batteries operate better at higher temperatures (room temperature) than at lower temperatures. A higher operating temperature facilitates the electrochemical reaction necessary for the production of current, and reduces the resistance and viscosity of the electrolyte, thus reducing the voltage drop within each cell. This allows post voltage to remain at a higher value longer.

The term final voltage refers to the minimum allowable voltage before possible battery damage may take place at various discharge rates. The nominally accepted minimum cell voltage is 1.75 V/cell.

Battery System Installation

Many factors must be considered for the proper location of batteries. The following are the most important of these:

1. Batteries should be in a warm, dry, well-ventilated room.
2. They should be located as conveniently as possible with respect to the electrical power source and usage point.
3. They should be located away from all sources of direct heat, flame or spark.

Temperature. Battery life, capacity and performance are functions of temperature and its relationship to specific gravity. As tempera-

Table 7-7. Specific Gravity for Storage Batteries.

HIGH SPECIFIC GRAVITY	LOW SPECIFIC GRAVITY
More capacity	Less capacity
Lower cost	Higher cost
Shorter life	Longer life
Less space required	More space required
Higher surge discharge rates	Lower surge discharge rates
More standby loss	Less standby loss
Daily use mode	No use standby mode
Usually 6- or 12-V battery	Usually 2-V battery

ture drops, the electrolyte viscosity increases. Viscosity will double as the temperature drops from 77°F (25°C) to 23°F (0°C), and below freezing, electrolyte viscosity increases even more dramatically, causing a rapid fall-off of discharge capacity. For instance, the capacity of a battery bank is normally rated at 77°F, but will be reduced to 89 percent of rated capacity at 55°F, 79 percent at 40°F and 73 percent at 32°F.

Conversely, at higher temperatures, the capacity of a battery bank will be increased: 103 percent at 85°F, 107 percent at 100°F and 111 percent at 120°F. However, continued operation of lead-acid batteries at temperatures greater than about 80°F is not recommended and will reduce the life expectancy of the set.

A second consideration in planning the temperature environment of your battery set is the point at which a battery will freeze and physically break its container. The freezing point of battery electrolyte is dependent on the acid concentration, and thus specific gravity and state of charge. A fully charged storage battery is virtually impossible to freeze at ordinary winter temperatures, while a fully discharged battery will freeze at only slightly less than 32°F.

Table 7-8 shows the relationship of specific gravity to freezing point. When specific gravity is measured with an ordinary hydrometer, the reading is often corrected to 77°F, thus, we also give these corrected values.

Ventilation. All lead-acid batteries produce certain amounts of free hydrogen gas which escapes through cell vents or ventilated filler caps. Hydrogen gas is colorless, odorless and extremely flammable. Since it is lighter than air, it will tend to rise to the highest point in the battery room or enclosure. The amount of hydrogen produced will vary with the type of service, but is greatest when rapid charging is taking place.

It is extremely important that adequate ventilation be provided to prevent the accumulation of hydrogen in the battery room. A large room which is not kept tightly closed will provide adequate ventilation without additional steps. On the other hand, if the battery room is small or tightly closed, one ventilator should be installed at or near the highest point, another near the lowest point.

In connection with the above, batteries should never be located near any source of heat, spark or flame, including any heaters, electrical switches or switching devices. Smoking should never be permitted near batteries; children should not be permitted near batteries; and batteries should simply not be placed in the ordinary living areas of a home.

Parallel and Series Battery System. Individual battery units may be interconnected to form battery systems in either of two ways (or a combination of both). The final system voltage and capacity desired will determine the way the units are interconnected.

Series connection means that the positive terminal of one battery is connected to the negative terminal of the next one, and so on down the set. When battery units are connected in series, the voltage of the set will equal the combined voltages of the units. For example, a common battery system for small

Table 7-8. Specific gravity vs. Freezing Point.

SPECIFIC GRAVITY		FREEZING POINT
CORRECTED TO 77°F	AT SAME TEMPERATURE	
1.080	1.00	+20
1.130	1.150	+10
1.160	1.185	0
1.180	1.210	−10
1.200	1.235	−20
1.215	1.250	−30
1.225	1.265	−40

wind systems consists of 19 units of 6-V batteries connected in series. The resultant system voltage will equal 114 V (6 times 19). If the rated capacity of each 6-V unit is "180 amp-hours at 6 V," the resultant system capacity will be rated at "180 amp-hours at 114 V."

Parallel connection means that the positive terminal of each battery unit is connected to the positive terminal of the next unit and the negative terminal is connected to the negative terminal of the next unit (and so on). When battery units are connected in parallel, the voltage of the resultant battery system will equal the voltage of each unit. For example, a common system consists of 6 units of 12-V batteries, connected in parallel. The voltage of the resultant system will still be 12 V, but if the capacity of each unit is rated at "220 amp-hours at 12 V," the system capacity will be rated at "1320 amp-hours at 12 V" (220 times 6).

An example of a combination of parallel and series connections might be the use of 38 units of 180 amp-hours at 114 V. This system would consist of two sets of 19 units. The 19 units in each set would be connected in series, which would result in 2 systems rated at 180 amp-hours at 114 V. Then the two sets are connected to each other in parallel and the result will be a combined system rated at 360 amp-hours at 114 V.

Battery Racks. Batteries are ordinarily placed on racks which are laid out to allow convenient hook-up. The rack should be made of wood, coated steel or other corrosion-proof material. Batteries should never be placed on concrete, concrete blocks, brick or other masonry material. Racks should be designed to support very heavy loads—50 to 100 lb/ft—and should have adequate clearance above the battery units for each access to the filler caps.

When arranging batteries on racks, the units should never by in physical contact with each other, and should have 1/2–1 in. of space left between the units. Interconnect cables should be of adequate size for the maximum charge or discharge current ever anticipated.

Battery Maintenance and Care. The following simple steps should always be followed for maximum life and performance:

1. Keep batteries clean and dry. Wash off any accumulated dirt or battery acid with clean water.
2. Check the electrolyte level occasionally and add only distilled water if required. Never add other chemicals to batteries.
3. Check that all cable connections are tight and clean. There are several good spray-on battery terminal protective compounds available which will prevent corrosion and poor connections.

Charging Rates. Since many users of wind systems wish to provide supplementary power for charging battery sets, it is important to note that it is quite possible to charge a battery set at a rate which is too high, with the result that the battery set can be severely damaged.

Battery sets provided with wind systems are designed to safely accept the maximum current that the wind generator can provide. If this charge rate is exceeded, the battery units will begin to heat, and if the condition persists, the batteries will eventually become hot enough to warp and deform the plates and the separators.

As a general rule of thumb, a charging rate (amps) equal to 10 percent of the total amp-hour rating of the battery set is safe for prolonged periods. If the charge rate greatly exceeds this value, it must not be continued for lengthy periods. In addition, as the batteries become charged, the rate should gradually be reduced to zero. This is called a "tapering" charge rate, as opposed to a constant charge rate. Most commercially available chargers are designed to automatically produce a tapering charge rate.

Testing and Monitoring. The best and simplest method for determining the state of charge of your battery set is by means of an ordinary hydrometer. This device measures the specific gravity of the electrolyte, and specific gravity is directly and linearly proportional to state of charge. The device works by drawing up a sample of the electrolyte out of

Table 7-9. Standing Voltages.

VOLTS PER CELL	PERCENT CHARGED (= PERCENT OF AMP-HOUR CAPACITY REMAINING)
2.00	100
1.99	90
1.98	80
1.97	70
1.96	60
1.94	50
1.93	40
1.91	30
1.87	20
1.83	10
1.75	0

the top of the battery, and it is usually necessary to pump liquid up and down the hydrometer four or five times in order to obtain an accurate reading, since the electrolyte becomes incompletely mixed after standing in a working battery.

The second method of determining state of charge is by reading the battery voltage. This is not an accurate measurement but may be used as an approximation. In order to use this method, the battery voltage should be measured while the battery is neither charging nor discharging. Table 7-9 gives "standing" voltages at different states of charge or capacity levels. Voltages are shown in terms of volts per cell. In order to convert volts per cell to system voltage, multiply by the number of cells which are connected in series in your system. For example, a 114-V lead-acid battery contains 57 cells in series.

UNIQUE STORAGE SYSTEMS

The battery is the primary storage system currently used for WECS. In the future, unique systems such as flywheel, hydrogen and pumped storage may be utilized.

Flywheel Storage.

The flywheel, which stores energy in mechanical form, is among the oldest of energy storage devices. As recently as only seven years ago, it was generally regarded as impractical for storing energy at any but low-mass and low-volume densities. In addition, safety problems associated with flywheels have limited their use. Nevertheless, experiments with small-scale energy storage are presently underway.

One real hope for energy storage at high mass and volume densities, and of achieving high power densities, lies in the "superflywheel" concept shown in Figure 7-17. The idea merits further research because it has the potential for revolutionizing power systems in general. The technical concepts appear sound, and developmental work is in progress by DOE.

The development of materials having unidirectional tensile strengths many times that of steel (termed unidirectional composite materials) has served to make the superflywheel concept a practical possibility. Two different structures have been championed by Dan Rabenhorst, of the Johns Hopkins University/Applied Physics Laboratory, and by Post and Post, respectively. The properties of the unidirectional composite materials are reported to be such that flywheel rupture, even at the highest rotational speeds, need not be a significant hazard. The preliminary calculations indicate energy densities of about 300 Wh/lb, and power densities of 300 W/lb are possible with presently available materials. Cost estimates, presumably for mass produced units, run about $35/kWh.

The advantages of flywheel energy storage over that provided by contemporary batteries are as follows:

1. Rapid charge/discharge capability.
2. Unlimited depth of discharge and number of cycles.
3. Simplicity, no maintenance, infinite shelf life.

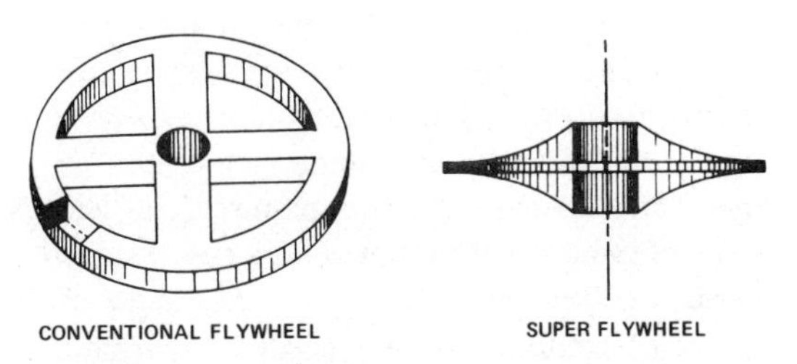

Figure 7-17. Conventional flywheel and super flywheel configurations.

4. Flexibilty—output/input can be electrical (ac or dc), hydraulic, mechanical, etc.

To these, the superflywheel would add the following:

6. A more favorable energy to mass ratio (possibly 15 to 1).
7. A more favorable energy to volume ratio.
8. Greater efficiency in charge/discharge cycling.

To reduce losses in a superflywheel, as shown in Figure 7-19, the high-speed disc spins in a vacuum chamber, and precision ball bearings, magnetic bearings or air bearings are used. These measures also enhance the safety of the flywheel by providing a housing that prevents damage should the wheel fracture or fail.

Various design studies confirm that a small superflywheel energy storage unit should be cost-competitive with an equivalent battery system and will most probably weigh less. Thus, this storage unit should see great potential in small WECS applications. As this is written, no small superflywheel is commercially available, and few are being tested. Research is expanding in this area, however, and a superflywheel energy storage package could become available.

Hydrogen Storage

Energy from the wind can be stored in chemical form by first converting to electricity and then producing hydrogen (and oxygen) in an electrolytic cell. This idea has been around for many years but thus far no way has been found to fabricate a total storage recovery system that can compete economically with power from conventional generating systems. The reasons for this situation are that: 1) electricity from WECS has been more expensive than from distribution systems; 2) several energy conversion steps are required, at least one of which has a low efficiency; and 3) several components of varying costs and degrees of complication are needed for a complete system, thus increasing initial costs, operational complexity and probability of breakdown. At least one expert has expressed the belief that electrolysis will never be a major hydrogen producer. With the increasing availability of petroleum products from the North Sea, Norway is reportedly switching from electrolytic production of hydrogen, even with .03 cent electricity available. Although the costs of electricity generated from WECS and conventional systems are becoming more comparable, it is difficult to believe the best economic use of wind in the near term will be to produce hydrogen either for use as a combustion fuel or in fuel cells. On the other hand, if hydrogen is required for chemical synthesis, hydrogen production from wind energy will stand a better chance of being economical, particularly if high-purity hydrogen is required.

The foregoing statements refer primarily to systems designed to convert all the energy from a WECS to hydrogen, prior to any subsequent use. It is more probably that such conversion will be economical when surplus electrical energy is available and whenever a good price can be obtained for the oxygen. Each situation will obviously require individual study. The economics will be improved by the fact that electrolytic cells become increasingly efficient as the rate of electrolysis decreases.

The considerable recent interest in a hydrogen energy economy evidences a general expectation for hydrogen production to occur on a major scale by at least one process. Some manner of interfacing WECS with this economy would be desirable. The wide range of hydrogen uses and the projected enormous increase in demand, coupled with diminishing natural gas availability, may drastically alter the current economic picture.

Fuel cells were not discussed earlier because their operational concept differs from that of ordinary batteries, since the fuel must be continuously provided from an external supply. The H^2-O^2 fuel cell is important, because hydrogen is the only fuel conveniently obtainable from wind energy. The Oklahoma State University group has been studying problems associated with the H^2-O^2 fuel cell and electrolysis cell operation for a number of years, and have constructed some small pilot plants. They report an overall efficiency of 0.60 for combined electrolysis cell-fuel cell operation at high temperature (~350°F) and pressure (~200 atm), with a satisfactory electrode life. The operating conditions do give rise to problems, however. Short operating lifetimes have been reported for fuel cells in general.

Water of high quality is required in the operation of the electrolysis cells commercially available, and it is difficult to believe this requirement can be modified if minimum down time and highest operational efficiencies are to be achieved. The region of the state lying within the highest wind energy regime is generally a water-deficient region, thus presenting an additional problem. Water resources for hydrogen production in the coastal region of high wind energy should be adequate.

The enormous quantity of hydrogen available on the globe assures a hydrogen supply. In contrast, chronic shortages of many metals needed for constructing other storage devices are expected to continue indefinitely. Despite the numerous problems clearly evident, the storage of some energy in the form of hydrogen may take place in time, particularly if the energy situation changes toward the direction of a hydrogen economy.

Pumped Storage

The term "pumped storage" referred for many years to the process of storing gravitational energy in water by pumping it from a lower to a higher level. Most of the stored energy could then be recovered as desired by controlling the release of the water through a turbine. More recently, the term has come to include the process of compressing air, which is later used in an indirect way to increase the efficiency of a heat engine operating on fossil fuel. In either case, the form of the stored energy is mechanical, despite the considerably different recovery techniques employed.

- *Hydrofirming*. Thomas appears to be among the first to have studied in some detail the aspects of storing wind energy by hydrofirming in conjunction with hydroelectric installation. The requirements for the presence of the latter severely limits the extent to which the concept can be successfully applied, despite its intrinsic simplicity. This manner of storing energy would be most economic, of course, in regions with highest wind energy regimes. (Several hydroelectric storage facilities currently exist within the U.S., but hydrofirming is accomplished at none of these with wind power. Some interest has developed in this direction in Michigan for WECS projects.)
- *Compressed air*. In recent years, a great deal of interest has developed around the idea of using compressed air as a means of storing energy in times of surplus power and releasing it in times when the demand is appropriately large. The principal application is in connection with turbines that burn fossil fuels (gas turbines). Ordinarily, about two-thirds of the power output of a gas turbine is required to drive the compressor supplying the turbine with air. During periods of high loads, the turbine output of electrical power could be increased by a factor of two to three if compressed air were independently available, as from a storage reservoir. A Swedish manufacturer of turbines is reported to be close to having a unit designed to operate in this manner. (With regard to this concept, Robinson states that "a hard engineering study, beyond exploratory design studies, has yet to be made in the United States." Southwest Research Institute personnel have, in a preliminary survey, looked into the possibility of using spent potash mines near Carlsbad, New Mexico for compressed air storage to supply nearby gas turbines operated by Southwestern Public Service. Wind power would be utilized for compressing the air in their plan.

It is generally agreed that large storage volumes enhance the economics of such projects, although only about 15 percent of the volume required to store an equivalent quantity of energy by hydrofirming is needed. Specially constructed storage systems are considered too expensive. This leaves underground storage in caverns, aquifiers, spent oil and gas fields and, of course, spent mines, for possibilities. For the latter, Southwest Research Institute estimates an initial cost of about one dollar per stored kW. It would also be of interest to determine whether such a charge of compressed air in a spent oil pool could simultaneously be used in secondary and tertiary recovery.

8
Wind Energy Conversion Systems

This chapter provides an in-depth summary of water-pumping and small-scale and large-scale WECS (Figure 8-1) which represent the state-of-the-art technology.

An exhaustive survey of WECS which have been produced or are in the process of development has been completed and the results are given in this chapter. This survey includes the assembly of key data from all known manufacturers, developers or distributors of WECS. The data sheets provided include information regarding each firm's address, model number(s) of WECS, rated output or rated power, output at 12 and 14 mph, rotor blade diameter, number of blades, machine description, cost, operating speed—rated/cut-in/cut-out/maximum, rpm at rated speed, blade materials, rotor weight, system weight, overspeed control, generator/alternator type, application, testing history, reliability prediction, warranty, maintenance requirements, country of origin, number produced, availability and comments. Accuracy of data rests with the specific vendors. Graphs of WECS performance are generalized since accurate data was not available. The DOE Rocky Flats Test Center can be contacted for limited performance test results on some WECS.

The WECS have been divided into two categories: 1) water-pumping windmills, and 2) electricity-producing WEC systems. The systems are listed alphabetically within each category by system configuration name.

This survey includes 49 manufacturers and more than 117 models of WECS. U.S. distributors of some WECS are listed in Table 8-1.

The following material represents responses to inquiries received from various vendors, public domain information sources, and related data research efforts. We do not assume any legal liability or responsibility for the ac-

Figure 8-1. Photograph of water-pumping windmill and DOE's MOD-0 large-scale wind energy conversion system at the Plumbrook site. (*Courtesy of DOE.*)

Table 8.1. Partial list of U.S. distributors of WECS.**

WECS MANUFACTURER	AUTOMATIC POWER PENNWALT	AERO-POWER	BOSTON WIND POWER	STEVE BLAKE	BUDGEN & ASSOCIATES	ENERTECH	ENERGY ALTERNATIVES	INDEPENDENT ENERGY SYSTEMS	NORTHWIND POWER CO.	PRAIRIE SUN AND WIND	REAL GAS & ELECTRIC	SENCENBAUGH ELECTRIC	WINDLITE ALASKA
Aeropower									X	X			
Aerowatt	X												X
Dempster		X											
Dunlite						X	X				X	X	
Elektro					X	X	X				X	X	
Jacobs*			X	X			X	X	X				
Lubing					X		X						
Sencenbaugh						X						X	X
Wincharger							X			X	X		

*Refurbished/modified.
**Distributorships are subject to change.

curacy, completeness or usefulness of any information, apparatus, product, or process disclosed, or represent that its use would not infringe privately owned rights. References herein to any specific commercial product, process, or service by trade name, mark, manufacturer or otherwise does not necessarily constitute or imply its endorsement, recommendation or favoring by the author or his representatives. This consolidation of vendor data must be reviewed with care by the audience.

The Department of Energy's Rocky Flats facility has conducted ongoing experimental data collection efforts for small Wind Energy Conversion Systems. Their first semiannual report is documented in RFP-2920/3533/78/6-2, dated September 28, 1978. Part of the consideration prior to purchasing a WECS should be a thorough evaluation by the buyer of all available test data. The executive summary from Volume II of the above report has been quoted for specific WECS.

WATER-PUMPING WINDMILLS

This section provides data on the Aeromotor, Dempster, Heller-Aller, KMP Parish and Sparco water-pumping windmills. The first four systems are similar multi-blade designs, and the Sparco is a Danish SWECS of two-blade configuration.

Joe Carter, in his series, "Using Water-pumping Windmills," in the *Wind Power Digest,* stated:

"American or American-based manufacturers of traditional windmills are listed (following data sheets), and upon request each may send you some of their literature and/or locate for you their nearest dealer. Unfortunately there really isn't any more detailed literature other than what the manufacturers offer, so we depend on their claims of windmill ouptut as the basis for designing a system. Which to choose? Is one make superior to another? It would be nice if the "you-pay-for-what-you-get" axiom could be applied here, but the most expensive line of windmills, the Aeromotor, is the one I would personally choose last, so that doesn't work. After working around different mills for a couple of weeks, listening to the testimonies of afficionados much more experienced than I, the quality highest, for my money, can be found in the Dempster line, if you are buying new . . ."

"The old Aeromotors are highly respected; they did, after all, account for 80–90% of all windmill sales for a long time. But something seems to have been lost in the move to the present manufacturing plant in Argentina. Bearings have a lot to do with it. Aeromotors have always run on babbitt, a rather soft metal that is poured into place, making replacement difficult, if not impossible. Heller-Aller Baker windmills run on cold rolled steel on machined cast iron, which is said to be a proven long-lived combination. Dempsters use two tapered roller bearings to carry the main shaft, which, as long as they stay oiled, develop the least friction."

By examining the data sheets which follow and carefully reviewing your application, you will be able to select the appropriate water-pumping windmill. (Refer to Chapter 9 for more detailed application information.)

The sale of water-pumping windmills is on the increase, partially due to higher cattle prices. These higher prices for cattle encourage larger herds, which require more water to support the animals. With the increasing cost of running electric lines to some rural areas reaching several thousand dollars a mile, windmills are becoming a realistic alternative to electricity-powered water pumps. Although the average water-pumping windmill costs around $3,000, it should be noted that this initial investment on a system that requires minimal investment is a sound one, based on the alternative of high electric power line costs plus increasing monthly electric bills.

Sales of water-pumping windmills such as the Dempster and Aeromotor systems have jumped 15 percent annually in the past several years, after being stagnant for the last four decades. For example, Aeromotor expects to sell 3500 units this year, up 15 percent from last year. Heller-Aller projects a 40 percent increase in sales.

AEROMOTOR 702-6

Manufacturer: Aeromotor (Division of Valley Industries)
Address: P.O. Box 1364, Conway, AR 72032
Contact: Mr. Stan Anderson
Telephone: (501) 329–9811
Model number: 702-6 (see Figure 8-2)
Rated output: See Table 8-2
Rotor blade diameter: 6 ft
Number of blades: Multi-blade
Machine type: Upwind, horizontal-axis water-pumping
Cost: Would not quote
Operating speed
  Rated speed (mph): 15–20
  Cut-in speed (mph): 9
  Cut-out speed (mph): 28
rpm at rated output: 125
Blade materials: Galvanized steel
System weight: 210 lb
Overspeed control: Rotor turns sideways to wind

Table 8-2. Aeromotor pumping capacities (15-20-mph wind). *(Courtesy of Aeromotor.)*

DIAMETER OF CYLINDER (in.)	CAPACITY* PER HOUR (gal)		TOTAL ELEVATION (ft) SIZE OF AERMOTOR					
	6 ft	8-16 ft	6 ft	8 ft	10 ft	12 ft	14 ft	16 ft
1¾	105	150	130	185	280	420	600	1,000
1⅞	125	180	120	175	260	390	560	920
2	130	190	95	140	215	320	460	750
2¼	180	260	77	112	170	250	360	590
2½	225	325	65	94	140	210	300	490
2¾	265	385	56	80	120	180	260	425
3	320	470	47	68	100	155	220	360
3¼		550			88	130	185	305
3½	440	640	35	50	76	115	160	265
3¾		730			65	98	143	230
4	570	830	27	39	58	86	125	200
4¼		940			51	76	110	180
4½	725	1050	21	30	46	68	98	160
4¾		1170				61	88	140
5	900	1300	17	25	37	55	80	130
5¾		1700				40	60	100
6		1875		17	25	38	55	85
7		2550			19	28	41	65
8		3300			14	22	31	50

*Capacities shown in the above table are approximate, based on the mill set on the long stroke, operating in a 15-20-mph wind. The short stroke increases elevation by one-third and reduces pumping capacity one-fourth. (Refer to Figure 8-3 for water-pumping terms.)

Figure 8-2. Aeromotor water-pumping windmill.

Application: Water-pumping
Testing history: 45 years of usage experience
Reliability prediction (MTBF): Unspecified
Warranty: One year, materials and workmanship
Maintenance requirements: Annual lubrication
Country of origin: U.S. (components manufactured in Argentina)
Number produced: 3500 per year planned sales
Availability: 60 days ARO
Comments: Valley Industries claims 75 percent of the windmill market and plan to sell 3500 units this year

AEROMOTOR 702-8

Manufacturer: Aeromotor (Division of Valley Industries)
Address: P.O. Box 1364, Conway, AR 72032
Contact: Mr. Stan Anderson
Telephone: (501) 329–9811
Model number: 702-8
Rated output: See Table 8-2
Rotor blade diameter: 8 ft
Number of blades: Multi-blade
Machine type: Upwind, horizontal-axis, water-pumping
Cost: Would not quote
Operating speed
  Rated speed (mph): 15–20
  Cut-in speed (mph): 9
  Cut-out speed (mph): 28
rpm at rated output: 105
Blade materials: Galvanized steel
System weight: 355 lb
Overspeed control: Rotor turns sideways to wind
Application: Water-pumping
Testing history: 45 years of usage experience
Reliability prediction (MTBF): Unspecified
Warranty: One year, materials and workmanship
Maintenance requirements: Annual lubrication
Country of origin: U.S. (components manufactured in Argentina)
Number produced: See 702-6 data sheet
Available: 60 days ARO

AEROMOTOR 702-10

Manufacturer: Aeromotor (Division of Valley Industries)
Address: P.O. Box 1364, Conway, AR 72032
Contact: Mr. Stan Anderson
Telephone: (501) 329–9811
Model number: 702-10
Rated output: See Table 8-2
Rotor blade diameter: 10 ft
Number of blades: Multi-blade
Machine type: Upwind, horizontal-axis, water-pumping
Cost: Would not quote
Operating speed
  Rated speed (mph): 15–20
  Cut-in speed (mph): 9
  Cut-out speed (mph): 28
rpm at rated output: 85
Blade materials: Galvanized steel
System weight: 645 lb
Overspeed control: Rotor turns sideways to wind
Application: Water-pumping
Testing history: 45 years of manufacturing experience
Reliability prediction (MTBF): Unspecified
Warranty: One year, materials and workmanship
Maintenance requirements: Annual lubrication

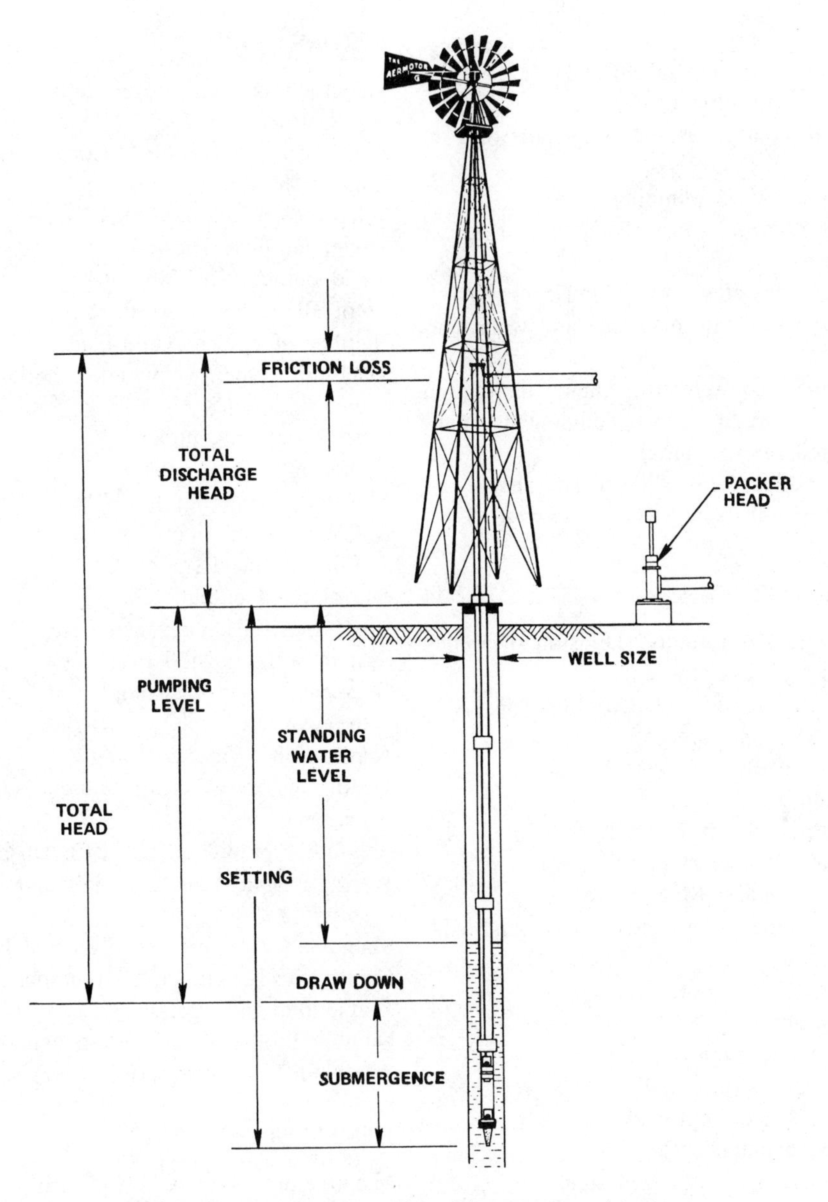

Figure 8-3. Aeromotor illustration of key water-pumping terms.

Country of origin: U.S. (components manufactured in Argentina)
Number produced: See 702-6 data sheet
Availability: 60 days ARO

AEROMOTOR 702-12

Manufacturer: Aeromotor (Division of Valley Industries)
Address: P.O. Box 1364, Conway, AR 72032
Contact: Mr. Stan Anderson
Telephone: (501) 329–9811
Model number: 702-12
Rated output: See Table 8-2
Rotor blade diameter: 12 ft
Number of blades: Multi-blade
Machine type: Upwind, horizontal-axis, water-pumping
Cost: Would not quote
Operating speed
  Rated speed (mph): 15–20
  Cut-in speed (mph): 9
  Cut-out speed (mph): 28

rpm at rated output: 73
Blade materials: Galvanized steel
System weight: 1090 lb
Overspeed control: Rotor turns sideways to wind
Application: Water-pumping
Testing history: 45 years of manufacturing experience
Reliability prediction (MTBF): Unspecified
Warranty: One year, materials and workmanship
Maintenance requirements: Annual lubrication
Country of origin: U.S. (components manufactured in Argentina)
Number produced: See 702-6 data sheet
Availability: 60 days ARO

AEROMOTOR 702-14

Manufacturer: Aeromotor (Division of Valley Industries)
Address: P.O. Box 1364, Conway, AR 72032
Contact: Mr. Stan Anderson
Telephone: (501) 329–9811
Model number: 702-14
Rated output: See Table 8-2
Rotor blade diameter: 14 ft
Number of blades: Multi-blade
Machine type: Upwind, horizontal-axis, water-pumping
Cost: Would not quote
Operating speed
 Rated speed (mph): 15–20
 Cut-in speed (mph): 9
 Cut-out speed (mph): 28
rpm at rated output: 62
Blade materials: Galvanized steel
System weight: 1695 lb
Overspeed control: Rotor turns sideways into wind
Application: Water-pumping
Testing history: 45 years of manufacturing experience
Reliability prediction (MTBF): Unspecified
Warranty: One year, materials and workmanship
Maintenance requirements: Annual lubrication
Country of origin: U.S. (components manufactured in Argentina)
Number produced: See 702-6 data sheet
Availability: 60 days ARO

AEROMOTOR 702-16

Manufacturer: Aeromator (Division of Valley Industries)
Address: P.O. Box 1364, Conway, AR 72032
Contact: Mr. Stan Anderson
Telephone: (501) 329–9811
Model number: 702-16
Rated output: See Table 8-2
Rotor Blade diameter: 16 ft
Number of blades: Multi-blade
Machine type: Upwind, horizontal-axis, water-pumping
Cost: Would not quote
Operating speed
 Rated speed (mph): 15–20
 Cut-in speed (mph): 9
 Cut-out speed (mph): 28
rpm at rated output: 53
Blade materials: Galvanized steel
System weight: 2450 lb
Overspeed control: Rotor turns sideways to the wind
Application: Water-pumping
Testing history: 45 years of manufacturing experience
Reliability prediction (MTBF): Unspecified
Warranty: One year, materials and workmanship
Maintenance requirements: Annual lubrication
Country of origin: U.S. (components manufactured in Argentina)
Number produced: See 702-6 data sheet
Availability: 60 days ARO

DEMPSTER 6 Ft

Manufacturer: Dempster Industries, Inc.
Address: P.O. Box 848, Beatrice, NB 68310
Contact: Mr. Keith Lynch
Telephone: (402) 223–4026
Model number: 6 ft
Rated output: See Table 8-3
Rotor blade diameter: 6 ft
Number of blades: Multi-blade
Machine type: Upwind, horizontal axis, water-pumping
Cost: $525, less tower
Operating Speed
 Rated speed (mph): 15
 Cut-in speed (mph): 5
 Cut-out speed (mph): 50

Table 8-3. Dempster pumping capacities (15-mph wind).*

CYLINDER SIZE	6 ft 5 in. STROKE		8 ft 7½ in. STROKE		10 ft 7½ in. STROKE		12 ft 12 in. STROKE		14 ft 12 in. STROKE	
	ELEV.	GPH	ELEV.	GPH	ELEV.	GPH	ELEV.	GPH	ELEV.	GPH
1⅞ in.	120	115	172	173	256	140	388	180	580	159
2 in.	95	130	135	195	210	159	304	206	455	176
2¼ in.	75	165	107	248	165	202	240	260	360	222
2½ in.	62	206	89	304	137	248	200	322	300	276
2¾ in.	54	248	77	370	119	300	173	390	260	334
3 in.	45	294	65	440	102	357	147	463	220	396
3¼ in.	39	346	55	565	86	418	125	544	187	465
3½ in.	34	400	48	600	75	487	108	630	162	540
3¾ in.	29	457	42	688	65	558	94	724	142	620
4	26	522	37	780	57	635	83	822	124	706

*GPH = gallons per hour.

Blade materials: Galvanized steel
Rotor weight: 100 lb
System weight: 280 lb
Overspeed control: Rotor turns sideways into wind
Application: Water-pumping
Testing history: Most reliable of field tested water pumpers
Reliability prediction (MTBF): Unspecified
Warranty: Limited five years, parts and workmanship
Maintenance requirements: Annual inspection and lubrication
Country of origin: U.S.
Number produced: Would not quote
Availability: Normally in stock

DEMPSTER 8 Ft

Manufacturer: Dempster Industries, Inc.
Address: P.O. Box 848, Beatrice, NB 68310
Contact: Mr. Keith Lynch
Telephone: (402) 223–4026
Model number: 8 ft
Rated output: See Table 8-3
Rotor blade diameter: 8 ft
Number of blades: Multi-blade
Machine type: Upwind, horizontal axis, water-pumping
Cost: $760, less tower
Operating Speed
  Rated speed (mph): 15
  Cut-in speed (mph): 5
  Cut-out speed (mph): 50
Blade materials: Galvanized steel
Rotor weight: 120 lb
System weight: 388 lb
Overspeed control: Rotor turns sideways into wind
Application: Water-pumping
Testing history: Most reliable of field tested water pumpers
Reliability prediction (MTBF): Unspecified
Warranty: Limited five years, parts and workmanship
Maintenance requirements: Annual inspection and lubrication
Country of origin: U.S.
Number produced: Would not quote
Availability: Normally in stock

DEMPSTER 10 Ft

Manufacturer: Dempster Industries, Inc.
Address: P.O. Box 848, Beatrice, NB 68310
Contact: Mr. Keith Lynch
Telephone: (402) 223–4026
Model number: 10 ft
Rated output: See Table 8-3
Rotor blade diameter: 10 ft
Number of blades: Multi-blade
Machine type: Upwind, horizontal axis, water-pumping
Cost: $1275, less tower
Operating Speed
  Rated speed (mph): 15
  Cut-in speed (mph): 5
  Cut-out speed (mph): 50
Blade materials: Galvanized steel
Rotor weight: 150 lb

System weight: 500 lb
Overspeed control: Rotor turns sideways into wind
Application: Water-pumping
Testing history: Most reliable of field tested water pumpers
Reliability prediction (MTBF): Unspecified
Warranty: Limited five years, parts and workmanship
Maintenance requirements: Annual inspection and lubrication
Country of origin: U.S.
Number produced: Would not quote
Availability: Normally in stock

DEMPSTER 12 Ft

Manufacturer: Dempster Industries, Inc.
Address: P.O. Box 848, Beatrice, NB 68310
Contact: Mr. Keith Lynch
Telephone: (402) 223–4026
Model number: 12 ft (See Figure 8-4)
Rated output: See Table 8-3
Rotor blade diameter: 12 ft
Number of blades: Multi-blade
Machine type: Upwind, horizontal axis, water-pumping
Cost: $2178, less tower
Operating Speed
  Rated speed (mph): 15
  Cut-in speed (mph): 5
  Cut-out speed (mph): 50
Blade materials: Galvanized steel
Rotor weight: 334 lb
System weight: 935 lb
Overspeed control: Rotor turns sideways into wind
Application: Water-pumping
Testing history: Most reliable of field tested water pumpers
Reliability prediction (MTBF): Unspecified
Warranty: Limited five years, parts and workmanship
Maintenance requirements: Annual inspection and lubrication
Country of origin: U.S.
Number produced: Would not quote
Availability: Special order only

DEMPSTER 14 Ft

Manufacturer: Dempster Industries, Inc.
Address: P.O. Box 848, Beatrice, NB 68310
Contact: Mr. Keith Lynch
Telephone: (402) 223–4026
Model number: 14 ft

Figure 8-4. Dempster 12-ft water-pumping windmill.

Rated output: See Table 8-3
Rotor blade diameter: 14 ft
Number of blades: Multi-blade
Machine type: Upwind, horizontal axis, water-pumping
Cost: $3291, less tower
Operating Speed
Rated speed (mph): 15
Cut-in speed (mph): 5
Cut-out speed (mph): 50
Blade materials: Galvanized steel
Rotor weight: 616 lb
System weight: 1450 lb
Overspeed control: Rotor turns sideways into wind
Application: Water-pumping
Testing history: Most reliable of field tested water pumpers
Reliability prediction (MTBF): Unspecified
Warranty: Limited five years, parts and workmanship
Maintenance requirements: Annual inspection and lubrication
Country of origin: U.S.
Number produced: Would not quote
Availability: Special order only

Figure 8-5. Heller-Aller Baker water-pumping windmill.

HELLER-ALLER BAKER 6 Ft

Manufacturer: Heller-Aller Company
Address: Perry and Oakwood Streets, Napoleon, OH 43545
Contact: Mr. Charles Buehrer
Telephone: (419) 592–1856
Model number: Baker 6 ft (See Figure 8-5)
Rated output: See Table 8-4
Rotor blade diameter: 6 ft
Number of blades: Multi-blade
Machine type: Upwind, horizontal axis, water-pumping
Cost: Contact factory
Operating Speed
Rated speed (mph): 15
Cut-in speed (mph): 7
Cut-out speed (mph): 25
rpm at rated output: 150
Blade materials: Galvanized steel
System weight: 220 lb
Overspeed control: Rotor turns sideways into wind
Application: Water-pumping
Testing history: Manufacturer since 1886, many years of field utilization
Reliability prediction (MTBF): Unspecified
Warranty: One year, parts and workmanship
Maintenance requirements: Inspection and lubrication twice yearly
Country of origin: U.S.
Number produced: Would not quote
Availability: Contact factory; varies from in-stock to 18 weeks
Comments: "With energy problems of today, wind power will become more than ever a way of life in this country."

HELLER-ALLER BAKER 8 Ft

Manufacturer: Heller-Aller Company
Address: Perry and Oakwood Streets, Napoleon, OH 43545
Contact: Mr. Charles Buehrer
Telephone: (419) 592–1856

Model number: Baker 8 ft
Rated output: See Table 8-4
Rotor blade diameter: 8 ft
Number of blades: Multi-blade
Machine type: Upwind, horizontal axis, water-pumping
Cost: Contact factory
Operating Speed
Rated speed (mph): 15
Cut-in speed (mph): 7
Cut-out speed (mph): 25
rpm at rated output: 150
Blade materials: Galvanized steel
System weight: 360 lb
Overspeed control: Rotor turns sideways into wind
Application: Water-pumping
Testing history: Manufacturer since 1886, many years of field experience
Reliability prediction (MTBF): Unspecified
Warranty: One year, parts and workmanship
Maintenance requirements: Inspection and lubrication twice yearly
Country of origin: U.S.
Number produced: Would not quote
Availability: Contact factory; varies from in-stock to 18 weeks

HELLER-ALLER BAKER 10 Ft

Manufacturer: Heller-Aller Company
Address: Perry and Oakwood Streets, Napoleon, OH 43545
Contact: Mr. Charles Buehrer
Telephone: (419) 592–1856
Model number: Baker 10 ft
Rated output: See Table 8-4
Rotor blade diameter: 10 ft
Number of blades: Multi-blade
Machine type: Upwind, horizontal axis, water-pumping
Cost: Contact factory
Operating Speed
Rated speed (mph): 15
Cut-in speed (mph): 7
Cut-out speed (mph): 25
rpm at rated output: 150
Blade materials: Galvanized steel
System weight: 475 lb
Overspeed control: Rotor turns sideways into wind
Application: Water-pumping
Testing history: Manufacturer since 1886, many years of field utilization
Reliability prediction (MTBF): Unspecified
Warranty: One year, parts and workmanship
Maintenance requirements: Inspection and lubrication twice yearly
Country of origin: U.S.
Number produced: Would not quote
Availability: Contact factory; varies from in-stock to 18 weeks

Table 8-4. Heller-Aller Baker pumping capacities (15-mph wind).*

MODEL	6-ft BAKER		8-ft BAKER		10-ft BAKER		12-ft BAKER	
TOTAL ELEVATION (ft)	DIAMETER OF CYLINDER (in.)	U.S. GPH**	DIAMETER OF CYLINDER (in.)	U.S. GPH	DIAMETER OF CYLINDER (in.)	U.S. GPH	DIAMETER OF CYLINDER (in.)	U.S. GPH
25	3	350	3½	900	4	1250	6	2400
	2½	240	3	720	3½	925	5	1625
	2¼	200	2½	450	3	700	4½	1425
75	2	160	2¼	350	2½	475	4	1125
100	2	150	2	250	2½	460	3	600
125	1⅝	120	1⅞	240	2	280	2½	525
150			1¾	220	2	280	2½	525
200					1⅞	260	2	325
250					1¾	215	2	325
300							1¾	200

*The capacities are approximate. By the total elevation in feet, we do not mean the depth of the well, but the distance to the cylinder. Do not use pipe smaller than that for which cylinders are fitted. Although we provide the above table as the nominal values, larger cylinders may, in many circumstances, be used with satisfaction.

**GPH = gallons per hour

HELLER-ALLER BAKER 12 Ft

Manufacturer: Heller-Aller Company
Address: Perry and Oakwood Streets, Napoleon, OH 43545
Contact: Mr. Charles Buehrer
Telephone: (419) 592–1856
Model number: Baker 12 ft
Rated output: See Table 8-4
Rotor blade diameter: 12 ft
Number of blades: Multi-blade
Machine type: Upwind, horizontal axis, water-pumping
Cost: Contact factory
Operating Speed
  Rated speed (mph): 15
  Cut-in speed (mph): 7
  Cut-out speed (mph): 25
rpm at rated output: 150
Blade materials: Galvanized steel
System weight: 800 lb
Overspeed control: Rotor turns sideways into wind
Application: Water-pumping
Testing history: Manufacturer since 1886, many years of field utilization
Reliability prediction (MTBF): Unspecified
Warranty: One year, parts and workmanship
Maintenance requirements: Inspection and lubrication twice yearly
Country of origin: U.S.
Number produced: Would not quote
Availability: Contact factory; varies from in-stock to 18 weeks

KMP PARISH WINDMILLS 8 Ft.

Manufacturer: KMP Lake Pump Mfg. Co., Inc.
Address: Box 441, Earth, TX 79031
Contact: Marketing Department
Telephone: (806) 257–3411
Model number: 8 ft (see Figure 8-6)
Rated output: See Table 8-5
Rotor blade diameter: 8 ft
Number of blades: 18 multi-blades
Machine type: Upwind, horizontal axis, water-pumping
Cost: $800
Operating Speed
  Rated speed (mph): Unspecified

Figure 8-6. KMP Parish water-pumping windmill.

  Cut-in speed (mph): Unspecified
  Cut-out speed (mph): Unspecified
  Maximum speed: Unspecified
Blade materials: Galvanized metal
Rotor weight: Unknown
System weight on tower: 90 lb
Overspeed control: Variable tail vane speed control to make high-wind runaway
Generator/alternator type: Unspecified
Application: Water-pumping
Testing history: Unknown
Reliability prediction (MTBF): Unknown
Warranty: Unspecified
Maintenance requirements: Unspecified
Country of origin: U.S.
Number produced: Unknown
Availability: Stock item
Comments: Tower (24' x 2" x 3/16") galvanized angle; Cost $1000

KMP PARISH WINDMILLS 10 Ft

Manufacturer: KMP Lake Pump Mfg. Co., Inc.
Address: Box 441, Earth, TX 79031
Contact: Marketing Department
Telephone: (806) 257–3411
Model number: Baker 10 ft (see Figure 8-6)
Rated output: See Table 8-5
Rotor blade diameter: 8 ft
Number of blades: 18 multi-blades

Table 8-5. KMP Parish pumping capacities (15-mph wind).

DIAMETER OF CYLINDER (in.)	8-ft MILL		10-ft MILL	
	TOTAL ELEVATION (ft)	U.S. GPH*	TOTAL ELEVATION (ft)	U.S. GPH
1⅞	182	173	270	140
2¼	120	248	180	202
2¾	100	370	130	300

*GPH = gallons per hour

Machine type: Upwind, horizontal axis, water-pumping
Cost: $1200
Operating Speed
Rated speed (mph): Unspecified
Cut-in speed (mph): Unspecified
Cut-out speed (mph): Unspecified
Maximum speed: Unspecified
Blade materials: Galvanized metal
Rotor weight: Unknown
System weight on tower: 120 lb
Overspeed control: Variable tail vane speed control to brake high-wind runaway
Application: Water-pumping
Testing history: Unknown
Reliability prediction (MTBF): Unknown
Warranty: Unspecified
Maintenance requirements: Unspecified
Country of origin: U.S.
Number produced: Unknown
Availability: Stock item
Comments: Tower (24' x 2" x 3/16") galvanized angle; cost: $1000

SPARCO D/P

Manufacturer: SPARCO, c/o Enertech Corporation
Address: P.O. Box 420, Norwich, VT 05055
Contact: Mr. Edmund Coffin
Telephone: (802) 649–1145
Model number: D-Diaphragm or P-Piston (see Figure 8-7)
Rated output: 58 GPH
Rotor blade diameter: 4.17 ft
Number of blades: 2
Machine type: Upwind, horizontal axis, water-pumping
Cost: D—$295, P—$349
Operating Speed
Rated speed (mph): Unknown
Cut-in speed (mph): 5
Cut-out speed (mph): Unknown
Maximum speed: Unknown
Blade materials: Cast aluminum
Rotor weight: 8 lb
System weight: 58 lb
Overspeed control: Mechanical, blade feathering
Application: Water-pumping
Testing history: Used in Denmark
Reliability prediction (MTBF): Unspecified
Warranty: One year, parts and workmanship
Maintenance requirements: Semi-annual inspection, lubrication
Country of origin: Denmark
Number produced: 30,000
Availability: Contact Enertech Corporation

ELECTRICITY PRODUCING WECS

This section provides data on a variety of electricity-producing WECS. A brief description of the systems which have or are currently being evaluated be DOE's Rocky Flats test site are summarized as follows:

Sencenbaugh 1000-14

"The Sencenbaugh is a lightweight machine manufactured by Sencenbaugh Wind Electric, Pao Alto, CA. The blades are manufactured from Sitka spruce, with a bonded copper leading edge and a polyurethane finish. The main generator casting is 356T6 aluminum alloy. The alternator is coupled through a 3:1 helical gearbox and is driven by a three-bladed propeller 3.65 m in diameter.

Overspeed control is provided by using the increasing wind pressure of the propeller (and the resultant propeller thrust) to swing the alternator assembly (which is offset from the bearing support column) out of the oncoming wind. The foldable tail automatically reopens as wind speed decreases due to gravitational forces on the tail assembly. Tail offset and inclination with respect to the rotor may be

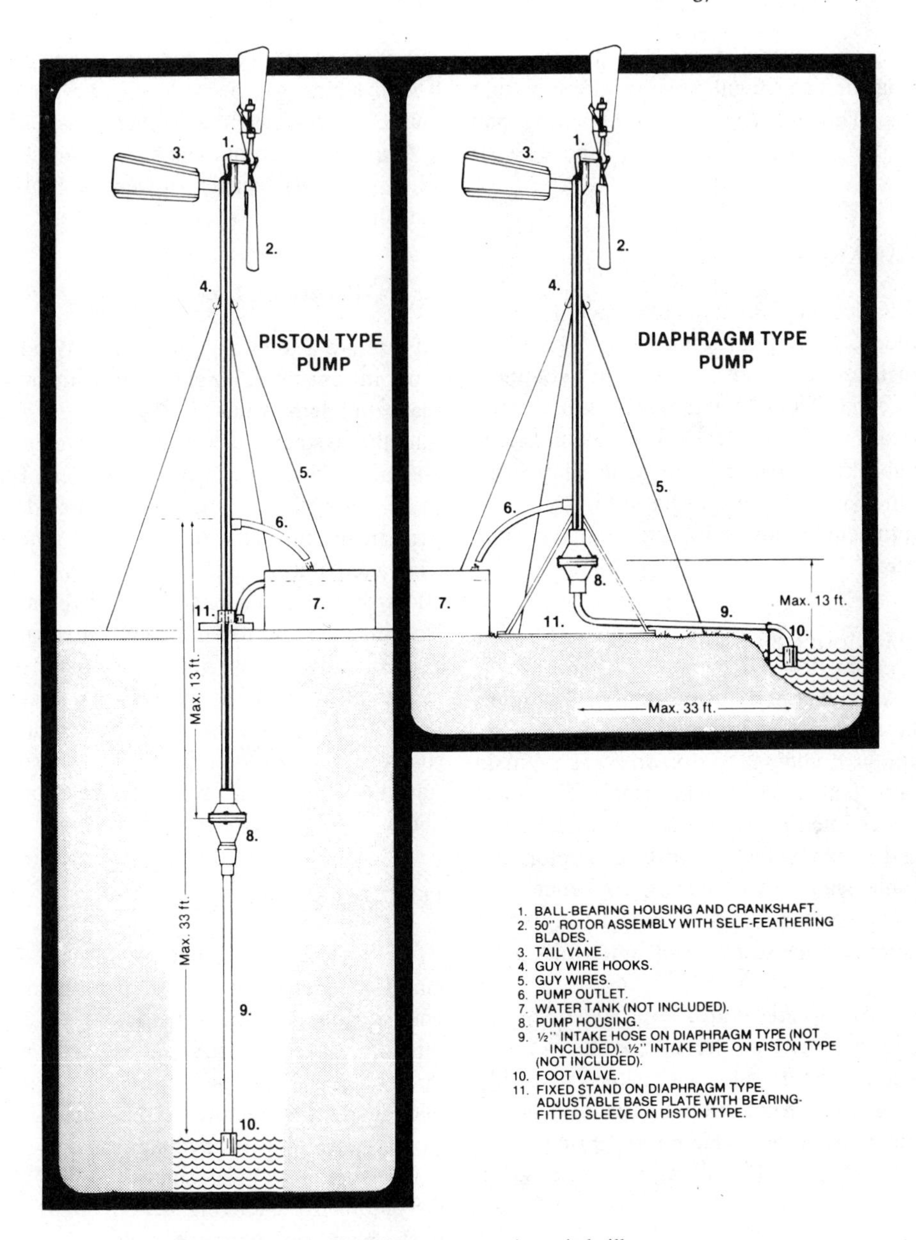

Figure 8-7. SPARCO water-pumping windmills.

varied, thus changing the cut-in and cut-out speeds of the machine."

Dunlite 81-002550

"The Dunlite is manufactured by Dunlite Electrical Company, a division of Pye Industrial Sales, Australia, and is distributed in the United States. The alternator is coupled through a 5:1 helical gearbox and is driven by a three-bladed propeller 4.10 m in diameter. The blades are made from galvanized steel sheets.

Propeller overspeed control is provided by the automatic feathering action of the blades. As speed of the blade increases, the centrifugal force on the governor weight on each of the blades overcomes the tension of a central spring and shock absorber unit and moves the blade to a coarser pitch. As the wind speed decreases, the propeller speed will slow, thus reducing the centrifugal force on the governor weights. This relaxes the central spring and returns the blades to maximum speed position."

KEDCO Model 1200

"The Kedco is a lightweight machine made by Kedco, Inc., Englewood, CA. The blade design incorporates an aircraft-type aluminum construction. The remaining structural components are welded steel. The main frame (strongback) is tubular, with main shaft bearings fitted into the tube. This design yields a very efficient, lightweight, and relatively stiff structure."

ALTOS Model 8B

"The Altos (formerly Amerenalt), is a lightweight machine manufactured by Altos Corporation, Boulder, Colorado. The main generator housing is welded steel. The alternator is coupled through a 11:1 cycloidal gearbox and is driven by a multi-bladed rotor, 2.4 m in diameter. The rotor is constructed of formed aluminum Clark Y airfoils supported by stainless steel spokes and inner and outer rims.

Rotor overspeed control is provided by utilizing increasing wind pressure on the rotor to swing the rotor and alternator assembly (which is offset from the tail assembly) out of the oncoming wind. The return spring automatically reopens the tail as the wind speed decreases."

GRUMMAN Windstream™ 25

"The Grumman Windstream 25 was designed as a stand-alone system for charging large battery banks. However, at the present time some are supplementing utilities through synchronous inverters.

No uncommon features have been incorporated in the design. For example, the rotor is controlled by blade pitch. There are three extruded aluminum blades. It is a downwind machine and the generator is driven through a standard shaft-mount reducer type gearbox. The machine has been designed with conservative safety factors wherever possible.

Grumman Corporation has recommended a 13.7 m (45 ft) concrete tower for most of its machines."

ZEPHYR Model 15kW

"There are a total of four Zephyr Wind Dynamos in existence. Zephyr has incorporated many modern wind system design features into the system, such as the downwind rotor and a positive yaw drive controller. But the most innovative concept incorporated is the permanent magnet, direct drive alternator. The alternator is approximately 0.74 m (2.4 ft) in diameter and only 0.20 m (8 in) wide. The magnets are fixed to the outer rim and, consequently, the relative circumferencial speeds of the magnets are high. A direct drive generator develops its rated power at a low rotational speed, eliminating the speed-up transmission that is usually needed for a machine of this size."

ELEKTRO WV50G

"The Elektro is manufactured by Elektro GmbH, Switzerland and is distributed in the United States. The alternator is coupled through a 4:1 gearbox and is driven by a three-bladed propeller 5.0 m in diameter. The blades are manufactured from spruce or redwood, depending on availability.

Propeller overspeed control is provided by the automatic feathering action of the blades. Each blade is equipped with a centrifugal weight, in a blade holder, which is preloaded by a spring. As the rotational speed of the blade increases, the centrifugal force exceeds the spring force and changes the pitch of the blade. An automatic control system protects the rotor from over-speeding, protects the machine from storms."

AMERICAN WIND TURBINE AWP-16

"The AWT is a lightweight machine manufactured by American Wind Turbine Co.,

Stillwater, OK; and is designed for use with an electric water pump. The alternator is rim-belt driven by a multi-bladed rotor, 4.88 m in diameter, with a 30:1 ratio. The rotor is constructed of formed aluminum airfoils supported by stainless steel spokes and inner and outer rims.

Rotor overspeed control is achieved by utilizing increasing wind pressure on the rotor to gradually swing the rotor and alternator assembly out of the oncoming wind. The return spring aligns the tail as the wind speed decreases."

NORTHWIND (EAGLE) 3kW-110

"The Eagle is sold by the North Wind Power Company, Warren, Vermont; however, the machine incorporates components from a unit originally manufactured by the Jacobs Wind Electric Co. While the Eagle tested at Rocky Flats may resemble the original machine in appearance, it cannot be considered a "Jacobs" unit and test results for the Eagle do not reflect upon the performance or reliability of any Jacobs-built machine or component. North Wind-manufactured components include the hub/blade pitch subsystem, blades and control box. The generator and slip rings were extensively rebuilt by North Wind and numerous bearings and fasteners were replaced.

A direct drive, shunt wound, dc generator is driven by a three-bladed variable blade-pitch rotor. The blades are pitched by centrifugal forces. In this rotor the centrifugal weights are the blades themselves. They are retained by springs and translate along the pitch axis under centrifugal force. As they translate, a linkage system pitches the blades."

MANUFACTURER'S SPECIFICATIONS

As in all emerging technologies, the age-old maxim "Caveat emptor" (Let the buyer beware) should be kept in mind when reviewing the following manuacturer data and claims. If you are interested in purchasing a WECS check with DOE's Rocky Flats Test Center for the latest system performance data.

AERO POWER SYSTEMS 1500

Manufacturer: Aero Power Systems, Inc.
Address: 2398 Forth Street, Berkeley, CA 94710
Contact: Mr. Mario Agnello
Telephone: (415) 848–2710
Model number: SL1500
Rated power (kW): 1.5 @ 24 mph (see Figure 8-8)

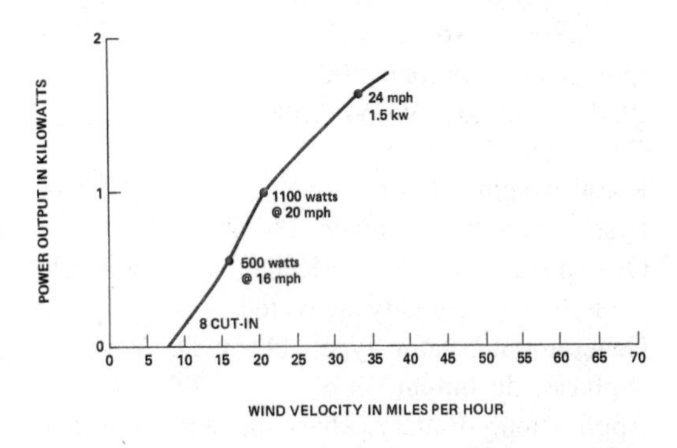

Figure 8-8. Aero Power Model SL 1500 power output curve.

Figure 8-8a. Photograph of Aero Power Model SL 1500 SWECS.

Power output at 12 mph: 200 W
Power output at 14 mph: 300 W
Rotor Blade Diameter: 12 ft
Number of blades: 3
Machine type: Upwind, horizontal axis
Cost: $3000 (12 VDC: $2995; 24 VDC: $3195)
Operating Speed
Rated speed (mph): 24
Cut-in speed (mph): 8
Cut-out speed (mph): 100
Maximum speed: 125
rpm at rated output: 500
Blade materials: Wood, Sitka spruce
Gear ratio: 2.5 to 1
Rotor weight: 50 lb
System weight on tower: 180 lb
Overspeed control: Mechanical, variable pitch, centrifugally activated
Generator/alternator type: 14 or 28 VAC, 3 phase, dc output
Application: Battery charging and interface with utility via synchronous inverter
Testing history: Field operation, seven years of WECS experience, SL1500, 18 months use
Reliability prediction (MTBF): Unknown
Warranty: One year, defects in workmanship and materials
Maintenance requirements: Semi-annual, grease hub, check blades
Country of origin: U.S.
Number produced: 160
Availability: In stock

AEROSPACE SYSTEMS DOE 1 KW

Manufacturer: Aerospace Systems, Inc.
Address: 1 Vinebrook Park, Burlington, MA 01803
Contact: Mr. John Zvara
Model number: DOE 1 kW (see Figure 8-9)
Rated power (kW): 1.2 @ 20 mph (see Figure 8-10)
Power output at 12 mph: Unknown
Power output at 14 mph: Unknown
Rotor Blade Diameter: 15 ft diameter; 8 ft high
Number of blades: 3
Machine type: Cycloturbine vertical-axis
Cost: $1500/kW

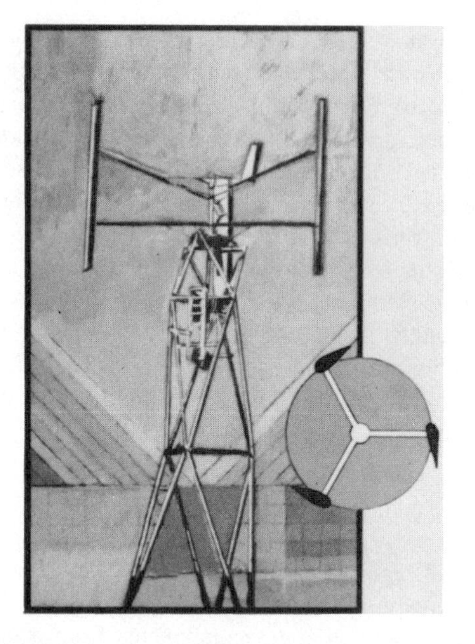

Figure 8-9. Aerospace Systems, Inc. Model 1 kW, WECS under development for DOE. (*Courtesy of DOE.*)

Operating Speed
Rated speed (mph): 20
Cut-in speed (mph): Unspecified (minimize with regard to power and system cost)
Cut-out speed (mph): 40
Maximum speed: 165
rpm at rated output: Unknown

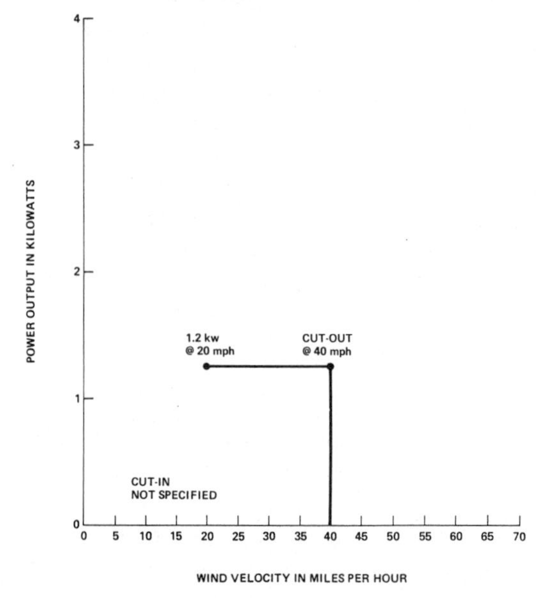

Figure 8-10. Aerospace Systems DOE 1-kW model power output curve.

Blade materials: Extruded D-shaped 6061-T6 aluminum
Rotor weight: Unknown
System weight on tower: 508 lb
Overspeed control: Unknown
Generator/alternator type: 24 Vdc natural power regulator, with regulation for battery charging
Application: For very small systems in rural and remote applications (such as pumping water into remote stock watering tanks), seismic monitoring stations and offshore navigation aids
Testing history: Under development
Reliability prediction (MTBF): High-reliability design (10 years MBTF)
Warranty: Unspecified
Maintenance requirements: Minimal
Country of origin: U.S.
Number produced: 1
Availability: DOE Development Project
Comments: System design based on cycloturbine design developed by Herman Dress of Pinson Energy Corporation.

AEROWATT 24 FP7

Manufacturer: Aerowatt S.A. 37 Rue Chanzy, 75-Paris 11e France, c/o Pennwalt Corporation—Automatic Power Division
Address: P.O. Box 18738, Houston, TX 77023
Contact: Mr. G. T. Priestly
Telephone: (713) 228–5208
Model number: 24 FP7 (see Figure 8-11)
Rated power: 28 W @ 15.7 mph (see Figure 8-12)
Power output at 12 mph: 15 W
Power output at 14 mph: 22 W
Rotor Blade Diameter: 3.3 ft
Number of blades: 2
Machine type: Horizontal-axis
Cost: $2735
Operation Speed: Efficiency of approximately 15 percent
  Rated speed (mph): 15.7
  Cut-in speed (mph): 6.7
  Cut-out speed (mph): 50
  Maximum speed: 150
rpm at rated output: 1150
Blade materials: Wood-polyurethane coated

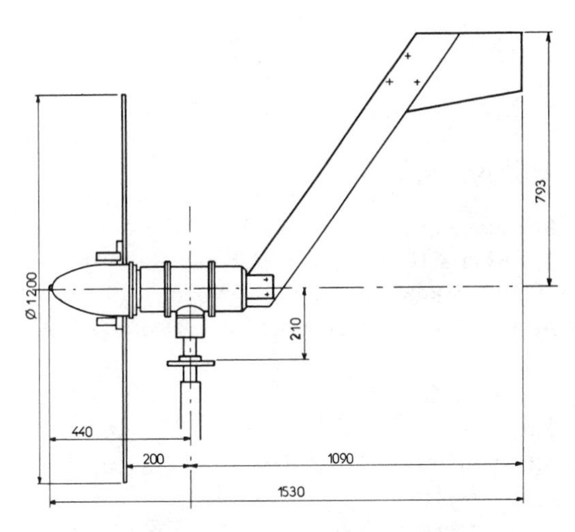

Figure 8-11. Aerowatt Model 24 FP7 WECS.

Rotor weight: 10 lb
System weight on tower: 32 lb
Overspeed control: Centrifugal pitch control
Generator/alternator type: ac permanent magnet alternator, 24 V
Application: Industrial SWECS
Testing history: In production for more than five years
Reliability prediction (MTBF): Unknown
Warranty: Five years on rotor
Maintenance requirements: Check slip rings and brushes every two years
Country of origin: France
Number produced: Over 100 systems
Availability: In stock
Comments: Concerning Aerowatt's views as to the future of wind power, "we feel it will continue to be useful as a source of electrical and mechanical energy in areas remote from commercially generated power, especially for industrial applications such as marine aids to navigation, data gathering, communication and cathodic protection. In particular areas, large size ma-

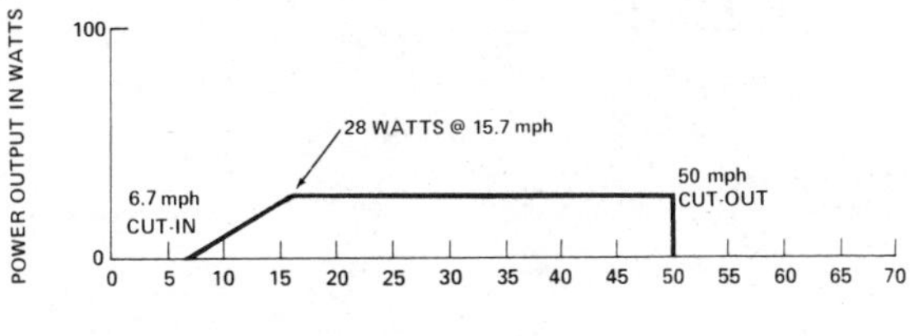

Figure 8-12. Aerowatt Model 24 FP7 power output curve.

chines may someday be useful to conserve fuel when operated in parallel with local electrical networks."

AEROWATT 150FRP7

Manufacturer: Aerowatt S.A. 37 Rue Chanzy, 75-Paris 11e France, c/o Pennwalt Corporation—Automatic Power Division
Address: P. O. Box 18738, Houston TX 77023
Contact: Mr. G. T. Priestly
Telephone: (713) 228–5208
Model number: 150 FRP7 (see Figures 8-13 and 8-14)
Rated power (kW): 130 W @ 15.7 mph (see Figure 8-15)
Power output at 12 mph: 75 W
Power output at 14 mph: 105 W
Rotor Blade Diameter: 6.7 ft
Number of blades: 2
Machine type: Horizontal-axis
Cost: $4945
Operating Speed
  Rated speed (mph): 15.7
  Cut-in speed (mph): 6.7
  Cut-out speed (mph): 50
  Maximum speed: 150

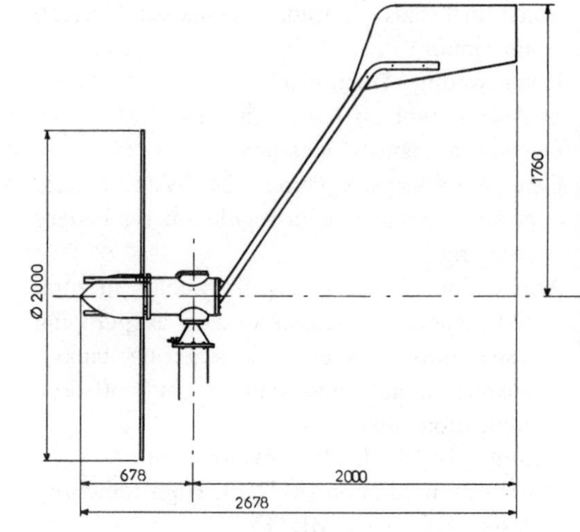

Figure 8-13. Aerowatt Model 150 FRP7 WECS.

rpm at rated output: 550
Blade materials: Aluminum
Rotor weight: 30 lb
System weight on tower: 110 lb
Overspeed control: Centrifugal pitch control
Generator/alternator type: 30 ac permanent magnet alternator, 24 V
Application: Industrial SWECS
Testing history: In production for five years

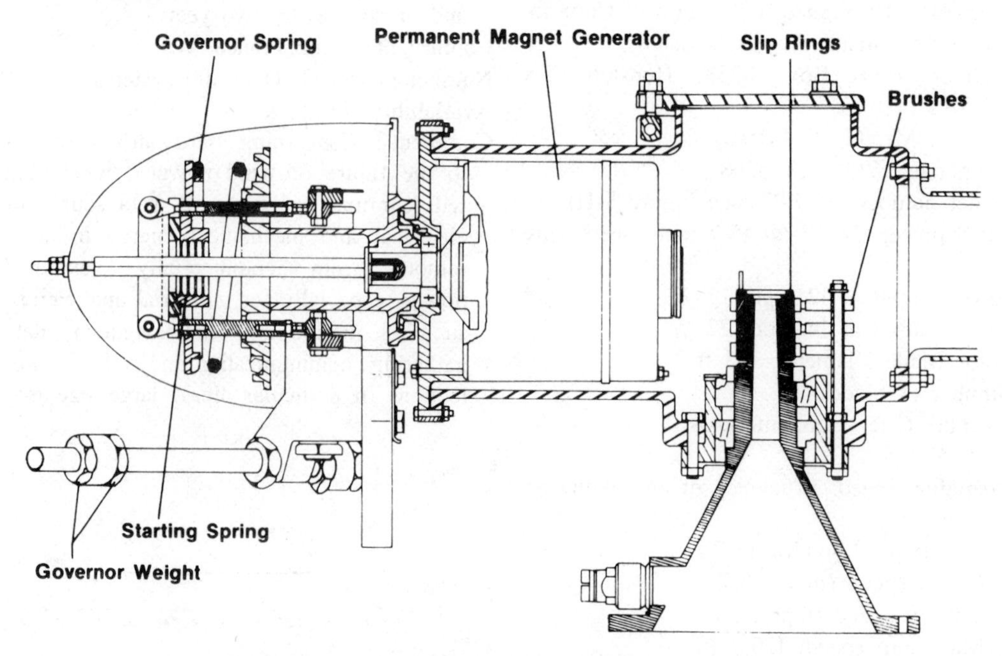

Figure 8-14. Aerowatt Model 150 FRP7 WECS configuration.

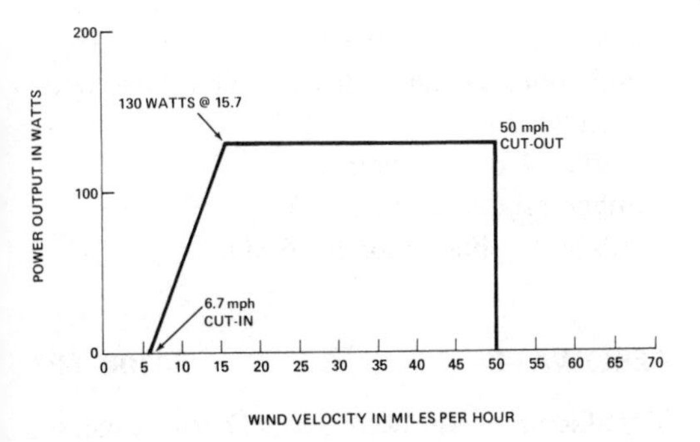

Figure 8-15. Aerowatt Model 150 FRP7 power output curve.

Reliability prediction (MTBF): Unknown
Warranty: Five years on rotor
Maintenance requirements: Slip ring and brushes should be checked every two years
Country of origin: France
Number produced: Over 100
Availability: In stock

AEROWATT 300 FP7

Manufacturer: Aerowatt S.A. 37 Rue Chanzy, 75-Paris 11e France, c/o Pennwalt Corporation—Automatic Power Division
Address: P.O. Box 18738, Houston, TX 77023
Contact: Mr. G. T. Priestly
Telephone: (713) 228–5208
Model number: 300 FP7 (see Figure 8-16)
Rated power: 350 W @ 15.7 mph (see Figure 8-17)
Power output at 12 mph: 200 W
Power output at 14 mph: 270 W
Rotor Blade Diameter: 10.7 ft
Number of blades: 2
Machine type: Horizontal-axis
Cost: $7400
Operating Speed
  Rated speed (mph): 15.7
  Cut-in speed (mph): 6.7
  Cut-out speed (mph): 50
  Maximum speed: 125
rpm at rated output: 550
Blade materials: Aluminum
Rotor weight: 100 lb
System weight on tower: 380 lb
Overspeed control: Centrifugal pitch control

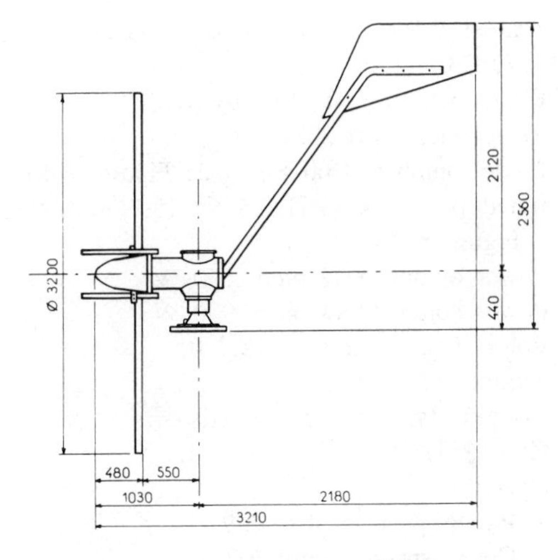

Figure 8-16. Aerowatt Model 300 FP7 WECS.

Generator/alternator type: 30 ac permanent magnet alternator, 110 V
Application: Industrial SWECS
Testing history: In production five years
Reliability prediction (MTBF): Unknown
Warranty: Five years on rotor
Maintenance requirements: Slip rings and brushes every two years
Country of origin: France
Number produced: Over 50
Availability: In stock

AEROWATT 1100 FP7

Manufacturer: Aerowatt S.A. 37 Rue Chanzy, 75-Paris 11e France, c/o Pennwalt Corporation—Automatic Power Division

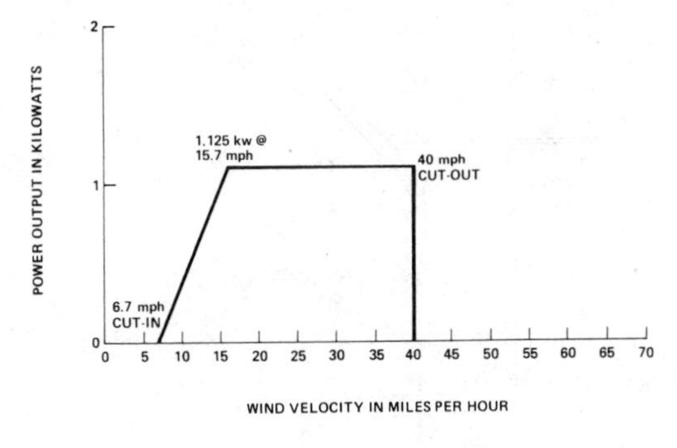

Figure 8-17. Aerowatt Model 300 FP7 power output curve.

Address: P.O. Box 18738, Houston, TX 77023
Contact: Mr. G. T. Priestly
Telephone: (713) 228–5208
Model number: 1100 FP7 (see Figure 8-18)
Rated power (kW): 1.125 @ 15.7 mph (see Figure 8-19)
Power output at 12 mph: 650 W
Power output at 14 mph: 900 W
Rotor Blade Diameter: 16.7 ft
Number of blades: 2
Machine type: Horizontal-axis
Cost: $15,475
Operating Speed
 Rated speed (mph): 15.7
 Cut-in speed (mph): 6.7
 Cut-out speed (mph): 40
 Maximum speed: 125
rpm at rated output: 178
Blade materials: Aluminum
Rotor weight: 220 lb
System weight on tower: 770 lb
Overspeed control: Centrifugal pitch control
Generator/alternator type: 30 ac permanent magnet alternator, 110 V
Application: Industrial SWECS
Testing history: Five years of production history
Reliability prediction (MTBF): Unknown
Warranty: Five years on rotor
Maintenance requirements: Check slip rings and brushes and check lubrication once each year
Country of origin: France
Number produced: Over 25
Availability: Four months ARO

AEROWATT 4100 FP7

Manufacturer: Aerowatt S.A. 37 Rue Chanzy, 75-Paris 11^{e} France, c/o Pennwalt Corporation—Automatic Power Division
Address: P.O. Box 18738, Houston, TX 77023
Contact: Mr. G. T. Priestly
Telephone: (713) 228–5208
Model number: 4100 FP7 (see Figure 8-20 and 8-21)
Rated power (kW): 4.1 @ 16.0 mph (see Figure 8-22)
Power output at 12 mph: 2.8 kW
Power output at 14 mph: 3.45 kW
Rotor Blade Diameter: 30.7 ft
Number of blades: 2
Machine type: Horizontal-axis

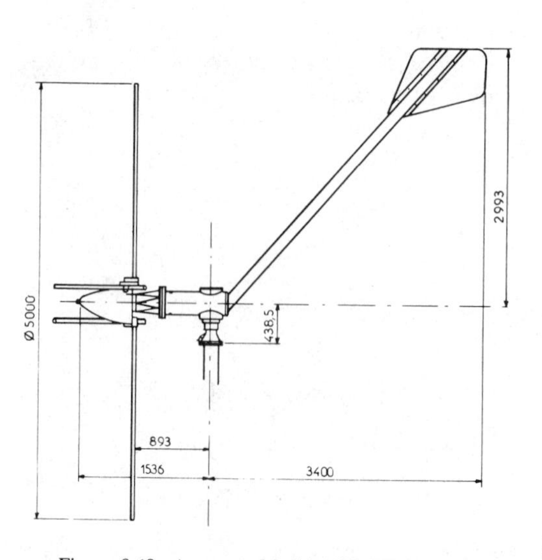

Figure 8-18. Aerowatt Model 1100 FP7 WECS.

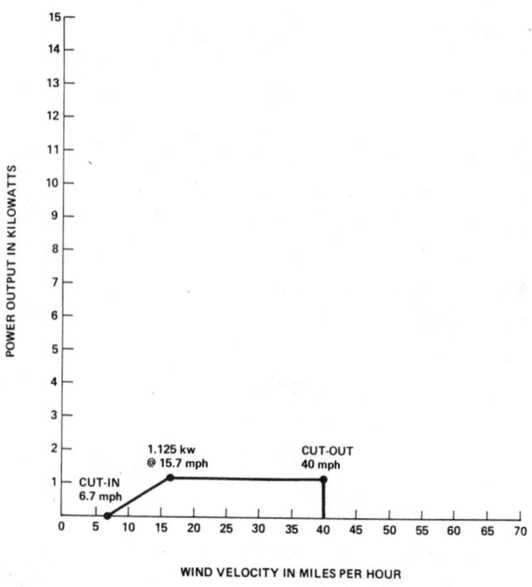

Figure 8-19. Aerowatt Model 1100 FP7 power output curve.

Cost: $35,350
Operating Speed
 Rated speed (mph): 16.0
 Cut-in speed (mph): 3.3
 Cut-out speed (mph): 55
 Maximum speed: 125
rpm at rated output: 152
Blade materials: Extruded aluminum (epoxy, neoprene or teflon coating available)
Rotor weight: 600 lb
System weight on tower: 2,100 lb
Overspeed control: Centrifugal pitch control
Generator/alternator type: 30 ac permanent magnet alternator, 110 V
Application: Industrial SWECS
Testing history: Five years of production
Reliability prediction (MTBF): Unknown
Warranty: Five years on rotor
Maintenance requirements: Check slip rings and brushes and check lubrication in gears once each year
Country of origin: France
Number produced: Over 25
Availability: Six months ARO

ALCOA DOE 8 kW ALVAWT

Manufacturer: Aluminum Company of America Labs
Address: Alcoa Technical Center, Alcoa Center PA 15069
Contact: Mr. Paul N. Vosburgh
Telephone: (412) 339–6651
Model number: DOE 8 kW ALVAWT (271806) (see Figure 8-23)
Rated power (kW): 11.0 @ 20 mph (see Figure 8-24)
Power output at 12 mph: Unspecified
Power output at 14 mph: Unspecified
Rotor Blade Diameter: 33 ft diameter, 34 ft high
Number of blades: 3
Machine type: Darrieus, vertical-axis wind turbine
Cost: $10,000; $750/kW design cost goal; cumulative development costs $305,000
Operating Speed
 Rated speed (mph): 20
 Cut-in speed (mph): Unknown
 Cut-out speed (mph): Unknown
 Maximum speed: Unknown
rpm at rated output: Unknown
Blade materials: Aluminum, NACA 0015 airfoil configuration
Rotor weight: Unknown
System weight on tower: 10,480 lb
Overspeed control: Unknown
Generator/alternator type: Unknown
Application: Produce high quality, 60-Hz ac electricity compatible and synchronized with grid energy
Testing history: Beginning demonstration phase
Reliability prediction (MTBF): Unknown
Warranty: Unspecified
Maintenance requirements: Not specified
Country of origin: U.S.
Number produced: 1
Availability: Demonstration design

Figure 8-20. Aerowatt Model 4100 FP7 WECS installed at DOE's Rocky Flats Test Site.

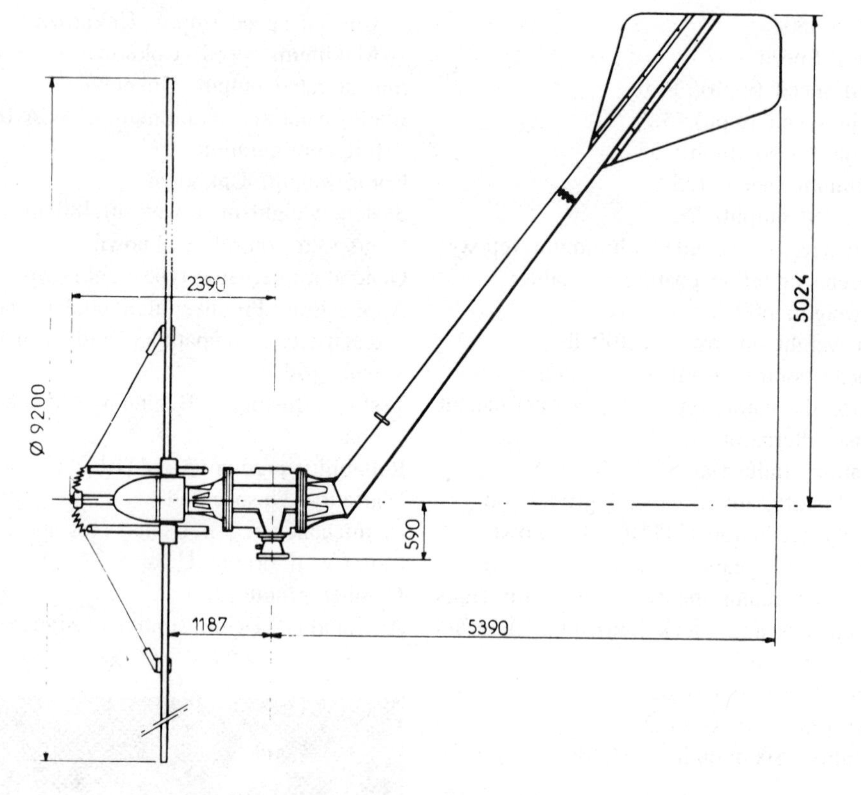

Figure 8-21. Aerowatt Model 4100 WECS mechanical configuration.

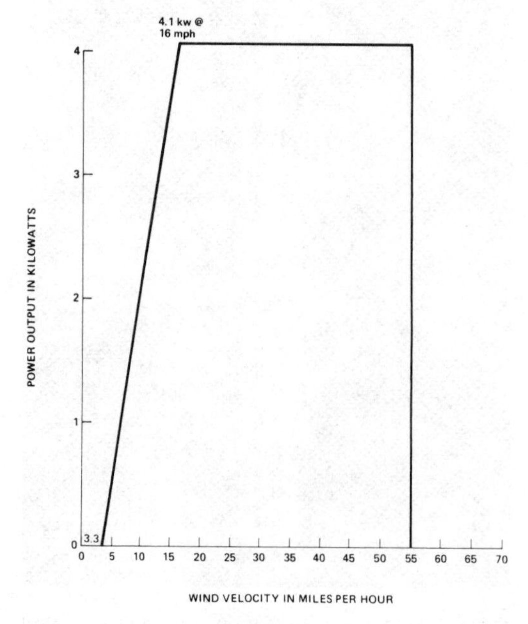

Figure 8-22. Aerowatt Model 4100 FP7 power output curve.

ALCOA — VARIOUS NEW WECS

"After investigating the history and the state-of-the-art of Wind Energy Conversion Systems, Alcoa decided in 1976 that the combination of advantages of Vertical Axis versus Horizontal Axis Wind Turbines, as well as a good fit between available VAWT technology and Alcoa resources, offered the best opportunity for Alcoa to develop a cost-effective, low-maintenance, performance-reliable line of wind conversion hardware.

VAWT ADVANTAGES

- ACCEPTS WIND FROM ALL DIRECTIONS—No Yawing
- SELF-REGULATING @ CONSTANT RPM—No Pitch Control
- 60 Hz—AC—GRID COMPATIBLE—No Power Conditioning or Storage
- WORKING PARTS @ GROUND LEVEL—No Weight or Bulk Limitations

Figure 8-23. Alcoa 8-kW WECS under development for DOE. (*Courtesy of DOE.*)

•NO TOWER—No Shadow, Energy at Base

•LOW COST—SIMPLE—RUGGED—Cost-Effective Energy

After exposure to the available Canadian (National Research Council of Canada) and United States (DOE/Sandia Laboratories) technology base, initial efforts were directed toward determining the most useful machine sizes and energy capacities and types. It was decided that the initial product design effort would be limited to machines larger than required for residential applications and with a maximum size limited by the largest aluminum blade chord that could be extruded in one piece. The initial product design effort was also limited to machines that could generate high quality 60 Hz, AC electricity compatible and synchronized with grid energy. Storage and power conditioning were specifically excluded from the design challenge as was mechanical energy output.

In support of the Alcoa-funded in-house design and development effort, Alcoa Laboratories responded to competitive requests for proposals by DOE and was awarded contracts for design and demonstration of a small machine (administered for DOE by Rockwell International at Rocky Flats, Colorado) and an intermediate-sized machine (administered for DOE by Sandia Laboratories at Albuquerque, New Mexico). Phase I of each of those efforts is complete and prototype fabrication was completed during the second quarter of 1979. Progress on the small machine was reported by Alcoa's project manager, Tom Stewart, at the DOE/Rockwell Workshop on Small Wind Energy Conversion Systems in Boulder, Colorado.

In addition to the two DOE-funded design efforts, Alcoa Laboratories was assigned the task of developing two additional small machines and one additional intermediate-sized unit to complete an introductory line of five basic models: #271806–08 kW, #453011–22 kW, #634214–57 kW, #835524–112 kW and #1238229–280 kW.

The model numbers simply indicate the rotor height (first two or three digits) and diameter (next two digits) in feet and the blade chord (last two digits) in inches along with a nominal kW rating for the machine when targeted for use at a site with a mean annual wind speed of 6.71 m/s (15 mph). For example: Model #634214–57 kW has a rotor 63′ high by 42′ in diameter with 14″ chord blades and a nominal 57 kW capacity induction generator. The largest of the five basic models has also been targeted for 18 mph wind regimes. It them becomes Model #1238229–500 kW.

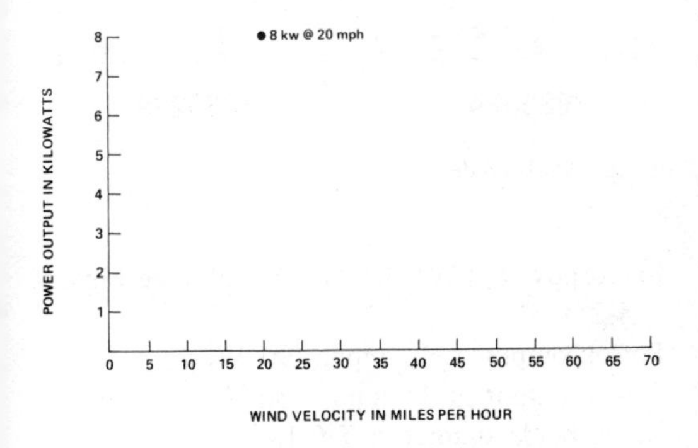

Figure 8-24. Alcoa DOE 8-kW model power output curve.

Following are installed cost estimates for the six introductory ALVAWT models:

ALVAWT MODEL	INSTALLED COST	ELECTRICITY COST* TYPICAL/kWh	ELECTRICITY COST* INVESTMENT/ ANNUAL kWh
#271806-08 kW	$ 10,000	$.06-.08	$.62
#453011-22 kW	$ 20,000	$.04-.06	$.44
#634214-57 kW	$ 40,000	$.03-.05	$.35
#835524-112 kW	$ 75,000	$.03-.05	$.32
#1238229-280 kW	$160,000	$.03-.05	$.26
#1238229-500 kW	$190,000	$.02-.04	$.17

All costs are in 1979 first-quarter dollars.

*Annual energy outputs for the first five machines are based on installation at a site with mean annual wind speeds of 6.71 m/s (15 mph). Annual energy output for Model #1238229-500 kW assumes installation at a site with mean annual wind speed of 8.05 m/s (18 mph).

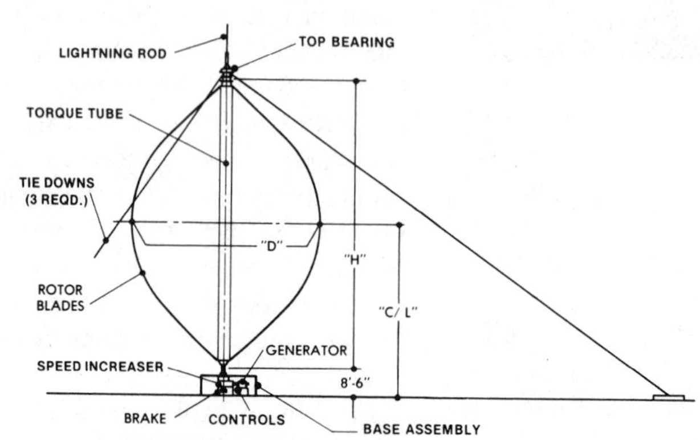

TYPICAL ALVAWT
ALCOA VERTICAL AXIS WIND TURBINE

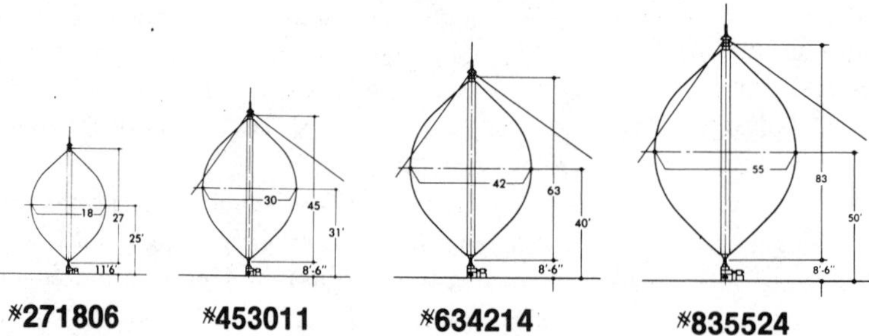

Figure 8-24a. Alcoa Basic ALVAWT Models.

ALTOS BWP-8B

Manufacturer: Altos—The Alternate Current
Address: P.O. Box 905, Boulder, CO 80302
Contact: Mr. Michael Blakely
Telephone: (303) 442-0855
Model number: BWP-8B

Rated power (kW): 1.5 @ 28 mph (see Figure 8-25)
Power output at 12 mph: 190 W
Power output at 14 mph: 350 W
Rotor blade diameter: 7.6 ft
Number of blades: Multi-blade

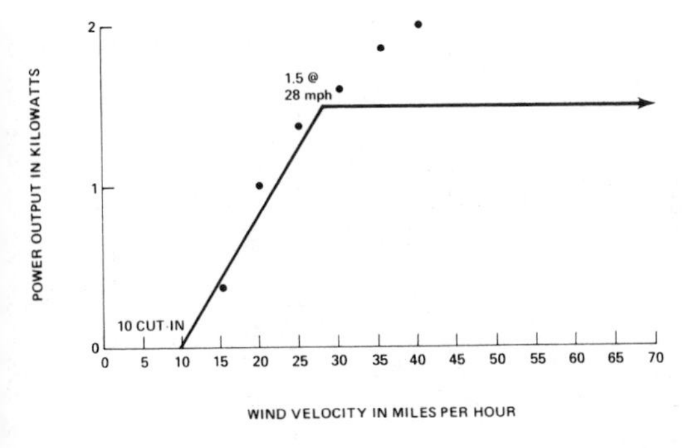

Figure 8-25. Altos Model BWP-8B power output curve.

Machine type: Upwind, horizontal-axis, multi-blade
Cost: Contact factory
Operating speed
  Rated speed (mph): 28
  Cut-in speed (mph): 10
  Cut-out speed (mph): 75
  Maximum speed (mph): Unspecified
rpm at rated output: 165
Blade materials: Aluminum, 5052
Rotor weight: 59 lb
System weight on tower: 250 lb
Overspeed control: Aerodynamic drag, rotor turns sideways
Generator/alternator type: 24 vdc, 0–70 A
Application: Small electric power applications
Testing history: Field testing
Reliability prediction (MTBF): Unspecified
Warranty: One year, parts and workmanship, 90 days on turbine
Maintenance requirements: Semi-annual fluid and system check, annual inspection
Country of origin: U.S.
Number produced: Unknown
Availability: Unknown
DOE Rocky Flats Tests: See DOE Report RFP-2920/3533/78/6-2 of Sept. 28, 1978 and later reports for more information.

ALTOS BWP-12A

Manufacturer: Altos—The Alternate Current
Address: P.O. Box 905, Boulder, CO 80302
Contact: Mr. Michael Blakely
Telephone: (303) 442-0855
Model number: BWP-12A
Rated power (kW): 2.2 @ 28 mph (see Figure 8-26)
Power output at 12 mph: 450 W
Power output at 14 mph: 650 W
Rotor blade diameter: 11.5 ft
Number of blades: Multi-blade
Machine type: Upwind, horizontal-axis, multi-blade
Cost: Contact factory
Operating speed
  Rated speed (mph): 28
  Cut-in speed (mph): 8
  Cut-out speed (mph): 60
  Maximum speed (mph): Unspecified
rpm at rated output: 116
Blade materials: Aluminum, 6061
Rotor weight: 111 lb
System weight on tower: 300 lb
Overspeed control: Aerodynamic drag, rotor turns sideways
Generator/alternator type: 115/200 vac, 3 phase
Application: SWECS
Testing history: Calculated performance data; Tests now underway
Reliability prediction (MTBF): Unspecified
Warranty: One year, parts and workmanship, 90 days on turbine

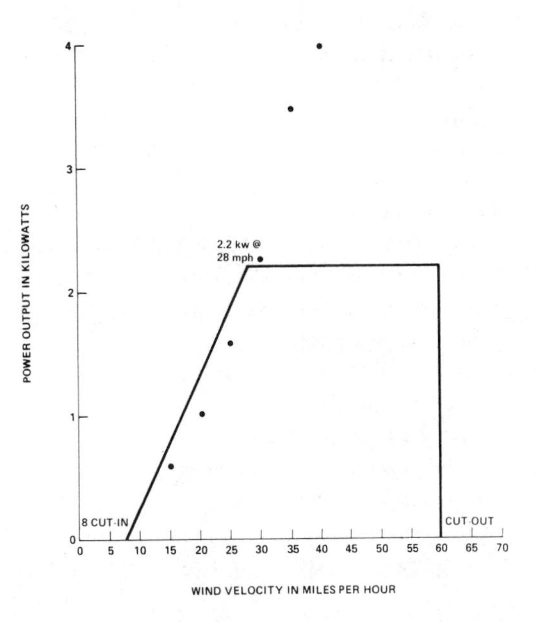

Figure 8-26. Altos Model BWP-12A power output curve.

Maintenance requirements: Semi-annual fluid and system check, annual inspection
Country of origin: U.S.
Number produced: Unknown
Availability: Contact factory

ALTOS BWP-12B

Manufacturer: Altos—The Alternate Current
Address: P.O. Box 905, Boulder, CO 80302
Contact: Mr. Michael Blakely
Telephone: (303) 442-0855
Model number: BWP-12B
Rated power (kW): 2.0 @ 28 mph (see Figure 8-27)
Power output at 12 mph: 400 W
Power output at 14 mph: 600 W
Rotor blade diameter: 11.5 ft
Number of blades: Multi-blade
Machine type: Upwind, horizontal-axis, multi-blade
Cost: Contact factory
Operating speed
  Rated speed (mph): 28
  Cut-in speed (mph): 8
  Cut-out speed (mph): 60
  Maximum speed (mph): Unspecified
rpm at rated output: 116
Blade materials: Aluminum, 6061
Rotor weight: 111 lb
System weight on tower: 300 lb
Overspeed control: Aerodynamic drag, rotor turns sideways
Generator/alternator type: 24 vdc alternator, trickle excitation
Application: SWECS
Testing history: Performance data calculated
Reliability prediction (MTBF): Unspecified
Warranty: One year, parts and workmanship, 90 days on turbine
Maintenance requirements: Semi-annual fluid and system check, annual inspection
Country of origin: U.S.
Number produced: Unknown
Availability: Contact factory

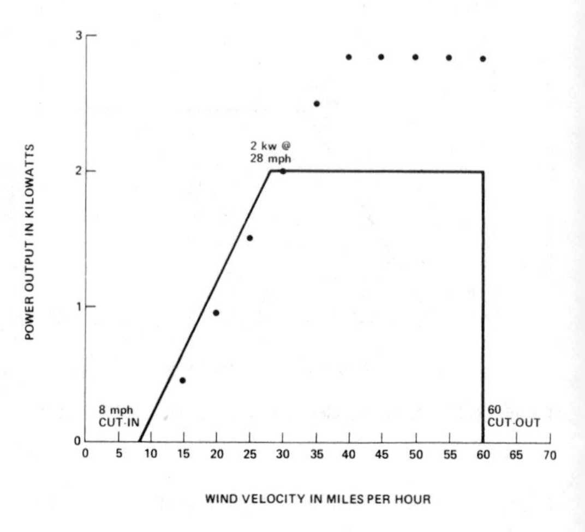

Figure 8-27. Altos Model BWP-12B power output curve.

AMERICAN WIND TURBINE SST-12

Manufacturer: American Wind Turbine Co.
Address: 1016 E. Airport Road, Stillwater, OK 74074
Contact: Ms. Nancy Thedfor
Telephone: (405) 377-5333
Model number: SST-12 (SST-Super Speed Turbine)
Rated power (kW): 0.900 @ 20 mph (see Figure 8-28)
Power output at 12 mph: 300 W
Power output at 14 mph: 450 W
Rotor blade diameter: 11.5 ft
Number of blades: Multi-blade, bicycle wheel
Machine type: Upwind, horizontal-axis, multi-blade, bicycle wheel
Cost: $1496
Operating speed
  Rated speed (mph): 20
  Cut-in speed (mph): 8
  Cut-out speed (mph): 35
  Maximum speed (mph): 125
rpm at rated output: 120
Blade materials: Aluminum
Rotor weight: 92 lb
System weight on tower: 328 lb
Overspeed control: Mechanical, tail turns, rotor vane deflects
Generator/alternator type: Permanent magnet, 3-phase alternator
Application: Low-wind speed areas
Testing history: Moving test bed, field testing
Reliability prediction (MTBF): Unspecified
Warranty: Limited 90 days, parts and workmanship
Maintenance requirements: Semi-annual, check belts, annual grease and inspect

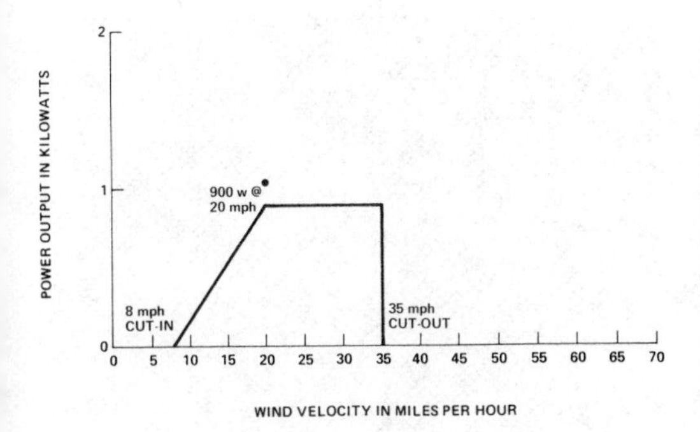

Figure 8-28. American Wind Turbine Model SST-12 power output curve.

Country of origin: U.S.
Number produced: Few
Availability: Contact factory

AMERICAN WIND TURBINE SST-16

Manufacturer: American Wind Turbine Co.
Address: 1016 E. Airport Road, Stillwater, OK 74074
Contact: Ms. Nancy Thedfor
Telephone: (405) 377-5333
Model number: SST-16 (SST-Super Speed Turbine) (see Figure 8-29)
Rated power (kW): 1.8 @ 20 mph (see Figure 8-30)
Power output at 12 mph: 600 W
Power output at 14 mph: 900 W
Rotor blade diameter: 15.3 ft
Number of blades: Multi-blade
Machine type: Upwind, horizontal-axis, multi-blade, bicycle wheel
Cost: $2420
Operating speed
 Rated speed (mph): 20
 Cut-in speed (mph): 8
 Cut-out speed (mph): 40
 Maximum speed (mph): Unspecified
rpm at rated output: 90
Blade materials: Multiple aluminum airfoils supported by spokes and rims
Rotor weight: 135 lb
System weight on tower: 520 lb
Overspeed control: Mechanical, tail turns, rotor vane deflects
Generator/alternator type: Permanent magnet, 3-phase alternator
Application: SWECS

Figure 8-29. American Wind Turbine Model SST-16 WECS. (*Courtesy of DOE.*)

Testing history: Moving test bed, field testing, being tested at DOE Rocky Flats test site
Reliability prediction (MTBF): Unspecified
Warranty: Limited 90 days, parts and workmanship

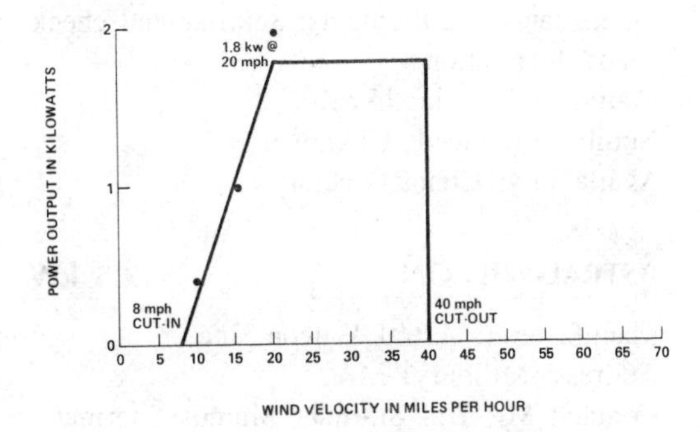

Figure 8-30. American Wind Turbine Model SST-16 power output curve.

Maintenance requirements: Semi-annual, check belts, annual inspection
Country of origin: U.S.
Number produced: Unknown
Availability: Contact factory

ASTRAL-WILCON AW 8-C

Manufacturer: Astral Wilcon, Inc.
Address: Millbury, MA
Contact: Mr. Bill Stern or Marcus Sherman
Telephone: (617) 865-9412
Model number: AW 8-C (see Figure 8-31)
Rated power (kW): 8.0 @ 22 mph (see Figure 8-32)
Power output at 12 mph: 900 W
Power output at 14 mph: 2.4 kW
Rotor blade diameter: 26 ft
Number of blades: 3
Machine type: Horizontal-axis
Cost: $7900
Operating speed
  Rated speed (mph): 22
  Cut-in speed (mph): 8
  Cut-out speed (mph): 50
  Maximum speed (mph): 130
rpm at rated output: 200
Blade materials: Reinforced plastic (fiberglass)
Rotor weight: 200 lb
System weight on tower: 535 lb
Overspeed control: Variable pitch rotor
Generator/alternator type: 3-phase, unregulated ac brushless alternator
Application: Water or space heat, battery charging, grid connected
Testing history: Operational prototype
Reliability prediction (MTBF): 10 years
Warranty: 1 year limited
Maintenance requirements: Semi-annual check and lubrication
Country of origin: U.S.
Number produced: Unknown
Availability: Contact factory

Figure 8-31. Astral-Wilcon Model AW 8-C SWECS.

ASTRAL-WILCON 2.5 kW

Manufacturer: Astral Wilcon, Inc.
Address: Millbury, MA
Contact: Mr. Bill Stern or Marcus Sherman
Telephone: (617) 865-9412
Model number: 2.5 kW
Rated power (kW): 2.5 @ 30 mph (see Figure 8-33)
Power output at 12 mph: 175 W (see Figure 8-34)
Power output at 14 mph: 275 W
Rotor blade diameter: 15 ft
Number of blades: 3
Machine type: Upwind, horizontal-axis

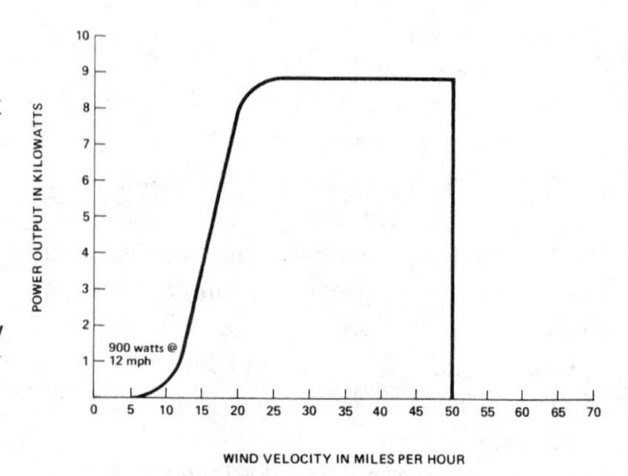

Figure 8-32. Astral-Wilcon Model AW 8-C power output curve.

Cost: $1989
Operating speed
  Rated speed (mph): 27
  Cut-in speed (mph): 9
  Cut-out speed (mph): Manual
  Maximum speed (mph): 130
rpm at rated output: 275
Blade materials: Polyurethane core and fiberglass skin
Rotor weight: 90 lb
System weight on tower: 275 lb
Overspeed control: Variable pitch rotor
Generator/alternator type: 3-phase unregulated ac brushless alternator
Application: Battery charging, water or space heat, grid connected
Testing history: 2 machines; 1½ years use each
Reliability prediction (MTBF): 10 years
Warranty: 1 year limited
Maintenance requirements: Semi-annual check, lubrication
Country of origin: U.S.
Number produced: 3
Availability: Contact factory

Figure 8-33. Astral-Wilcon Model 2.5 kW SWECS.

DAF VAWT-50

Manufacturer: DAF, INDAL, Ltd.
Address: 3570 Hawkestone Road, Mississauga, Ontario, L5C2V8, Canada
Contact: Mr. C. F. Wood
Telephone: (416) 275-5300
Model number: VAWT-50
Rated power (kW): 50
Power output at 12 mph: 0
Power output at 14 mph: Unspecified
Rotor blade diameter: 37 ft
Number of blades: 2
Machine type: Darrieus, vertical-axis, 37 ft x 55 ft
Cost: $90,000 prototype
Fixed speed: 83 rpm
Operating speed
  Rated speed (mph): Unspecified
  Cut-in speed (mph): Unspecified
  Cut-out speed (mph): Unspecified
  Maximum speed (mph): Unspecified
Blade materials: Aluminum
Rotor weight: Unspecified
System weight on tower: Unspecified
Overspeed control: Spoilers and brakes
Generator/alternator type: Induction
Application: Electricity production for grid-connected application
Testing history: 1500 hours
Reliability prediction (MTBF): Unspecified
Warranty: Unspecified
Maintenance requirements: Unspecified
Country of origin: Canada
Number produced: 7
Availability: Unspecified

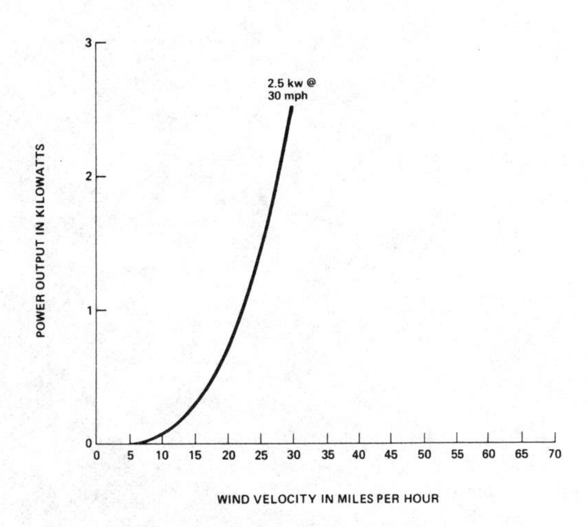

Figure 8-34. Astral-Wilcon Model 2.5 kW power output curve.

DAF **VAWT-200**

Manufacturer: DAF, INDAL, Ltd.
Address: 3570 Hawkestone Road, Mississauga, Ontario, L5C2V8, Canada
Contact: Mr. C. F. Wood
Telephone: (416) 275-5300
Model number: VAWT-200 (see Figure 8-35)
Rated power (kW): 200
Power output at 12 mph: 0
Power output at 14 mph: Unspecified
Rotor blade diameter: 80 ft
Number of blades: 2
Machine type: Darrieus, vertical-axis, 80 ft x 120 ft
Cost: $350,000 prototype
Fixed speed: 36 rpm
Operating speed
  Rated speed (mph): Unspecified
  Cut-in speed (mph): Unspecified
  Cut-out speed (mph): Unspecified
  Maximum speed (mph): Unspecified
Blade materials: Aluminum
Rotor weight: Unspecified
System weight on tower: Unspecified
Overspeed control: Spoilers and brakes

Figure 8-35. DAF Model 200-kW vertical axis wind turbine.

Generator/alternator type: Induction
Application: Electricity production for grid-connected application
Testing history: 1500 hours
Reliability prediction (MTBF): Unspecified
Warranty: Unspecified
Maintenance requirements: Unspecified
Country of origin: Canada
Number produced: 1
Availability: Unspecified

DAKOTA BC 4

Manufacturer: Dakota Wind and Sun, Ltd.
Address: P.O. Box 1781, Aberdeen, SD 57401
Contact: Mr. R. L. Kirchner
Telephone: (303) 632-0590
Model number: BC 4 (see Figure 8-36)
Rated power (kW): 4.0 @ 27 mph (see Figure 8-37)
Power output at 12 mph: 0.7 kW
Power output at 14 mph: 1.15 kW
Rotor blade diameter: 14 ft
Number of blades: 3
Machine type: Upwind, horizontal-axis
Cost: $7350
Operating speed
  Rated speed (mph): 27
  Cut-in speed (mph): 8–9
  Cut-out speed (mph): None
  Maximum speed (mph): 150
rpm at rated output: 300
Blade materials: Wood (Sitka Spruce), epoxy paint
Rotor weight: 75 lb
System weight on tower: 550 lb
Overspeed control: Centrifugal blade feathering
Generator/alternator type: 110-V dc generator
Application: Electricity, SWECS
Testing history: Field tests
Reliability prediction (MTBF): No data available
Warranty: Three-year replacement of defective parts
Maintenance requirements: Semi-annual, inspect, lubricate system
Country of origin: U.S.
Number produced: 105
Availability: Six weeks
Note: Dakota plans to produce a 10 kW @ 20 mph SWEC

Figure 8-36. Dakota Wind and Sun Model BC 4 WECS.

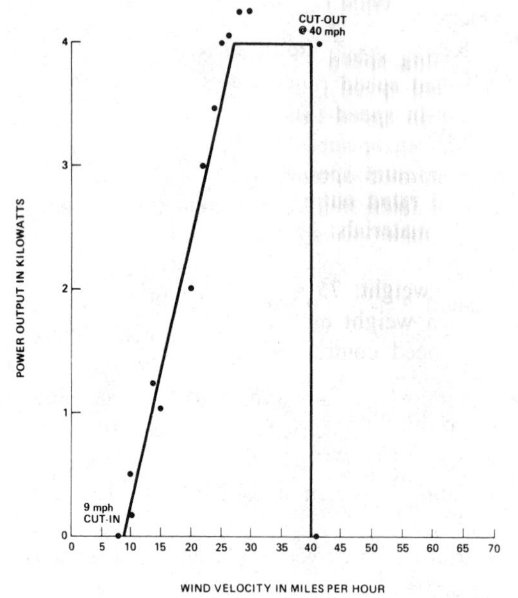

Figure 8-37. Dakota Model BC 4 power output curve.

DOE/NASA-LEWIS DOE MOD-0

Manufacturer: NASA-Lewis Research Center
Address: Cleveland, OH 44135
Contact: Richard Thomas
Telephone: (216) 433-4000 extension 6134
Model number: MOD-0 (see Figure 8-38 and 8-39)
Rated power (kW): 100 (see Figure 8-40)
Power output at 12 mph: 0
Power output at 14 mph: Did not specify
Rotor blade diameter: 125 ft uncovered
Number of blades: 2
Machine type: Large horizontal-axis, downwind, rigid nob
Cost: $1,000,000
Operating speed
  Rated speed (mph): 21.7 @ hub height
  Cut-in speed (mph): 13 @ hub height
  Cut-out speed (mph): 40 @ hub height
  Maximum speed (mph):
rpm at rated output: 40
Blade materials: Aluminum
Rotor weight: 2300 lb/blade
Cone angle: 7°
System weight on tower: 45,000 lb
Overspeed control: Computer-controlled hydraulics
Generator/alternator type: Sync ac, 250 KVA rating, 480 V, 1800 rpm
Application: Research Wind Energy Conversion System
Testing history: Currently under test at Plumbrook, Ohio test site (NASA-Lewis Research Center)
Reliability prediction (MTBF): Unknown
Warranty: None
Maintenance requirements: Unspecified
Country of origin: U.S.
Number produced: 1
Availability: DOE demonstration WECS
Description: Figure 8-38 shows a drawing of the MOD-0 WECS in operation at the NASA Plumbrook site near Sandusky, Ohio. The WECS has a 2-bladed constant

Figure 8-38. DOE/NASA-Lewis MOD-0 LWECS. (*Courtesy of DOE/NASA LRC.*)

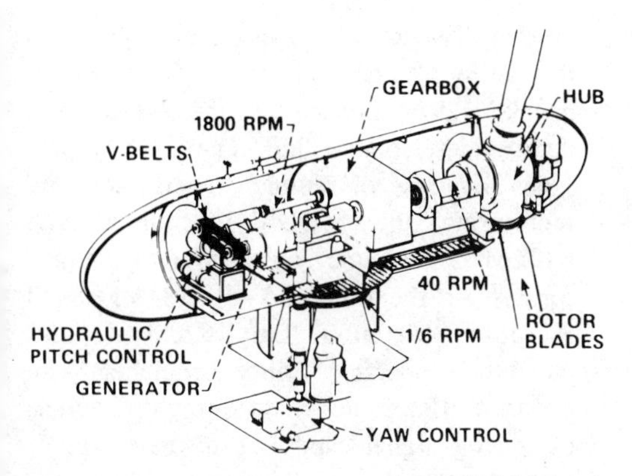

Figure 8-39. Key components of DOE/NASA-Lewis MOD-0 LWECS. (*Courtesy of DOE/NASA LRC.*)

40 rpm, 125-foot diameter rotor located downwind of the tower. The rotor drives a 100-kW synchronous alternator through a step-up gear box. The drive train and rotor are located in a nacelle with a centerline 100 feet above ground. The nacelle sits on top of a 4-legged steel truss tower. Wind direction is sensed by a wind vane on top of the nacelle and is used as a signal for the yaw control for keeping the wind turbine aligned in the direction of the wind. Details of the drive train system and the yaw system are shown in Figure 8-39. The wind turbine including the yaw drive, drive train and rotor blades was lifted to the top of the tower in one operation.

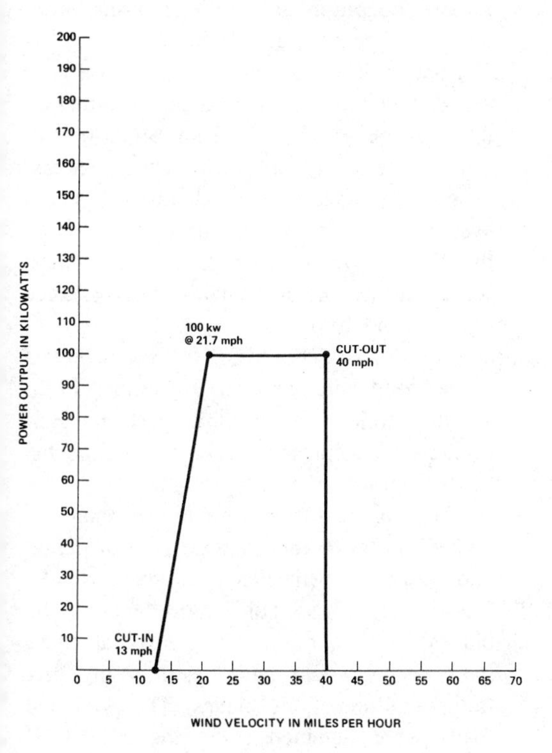

Figure 8-40. DOE/NASA-Lewis MOD-0 power output curve.

Operation: The operation of the MOD-0 WECS consists primarily of startup, normal operations connected to the utility network, shutdown and standby. In addition to connection to the Ohio Edison utility system, MOD-0 can be connected to: (1) a load bank; (2) a diesel generator of approximately 160 kW; and (3) the Plumbrook network. The Plumbrook network can be disconnected from Ohio Edison to provide a good simulation of a small utility network with several small generators and real load characteristics.

The basic controls are: (1) the yaw control for aligning the wind turbine with the wind direction; and (2) the blade pitch control used for startup, shutdown and power control. All normal operation functions are programmed into a microprocessor which provides the supervisory control. A safety shutdown system is wired into the MOD-0 controls to automatically and safely shut WECS down in the event any key parameters are out-of-tolerance. In addition to the on-site microprocessor and safety systems, a remote control and monitoring system is designed to allow the Wind Energy Project Office (or a local utility dispatcher in the case of follow-on wind turbines) to monitor and control the operation.

Normal Control: When WECS is shutdown, the blades are feathered and are free to slowly rotate. For wind speed at cut-in (13 mph) or greater, the yaw control is activated and the WECS aligned with the wind. The blades are then pitched at a programmed rate and the rotor speed is brought to about 41 rpm. At this time the automatic synchronizer is activated and WECS is synchronized with the utility network.

For wind speeds between cut-in and rated (22 mph), the blade pitch angle is held constant at 0° (at 3/4 of the blade radius).

For wind speeds above rated, the blade pitch angle is automatically controlled to limit the wind turbine power to 100 kW. For winds above cut-out (40 mph), the blades are feathered and the system is shut down. The yaw control, hydraulic systems and electrical systems are all turned off when the system is shutdown.

Microprocessor: For utility operation, it is planned that WECS will be located some distance from the control power station or the dispatcher's control room. WECS are also planned to be completely automatic and to deliver power to the network whenever the wind speeds are between cut-in and cut-out.

To accomplish this, the control logic for achieving proper startup, synchronization, power control, and shutdown is all contained in a supervisory control system utilizing a microprocessor. The microprocessor is located at the WECS site with the other electrical controls and is programmed to automatically execute all the normal wind turbine operation and control sequences. For example, the microprocessor tracks the wind speed and initiates the startup sequence when wind speed reaches cut-in and shutdown when the wind speed goes above cut-out.

Safety Shutdown System: The safety shutdown system for MOD-0 is an emergency shutdown system which operates independently of all other controls. The safety system uses a set of sensors to monitor potential problem areas to protect the wind turbine from a catastrophic failure. These sensors monitor key parameters such as: rotor overspeed, overcurrent or reverse current, vibration, yaw error, pitch system hydraulic fluid level, key temperatures and microprocessor failure. If any sensor detects an out-of-tolerance signal, the safety system initiates shutdown of the WECS. Critical sensors and circuits such as those for monitoring speed are redundant. A failure of the safety system will also initiate a shutdown.

Remote Control and Monitoring System: A simple remote control and monitoring system (RC&M) is incorporated in the MOD-0 system. The RC&M is located at the Lewis Research Center, 50 miles from Plumbrook, and is connected to the WECS by a dedicated telephone line. The RC&M provides a means of remotely controlling and monitoring the MOD-0 operation. The RC&M is the same control system planned for use on the MOD-0A WECS scheduled for utility operation. The RC&M provides the remote operator (utility dispatcher) with a start and stop mode, up to eight channels of analog signals and key discrete signals that can shut the wind turbine down.

Operations Summary: The MOD-0 operations to date have shown that the basic controls for speed, power and yaw work satisfactorily. Synchronization to the utility network has been demonstrated routinely. The normal operations for startup, utility operation, shutdown and standby have all been demonstrated and performance is satisfactory. The MOD-0 has also been used to check out remote operation planned for follow-on MOD-0A and MOD-1 WECS. In summary, the MOD-0 has exhibited satisfactory operation and no operations problems are apparent at this time.

Performance: The purpose of this section is to summarize the performance information that has been obtained from MOD-0 testing. Three areas of performance are discussed: (1) blade dynamic loads; (2) power system dynamics in the drive train when the WT is synchronized to the utility network; and (3) the aerodynamic performance of the wind turbine.

Dynamic Loads: The MOD-0 first achieved rated speed and power in December 1975. At that time, the machine performed as predicted except for larger than expected blade bending moments.

These blade loads were higher than expected for both the flatwise (out-of-plane) and edgewise (in-plane) moment loads. These high loads did not damage the blades, but continuous operation at these load levels would have resulted in early fatigue failure of the blades. The predicted loads were obtained using the MOSTAB rotor analysis code. The cyclic moments plotted are at station 40 in the blade shank

(40 in. from the rotor axis) and are plotted versus nominal wind speed. Cyclic moment is equal to one-half the difference between the maximum and minimum values of moment during one revolution of the rotor. The data are represented by mean values with the bars representing $\pm 1\ \sigma$ ($\pm$ 34% of the data about the mean). The variations about the mean are caused by such things as variations in wind direction and velocity and control changes.

As a result of the high blade loads, an intensive study was undertaken to analyze the loads data, to determine the causes of the high loads, and to recommend modifications to reduce the loads. This study showed that the flatwise bending moments were primarily caused by the impulse applied to the blade each time it passed through the wake of the tower. It was concluded that the tower was blocking the airflow much more than had been expected. The higher blockage was confirmed by site wind measurements and wind tunnel tower model tests. To reduce the tower blockage and increase the airflow through the tower, it was decided to remove the stairways from the tower. This modification reduced the tower blockage from 0.64 to 0.35 (a blockage of 1.0 meaning that the airflow through the tower is zero).

The study of the edgewise blade loads, particularly their harmonic content, led to the conclusion that these high loads were caused by excessive nacelle yawing motion. To reduce these loads, it was recommended that the single yaw drive be replaced by a dual yaw drive. This dual yaw drive was expected to help by (1) changing the torsional frequency of the system and moving it away from the 2P (two cycles per rotor revolution) resonance; and (2) eliminating the free-play present in the single yaw drive. It was also decided to add three brakes to the yaw system to provide additional stiffness. The result of these modifications on the blade moments was that both flatwise and edgewise bending moments were reduced below the values predicted by MOSTAB.

It should be noted that tower shadow and yaw stiffness are examples of why the MOD-0 project was initiated. The early higher than expected loads led to extensive re-evaluation of the analytical tools and subsequent redesign of the WECS. This information has been important in the design of the follow-on large-scale WECS.

Power System Dynamics: Operation of the WECS synchronized to the utility network showed two types of power variations. The mean power was found to vary slowly with changes in wind speed and a higher frequency variation about the mean was also observed. A study of the data was initiated to examine the source of the variations and to determine means of reducing them.

Strip chart data traces of the blade pitch angle, alternator power and wind speed, indicate variations of the mean power is evident and occurs as the wind speed increases from 20 to 30 mph. Also, the 2P power variation about the mean has been recorded. The following describes the control and design changes that have been made to the MOD-0 WECS to reduce these power variations to acceptable levels. Again, it is noted that operation of the MOD-0 WECS has led to an early discovery of potential problems and their solutions. This information is also directly applicable to the follow-on large-scale WECS.

Elimination of the varying mean power revealed a problem in the blade pitch control. This control is an integral control that adjusts blade pitch angle to control the alternator power to the selected power set point. Increasing the gain of this control results in instability while reducing the gain results in a very slow responding control. Analysis and tests showed that the addition of proportional control to the system was unsatisfactory, primarily because of the power drive train resonance occurring at less than 1 Hz.

To correct this problem, it was necessary to go to a feed-forward control using wind speed as the feed forward signal. The control essentially performs by sensing wind speed changes on the WECS nacelle and using this signal to make blade pitch changes. The closed loop control is still active and controls to the required power setting. The feed forward allows more rapid control to follow wind gust changes. A comparison of the power output with and without feed forward control appears to have reduced the variation of the mean power; this control is still under study.

The downwind rotor causes a pulse to be put on the power drive train everytime a blade passes through the tower wake. The shaft torque is a function of wind speed and blade position. The torque pulse is transmitted through the drive train twice per rotor revolution. Because of the power train dynamics, the magnitude of the torque pulse is amplified by about a factor of 2 at the alternator output.

The power train torque variations can be reduced by reducing the tower shadow or increasing the damping in the power train. Since the tower shadow was already reduced by removing the stairs, the power train damping was increased by adding a fluid coupling in the high speed shaft. Data shows that for wind speeds of 30 mph, the power oscillations were reduced from ±30 kW to ±12 kW by the fluid coupling. As a result, the power oscillations are not considered to be a problem for utility operation.

Aerodynamic Performance: The MOD-0 was designed to deliver 100 kW of the alternator output for hub height wind speeds of 18 mph.

Tests have been conducted at Plumbrook to determine the MOD-0 performance. The performance of the MOD-0 was determined by using data from the WECS, the nacelle anemometer and the meteorological (met) anemometer. The alternator power, nacelle wind speed and met tower wind speed data were recorded simultaneously with the WECS synchronized to the utility network. Approximately two hours of data were taken with wind speeds in the range of 10 mph to 35 mph.

To obtain the WECS power versus met tower wind speed, three steps were taken. First, the alternator power was plotted versus nacelle wind speed. The scatter in these data is probably due to wind speed and wind direction changes. The data were then averaged over regions of 1 mph to obtain the plot which shows a leveling off of the averaged power at less than 100 kW in high winds which was due to a power control set point of 90 kW. It also shows the lack of convergence at the extremes of the wind speed range, which results from having too few data points. Next, the wind shear was calculated from four simultaneous readings of the anemometers on the meteorological tower, and the data interpolated to obtain the wind speed for the 100-foot level. The nacelle wind speed was then correlated with the 100-foot wind speed at the meteorological tower. These data were averaged over two-minute intervals to obtain average nacelle wind speed for an average met tower wind speed. Analysis was performed that showed two minutes was sufficient for obtaining correlation between two sites located 650 feet apart. The third step was to cross-plot the results of the first two steps to obtain a plot of WECS power versus met tower wind speed. This result was superimposed on the plot of predicted performance. The results show that the WECS performs better than predicted at these wind speeds. A possible explanation for this result is the fact that the MOD-0 was sized assuming a rough airfoil and the actual airfoil is more like a smooth airfoil.

Summary: The DOE/NASA 100 kW MOD-0 was first operated at rated speed and power in December 1975 in winds of 25–35 mph. These tests showed higher than expected loads, particularly blade loads. As a result, WECS was modified to reduce these loads. These modifications included removing the stairways in the tower to reduce the tower blockage of wind and stiffening the yaw drive system. These modifications resulted in reducing the mean loads to levels below those predicted for the wind turbine by MOSTAB analysis.

General operation of a large-scale WECS has been fully demonstrated by the MOD-0. These operations include startup, synchronization to a utility network, blade pitch control for control of power level, and shutdown. Limited operations have been demonstrated in a fully automatic mode and by use of a remote control and monitoring panel, 50 miles from the site, similar to what a utility dispatcher would use.

Early operation of the MOD-0 connected to the utility network revealed oscillations in the alternator power output of ±30 kW. The oscillations were due to the reduced torque impulse caused by each blade passing through the tower wake. The oscillations were reduced to less than ±12 kW by adding a commercially available fluid coupling to the high speed shaft.

Aerodynamic performance of the MOD-0 was determined to be better than predicted.

In summary, the MOD-0 WECS has been meeting its primary objective of providing the wind energy program with early operations and performance data for large wind turbines. The engineering data on dynamic loads, utility operation, aerodynamic performance, etc., has contributed to increasing the probability of success for the follow-on large-scale WECS.

Figure 8-41. DOE/NASA-Lewis MOD-0A LWECS. (*Courtesy of DOE/NASA LRC.*)

DOE/NASA-Lewis DOE MOD-0A

Manufacturer: NASA-Lewis Research Center (Schenectady-Westinghouse/Lockheed)
Address: Cleveland, OH 44135
Contact: Richard Thomas
Telephone: (216) 433-4000, extension 6134
Model number: MOD-0A (see Figure 8-41)
Rated power: 200 kW @ 19 mph (see Figure 8-42)
Power output at 12 mph: 0
Power output at 14 mph: 26 kW
Rotor blade diameter: 125 ft uncovered
Number of blades: 2
Machine type: Large horizontal-axis, downwind, rigid hub
Cost: $1,738,000 total installed
- Rotor (blades, hub, PCM) $814,000
- Mechanical equipment $337,000
- Electrical equipment $70,000
- Tower assembly $150,000
- Control equipment $91,000
- Shipping $18,000
- Installation, start-up $250,000

Operating speed
- Rated speed (mph): 21.7 hub height
- Cut-in speed (mph): 13 @ hub height
- Cut-out speed (mph): 40 @ hub height

rpm at rated output: 40
Blade materials: Aluminum, NASA 23000 airfoil
Rotor weight: 12,200 lb
Cone angle: 7°
System weight on tower: 45,000 lb
Overspeed control: Computer-controlled hydraulics
Generator/alternator type: Sync ac, 250 KVA rating, 480 V, 1800 rpm
Application: Test program to determine WECS operating, performance and dynamic characteristics and the economics of large utility-based wind systems
Testing history: Currently under test utilization at Clayton, New Mexico, Block Island, Rhode Island, and the Island of Culebra, Puerto Rico
Reliability prediction (MTBF): Unknown; design life 30 years
Warranty: None

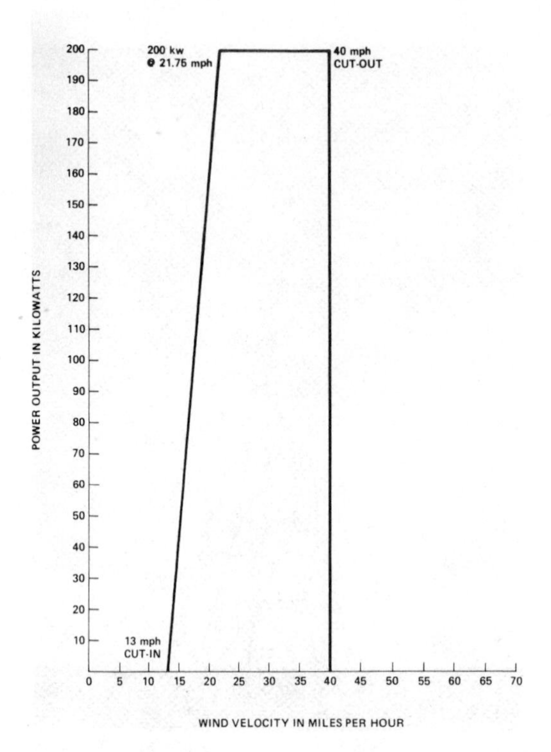

Figure 8-42. DOE/NASA-Lewis MOD-0A power output curve.

Maintenance requirements: Unspecified
Country of origin: U.S.
Number produced: 3
Availability: DOE development and demonstration project
Description: The MOD-0A WECS consists of all the elements required to generate electrical power from the wind. The machine is a horizontal axis wind turbine and consists of a rotor (blades, hub and pitch change mechanism), the drive train (including the shaft, speed increaser and generator), the nacelle (consisting of a shroud, bedplate and yaw mechanism), the tower, and the electrical and controls system. A picture of the machine without blades is shown in Figure 8-41.

The 200-kW MOD-0A WECS has a two-bladed, 125 ft diameter rotor driving a 200-kW synchronous alternator through a speed increaser gear box. The rotor is located downwind of the tower and rotates at a constant speed of 1800 rpm and delivers 60 Hz three-phase power. The wind turbine starts rotating at a wind speed of 5 mph and begins to deliver power at 13 mph. Its design and operating speed is 40 rpm.

Development Program: The MOD-0A development program is proceeding in a series of steps starting with the definition of system requirements to the final operation of the wind turbine at the utility location. These steps are described below:

The system requirements are used as the basis for the machine design. With the definition of wind regime and power output, the mechanical design of the machine can proceed. The first and critical step is to define the steady and cyclic loads on the machine. Loads definition is critical throughout the machine and the loads on the rotor are especially critical. For the wind turbines currently in the development process (including MOD-0A), extremely careful attention is paid to the loads definition in the rotor area because of the critical nature of the blade design. All computer codes used for loads definition should be validated by the use of actual measured loads on the MOD-0 machine. Results can then be accurately extrapolated to similar machines.

The machines are designed so that they can be easily synchronized to an electrical power network with a control system that automatically starts the machine at the proper wind conditions, brings the machine to the design speed (40 rpm), synchronizes the wind turbine with the utility grid, controls the power level under varying wind conditions and shuts the machine down.

The MOD-0A wind turbine has no formal qualification program for design verification. The mechanical design of the machine is validated by design analysis with the analysis tools verified by MOD-0 test results. The MOD-0A electrical and control systems are duplicates of the MOD-0 systems and are validated by MOD-0 testing prior to utility operation.

The assembly and acceptance testing of the wind turbine is accomplished in three phases. The first phase consists of the assembly and testing of the drive train. Initially, the 200-kW alternator, high speed shaft, gear box, and low speed shaft are assembled on the machine bedplate and connected to a dynamometer and speed reducer used to power the machine. The mechanical alignments and system balance are then checked. The power train is then run for approximately 40 hours under load and the mechanical and electrical performance of the system is evaluated. Following the successful completion of the drive train evaluation, Phase II assembly and testing proceeds. In Phase II, the yaw drive assembly, rotor hub, and pitch change hydraulic system are added to the power train components. This is the complete wind turbine except for the blades. In this serves of tests, the machine instrumentation is calibrated, the mechanical balance of the system is verified, and the primary controls (pitch and yaw) are evaluated. Upon the successful completion of these tests, the machine is ready for assembly and checkout at a utility site.

The third and final phase of the testing occurs at the utility site just prior to on-line operation with the public utility. In this testing, a thorough mechanical, electrical and controls evaluation of the wind turbine is made after it is mounted on the tower. Extensive instrumentation is requird for these tests. In order to accommodate these instrumentation requirements, a portable instrumentation and data system was designed and built for the initial evaluation of these wind turbines. The data system has the capability for on-line analog recording and digital data processing. Since the data system is portable, it will be used to conduct startup operation of all MOD-0A machines.

Program Status: The MOD-0A unit on Block Island, a resort area off the coast of Rhode Island and Connecticut, will generate a peak output of approximately 200 kW. This is enough power to serve 35 to 50 of the island's 200–300 year-round residences. The units installed are running well and have generated nearly 500,000 kWh of energy. This DOE program has two goals: to test the design of the machine, and to answer questions about the economical and technical aspects of their operation by a utility.

DOE/NASA-Lewis — DOE MOD-1

Manufacturer: NASA-Lewis Research Center (Subcontractor—GE)
Address: Cleveland, OH 44135
Contact: Richard Thomas
Telephone: (216) 433-4000, extension 6134
Model number: MOD-1 (see Figure 8-43)
Rated power: 2 mW @ 18 mph (see Figure 8-44)
Power output at 12 mph: 0
Power output at 14 mph: 0
Rotor blade diameter: 201.7 ft
Number of blades: 2
Machine type: Large horizontal-axis, downwind, rigid hub
Cost: $5,770,000 total installed
  Rotor (blades, hub, PCM) $2,120,000
  Nacelle structure $780,000
  Power-generating equipment $360,000
  Controls $230,000
  Yaw drive system $330,000
  Tower assembly $420,000
  Assembly, testing $750,000
  Site preparation, installation and checkout $750,000
Operating speed
  Rated speed (mph): 18
  Cut-in speed (mph): 15.7
  Cut-out speed (mph): 42.5
  Maximum speed (mph): 33
rpm at rated output: 35
Blade materials: Steel/foam
Rotor weight: 108,000 lb
Cone angle: 9°
System weight on tower: 335,000 lb
Overspeed control: Microprocessor-controlled hydraulics
Generator/alternator type: Sync ac, 2225 KVA, 0.9 power factor, 4160 V, 1800 rpm
Application: Small community power source: the power generated by this experimental SWECS will be fed into local utilty power grids
Testing history: Howards Knob, Boone, North Carolina undergoing field test
Reliability prediction (MTBF): Unknown
Warranty: None
Maintenance requirements: Unspecified
Country of origin: U.S.
Number produced: 1

Figure 8-43. DOE/NASA-Lewis MOD-1 LWECS. (*Courtesy of DOE/NASA LRC.*)

Availability: Boone, North Carolina

Description: As shown in Figure 8-43, the MOD-1 WECS incorporates a two-bladed, downwind rotor driving an alternating current generator through a speed increaser atop a steel, truss-type tower. The major characteristics of the MOD-1 WECS are briefly described as follows:

Rotor: Two variable-pitch, steel blades are attached to the hub barrel via three-row roller bearings which permit about 105° pitch excursion from full feather to maximum power. Blade pitch is controlled by hydraulic actuators which provide a maximum pitch rate of 14°/sec.

The hub tailshaft provides the connection to the low-speed shaft and to the dual-tapered-roller main bearing, which supports the rotor and one end of the low-speed shaft.

Drive Train: Comprising the drive train are the low-speed shaft and couplings, the gearbox, and the high-speed shaft/slip-clutch which drives the generator. Cooling of the gearbox and the main rotor bearing is through a pressured oil lubrication system. Waste heat is rejected by an oil cooler suspended below the nacelle. The slip-clutch precludes excessive torques from developing in the entire drive train due to extreme wind gusts and/or faulty synchronization.

Power Generation/Control: A General Electric synchronous ac generator is driven at 1800 rpm through the high-speed shaft. A shaft-mounted, brushless exciter controlled by a solid-state regulator and power stabilizer inputs provides voltage control. The generator output at 4160 volts is brought by cables and slip-rings to the control enclosure. After routing through a circuit breaker, the output is connected to the utility via a

voltage-matching step-up transformer.
Nacelle Structure: The core of the nacelle structure is the welded steel bedplate. All nacelle equipment and the rotor are supported by the bedplate, which provides the load path from the rotor to the yaw structure. Other equipment supported by the bedplate includes the pitch control and yaw drive hydraulic packages, walkways, oil coolers, heaters, hydraulic plumbing, electronics boxes, cabling and the fairing. The entire fairing is removable for crane access into the nacelle area. Redundant insrumentation booms, with wind speed, temperature and direction sensors are mounted on the upwind end of the fairing.
Yaw Drive: Unlimited yaw rotation is provided by the yaw drive system, comprising the upper and lower yaw structures, the two-motor hydraulic drive, the hydraulic yaw brake, and the large cross-roller yaw bearing. Power and signal data are transferred to the tower-mounted cabling by slip-rings. The yaw drive is capable of yawing the rotor/nacelle at 0.25°/sec. To provide adequate yaw drive stiffness, the yaw brake is fully activated when not in a yaw maneuver and partially activated during the maneuver to avoid backlash in the yaw drive gear train.
Data Acquisition: Two independent data acquisition systems are provided: an Engineering Data Acquisition System (EDAS) and the Control/Operational Data Acquisition System (CODAS). EDAS includes a collection of rotor, nacelle, tower and surface-mounted sensors with data being routed through three NASA-provided multiplexers to a NASA-provided Portable Instrumentation Van (PIV) located near the base of the tower. CODAS includes sensors, electronics and software as necessary for eventual unattended operation of the WTG.
Tower: The truss tower is made up of seven vertical bays, including the base and top (pintle) sections. Tubular steel columns are used at the four corners to carry the main loads. Back-to-back channels serve as cross-members where loads permit. However, in most bays, tube-section cross-members are still required because of high loads and to reduce "tower shadow." Access to the yaw drive and nacelle area is provided via a cable-guided, gondola-type elevator.
Surface Installations: The major surface elements of the WECS are the ground enclosure, the backup battery system, the step-up transformer, cabling, security fencing, area lighting, and the tower foundation. A prefabricated, air conditioned/heated, steel structure, about 10 ft. x 12 ft. x 28 ft. long, which contains switchgear and CODAS/EDAS equipment such as recorders, computer, etc., provides protection for the sensitive ground equipment.

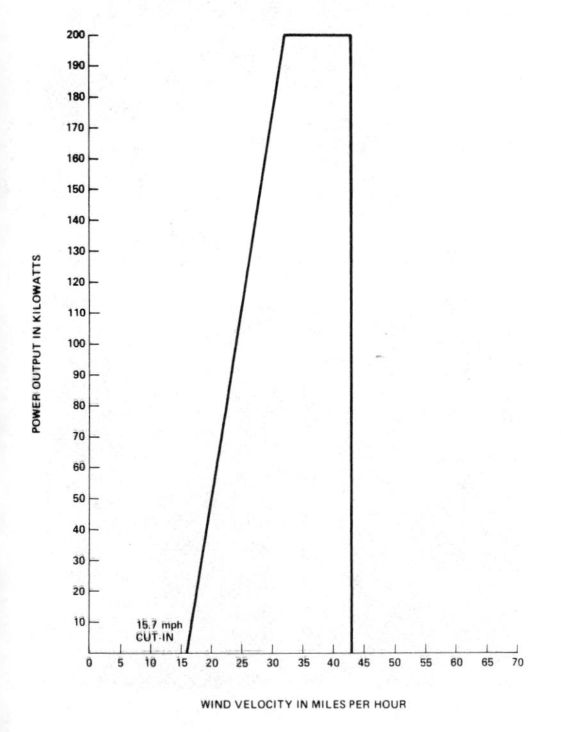

Figure 8-44. DOE NASA/Lewis MOD-1 power output curve.

WECS Weight: The final system weight (rotor, nacelle and tower) is expected to be about 650,000 lbs.

WECS Installation: To reduce shipping and erection costs, WECS was designed to permit transport and assembly with a 50-ton limit on any one subassembly. This dictates an "on-tower" final assembly, but eliminates the requirement for expensive, difficult-to-obtain cranes with adequate single-lift capability.

Performance: The WECS will be capable of delivering 2000 kW(e) to the utility grid in a windspeed of about 25 mph. Power output will vary from cut-in wind velocity (11 mph) up to rated power (2000 kW). As wind speed increases above 25 mph, power output is maintained at 2000 kW by varying the blade pitch. At steady wind speeds above the cut-out speed of 35 mph, WECS is shut-down via a rotor feathering maneuver. Note that all the wind speeds are referenced to 30 ft. elevation above the surface.

Economics: Production cost estimates for the MOD-1 WECS will be generated late this year for production rates which may be realizable in the early-to-mid-1980's. Until those data are available, recurring (second-unit) costs are used as a reasonable starting-point for the application of standard learning-curve techniques. The initial recurring unit cost, based on the present General Electric/subcontractor/supplier structure for a 2000-kW system, is expected to be $3.3M, including all burdens.

The specific installed cost (at 2000 kW) is then $1650/kW, and the cost of energy is calculated at 8.1¢/kWh using an 18% fixed charge rate, and an annual energy output of 7.4 x 10^6 kW-hrs which assumes 90% WECS availability.

DOE/NASA-Lewis — DOE MOD-2

Manufacturer: NASA-Lewis Research Center, Cleveland, OH 44135

Address: Subcontractor, Boeing Engineering and Construction, P.O. Box 3999, Seattle, WA 98124

Contact: Richard Thomas

Telephone: (216) 433-4000, extension 6134

Model number: MOD-2 (see Figure 8-45)

Rated power: 2.5 mW @ 18 mph (see Figure 8-46)

Power output at 12 mph: 0

Power output at 14 mph: 200 kW

Rotor blade diameter: 300 ft

Number of blades: 2

Machine type: Large horizontal-axis, downwind, teetered hub

Cost: $3,540,000 total installed

- Rotor (blades, hub, PCM) $1,040,000
- Drive Train $630,000
- Nacelle $190,000
- Electrical/electronics $200,000
- Tower assembly $200,000
- Factory checkout $490,000
- Site preparation $790,000

Operating speed

- Rated speed (mph): 28
- Cut-in speed (mph): 14
- Cut-out speed (mph): 60
- Maximum speed (mph): Unspecified

rpm at rated output: 17.5

Figure 8-45. DOE/NASA-Lewis MOD-2 LWECS. (*Courtesy of DOE.*)

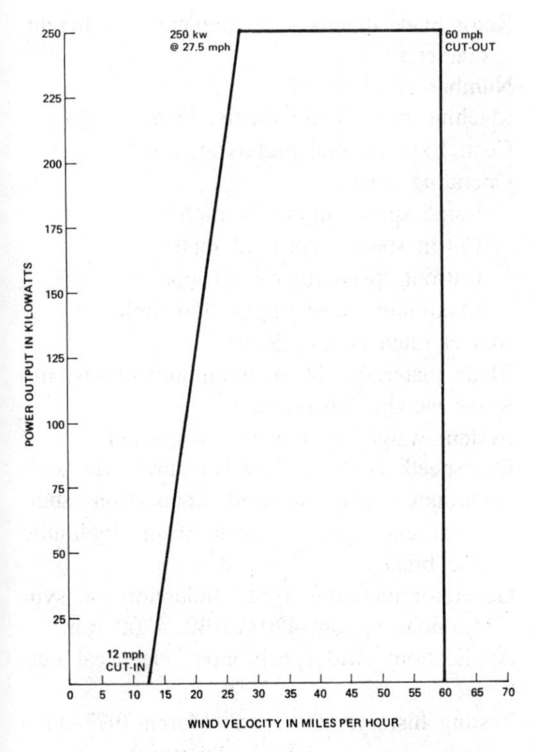

Figure 8-46. DOE/NASA-Lewis MOD-2 power output curve.

Blade materials: Steel; paper honeycomb
Rotor weight: 173,000 lb
System weight on tower: 348,000 lb
Overspeed control: Microprocessor-controlled hydraulics
Generator/alternator type: Synchronous ac, 3125 KVA, 0.8 power factor, 4160 V, 1800 rpm
Application: Utility power grid interconnection
Testing history: Unspecified
Reliability prediction (MTBF): Availability 0.92
Warranty: None
Maintenance requirements: Unspecified
Country of origin: U.S.
Number produced: 1
Availability: Demonstration unit
Description: The Boeing Engineering and Construction Company, a division of the Boeing Company, was selected to develop the MOD-2 WECS. The following briefly summarizes the major features of the MOD-2 WECS as presently envisioned by Boeing. As the project proceeds into its analysis and design tasks, many of those features will be reexamined and may undergo some changes before a final design is selected.

The general arrangement of the MOD-2 WECS is shown in Figure 8-45. WECS begins to generate power when the wind speed at the hub reaches 14 mph. Rated power of 2500 kW is reached at a wind speed of 28 mph at the hub, and the wind turbine is shut down at wind speeds above 60 mph at the hub. These design parameters were selected to convert the largest amount of wind energy to electricity when the system is operating in a wind regime with a mean speed of 14 mph at 30 feet. Preliminary cost analysis indicates that the cost of electricity for the one-hundredth wind turbine will be about 3.7¢/kWh.

There are several unique features to this design. The blades, are constructed of steel and paper honeycomb. The primary load-carrying structure of the blade is a "D" spar constructed of a steel plate weldment. The spar is carried inboard past the 20% radius to provide structure to connect to the hub. The spar is connected to the hub through two door-hinge-type bearings. These bearings are spaced about 16 feet apart to spread out the large blade bending loads. The trailing edge of the blade is constructed from thin sheet steel bonded to resin impregnated Kraft paper honeycomb. This assembly is then bonded to the "D" spar leading edge. The completed blade has an NACA 23000 airfoil shape. A linear twist of 8° is built into the blade. The rotor subassembly is connected to the gear box via a long, torsionally "soft" quill shaft.

Another unique feature of this baseline design is the tower. The tower is a 187-foot steel shell configuration with four tubular strut braces at the 50-foot level. The tower is made in two sections, the first being 150-feet tall. This locates the field splice in a low bending moment area. The tower sections are fabricated using manufacturing techniques developed for making utility cantilever power poles. The tower is designed to have a bending frequency which is substantially lower than those frequencies which would couple with the rotor.

No special materials, tools, or fabrication techniques are used in any part of the MOD-2 WECS, in keeping with the low cost goals of the project.

eneral Approach: This project is planned to be a 30-month effort for the analysis, design, fabrication, assembly and installation at a given utility site of a minimum 300-foot diameter WECS. At this time it is not certain how many MOD-2 machines will be built. The 30-month period applies to installation of one WECS.

Conceptual design and cost assessment analyses will be conducted to establish the design specifications of the WECS that produces cost-effective power. The sensitivity of energy costs will be determined as a function of rotor diameter electrical power output, as well as selected variations in wind turbine design. Following these analyses, the preliminary design will be established in sufficient detail to permit accurate cost estimates of all components. A key factor in the preliminary design will be that the MOD-2 WECS should meet its goal of producing power at a cost that is attractive to industry. Following the preliminary design, a design review will be held to evaluate and assess all results in addition to the design itself.

Following the approval of the results of the preliminary design and cost optimization tasks, Boeing may proceed with detailed design, fabrication, assembly, and installation tasks.

DOE/SANDIA 17-Meter

Manufacturer: DOE/Sandia Laboratories
Address: Advanced Energy Projects, Division 4715, Albuquerque, NM 87185
Contact: Richard H. Braasch
Telephone: (505) 264-3850
Model number: 17-Meter Experimental VAWT (see Figure 8-47)
Rated power: 60 kW (see Figure 8-48)
Power output at 12 mph: 5.5 kW*
Power output at 14 mph: 10.3 kW*
Rotor blade diameter: 17-diameter; 1.0 height/diameter
Number of blades: 2
Machine type: Vertical-axis, Darrieus design
Cost: Experimental prototype, not for sale
Operating speed
  Rated speed (mph): 31 mph*
  Cut-in speed (mph): 12 mph*
  Cut-out speed (mph): 60 mph
  Maximum speed (mph): 150 mph
rpm at rated output: 50.6*
Blade materials: 24 in. aluminum extrusion
Rotor weight: Unspecified
System weight on tower: Unspecified
Overspeed control: Speed control via synchronous generator-grid connection shutdown and park via dual 36-in. hydraulic disc brakes
Generator/alternator type: Induction or synchronous option; 480 V, 30, 1800 rpm
Application: Grid synchronous electrical output
Testing history: 600 hours; March 1977–June 1979
Reliability prediction (MTBF): Insufficient data; three unscheduled maintenance periods on operating system through June 1979
Warranty: Unspecified
Maintenance requirements: Visual inspection bi-weekly, semi-annual guy tensioning and gearbox lube
Country of origin: U.S.
Number produced: 1
Availability: Experimental prototype
*System configuration is variable. Data given is for 24-in. NACA 0015 blade without struts operating at 50.6 rpm. Wind speed measurements are referenced to 30 ft height by 1/7 power law.

DUNLITE MODEL L-HIGH WIND SPEED

Manufacturer: Dunlite Electrical Products, Division of Pye, Ltd., 21 Frome St., Adelaide 5000, Australia
Address: c/o Enertech, P.O. Box 420, Norwich, VT 05055
Contact: Mr. Ned Coffin
Telephone: (802) 649-1145
Model number: L-High Wind Speed

Figure 8-47. DOE 17-Meter vertical-axis WECS. (*Courtesy of DOE/Sandia.*)

Rated power (kW): 1.0 @ 25 mph (see Figure 8-49)
Power output at 12 mph: 150 W
Power output at 14 mph: 260 W
Rotor blade diameter: 12 ft
Number of blades: 3
Machine type: Horizontal-axis
Cost: Contact Enertech for current price
Operating speed
  Rated speed (mph): 25
  Cut-in speed (mph): 10

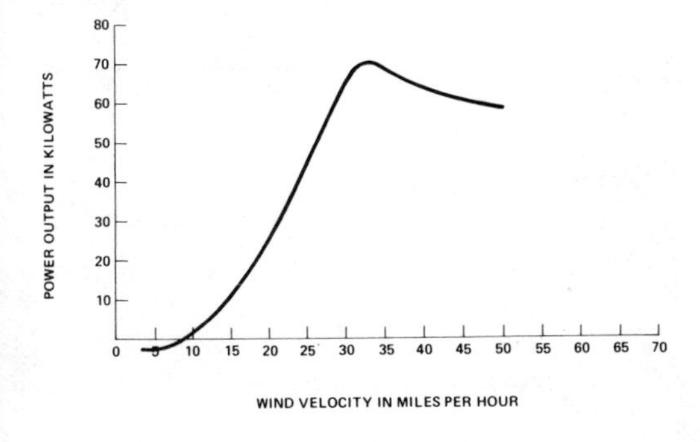

Figure 8-48. DOE/Sandia Laboratories vertical-axis power output curve.

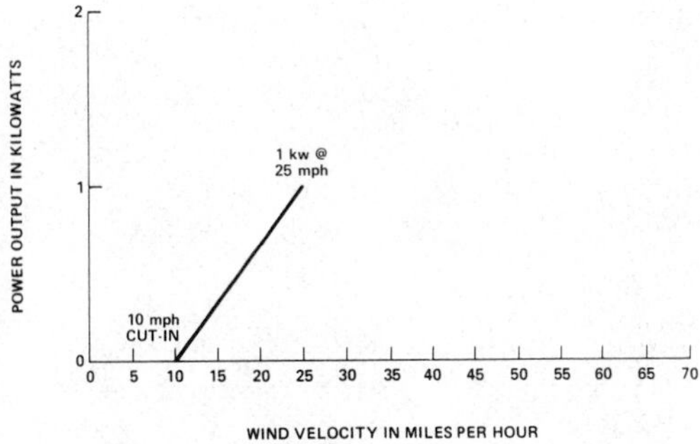

Figure 8-49. Dunlite L-High Wind Speed power output curve.

  Cut-out speed (mph): Unspecified
  Maximum speed (mph): 80
rpm at rated output: Unspecified
Blade materials: Galvanized sheet steel
Rotor weight: 130 lb
System weight on tower: 500 lb
Overspeed control: Centrifugally 80° activated blade full feathering
Generator/alternator type: 12, 24, 32, 36, 110 V alternator
Application: Electric power SWECS
Testing history: Wind tunnel, field testing, 40 years of experience in WECS
Reliability prediction (MTBF): Unspecified
Warranty: One year, parts and workmanship
Maintenance requirements: Annual, check oil and inspect system
Country of origin: Australia
Number produced: 1000's
Availability: Contact Enertech for delivery

DUNLITE MODEL M STANDARD

Manufacturer: Dunlite Electrical Products, Division of Pye, Ltd., 21 Frome St., Adelaide 5000, Australia
Address: c/o Enertech, P.O. Box 420, Norwich, VT 05055
Contact: Mr. Ned Coffin
Telephone: (802) 649-1145
Model number: M Standard 81 002550 (see Figure 8-50)
Rated power (kW): 2.0 @ 25 mph (see Figure 8-51)
Power output at 12 mph: 0.25 kW
Power output at 14 mph: 0.75 kW

Figure 8-50. Dunlite Model M Standard WECS.

Rotor blade diameter: 13 ft
Number of blades: 3
Machine type: Upwind, horizontal-axis
Cost: $6350
Operating speed
Rated speed (mph): 25
Cut-in speed (mph): 9–10
Cut-out speed (mph): 80
Maximum speed (mph): 80
rpm at rated output: 200
Blade materials: Galvanized sheet steel variable-pitch blades
Rotor weight: 130 lb
System weight on tower: 350 lb
Overspeed control: Mechanical, centrifugal, 80° activated blade feathering governor
Generator/alternator type: 12, 24, 32, 48, 110 vdc 3-phase alternator
Application: Electric power, battery charging
Testing history: Wind tunnel tests, field testing, 30 years of experience
Reliability prediction (MTBF): Unspecified
Warranty: One year, parts and workmanship
Maintenance requirements: Annual, check oil and inspect system
Country of origin: Australia
Number produced: 1000's
Availability: Contact Enertech
Comments: Efficiency of approximately 18 percent
DOE Rocky Flats Tests: "The 3-bladed horizontal-axis Dunlite machine became fully operational in February 1977. In 16 months of testing, the Dunlite experienced two failures. The machine produced a significant amount of useful data. In November 1977, the machine was nearly destroyed in a windstorm. After repairs were made to the blades and tail, a loose roller in the hub fork-end assembly required additional repair in April 1978. No manufacturer power curve suitable for comparison is available due to the fact that the RF machine does not have the magnetic latch in the hub (which delays blade pitch-action) for which manufacturer data is available. However, when charging batteries, the Dunlite produced approximately 1900 watts at 11 m/s compared with a manufacturer data point of 2000 watts at 11 m/s."
See DOE Report RFP-2920/3533/78/6-2 of September 28, 1978 and new reports for more information.

DYNERGY 5-METER

Manufacturer: Dynergy Corporation
Address: P.O. Box 428, 1269 Union Ave., Laconia, NH 03246
Contact: Mr. Robert B. Allen

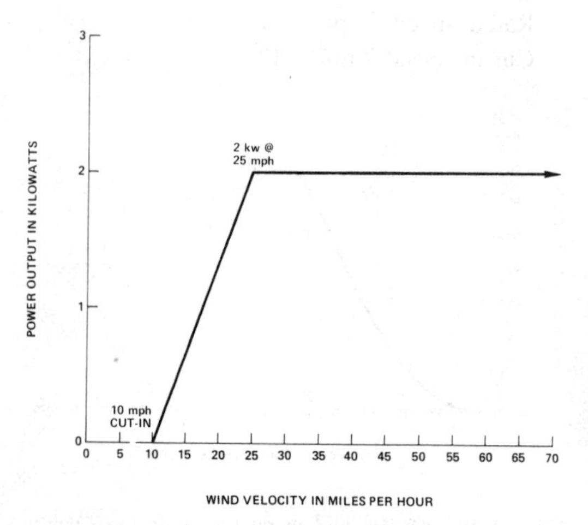

Figure 8-51. Dunlite Model M-Standard power output curve.

Telephone: (603) 524-8313
Model number: 5-Meter (see Figure 8-52)
Rated power (kW): 3.3 @ 24 mph (see Figure 8-53)
Power output at 12 mph: 500 W
Power output at 14 mph: 950 W
Rotor blade diameter: 15 ft
Number of blades: 3
Machine type: Vertical-axis
Cost: $6975
Operating speed
  Rated speed (mph): 24
  Cut-in speed (mph): 10
  Cut-out speed (mph): 80
  Maximum speed (mph): 120
rpm at rated output: 243
Blade materials: Extruded aluminum 6061-T6, with cast aluminum fittings
Rotor weight: 442 lb
System weight: 850 lb
Overspeed control: Disc brake, aerodynamic stall load-controlled
Generator/alternator type: Unknown
Application: Electricity SWECS
Testing history: Field tests
Reliability prediction (MTBF): Unspecified

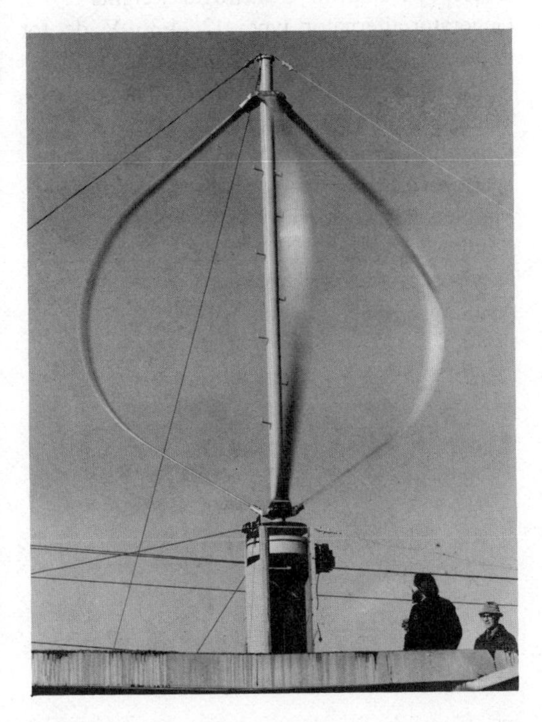

Figure 8-52. Dynergy 5-Meter (3-blade) Darrieus WECS.

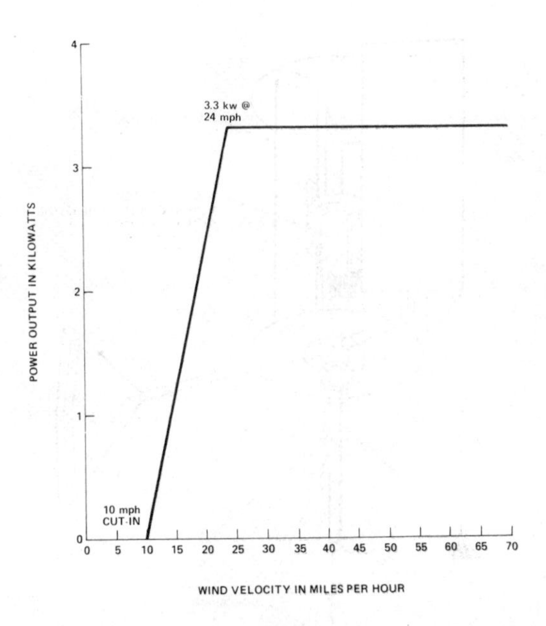

Figure 8-53. Dynergy 5-Meter power output curve.

Warranty: Unspecified
Maintenance requirements: Semi-annual inspection of blades, annual greasing of fittings
Country of origin: U.S.
Number produced: 1
Availability: Contact factory

ELEKTRO G.m.b.h. W50

Manufacturer: Elektro G.m.b.h., St. Gallerstrasse 27, Winterthur, Switzerland
Address: c/o Budgen & Associates and Real Gas & Electric Co.
Model Number: W50 (see Figure 8-54)
Rated power: 50 watts (see Figure 8-55)
Power output at 12 mph: 10 W
Power output at 14 mph: 12.5 W
Rotor blade diameter: 1.47 ft
Number of blades: 1 (Savonius rotor with 6 blades)
Machine type: Savonius vertical-axis wind turbine with ac generator
Cost: $1941 for WECS
Operating speed
  Rated speed (mph): 39
  Cut-in speed (mph): 7
  Cut-out speed (mph): Unspecified
  Maximum speed (mph): 100
rpm at rated output: 120–500

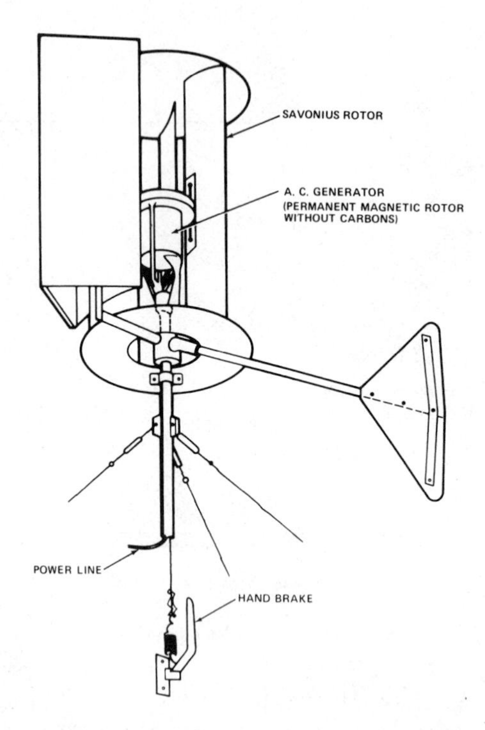

Figure 8-54. Elektro Savonius Wind Turbine Models W50 and W250.

Blade materials: Wood
Rotor weight: Unknown
System weight on tower: 77 lb
Overspeed control: Centrifugal weights
Generator/alternator type: 6/12/24 V dc for battery charging
Application: Battery charging, lighting
Testing history: 37 years of manufacturing and utilization experience
Reliability prediction (MTBF): Unspecified
Warranty: Unknown
Maintenance requirements: Unknown
Country of origin: Switzerland
Number produced: Unknown
Availability: Six months ARO

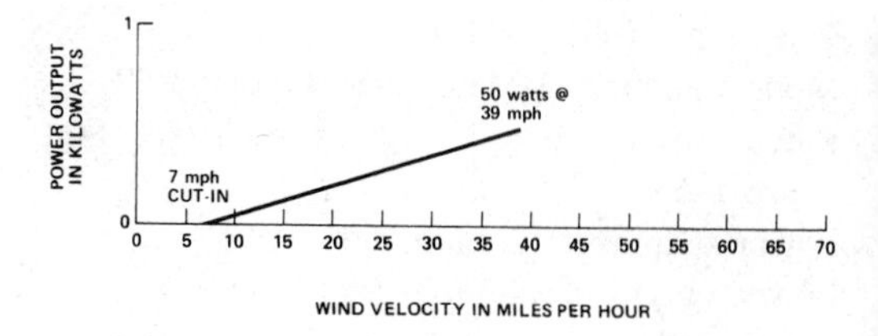

Figure 8-55. Elektro Model W50 power output curve.

ELEKTRO G.m.b.h. W250

Manufacturer: Elektro G.m.b.h., St. Gallerstrasse 27, Winterthur, Swtizerland
Address: c/o Budgen & Associates and Real Gas & Electric Co.
Model Number: W250 (see Figure 8-54)
Rated power: 150 W @ 23 mph (see Figure 8-56)
Power output at 12 mph: 45 W
Power output at 14 mph: 65 W
Rotor blade diameter: 2.16 ft
Number of blades: 1 (Savonius rotor with 6 blades)
Machine type: Savonius vertical-axis wind turbine with ac generator
Cost: $2364 turbine
Operating speed
- Rated speed (mph): 23
- Cut-in speed (mph): 7
- Cut-out speed (mph): Unspecified
- Maximum speed (mph): 100

rpm at rated output: 80/400
Blade materials: Wood
Rotor weight: Unknown
System weight on tower: 154 lb
Overspeed control: Centrifugal weights
Generator/alternator type: 12/24/36 V dc for battery charging
Application: Battery charging, lighting
Testing history: 37 years of experience
Reliability prediction (MTBF): Unspecified
Warranty: Unknown
Maintenance requirements: Unknown
Country of origin: Switzerland
Number produced: Unknown
Availability: Six months ARO

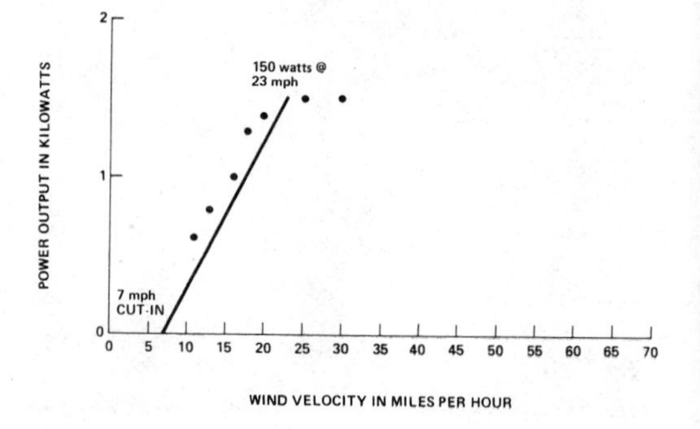

Figure 8-56. Elektro Model W250 power output curve.

ELEKTRO G.m.b.h. WV05

Manufacturer: Elektro G.m.b.h., St. Gallerstrasse 27, Winterthur, Switzerland
Address: c/o Budgen & Associates and Real Gas & Electric Co.
Model number: WV05
Rated power: 500 W @ 20 mph (see Figure 8-57)
Power output at 12 mph: 150 W
Power output at 14 mph: 300 W
Rotor blade diameter: 8.2 ft
Number of blades: 2
Machine type: Horizontal-axis
Cost: $3871 turbine only
Operating speed
  Rated speed (mph): 20
  Cut-in speed (mph): 7
  Cut-out speed (mph): 50
  Maximum speed (mph): 80
rpm at rated output: 250/700
Blade materials: Wood
Rotor weight: Unknown
System weight on tower: 143 lb
Overspeed control: Full feathering
Generator/alternator type: 12/24/36/48 V dc for battery charging
Application: Battery charging, lighting
Testing history: 37 years of experience
Reliability prediction (MTBF): Unspecified
Warranty: Unknown
Maintenance requirements: Unknown
Country of origin: Switzerland
Number produced: Unknown
Availability: Six months ARO

ELEKTRO G.m.b.h. WV15G

Manufacturer: Elektro G.m.b.h., St. Gallerstrasse 27, Winterthur, Switzerland
Address: c/o Budgen & Associates and Real Gas & Electric Co.

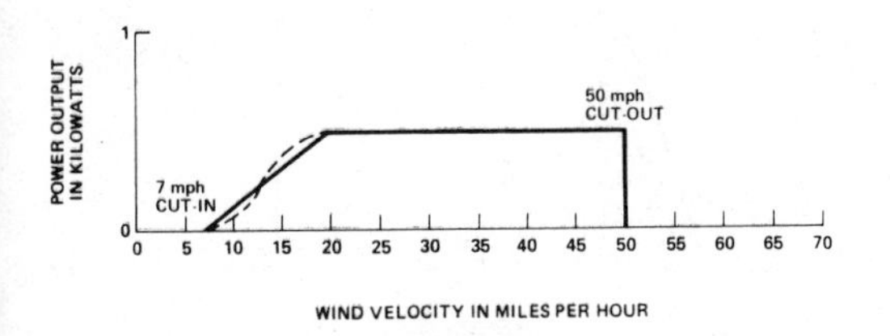

Figure 8-57. Elektro Model WV05 power output curve.

Model number: WV15G
Rated power: 1.2 @ 2.3 mph (see Figure 8-58)
Power output at 12 mph: 300 W (with field); 265 W (with magnet)
Power output at 14 mph: 500 W (with field); 463 W (with magnet)
Rotor blade diameter: 9.84 ft
Number of blades: 2
Machine type: Horizontal-axis
Cost: $4402 turbine only
Operating speed
  Rated speed (mph): 23
  Cut-in speed (mph): 7
  Cut-out speed (mph): 50
  Maximum speed (mph): 80
rpm at rated output: 220/700
Blade materials: Wood
Rotor weight: Unknown
System weight on tower: 297 lb
Overspeed control: Full feathering
Generator/alternator type: 12/24/36/48 V dc for battery charging
Application: Battery charging, lighting
Testing history: 37 years of experience
Reliability prediction (MTBF): Unspecified
Warranty: Unknown
Maintenance requirements: Unknown
Country of origin: Switzerland
Number produced: Unknown
Availability: Six months ARO

ELEKTRO G.m.b.h. WV25D

Manufacturer: Elektro G.m.b.h., St. Gallerstrasse 27, Winterthur, Switzerland
Address: c/o Budgen & Associates and Real Gas & Electric Co.

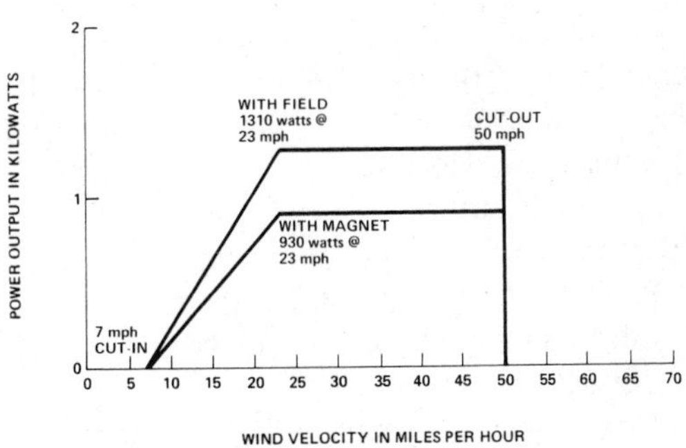

Figure 8-58. Elektro Model WV15G power output curve.

Model number: WV25D
Rated power (kW): 2.0 (see Figure 8-59)
Power output at 12 mph: Unknown
Power output at 14 mph: Unknown
Rotor blade diameter: 11.81 ft
Number of blades: 2
Machine type: Horizontal-axis
Cost: $5681 turbine only
Operating speed
 Rated speed (mph): Unknown
 Cut-in speed (mph): Unknown
 Cut-out speed (mph): Unknown
 Maximum speed (mph): 80
rpm at rated output: 40/90
Blade materials: Wood
Rotor weight: Unknown
System weight on tower: 396 lb
Overspeed control: Centrifugal weights
Generator/alternator type: 110/190 V, 3-phase VAC
Application: 3-phase ac for heating
Testing history: 37 years of experience
Reliability prediction (MTBF): Unspecified
Warranty: Unknown
Maintenance requirements: Unknown
Country of origin: Switzerland
Number produced: Unknown
Availability: Six months ARO

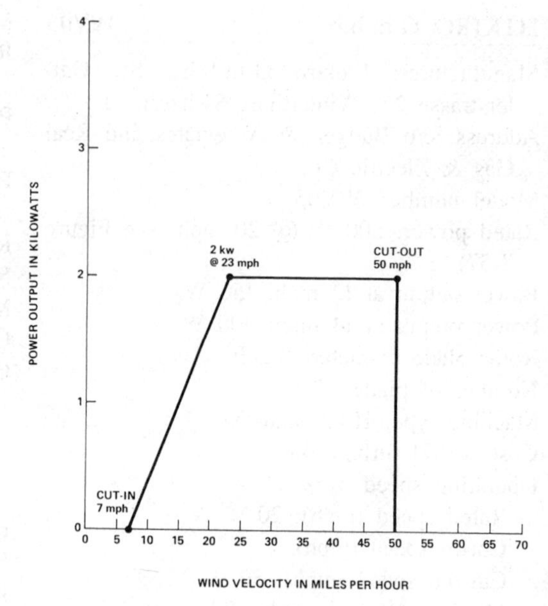

Figure 8-59. Elektro Model WV25D power output curve.

ELEKTRO G.m.b.h. WV25G

Manufacturer: Elektro G.m.b.h., St. Gallerstrasse 27, Winterthur, Switzerland
Address: c/o Budgen & Associates and Real Gas & Electric Co.
Model number: WV25G
Rated power (kW): 2.2 @ 22 mph (see Figure 8-60)
Power output at 12 mph: 380 W
Power output at 14 mph: 575 W
Rotor blade diameter: 11.81 ft
Number of blades: 2
Machine type: Horizontal-axis
Cost: $5020 turbine only
Operating speed
 Rated speed (mph): 22
 Cut-in speed (mph): 7
 Cut-out speed (mph): 50
 Maximum speed (mph): 80
rpm at rated output: 200/600
Blade materials: Wood
Rotor weight: Unknown
System weight on tower: 396 lb
Overspeed control: Full feathering
Generator/alternator type: 24/36/48/110 V dc for battery charging
Application: Battery charging, lighting
Testing history: 37 years of experience
Reliability prediction (MTBF): Unspecified
Warranty: Unknown
Maintenance requirements: Unknown
Country of origin: Switzerland
Number produced: Unknown
Availability: Six months ARO

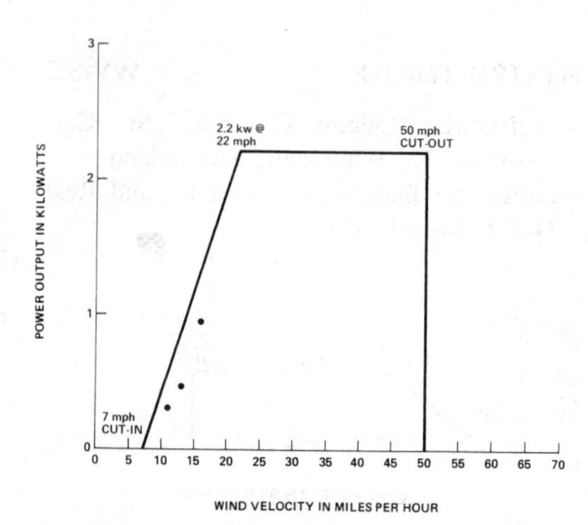

Figure 8-60. Elektro Model WV25G power output curve.

ELEKTRO G.m.b.h. WV25/3G

Manufacturer: Elektro G.m.b.h., St. Gallerstrasse 27, Winterthur, Switzerland
Address: c/o Budgen & Associates and Real Gas & Electric Co.
Model number: WV25/3G
Rated power (kW): 2.5 @ 23 mph (see Figure 8-61)
Power output at 12 mph: 750 W
Power output at 14 mph: 1.1 kW
Rotor blade diameter: 12.5 ft
Number of blades: 3
Machine type: Horizontal-axis
Cost: Contact distributor
Operating speed
Rated speed (mph): 23
Cut-in speed (mph): 7
Cut-out speed (mph): 50
Maximum speed (mph): 80
rpm at rated output: Unknown
Blade materials: Wood
Rotor weight: Unknown
System weight on tower: Unknown
Overspeed control: Full feathering
Generator/alternator type: Unknown
Application: SWECS
Testing history: 37 years of experience
Reliability prediction (MTBF): Unspecified
Warranty: Unknown
Maintenance requirements: Unknown
Country of origin: Switzerland
Number produced: Unknown
Availability: Six months ARO

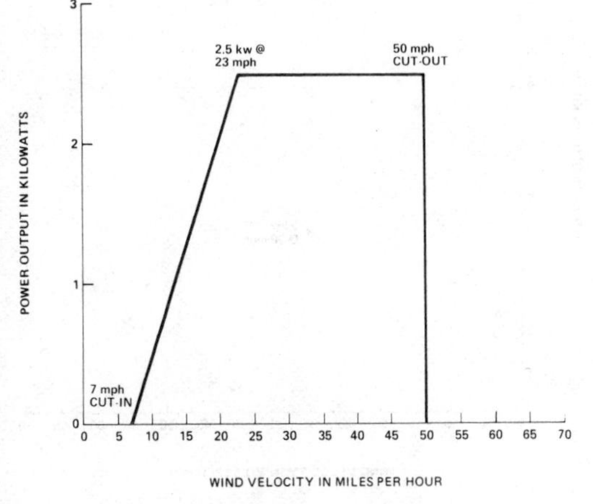

Figure 8-61. Elektro Model WV25/3G power output curve.

ELEKTRO G.m.b.h. WV35D

Manufacturer: Elektro G.m.b.h., St. Gallerstrasse 27, Winterthur, Switzerland
Address: c/o Budgen & Associates and Real Gas & Electric Co.
Model number: WV35D
Rated power (kW): 3.5 (see Figure 8-62)
Power output at 12 mph: Unknown
Power output at 14 mph: Unknown
Rotor blade diameter: 14.43 ft
Number of blades: 3
Machine type: Horizontal-axis
Cost: $6738 turbine only
Operating speed
Rated speed (mph): Unknown
Cut-in speed (mph): Unknown
Cut-out speed (mph): Unknown
Maximum speed (mph): 80
rpm at rated output: 35/70
Blade materials: Wood
Rotor weight: Unknown
System weight on tower: 539 lb
Overspeed control: Unknown
Generator/alternator type: 125/220 V, 3-phase VAC
Application: 30 V ac for heating
Testing history: 37 years of experience
Reliability prediction (MTBF): Unspecified
Warranty: Unknown

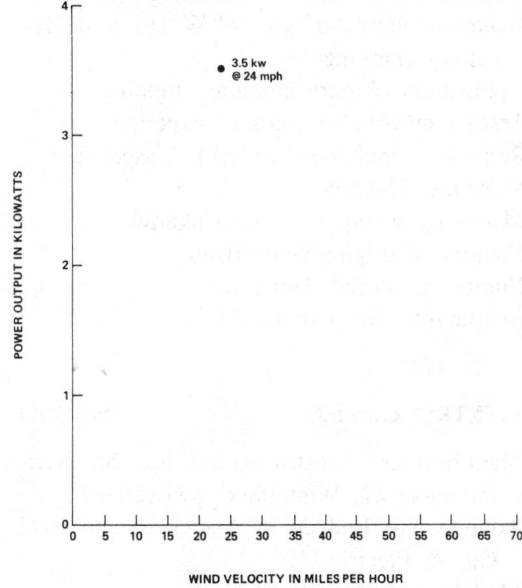

Figure 8-62. Elektro Model WV35D power output curve.

Maintenance requirements: Unknown
Country of origin: Switzerland
Number produced: Unknown
Availability: Six months ARO

ELEKTRO G.m.b.h. WV35G

Manufacturer: Elektro G.m.b.h., St. Gallerstrasse 27, Winterthur, Switzerland
Address: c/o Budgen & Associates and Real Gas & Electric Co.
Model number: WV35G
Rated power (kW): 4 @ 24 mph (see Figure 8-63)
Power output at 12 mph: 585 W
Power output at 14 mph: 900 W
Rotor blade diameter: 14.43 ft
Number of blades: 3
Machine type: Horizontal-axis
Cost: $6086, turbine; $16,299, complete in U.S.
Operating speed
  Rated speed (mph): 24
  Cut-in speed (mph): 7
  Cut-out speed (mph): 50
  Maximum speed (mph): 80
rpm at rated output: 160/450
Blade materials: Wood
Rotor weight: Unknown
System weight on tower: 517 lb
Overspeed control: Full feathering
Generator/alternator type: 48/60/110 V dc for battery charging
Application: Battery charging, lighting
Testing history: 37 years of experience
Reliability prediction (MTBF): Unspecified
Warranty: Unknown
Maintenance requirements: Unknown
Country of origin: Switzerland
Number produced: Unknown
Availability: Six months ARO

ELEKTRO G.m.b.h. WV50D

Manufacturer: Elektro G.m.b.h., St. Gallerstrasse 27, Winterthur, Switzerland
Address: c/o Budgen & Associates and Real Gas & Electric Co.
Model number: WV50D

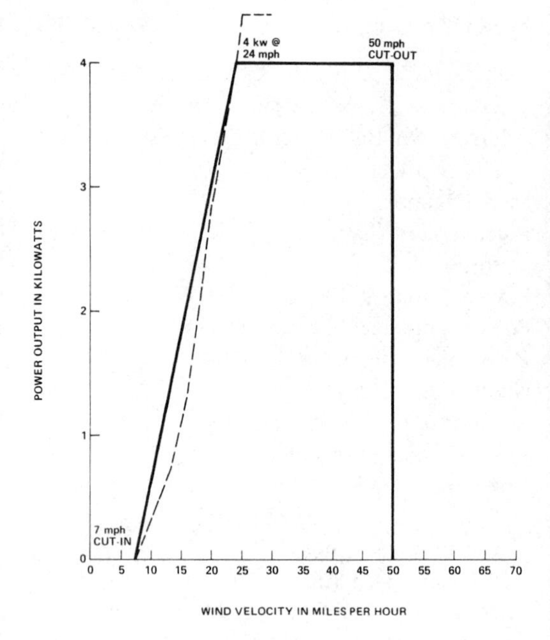

Figure 8-63. Elektro Model WV35G power output curve.

Rated power (kW): 5.0 (see Figure 8-64)
Power output at 12 mph: Unknown
Power output at 14 mph: Unknown
Rotor blade diameter: 16.4 ft
Number of blades: 3
Machine type: Horizontal-axis

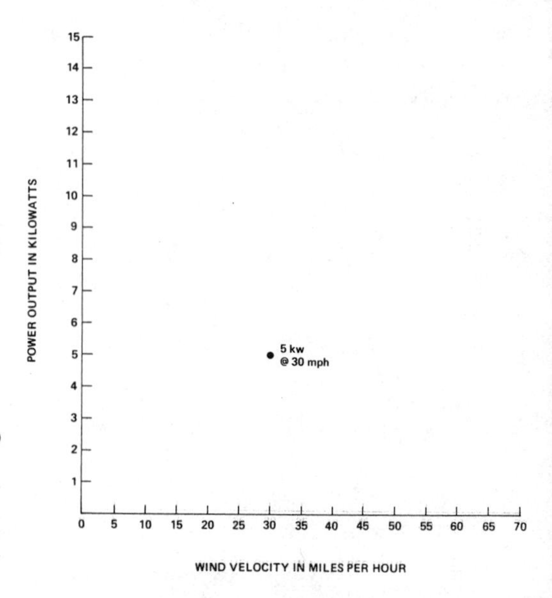

Figure 8-64. Elektro Model WV50D power output curve.

Cost: $7608, turbine only
Operating speed
Rated speed (mph): Unknown
Cut-in speed (mph): Unknown
Cut-out speed (mph): Unknown
Maximum speed (mph): 80
rpm at rated output: 60/30
Blade materials: Wood
Rotor weight: Unknown
System weight on tower: 605 lb
Overspeed control: Unknown
Generator/alternator type: 125/220 V, 3-phase VAC
Application: 30 V ac for heating
Testing history: 37 years of experience
Reliability prediction (MTBF): Unspecified
Warranty: Unknown
Maintenance requirements: Unknown
Country of origin: Switzerland
Number produced: Unknown
Availability: Six months ARO

Figure 8-65. Elektro Model WVG 50G WECS.

ELEKTRO G.m.b.h. WVG 50G

Manufacturer: Elektro G.m.b.h., St. Gallerstrasse 27, Winterthur, Switzerland
Address: c/o Budgen & Associates and Real Gas & Electric Co.
Model number: WVG 50G (see Figure 8-65)
Rated power (kW): 6 @ 30 mph (see Figure 8-66)
Power output at 12 mph: 800 W
Power output at 14 mph: 1375 W
Rotor blade diameter: 16.42 ft
Number of blades: 3
Machine type: Horizontal-axis
Cost: $5523, turbine; $14,874, complete in U.S.
Operating speed
Rated speed (mph): 30
Cut-in speed (mph): 7
Cut-out speed (mph): 50
Maximum speed (mph): 80
rpm at rated output: 120/250
Blade materials: Wood
Rotor weight: Unknown
System weight on tower: 583 lb
Overspeed control: Full feathering
Generator/alternator type: 60/110 V dc for battery charging
Application: Battery charging
Testing history: 37 years of experience
Reliability prediction (MTBF): Unspecified
Warranty: Unknown
Maintenance requirements: Unknown
Country of origin: Switzerland
Number produced: Unknown
Availability: Six months ARO
DOE Rocky Flats Tests: See DOE Report RFP–2920/3533/78/6-2 of September 28, 1978 and later reports for more information.

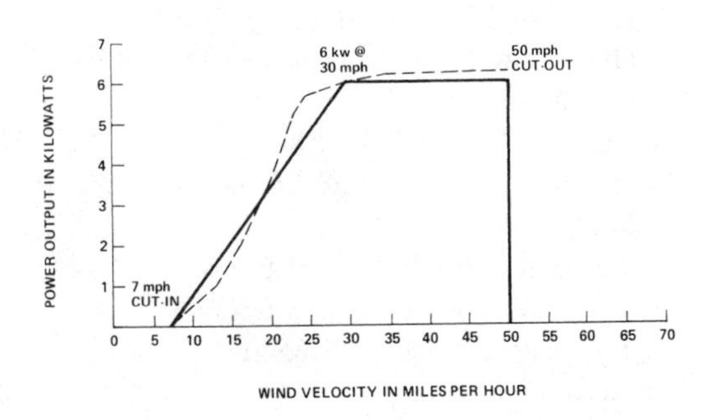

Figure 8-66. Elektro Model WVG 50G power output curve.

ELEKTRO G.m.b.h. WVG 120D

Manufacturer: Elektro G.m.b.h., St. Gallerstrasse 27, Winterthur, Switzerland
Address: c/o Budgen & Associates and Real Gas & Electric Co.
Model number: WVG 120D
Rated power (kW): 8.0 @ (see Figure 8-67)
Power output at 12 mph: Unknown
Power output at 14 mph: Unknown
Rotor blade diameter: 21.65 ft
Number of blades: 3
Machine type: Horizontal-axis
Cost: $10,055, turbine only
Operating speed
  Rated speed (mph): Unknown
  Cut-in speed (mph): Unknown
  Cut-out speed (mph): Unknown
  Maximum speed (mph): 80
rpm at rated output: 60/130
Blade materials: Wood
Rotor weight: Unknown
System weight on tower: 704 lb
Overspeed control: Unknown
Generator/alternator type: 125/200 V, 3-phase VAC
Application: 3-phase ac for heating
Testing history: 37 years of experience
Reliability prediction (MTBF): Unspecified
Warranty: Unknown
Maintenance requirements: Unknown
Country of origin: Switzerland
Number produced: Unknown
Availability: Six months ARO

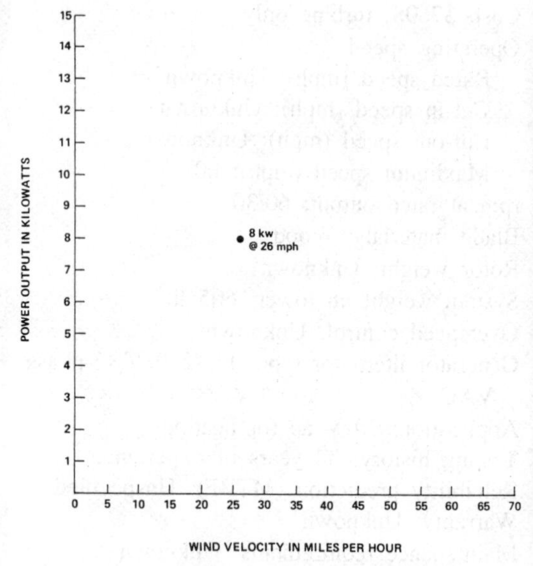

Figure 8-67. Elektro Model WVG 120D power output curve.

ELEKTRO G.m.b.h. WVG 120G

Manufacturer: Elektro G.m.b.h., St. Gallerstrasse 27, Winterthur, Switzerland
Address: c/o Budgen & Associates and Real Gas & Electric Co.
Model number: WVG 120G
Rated power (kW): 10 @ 26 mph (see Figure 8-68)
Power output at 12 mph: 2.6 kW
Power output at 14 mph: 3.6 kW
Rotor blade diameter: 21.65 ft
Number of blades: 3
Machine type: Horizontal-axis
Cost: $9589, turbine; $23,378, complete in U.S.
Operating speed
  Rated speed (mph): 26
  Cut-in speed (mph): Unknown
  Cut-out speed (mph): Unknown
  Maximum speed (mph): 80
rpm at rated output: 120/250
Blade materials: Wood
Rotor weight: Unknown
System weight on tower: 693 lb
Overspeed control: Unknown
Generator/alternator type: 110 V dc for battery charging
Application: SWECS
Testing history: 37 years of experience
Reliability prediction (MTBF): Unspecified
Warranty: Unknown
Maintenance requirements: Unknown
Country of origin: Switzerland
Number produced: Unknown
Availability: Six months ARO

ELTEECO LTD. 30-kW

Manufacturer: Elteeco Ltd. (Sir Henry Lawson-Tancred, Sons & Co. Ltd.)
Address: Aldborough Manor, Boroughbridge, North Yorks, England, Y05 9EP
Contact: J. R. Thompson

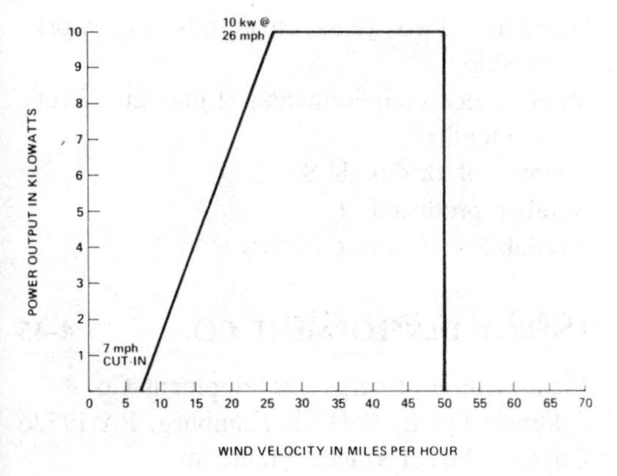

Figure 8-68. Elektro Model WVG 120G power output curve.

Telephone: (090-12) 3223 and 2716
Model number: Type 1 and 2
Rated power (kW): 30 @ 20 mph (see Figure 8-69)
Power output at 12 mph: Unknown
Power output at 14 mph: Unknown
Rotor blade diameter: 56 ft
Number of blades: 3
Machine type: Horizontal-axis, down-wind

Figure 8-68A. Elteeco (Aldborough 56 ft WEC.)

Cost: $19,200
Operating speed
  Rated speed (mph): 20
  Cut-in speed (mph): Unknown
  Cut-out speed (mph): 35
  Maximum speed (mph): 90
rpm at rated output: Unspecified
Blade materials: Steel spars supporting fiber-glass-molded blade
Rotor weight: Unknown
System weight on tower: Unknown
Overspeed control: Pressure-reducing valve operates centrifugally-operated blade flaps
Generator/alternator type: Unknown
Application: Unknown
Testing history: Unknown
Reliability prediction (MTBF): Unknown
Warranty: Unknown
Maintenance requirements: Unknown
Country of origin: England
Number produced: 1
Availability: Special order

ENERGY DEVELOPMENT CO. 4-40

Manufacturer: Energy Development Co.
Address: 179 E. R.D. 2, Hamburg, PA 19526
Contact: Mr. Terrance Mehrkam
Telephone: (215) 562-8856
Model number: 4-40 (see Figure 8-70)
Rated power: 40.0 @ 25 mph (see Figure 8-71)
Power output at 12 mph: 14 kW
Power output at 14 mph: 18 kW
Rotor blade diameter: 38 ft
Number of blades: 4
Machine type: Downwind, horizontal-axis
Cost: Contact Energy Development Co.

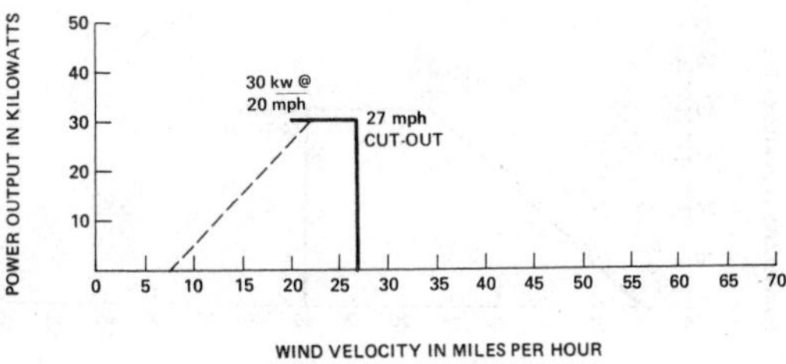

Figure 8-69. Elteeco Ltd. 30-kW Model power output curve.

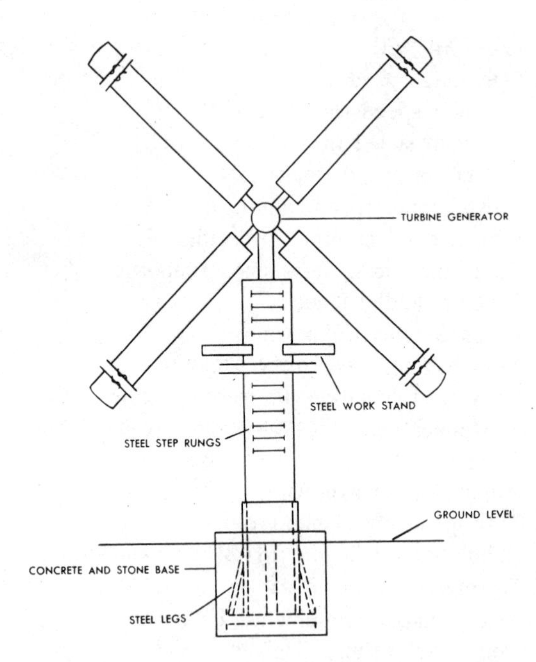

Figure 8-70. Energy Development Co. Model 440 40-kW WECS.

Operating speed
- Rated speed (mph): 27
- Cut-in speed (mph): 5
- Cut-out speed (mph): 40
- Maximum speed (mph): 60; maximum survival speed is 120 mph

rpm at rated output: 60
Blade materials: Aluminum T-6
Rotor weight: 925 lb
System weight on tower: 7000 lb
Overspeed control: Mechanical brake
Generator/alternator type: Synchronous, 3-phase vac
Application: Electricity for residential and commercial use
Testing history: Monitored during operation
Reliability prediction (MTBF): Unspecified

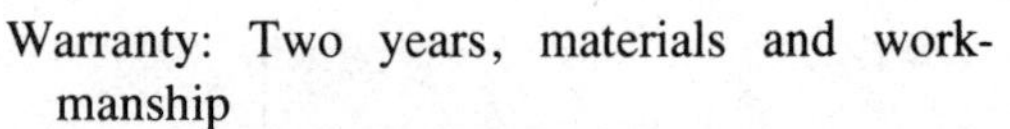
Warranty: Two years, materials and workmanship
Maintenance requirements: Lubricate every six months
Country of origin: U.S.
Number produced: 1
Availability: Contact factory

ENERGY DEVELOPMENT CO. 4-45

Manufacturer: Energy Development Co.
Address: 179 E. R.D. 2, Hamburg, PA 19526
Contact: Mr. Terrance Mehrkam
Telephone: (215) 562-0856
Model number: 4-45
Rated power (kW): 45.0 @ 25 mph (see Figure 8-72)
Power output at 12 mph: 16 kW
Power output at 14 mph: 20 kW
Rotor blade diameter: 40 ft (12.2 meters)
Number of blades: 4
Machine type: Downwind, horizontal-axis
Cost: Contact Energy Development Co.
Operating speed
- Rated speed (mph): 27
- Cut-in speed (mph): 5
- Cut-out speed (mph): 40
- Maximum speed (mph): Unknown

rpm at rated output: 60
Blade materials: Aluminum T-6
Rotor weight: 1250 lb
System weight on tower: 9500 lb
Overspeed control: Mechanical brake
Generator/alternator type: Synchronous
Application: Electricity for residential and commercial use
Testing history: Field test
Reliability prediction (MTBF): Unspecified
Warranty: Two years, materials and workmanship

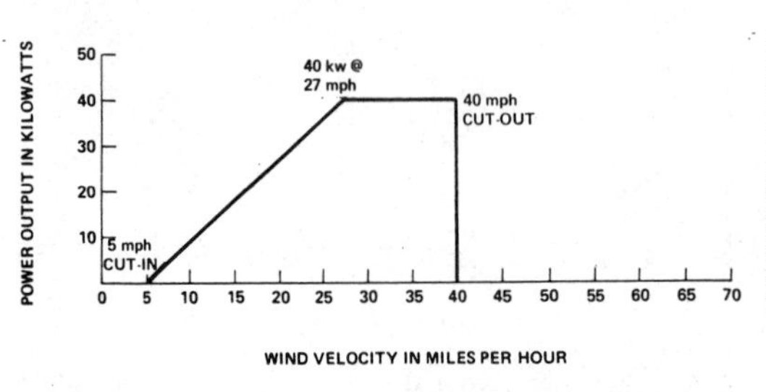

Figure 8-71. Energy Development Co. Model 440 power output curve.

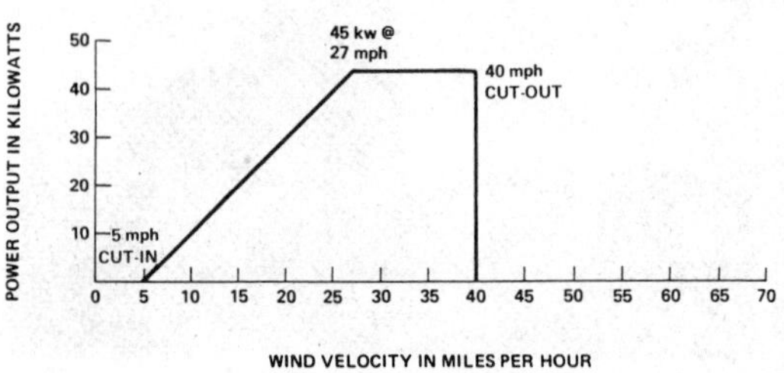

Figure 8-72. Energy Development Co. Model 445 power output curve.

Maintenance requirements: Lubricate every six months
Country of origin: U.S.
Number produced: 1
Availability: Contact factory

ENERGY DEVELOPMENT CO. 4-100

Manufacturer: Energy Development Co.
Address: 179 E. R.D. 2, Hamburg, PA 19526
Contact: Mr. Terrance Mehrkam
Telephone: (215) 562-8856
Model number: 4-100
Rated power (kW): 100
Power output at 12 mph: Unspecified
Power output at 14 mph: Unspecified
Number of blades: 4
Machine type: Downwind, horizontal-axis
Cost: $55,000
Operating speed
  Rated speed (mph): 27
  Cut-in speed (mph): 5
  Cut-out speed (mph): 60
  Maximum Speed: 60
rpm at rated output: 60
Blade materials: Aluminum T-6, NACA 441B configuration
Rotor weight: Unspecified
System weight on tower: Unspecified
Overspeed control: Automatic shutdown @ 60 rpm
Generator/alternator type: AC three-phase alternator
Application: Small community electric power
Testing history: Unspecified
Reliability prediction (MTBF): Unspecified
Warranty: Two years, materials and workmanship
Maintenance requirements: Lubricate every six months
Country of origin: U.S.
Number produced: 1
Availability: Contact factory

ENERGY DEVELOPMENT CO. 4-225

Manufacturer: Energy Development Co.
Address: 179 E. R.D. 2, Hamburg, PA 19526
Contact: Mr. Terrance Mehrkam
Telephone: (215) 562-8856
Model number: 4-225
Rated power (kW): 225
Power output at 12 mph: Unspecified
Power output at 14 mph: Unspecified
Number of blades: 4
Machine type: Downwind, horizontal-axis
Cost: $115,000
Operating speed
  Rated speed (mph): 27
  Cut-in speed (mph): 5
  Cut-out speed (mph): 60
  Maximum Speed: 60
rpm at rated output: 60
Blade materials: Aluminum T-6, NACA 441B configuration
Rotor weight: Unspecified
System weight on tower: Unspecified
Overspeed control: Automatic shutdown @ 60 rpm
Generator/alternator type: AC three-phase alternator
Application: Small community electric power
Testing history: Unspecified
Reliability prediction (MTBF): Unspecified
Warranty: Two years, materials and workmanship
Maintenance requirements: Lubricate every six months
Country of origin: U.S.
Number produced: 1
Availability: Contact factory

ENERGY DEVELOPMENT CO. 2 mW

Manufacturer: Energy Development Co.
Address: 179 E. R.D. 2, Hamburg, PA 19526
Contact: Mr. Terrance Mehrkam
Telephone: (215) 562-8856
Model number: 2 megawatt (see Figure 8-72A)
Rated power: 2 megawatt
Power output at 12 mph: Unspecified
Power output at 14 mph: Unspecified
Rotor blade diameter: 160 ft
Number of blades: 6
Machine type: Downwind, horizontal-axis
Cost: Unspecified
Operating speed
  Rated speed (mph): 5
  Cut-in speed (mph): 35
  Cut-out speed (mph): 40
  Maximum Speed: 60

Figure 8-72A. Mehrkam 2 megawatt Wind Turbine.

rpm at rated output: 36
Blade materials: Aluminum alloy, NACA 441B airfoil
Rotor weight: 2400 lb
System weight on tower: Unspecified
Overspeed control: Failsafe brake
Generator/alternator type: Three-phase induction generator
Application: Community electric power
Testing history: Unspecified
Reliability prediction (MTBF): Unspecified
Warranty: Unspecified
Maintenance requirements: Unspecified
Country of origin: U.S.
Number produced: Unspecified
Availability: Contact factory

ENERTECH CORPORATION 1500

Manufacturer: Enertech Corporation
Address: P.O. Box 420, Norwich, VT 05055
Contact: Mr. Ned Coffin
Telephone (802) 649-1145
Model number: 1500 (see Figure 8-73)
Rated power (kW): 1.5 @ 22 mph (see Figure 8-74)
Power output at 12 mph: 350 W
Power output at 14 mph: 600 W
Rotor blade diameter: 13.2 ft
Number of blades: 3
Machine type: Downwind, horizontal-axis
Cost: $2900
Operating speed
  Rated speed (mph): 22
  Cut-in speed (mph): 9
  Cut-out speed (mph): 40
  Maximum speed (mph): 100
rpm at rated output: 170
Blade materials: Wood
Rotor weight: 48 lb
System weight on tower: 185 lb
Overspeed control: Automatic brake
Generator/alternator type: Induction generator, 115 vac
Application: Electricity
Testing history: Truck tested, field tested
Reliability prediction (MTBF): Unspecified
Warranty: One year, parts and workmanship

Figure 8-73. Enertech Model 1500 SWECS. (*Courtesy of Enertech.*)

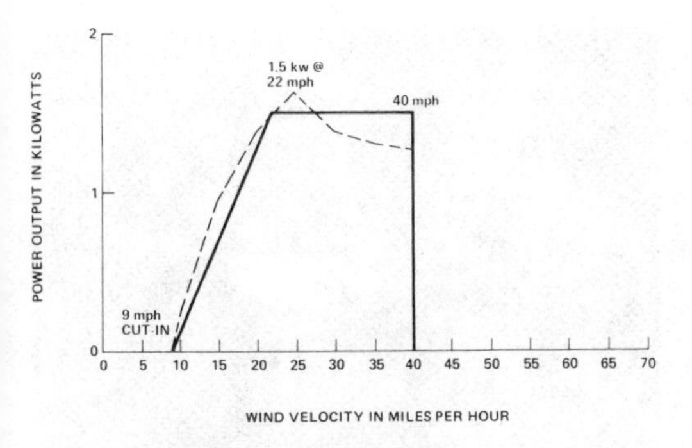

Figure 8-74. Enertech Model 1500 power output curve.

Maintenance requirements: Annual oil change and systems inspection
Country of origin: U.S.
Number produced: Unknown
Availability: Contact Enertech

ENERTECH CORPORATION DOE 1-kW

Manufacturer: Enertech Corporation
Address: P.O. Box 420, Norwich, VT 05055
Contact: Mr. Bill Drake
Telephone (802) 649-1145
Model number: DOE 1-kW (see Figure 8-75)

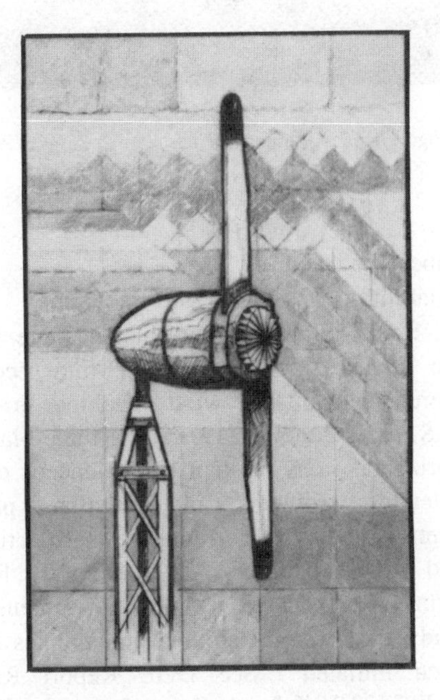

Figure 8-75. Enertech 1-kW WECS under development for DOE. (*Courtesy of DOE.*)

Rated power (kW): 2.1 @ 20 mph (see Figure 8-76)
Power output at 12 mph: Unknown
Power output at 14 mph: Unknown
Rotor blade diameter: 16.4 ft
Number of blades: 2
Machine type: Downwind, horizontal-axis
Cost: $6500. $1500/kW cost goal; development costs: $150,000
Operating speed
  Rated speed (mph): 20
  Cut-in speed (mph): Unspecified (minimum)
  Cut-out speed (mph): Unspecified (maximum)
  Maximum speed (mph): 165
rpm at rated output: Unknown
Blade materials: Composite blade covered with polyurethane
Rotor weight: Unknown
System weight on tower: 695 lb
Overspeed control: Stall
Generator/alternator type: 24 vdc, with regulation for battery charging. Dual output, Maremont
Application: For very small-scale uses such as seismic monitoring power and offshore navigation station
Testing history: Under development
Reliability prediction (MTBF): High reliability design (10-year MBTF).
Warranty: Unspecified

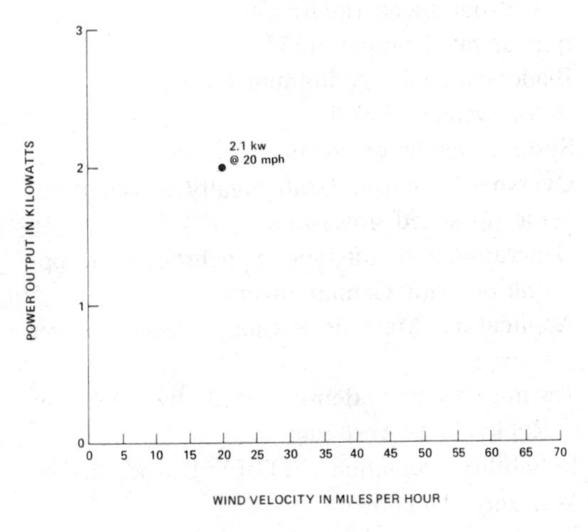

Figure 8-76. Enertech DOE 1-kW Model power output curve.

Maintenance requirements: Minimal
Country of origin: U.S.
Number produced: 1
Availability: DOE development project
Comments: Uses a Moremont Corp. attenuator and electrical sybsystem to provide a valid power of 2.1 kW. There are two isolated attenuator sections which operate in parallel to meet the high reliability design goals.

GRUMMAN WINDSTREAM™ 25

Manufacturer: Grumman Energy Systems
Address: 4175 Veterans Memorial Highway, Ronkonkoma, NY 11779
Contact: Mr. Paul Henton
Telephone: (516) 575-7261
Model number: Windstream™ 25 (see Figures 8-77 and 8-78)
Rated power (kW): 20 @ 28 mph (see Figure 8-79)
Power output at 12 mph: 1.2 kW
Power output at 14 mph: 2 kW
Rotor blade diameter: 25 ft
Number of blades: 3
Machine type: Downwind, horizontal-axis, variable pitch rotor
Cost: $19,900 Windstream™ 25; $4715 for work-stand rental and tower
Operating speed
Rated speed (mph): 28
Cut-in speed (mph): Unknown
Cut-out speed (mph): 50
rpm at rated output: 125
Blade materials: Aluminum blades
Rotor weight: 750 lb
System weight on tower: 1250 lb
Overspeed control: Centrifugally activited rotor tip speed governors
Generator/alternator type: Synchronous ac operation with Gemini inverter
Application: Main or auxiliary electric power source
Testing history: Being tested by DOE at Rocky Flats Test Site
Reliability prediction (MTBF): Unspecified
Warranty: Unknown
Maintenance requirements: Unknown
Country of origin: U.S.

Figure 8-77. Grumman Windstream™ 25 WECS.

Number produced: 10
Availability: 120 days ARO
DOE Rocky Flats Tests: "This machine was the largest (25-foot diameter rotor) commercially available wind machine in the U.S. as of June 30, 1978. The three-bladed horizontal axis system experienced considerable problems, in particular, pitch control and voltage control malfunctions and stiffness in yaw. In addition to long term performance testing, an intensive study of tower dynamic characteristics has been initiated." See DOE Report RFP-2920/3523/78/6-2 of September 28, 1978 and later reports for more information."

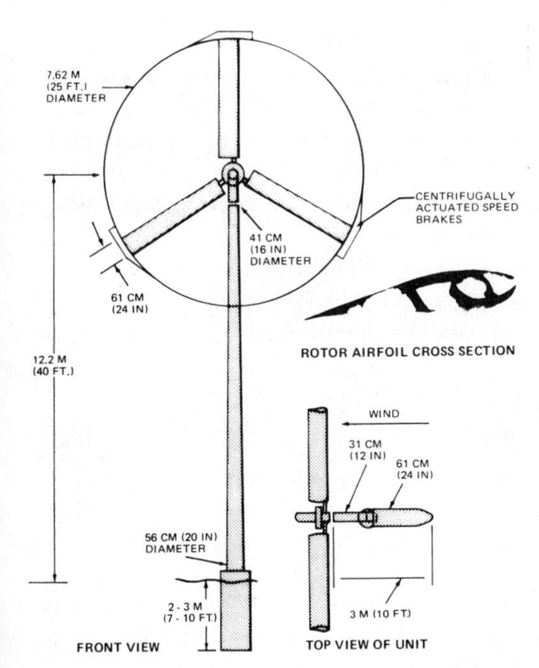

Figure 8-78. Grumman Windstream™ 25 WECS mechanical configuration.

GRUMMAN DOE 8-kW

Manufacturer: Grumman Energy Systems
Address: 4175 Veterans Memorial Highway, Ronkonkoma, NY 11779
Contact: Mr. Frank Adler
Telephone: (516) 575-7261
Model number: DOE 8-kW (see Figure 8-80)
Rated power (kW): 11.0 @ 20 mph (see Figure 8-81)
Power output at 12 mph: Unspecified
Power output at 14 mph: Unspecified
Rotor blade diameter: 33.25 ft
Number of blades: 3
Machine type: 3-blade horizontal-axis, downwind
Cost: $8250 ($750/kW design cost goal); development costs: $310,000
Operating speed
  Rated speed (mph): 20
  Cut-in speed (mph): Unspecified
  Cut-out speed (mph): Unspecified
  Maximum speed (mph): Unspecified
rpm at rated output: Unspecified
Blade materials: Unspecified
Rotor weight: Unspecified
System weight on tower: 2590
Overspeed control: Feather
Generator/alternator type: Induction
Application: Provide power for homes and farm buildings
Testing history: Under development

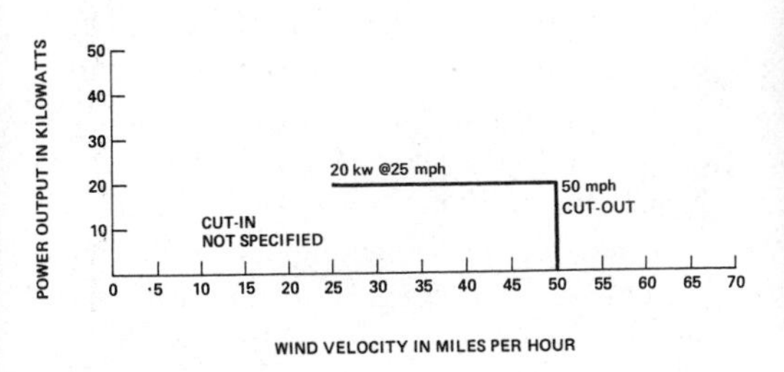

Figure 8-79. Grumman Windstream™ 25 power output curve.

Figure 8-80. Grumman 8-kW Model WECS under development for DOE.

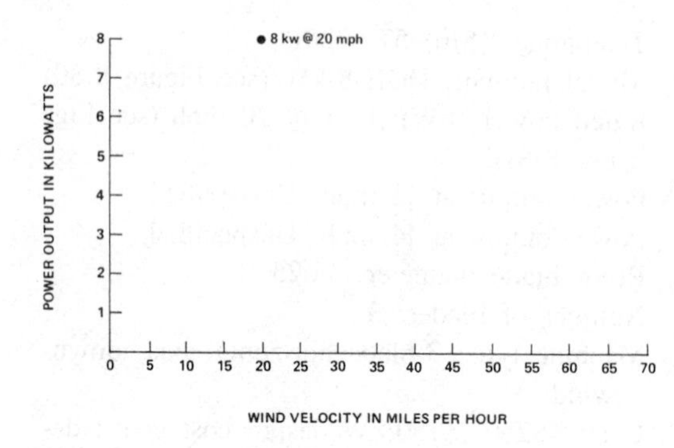

Figure 8-81. Grumman DOE 8-kW Model power output curve.

Reliability prediction (MTBF): Unspecified
Warranty: Unspecified
Maintenance requirements: Unspecified
Country of origin: U.S.
Number produced: 1
Availability: DOE development project

HUMMING BIRD 4000/22

Manufacturer: Power Group International Corp.
Address: 13315 Stuebner-Airline Road, Suite 106, Houston Texas 77014
Contact: Sales Manager
Model number: 4000/22 (see Figure 8-82)
Rated power (kW): 4 @ 22 mph
Power output at 12 mph: Unspecified
Power output at 14 mph: Unspecified
Rotor blade diameter: 14 ft
Number of blades: 3
Machine type: Upwind horizontal axis, fixed pitch blades
Cost: Unspecified
Operating speed
  Rated speed (mph): 22
  Cut-in speed (mph): 8
  Cut-out speed (mph): 26–28
  Maximum speed (mph): Unspecified
rpm at rated output: 266
Blade materials: Epoxy resin, polyurethane foam core
Rotor weight: 25 lb
System weight on tower: 500 lb
Overspeed control: Automatic dynamic braking
Generator/alternator type: Direct drive, brushless, synchronous, 3 phase
Application: SWECS
Testing history: Three years field experience and wind tunnel test
Reliability prediction (MTBF): Unspecified
Warranty: Unspecified
Maintenance requirements: Annual maintenance
Country of origin: U.S.
Number produced: Unspecified
Availability: Unspecified

JACOBS J8KVA

Manufacturer: Jacobs Wind Electric Company, Inc.
Address: Route 13, Box 722, Fort Myers, FL 33908
Contact: M. L. Jacobs
Telephone: (813) 481-3113, 334-0339
Model number: Jacobs 8-KVA (see Figure 8-83)
Rated power (kW): 8
Rotor blade diameter: 7 m
Number of blades: 3
Machine type: Horizontal-axis
Cost: Contact factory
Overspeed control: Blade-actuated system of speed control
Generator/alternator type: Jacobs generator
Application: Home electricity generation
Testing history: More than 50 years experience, system under design and test for several years
Maintenance requirements: Inspection
Country of origin: U.S.
Availability: Contact factory
Comments: Figure 8-83 shows construction of Jacobs new 8 KVA WECS. "Note drive pinion at top of gear case, no heavy oil drag power loss. The offset Hypoid Drive System (patented) balances gear torque against propeller back-thrust pressure to give a steady equalized power delivery to the alternator. Pinion gear shaft operates in a sealed tube, eliminating any oil seal at bottom of gear case. Thus, oil can never leak out of the gear case and destroy gears. Brushless alternator is accessible in tower, with no collector rings to ever give trouble."

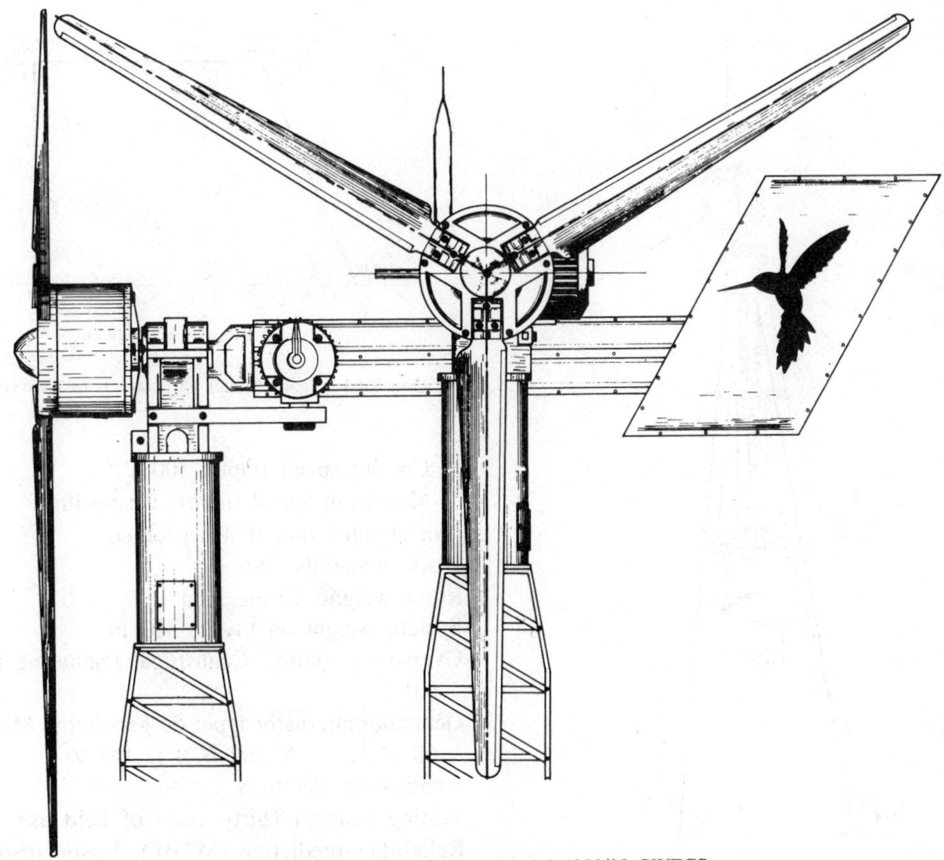

Figure 8-82. Hummingbird Model 4000/22 SWECS.

JACOBS Model 45

Manufacturer: Jacobs Wind Electric Company, Inc.*
Address: Route 13, Box 722, Fort Myers, FL 33901
Contact: Mr. M. L. Jacobs
Telephone: (813) 481-3113
Model number: 45
Rated power (kW): 1.8 @ 23 mph (see Figure 8-84)
Power output at 12 mph: 550 W
Power output at 14 mph: 760 W
Rotor blade diameter: 14 ft
Number of blades: 3
Machine type: Horizontal-axis
Cost: $1800
Operating speed
  Rated speed (mph): 23
  Cut-in speed (mph): 7
  Cut-out speed (mph): 100
  Maximum speed (mph): Unspecified
rpm at rated output: Unspecified
Blade materials: Wood
Rotor weight: Unspecified
System weight on tower: 500 lb
Overspeed control: Centrifugal feathering flyball
Generator/alternator type: dc generator, 32 V
Application: SWECS
Testing history: Thirty years of sales experience
Reliability prediction (MTBF): Unspecified
Warranty: Five years original
Maintenance requirements: Annual check, semiannual lubrication and cleaning
Country of origin: U.S.
Number produced: Thousands
Availability: Refurbished

*The Model 45 was manufactured by Jacobs Wind Electric Co. from 1930 to 1960 in Minneapolis, Minn. These units are available only as refurbished systems. They are not available from this address.

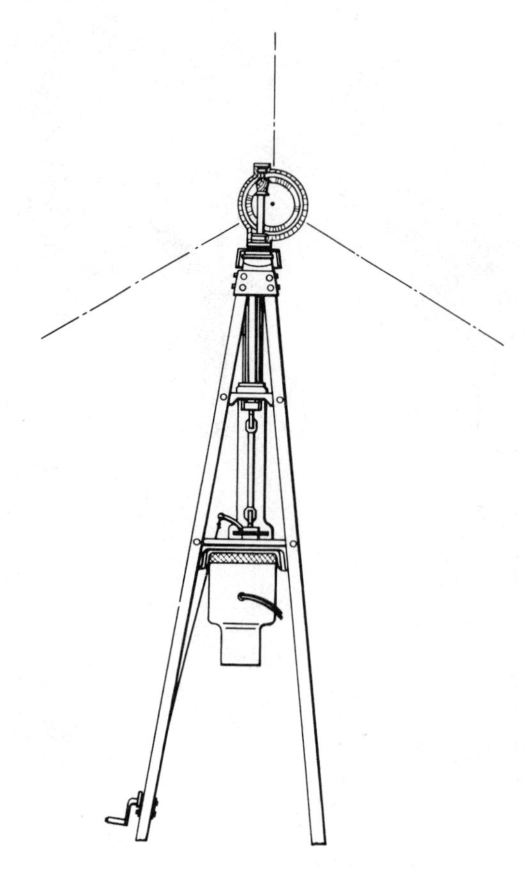

Figure 8-83. Jacobs new 8-KVA WECS.

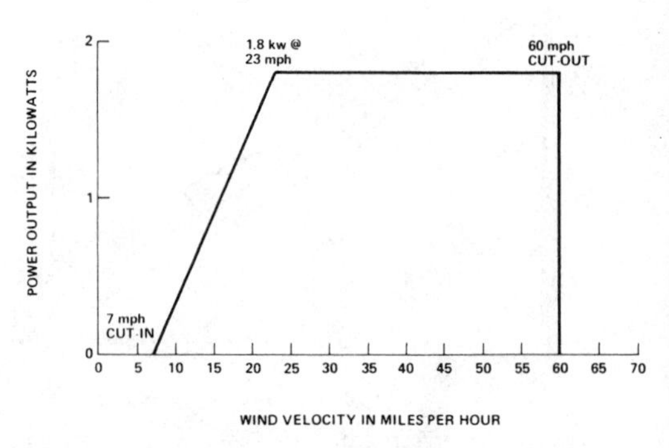

Figure 8-84. Jacobs Model 45 power output curve.

JACOBS Model L 60/60B

Manufacturer: Jacobs Wind Electric Company, Inc.*
Address: Route 13, Box 722, Fort Myers, FL 33908
Contact: Mr. M. L. Jacobs
Telephone: (813) 481-3113
Model number: 60/60B
Rated power (kW): 2.5 @ 23 mph (see Figure 8-85)
Power output at 12 mph: 800 W
Power output at 14 mph: 1.1 kW
Rotor blade diameter: 14 ft
Number of blades: 3
Machine type: Horizontal-axis
Cost: Unspecified
Operating speed
Rated speed (mph): 23
Cut-in speed (mph): 7
Cut-out speed (mph): 100
Maximum speed (mph): Unspecified
rpm at rated output: Unspecified
Blade materials: Wood
Rotor weight: Unspecified
System weight on tower: 500 lb
Overspeed control: Centrifugal feathering flyball
Generator/alternator type: dc generator, Model 60 is 32 V, Model 60B is 110 V
Application: SWECS
Testing history: Thirty years of field use
Reliability prediction (MTBF): Unspecified
Warranty: Five years original
Maintenance requirements: Annual inspection, semiannual lubrication
Country of origin: U.S.
Number produced: Thousands
Availability: Refurbished
Comments: Unspecified

*The Model 60/60B was manufactured by Jacobs Wind Electric Co. from 1930 to 1960 in Minneapolis, Minn. These units are available only as refurbished systems. They are not available from this address.

KAMAN DOE 40 kW

Manufacturer: Kaman Aerospace Corporation
Address: Old Windsor Road, Bloomfield, CT 06002
Contact: H. Howes
Model number: DOE 40 kW
Rated power (kW): 40 @ 20 mph (see Figure 8-87)

Figure 8-85. Jacobs Model 60 WECS.

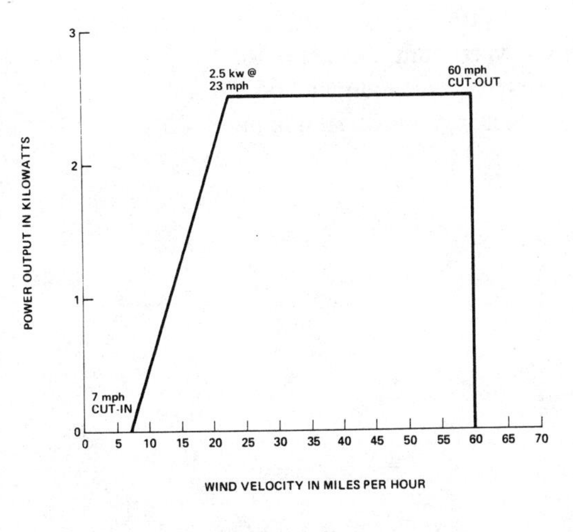

Figure 8-86. Jacobs Model 60 power output curve.

Power output at 12 mph: Unspecified
Power output at 14 mph: Unspecified
Rotor blade diameter: 64 ft
Number of blades: 2
Machine type: Downwind, horizontal-axis
Cost: $500/kW, cost design goal; development costs: $370,000
Operating speed
  Rated speed (mph): 20
  Cut-in speed (mph): Unspecified
  Cut-out speed (mph): Unspecified
  Maximum speed (mph): Unspecified
rpm at rated output: Unspecified
Blade materials: Unspecified
Rotor weight: Unspecified
System weight on tower: 4900 lb

Figure 8-87. Kaman DOE 40-kW Model WECS.

Figure 8-88. Kedco Model 1200 WECS.

Overspeed control: Unspecified
Generator/alternator type: Unspecified
Application: Deep well irrigation, small community electricity
Testing history: Under development
Reliability prediction (MTBF): Unspecified
Warranty: Unspecified
Maintenance requirements: Unspecified
Country of origin: U.S.
Number produced: 1
Availability: DOE development project

KEDCO, INC. 1200

Manufacturer: Kedco, Inc.
Address: 9016 Aviation Blvd. Inglewood, CA 90301
Contact: Mr. Terry Rainey
Telephone: (213) 776-6636
Model number: 1200 (see Figure 8-88)
Rated power (kW): 1.2 @ 21 mph (see Figure 8-89)
Power output at 12 mph: 450 W
Power output at 14 mph: 620 kW
Rotor blade diameter: 12 ft
Number of blades: 3
Machine type: Downwind, horizontal-axis
Cost: $2300
Operating speed
Rated speed (mph): 21
Cut-in speed (mph): 7
Cut-out speed (mph): 70
Maximum speed (mph): Unspecified
rpm at rated output: 300
Blade materials: Aluminum, 2024-T3

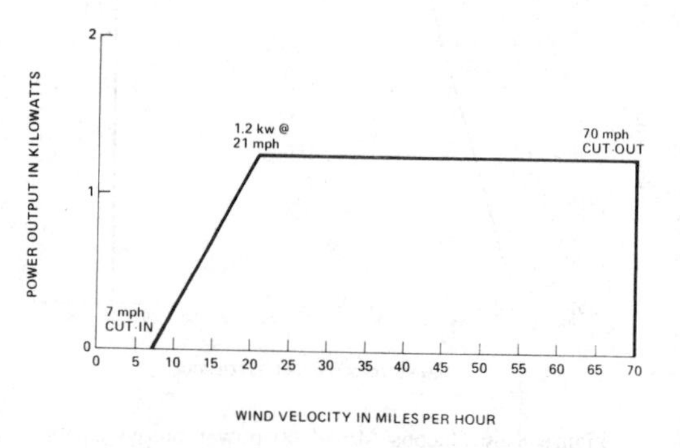

Figure 8-89. Kedco Model 1200 power output curve.

Rotor weight: 7 lb
System weight on tower: 202 lb
Overspeed control: Mechanical, centrifugal blade feathering
Generator/alternator type: 14.4 vdc alternator
Application: SWECS
Testing history: Moving test bed
Reliability prediction (MTBF): Unspecified
Warranty: One year, parts and labor
Maintenance requirements: Annual, check fluid levels, inspect system
Country of origin: U.S.
Number produced: Unknown
Availability: Contact Kedco
DOE Rocky Flats Tests: See DOE Report RFP-2920/3533/78/6-2 of September 28, 1978 and later reports for more information.

KEDCO, INC. 1205

Manufacturer: Kedco, Inc.
Address: 9016 Aviation Blvd. Inglewood, CA 90301
Contact: Mr. Terry Rainey
Telephone: (213) 776-6636
Model number: 1205
Rated power (kW): 1.2 @ 22 mph (see Figure 8-90)
Power output at 12 mph: 300 W
Power output at 14 mph: 520 W
Rotor blade diameter: 12 ft
Number of blades: 3
Machine type: Downwind, horizontal-axis
Cost: Contact Kedco
Operating speed
  Rated speed (mph): 22
  Cut-in speed (mph): 8
  Cut-out speed (mph): 70
  Maximum speed (mph): Unspecified
rpm at rated output: 300
Blade materials: Aluminum, 2024-T3
Rotor weight: 7 lb
System weight on tower: 202 lb
Overspeed control: Mechanical, centrifugal blade feathering
Generator/alternator type: 28.4 vdc
Application: SWECS
Testing history: Moving test bed
Reliability prediction (MTBF): Unspecified
Warranty: One year, parts and labor
Maintenance requirements: Annual, check fluid levels, inspect system
Country of origin: U.S.
Number produced: Unknown
Availability: Contact Kedco

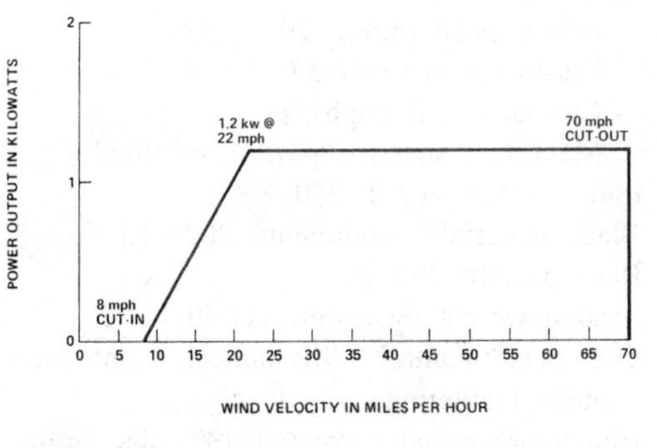

Figure 8-90. Kedco Model 1205 power output curve.

KEDCO, INC. 1210

Manufacturer: Kedco, Inc.
Address: 9016 Aviation Blvd. Inglewood, CA 90301
Contact: Mr. Terry Rainey
Telephone: (213) 776-6636
Model number: 1210
Rated power (kW): 2 @ 25 mph (see Figure 8-91)
Power output at 12 mph: 150 W
Power output at 14 mph: 450 W
Rotor blade diameter: 12 ft
Number of blades: 3
Machine type: Downwind, horizontal-axis
Cost: $2595
Operating speed
  Rated speed (mph): 25
  Cut-in speed (mph): 11
  Cut-out speed (mph): 70
  Maximum speed (mph): Unspecified
rpm at rated output: 300
Blade materials: Aluminum, 2024-T3
Rotor weight: 7 lb
System weight on tower: 252 lb
Overspeed control: Mechanical, centrifugal blade feathering
Generator/alternator type: 28.4 vdc
Application: SWECS
Testing history: Moving test bed
Reliability prediction (MTBF): Unspecified

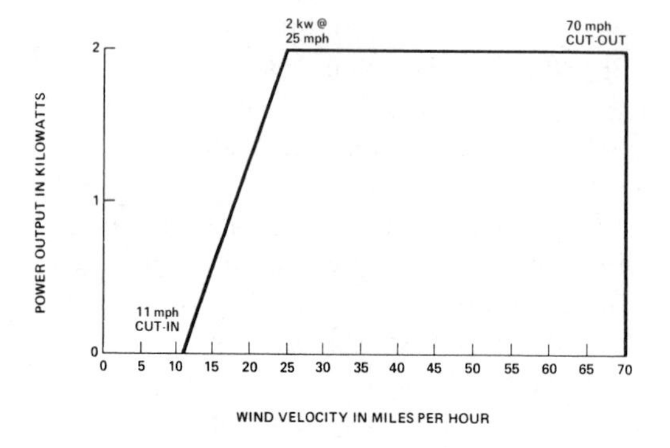

Figure 8-91. Kedco Model 1210 power output curve.

Warranty: One year, parts and labor
Maintenance requirements: Annual, check fluid levels, inspect system
Country of origin: U.S.
Number produced: Unknown
Availability: Contact Kedco

KEDCO, INC. 1600

Manufacturer: Kedco, Inc.
Address: 9016 Aviation Blvd. Inglewood, CA 90301
Contact: Mr. Terry Rainey
Telephone: (213) 776-6636
Model number: 1600
Rated power (kW): 1.2 @ 17 mph (see Figure 8-92)
Power output at 12 mph: 400 W
Power output at 14 mph: 600 W
Rotor blade diameter: 16 ft
Number of blades: 3
Machine type: Downwind, horizontal-axis
Cost: $2895

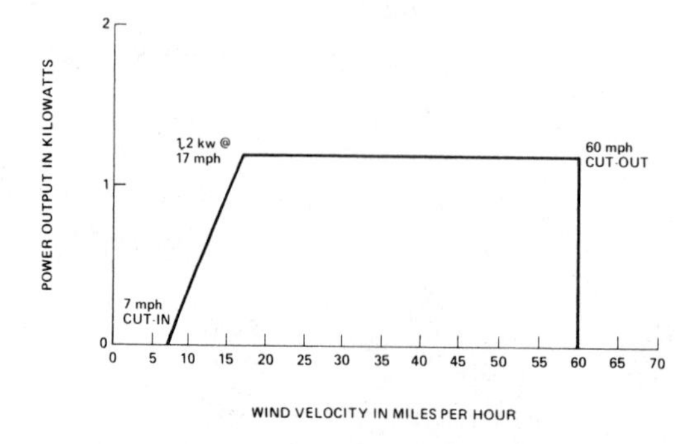

Figure 8-92. Kedco Model 1600 power output curve.

Operating speed
  Rated speed (mph): 17
  Cut-in speed (mph): 7
  Cut-out speed (mph): 60
  Maximum speed (mph): Unspecified
rpm at rated output: 250
Blade materials: Aluminum, 2024-T3
Rotor weight: 8.5 lb
System weight on tower: 217 lb
Overspeed control: Mechanical, centrifugal blade feathering
Generator/alternator type: 14.4 vdc
Application: SWECS
Testing history: Moving test bed
Reliability prediction (MTBF): Unspecified
Warranty: One year, parts and labor
Maintenance requirements: Annual, check fluid levels, inspect system
Country of origin: U.S.
Number produced: Unknown
Availability: Contact Kedco

KEDCO, INC. 1605

Manufacturer: Kedco, Inc.
Address: 9016 Aviation Blvd. Inglewood, CA 90301
Contact: Mr. Terry Rainey
Telephone: (213) 776-6636
Model number: 1605
Rated power (kW): 2 @ 20 mph (see Figure 8-93)
Power output at 12 mph: 750 W
Power output at 14 mph: 1060 W
Rotor blade diameter: 16 ft
Number of blades: 3
Machine type: Downwind, horizontal-axis
Cost: Contact factory
Operating speed
  Rated speed (mph): 20
  Cut-in speed (mph): 7
  Cut-out speed (mph): 60
  Maximum speed (mph): Unspecified
rpm at rated output: 250
Blade materials: Aluminum, 2024-T3
Rotor weight: 8.5 lb
System weight on tower: 217 lb
Overspeed control: Mechanical, centrifugal blade feathering
Generator/alternator type: 0–180 vdc, permanent magnet

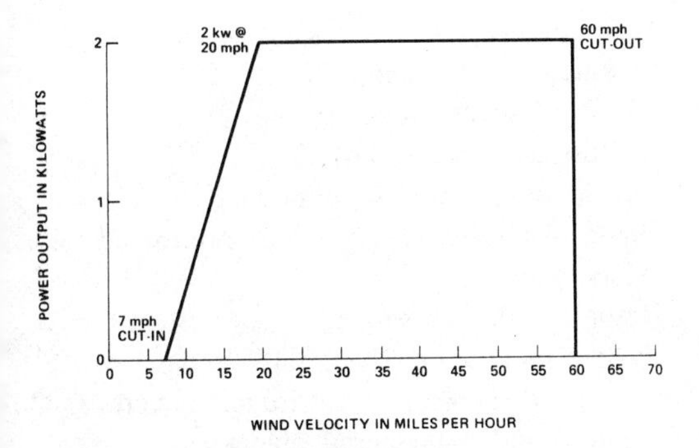

Figure 8-93. Kedco Model 1605 power output curve.

Application: SWECS
Testing history: Moving test bed
Reliability prediction (MTBF): Unspecified
Warranty: One year, parts and labor
Maintenance requirements: Annual, check fluid levels, inspect system
Country of origin: U.S.
Number produced: Unknown
Availability: Contact Kedco

KEDCO, INC. 1610

Manufacturer: Kedco, Inc.
Address: 9016 Aviation Blvd. Inglewood, CA 90301
Contact: Mr. Terry Rainey
Telephone: (213) 776-6636
Model number: 1610
Rated power (kW): 2.0 @ 22 mph (see Figure 8-94)
Power output at 12 mph: 350 W
Power output at 14 mph: 675 W
Rotor blade diameter: 16 ft

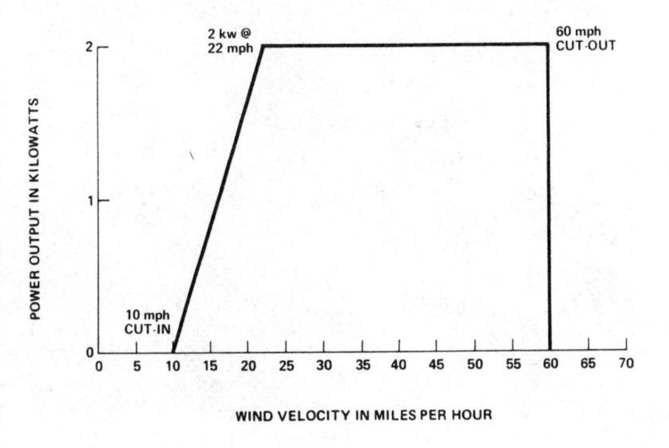

Figure 8-94. Kedco Model 1610 power output curve.

Number of blades: 3
Machine type: Downwind, horizontal-axis
Cost: $3195
Operating speed
  Rated speed (mph): 22
  Cut-in speed (mph): 10
  Cut-out speed (mph): 60
  Maximum speed (mph): Unspecified
rpm at rated output: 250
Blade materials: Aluminum, 2024-T3
Rotor weight: 8.5 lb
System weight on tower: 267 lb
Overspeed control: Mechanical, centrifugal blade feathering
Generator/alternator type: 0–180 vdc, permanent magnet
Application: SWECS
Testing history: Moving test bed
Reliability prediction (MTBF): Unspecified
Warranty: One year, parts and labor
Maintenance requirements: Annual, check fluid levels, inspect system
Country of origin: U.S.
Number produced: Unknown
Availability: Contact Kedco

KEDCO, INC. 1620

Manufacturer: Kedco, Inc.
Address: 9016 Aviation Blvd. Inglewood, CA 90301
Contact: Mr. Terry Rainey
Telephone: (213) 776-6636
Model number: 1620
Rated power (kW): 3.0 @ 25 mph (see Figure 8-95)
Power output at 12 mph: 250 W
Power output at 14 mph: 650 W
Rotor blade diameter: 16 ft
Number of blades: 3
Machine type: Downwind, horizontal-axis
Cost: Contact Kedco
Operating speed
  Rated speed (mph): 25
  Cut-in speed (mph): 11
  Cut-out speed (mph): 60
  Maximum speed (mph): Unspecified
rpm at rated output: 250
Blade materials: Aluminum, 2024-T3
Rotor weight: 75.5 lb
System weight on tower: 293 lb

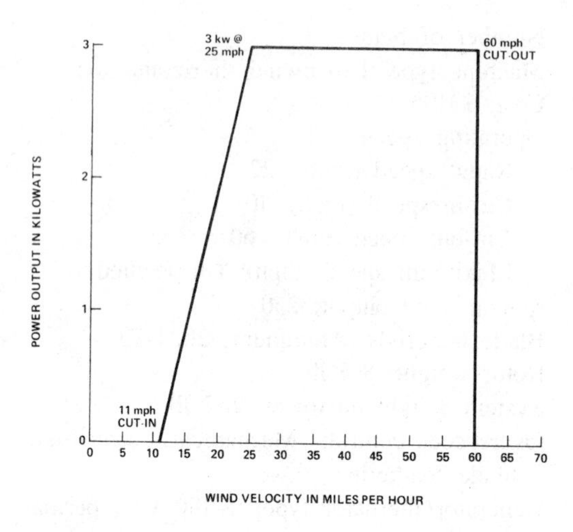

Figure 8-95. Kedco Model 1620 power output curve.

Overspeed control: Mechanical, centrifugal blade feathering
Generator/alternator type: 0–180 vdc, permanent magnet
Application: SWECS
Testing history: Moving test bed
Reliability prediction (MTBF): Unspecified
Warranty: One year, parts and labor
Maintenance requirements: Annual, check fluid levels, inspect system
Country of origin: U.S.
Number produced: Unknown
Availability: Contact Kedco

LUBING M022-3-G024-400

Manufacturer: Lubing Maschinenfabrik, Ludwig Bening, 2847, Barnsdorf, P.O. Box 171, Germany
Address: c/o Budgen & Associates, 72 Broadview Ave., Pointe Claire, Quebec, Canada
Contact: Dr. Budgen
Telephone: (514) 695-4073
Model number: M022-3-G024-400 (see Figure 8-96)
Rated power: 400 W @ 22 mph (see Figure 8-97)
Power output at 12 mph: 70 W
Power output at 14 mph: 155 W
Rotor blade diameter: 7 ft
Number of blades: 6
Machine type: Downwind, horizontal-axis
Cost: $2304
Operating speed
  Rated speed (mph): 22
  Cut-in speed (mph): 9
  Cut-out speed (mph): 35
rpm at rated output: Unknown
Blade materials: Epoxy resins reinforced with fiberglass
Rotor weight: Unknown
System weight on tower: 309 lb
Overspeed control: Centrifugal governor, 3 flared and 3 feathered blades
Generator/alternator type: 3-phase permanent magnet alternator, 400 V rectified to 24 or 12 V dc
Application: Battery charging, lighting
Testing history: 25 years of experience, highest order of workmanship
Warranty: Unknown
Maintenance requirements: Annual oil change of two-step gear system
Country of origin: France
Number produced: Several thousand
Availability: Four months ARO

Figure 8-96. Lubing 400-W SWECS.

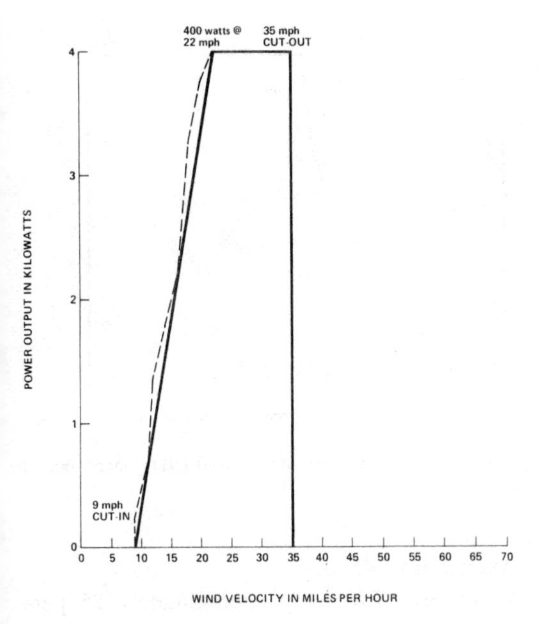

Figure 8-97. Lubing Model M022-3-C024-400 power output curve.

McDONNELL DOUGLAS DOE 40 kW

Manufacturer: McDonnell-Douglas Aircraft Corporation
Address: Box 516, St. Louis, MO 63166
Contact: John Anderson
Model number: DOE 40 kW
Rated power (kW): 40 @ 20 mph (see Figure 8-98)
Power output at 12 mph: Unspecified
Power output at 14 mph: Unspecified
Rotor blade diameter: 32.5 ft by 65 ft diameter
Number of blades: 3
Machine type: Vertical-axis gyromill
Cost: $20,000; $500/kW design goal; $10,000 development cost
Operating speed
  Rated speed (mph): 20
  Cut-in speed (mph): Minimum
  Cut-out speed (mph): Maximum
  Maximum speed (mph): 125
rpm at rated output: Unspecified
Blade materials: Unspecified
Rotor weight: Unspecified
System weight on tower: Unspecified
Overspeed control: Unspecified
Generator/alternator type: Unspecified
Application: Deep well irrigation, small community power, small factories
Testing history: Under development
Reliability prediction (MTBF): Unspecified
Warranty: Unspecified
Maintenance requirements: Unspecified
Country of origin: U.S.
Number produced: 1
Availability: DOE development project

MILLVILLE WINDMILLS 10-3-IND

Manufacturer: Millville Windmills and Solar Equipment Co.
Address: P.O. Box 32, 10335 Old 44 Drive, Millville, CA 96062
Contact: Mr. Devon Tassen
Telephone: (916) 547-4302
Model number: 10-3-IND (see Figure 8-99)
Rated power (kW): 10.0 @ 25 mph (see Figure 8-100)
Power output at 12 mph: 0.9 kW
Power output at 14 mph: 1.9 kW
Rotor blade diameter: 24.3 ft

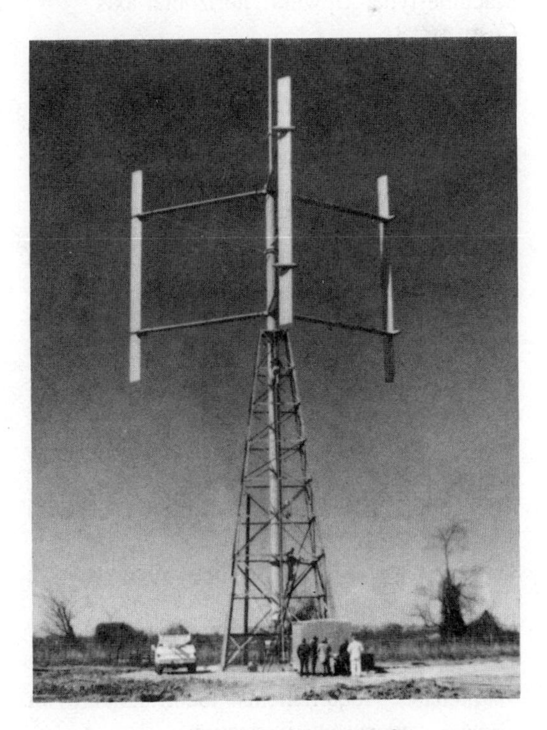

WIND VELOCITY IN MILES PER HOUR

Figure 8-98. McDonnell-Douglas Model DOE 40-kW Model wind energy conversion system.

Figure 8-99. Millville Model 10-3-IND downwind 25-ft, diameter WECS.

Number of blades: 3
Machine type: Upwind, horizontal-axis
Cost: $6250
Operating speed
  Rated speed (mph): 25
  Cut-in speed (mph): 9
  Cut-out speed (mph): 60
  Maximum speed (mph): 80
rpm at rated output: 80
Blade materials: Aluminum and stainless steel
Rotor weight: 230 lb
System weight on tower: 1000 lb
Overspeed control: Mechanical; blade feathering, rotor turns edgewise to excessive winds
Generator/alternator type: 220 V, induction generator, 42 A, 60 Hz ac
Application: Electricity production for utility tie-in
Testing history: Field testing, operated on tower
Reliability prediction (MTBF): Unknown
Warranty: One year
Maintenance requirements: Semi-annual, oil check and system inspection
Country of origin: U.S.
Number produced: 10

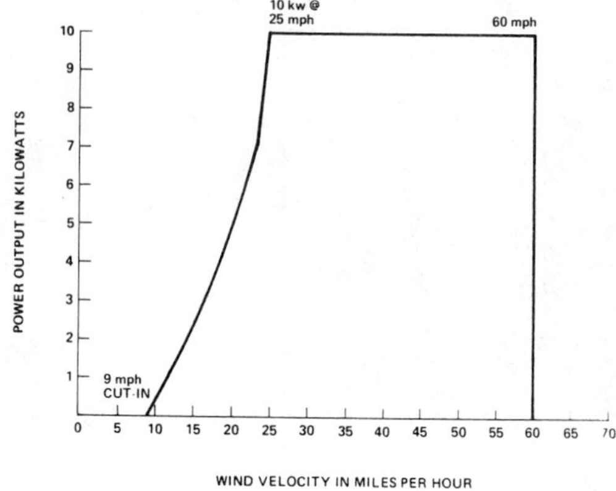

Figure 8-100. Millville Model 10-3-IND power output curve.

Availability: 45 days
Comments: Efficiency approximately 25 percent

NOAH 55 kW

Manufacturer: NOAH
Model number: 55kW
Rated power (kW): 55.0 (see Figure 8-101)
Power output at 12 mph: 22 kW
Power output at 14 mph: 30 kW
Rotor blade diameter: 36 ft
Number of blades: 10, double rotor
Operating speed
  Rated speed (mph): 20
  Cut-in speed (mph): 6.7
  Cut-out speed (mph): 45
  Maximum speed (mph): Unspecified
rpm at rated output: 71.4
Blade materials: Unspecified
Rotor weight: Unspecified

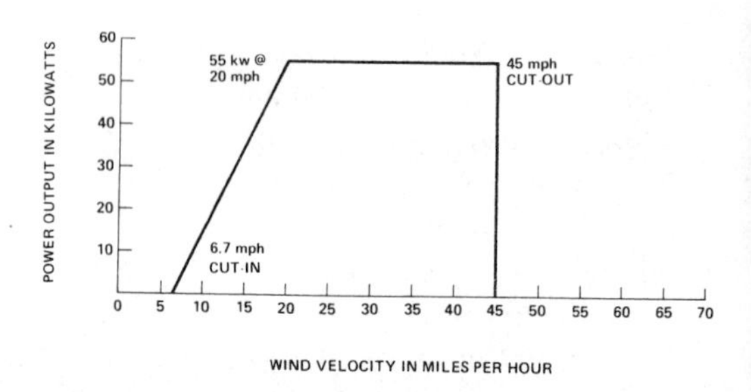

Figure 8-101. Noah 55-kW Model power output curve.

System weight on tower: Unspecified
Overspeed control: Electronic
Generator/alternator type: 28 permanent magnet, ac generator
Application: Electricity production
Testing history: Unspecified
Reliability prediction (MTBF): Unspecified
Warranty: Unspecified
Maintenance requirements: Unspecified
Country of origin: Switzerland
Number produced: Unspecified
Availability: Unspecified

NORTH WIND — DOE 1 kW

Manufacturer: North Wind Power Co.
Address: Box 315, Warren, VT 05674
Contact: Mr. Don Mayer
Telephone: (802) 496-2995
Model number: DOE 1 kW (see Figure 8-102)
Rated power (kW): 2.2 kW @ 20 mph (1 kW @ 20 mph specified) (see Figure 8-103)
Power output at 12 mph: Unknown
Power output at 14 mph: Unknown
Rotor blade diameter: 16.4 ft
Number of blades: 3
Machine type: Upwind, horizontal-axis
Cost: $1500/kW; development costs: $260,000
Operating speed
  Rated speed (mph): 20
  Cut-in speed (mph): Unspecified (Minimum)
  Cut-out speed (mph): Unspecified (Maximum)
  Maximum speed (mph): 165
rpm at rated output: 250
Blade materials: Sitka Spruce Wood
Rotor weight: Unknown
System weight on tower: 625 lbs
Overspeed control: Based on Paris-Dunn Co. system of the 1930's; rotor tilts to 90° horizontal position at 105 mph, rotor, tilt-back
Generator/alternator type: 3-phase, 12 pole, direct drive Lundell rotor
Application: For small-scale rural and remote applications
Testing history: Under development
Reliability prediction (MTBF): High-reliability design (10 years MBTF) proposed
Warranty: Unspecified

Figure 8-102. North Wind 1-kW WECS under development for DOE.

Maintenance requirements: Minimal
Country of origin: U.S.
Number produced: 1
Availability: DOE development project

NORTH WIND — 2 kW, 32V

Manufacturer: North Wind Power Co.
Address: P.O. Box 315, Warren, VT 05674
Contact: Mr. Don Mayer
Telephone: (802) 496-2995
Model number: 2kW, 32V

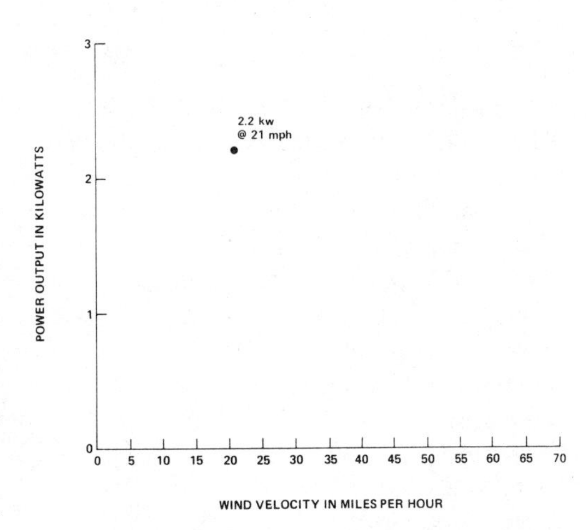

Figure 8-103. North Wind Power DOE 1-kW Model power output curve.

Rated power (kW): 2.0 @ 22 mph (see Figure 8-104)
Power output at 12 mph: 350 W
Power output at 14 mph: 650 W
Rotor blade diameter: 13.6 ft
Number of blades: 3
Machine type: Upwind, horizontal-axis
Cost: $2200
Operating speed
  Rated speed (mph): 22
  Cut-in speed (mph): 8
  Cut-out speed (mph): None
  Maximum speed (mph): Unspecified
rpm at rated output: 265
Blade materials: Wood, Sitka Spruce, fiberglass coating
Rotor weight: 70 lb
System weight on tower: 480 lb
Overspeed control: Mechanical, centrifugal blade pitching
Generator/alternator type: 32 vdc generator, direct drive
Application: SWECS
Testing history: Not available
Reliability prediction (MTBF): Unspecified
Warranty: Contact North Wind Power Co.
Maintenance requirements: Grease, clean or replace brushes, refurbish blades every five years
Country of origin: U.S.
Number produced: 150
Availability: Contact North Wind Power Co.

NORTH WIND 2 kW, 110V

Manufacturer: North Wind Power Co.
Address: P.O. Box 315, Warren, VT 05674
Contact: Mr. Don Mayer
Telephone: (802) 496-2995
Model number: 2 kW, 110 V
Rated power (kW): 2.0 @ 22 mph
Power output at 12 mph: 400 W
Power output at 14 mph: 650 W
Rotor blade diameter: 13.6 ft
Number of blades: 3
Machine type: Upwind, horizontal-axis
Cost: $3500
Operating speed
  Rated speed (mph): 22
  Cut-in speed (mph): 8

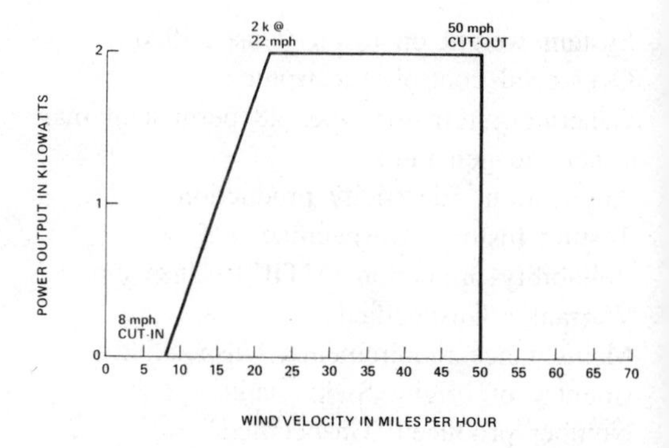

Figure 8-104. North Wind Power 2-kW, 32 V Model power output curve.

  Cut-out speed (mph): None
  Maximum speed (mph): Unspecified
rpm at rated output: 265
Blade materials: Wood, Sitka Spruce, fiberglass coating
Rotor weight: 70 lb
System weight on tower: 480 lb
Overspeed control: Mechanical, centrifugal blade pitching
Generator/alternator type: 110 vdc generator, direct drive
Application: SWECS
Testing history: Not available
Reliability prediction (MTBF): Unspecified
Warranty: One year unconditional, generator and main components
Maintenance requirements: Grease, clean or replace brushes, refurbish blades every five years
Country of origin: U.S.
Number produced: Several hundred
Availability: Contact North Wind Power Co.

NORTH WIND 3 kW, 32V

Manufacturer: North Wind Power Co.
Address: P.O. Box 315, Warren, VT 05674
Contact: Mr. Don Mayer
Telephone: (802) 496-2995
Model number: 3 kW, 32 V (see Figure 8-105)
Rated power (kW): 3.0 @ 27 mph (see Figure 8-106)
Power output at 12 mph: 650 W
Power output at 14 mph: 950 W

Figure 8-105. North Wind Power Co. refurbished Jacobs WECS. (This photo includes the original generator tail vane. A different governor and propeller are shown.)

Rotor blade diameter: 13.6 ft
Number of blades: 3
Machine type: Upwind, horizontal-axis
Cost: $3200
Operating speed
  Rated speed (mph): 27
  Cut-in speed (mph): 8
  Cut-out speed (mph): None
  Maximum speed (mph): None
rpm at rated output: 265
Blade materials: Wood, Sitka Spruce, fiberglass coating
Rotor weight: 70 lb
System weight on tower: 480 lb
Overspeed control: Mechanical, centrifugal blade pitching
Generator/alternator type: 32 vdc generator, direct drive
Application: SWECS
Testing history: Not available
Reliability prediction (MTBF): Unspecified
Warranty: One year unconditional, generator and main components
Maintenance requirements: Grease, every three to five years, clean or replace brushes, refurbish blades every five years
Country of origin: U.S.
Number produced: 1000
Availability: Contact North Wind Power Co.

NORTH WIND 3 kW, 110V

Manufacturer: North Wind Power Co.
Address: P.O. Box 315, Warren, VT 05674
Contact: Mr. Don Mayer
Telephone: (802) 496-2995
Model number: 3 kW, 110 V
Rated power (kW): 3.0 @ 25 mph
Power output at 12 mph: 650 W
Power output at 14 mph: 950 W
Rotor blade diameter: 13.6 ft
Number of blades: 3
Machine type: Upwind, horizontal-axis
Cost: $4600
Operating speed
  Rated speed (mph): 25
  Cut-in speed (mph): 8
  Cut-out speed (mph): None
  Maximum speed (mph): None
rpm at rated output: 265
Blade materials: Wood, Sitka Spruce, fiberglass coating
Rotor weight: 70 lb
System weight on tower: 480 lb

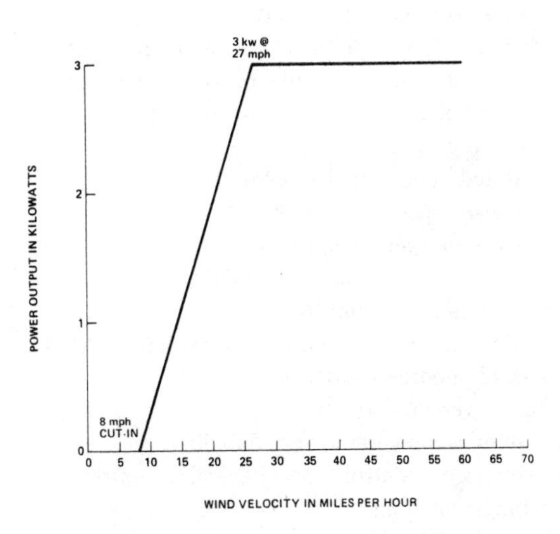

Figure 8-106. North Wind Power, 3-kW, 32V Model power output curve.

Overspeed control: Mechanical, centrifugal blade pitching
Generator/alternator type: 110 vdc generator, direct drive
Application: SWECS
Reliability prediction (MTBF): Unspecified
Warranty: One year unconditional, generator and main components
Maintenance requirements: Grease, clean or replace brushes, refurbish blades every five years
Country of origin: U.S.
Number produced: Thousands
Availability: Contact North Wind Power Co.
DOE Rocky Flats Tests: See DOE Report RFP-2920/3533/78/6-2 of September 28, 1978 and later reports for more information.

PINSON CYCLOTURBINE C2E3

Manufacturer: Pinson Energy Corporation
Address: P.O. Box 7, Marston Mills, MA 02648
Contact: Mr. Tom Sadler
Telephone: (617) 428-8535
Model number: C2E 3 (see Figure 8-107)
Rated power (kW): 2.2 @ 22.4 mph (see Figure 8-108)
Power output at 12 mph: 336 W
Power output at 14 mph: 533 W
Rotor blade diameter: 15.6 ft
Number of blades: 3, vertical
Machine type: Self-starting and governing, straight-bladed variable pitch vertical axis.
Cost: $4500 (rotor and transmission)
Operating speed
  Rated speed (mph): 22.4
  Cut-in speed (mph): 8
  Cut-out speed (mph): 50
  Maximum speed (mph): 120
rpm at rated output: 200
Blade materials: Aluminum 6061-T6, NACA 0015 modified airfoil
Rotor weight: 275 lb
System weight on tower: 533 lb
Overspeed control: Mechanical, centripetal blade pitching
Generator/alternator type: 120V-33A or 240V-16A internal excitation

Figure 8-107. Pinson Cycloturbine Model C2E WECS.

Application: Electrical restrictive water heating, compressing air, utility grid interfacing with synchronous inverter and direct water heating via stirring, battery charging
Testing history: Tachometer, anemometer, voltmeter, resistive load bank
Reliability prediction (MTBF): Unspecified
Warranty: Ninety days labor, one year parts
Maintenance requirements: Semi-annual lubrication bearings, tighten hardware, replace drive belts, visual inspection

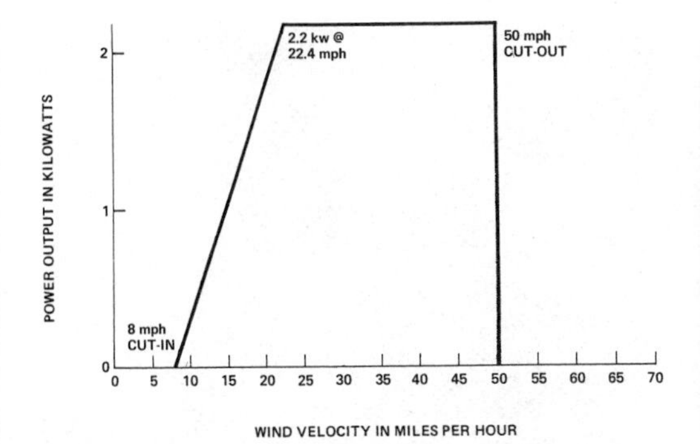

Figure 8-108. Pinson Cycloturbine Model C3E power output curve.

Country of origin: U.S.
Number produced: 20
Availability: Contact factory
Comments: The cycloturbine is an innovative vertical-axis wind machine that overcomes the basic problem associated with vertical-axis designs—the inability to self-start without complex and expensive logic systems. Prototype cycloturbines have performed very well in testing over the past year, and the first production run of machines are now being installed in the Cape Cod area.

SENCENBAUGH WIND ELECTRIC 24-14

Manufacturer: Sencenbaugh Wind Electric
Address: P.O. Box 1174, Palo Alto, CA 94306
Contact: Mr. Jim Sencenbaugh
Telephone: (415) 964-1593
Model number: 24-14 (see Figure 8-109)
Rated power: 24 W @ 21 mph (see Figure 8-110)
Power output at 12 mph: 5 W
Power output at 14 mph: 10 W
Rotor blade diameter: 20 in.
Number of blades: 3
Machine type: Horizontal-axis
Cost: $485
Operating speed
- Rated speed (mph): 21
- Cut-in speed (mph): 8–9
- Cut-out speed (mph): Unspecified
- Maximum speed (mph): 100

Figure 8-109. Sencenbaugh Model 24-14 SWECS.

rpm at rated output: 3200
Blade materials: Aluminum 356-T6
Rotor weight: 2.5 lb
System weight on tower: 18 lb
Overspeed control: None required
Generator/alternator type: dc generator provides 1.7 A output at 14.4 vdc in 21-mph wind
Application: Trickle charge 12-V battery
Testing history: Two years in-use experience
Reliability prediction (MTBF): Unspecified
Warranty: One year, parts and labor
Maintenance requirements: Grease bearings once a year
Country of origin: U.S.

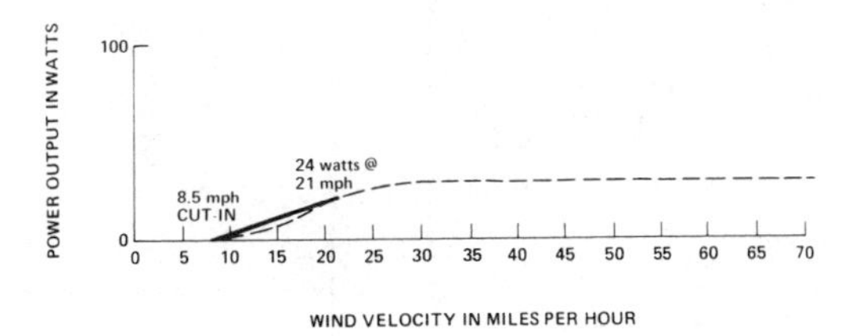

Figure 8-110. Sencenbaugh Model 24-14 power output curve.

Number produced: 20
Availability: 4–6 weeks ARO
Comments: Sencenbaugh provides an excellent user-oriented catalog for SWECS. Send $2 for Catalog 1078 to above address.

SENCENBAUGH WIND ELECTRIC 400-14HDS

Manufacturer: Sencenbaugh Wind Electric
Address: P.O. Box 1174, Palo Alto, CA 94306
Contact: Mr. Jim Sencenbaugh
Telephone: (415) 964-1593
Model number: 400-14HDS (see Figures 8-111 and 8-112)
Rated power: 400 W @ 20 mph (see Figure 8-113)
Power output at 12 mph: 75 W
Power output at 14 mph: 145 W
Rotor blade diameter: 7 ft
Number of blades: 3
Machine type: Upwind, horizontal-axis
Cost: $1250
Operating speed
  Rated speed (mph): 20
  Cut-in speed (mph): 9–9
  Cut-out speed (mph): Unspecified
  Maximum speed (mph): 120
rpm at rated output: 1000

Figure 8-111. Sencenbaugh Model 500-14 HDS small-scale WECS. (*Courtesy of Sencenbaugh Wind Electric.*)

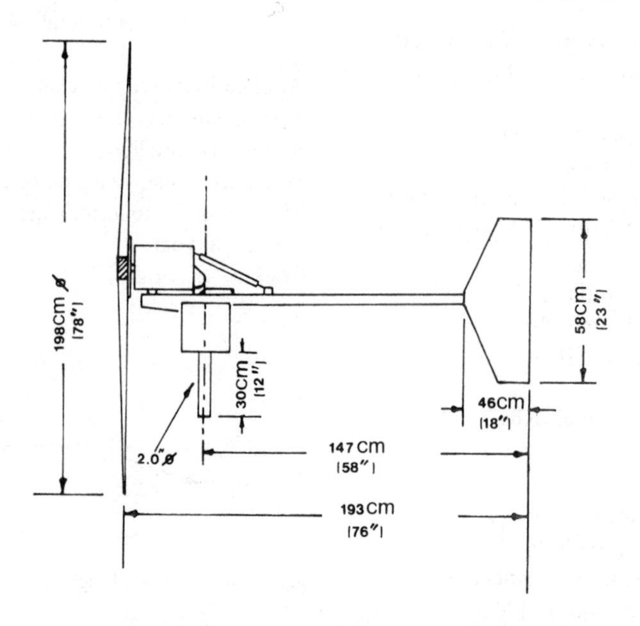

Figure 8-112. Sencenbaugh Model 500-14 HDS SWECS mechanical configuration.

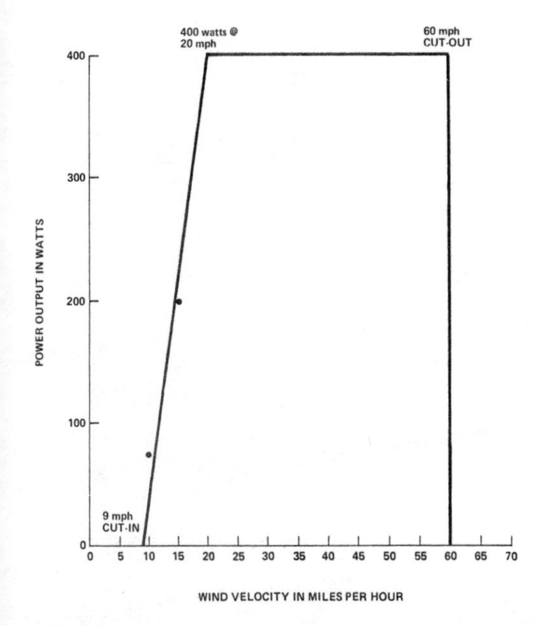

Figure 8-113. Sencenbaugh Model 500-14 HDS power output curve.

Blade materials: Sitka Spruce (wood), polyurethane finish
Rotor weight: 3 lb
System weight on tower: 65 lb
Overspeed control: Rotor tilts upward in excessive winds
Generator/alternator type: 14 VDC, self-excitation, 3-phase alternator
Application: 12-V battery storage system
Testing history: Field test, firm started in 1972
Reliability prediction (MTBF): 15 years
Warranty: One year, parts and labor
Maintenance requirements: Annual inspection
Country of origin: U.S.
Number produced: 29
Availability: Four weeks ARO

SENCENBAUGH WIND ELECTRIC

500-14

Manufacturer: Sencenbaugh Wind Electric
Address: P.O. Box 1174, Palo Alto, CA 94306
Contact: Mr. Jim Sencenbaugh
Telephone: (415) 964-1593
Model number: 500-14 (see Figures 8-114 and 8-115)
Rated power: 500 W @ 25 mph (see Figure 8-116)
Power output at 12 mph: 50 W
Power output at 14 mph: 125 W
Rotor blade diameter: 6 ft
Number of blades: 3
Machine type: Upwind horizontal-axis
Cost: $2250
Operating speed
  Rated speed (mph): 25
  Cut-in speed (mph): 9–10
  Cut-out speed (mph): Unspecified
  Maximum speed (mph): 140
rpm at rated output: 1000
Blade materials: Sitka Spruce (wood), epoxy finish
Rotor weight: 10 lb
System weight on tower: 243 lb
Overspeed control: Rotor turns sideways to the wind
Generator/alternator type: 14 or 28 vdc, self-excitation
Application: 12 or 24 vdc battery storage system
Testing history: Field test, firm started in 1972
Reliability prediction (MTBF): 15 years
Warranty: One year, parts and labor
Maintenance requirements: Annual inspection and lubrication
Country of origin: U.S.
Number produced: 10
Availability: Four to five weeks ARO
Comments: Efficiency of 500-14 is approximately 24 percent

SENCENBAUGH WIND ELECTRIC

1000-14

Manufacturer: Sencenbaugh Wind Electric
Address: P.O. Box 1174, Palo Alto, CA 94306
Contact: Mr. Jim Sencenbaugh
Telephone: (415) 964-1593
Model number: 1000-14 (see Figures 8-117 and 8-118)
Rated power (kW): 1.0 @ 23 mph (see Figure 8-119)
Power output at 12 mph: 300 W
Power output at 14 mph: 450 W

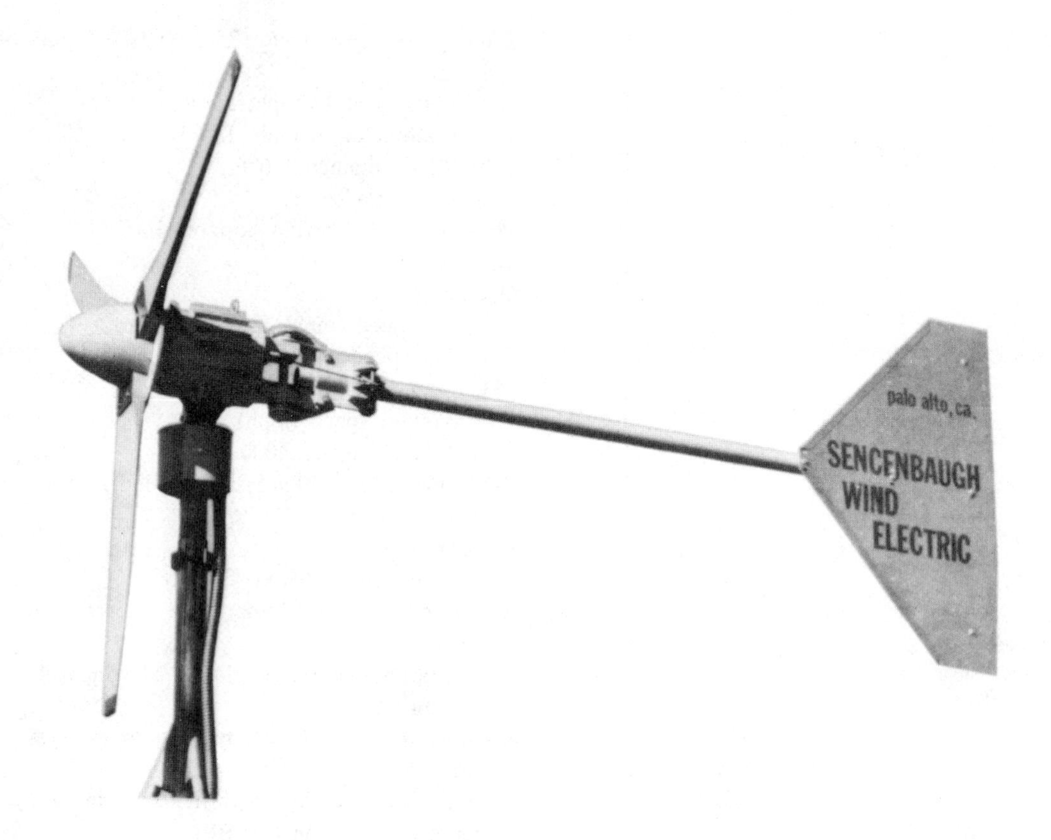

Figure 8-114. Sencenbaugh Model 500-14 SWECS.

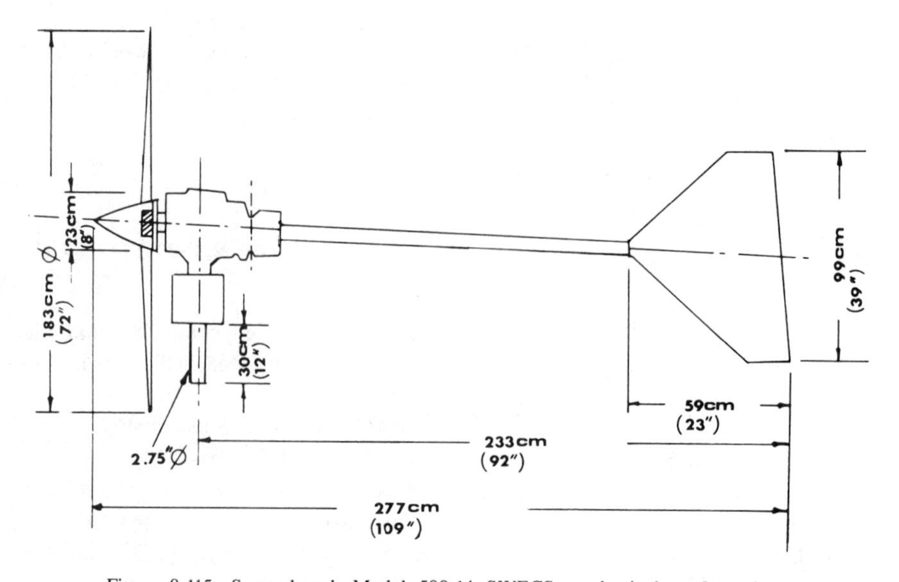

Figure 8-115. Sencenbaugh Model 500-14 SWECS mechanical configuration.

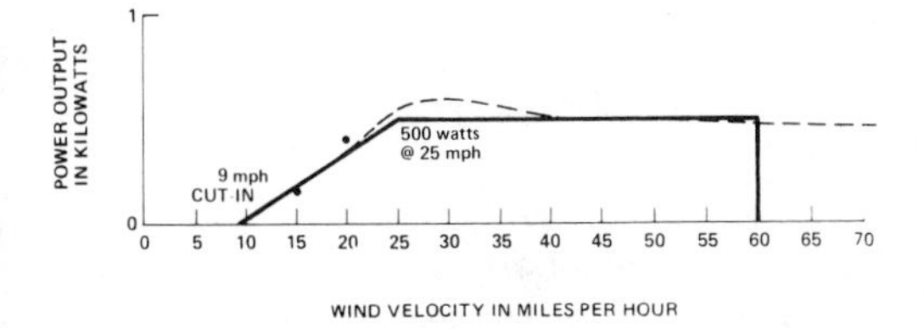

Figure 8-116. Sencenbaugh Model 500-14 power output curve.

Rotor blade diameter: 12 ft
Number of blades: 3
Machine type: Upwind horizontal-axis, propeller type, fixed pitch
Cost: $2950
Operating speed
 Rated speed (mph): 22–23
 Cut-in speed (mph): 6–8
 Cut-out speed (mph): Unspecified
 Maximum speed (mph): 80
rpm at rated output: 290
Blade materials: Sitka Spruce (wood), epoxy finish, bonded copper leading edge
Rotor weight: 16 lb
System weight on tower: 300 lb

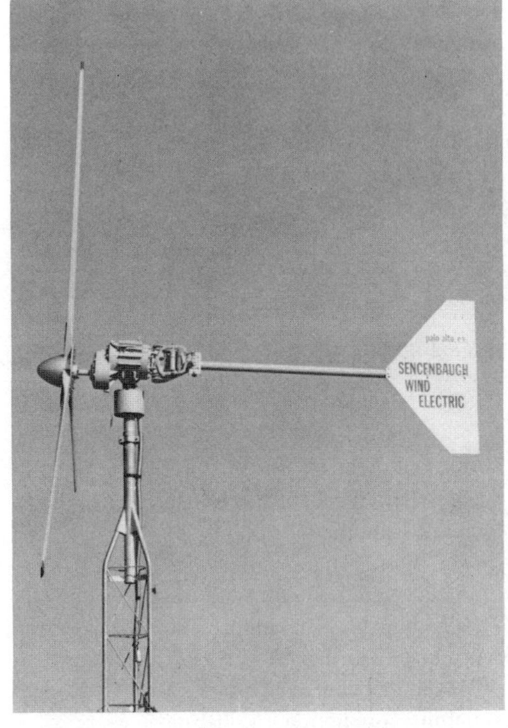

Figure 8-117. Sencenbaugh Model 1000-14 SWECS.

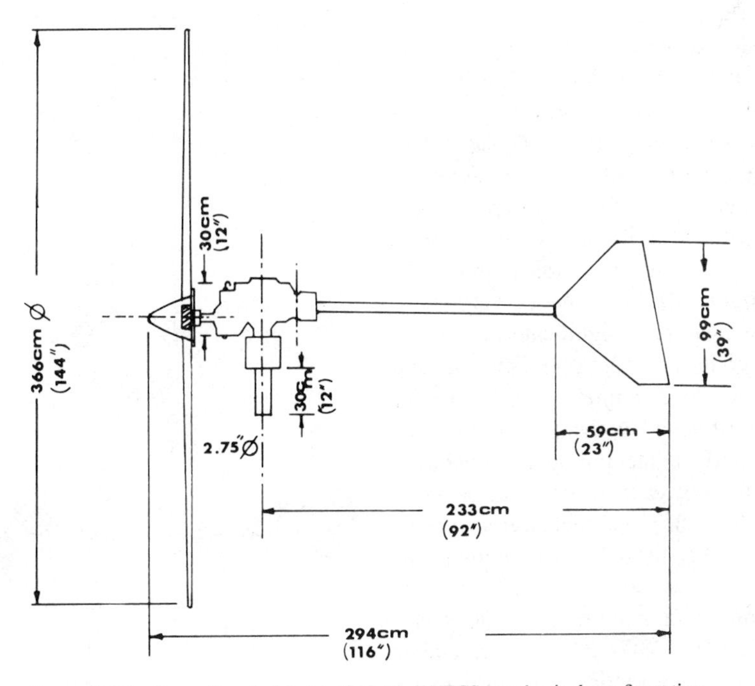

Figure 8-118. Sencenbaugh Model 1000-14 SWECS mechanical configuration.

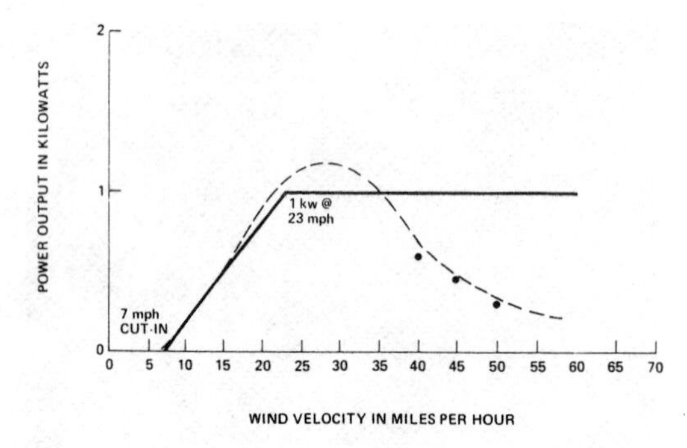

Figure 8-119. Sencenbaugh Model 1000-14 power output curve.

Overspeed control: Rotor turns edgewise to the wind, mechanical; foldable tail
Generator/alternator type: 14 or 28 vdc, self-excitation, 3-phase alternator
Application: 12-V battery storage system
Testing history: The model 1000-14 has been produced since January 1977; has more than 240,000 operational hours.
Reliability prediction (MTBF): 15 years
Warranty: One year, parts and labor
Maintenance requirements: Annual inspection and lubrication
Country of origin: U.S.
Number produced: 45
Availability: Four weeks ARO
DOE Rocky Flats Tests: "This three-bladed, horizontal-axis upwind machine was extensively tested for 8 months prior to June 30, 1978 without a significant failure. Minor problems with oil seal leakage did develop. However, the oil leakage was found to be peculiar to the RF machine (a very early production model). A significant amount of data was collected from the Sencenbaugh under various load configurations and with various adjustments to the machine's foldable tail. Only in the tail-locked operation mode did the power produced matched that specified by the manufacturer's power curve. However, testing is continuing and these data should be considered reliable, but preliminary in nature." See DOE Report RFP-2920/3533/78/6-2 of September 28, 1978 and later reports for more information.

SKYHAWK II

Manufacturer: Independent Energy Systems, Inc.
Address: 6043 Sterrettania Road (Rt. 832), Fairview, PA 16415
Contact: Mr. Bob Hauser or John D'Angelo
Telephone: (814) 833-3567
Model number: II (see Figure 8-120)
Rated power (kW): 2.0 @ 23 mph (see Figure 8-121)
Power output at 12 mph: 600 W
Power output at 14 mph: 850 W
Rotor blade diameter: 13.5 ft
Number of blades: 3
Machine type: Upwind, horizontal-axis
Cost: $3495
Operating speed
  Rated speed (mph): 23
  Cut-in speed (mph): 7–8
  Cut-out speed (mph): 90
  Maximum speed (mph): None
rpm at rated output: 265
Blade materials: Aircraft quality Sitka Spruce (wood) with protected leading edge
Rotor weight: 45 lb
System weight on tower: 485 lb
Overspeed control: Centrifugally activated variable blade pitching
Generator/alternator type: Direct-drive shunt-wound, 6 pole dc generator, 20 A @ 140 vdc; available in 24, 32, 48, 120, and 200 V models

Figure 8-120. Skyhawk II SWECS using portions of Jacobs design. (*Courtesy of Independent Energy Systems.*)

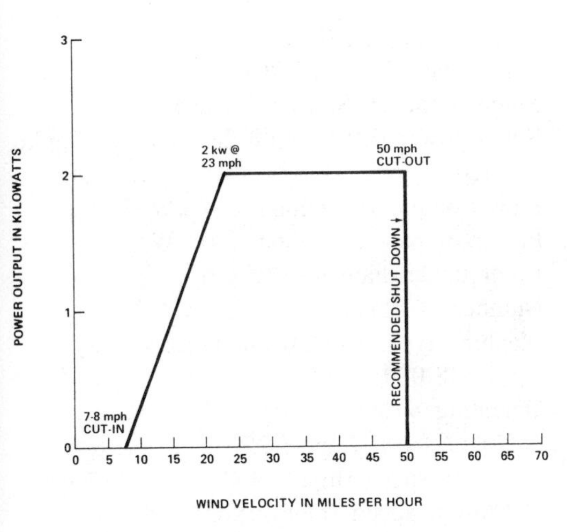

Figure 8-121. Skyhawk Model II power output curve.

Application: SWECS for electricity generation
Testing history: Bench testing, field testing
Reliability prediction (MTBF): Unspecified
Warranty: Material and workmanship to two years; lighting, two years

Maintenance requirements: Grease twice each year, refurbish blades every five years
Country of origin: U.S.
Number produced: Refurbished unit
Availability: In stock

SKYHAWK IV

Manufacturer: Independent Energy Systems, Inc.
Address: 6043 Sterrettania Road (Rt. 832), Fairview, PA 16415
Contact: Mr. John D'Angelo
Telephone: (814) 833-3567
Model number: IV (see Figure 8-122)
Rated power (kW): 4.0 @ 23 mph (see Figure 8-123)
Power output at 12 mph: 1.175 kW
Power output at 14 mph: 1.7 kW
Rotor blade diameter: 15.0 ft
Number of blades: 3
Machine type: Upwind, horizontal-axis
Cost: $4795

Figure 8-122. Skyhawk Model IV WECS.

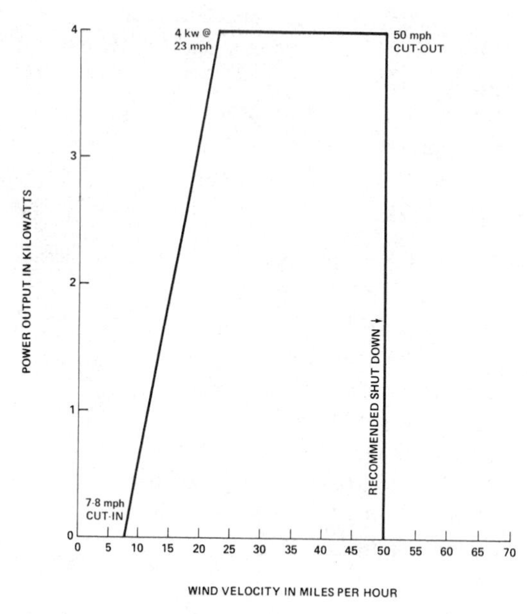

Figure 8-123. Skyhawk Model IV power output curve.

Operating speed
  Rated speed (mph): 23
  Cut-in speed (mph): 7–8
  Cut-out speed (mph): 90
rpm at rated output: 285
Blade materials: Aircraft quality Sitka Spruce (wood) with protected leading edge
Rotor weight: 60 lb
System weight on tower: 600 lb
Overspeed control: Centrifugally activated variable blade pitching
Generator/alternator type: Direct-drive dc generator, 32/120/200 V; 32, 48, 120, 200 V models available
Application: SWECS for electricity generation
Testing history: Bench testing, field testing
Reliability prediction (MTBF): Unspecified
Warranty: Material and workmanship to two years
Maintenance requirements: Grease twice each year, refurbish blades every five years
Country of origin: U.S.
Number produced: 25
Availability: In stock

STORM MASTER — STORM MASTER 10

Manufacturer: Wind Power Systems, Inc.
Address: P.O. Box 17323, San Diego, CA 92117
Contact: Mr. Ed Selter
Telephone: (714) 452-7040
Model number: Storm Master 10
Rated power (kW): 18 @ 24 mph (see Figure 8-124)
Power output at 12 mph: 2.4 kW
Power output at 14 mph: 3.6 kW
Rotor blade diameter: 32.8 ft
Number of blades: 3
Machine type: Downwind, horizontal-axis
Cost: $18,000
Operating speed
  Rated speed (mph): 25
  Cut-in speed (mph): 9–8
  Cut-out speed (mph): None
  Maximum speed (mph): None
rpm at rated output: 150
Blade materials: Fiberglass shell, foam core
Rotor weight: 285 lb
System weight on tower: 950 lb
Overspeed control: Brake, variable pitch, centrifugally controlled governor
Generator/alternator type: 3-phase alternator (permanent magnet), 60 Hz, 240 or 48 vac
Application: Electricity production for home use, 3-phase parallel generation

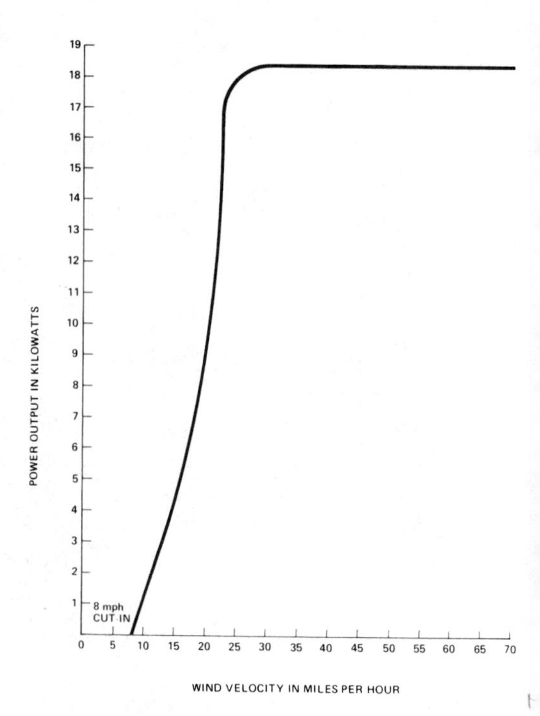

Figure 8-124. Storm Master Model 10 power output curve.

Testing history: Data calculated, field test
Reliability prediction (MTBF): Unknown
Warranty: Two years, materials and workmanship
Maintenance requirements: Unspecified
Country of origin: U.S.
Number produced: 4
Availability: Mid-1980

TETRAHELIX S

Manufacturer: Zephyr Wind Dynamo Company
Address: P.O. Box 241, 21 Stanwood Street, Brunswick, ME 04011
Contact: Mr. Willard Gillette
Telephone: (207) 725-6534
Model number: Tetrahelix S (see Figures 8-125 and 8-126)
Rated power: 7 W @ 25 mph
Power output at 12 mph: 1.3 W
Power output at 14 mph: Unknown
Rotor blade diameter: 8 ft long, 2 ft diameter
Number of blades: 2
Machine type: Vertical-axis, gyromill, self-starting, omni-direction
Cost: $285
Operating speed
  Rated speed (mph): 25
  Cut-in speed (mph): 8–8
  Cut-out speed (mph): Unspecified
  Maximum speed (mph): 50
rpm at rated output: 360
Blade materials: Dacron, nylon, aluminum, kevar ties
Rotor weight: 5 lb
System weight on tower: 5 lb

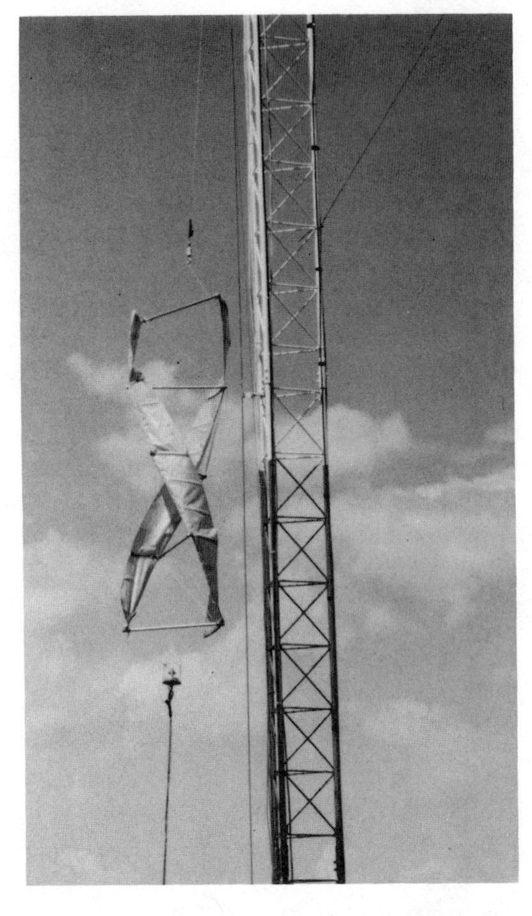

Figure 8-125. Zephyr Tetrahelix Model "S" SWECS.

Overspeed control: Non-destructive collapse in high winds
Generator/alternator type: 14 vdc, permanent magnet
Application: Very small battery charging applications

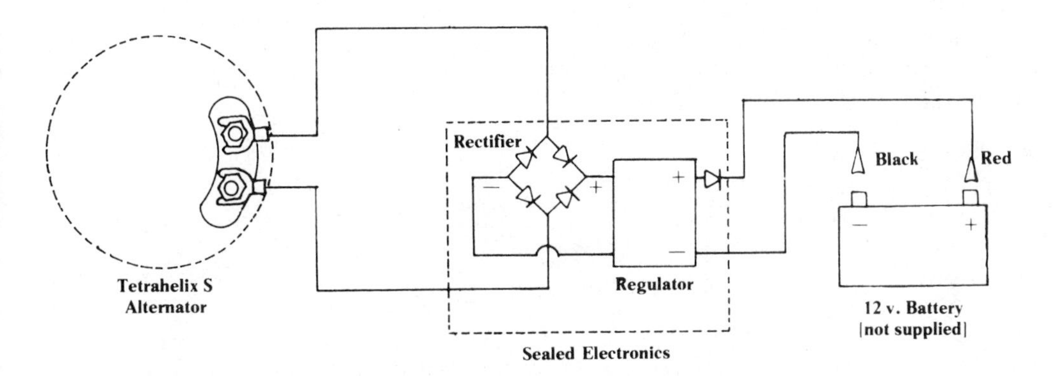

Figure 8-126. Schematic of Zephyr Tetrahelix SWECS.

Testing history: Moving test bed, field testing
Reliability prediction (MTBF): Unspecified
Warranty: 30 days, parts and workmanship
Maintenance requirements: Nominal
Country of origin: U.S.
Number produced: Several
Availability: Four to six weeks ARO

TOPANGA

Manufacturer: Topanga Power
Address: P.O. Box 712, Topango, CA 90290
Rotor blade diameter: 12 or 16 ft
Cost: $1400
rpm at rated output: 300
System weight on tower: 200 lb

TWR I, II, III

Manufacturer: TWR Enterprises/Sun Wind Home Concepts Division
Address: 72 W. Meadows Lane, Sandy UT 84070
Contact: Tom W. Rentz
Model number: Development Model and WINDTITAN I, II, III
Rated power: 800 @ 24 mph (see Figure 8-127)
Power output at 12 mph: 230 W
Power output at 14 mph: 320 W
Rotor blade diameter: 6, 10, 14 and 16 ft
Number of blades: 3
Machine type: Horizontal-axis, high-efficiency props
Cost: $535
Operating speed
Rated speed (mph): 24
Cut-in speed (mph): 7
Cut-out speed (mph): 45
Maximum speed (mph): 90
rpm at rated output: 275–195 at 24 mph
Blade materials: Spruce and aluminum

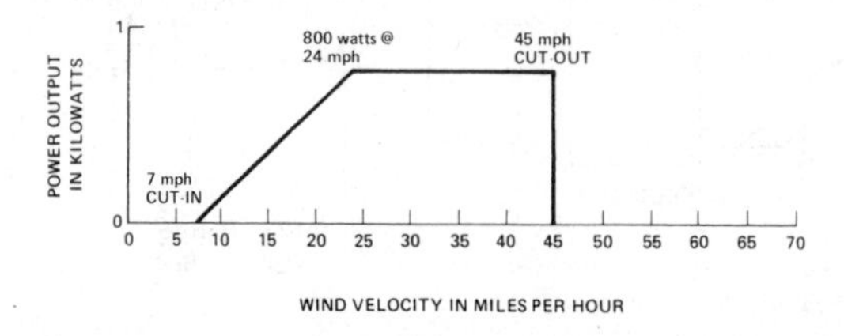

Figure 8-127. TWR 6-ft Model power output curve.

Rotor weight: Unspecified
System weight on tower: 225–370 lb
Overspeed control: Centraxis governor, parawind rudder
Generator/alternator type: 3-phase, slow speed
Application: Home power, electric bicycle and auto charging
Testing history: Evolution from seven previous designs
Reliability prediction (MTBF): Unknown
Warranty: One year, all parts
Maintenance requirements: Lubrication every six months; check field brushes once a year; replace bearings 5–10 years
Country of origin: U.S.
Number produced: 10
Availability: 45–90 days

UNITED TECHNOLOGIES DOE 8 kW

Manufacturer: United Technologies Research Center
Address: Silver Lane, East Hartford, CT 06108
Contact: Mr. M. C. Cheney, Jr.
Telephone: (203) 727-7536
Model number: DOE 8 kW (see Figure 8-128)
Rated power (kW): 9.0 @ 20 mph (see Figure 8-129)
Power output at 12 mph: 2 kW
Power output at 14 mph: 3.2 kW
Rotor blade diameter: 31 ft
Number of blades: 2
Machine type: Downwind, self-adjusting, flex-beam rotor, horizontal-axis
Cost: $6750. Design cost goal is $750/kW. Development cost $38,000
Operating speed
Rated speed (mph): 20
Cut-in speed (mph): 9
Cut-out speed (mph): 100
Maximum speed (mph): 165
rpm at rated output: 108
Blade materials: Fiberglass
Rotor weight: 200 lb
System weight on tower: 1875 lb
Overspeed control: "Flex beam" between the hub and airfoil will allow twisting and feathering motion by simple centrifugal pendulum device; stall technique
Generator/alternator type: Induction generator

Figure 8-128. United Technologies Research Center 8-kW WECS under development for DOE.

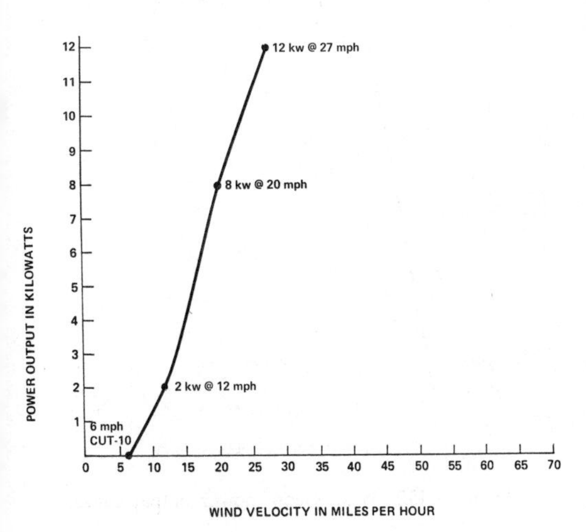

Figure 8-129. United Technologies DOE 8-kW Model power output curve.

Application: Electricity for farm and home use
Testing history: Prototype shakedown
Reliability prediction (MTBF): Unspecified
Warranty: Unspecified
Maintenance requirements: Unspecified
Country of origin: U.S.
Number produced: 1
Availability: DOE development project

WINCO-WINCHARGER 1222 H

Manufacturer: Dyna Technology, Inc.
Address: E. 7th at Division St., P.O. Box 3253, Sioux City, IO 51102
Telephone: (712) 252-1821
Model number: 1222 H (see Figure 8-130)
Rated power: 200 W @ 23 mph (see Figure 8-131)
Power output at 12 mph: 26 W

Figure 8-130. Winco-Wincharger Model 1222H small-scale WECS.

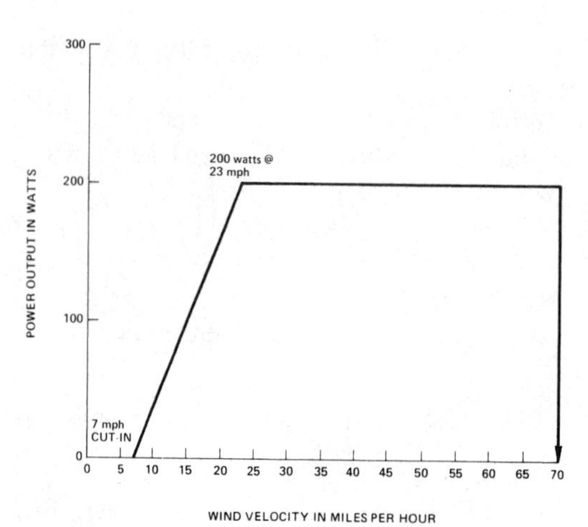

Figure 8-131. Winco-Wincharger Model 1222H power output curve.

Power output at 14 mph: 75 W
Rotor blade diameter: 6 ft
Number of blades: 2
Machine type: Upwind, horizontal-axis
Cost: $575
Operating speed
  Rated speed (mph): 23
  Cut-in speed (mph): 7–8
  Cut-out speed (mph): Unspecified
  Maximum speed (mph): 70
rpm at rated output: 900
Blade materials: Wood, copper leading edge
Rotor weight: 25 lb
System weight on tower: 134 lb
Overspeed control: Air brake governor
Application: Battery charger
Testing history: America's oldest continuous manufacturer of small wind plants
Reliability prediction (MTBF): Unspecified
Warranty: One year
Maintenance requirements: Annual lubrication
Country of origin: U.S.
Number produced: Thousands
Availability: 60 days ARO

WIND GENNI

Manufacturer: Product Development Institute
Address: 508 S. Byrne Rd., Toledo, OH 43609
Telephone: (419) 382-3423
Rated power (kW): 3.0 @ 20 mph (see Figure 8-132)
Power output at 12 mph: 850 W
Power output at 14 mph: 1.4 W
Number of blades: 3
Machine type: Upwind, horizontal-axis
Cost: $3595
Operating speed
  Rated speed (mph): 20
  Cut-in speed (mph): 9
Blade materials: Fiberglass
Overspeed control: Automatic feather
Generator/alternator type: Base load injector for power line tie-in
Country of origin: U.S.
Number produced: 1

WINDWIZARD C9D

Manufacturer: Aero Lectric Co.
Address: 1357 Winters Ave., Creasptown, MD 21502
Contact: Mr. Michael Glick
Telephone: (301) 724-9165
Model number: C9D
Rated power: 600 W @ 26 mph (see Figure 8-133)

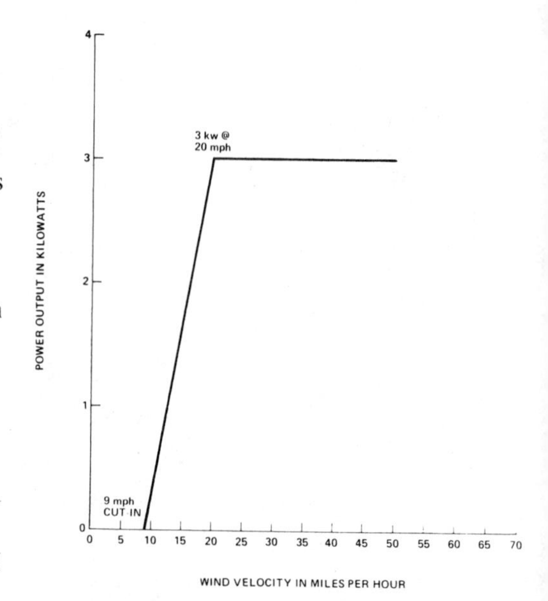

Figure 8-132. Wind Genni power output curve.

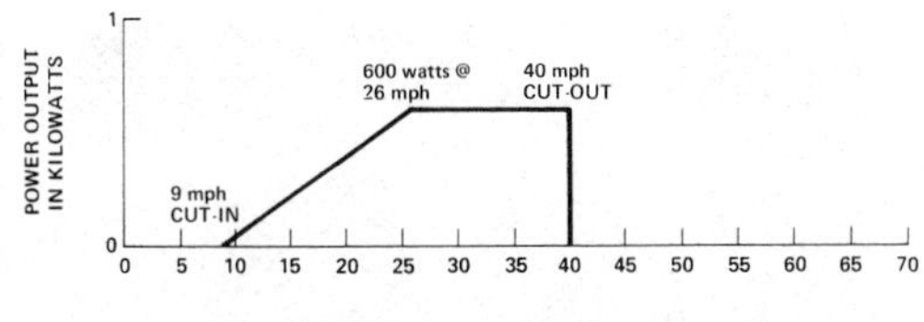

Figure 8-133. Wind Wizard Model C9D power output curve.

Power output at 12 mph: 100 W
Power output at 14 mph: 175 W
Rotor blade diameter: 9 ft
Number of blades: 3
Machine type: Upwind, horizontal-axis
Cost: $995
Operating speed
  Rated speed (mph): 26
  Cut-in speed (mph): 9
  Cut-out speed (mph): 40
  Maximum speed (mph): Unspecified
rpm at rated output: 337
Blade materials: Wood, urethane, and fiberglass
Rotor weight: 22 lb
System weight on tower: 50 lb
Overspeed control: Rotor turns sideways
Generator/alternator type: Alternator adjustable voltage 12–18 V
Application: SWECS electricity generation
Testing history: Field test
Reliability prediction (MTBF): Unspecified
Warranty: One year, parts and workmanship limited
Maintenance requirements: Semi-annual inspection of system
Country of origin: U.S.
Number produced: Unknown
Availability: Four to six weeks ARO

WINDWORKS DOE 8 kW

Manufacturer: Windworks
Address: Box 329, Route 3, Mukwonago, WI 53149
Contact: Mr. Hans Meyer
Telephone: (414) 363-4088
Model number: DOE 8 kW (see Figure 8-134)

Figure 8-134. Windworks 8-kW Model WECS under development for DOE.

Rated power (kW): 8.0 @ 20 mph (see Figure 8-135)
Power output at 12 mph: Unspecified
Power output at 14 mph: Unspecified
Rotor blade diameter: 32.8 ft
Number of blades: 3
Machine type: Downwind, horizontal-axis propeller type
Cost: $6000; desired cost goal: $750/kW

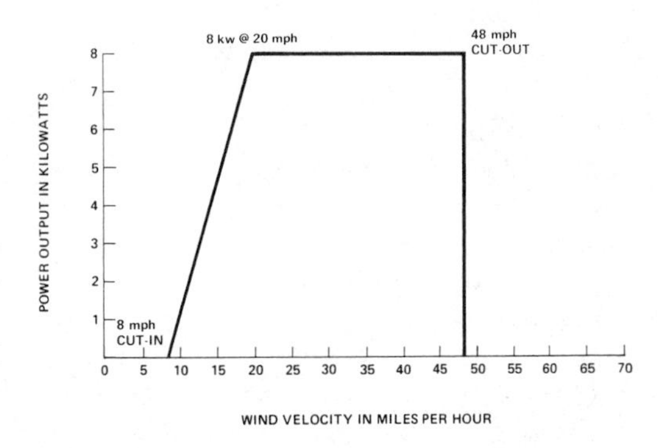

Figure 8-135. Windworks Model DOE 8-kW Model power output curve.

Operating speed
  Rated speed (mph): 20
  Cut-in speed (mph): 8
  Cut-out speed (mph): 48
  Maximum speed (mph): 165
rpm at rated output: 145
Blade materials: Aluminum, extruded, 6063T6
Rotor weight: 400 lb
System weight on tower: 1620 lb
Overspeed control: Blade feathering by hudraulic blade-pitch control system
Generator/alternator type: Permanent magnet samarium cobalt direct drive
Application: Tie-in with utility or gasoline-fueled generator, independent of back-up power source
Testing history: Under development
Reliability prediction (MTBF): Unspecified
Warranty: Unspecified
Maintenance requirements: Unspecified
Country of origin: U.S.
Number produced: 1
Availability: DOE development project

WHIRLWIND POWER CO. A

Manufacturer: Whirlwind Power Co.
Address: 2458 W. 29th Ave., Denver, CO 80211
Contact: Mr. Elliott Bayly
Telephone: (303) 477-6436
Model number: A (see Figure 8-136)
Rated power (kW): 2.0 @ 25 mph (see Figure 8-137)
Power output at 12 mph: 200 W
Power output at 14 mph: 400 W
Rotor blade diameter: 10 ft
Number of blades: 2
Machine type: Downwind, horizontal-axis
Cost: $2995
Operating speed
  Rated speed (mph): 25
  Cut-in speed (mph): 10
  Cut-out speed (mph): 50
  Maximum speed (mph): 80
rpm at rated output: 900
Blade materials: Sitka Spruce wood
Rotor weight: Unspecified
System weight on tower: 71 lb

Figure 8-136. Whirlwind Power Model "A" WECS. (*Courtesy of Whirlwind Power Company.*)

Overspeed control: Electro magnetic brake
Generator/alternator type: 12, 24, 32, 48, 120, 240 vac permanent magnet
Application: SWECS electricity generation, battery charging, water or space heating
Testing history: Field test
Reliability prediction (MTBF): Unspecified
Warranty: One year, parts and labor
Maintenance requirements: Every five years lubricate bearings
Country of origin: U.S.

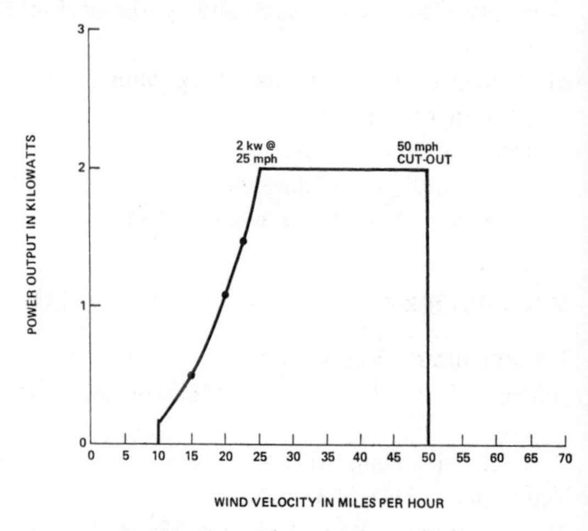

Figure 8-137. Whirlwind Power Model "A" power output curve.

Number produced: 30
Availability: Stock to 120 days

WIND POWER PRODUCTS 3MW

Manufacturer: Wind Power Products Co.
Address: 213 Boeing Field Terminal, Seattle, Washington 98108
Model number: 3MW
Rated power: 3 MW @ 40 mph
Rotor blade diameter: 165 ft
Number of blades: 3
Machine type: Horizontal-axis
Operating speed
  Rated speed (mph): 40
  Cut-in speed (mph): Unknown
  Cut-out speed (mph): Unknown
  Maximum speed (mph): Unknown
Application: Large-scale WEC
Testing history: System under private development
Reliability prediction (MTBF): Unknown
Warranty: Unspecified
Country of origin: U.S.
Number produced: 1
Availability: Contact factory

Figure 8-138. WTG Model MP1-200 200-kW LWECS.

WTG MP1-200

Manufacturer: WTG Energy Systems, Inc.
Address: 251 Elm Street, Buffalo, NY 14203
Contact: Mr. Allen Spaulding, President
Telephone: (716) 856-1620
Model number: MP1-200 (see Figures 8-138 and 8-139)
Rated power (kW): 200 @ 28 mph (see Figure 8-140)
Power output at 12 mph: 15 kW
Power output at 14 mph: 30 kW
Rotor blade diameter: 80 ft
Number of blades: 3
Machine type: Upwind, horizontal-axis, fixed pitch blades, design based on Dutch Gedser WECS
Cost: $226,000 F.O.B. Buffalo, NY; includes 80 ft tower
Operating speed
  Rated speed (mph): 28
  Cut-in speed (mph): 8
  Cut-out speed (mph): 60
  Maximum speed (mph): 150
rpm at rated output: 30 rpm (synchronous speed)
Blade materials: Steel tubing spar, galvanized steel skin, fixed pitch
Rotor weight: 15,000 lb
System weight on tower: 85,000 lb with tower
Overspeed control: Hydraulic held, spring activated drag flaps at blade tips, interval disk brake, stall effect of airfoil, load on generator
Generator/alternator type: ac synchronous generator
Application: Small community electricity production, interface with continuous duty diesel utilities

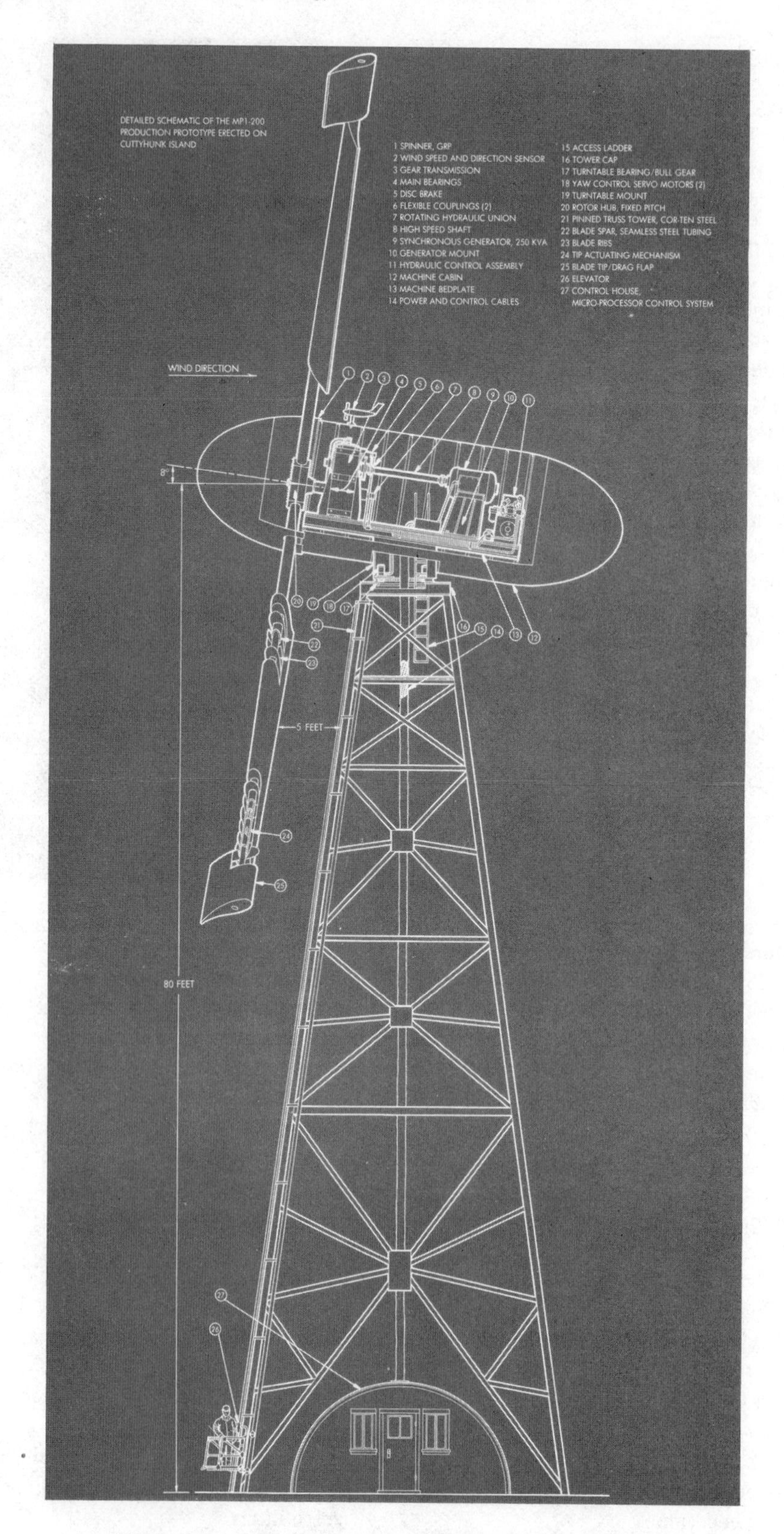

Figure 8-139. Major components of WTG MP-1-200 LWECS.

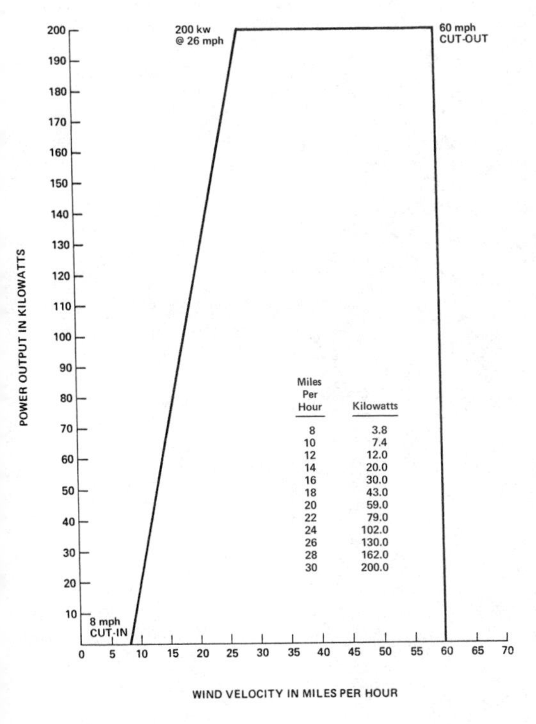

Figure 8-140. WTG Model MP-1-200 power output curve.

Testing history: Field tested since 1977
Reliability prediction (MTBF): Thirty year design life
Warranty: Twelve months
Maintenance requirements: Lubrication and inspection per schedule
Country of origin: U.S.
Number produced: 1
Availability: Six months

ZEPHYR 647 VLS-PM

Manufacturer: Zephyr Wind Dynamo
Address: P.O. Box 241, Brunswick, ME 04011
Contact: Mr. Willard Gillette
Telephone: (207) 725-6534
Model number: 647 VLS-PM (see Figure 8-141)
Rated power (kW): 15 @ 30 mph (see Figure 8-142)
Power output at 12 mph: 2.8 kW
Power output at 14 mph: 4.2 kW
Rotor blade diameter: 20 ft
Number of blades: 3
Machine type: Downwind, horizontal-axis
Cost: $12,000 for turbine, $25,000 installed
Operating speed
  Rated speed (mph): 30
  Cut-in speed (mph): 8
  Cut-out speed (mph): 45
  Maximum speed (mph): Unspecified
rpm at rated output: 300
Blade materials: Urethane mold covered with Kevlar skin
Rotor weight: 225 lb
System weight on tower: 600 lb
Overspeed control: Glide out spoilers and servo control
Generator/alternator type: Alternator
Application: SWEC electricity generation
Testing history: Limited field test
Reliability prediction (MTBF): Unspecified
Warranty: Unspecified
Maintenance requirements: Unspecified

Figure 8-141. Zephyr Model 647 VLS-PM WECS.

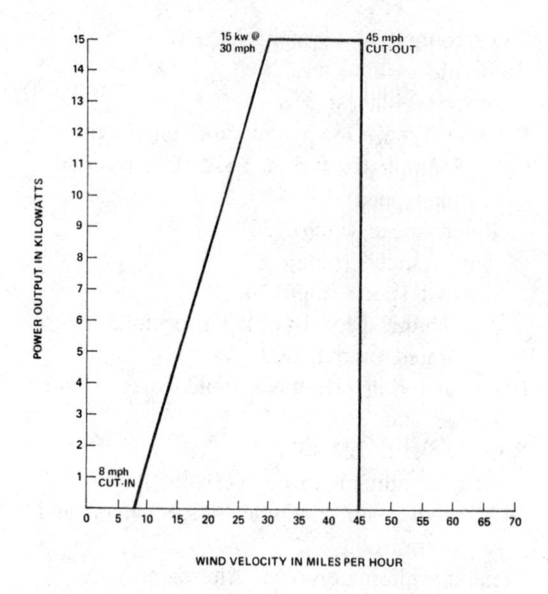

Figure 8-142. Zephyr Model 647 VLS-PM power output curve.

Country of origin: U.S.
Number produced: 3
Availability: Four to six weeks ARO
DOE Rocky Flats Tests: See DOE Report RFP-2920/3533/78/6-2 of September 28, 1978 and later reports for more information.

9 Applications

POTENTIAL APPLICATIONS OF WIND ENERGY CONVERSION SYSTEMS

In general, wind energy conversion, as has been discussed in this book, is the process by which the kinetic energy in the wind is initially converted into mechanical and (usually) electrical energy. Possible final uses of the energy are the following:

- Electrical energy.
- Chemical energy (production of hydrogen and oxygen, production of ammonia, etc.).
- Thermal energy (compression, heat pumps, etc.).
- Potential energy (hydrostorage, pumping water, etc.).

Wind energy conversion systems designed to produce 60-Hz power at a standard voltage (110, 220, etc.) have an immediate and widespread utility. Unfortunately, the nature of the wind and of energy conversion processes causes systems of this type to be penalized with lowered efficiencies, regardless of whether the 60-Hz frequency is obtained directly from synchronous alternators operated at constant speed, or indirectly from battery storage followed by reconversion to ac by means of inverters. Certain systems require electrical power at voltages and frequencies different from those available directly from distribution systems. For example, electrolytic cells require direct current, and at voltages less than 110. In these cases, the output of a WECS might be designed to supply power in directly usable form, but the amount of power used in the totality of all such systems is minute compared to the total electrical power consumed in the U.S. Either the output of WECS power must be in one of the forms currently present in distribution systems, and/or a new technology must evolve to utilize WECS power from its most efficient mode of operation, if the extensive utilization of wind energy is to become attractive. A second problem of comparable magnitude arises in connection with matching the available power to the demand of the user. Wind energy must be captured from moment to moment at a given site. Furthermore, its availability is in no way related to the demands of the consumer. This varying availability must be given prime consideration in the planning and design of any WECS if it is to be practical and to provide reasonably convenient power. In situations for which the wind is the only energy source, energy storage is likely to be a necessity, and the specifications it must meet are rather inflexible. Whenever wind energy is to be supplemented by energy from other sources, storage requirements become less stringent. Situations can be envisioned in which WECS without storage may be practical. In any serious effort to expedite the practical use of the wind energy resource, several reasonable combinations of wind and auxiliary energy sources must be considered, and for a wide range of relative contributions from each.

Storage systems are discussed in Chapter 7; however, the storage problem is nonetheless present, if only tacitly, and must be dealt with in almost every practical scheme for wind energy utilization.

Table 9-1. Different possible ways to pump water with a wind system.

	METHOD	ADVANTAGES	DISADVANTAGES
A	Windmill direct over well driving pump at well with drive shaft or push rod.	Simple. Possibly relatively low cost. Equipment has long history of availability.	Well site may not be a good wind site.
B	Drive shaft to remote well site.	Allows some flexibility in WECS siting. Allows power take off for other requirements.	Safety hazard of drive shaft. Relatively high cost of drive components.
C	Wind generator electric pump.	Allows energy storage (in batteries, etc.) Electricity for other requirements. Relatively high efficiency. Equipment has long history of availability. Allows best flexibility in WECS siting.	Energy loss in long wire runs. Relatively high cost. Safety requirements of electric wire runs.
D	Windmill – hydraulic/pneumatic system. Water can be pumped directly by bubbled air, or by pneumatic pump.	Allows greater flexibility of WECS siting than B. Hydraulic power or compressed air available for other requirements.	Energy loss in long fluid pipes. Safety considerations of compressed fluids or air. Relatively high cost.
E	Jet pump geared directly to windmill. Two or three pipes come down the tower from the pump.	Allows greater flexibility of WECS siting than B. Common, relatively inexpensive pump, which is relatively efficient. Avoids generator and motor losses of the electrical system.	Minimum wind speed for developing minimum head for pumping may be quite high. Must prime the pump at the top of the tower.

Small-scale WECS

In the past, the relative abundance of fossil fuels has restricted wind generated electrical power to only a few unique situations. These have involved relatively isolated locations, too far from electrical distribution systems to justify the line extensions necessary to supply the relatively small demand. In some of these locations—the American Great Plains before the coming of rural electrification; the Australian Outback; offshore drilling platforms, as shown in Figure 9-1; and certain regions of Argentina—it has proved cost-competitive to erect SWECS to produce electricity. Thus, it may be safely said that the most extensive utilization of wind energy up to the present time has been accomplished with the use of small systems.

In general, the energy capture sections of these small systems benefited from much in-situ "engineering" with the result that the towers, rotors and generators came to be reliable in operation. The more serious problems with the systems, at least in the Great Plains, were the following:

- Adequate storage systems were too expensive. Systems could not supply electrical power without some interruptions because of discharged batteries and no wind.
- The lead-acid battery storage systems required more accommodation in use and more careful servicing than could be given under the circumstances.
- The capture unit was too small for the user demand and there was no economical way to obtain incremental increases in its size.
- The system output was low-voltage dc, so that all electrical appliances, etc., were of a special kind, were not always easy to obtain and were more expensive than the 110-V, 60-Hz items commonly available.

Figure 9-1. Automatic Power SWECS on at-sea-pumping platform.

Despite these drawbacks, a large number of small units brought electricity to rural families in America during the 1930's and 1940's. A return to SWECS of a similar type in significant numbers is currently proposed by some as the "decentralized" concept for individual self-sufficiency. Some design modifications are appearing in systems currently offered, but the cost of power from these new systems is still not competitive with that from the large distribution systems, except in unique applications. Nevertheless, if large numbers of small users can obtain a large fraction of their energy requirements from wind-driven generating systems at satisfactory cost, then a considerable saving of fossil fuel would result.

It is becoming more apparent that wind energy utilization in small producer-consumer units is likely to be more feasible in the economic sense if: 1) its use can be coordinated with other energy sources that may be present, and 2) it can be used, at least in part, to supply energy needs other than electrical.

Total Electrical Supply for a (Rural) Home, Farm, Ranch, Small Business (etc.) A listing of the currently available commercial systems in the SWECS classification has been provided in Chapter 7. These units make use of horizontal-axis rotors, as shown in Figure 9-2, with two or three blades, driving specially designed dc generators through a system of gears increasing the armature rotation rate by a factor of four or more. Descriptions of other vertical-axis machines, such as the cycloturbine, Darrieus and unique configurations, are included in Chapter 8. The dc power is used directly to charge a bank of lead-acid storage batteries. If the user is to operate devices requiring ac, either a motor-generator unit or an inverter is required. Because the dc to ac conversion process is usually 65 percent efficient, the user is advised to operate as much of his electrical equipment as possible directly on the dc from the system of storage batteries. Some dual wiring is required, and a more complicated and expensive installation results.

For the average home, a 6-kW WECS would probably be the minimum needed for uninterrupted service. Many factors operate to determine the exact requirement, some of which are difficult—or take a long period of time—to assess. (Refer to Chapter 3 for wind characteristics and siting information.)

Supplementary Electrical Supply for a (Rural) Home, Farm, Ranch, Small Business (etc.) Under certain circumstances, it could be worthwhile for the small-scale consumer to produce only part of his power from the wind and to depend upon utility power for the rest. By planning to supply only part of the power needed, the storage system cost can be reduced significantly because the cost of storage per unit of electrical energy is relatively great.

The same commercially available units used in total supply systems can likewise be adapted for use in supplementary supply systems.

Asynchronous Alternator for Supplementary Supply of a Small-scale Consumer Demand. A 6-kW asynchronous alternator producing 60-Hz constant ac voltage power over a wide range of rotational speeds has been developed by Dr. Hughes and his group at Oklahoma State University. The variable rotational speed feature permits an important increase in rotor capture efficiency over the wide range of wind speeds encountered in practice. Moreover, the power produced would be directly usable by a customer.

It is apparent that the maximum economic and technological potential of the asynchronous alternator can be achieved in phase operation with a utility grid. Because the instantaneously available wind power and the consumer demand vary independently, the WECS would only on occasion match the demand. When the wind power is deficient, the difference would be made up from the utility line, and when the wind power is excessive, the excess power would be returned to the utility line for redistribution. Total usage of the wind power would thus be achieved without any provision for storage. Agreements with the power company involved are still in the process of being worked out in the operation of such systems.

Synchronous alternators were used at Grandpa's Knob (1250 kW), in Denmark (200 kW) and in the prototype (100 kW) built in Germany by Hutter. Synchronous operation may be uneconomical for some small-scale systems, but the idea has persisted for several years that phased operation of small asynchronous alternators may be justifiable.

Heating and Cooling. Increasing costs of LPG and utilities in general may make feasible the use of wind power for the heating and cooling of structures. Comparisons of the theoretical performance of three wind-driven systems have been completed: mechanical energy to heat; closed Brayton cycle with air as a working fluid; and a Freon-12 heat pump. The calculated minimum wind speed for the last two systems was 8 mph, while the first would operate at any wind speed. The Freon-12 heat pump was the best system; with a suitable reservoir, such a system would be autonomous.

The most efficient manner of utilizing the available energy would be with a refrigerator compressor (heat pump), probably of the automotive type, which is designed to operate at

Figure 9-2. Residential installation of Enertech Model 1500 WECS.

acceptable efficiencies over a fairly large range of rotational speeds. The systems could be designed to do nothing more than transfer heat from a cold to a hot reservoir, from which arrangements are made to transfer heat from the hot reservoir to the structure in cold weather, and from the structure to the cold reservoir in hot weather, thus diminishing the use of fuel and electricity. Such a system would be without most of the problems associated with the use of wind energy in the electrical form, and, in particular, would not be involved with interfacing electrical systems.

The heat storage characteristics of several commonly available materials, as well as those of other materials showing promise in this type of application, have already been evaluated. With manufacturers' information on refrigeration unit performance, wind behavior characteristics, heat conductivities and capacities of soils, new insulating materials, etc., it should be possible to estimate the economic feasibility of such a system. WECS of this type would clearly have their best chance for economic success when designed for a structure in advance of its actual construction.

The problems of heat storage and release have been investigated in a number of studies on the utilization of solar energy for heating houses. For the most part, the hope has been to bring maximum benefit to residents of regions experiencing prolonged cold seasons.

Pumping Water. The use of the Dempster-type windmill shown in Figure 9-3 to pump water has been the only continuously successful utilization of wind power in the U.S. Its use at many homes, farms and ranches has persisted despite extensive inroads made by the electrical pump. It is well known that the windmill made the settlement of many western states possible. In addition to the use of water-pumping windmills to provide small quantities of water for stock, irrigation and personal needs, there is a need to expand the potential of water-pumping windmill applica-

Figure 9-3. Dempster water-pumping wind energy conversion system used to support remote cattle watering application.

tions. These new applications can be met by larger WECS of different designs, such as the Darrieus vertical-axis wind turbine.

An obvious application would be the pumping of irrigation water for larger farms. These wells are presently powered by internal combustion engines or by electrical motors. Wind powered units should experience increasing appeal if natural gas availability continues to be unsure.

The power required for individual wells depends primarily on water lift and output, and in some situations exceeds 100 hp, but a large demand exists for power less than 50 kW (67 hp). Irrigation on a large scale is practiced in several areas of the Southwest as well as in certain other regions of the U.S. Should it be possible to design and produce WECS to satisfy a substantial fraction of this need, an important sector of agricultural industry would benefit, and the use of non-renewable energy resources would be decreased by that amount.

Many problems will have to be confronted in such an endeavor. The seasonal nature of irrigation, the need and the variability of the wind are among the primary ones, even when adequate water is present in storage. An off-season usage of power from the WECS would likely be necessary to achieve an economically satisfactory operation.

Experimentation along this line in recent years has been conducted on the island of

Barbados by the Brace Research Institute and by the USDA in Texas. A hydromechanical system of 35 hp (26 kW) rated output, constructed largely from readily obtainable automotive parts, is used to lift water 100 ft to a sprinkler system for irrigating an area of about 10 acres in the Brace System. (The USDA approach is covered in detail later in this chapter.)

A second water-pumping application for small units is concerned with the water systems of small municipalities. Cities with populations of 10,000 or less, dependent upon wells for their water, often have several pumps, both on the wells themselves and in booster service driven by electric motors rated at 50 kW or less. Certain situations undoubtedly exist wherein wind power units might be profitably brought into service. For example, such municipalities often find it necessary to add pumping capacity, or to replace obsolete or worn out equipment. The prospect of zero fuel costs from a non-depletable resource should appeal to those buying a new pumping system.

In many respects, the technical problems involved in city water supplies are similar to those in crop irrigation. The principal difference is that the demand made of municipal systems is much less seasonal, although the use of water in the summer is greater.

Possible Future Developments in Vehicle Propulsion, and Related Small-scale Storage. Several research programs are under way which have the aim of developing satisfactory propulsion systems not involving the combustion of fossil fuels. Related research is being directed toward improved battery technology, performance and economics, and energy storage in superflywheels. Should significant breakthroughs occur, the usefulness of prime energy sources such as small-type WECS would increase markedly. Whether any of these ideas will bear fruit is a speculative matter at present, and any associated technology is at least several years in the future. Nevertheless, any experience gained in producing an economically successful WECS in the near future would prove of great value in the planning of new systems.

Pumping Oil. Shallow oil wells (~150 ft) in northwestern New Mexico are being pumped with ordinary windmills. Several oil fields in Texas are located in areas having good energy regimes. Although most oil wells are much deeper than the New Mexico wells, it may be possible to use wind power to pump these wells. If a large number of wells are located relatively close to one another, it may be more economical to erect one medium or large WECS to generate electricity for pumping these wells.

The practical applications discussed above probably have the greatest potential for directly affecting large numbers of citizens. Moreover, practical systems of these types could undoubtedly be developed in a shorter time span than those considered above. Combinations of systems, such as wind-solar energy systems, have been proposed and operated. The combination of wind and solar energy are compatible in part due to their synergistic meteorological overlap.

It should be kept in mind that the total situation must be viewed against a background of gradually diminishing availabilities of conventional fuels, accompanied by increasing prices. On occasion, certain fuels have been unavailable because of distribution problems, and these problems are not likely to diminish. Fuel allocation procedures invariably operate to the great inconvenience, discomfort and displeasure of most consumers. Finally, what may be slightly uneconomical one year could become economical the next year.

Intermediate-scale WECS

Total Electical Supply for Communities or Industry. Few, if any, situations exist today where this concept can be applied. The cost of storage using presently available devices would be excessive, and few industries could adjust to the intermittent delivery of power that would occur without storage.

Supplementary Electrical Supply for a Utility Grid, Small Community Group or Small Industry. This concept has been the most extensively accepted and tested of all concepts relating to the intermediate- to large-scale WECS utilization of wind power. As commonly visualized, it consists of a wind-driven synchronous generator having its regulated output phased into a utility grid, as shown in Figure 9-4. As such, it does not supply the consumer directly, but does so through the local utility grid. It provides power to supplement that generated in the conventional power plants. A variation of this concept has the wind-driven unit as the direct supplier of power to the consumer (etc.), with power from the utility grid as the supplement. The essential differences are determined by the ownership of the WECS and the point at which it delivers its output.

DOE is currently utilizing the MOD-OA WECS to provide this type of utility-grid connection needed for Block Island, Rhode Island and similar sites.

The energy storage devices or systems currently in use do not offer economically reasonable and generally applicable possibilities for intermediate systems. Systems without electrical storage seem to be all that warrant consideration at the present time. A requirement for optimizing efficiency in such cases is that all power be consumed as it is produced, and this condition will be assured if minimum demand always exceeds the maximum WECS output. The greater the ratio of minimum demand to maximum WECS output, the smaller will be the relative effect of the WECS power on the demand of the total system on the conventional generating station.

The system of rural electric cooperatives could provide nearly ideal situations in which to investigate this concept. Their rural locations allow considerable choice in siting wind generators; minimum demand (_|5000 kW) is considerably greater than the maximum output of currently contemplated WECS; the rural electric cooperative owns the local system of transmission lines; and other than matters relating to long-term contracts and agreements with utility companies, these organizations have a virtually autonomous operation.

Certain combinations of circumstances may exist under which small, individual industrial plants or larger farms could make economical use of wind generated electrical power from intermediate-scale units.

Pumping Water. The same general considerations apply with intermediate-scale units used in municipal and other public water supplies as apply for small-scale units. Moreover, intermediate-scale units could prove useful in pumping water from surface reservoirs to higher storage levels. Many factors combine to make use of intermediate-scale units generally less feasible in pumping irrigation water, except where satisfactory storage facilities are

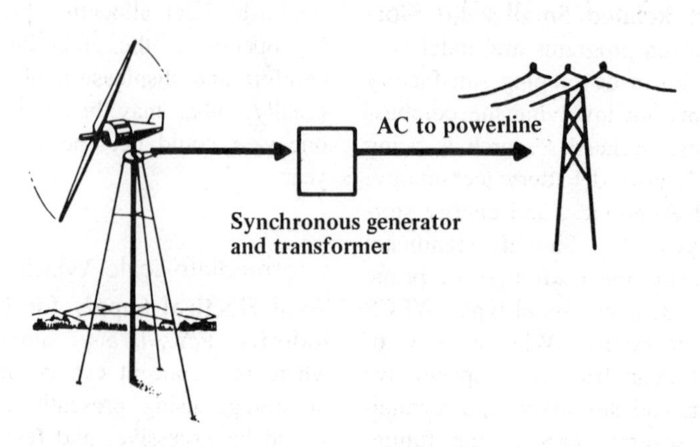

Figure 9-4. Interface between WECS and utility powerline. (*Courtesy of DOE/NASA LRC.*)

available from which gravity or other low-energy flow is possible.

Grain Drying and Elevator Operation. Grain drying and elevator operations constitute an important link in the chain that channels much of agricultural production into the marketplace. Because large numbers of these operations occur in rural settings, as shown in Figure 9-5, it is worthwhile to consider the benefits of wind power in this application. Moreover, all elevators project a solid structure upward of 100 ft or more, thus offering a ready-made "tower" capable of supporting a relatively large wind rotor.

The "typical" grain elevator in the Southwest region uses 80–200 hp (60–150 kW) in units up to 25 hp (19 kW) or so. Most of these motors drive blowers that force air through the bins, mostly for drying, but in some cases for cooling. Their operation is highly intermittent in the long term, because they are usually operated no more than a total of two months out of the year. Heat is used for drying grain sorghums and corn, but not for wheat. This drying operation takes place during October and November. Elevator operations are the most seasonal of all considered, because they are intimately tied to the local harvest.

This great irregularity in power demand poses formidable problems in designing a WECS for economically feasible operation.

Figure 9-5. Grumman WECS used to provide power for agricultural storage facility. (*Courtesy of DOE.*)

The fact of the ready-made towers can scarcely be ignored, however, in a general plan to investigate the technology and economics of wind power utilization. A generally applicable procedure now apparent would involve a system wherein the WECS could return power to the local distribution system during the times when the demand for power for elevator operations is low.

Large-scale WECS

Total Electrical Supply for an Organized Consumer Group or an Industry. Although future technological developments probably will render this concept more practical, the same considerations discussed for intermediate-scale systems generally apply for large-scale systems as well.

Supplementary Electrical Supply for a Utility Grid or an Organized Consumer Group. The potentially most extensive use of large-scale WECS is operation into grids (Figure 9-6) of electric companies. The operation and maintenance of such large systems would probably integrate into the electric power company's systems more smoothly than into that of any other organized business. Nevertheless, the operation of large WECS might be feasible for some of the larger rural electric cooperatives.

Two units in the large class have been successfully operated into utility grids. The 1250-kW Putnam LWECS at Grandpa's Knob in Vermont operated into a utility grid for a period of 16 months during 1941–1943 before a bearing failed for reasons apparently unrelated to wind turbine peculiarities. Repair under wartime conditions took more than two years. Shortly after operation resumed, a blade failed. No phasing problems were encountered. The Bureau d'Etudes Scientifiques et Techniques of France engineered a 800-kVA wind generator for the French Electricity Authority. This unit was erected near Nogent Le Roi, 120 km southwest of Paris, and was operated into a utility network over an 18-month period, from 1958 to 1960. In an effort to improve operational characteristics, the original rigid blade was replaced by a flexible one (which, however, had exhibited a flutter in wind tunnel tests). One blade broke off and the unbalance destroyed the hub. The unit was not repaired and the experiment ended.

Refer to Chapter 10 for a detailed presentation of the current Federal Wind Energy Program. DOE has provided financial support for

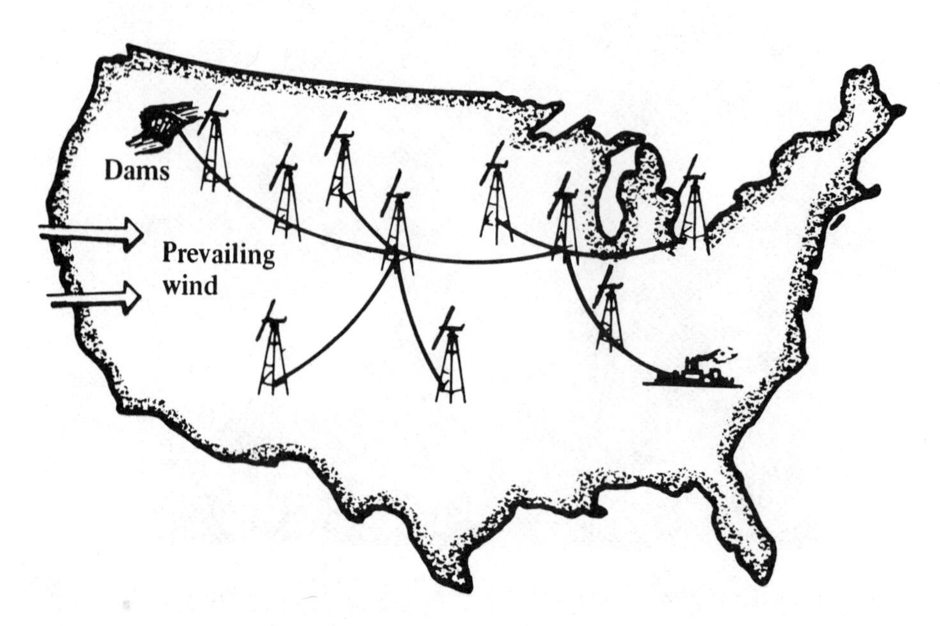

Figure 9-6. Large-scale WECS operating as part of the utility grid. (*Courtesy of DOE/NASA LRC.*)

the development of large-scale WECS such as the MOD-O, MOD-OA, MOD-1 and MOD-2 WECS. These systems are currently under experimental test and evaluation.

Indirect Supplementary Power in Certain Conventional Power Generating Systems. The technique of using a supplementary power source to pump water from a low level into a reservoir (as shown in Figure 9-7) at a higher level is termed "hydrofirming." When done in conjunction with a hydroelectric generating plant, a large fraction of the gravitational energy stored in the water is recovered when it cycles through the turbine. It is obvious that the process provides storage for energy available from the auxiliary power source and that wind energy can be stored in this manner when the power for pumping is derived from a WECS.

The hydrofirming concept is one of the earliest to be suggested in considerations of wind power utilization because it makes use of a familiar type of energy storage and a directly usable procedure for energy recovery. An operating hydroelectric plant is required for application of this concept, and a high wind energy regime increases chances for an economically justifiable operation.

The power demand on a gas turbine to drive its supercharger (70 percent) could be available for generating electricity if compressed air were available from an independent source such as wind power (as shown in Figure 9-8). Wind energy can be used to compress air, and if adequate storage facilities exist, or are not too costly to construct, the wind can be used indirectly to boost output or save fuel.

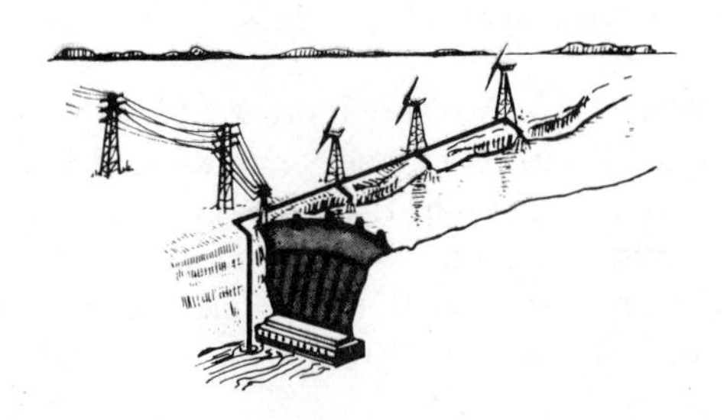

Figure 9-7. Effective combination of wind and water power through hydrofirming. (*Courtesy of DOE/NASA LRC.*)

Indirect Supplementary Power for Energy Intensive Industries. Certain industries require large amounts of energy in the manufacture of their product(s). Examples are the cement, glass and steel industries, and there are the most likely to be seriously affected by shortages of energy resources. The possibility exists that wind energy can be utilized in some special way to benefit these types of industrial operations.

In particular, Southwestern Portland Cement Company has found their kiln production to be increased by 25 percent if the oxygen content of the air for combustion is increased by only 2 percent. No increase in fuels is required. Oxygen can be produced in at least three ways from wind energy, and stored as compressed gas or as a liquid. This storage can be used as a means to minimize the problems associated with wind intermittency.

In general, if an industry consumes a raw material requiring substantial quantities of energy to produce, and wind energy can be used conveniently to produce that raw material, a situation may exist wherein wind power utilization is economical in that industry. The raw material would thus become the means of storing wind energy, thereby solving a primary problem in its utilization. The production of ammonia might be boosted with the assistance of wind power.

RESIDENTIAL APPLICATIONS

Wind systems have caught the public imagination as an alternative energy source. The idea of installing a SWECS which produces power out of thin air and allows its owner to be self-sufficient in producing his electric power and be insulated from recurring utility bills is appealing to many. Those who, in increasing numbers, have sought to make this idea a reality have found that harnessing the wind is usually neither as inexpensive nor as easy as it sounds. Unless you build your own, the initial cost of SWECS can be high. In most cases, utility power is still needed during windless periods and times of high power demand. And even with storage batteries,

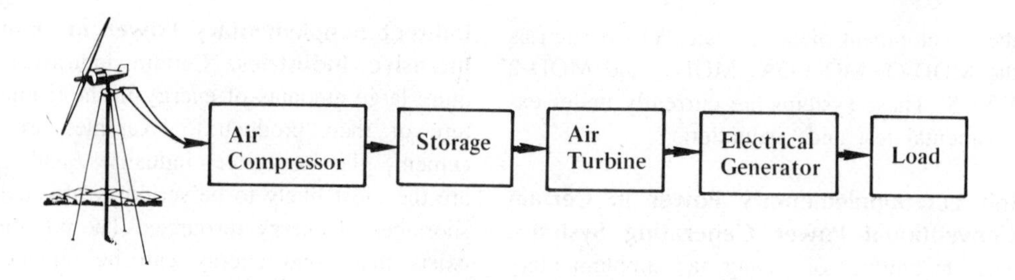

Figure 9-8. WECS compressed air schematic. (*Courtesy of DOE/NASA LRC.*)

many wind system owners have found that wind power can meet only part of their energy requirement.

But despite these drawbacks and limitations, SWECS ownership can still be a satisfying experience. If your power costs are high, if you need mechanical or electrical power in a remote location (Figure 9-9) away from existing utility lines, if you live in an area with documented high annual average winds, or if the use of alternate energy sources makes practical and philosophical sense to you (Figure 9-10), installing a SWECS may offer definite economical and operational, as well as emotional, rewards.

This section is intended to provide a basis for determining the practicality of applying wind energy to your particular situation. Whether or not it is a practical solution depends upon your specific energy needs and a variety of other considerations. If you decide that the wind is a practical energy source for you, other decisions are required, such as the type and amount of equipment you will need and whether to buy a SWECS or build your own.

To decide if wind energy is practical for you, you will want to determine your energy requirements, your available wind energy resource and the equipment needed to convert and use the available energy. You will have to consider the cost of energy obtained from a wind system and decide if the wind is a practical source of power for you. This section will provide you with the basic information or methods you need to make these decisions.

Power and Energy Requirements

The entire process of selecting a suitable SWECS involves determining what power is available from the wind at your site, knowing what you need in the way of energy and power, then matching these to arrive at a SWECS that will do the job. This discussion of power and energy requirements is separated into two parts. The first will discuss electric load estimation, and the second will discuss mechanical load estimation, particularly for water-pumping.

Before the discussion of power requirements, let us keep in mind that while it is well to calculate how much power you need, some consideration must be given to losses or inefficiency. That is, if you figure how much electric power you need to run a light bulb, it will take a little extra, over and above the amount needed by the light bulb alone. The wires running from the generator to the light will waste some power because of electrical resistance. Friction is another example of inefficiency. This waste ends up as a power loss in the form of heat. Estimation of losses is an important part of your calculation of your power and energy needs.

Electric Load Estimation. Two different numbers will result from performing the simple calculations—first, the power load, which is expressed as the number of watts (W) or kilo (thousand) watts (kW), and second, the energy requirement, which is expressed in kilowatt-hours (kWh).

Figure 9-9. Small-scale Dunlite WECS installed in remote location.

Electric power is a product of electric pressure, called volts (V), and current flow, called amperes amps (A). Just as force times rate of motion equals power, usually expressed as horsepower (hp), volts times amps equals power, expressed in watts. For example, a 12-

Figure 9-10. Independent Energy SWECS installed at modern home site.

V battery that pushes 10 A through a light uses electric power equal to 12 × 10 = 120 W. Now suppose that the 120 W light is left on for 10 hours. Then, the electric energy consumer will equal 120 W times 10 hours, which equals 1200 W-hours, or 1.2 kWh.

If a battery could store 2400 W-hours of energy, then it would have a capacity to produce 120 W of power for 20 hours. That is, 2400 W-hours divided by 120 W equals 20 hours.

In order to estimate your electrical load requirements, it is necessary to determine two types of information:

- Which electrical devices you will use and how much power, in watts, they will draw.
- How long these devices will operate (say, in hours per month).

A third item of information—at what time of the day the devices operate—will also be discussed.

If it turns out that your WECS will provide power to electrical devices that you already use, and perhaps have used for some time, then load analysis becomes a simple task of checking all your electric bills for the last dozen or so months. It is a good idea to know how your electric bill changes with each month of the year to see what seasonal changes look like. In many cases, changes in the weather affect your energy use patterns. Here, you are not concerned with the dollar figure of the bill, but instead the actual demand figure expressed as kWh. Make sure that your utility company meter-reader has really read your meter, as occasionally utilities will estimate your use. Estimated kWh figures will not help you at all.

If you have not saved enough bills to check the demand, you can usually obtain a summary from your electric utility company. In either case, the utility bill will give you the monthly energy demand. It will not, however, give you the power demand in W or kW. For that, you will have to list the devices you use and determine the power requirement of each. Figure 9-11 will assist you. This figure will also serve to assist in estimation of energy demand if you do not have your utility bills.

We will follow the steps in Figure 9-12 to apply a logical sequence to the following discussion.

NAME	WATTS	HRS/MO	KWHRS/ MO
Air conditioner, central			620*
Air conditioner, window	1566	74	116*
Battery charger			1*
Blanket	190	80	15
Blanket	50-200		15
Blender	350	3	1
Bottle sterilizer	500		15
Bottle warmer	500	6	3
Broiler	1436	6	8.5
Clock	1-10		1.4*
Clothes drier	4600	20	92*†
Clothes drier, electric heat	4856	18	86*†
Clothes drier, gas heat	325	18	6*†
Clothes washer			8.5*
Clothes washer, automatic	250	12	3*
Clothes washer, conventional	200	12	2*†
Clothes washer, automatic	512	17.3	9*
Clothes washer, ringer	275	15	4*†
Clippers	40-60		½
Coffee maker	800	15	12
Coffee maker, twice a day			8
Coffee percolator	300-600		3-10
Coffee pot	894	10	9
Cooling, attic fan	1/6-3/4HP		60-90*†
Cooling, refrigeration	3/4-1½ ton		200-500*
Corn popper	460-650		1
Curling iron	10-20		½
Dehumidifier	300-500		50*
Dishwasher	1200	30	36*
Dishwasher	1200	25	30*
Disposal	375	2	1*
Disposal	445	6	3*
Drill, electric, ¼"	250	2	5
Electric baseboard heat	10,000	160	1600
Electrocuter, insect	5-250		1*
Electronic oven	3000-7000		100*
Fan, attic	370	65	24*†
Fan, kitchen	250	30	8*†
Fan, 8"-16"	35-210		4-10*†
Food blender	200-300		½
Food warming tray	350	20	7
Footwarmer	50-100		1
Floor polisher	200-400		1
Freezer, food, 5.30 cu.ft.	300-800		30-125*
Freezer, ice cream	50-300		½
Freezer	350	90	32*
Freezer, 15 cu.ft.	440	330	145*
Freezer, 14 cu.ft.			140*
Freezer, frost-free	440	180	57*
Fryer, cooker	1000-1500		5
Fryer, deep fat	1500	4	6
Frying pan	1196	12	15
Furnace, electric control	10-30		10*
Furnace, oil burner	100-300		25-40*
Furnace, blower	500-700		25-100*†
Furnace, stoker	250-600		3-60*†
Furnace, fan			32*†
Garbage disposal equipment	1/4-1/3 HP		½*
Griddle	450-1000		5
Grill	650-1300		5
Hair drier	200-1200		½-6*
Hair drier	400	5	2*
Heat lamp	125-250		2
Heater, aux.	1320	30	40
Heater, portable	660-2000		15-30
Heating pad	25-150		1
Heating pad	65	10	1
Heat lamp	250	10	3
Hi Fi Stereo			9*
Hot plate	500-1650		7-30
House heating	8000-15,000		1000-2500
Humidifier	500		5-15*
Iron	1100	12	13
Iron			12
Iron, 16 hrs/month			13
Ironer	1500	12	18
Knife sharpener	125		¼*
Lawnmower	1000	8	8*†
Lighting	5-300		10-40
Lights, 6 room house in winter			60
Light bulb, 75	75	120	9
Light bulb, 40	40	120	4.8
Mixer	125	6	1
Mixer, food	50-200		1
Movie projector	300-1000		
Oil burner	500	100	50*
Oil burner			50*
Oil burner, 1/8 HP	250	64	16*
Pasteurizer, ½ gal.	1500		10-40
Polisher	350	6	2
Post light, dusk to dawn			35
Power tools			3
Projector	500	4	2*
Pump, water	450	44	20*†
Pump, well			20*†
Radio			8
Radio, console	100-300		5-15*
Radio, table	40-100		5-10*
Range	8500-1600		100-150
Range, 4 person family			100
Record player	75-100		1-5
Record player, transistor	60	50	3*
Record player, tube	150	50	7.5*
Recorder, tape	100	10	1*
Refrigerator	200-300		25-30*
Refrigerator, conventional			83*
Refrigerator-freezer	200	150	30*
Refrigerator-freezer 14 cu. ft.	326	290	95*
Refrigerator-freezer, frost-free	360	500	180*
Roaster			40
Rotisserie			42*
Sauce pan	300-1400		2-10
Sewing machine	30-100		½-2
Sewing machine	100	10	1
Shaver	12		1/10
Skillet	1000-1350		5-20
Skil Saw	1000	6	6
Sunlamp	400	10	4
Sunlamp	279	5.4	1.5
Television	200-315		15-30*
TV, BW	200	120	24*
TV, BW	237	110	25*
TV, color	350	120	42*
TV, color			100*
Toaster	1150	4	5
Typewriter	30	15	.5*
Vacuum cleaner	600	10	6
Vacuum cleaner, 1 hr/wk			4

(continued)

Figure 9-11. Power and energy requirements of appliances and farm equipment.

Figure 9-11. Power and energy requirements of appliances and farm equipment—continued

NAME	WATTS	HRS/MO	KWHRS/ MO
Vaporizer	200-500		2-5
Waffle iron	550-1300		1-2
Washing machine, 12 hrs/mo			9*
Washer, automatic	300-700		3-8*
Washer, automatic	100-400		2-4*
Washer, conventional	4474	89	400
Water heater	1200-7000		200-300
Water pump (shallow)	½ HP		5-20*†
Water pump (deep)	1/3-1 HP		10-60*†

AT THE BARN

NAME	CAPACITY HP OR WATTS	EST. KWHR
Barn cleaner	2-5 HP	120/yr.*
Clipping	fractional	1/10 per hr.
Corn, ear crushing	1-5 HP	5 per ton*
Corn, ear shelling	¼-2	1 per ton*†
Electric fence	7-10 watts	7 per mo.*†
Ensilage blowing	3-5	½ per ton
Feed grinding	1-7½	½-1½ per 100 lbs.*†
Feed mixing	½-1	1 per ton*†
Grain cleaning	¼-½	1 per ton bu*†
Grain drying	1-7½	5-7 per ton*†
Grain elevating	¼-5	4 per 1000 bu*†
Hay curing	3-7½	60 per ton*
Hay hoisting	½-1	1/3 per ton*†
Milking, portable	¼-½	1½ per cow/mo.*†
Milking, pipeline	½-3	2½ per cow/mo.*†
Sheep shearing	fractional	1½ per 100 sheep
Silo unloader	2-5 HP	4-8 per ton*
Silage conveyor	1-3 HP	1-4 per ton*
Stock tank heater	200-1500 watts	varies widely
Yard lights	100-500 watts	10 per mo.
Ventilation	1/6-1/3 HP	2-6 per day*† per 20 cows
IN THE MILKHOUSE		
Milk cooling	½-5 HP	1 per 100 lbs. milk*
Space heater	1000-3000	800 per year
Ventilating fan	fractional	10-25 per mo.*†
Water heater	1000-5000	1 per 4 gal
FOR POULTRY		
Automatic feeder	¼-½ HP	10-30 KWHR/mo*†
Brooder	200-1000 watts	½-1½ per chick per season
Burglar alarm	10-60 watts	2 per mo.*
Debeaker	200-500 watts	1 per 3 hrs.
Egg cleaning or washing	fractional HP	1 per 2000 eggs*†
Egg cooling	1/6-1 HP	1¼ per case*

NAME	CAPACITY HP OR WATTS	EST. KWHR
Night lighting	40-60 watts	10 per mo. per 100 birds
Ventilating fan	50-300 watts	1-1½ per day*† per 1000 birds
Water warming	50-700 watts	varies widely
FOR HOGS		
Brooding	100-300 watts	35 per brooding period/litter
Ventilating fan	50-300 watts	¼-1½ per day*†
Water warming	50-1000 watts	30 per brooding period/litter
FARM SHOP		
Air compressor	¼-½ HP	1 per 3 hr.*
Arc welding	37½ amp	100 per year*
Battery charging	600-750 watts	2 per battery charge*
Concrete mixing	¼-2 HP	1 per cu. yd.*†
Drill press	1/6-1 HP	½ per hr.*†
Fan, 10"	35-55 watts	1 per 20 hr.*†
Grinding, emergy wheel	1/4-1/3 HP	1 per 3 hr.*†
Heater, portable	1000-3000 watts	10 per mo.
Heater, engine	100-300 watts	1 per 5 hr.
Lighting	50-250 watts	4 per mo.
Lathe, metal	¼-1 HP	1 per 3 hr.
Lathe, wood	¼-1 HP	1 per 3 hr.
Sawing, circular 8"-10"	1/3-1/2 HP	1/2 per hr.
Sawing, jig	1/4-1/3 HP	1 per 3 hr.
Soldering, iron	60-500 watts	1 per 5 hr.
MISCELLANEOUS		
Farm chore motors	½-5	1 per HP per hr.
Insect trap	25-40 watt	1/3 per night
Irrigating	1 HP up	1 per HP per hr.
Snow melting, sidewalk and steps, heating–cable imbedded in concrete	25 watts per sq. ft.	2.5 per 100 sq. ft. per hr.
Soil heating, hotbed	400 watts	1 per day per season
Wood sawing	1-5 HP	2 per cord

Symbol Explanation
*AC power required
†Normally AC, but convertible to DC

Notes: Lighting in this table is assumed to be incandescent–if fluorescent, the wattage bulbs consume the same power but deliver 3 times as much light–fluorescent bulbs also require AC, but can be converted to DC.

These figures can be cut by 50% with conservation of electricity.

Step 1. Determine the appliance load rating, expressed in W or kW. *Example:* Brand C electric motor is a 1-hp motor. Its electrical load when operating is 860 W, and when starting, 1400 W for 1 second. These data are found on the appliance data plate, by writing to the manufacturer, by testing an appliance yourself or from Figure 9-11. You can easily test the appliance if you presently have electric service. Watch your electric utility meter, which measures energy usage. Usually it contains a slowly spinning disc, and some number of revolutions of it indicates that 1 kWh has been consumed. Ask your power company what each revolution means. Turn off all other appliances so the meter stops. Then turn on the appliance you wish to rate and time the spinning disc.*

Step 2. Determine the load cycle time and the number of hours the load will operate on

*This will not be adequate for obtaining the starting load, but that is not necessary for these load calculations.

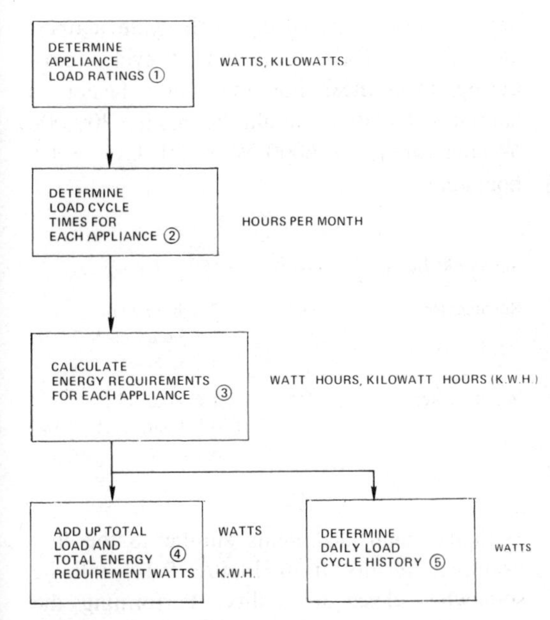

Figure 9-12. Estimating residential energy requirements. (*Courtesy of DOE.*)

a monthly basis. *Example:* Brand D refrigerator will operate an average of 15 hours per month. Note that this information depends on how well insulated the refrigerator is, the number of times the door is opened, how much bulk will be stored, and the room temperature.

Step 3. Determine the appliance's monthly energy requirement. This is calculated by multiplying together the data from Steps 1 and 2 above. The result is in W-hours or kWh. *Example:* A color television is determined in Step 1 to require 350 W. You know that it will be used for 150 hours per month, so:

350 W × 150 hours = 52,500 W-hours.

To get kWh: 52,500 ÷ 1000 = 52.5 kWh.

Step 4. Determine maximum load. This calculation will result in your maximum power demand, expressed as kW, which would occur if all of your appliances were operating at the same time. *Example:*

Item		*Power (W)*
TV (B&W)		200
Coffee pot		894
Dishwasher		1200
Refrigerator		300
Water heater (custom)		6000
Maximum load	=	8594 W
	=	8.59 kW.

Step 5. Determine the total monthly energy requirement. This is the energy which must be provided by the SWECS each month, or which you pay for from your utility company. *Example:*

*Item**		*Energy Requirement (kWh)*
TV		24
Coffee pot		9
Dishwasher		36
Refrigerator		30
Water heater		300
Total energy requirement	=	399 kWh/month.

(A daily breakdown of your electrical load is described in Step 6.)

Step 6. Determine the daily load cycle history. This calculation will provide a much better estimate of the actual electric load demand. Simply adding up all of the loads, as in the above example, assumes that all appliances will be on at the same time and gives a worst case figure, but does not reflect a real case. To arrive at a load cycle history, you must make estimates of the time of day your devices will be on and for how long. This estimate may be as accurate or as rough as necessary. How accurate you decide to be in making estimates will depend entirely on your assessment of the importance of this calculation. For an accurate estimation, it will be necessary to actually monitor any items that operate on a cyclic basis, such as refrigerators. For less accuracy, it may be reasonable to assume such loads "on" continuously. As a first example, we shall separate items by their nature: those which you control, and those that operate automatically. For example, items that operate automatically shall be assumed to operate continuously. The other loads will require estimation of operating cycles, as listed below.

Automatic Items	*Load (W)*	*Time*
Refrigerator	300	Continuous
Water heater	6000	Continuous

*The above examples are selected at random from Figure IX-11 and do not necessarily represent a typical household load.

User-controlled Items		
TV	200	4 hours/day as follows: 1 hour: 8AM–9AM 3 hours: 6PM–9PM
Coffee pot	894	20 minutes/day 7:30AM–7:50AM
Dishwasher	1200	1 hour/day: 5PM–6PM

From this table, we can see a *base load,* that is, a continuous load equal to 6300 W, with peak loads going as much as 1200 W higher. If we made a simple graph of this load, it would look like Figure 9-13. For a second example, let us be more realistic. The assumption that all automatic items are continuous loads should be adjusted. Refer to Figure 9-11 for the basic load data. Notice that the refrigerator is listed as 200–300 W, for 25–30 kWh/month. Using 300 W and 30 kWh (or 30,000 W-hours)/month, we can calculate the hours/month this device operates:

$$30{,}000 \text{ W-hours/month} \div 300 \text{ W} = 100 \text{ hours/month.}$$

Now, assume a 30-day month, 100 hours/month ÷ 30 days = 3.3 hours/day. This is the estimated number of hours/day this refrigerator will operate. Now estimate when, and for how long, during each cycle it operates. A safe estimate is that it cycles most during mealtimes. For the water heater, a similar calculation should be made: 300,000 W-hours/month ÷ 6000 W ÷ 30 days = 1.6 hours/day.

Automatic Items	*Load*	*Time*
Refrigerator	300	3.3 hours/day: 1.1 hours each 7 AM, noon, 5 PM
Water heater	6000	1.6 hours/day: 0.8 hours each 8 AM, 6 PM

With use-controlled items similar to the first example, results from Figure 9-14 will be somewhat closer to reality. Performing the type of analysis in figure 9-14 may not actually be necessary for your energy requirement estimates, but it is a good way to understand the nature and characteristics of the electric load you expect.

Step 7. Determining your monthly energy requirements. A previous example illustrated how a monthly energy requirement of 399 kWh was established. This could have been any month. For some months, a heavy demand for heating may raise electrical con-

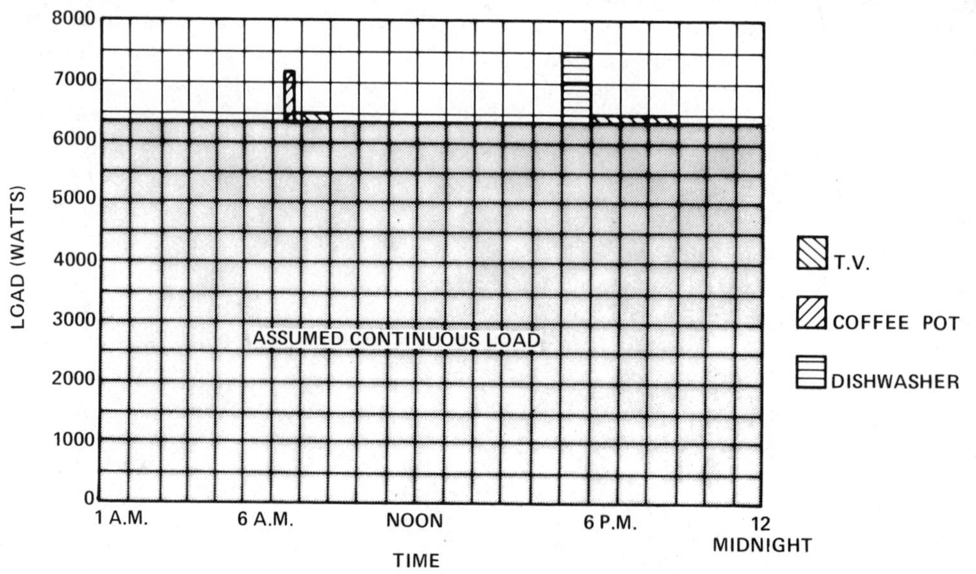

Figure 9-13. Graph of example load history. *(Courtesy of DOE.)*

sumption, while in others, air conditioning will prevail. Thus, you must complete a demand analysis for each month of the year. A graph plotted from your totals would be similar to Figure 9-15.

By following these seven steps the electric load can be estimated for SWEC applications.

Mechanical Load Estimation. Estimation of mechanical load may be as simple as reading the data plate on a device you expect to drive mechanically by wind power, and can be as complex as calculating the horsepower required to pump water through some pipes. This section will deal primarily with the estimation of power required to pump water.

Figure 9-16 illustrates a complex water pump and storage system. In any system such as this one, you are expected to know the well depth and tank height. Add these two together and you get water head—the maximum height to which the pump must lift water (we assume the tank is not pressurized).*

Water head = well depth + storage tank height.

(Water head is measured in ft, well depth is measured in ft from ground surface [minimum allowable water height] and tank height is measured from ground surface to top of water outlet at tank.) *Example:*

Well depth	200 ft
Tank height	70 ft
Water head	= 200 ft + 70 ft = 270 ft

Notice that while the water is being lifted to the total height of 270 ft as in the examples, it must pass through pipes that are considerably longer, unless the tank is directly on top of the well. The loss of pressure (head loss) from water flowing through the pipes will increase the amount of load on the pump. We calculate the head loss using the head loss factor from the graph on Figure 9-17. This factor is the head loss (ft) per 100 ft of pipe run. Thus, you need to measure total length of pipe run and know flow rate measured in gallons/minute (gpm) and the pipe diameter.*

From Figure 9-17, a value of head loss factor can be determined after you have meas-

*For a pressurized tank, add 2.31 ft of water head for each psi.

*Figure 9-17 assumes standard ("schedule 40") steel pipe. For other diameters than those listed, the head loss at the same flow rate is proportional to the fifth power of the ratio of the pipe diameters. Pipe diameter is approximately the internal diameter. The outside diameter will be 1/4–1/2 in. larger.

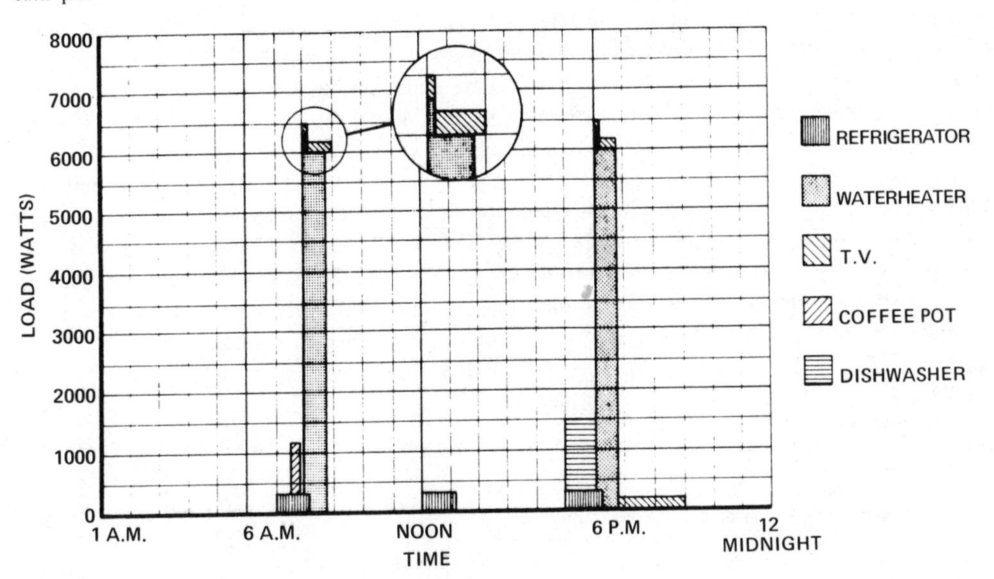

Figure 9-14. Graph of load history. (*Courtesy of DOE.*)

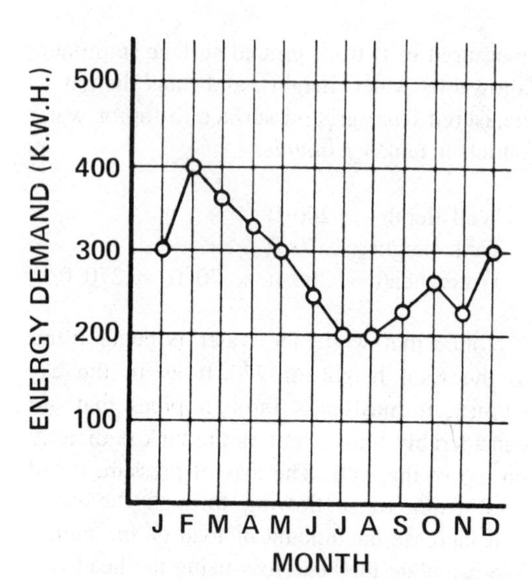

Figure 9-15. Typical monthly energy demand. (*Courtesy of DOE.*)

ured pipe run, which can then be converted to head loss.*

Head loss = head loss factor × pipe run ÷ 100 (when pipe run measured in ft).

Example:

Maximum pump capacity—5 gpm
Pipe run—250 ft
Water head—250 ft
Pipe diameter—3/4 in.
Two 90° elbow fittings installed.

From Figure 9-17, find the 5-gpm flow rate on the horizontal line. From this point, go straight up to the 3/4-in. pipe curve line. From there, look straight to the left to read the head loss factor = 6.0 on the vertical scale. This is 6 ft per 100 ft of pipe run. Then total head loss factor = 6 × 250 ÷ 100 = 15 ft. For the two elbow fittings add 2.3 ft each. Then head loss = 15 + 4.6 = 19.6 ft. From here, we calculate total head, which is the actual load presented to the pump.

*This value is for steel pipe and does not include valve and fitting losses. If several hundred feet of pipe length is involved, fitting losses can be neglected. To include them in an approximately way, count up the number of tees and elbows. For 1-in. pipe add 3 ft of pipe length for each fitting, and for fully open valves, add none for a gate, 12 ft for an angle, and 30 ft for a globe valve (the usual spigot-type valve). This is the total equivalent pipe length. For other pipe sizes, proportion these lengths to the pipe size (i.e., twice these values for 2-in. pipe). For smooth plastic pipe, reduce the head loss factor by 40 percent.

Total head = water head + head loss (where all factors are measured in ft).

Continuing the example, total head loss = 250 + 19.6 = 269.6 ft, or 270 ft.

Because of head loss, the pump is loaded as if it has to pump water 19.6 ft higher than it really does.

Now we calculate horsepower required by the pump and supplied by the WECS. From Figure 9-18, you can read the theoretical horsepower (no losses) to know the total head and flow rate. Continuing our example, for total head = 270 ft, and 5 gpm, the theoretical horsepower = 0.3. This is the horsepower supplied by the pump. Now you must calculate horsepower supplied to the pump by the WECS. For most well-maintained or new piston pumps installed on WECS, assume a pump efficiency of 70 percent. Then: wind turbine horsepower = pump horsepower ÷ 0.7.

Example: Pump horsepower calculated in the previous example = 0.3. Then wind turbine horsepower = 0.3 ÷ 0.7 = 0.43. Thus, a 1/2 hp wind turbine can pump water at a little more than 5 gpm up to a total height of 250 ft, through 250 ft of 3/4-in. iron pipe with two elbow fittings installed.

At this point in our calculations, we have developed a method to predict how much horsepower is needed. This is a power requirement, but, as with electrical systems, we need to know hp-hours, the energy requirement.

For this, you must estimate your daily (or monthly) water requirements in gallons, just as you would estimate electrical requirements.

Use Figure 9-19 to add up the gallons of water you need daily (adjust these according to your experience). Multiply values by 30 for monthly calculations. *Example:*

(2) Persons:	75 gallons/day × 30 = 2250 gallons/month
(1) Beef cow:	12 gallons/day × 30 = 360 gallons/month
(20) Chickens:	(4 gallons/day ÷ 100) × 20 ÷ 100 × 30 = 24 gallons/month

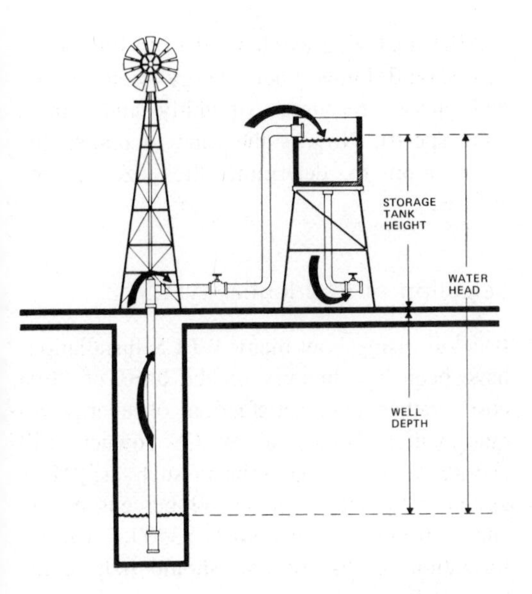

Figure 9-16. Water pump system diagram. (*Courtesy of DOE.*)

Lawn (1000 ft^2): 160 × 15 = 2400 gallons/month

Total: 5034 gallons/month

Using the water requirement data, you can calculate hp hours. *Example:* Assume the 1/2-hp pump of previous examples pumps at an average rate of 5 gpm (this rate varies with the changing wind speed). Calculate hp hours/month: 5034 gallons/month ÷ 5 gpm = 1007 minutes/month, or 1007 ÷ 60 = 16.8 hours/month required at an average flow rate of 5 gpm. Then, energy required (monthly hp-hours) = 16.8 × 1/2 = 8.4 hp-hours.

Following these steps, one can effectively determine the mechanical load estimates for a SWECS.

Wind Energy Resource Estimation

Chapter 3 provides an in-depth discussion of basic meteorology and site selection. This section will use that information as it relates to the application of installing a small WECS.

The evolution of the wind energy available at a SWECS site requires more than obtaining the local advice of the neighbors about the weather. Most people tend to overestimate wind speeds. The amount of time and money spent on this task is directed by the need for accuracy and by the size and importance of the wind system (from the standpoint of safety and energy production). The options are hiring a consultant to perform a wind site survey or doing the survey yourself.

Techniques used in a site survey have varying degrees of effectiveness. You might contact a weather service for climatic data for your area, or perhaps a local airport for its average wind speed. It is unlikely, however, that either source will have information pertinent to the exact area you have in mind. Average monthly wind power calculations for over 700 locations in the U.S. are included in the reference material at the back of this book. The most effective approach is using instruments—the placement of wind speed and wind direction equipment—to get actual site readings.

During your intuitive exploration and/or site instrumentation, you should have an awareness of the effects on wind speed by man-made and natural blockages as noted in Chapter 3. Examples are buildings, trees and other features of the terrain. These obstructions interfere with normal wind speeds at given

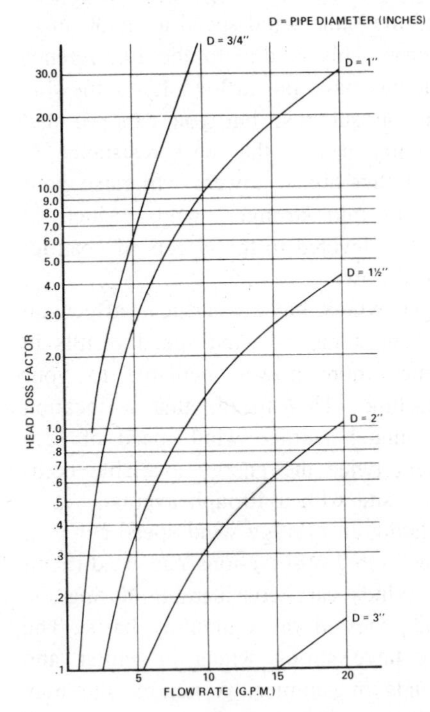

Figure 9-17. Head loss factor. (*Courtesy of DOE.*)

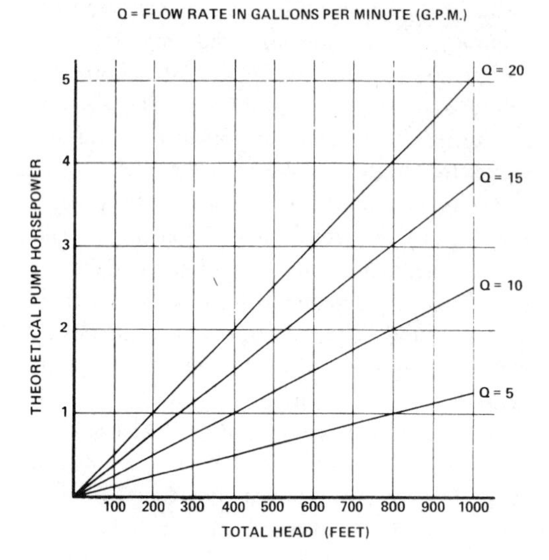

Figure 9-18. Theoretical pump horsepower. (*Courtesy of DOE.*)

heights and thus will influence your eventual tower height.

In the field, you might measure a wind speed of 20 mph at about 30 m high (98.5 ft, which is very high for a tower intended for a home or farm wind system). Back in the city, however, you may have to go as high as 200 m to find the same wind speed as at 30 m in the country. This is due to the interference that buildings have on airflow. It's somewhat better in the suburbs, but you can see that open country causes the least resistance to wind, and that higher towers will raise your machine up into stronger winds. Much of Chapter 3 is devoted to the effects of obstructions on airflow.

Stronger winds have a critical effect on WECS: a doubling of wind speed results in eight times more power available to your wind machine. This means that a location with an annual average wind speed of 12.6 mph offers twice the energy available compared to a site with a 10-mph average.

Calculating an average wind speed per year based on limited observations can be difficult at best. Winds vary considerably during the year when viewed on a monthly basis. The site may have strong winds in winter and weak winds in summer. A nearby site may have little or no wind at all.

After analyzing available data to this point, you should know your energy requirements and energy resources (monthly and annual wind speed). Now is the time to match this information to determine the size of the WECS.

Selection of Residential WECS

It is surprising how many WECS installations have been bought only on the basis of "first cost" rather than satisfaction of energy requirements. Often, a WECS owner will choose a system on a basis such as: "This system will supply about 60 percent of my energy needs and only costs $3400." The information in this section should help guide you in determining the best system for your needs, the resulting actual cost of power and the likely cost of power from the utility company in the future.

"First cost" and emotional factors such as habits and desires often have a strong effect on one's estimate of energy requirements. "First cost" of a wind system has more than once convinced a family moving to a remote location that two television sets operating off a wind-charged battery bank is an unacceptably high level of luxury.

Figure 9-20 can focus your thinking on the steps required to accomplish a rational selection of the most appropriate WECS.

The demand will establish whether the rotor will drive mechanical devices, such as pumps, compressors or grinding wheels; or electrical devices, such as generators or alternators. Mechanical devices demand a rotor design of relatively high solidity, whereas electric generators, for reasons mentioned earlier, tend to be equipped with relatively low-solidity rotors.

Once established, the type of rotor work performed enables you to evaluate the devices on which the work is performed, such as pumps and generators.

To visualize some of the alternative WECS applications, refer to Figure 9-21. Follow any path on this diagram that leads from top to bottom, from WECS to user. You will note most of the practical energy devices and proc-

Effect of External Temperature on Water Consumption

Hogs Temperature (°F.)	Water Consumption Pounds Per Hog Per Hour		
	75-125 lb. hogs	275-380 lb. hogs	Pregnant Sows
50	0.2	0.5	0.95
60	.25	.5	.85
70	.30	.65	.80
80	.30	.85	.95
90	.35	.65	.90
100	.60	.85	.80

Dairy Cows	Gallons Per Day Per Cow		
Temperature	Lactating Jerseys	Lactating Holsteins	Dry Holsteins
50	11.4	18.7	10.4
50-70	12.8	21.7	11.5
75-85	14.7	21.2	12.3
90-100	20.1	19.9	10.7

	Milliliter Per Bird Per Day	
Temperature	White Leghorn	Rhode Island Red
70	286	294
80	272	321
90	350	408
100	392	371
70	222	216
70	246	286

Water Consumption of Sheep
(Pounds of Water per Day)

On range or dry pasture	5-13
On range (salty feeds)	17
On rations of hay and grain or hay, roots and grain	0.3-6
On good pasture	Very little (if any)

In these experiments water was available for consumption.

Water Consumption of Pigs
(Pounds of Water per Day)

Conditions	
Body Weight = 30 lbs.	5-10
Body Weight = 60-80 lbs.	7
Body Weight = 75-125 lbs.	16
Body Weight = 200-380 lbs.	12-30
Pregnant Sows	30-38
Lactating Sows	40-50

Water Consumption of Chickens
(Gallons per 100 Birds per Day)

Conditions	
1-3 weeks of age	0.4-2.0
3-6 weeks of age	1.4-3.0
6-10 weeks of age	3.0-4.0
9-13 weeks of age	4.0-5.0
Pullets	3.0-4.0
Nonlaying Hens	5.0
Laying Hens (moderate temperatures)	5.0-7.5
Laying Hens (temperature 90°F)	9.0

Water Consumption of Growing Turkeys
(Gallons per 100 Birds per Week)

Conditions	
1-3 weeks of age	8-18
4-7 weeks of age	26-59
9-13 weeks of age	62-100
15-19 weeks of age	117-118
21-26 weeks of age	95-105

Water Consumption of People

Average Person: 75 gallons per day

Lawn: 0-200 gallons per 1000 square feet, every other day.

Water Consumption of Cattle

Class of Cattle	Conditions	(Pounds per Day)
Holstein calves (liquid milk or dried milk and water supplied)	4 weeks of age	10-12
	8 weeks of age	13
	12 weeks of age	18-20
	16 weeks of age	25-28
	20 weeks of age	32-36
	26 weeks of age	33-48
Dairy Heifers	Pregnant	60-70
Steers	Maintenance ration	35
	Fattening ration	70
Range Cattle		35-70
Jersey Cows	Milk Production 5-30 lbs/day	60-102
Holstein Cows	Milk Production 20-50 lbs/day	65-182
	Milk Production 80 lbs/day	190
	Day	90

From: Water, Yearbook of Agriculture, 1955
U.S. Department of Agriculture

Figure 9-19. Water requirements.

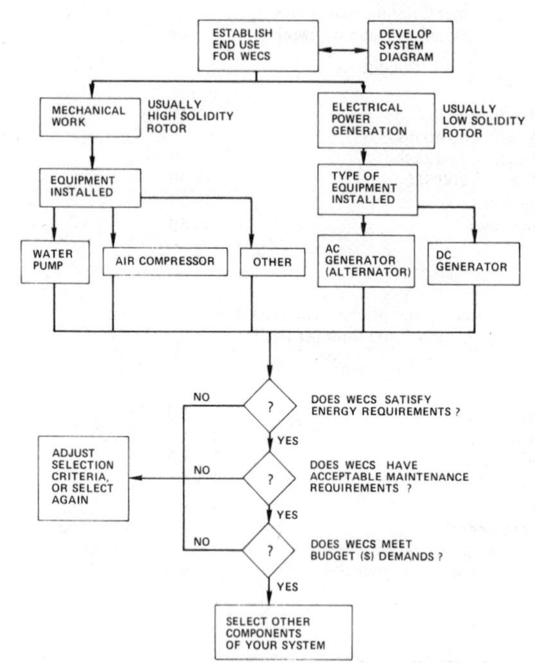

Figure 9-20. System design steps. (*Courtesy of DOE.*)

esses along each path. The diagram illustrates the most common system, as well as some systems being developed.

A basic system consists of the energy source (wind), a conversion device (wind turbine), energy storage (batteries, pumped water, etc.), energy use (such as heaters, motors, TV sets) and a backup source of energy (such as gasoline generators or solar cells).

Some of the options, as shown in Figure 9-22, to select the best system to pump water are: a wind turbine mounted directly over the pump using a push rod, a drive shaft to a remotely located pump, an electric pump system, a hydraulic or pneumatic system or a jet pump geared directly to the wind turbine rotor.

Table 9-1 is a chart that presents many of the factors you would consider in the selection of the best water-pumping system for your needs. You would normally evaluate each option by following the steps of Figure 9-20. It may be that the step regarding cost and your budget will eliminate several of the options from your list.

Another example of widely different choices available for accomplishing a specific task is depicted in Figure 9-23. Two methods are illustrated for preventing the freezing of a stock water pond and reducing fish kill caused by ice blocking the absorption of oxygen into the water. Either method works, and there are many other possibilities. One method uses a small Savonious rotor, mounted above the pond and driving a propeller that churns the water. This circulates warmer water to the surface, preventing ice formation and adding oxygen to the water. The air pump method bubbles air into the water and also causes the warmer water to rise to the surface.

By following the steps shown in Figure 9-20, you can effectively evaluate the potential for utilizing residential SWECS.

In preparing a system plan, you may discover it sometimes is more desirable to reduce your energy usage than to buy a larger wind machine. It is almost always prudent to evaluate your options in the area of conservation. It also is well to allow for future growth in energy needs, but many times you will still find that a good system plan can be vastly improved by saving more power.

System Cost/Benefit Analysis. All too often, a WECS installation is purchased according to first cost rather than performance. Since "the cheapest windmill will do" is never really an appropriate criterion, you should understand what "the cheapest windmill" really is.

The analysis of economic factors can be as complex or as simple as you wish to make it. For a simple analysis, you need to know the following:

1. Total installed cost (dollars).
2. Expected system life (years).
3. Total energy yield over the entire system life (kWh, hp-hours, etc.).
4. Annual maintenance and repair costs (dollars).
5. Other annual costs and savings (dollars).
6. Expected resale value at end of service life (dollars).
7. Other factors.

Generally, the bigger a WECS system is, the less it will cost per unit of rated output.

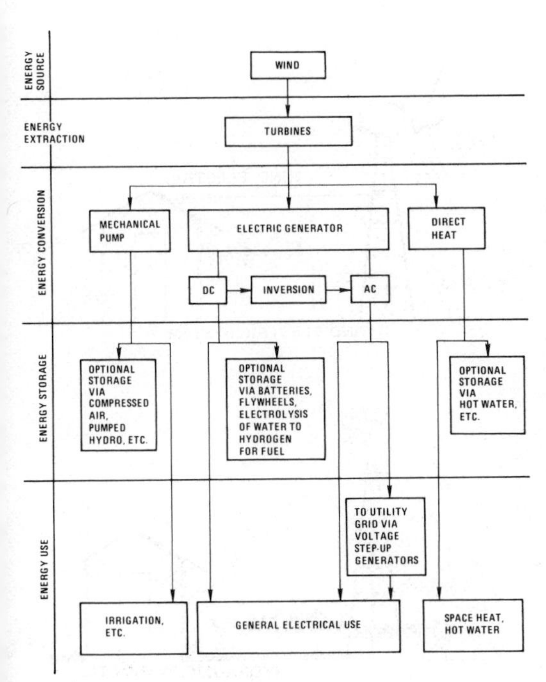

Figure 9-21. Alternative WECS applications. (*Courtesy of DOE.*)

The installed costs of WECS, measured in dollars per kW of rated power, tend to decrease with increasing rated power, as shown in Figure 9-24.

This total cost is broken down into component costs (Figure 9-25). These pie charts show the relative costs by the size of the pie slice. For the WECS, batteries may cost as much as the wind turbine, as illustrated. The graph may not actually reflect the system you are planning, but it is a good idea to look at the relative costs.

Costs normally include everything you must purchase and install to provide normal or desired operation of the system. This would include:

1. Wind turbine (WECS).
2. Tower, footing, guy wires, etc.
3. Batteries.
4. Pumps.
5. Storage sheds for batteries or other equipment.
6. Storage tanks.
7. Wires/electrical.
8. Plumbing.
9. All installation costs, such as delivery, plumbing, electrical and building permits.

When simply comparing system costs, some costs are often omitted, like the wiring and water heater unit. These are assumed to be already available. Such omissions are not always valid, though, as in a case where the wind-electric water heater is being compared with a propane gas or solar-powered water heater. Here, water heater costs may be very different.

Expected System Life. The Jacobs WECS has been installed at the South Pole and has performed for more than 20 years. Wind turbines have been installed in the Rocky Mountains and have been destroyed, or damaged, in just a few months. Both locations are windy and both are subject to severe weather conditions. In trying to analyze the expected life of your system, you are confronted with several problems. A 20-year life rating of WECS by some consumer-oriented organization is not available. It would be nice to simply assume that the manufacturer's or dealer's statements concerning expected life are valid for your site. To do so will require intuition and an unemotional, scientific guess of the relative credibility of such statements. Most manufacturers tend to shy away from making claims on the life of their units, as noted in Chapter 8.

Wind system designers, however, tend to plan their equipment for useful lifetimes of much longer than 20 years. Bearings, belts and some other parts may have to be replaced every several years, but the basic machinery—if well designed—should last. The DOE Small Wind Systems Test Center at Rocky Flats near Golden, Colorado (Figure 9-26), is testing designs and publishing results that will help answer this question. Eventually, dealers and manufacturers may publish such test data in their product literature.

The logical first estimate in any case comes from:

- Your needs.
- Dealer's estimate.
- Interviews and opinions of other WECS owners.

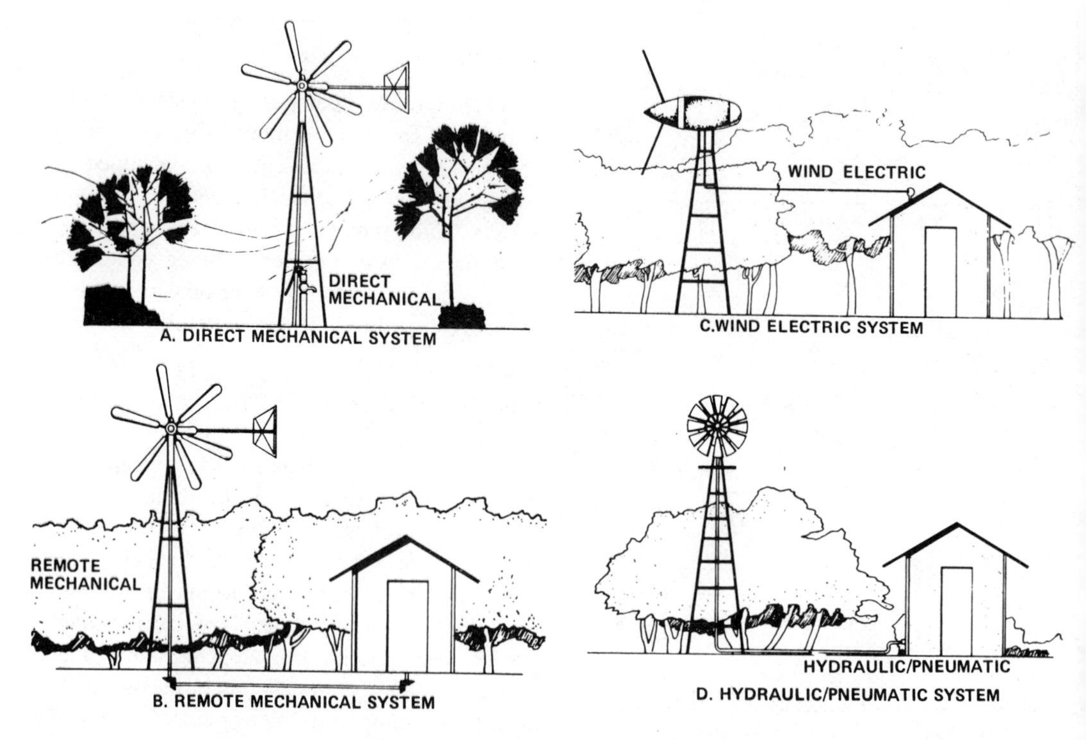

Figure 9-22. Water-pumping options. (*Courtesy of DOE.*)

Perhaps you intend to own a WECS for a limited time; use this time value in your cost study. Maybe a nearby WECS owner has had good service for 10 years; see if you can find out his expectations for continued service life (and his technique for getting such good service). Start with these and the dealer's comments. Probably you will not be far off.

Total Energy Yield. This value, expressed in kWh, or hp-hours, is the result of the energy resource study and site analysis you performed. Total energy yield represents the work of your entire system planning process.

You will likely estimate or calculate energy requirements on a monthly basis. Simple add these together for all the months of a year, and multiply the annual total by the expected life. This results in the total energy yield you can expect from your system.

Annual Maintenance and Repair Costs. It is possible to purchase a maintenance contract from some WECS dealers. Depending on the terms of such a contract, it might be possible to use the cost of the contract, plus a small contingency cost, for replacement of broken parts, as the annual maintenance cost.

Another approach is the "other owner interview." Find out what everybody else is paying to keep their machine operating. At the same time, try to evaluate the maintenance practices of the owners relative to the manufacturers' recommendations. Some owners of water-pumpers will report 20 or more years of good service and say that the only maintenance performanced was an occasional topping of transmission oil. One such performance was obtained from a machine whose manufacturer recommended an annual oil change. From these interviews and discussions with the dealers, form an estimate of the annual maintenance costs.

Expected Resale Value. Some farmers who bought wind electric systems prior to the advent of the Rural Electrification Act are reselling their old machines for prices varying from scrap iron rates to the original price. Allowing for inflation, this would indicate that these

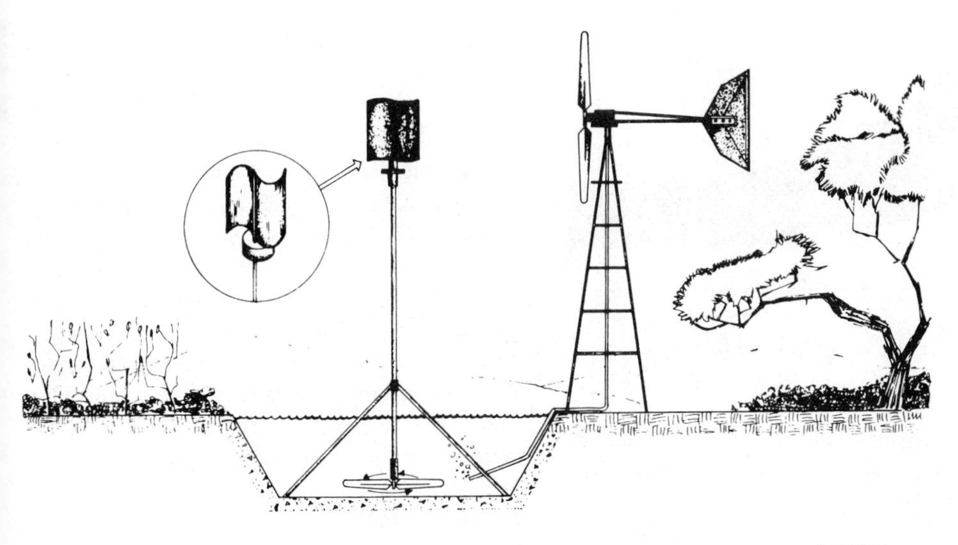

Figure 9-23. Two methods for wind-powered pond aeration. (*Courtesy of DOE.*)

WECS owners are enjoying as little as 50 percent depreciation in value over a 40-year life-span.

Other folks sometimes purchase these machines, rebuild them, and resell them at prices reflecting inflation. An old Jacobs SWECS may have been purchased for $900, then be sold 25 years later for between $100 and $1000, and, finally, be repaired and restored to its original condition and resold for $2000 to $3000. History shows that wind turbines, if properly maintained, can sell for their original cost plus an addition for inflation.

The resale price of old machines has risen rapidly in recent years. Greatly increased demand, coupled with the availability of old machines at reasonable cost, has contributed to the resale value trend. Introduction of new wind turbines from more manufacturers could soften the resale price structure, but this depends upon the ability of the WECS industry to satisfy the demand. It seems reasonable to expect resale value.

Other Annual Costs and Savings. Added to the list of costs are the bank interest you pay for money you borrow to purchase a WECS (or money you do not earn if you withdraw from savings), insurance and taxes.

Taxes, as inevitable as the wind, work both ways. First, the bad news—your tax assessor may be delighted to see you erect that permanent looking structure! Rather than bothering to figure the property tax rate, percentage of assessed value, homeowner's tax rebate and all the other present-day gimmicks, you can calculate your tax rate by dividing your total annual real estate taxes by the true estimated value of your property. For instance, if your "spread" is worth $50,000, and you pay $500 in taxes, your tax rate is really $500 ÷ $50,000 = 0.01 = 1 percent. So you would expect to pay about 30 dollars tax on a 3000 dollar wind system.

Next, think about some income tax angles. If your wind system is used for your farm or business, you can depreciate it a certain amount each year; that is, you include in your cost of doing business part of the original purchase price until it has been charged off to a preset salvage value. A reasonable lifetime for a wind turbine for tax purposes is expected to be 10 years for wind-electric types and 15 years for water-pumpers. The salvage value at that time may be 10 percent of the original cost. Such values are always conservative—less than the actual life if the device receives reasonable maintenance, and less than the actual resale value. If only part of the energy or water produced is used for the farm or business and the rest is for personal use, depreciation may be applied only to the

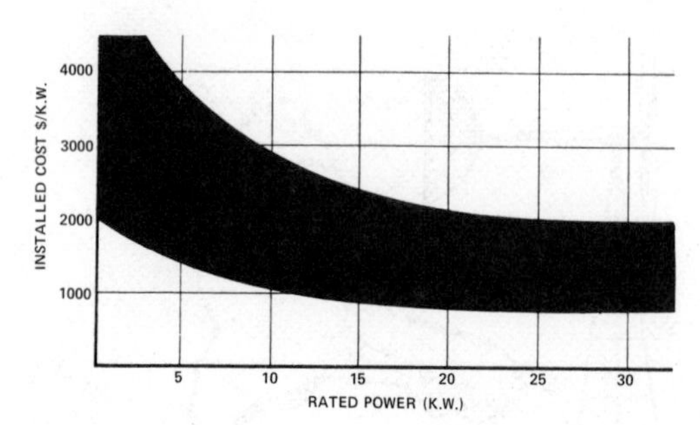

Figure 9-24. Costs of complete WECS in the 1970's. (*Courtesy of DOE.*)

farm and business portions. If you sell the wind turbine for more than its depreciated value, the excess is taxed as capital gains. Note that the cost of utility power that is no longer needed for the farm or business is a lost expense item of your tax form. Finally, if you borrow money for your wind turbine, the interest is tax-deductible.

New solar and alternative energy tax laws have been drawn up in various states and at the federal level to provide tax relief and incentives which would reduce the total cost of WECS ownership. Your local congressperson, the Internal Revenue Service, AWEA or your State Energy Office can give you the details on the legislation.

Additional homeowner's insurance will be another added cost. You may not feel a need for fire insurance, but liability insurance is a must. Depending on the cost, local windstorm conditions and dealer's warranties, wind damage insurance, if offered, is desirable. This is sometimes difficult to obtain.

The cost of your investment is a very important item to consider. If you plan to have a long-term loan on the WECS, you may select

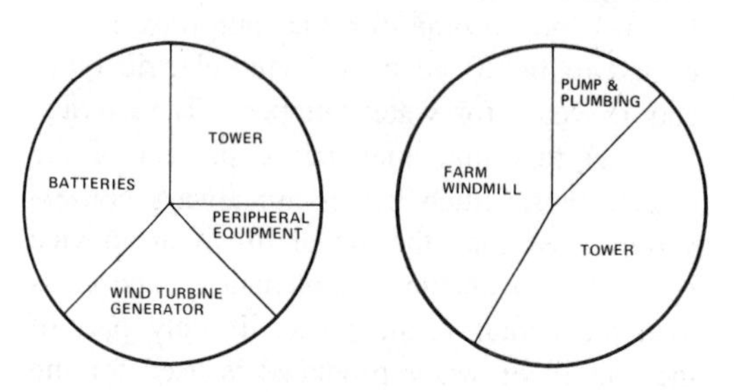

Figure 9-25. Typical relative costs for small-scale WECS. (*Courtesy of DOE.*)

the lifetime of your WECS, for analysis purposes, to be the same as the bank loan (this does not have to be the 10 years used for tax purposes). The annual cost of your investment is then the same as your annual payments.

If you take money from your savings account to buy a wind turbine, you could just take the interest you lose on that money, minus the income tax you would have paid on the interest, as the cost of your investment. This is not the true rate, since you cannot count on anyone returning your investment to you at the end of the life of the WECS. Rates set up just like loan payments are the desired ones for a correct analysis—equal, regular payments that include both capital payback and interest. Table 9-2 gives annual payback rates for various interest rates and lifetimes. The most appropriate value for you to use is the one you have obtained on your invested money in the past, minus taxes.

Other factors deserve consideration. Your estimation of inflation rates for utility power and WECS can greatly influence your final decision on wind power. A 9-percent long-term rate increase for utility power is one estimate at this time. How your electrical bill could increase (for the same power consumption) is indicated in the following table for several rates of inflation. Table 9-3 can be used for the price change of any other service or product, or annual interest on money.

Effect of Inflation of Future Value, Such as Cost of Energy. Table 9-3 shows, for instance, that a 9-percent annual inflation rate leads to a new value 5.14 times as much in 20 years. If your present electrical rate is 5¢/kWh, it would then be 5 × 5.14 = 25.7¢. If propane or heating oil costs 40¢/gallon now, it would be $2.06/gallon in 20 years.

The average cost of these items over the years (instead of the final value) is shown in Table 9-4.

If occasional utility power outages require you to have an alternative energy supply, you may wish to compare the relative costs of a gasoline-driven portable or stationary power unit and a WECS. The initial cost of the WECS and storage system is probably consid-

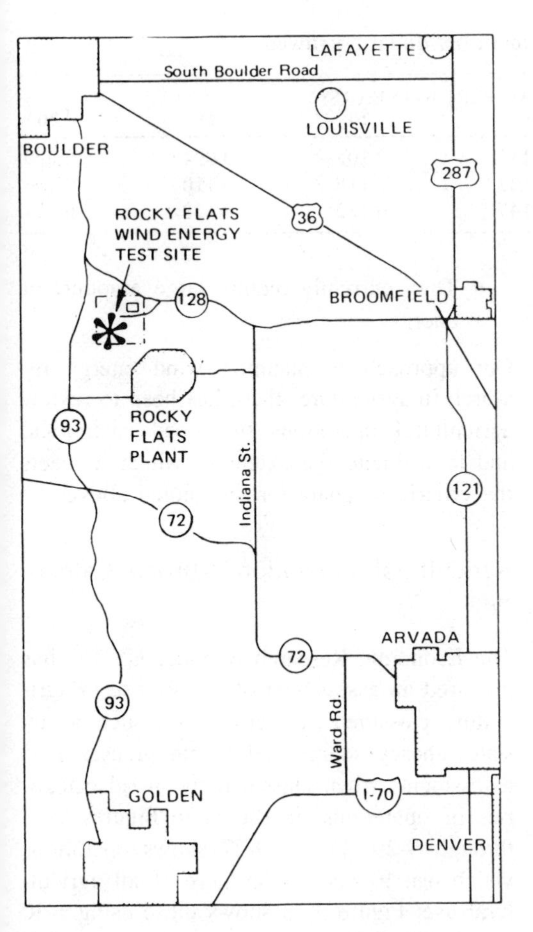

Figure 9-26. Small wind test center at Rocky Flats, Colorado. (*Courtesy of DOE.*)

erably greater than the cost of the power unit, but the WECS might more than make up for this difference by producing usable power much of the time.

Purchase of WECS. Selection of the type and brand of WECS, tower and other components will require careful consideration. There are various types of wind machines from which to choose. Chapter 8 provides a description of each of the currently available WECS and provides a theoretical power curve.

After deciding what equipment is wanted, you have three options as to how the equipment will be purchased: 1) buy directly from the manufacturer; 2) buy new or used equipment from a dealer; or 3) buy used equipment from someone else. Purchasing is important enough to warrant a brief comment on at least the first two alternatives.

Buying factory-direct may seem to be a way to save time or money, and for some it may be; but is the factory fully equipped to come out and install your machine and maintain it? If it isn't, are you? Many wind turbine manufacturers simply do not offer such services and you should check before you buy. If you hire someone to do such work, don't overlook the liability aspects that might be involved.

Dealers are usually organized and staffed to provide all of the services you need, in addition to offering the products needed to fully equip your system. Again, it is essential to ask if the dealer provides planning, installation and maintenance services. Generally, they will help you perform all of the system planning tasks discussed here.

WIND POWER USES IN AGRICULTURE

Dr. L. A. Liljedahl of the U.S. Department of Agriculture, Agricultural Research Service, provided the following overview of wind power in agriculture.

Agriculture presents a number of interesting and potentially valuable opportunities for the use of wind power. This is true not only because agriculture once was the principal user of wind power in the U.S. and thus might be better prepared to return to it. Nor is it due to the fact that agriculture is, like wind power itself, spread out over the country rather widely, although this does tend to suggest agriculture as a more likely industry to use wind power than more concentrated industries such as, say, aluminum production.

The opportunity for wind power use comes from the fact that several operations in agriculture which use appreciable amounts of energy are interruptible for short periods of time and a number of others are capable of storing energy in inexpensive forms, or can store work done for future use. This capability, then, accommodates a major characteristic of wind power; i.e., its variability and discontinuity.

From all of these potential applications,

Table 9-2. Annual loan payment per dollar borrowed.

INTEREST RATE (%)	LOAN PERIOD (YEARS) 5	10	15	20	25	30
8	0.251	0.149	0.117	0.102	0.094	0.089
10	0.264	0.163	0.132	0.118	0.110	0.106
12	0.277	0.177	0.147	0.132	0.128	0.124

however, those which will be most economical are those which will be used for the greatest portion of the year. These will produce the greatest yearly amount of energy per unit of power, and in turn yield the fastest economic return for the system. Because of the capital intensiveness of WECS, applications having high annual usage are more likely to be economical, other factors being equal.

Other applications which are more likely to be economical are those which are used predominately in areas of high average wind or those used predominantly in seasons of high average wind. In either case, the needed energy can be captured from the wind with a smaller machine, which will reduce the capital investment required and reduce the breakeven cost of conventional energy with which it can compete.

Therefore, we have reviewed operations in agriculture which may be good candidates for wind energy based upon the extent to which they are:

- Interruptible or use inexpensive energy storage.
- Used a large number of weeks per year.
- Used in areas or seasons of high wind.

To maximize replacement of conventional energy sources, also, publicly-supported research should, where a choice is necessary, concentrate on operations which also have the characteristic that:

- They currently require large amounts of energy.

Our approach to planning wind energy research in agriculture, then, has been to review agricultural operations by broad categories, and to estimate the extent to which it meets the criteria of characteristics noted above.

Agricultural Operations—Broad Categories

The Economic Research Service, USDA, has prepared an assessment of energy use in agriculture classified in many ways, such as by state, energy source and month of use. This assessment, when classified by broad categories of operations, is shown in Figures 9-27 through 9-29. Figure 9-27 shows operations which use 10 percent or more of all agricultural use; Figure 9-28 shows those using 1–10 percent; and Figure 9-29 shows those using 0.1–1 percent.

It is then obvious that there is a range of several orders of magnitude in the energy used annually for these operations, even when classified by broad categories. Under such circumstances, there is greater than usual reason to give importance to the amount of energy used annually in the operation.

Fertilizer Production. Although fertilizer is used on the farm at relatively few times of the year, it does not need to be made just when it

Table 9-3. Estimated final value inflation rates.

RATE OF INFLATION (%)	VALUE OR COST AT (YEARS): 1	5	10	15	20	25	30
5	1.00	1.22	1.55	1.98	2.53	3.23	4.12
7	1.00	1.31	1.84	2.58	3.62	5.07	7.11
9	1.00	1.41	2.17	3.34	5.14	7.91	12.17
11	1.00	1.52	2.56	4.31	7.26	12.24	20.62

Table 9-4. Average cost increase due to inflation.
Average Value or Cost Over (years):

RATE OF INFLATION (%)	5	10	15	20	25	30
5	1.11	1.27	1.46	1.76	1.91	2.21
7	1.15	1.42	1.79	2.31	2.53	3.15
9	1.20	1.58	2.17	3.07	3.39	4.54

is used. As is now the case with present processes and energy sources, fertilizer could be made the entire year by wind power when it is available, storing the results of the work done until needed (see Figure 9-30). The processes now used are not readily started and stopped, but this may not constitute a fundamental barrier. The current process is one of a dozen or more which have been used or previously considered for use in fertilizer manufacture (actually, nitrogen fixing), and it has come into use primarily because of its efficiency when used with present energy sources from petroleum, rather than because of any basic value. Thus, the feasibility of manufacturing fertilizer using wind generated energy cannot be immediately assessed without a thorough review of all known nitrogen fixing processes, with specific attention to the particular advantages, or disadvantages, or using wind energy as an input.

Vehicle Operation. A similar situation exists with regard to vehicular operation. There does not appear to be any way for these operations to be directly wind powered, and they are not likely to be considered interruptible. Furthermore, the prime movers now used have come into use primarily because of their efficiency and convenience when used with present energy sources from petroleum, and not because of any basic intrinsic value. Thus, the feasibility of using wind power indirectly for vehicular operation cannot be immediately assessed without a broad ranging review of all conceivable prime movers and intermediate fuels, as well as the characteristic of the processes for making the fuels, with attention to the advantages and disadvantages of each when wind energy is used as an input.

Stationary, On-farm Operations. Very few of the other operations in farming require such basic, long-term reassessments to yield a practical scenario (conceptual plan) for the use of wind to power the operation. Scenarios have been developed for the use of wind power for these operations, and the four previously described criteria for the use of wind power were applied to the operation and scenario. The results of this, summarized in Figure 9-31, were the first overall assessment of the wind power use in agriculture.

Any conclusions to be drawn from this assessment must be tentative, as the groupings shown are rather coarse and the quantitative knowledge is rather limited concerning most of these characteristics.

Not only are there wide variations in the annual operating usage of many of these operations, but there is a lack of precision about

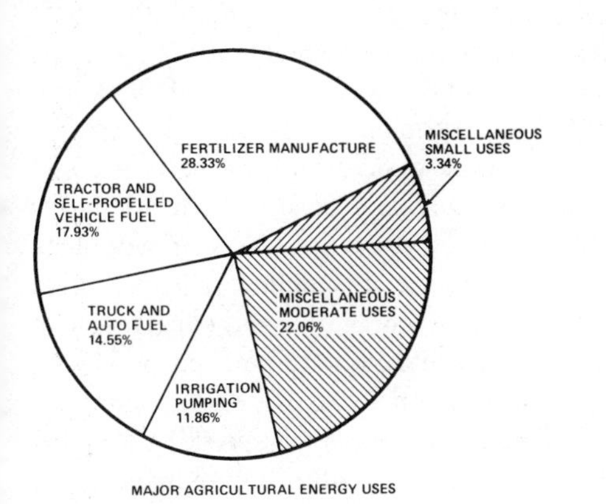

Figure 9-27. Major agricultural energy uses. (*Courtesy of USDA.*)

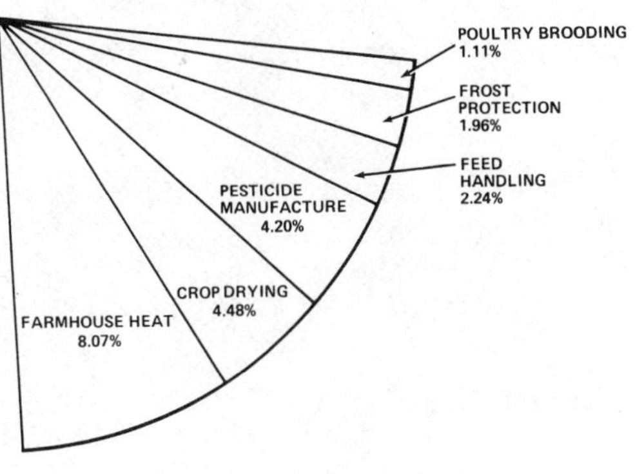

Figure 9-28. Moderate agricultural energy uses. (*Courtesy of USDA.*)

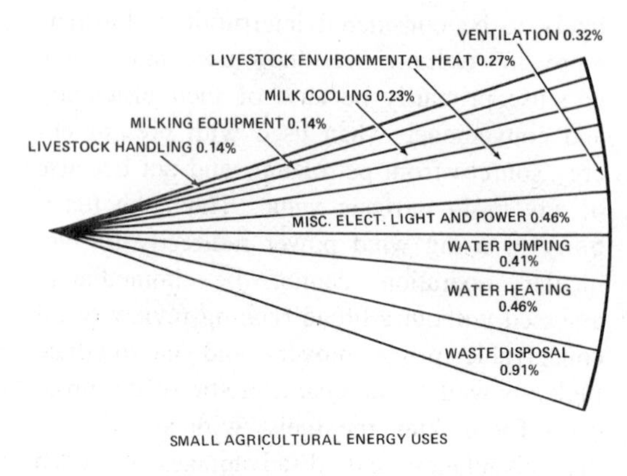

Figure 9-29. Small agricultural energy uses. (*Courtesy of USDA.*)

geographical distribution of wind power (nationally and locally) and about the cost and practicality of storing work done or storing energy in an intermediate medium; these factors are highly dependent on future technological developments, which are uncertain.

This assessment also presents these char-

Figure 9-30. Automatic plant powered by the wind to produce nitrogen fertilizer from air and water. (*Courtesy of DOE.*)

acteristics with equal weight. This may suffice as a first step, but it is recognized that a more critical assessment might give greater weight to some characteristics than to others.

From the previous assessment, irrigation and feed processing appear to have characteristics best suited for wind power use. Building heating and farm product processing and storage, such as milk cooling, grain drying and potato storage ventilation, also appear to be fairly well suited for wind power use. Domestic water-pumping, general purpose electric usage and frost protection appear to be less well suited, primarily due to the noninterruptibility of the demand, and to the more expensive energy storage required for continuous service.

Irrigation Pumping. The pumping required for irrigation lifts water from the supply source to the distribution system, and provides additional pressure for operation of the distribution system. The scenario for using wind energy uses WECS to pump water into a storage device of sufficient capacity (1–4 acre ft) to provide water to the distribution system under a constant pressure for a single watering period, commonly 8–12 hours. For gravity distribution systems such as gated pipe-furrow or header ditch-siphon tube furrow, the pressure required is no more than 2–3 psi, which can be economically provided with a small surface reservoir a few feet (vertically) upgrade of the field. Sprinkler distribution, by contrast, would require a pressure vessel of 10^5–10^6 gallons, or a standpipe of the same capacity, to provide water at 30–45 psi, the pressure presently needed for satisfactory operation of sprinkler systems. Exceptions would be those cases where the terrain could provide an equivalent head, which are infrequent. As the cost of the reservoir would be several orders of magnitude cheaper than pressure vessel storage, it is expected that wind power would initially be used only with gravity distribution systems. Fortunately, despite popular awareness of sprinkler systems and their recent growth, approximately 70 percent of all irrigation energy is consumed in systems using gravity distribution. Thus, this

SUMMARY OF CHARACTERISTICS RELEVANT TO WIND ENERGY USE IN AGRICULTURE

	MAJOR USE OF ENERGY	ANNUAL OPERATING USAGE	LOW COST STORAGE OR INTER-RUPTIBLE	USED IN HIGH WIND AREAS
IRRIGATION	GOOD	GOOD	GOOD	GOOD
RESIDENTIAL HEAT	GOOD	GOOD	AVERAGE	AVERAGE
POULTRY BROODING	GOOD	GOOD	AVERAGE	POOR
LIVESTOCK ENVIRONMENT	AVERAGE	AVERAGE	AVERAGE	AVERAGE
WATER HEATING	AVERAGE	GOOD	GOOD	AVERAGE
CROP DRYING	GOOD	POOR	GOOD	AVERAGE
FEED HANDLING	GOOD	GOOD	GOOD	AVERAGE
MILK COOLING	AVERAGE	GOOD	AVERAGE	AVERAGE
REFRIGERATED STORAGE	AVERAGE	GOOD	GOOD	AVERAGE
FROST PROTECTION	GOOD	POOR	POOR	POOR
WATER PUMPING	AVERAGE	GOOD	POOR	AVERAGE
MISCELLANEOUS ELECTRIC POWER AND LIGHTS	AVERAGE	GOOD	POOR	AVERAGE
VENTILATION	AVERAGE	GOOD	POOR	AVERAGE
MILK EQUIPMENT	POOR	GOOD	POOR	AVERAGE
WASTE DISPOSAL	AVERAGE	GOOD	AVERAGE	AVERAGE
LIVESTOCK HANDLING	POOR	POOR	POOR	AVERAGE
GRAIN HANDLING	POOR	POOR	GOOD	AVERAGE

LEGEND

VALUE	SYMBOL	PROPORTION OF TOTAL AGRICULTURAL ENERGY USE	ANNUAL OPERATING USAGE WEEKS/YEAR	INTERRUPTIBILITY OF ENERGY STORABILITY	AVERAGE WIND IN PREDOMINANT USE AREA W/m²
GOOD	(crosshatch)	>5%	>26	WORK INTERRUPTIBLE AND WORK STORABLE	>150
AVERAGE	(stipple)	<5% >0.5%	<26 >5	LOW COST INTERMEDIATE ENERGY STORAGE	<150 >50
POOR	(diagonal lines)	<0.5%	<5%	HIGH COST INTERMEDIATE ENERGY STORAGE	<50

Figure 9-31. Summary of characteristics relevant to wind energy use in agriculture. (*Courtesy of USDA.*)

scenario has considerable potential for energy savings, even though it would not be used with one of the major categories of distribution systems.

Longer-term energy storage would be achieved in such systems by storage of water in the soil root zone in excess of short-term plant needs. This, strictly speaking, would not actually be energy storage, but rather a storage of work done for later use.

Typical power requirements for irrigation would be 5–15 kW for pumping from surface water and shallow wells, and 40–60 kW for deep wells in sub-surface aquifers. Power transmission from the WECS to the pump might be either mechanical or electrical, as shown in Figures 9-32 and 9-33.

Building Heat. The farmhouse, poultry brooding, milking parlor and greenhouse space heat require the bulk of building heat in agriculture. A scenario for wind power use in all of

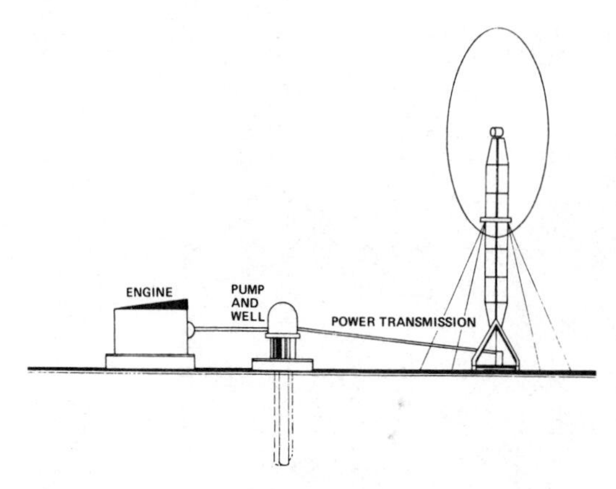

Figure 9-32. Scheme for wind power use in irrigation pumping with mechanical transmission. (*Courtesy of USDA.*)

these operations involves the use of intermediate heat storage in a container of material having a high specific heat, such as water or crushed stone. The storage media would be heated by the wind turbine by means of electrical or mechanical dissipation, or by the use of a heat pump. Electrical dissipation would use an alternator or generator driven by the WECS, the output of which would be loaded by an appropriate group of electrical resistance heating elements, as shown in Figure 9-34. Mechanical dissipation would likely use a churn or turbine to drive a fluid whose kinetic energy could then be dissipated hydraulically. Heat storage at relatively high temperatures might be possible with such systems, as the outlet temperature has no effect on the general system performance. If a heat pump were used, as shown in Figure 9-35, the compressor would be driven either mechanically or electrically, by the turbine. The coefficient of performance of the heat pump would permit use of a smaller wind turbine, but the limited condenser temperature of the heat pump would restrict the maximum temperature at which the energy could be stored.

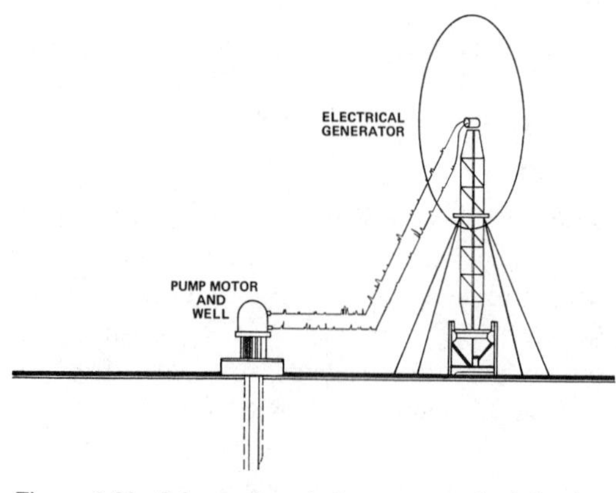

Figure 9-33. Scheme for wind power use in irrigation pumping with electrical transmission. (*Courtesy of USDA.*)

The amount of heat storage required would depend upon the recurrence frequency and duration of low wind periods, for which there has been only limited summarization and generalization of recorded data, so far. If recurrence frequency of major weather patterns can be used as a guide, this is likely to be of the order of four to six days. Transfer of the heat from the storage media would be similar to the systems now used for solar heating systems in the same application.

Product Processing and Storage. These applications of wind power are a more diverse category, and the scenario of wind power use is distinctly different for each application. For example, the simplest use of wind power for crop drying would be to drive fans used for unheated air-drying with WECS, with the power transmitted either mechanically or electrically, as shown in Figure 9-36. It is possible that we will find an inverse correlation between wind speed and drying capacity of the air, however. If this is appreciable, the simplicity and economy of this approach will be offset by its decreased effectiveness. An alternative to this would be the use of a wind-driven heat pump to dehumidify air used for crop-drying, using a smaller fraction of the energy available for moving of the air. If drying air dehumidification should be too complex, another alternative would be the use of wind power for refrigerated storage/drying of the crop. This cools the moist crop to a temperature where deterioration is reduced for a long period, during which the crop can be dried at a slower rate.

Milk and stored apples could be refrigerated with machinery driven by WECS, using storage of cooling capacity in the form of frozen water or solutions to accommodate periods without wind.

Obviously, some feed processing equipment, such as grinders, can be readily driven with WECS, with automatic control equipment and storage bins permitting storage of work done during period of high wind.

Deployment Modes. The scenarios described above for the agricultural applications might be used with several types of WECS installations. The simplest of these would be dedicated application, where the WECS drives only the load of one specific application, as shown in Figures 9-34 to 9-37. In this case, the WECS would be installed close to the application, and if possible, would remain only at that location, and power would be transmitted to that load. While this would be simplest, it would require an application having a high number of hours per year of usage to provide maximum economic return.

A second alternative would be use of a WECS to drive several loads whose use does not coincide with other loads during the year, so that no more than one load is driven at a time. An example of this would be a single WECS which drives an irrigation pump during the summer, a drying system in the fall and a heating system during the winter. If a mechanical drive system were used in this case, the WECS would have to be portable to permit moving from site to site during the year. Although this may appear to be cumbersome, farmers do routinely move large devices, such as center-pivot irrigation machines, from site to site. The time required for such moves and the hazard of damage to the machine, however, may limit the practicality of this approach. An electrical transmission system on the farm would permit a stationary WECS to supply power to several loads without moving the machine, as shown in Figure 9-37, at the cost of the additional investment for the electrical transmission equipment. The cost of such equipment might be reduced, however, by use of higher voltages and frequencies than normal utility service, if compatibility with utility service were not required.

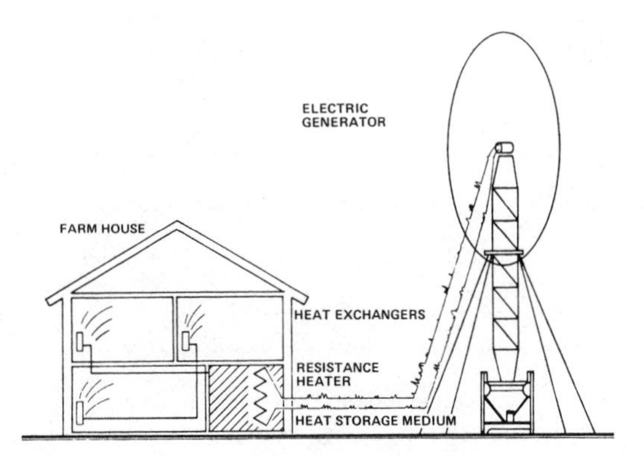

Figure 9-34. Scheme for wind power use in building heating using dissipative heating. (*Courtesy of USDA*.)

Another obvious alternative is the transmission of power from the wind turbine to the major loads as utility-compatible electric power. Such a scheme would permit the use of more conventional electric motors; the use of electric utility power as a backup or emergency power source; and, possibly, arrangement for sale of excess power back to the utility when the generated power exceeds the needs of the farm enterprise, as shown in Figure 9-38.

Current USDA Application Programs

Based upon the foregoing qualitative analysis of the potential for agricultural use of wind power, the Agricultural Research Service, in cooperation with the Department of Energy, has undertaken a number of detailed studies.

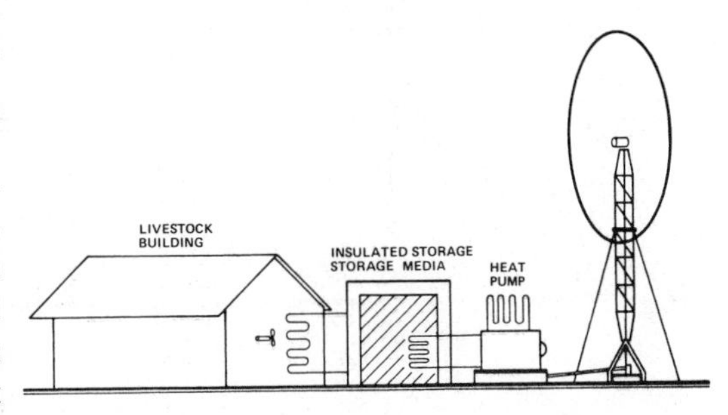

Figure 9-35. Scheme for wind power use in building heating using heat pump. (*Courtesy of USDA*.)

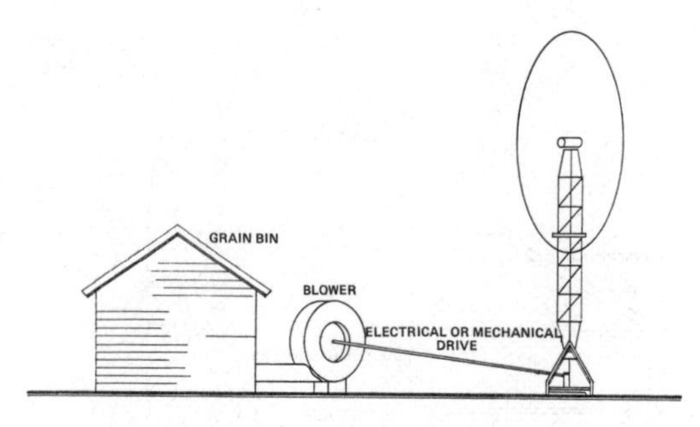

Figure 9-36. Scheme for wind power use in crop drying. (*Courtesy of USDA*.)

Analytical Studies. The Agricultural Research Service at Manhattan, Kansas, has made a systems analysis of the use of wind power for irrigation pumping, using known general characteristics of wind power, wind machines, pumps, irrigation wells and crop water requirements. This study has shown that pumps driven only by wind power could not irrigate the present area of land now being irrigated with continuously available power unless cropping patterns were changed or the number of wells were increased.

Development Planning and Research Associates, Manhattan, Kansas, have conducted a quantitative review of possible wind energy applications in agriculture using published data on power demand in agriculture, wind energy distribution and distribution of types and sizes of farms in the U.S. It has estimated the cost of conventional energy against which wind power economically competes, and identified the distribution of sizes of wind powered units which would be most economical, and their total number, nationally.

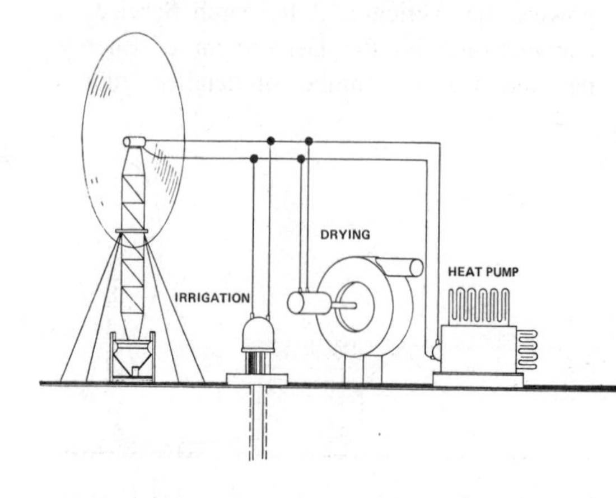

Figure 9-37. Scheme for multiple wind power use. (*Courtesy of USDA*.)

Experimental Studies. R. Nolan Clark of the Agricultural Research, Science and Education Administration of USDA is the principal investigator for wind energy irrigation pumping, and he has provided the following summary of their program, involving a 40-kW WECS.

A 40-kW, vertical-axis WECS was erected at the USDA Southwestern Great Plains Research Center, Bushland, Texas. Objectives of this project were to assemble a complete wind power pumping system, adapt or modify existing pumping equipment so that it would be effectively powered by a wind turbine and make economic analyses of wind-pumping systems.

The pumping system used both a WECS and an electrical motor to power an existing deep-well irrigation pump in a wind-assist concept, as shown in Figure 9-39. The system delivered 104m^3/hour against a head of 105 m. An overrunning clutch was used to synchronize the power sources.

All components of the wind-assisted pumping system worked satisfactorily. The power produced by the wind turbine was greater than predicted by the computer model. The power produced by the wind turbine averaged 36 kW and provided an energy savings of 65 percent for the pumping system.

Another study, by the Agricultural Research Service at Manhattan, Kansas, involves the installation of a 20-ft-diameter, 30-ft-high Darrius rotor wind turbine coupled to a shallow-well irrigation pump for use in installations where pumping is from surface water supplies. The unit is tested in an experimental sump and reservoir which will be followed by installation in a field site for irrigation operation during the irrigation season.

The Agricultural Research Service at Ames, Iowa, has installed a 25-ft-diameter horizontal-axis WECS to supply supplementary heat at a rural residence. Power transmission is three-phase electric ac, and a water tank pro-

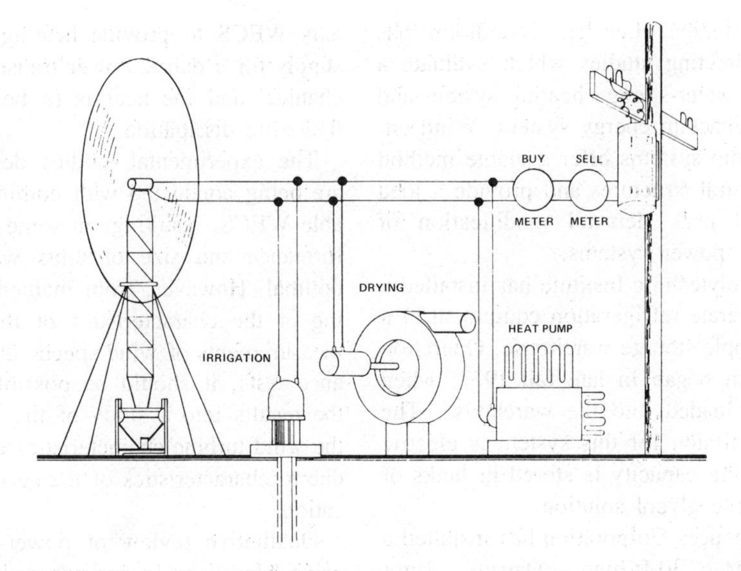

Figure 9-38. Multiple wind power use with utility power compatibility. (*Courtesy of USDA.*)

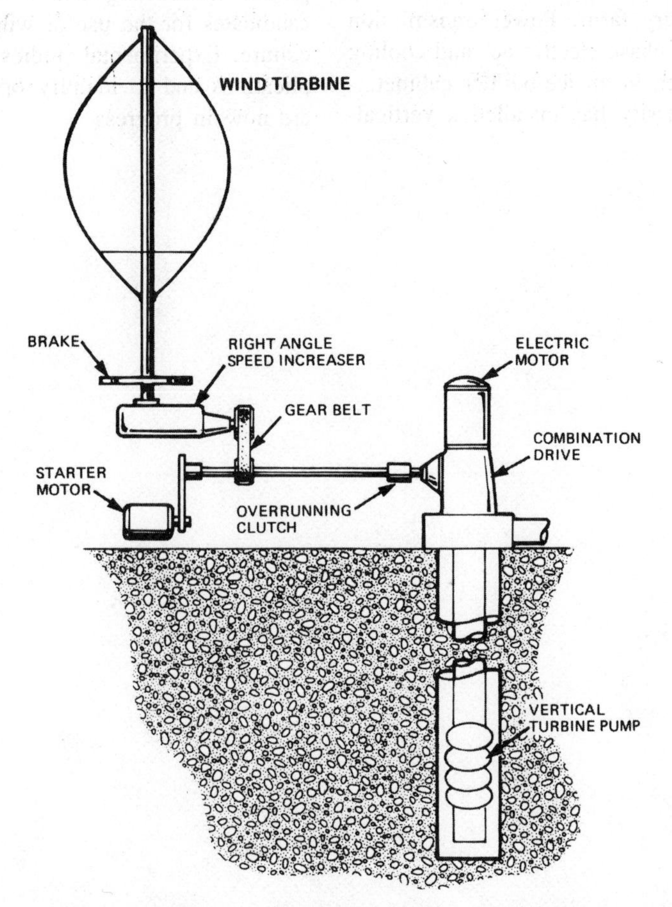

Figure 9-39. Schematic of wind-assisted irrigation pumping system. (*Courtesy of USDA.*)

vides heat storage. Leo H. Soderholm has also been directing studies which evaluate a wind energy/water-storage heating system and a heat-pump/backup energy system. Wind energy/heat-pump systems offer a viable method for heating rural structures and provide a load leveling and peak demand modification of rural electric power systems.

Virginia Polytechnic Institute has installed a WECS to operate refrigeration equipment in a pilot-scale apple storage warehouse. Operation of the system began in late fall 1977, when apples were loaded into the warehouse. The power transmission for this system is electric dc, and cooling capacity is stored in tanks of frozen ethylene glycol solution.

Kaman Sciences Corporation has installed a 20-ft-diameter, 30-ft-high Darrius rotor WECS to drive refrigeration equipment for milk cooling at the Colorado State University experimental dairy farm. Power transmission is through three-phase electric ac, and cooling capacity is stored in an ice-builder cabinet.

Cornell University has installed a vertical-axis WECS to provide heating of the water supply for a dairy. Power transmission is mechanical and the heat is to be generated by hydraulic dissipation.

The experimental studies described above are being conducted with commercially available WECS, resulting, in some cases, in performance and size of units which are suboptimal. However, from mathematical modeling of the characteristics of the WECS and measurements of wind speeds during performance tests, it should be possible to partition the results into a study of the effects due to the wind turbine characteristics and the effects due to characteristics of the agricultural application.

Qualitative review of power- and energy-using operations in agriculture have identified irrigation pumping, building, heating and product processing and storage as possible candidates for the use of wind power in agriculture. Experimental studies to test these applications and to identify operation problems are now in progress.

10
Federal Wind Energy Program

The Federal Wind Energy Program will provide approximately $67 million of research, exploratory development, technology development, engineering development and demonstration funds in FY 1980. These funds represent a significant investment in commercializing wind power. In comparison, the 1945 engineering costs to build the Smith-Putnam wind turbine were estimated to be $116,560.

This chapter provides a review of the purpose, strategy, significant programs, funding, technical status, program summaries and organization of the current Federal Wind Energy Program. The program consists of five major efforts, which include Program Development and Technology, Farm and Rural Use (Small) Systems, 100-kW-scale Systems, MW-scale Systems, and Large Multi-unit Systems.

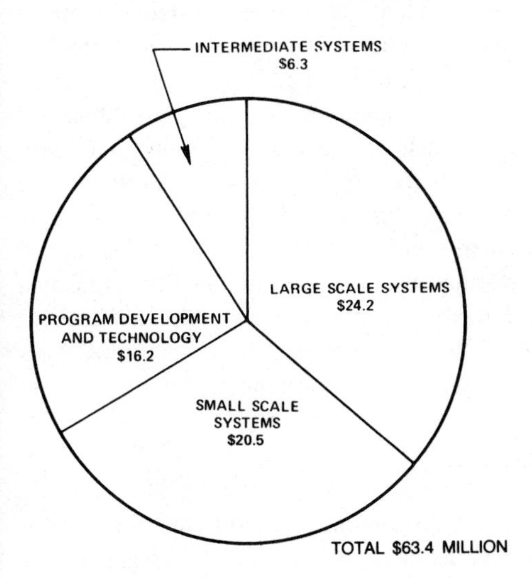

Figure 10-1. Federal Wind Energy Program Budget Allocation.

These five programs will utilize $67 million to accomplish the Federal Wind Energy program goals in FY 1980, as shown in Figure 10-1.

PURPOSE

The long-term objective of the Federal Wind Energy Program is to accelerate the development and commercialization of reliable, safe and economically viable wind energy systems. It has long been recognized that wind energy is a potentially abundant source of clean and renewable mechanical and electrical power. With the advent of the Rural Electrification Administration and inexpensive electrical power in the 1930's, the wind industry in effect ceased to function in the U.S. The increased uncertainty of future domestic energy supplies, with concurrent increased reliance on foreign sources of some fossil fuels, has led to renewed interest in wind systems. Cost remains the greatest barrier to the widespread use of wind systems. Thus, the purposes of the program are to reduce the cost to a competitive level with other power sources while resolving the technical, environmental and social issues, and to provide an environment conducive to the business development of wind systems so that wind systems may be used to contribute significantly to the nation's energy needs.

The ultimate goal of the Federal Wind Energy Program is an economically competitive U.S. industry devoted to commercial generation of mechanical and electric power using wind energy. Not only would electric utilities augment their generating capabilities through large-scale wind energy systems, but smaller (mostly farm and rural) consumers would be able to displace significant amounts of utility generated electricity and on-site use of fossil fuels by using wind power.

It is anticipated that an initial market will begin to form at about 4–5¢/kWh (1978 dollars), which is sufficient to support early production quantities.

To achieve the long-range program objectives and the quantity market necessary to realize maximum production efficiencies, an energy cost goal of 2¢/kWh has been established for both small and large wind systems. For small systems, this goal represents a "typical" midwest farm site interconnected to an electric utility for backup power for which the utility imposes a demand charge. For large systems, the goal assumes that they are deployed in sufficient numbers over a utility grid to receive appropriate capacity credit, in addition to saving the cost of conventional fuels.

STRATEGY

The Federal Wind Energy Program strategy has evolved from the principal barriers to commercialization of wind systems and represents a balance between time, risk and cost to the government. The three immediate requirements of the program are to:

1. Lower the cost of wind power while achieving high reliability and safety.
2. Increase user awareness and acceptance of wind systems (including user education designed to expedite use of wind energy).
3. Promote an industrial wind system manufacturing capability and supporting infrastructure.

The strategy of the wind energy program in addressing these key requirements can be summarized as follows:

- Continue research and technology development leading to progressively more advanced systems with lower costs.
- As cost reductions reach competitive levels for high-potential/high-cost applications, phase in demonstrations (including Federal sector) to foster user awareness and to build industrial production capacity whose efficiencies can contribute to further cost reductions.
- As costs approach the levels of the broader general market, replace the technology development and demonstrations with user incentives to stimulate market penetration and manufacturer's incentives to realize the enhanced efficiencies of quantity production.
- Provide convenient tools for siting, system selection and economic analysis to the various potential wind system users to reduce risks and to promote user confidence.

These key elements of the strategy satisfy the program requirements in a variety of ways, as discussed below.

Systems Development

In addition to its basic objective of lowering system cost, the technology development activities permit industry to gain experience in the design and fabrication of systems. By emphasizing a series of advanced systems with integrated technology development instead of predominately separate research projects, the industry is able to maintain continuity of experience. This aspect, plus the frequent use of parallel development contracts, could also result in a broad industrial base with its resulting price competition. Usually, more than one unit of each design will be produced to provide fabrication experience. Furthermore, these units will be field tested by typical users (farmers, utilities, etc.) to determine actual performance and costs, to identify and solve technical and operational problems and to generally guide the efforts to minimize costs and promote user awareness.

Systems Demonstrations

The demonstrations are primarily an interim market development activity. Costs are further reduced through advanced systems development. The increased number of systems is intended to provide a market for industry and thereby permit the establishment of reasonable production facilities. These facilities, with their increased efficiencies, are considered a vital element of the cost reduction strategy. The demonstrations are also expected to stimulate market development, including the Federal sectors, such as DOE, FPA and TVA. After the first few units, users will select their own systems, including privately developed systems that are subject to some minimum criteria. Cost-sharing will be an integral part of the program. Thus, user familiarity will increase and the competitive industry will be stimulated.

Incentives Program

As the broader market begins to develop, the systems development and demonstration activities will be replaced by a higher leverage incentives program. The most advantageous combination of incentive options is uncertain; however, studies are under way that will permit the government to develop appropriate recommendations. It is expected that the recommendations will include predominately user incentives to stimulate market growth, as well as some manufacturer incentives to assist in initial market entry and expanded production capacity.

Supporting Activities

Cost-competitiveness and production capacity are not sufficient to ensure the commercial deployment of a technology. There are a number of other issues which collectively affect market risk and firmness but are too expensive and too broad in scope to be addressed by the private sector. Thus, the wind program must also:

- Resolve and clarify, in the near-term, legal, environmental and aesthetic uncertainties and fulfill environmental impact assessment requirements.
- Establish a broad-based user awareness and information dissemination program involving various Federal and state agencies.
- Develop wind survey techniques and conduct national and regional wind resource assessments.
- Develop cost-effective siting techniques and handbooks for use by individuals, utilities and state and local agencies.
- Conduct applications studies and economic analyses to determine operational requirements and strategies for implementing wind systems in particular applications.
- Provide economic planning and applications handbooks designed to simplify site-specific selection and use of wind systems.
- Perform market assessments at the national and regional levels to firm estimates of market size as a function of wind system costs.
- Develop and test advanced components and innovative systems in an effort to increase the performance and/or lower the cost of wind systems and investigate the feasibility of newly proposed concepts.

An important aspect of this strategy is the control available at each phase. If the cost goals are not met at any stage, the level and thrust of program activities can be adjusted due to the relatively modular nature of the system. If an additional development cycle is required, it can be accomplished for approximately \$20–40 million. Conversely, if the cost goals are met, the demonstration activity can be accelerated for about \$2–4 million per machine. These features of the wind energy program permit reassessment and, if needed, restructuring of the commercialization activities at defined decision points.

SIGNIFICANT STRATEGIC PROGRAMS

The Federal Wind Energy Program has been structured to be responsive to the strategy elements defined in the previous material. In general terms, the program structure can be summarized as:

- Program development and technology, including the general market and applications studies and analyses; legal/-

environmental/social issues; wind characteristic program; research and development; and investigation of advanced/innovative concepts.

- Systems development elements, including small systems (less than 100 kW), intermediate systems (100–1000kW) and large systems (greater than 1 MW). These elements also include R&D and various analyses related to specific development efforts.
- Systems demonstrations elements under this task regarding the planning of demonstrations, site qualification and construction of wind systems are being addressed.

Later in this chapter, the individual program elements, their schedules and their relationships to the program objectives and goals are fully described. The key activities in each element are discussed below.

Program Development and Technology

These activities are directed towards providing users and manufacturers with the data and tools necessary to support the commercialization of wind systems. They also provide the information vital to effective program planning and Federal and non-Federal decision-makers. These activities have been divided into six sub-elements, which are described as follows.

Mission Analysis. This sub-element, shown in Figure 10-2, addresses the broad studies of mission analysis, marketing assessment, venture analysis, incentives studies and economic analysis. The ultimate goal of these activities is a firm estimate of market size as a function of system cost, now scheduled for completion for FY 1981. Toward this goal, initial mission analyses, incentives studies and venture analyses have been completed. In FY 1979, the activities included detailed incentives studies and ventures analysis, as well as selected regional and sector market assessments. Integrating the results of these activities and those from the application/system studies element provides a refined definition of the cost goals required to penetrate the various markets.

Applications/System Studies. These studies investigate the complex operational and economic questions associated with the use of wind power intermixed with existing conven-

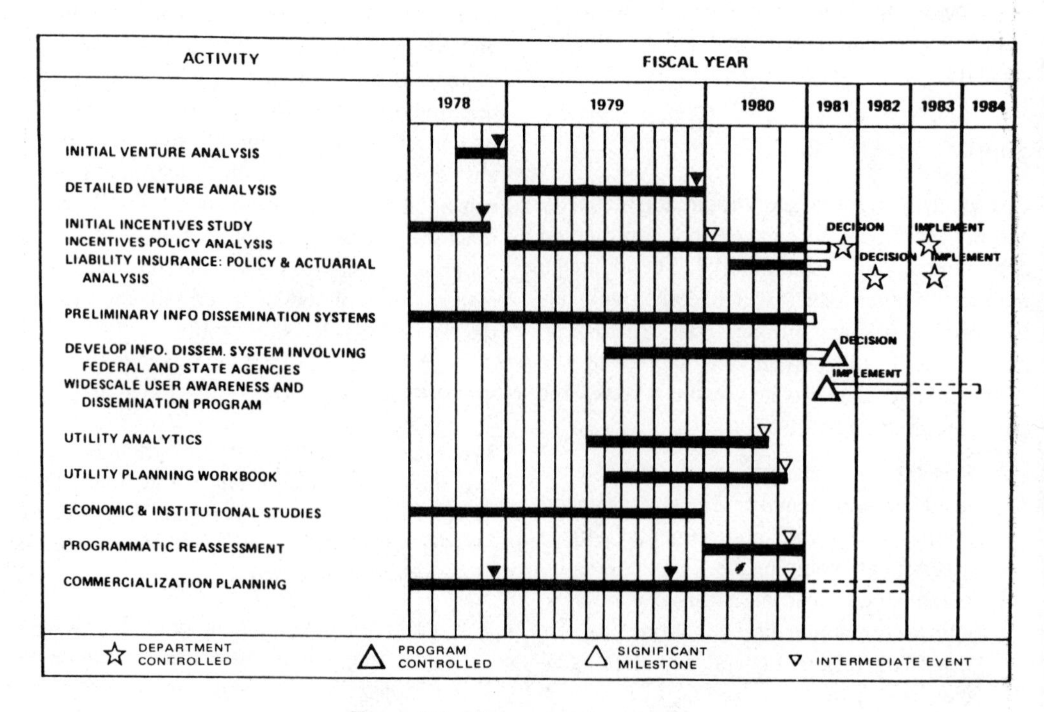

Figure 10-2. Mission analysis activities.

tional power systems. As shown in Figure 10-3, specific and genuine applications are being analyzed to provide a general economic model of wind systems interconnected to existing power grids. The key products of these activities will be planning workbooks and analytical tools to permit individual users and utilities to effectively evaluate and plan for the use of wind energy systems.

Legal/Environment/Social Issues. General and specific assessments which established that impacts from most legal, environmental and social issues associated with wind energy are likely to be minimal have been completed. The operation of field tests, as shown in Figure 10-4, will be monitored to confirm these assessments. Studies are continuing in selected areas of concern to further refine the assessments and to identify mitigating actions where necessary. Particular study areas started in FY 1979 include television interference, noise, general environmental studies, insurance and standards related to performance, testing, safety labeling and nomenclature.

Wind Characteristics Program.

The wind characteristics activities shown in Figure 10-5 are directed towards the following products:

- Large-area wind survey techniques.
- National and regional wind resource assessments.
- Siting techniques and handbooks appropriate for small and large systems.
- Wind characteristics data for use in system design.
- Standardized data collection and reduction methods.
- Wind forecasting techniques of operational use.
- Candidate site selection for use in field testing of new systems.

To date, significant progress has been made in these areas. An initial national resource assessment has been completed and large-area survey techniques were developed for testing in the Pacific Northwest during FY 1979. The first siting handbook has been published. Significant new and available wind data have been collected and standardized for use in site qualification and design efforts. An original set of 17 candidate field test sites was selected and a new Program Opportunity Notice

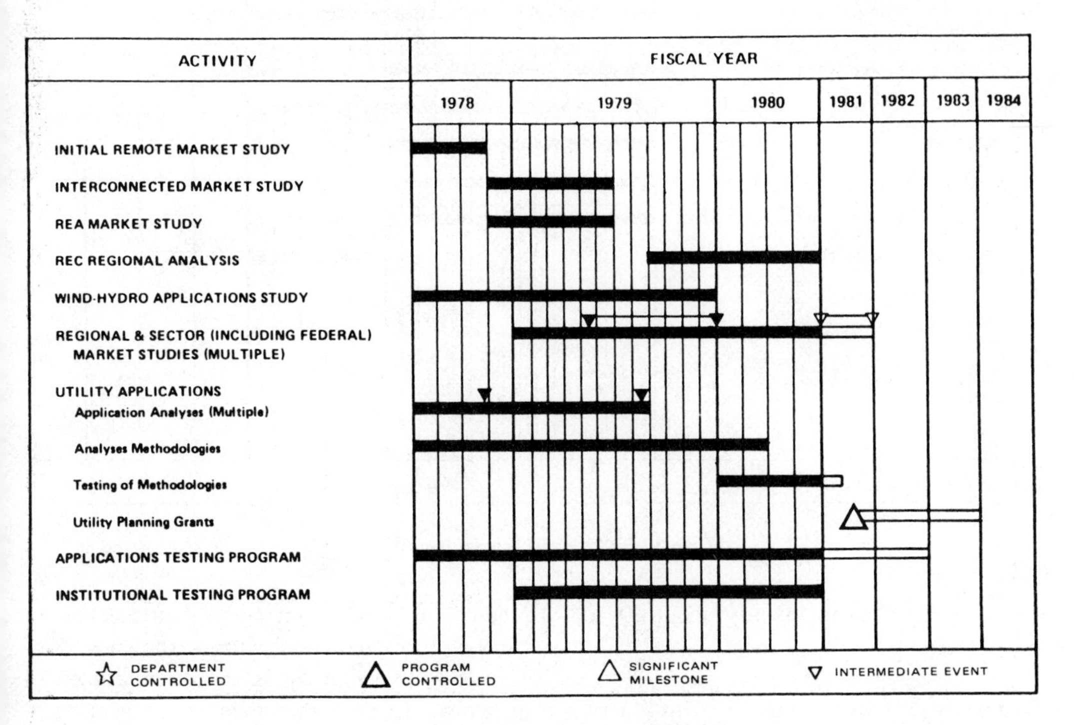

Figure 10-3. Applications/systems studies activities.

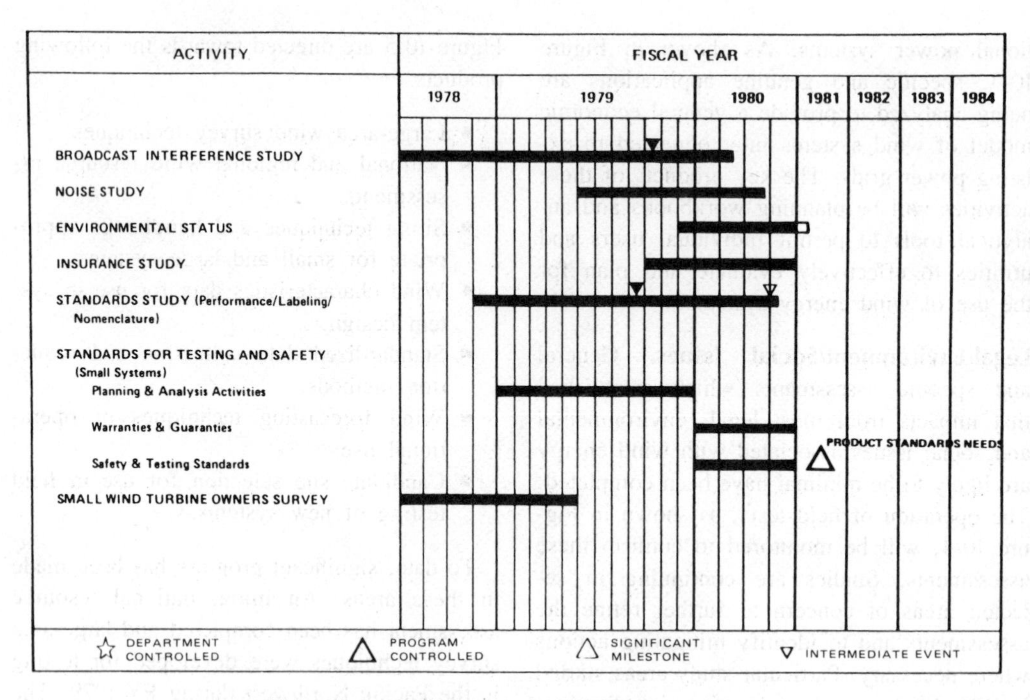

Figure 10-4. Legal/social/environmental issues activities.

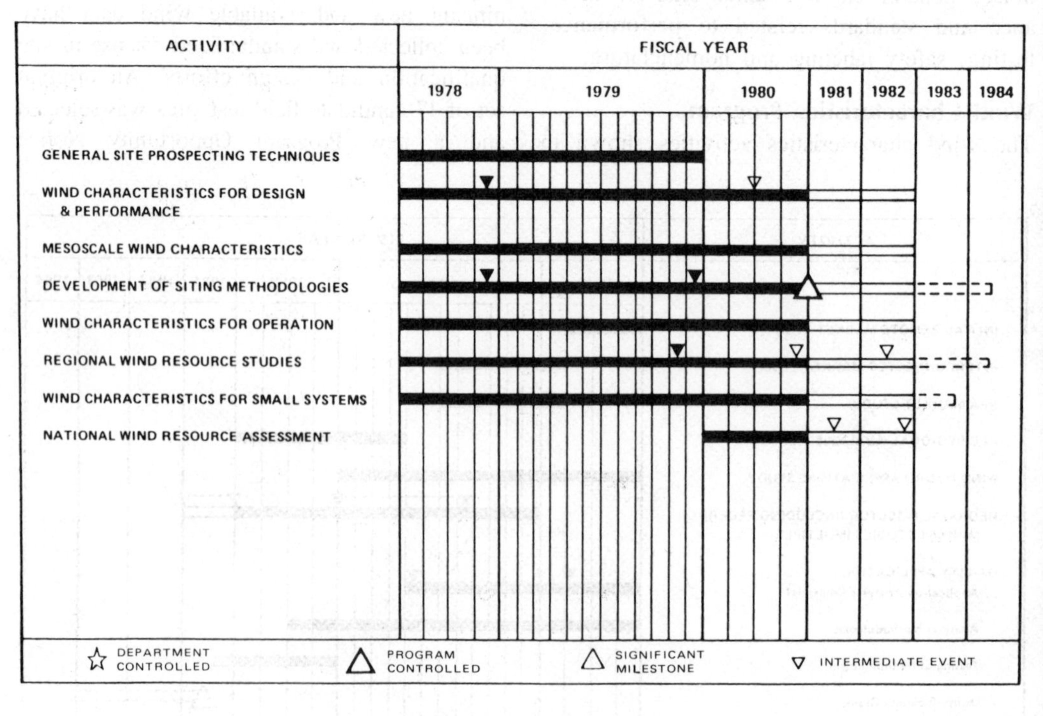

Figure 10-5. Wind characteristics activities.

(PON) is expected to identify about 20 additional sites.

Technology Development. This activity seeks to improve the performance and lower the cost of wind systems and their mechanical and electrical components. As shown in Figure 10-6, the emphasis is on system aspects of machine design rather than component technology. Research is continuing on utility in-

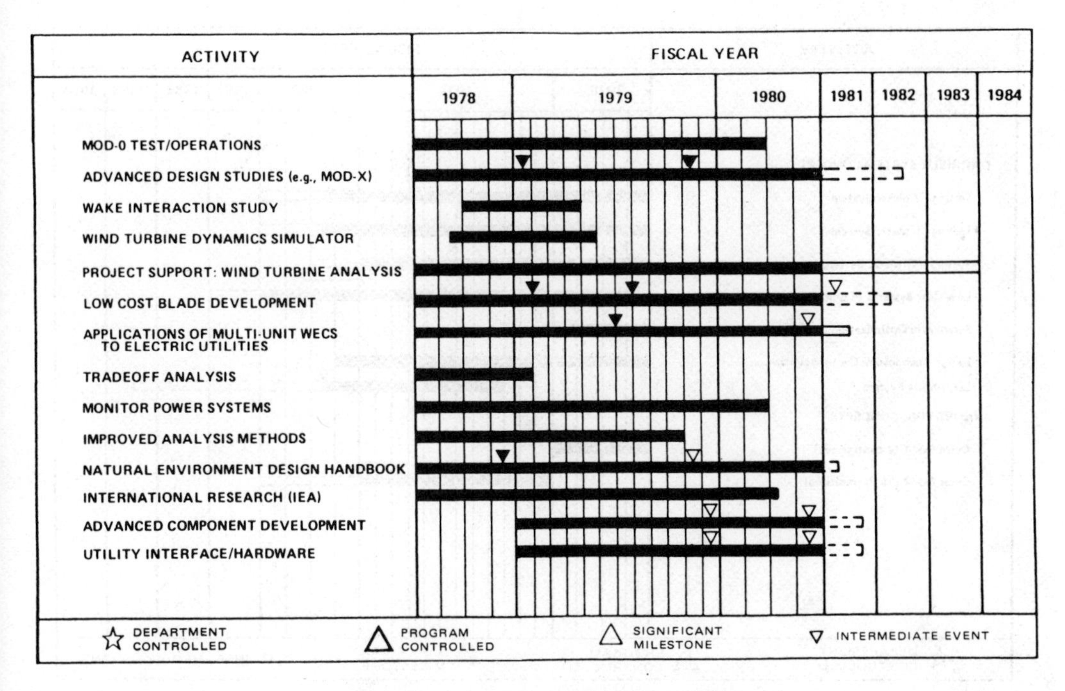

Figure 10-6. Technology development activities.

terface requirements and electrical stability problems. As a major cost component with major technical unknowns, research is also focusing on rotor development, where investigations are directed towards reducing rotor costs, determining the structural dynamics and fatigue characteristics of large rotors, and predicting machine performance and loads for use by designers. Simplified design techniques and tools are being developed for use by small and large machine designers/manufacturers.

Advanced System Concept. The purpose of the element is to determine the potential of alternative or innovative wind system configurations to offer increased energy output per unit cost over conventional (propeller-type, horizontal-axis) systems. A wide range of concepts, as shown in Figure 10-7, have been or are now being evaluated. Some of these concepts have failed to demonstrate the required potential and have been eliminated from the program. Other concepts shown in Figure 10-7 have been or are now being evaluated. Some of these concepts have failed to demonstrate the required potential and have been eliminated from the program. Other concepts have been more promising and have now entered the normal competitive system development cycle. For example, parallel awards in Small Scale systems competition include development projects for Darrieus, gyromill/cycloturbine and self-twisting blade concepts. New innovative concepts can continually be examined to determine their potential for significant advances in performance per unit cost.

Small Systems Development

Under this element, tests are currently being performed on commercially available small wind systems at the DOE Rocky Flats Plant in Golden, Colorado. In addition, the first cycle of new systems development is under way, with nine contractors designing systems with 1-kW, 8-kW and 40-kW power ratings for farm and rural residential applications (see Figure 10-8). Prototypes of these systems are scheduled to begin testing in CY 1980, as is the initiation of a second cycle of new systems development. Third and fourth cycles are scheduled for FY 1980 and FY 1981, respectively.

Intermediate Systems Development

The MOD-OA, first-generation technology, with a 125-ft diameter and a 200-kW power

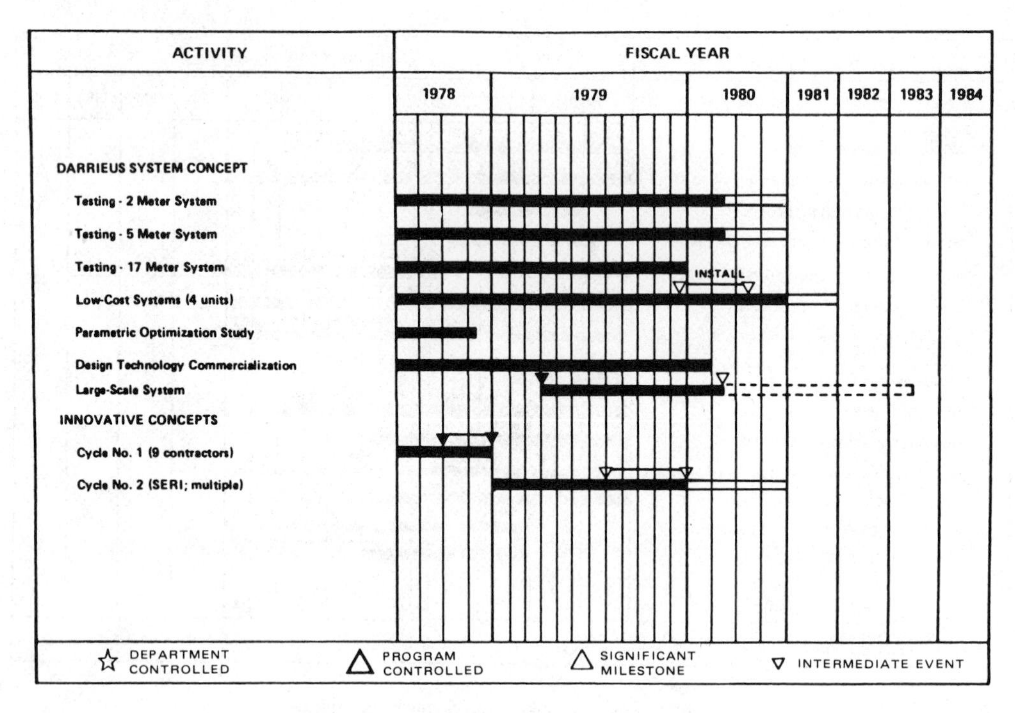

Figure 10-7. Advanced system concept activities.

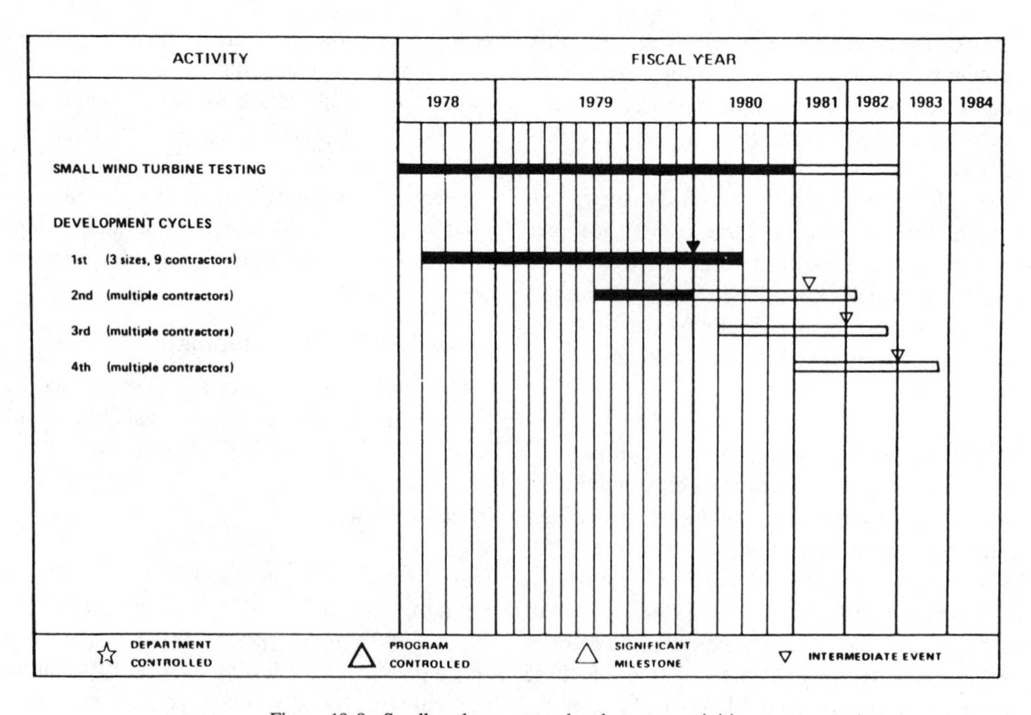

Figure 10-8. Small-scale systems development activities.

rating, is undergoing operational tests with local utilities at Clayton, New Mexico, and Culebra, Puerto Rico. A third unit was installed at Block Island, Rhode Island, in FY 1979; a fourth unit will be installed on Oahu, Hawaii, in FY 1980. Concurrently, with the testing of MOD-OA, the two parallel contracts, as shown in Figure 10-9, were awarded in FY 1979 for the design of second-generation systems in the same size and power

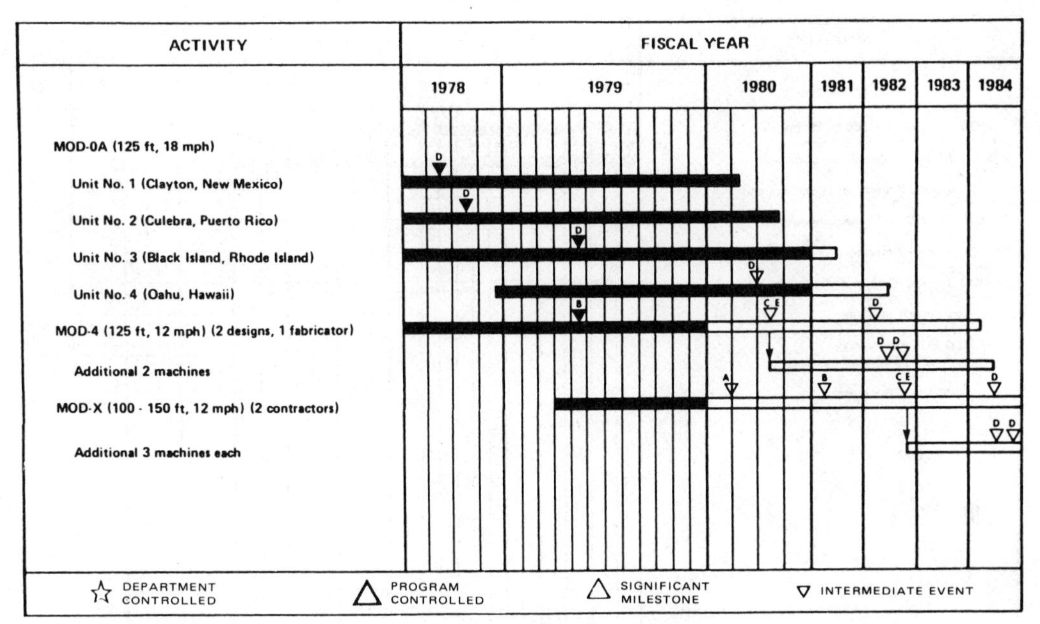

Figure 10-9. Intermediate-scale systems development activities.

range. A third generation of this general size (MOD-X) is scheduled to begin in FY 1981.

Large Systems Development

The MOD-1, first-generation large-scale technology, with a 200-ft diameter and a 2-MW power rating, was fabricated for installation at Boone, North Carolina, during FY 1979, as shown in Figure 10-10. The larger MOD-2, with a 300-ft diameter and a 2.5-MW power rating, is being designed. The MOD-2 represents second-generation technology, although it is the first system of this size. Plans call for four units to be installed, forming the first multi-unit system with 10 MW of power. Second-generation systems at the 200-ft-diameter size are planned with two parallel development contracts. Site selection for these systems was completed in FY 1979. Third-generation systems (MOD-Q) with optimized sizing are now planned for initiation in FY 1981.

Systems Demonstrations

Systems demonstrations are key elements in the Wind Energy Program to achieve user awareness and familiarity and to provide an interim market for early systems that are expected to be cost-effective in high wind/high potential applications. Current plans include the demonstration of 400 small wind systems in three cycles, beginning in FY 1981, and the demonstration of about 70 large systems in two cycles, beginning in FY 1982 (see Figure 10-11). Activities begun in FY 1979 and FY 1980 will address the numerous issues involved in planning an effective demonstration program; identifying criteria for inclusion of privately developed machines; site qualification; studies of multi-unit systems interconnected to utility grids; and the testing of components (especially blades) to ensure safety of the systems demonstrated.

FUNDING REQUIREMENTS

Funding requirements (budget authority) for fiscal years 1979–1983 are shown by program elements in Table 10-1. The estimated funding levels are predicated upon the assumptions that each successive generation of wind system performs as expected; that the optional duplicate machines under each development contract are purchased for experimentation;

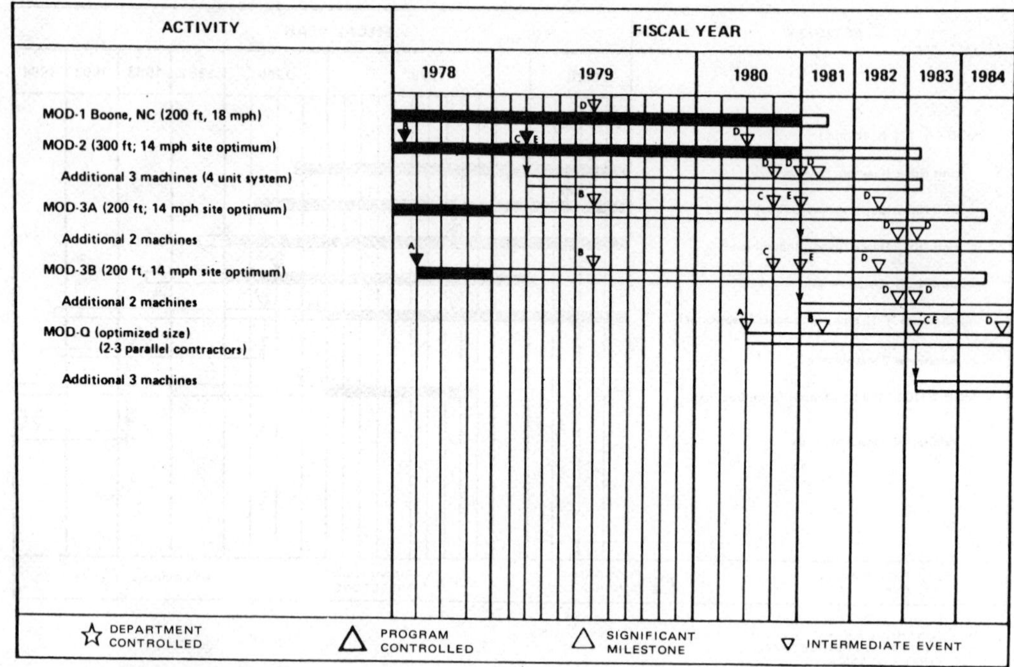

Figure 10-10. Large-scale systems development activities.

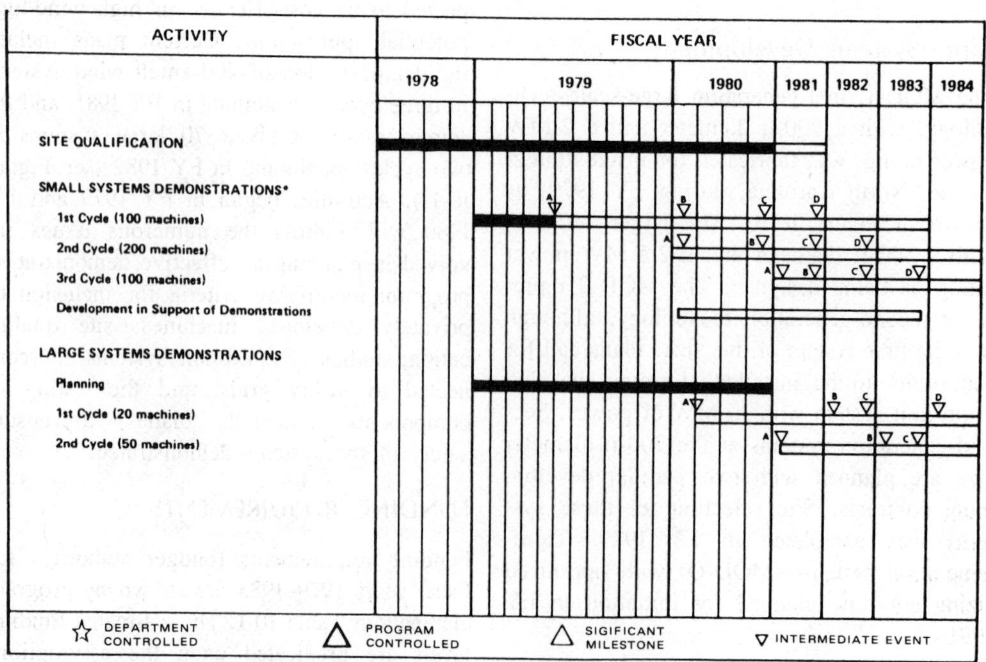

Figure 10-11. Systems demonstration activities.

Table 10-1. Wind Energy Program multi-year funding requirements by program element (budget authority) in millions of dollars.

PROGRAM ELEMENTS/SUB-ELEMENTS	FY 73-74[B]	FY 75[B]	FY 76	TQ[C]	FY 77	FY 78	FY 79[E]	FY 80[E] (EST)	FY 81 (EST)	FY 82 (EST)	FY 83 (EST)
1. PROGRAM DEVELOPMENT AND TECHNOLOGY	429	5,605	7,698	2,987	7,925	6,762	14,400	16,200	15,000	15,000	13,700
a. MISSION ANALYSIS	12	1,048	831	371	1,272	745	1,750	2,800	3,600	4,000	4,500
b. APPLICATIONS OF WIND ENERGY	–	579	556	230	160	25	700	1,800	2,500	3,800	2,700
c. LEGAL/SOCIAL/ENVIRONMENTAL ISSUES	–	422	200	300	199	95	420	800	600	400	200
d. WIND CHARACTERISTICS	14	399	992	669	1.900	2,320	2,800	1,400	4,200	4,800	5,800
e. TECHNOLOGY DEVELOPMENT	279	2,281	3,069	364	2,650	2,707	4,250	4,900	3,000	1,500	500
f. ADVANCED SYSTEMS	124	881	2,050	1,053	1,744	870	880	1,300	1,100	500	D
2. FARM AND RURAL USE (SMALL) SYSTEMS	–	736	1,507	231	2,641	7,669	19,700	20,500	22,000	22,000	10,000
3. 100 KW SCALE SYSTEMS	865	970	1,439	628	1,444	3,573	7,300	6,200	17,200	17,200	14,600
4. MW SCALE SYSTEMS	500	600	3,140	832	12,465	14,450	19,400	24,200	47,000	64,700	43,900
5. LARGE MULTI-UNIT SYSTEMS	–	–	582	292	25	1,497	4,200	16,000	25,000	30,000	40,000
TOTALS	1,794	7,911	14,366	4,970	24,500	34,036	65,000	63,400	126,200	148,900	253,600
		CAPITAL EQUIPMENT AND CONSTRUCTION			3,100	1,421	1,400	18,140	D	D	D
		TOTAL			27,600	35,457	59,600	63,400	126,200	148,900	253,600

A–IN OBLIGATIONS; OTHER FEDERAL DOCUMENTS MAY LIST OUTLAYS OR COST INCURRED
B–INCLUDES NSF FUNDING
C–TRANSITION QUARTER (JULY-SEPTEMBER 1976)
D–NOT SPECIFIED
E–FY 79 FUNDING FOR SUB-ELEMENTS IS NOT FIRM

NOTES: THE OUTYEAR BUDGET FIGURES FOR FY 1981-1983 WERE PRESENTED IN THE SOLAR MULTI-YEAR PLANS. HOWEVER MOST OF THESE NUMBERS ARE BASED UPON LITTLE REALITY. MOST DOE FINANCIAL PLANNING IS FROM YEAR TO YEAR.

and that small and large system demonstrations are initiated as currently planned. Under these assumptions, the funding level peaks in FY 1982, at about $150 million, when the major commitments for systems demonstration are scheduled. Thereafter, the funding levels decrease, reflecting the anticipated commercial viability of wind systems and a phasing out of Federally-sponsored developmental efforts.

In the current planning, there are situations which cannot reasonably be foreseen and, hence, for which there is presently no provision for resource requirements. For example, if some major breakthrough should occur under the "Advanced and Innovative Concepts" activity, revision of the entire Federal plan or sections of it could become necessary; also, depending on many evolving variables, government incentives may or may not be necessary in the FY 1983-84 time period. Costs associated with future incentives programs (e.g., accelerated depreciation, tax incentives) are not included at this time.

In addition to continuously ongoing program element assessments, major assessments and decision points are built into the program.

In Figure 10-12 the estimated total program funding is broken down into "hard-rock" baseline commitments and a planning wedge. The hard-rock commitment is based on a cost run-out of the systems development and testing underway in FY 1980, together with their supporting analyses and planning activities. The planning wedge reflects required activities now scheduled to start in FY 1981 and thereafter. In general, the wedge includes the development and testing of third-generation systems (all sizes), the planned systems demonstrations and their specific analyses and planning activities. The wedge does not include the potential cost of any incentives programs that may be recommended by OMB or the Congress.

The "hard rock" resources shown in Figure 10-12 would produce second-generation wind systems which would be effective in the relatively small high-wind/high-energy cost and remote power markets. Succeeding generations of machines will be required for major energy markets. The planning wedge reflects

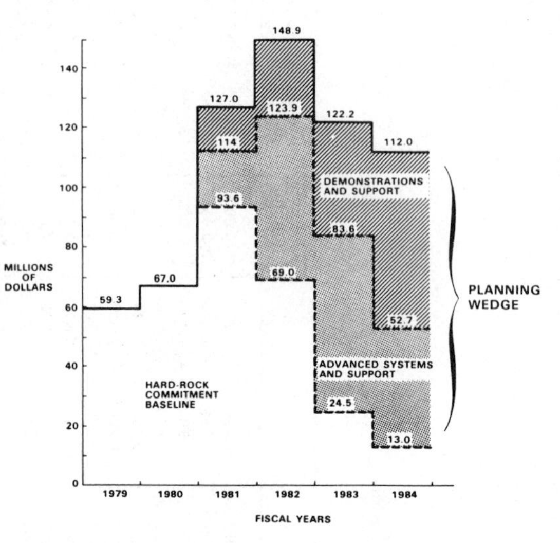

Figure 10-12. Wind Energy Program funding requirements (hard-rock commitments plus planning wedge).

the cost of triggering development of an industry, with DOE playing the part of a catalyst, that will serve the broadest possible market. A more modest financial resource commitment than that shown by the planning wedge would produce machines capable of competing in some limited markets, but not the required self-sustaining industry. It is expected that DOE must continue to accept development "risks" and encourage wind industry growth until wind energy systems can compete in the larger markets and are attractive to industry. The high-wind/high-energy cost markets open to second generation wind system technology provide an important entry point, particularly for small wind systems, and will build manufacturing and market infrastructure as well as product acceptance.

PROGRAM TECHNICAL STATUS

Overview

The goal of rapid commercialization of wind energy requires intensive planning and analysis efforts as well as development of reliable, cost-effective machines. In FY 1978, the Federal Wind Energy Program placed additional emphasis on studies to identify both good potential areas and applications for wind systems and possible legal, social, environmental, and institutional barriers. Applications of wind energy for uses ranging from small and large utilities to water-pumping and apple cooling were studied. Techniques and methodologies for identifying sites with good wind resources were assessed, and wind resource data were recorded and analyzed for candidate sites.

Economic analyses continued to provide useful information to potential users, manufacturers and planners. Further impetus toward commercialization is being considered with the assistance of a recently completed economic incentives study. Environmental assessments and studies of social, legal and institutional factors have identified potential barriers to widespread use of wind energy and provided planning information for successful coordination with existing institutions.

Technology development and wind turbine engineering development programs continued, with important advances in small and large machine development and testing. The small machine test facility at the DOE Rocky Flats Plant in Golden, Colorado, is gathering data on commercial machine performance and reliability, and recent contract awards for 1-,8- and 40-kW machines will support development of advanced systems to fulfill requirements for such diverse applications as remote communication stations, homes and agriculture. Data concerning wind resources, wind systems and commercial manufacturers of WECS are now available to potential users as a result of recent publications. The MOD-O experimental 100-kW machine continued to operate as a testbed for new rotor blades and other turbine components, as well as to test various configurations to provide additional technical information for both present and future generations of large wind turbines.

The first experimental MOD-OA (200-kW) wind turbines began field testing at sites in Clayton, New Mexico, and Culebra, Puerto Rico, and Block Island, Rhode Island. A fourth MOD-OA site, on the island of Oahu in Hawaii, will further diversify the testing environments for these utility-operated turbines.

Recognizing that alternative energy sources may be exploited through radically new and different concepts as well as established technologies, the program supported investigations of several advanced and innovative systems during FY 1978. Development of the Darrieus vertical-axis wind turbine continued, with construction and performance testing of a 17-m prototype. Other investigations, such as a tornado-type turbine and a polarized generator driven directly by the wind, are being supported to assess the potential of unique inventions to fulfill present and future energy requirements.

Program Elements

The Federal Wind Energy Program is designed to allow the earliest possible commercialization of wind power by simultaneously developing the WECS; assessing the technical, economical and institutional requirements for their widespread use; and stimulating their commercial utilization. At the same time, the program's iterative structure allows activities performed under discrete work elements to advance general and technical knowledge and the state-of-the-art in a systematic manner. The goal of this approach is to establish a firm base for wind systems by 1) performing research and development to build durable and economical wind systems; 2) performing field tests and testing applications to show that wind power can be implemented on both a small and a large scale on a widespread basis; and 3) developing the technological capability of private industry to ensure that commercialization can be accomplished.

As shown in Figure 10-13, the program is organized into the following five discrete, yet interrelated, program elements, each directed toward specific program objectives.

- Program development and technology
  - -Mission analysis
  - -Applications of wind energy
  - -Legal, social, environmental issues
  - -Wind characteristics
  - -Technology development
  - -Advanced and innovative concepts
- Farm and rural use (small) systems
- 100-kW-scale systems
- MW-scale systems
- Large-scale multi-unit systems

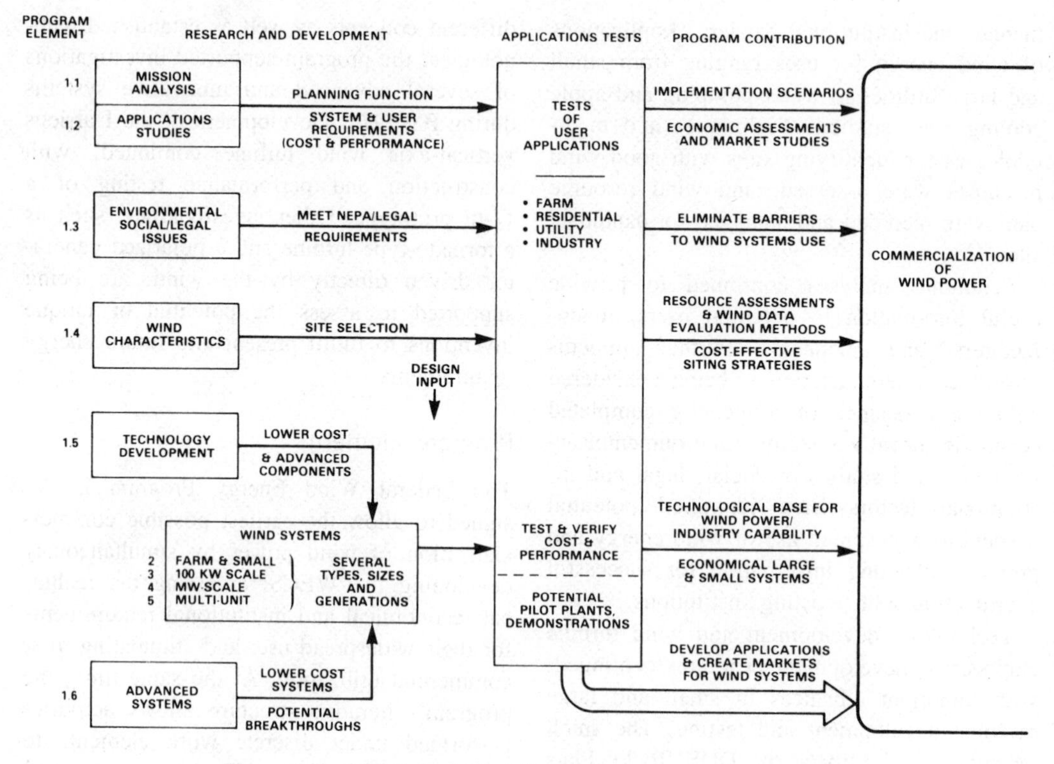

Figure 10-13. DOE Federal Wind Energy Program. *Courtesy of DOE.)*

The following program highlights further describe significant activities by program element areas.

Current Program Highlights

Program Development and Technology

Mission Analysis and Regional Applications Studies. Economic and market analyses provide program planners, industry and potential users with vital information and forecasts for large and small systems. A study of economic incentives has provided a basis for program planners to consider government action required to stimulate WECS commercialization. The study analyzed the potential impact of such incentives as direct cash subsidies, tax credits, loans and loan guarantees on WECS manufacturers and utility, residential, agricultural and industrial users.

Regional application studies were expanded to include assessment of potential markets in Hawaii and New England. The studies are identifying high-potential applications, analyzing costs and cost-sensitivity, determining wind resources and defining operational, maintenance and interface requirements related to each application. Results of the studies will provide information about specific regional requirements and applications, as well as a data base for future, expanded large and small machine applications studies.

Legal/Social/Environmental Issues. Studies of the potential environmental, legal and social impact of wind turbines identify issues which may impede rapid development of the wind energy resource. Field studies conducted by Battelle Memorial Institute in Columbus, Ohio indicate that an operating wind turbine generator does not pose a high risk to airborne fauna and is not likely to result in any secondary effects to vegetation, including crops. University of Michigan studies used the prototype MOD-O wind turbine at the National Aeronautics and Space Administration's (NASA) Plum Brook Station in Sandusky, Ohio to measure the effects of turbine operation on electromagnetic conditions and the biophysical environment. No interference to FM transmission or television audio signal was observed in the study. Video distortion was marked at close ranges and at high frequencies; however, this is a site- and machine-specific problem which will require evaluation for each individual machine site.

Public acceptance of wind systems will have an important effect on commercialization efforts. To measure public attitudes, a recent study surveyed knowledge of and reaction to wind and solar energy generally, and wind turbine designs in a variety of settings specifically. Results of the study indicated a positive correlation between knowledge about wind energy and attitudes toward the use, location and design of wind turbines.

Wind Characteristics. Pacific Northwest Laboratory (PNL) is responsible for the technical direction of the wind characteristics sub-element, providing meteorological work and background useful in site determination. Research projects in this field in 1978 focused primarily on the investigation of mesoscale, or large-area techniques, which involved gaining more detailed information on wind regions to provide more information to users. *A Handbook of Wind Turbine Siting Techniques Relative to Two-dimensional Terrain Features* was published in 1978 as a user's manual for utility and civil engineers, city planners and architects. The study covers general siting practices; hazards; where to find wind power at its maximum; and the optimum site relative to specific terrain features—i.e., bluffs, cliffs and shelterbelts. In a similar effort for small turbine users, PNL has produced *A Siting Handbook for Small Wind Energy Conversion Systems*. This report was designed for the layman interested in a wind turbine with a power output of less than 100 kW. A significant portion of this material is presented in Chapter 3.

Turbine Component and Systems Research. Efforts to lower system costs and improve performance continued during 1978, with a significant achievement recorded in blade technology. The largest blade ever fabricated, a 150-ft composite fiberglass-reinforced plastic structure, has now undergone static and dynamic tests. The effort to develop the unique design, fabrication and testing techniques for exceptionally large, lightweight blades will provide important data on structural characteristics, size limitations and costs for rotor designs. Sub-components, including joints and materials, have been separately tested for fatigue and structural properties.

The search for appropriate low-cost blade materials has been expanded through feasibility studies of wood, urethane and prestressed concrete structures. The studies include detailed cost estimates for blades using these materials, as well as an assessment of the effect of blade weight on wind turbine system design and cost.

Internationally, a joint effort is under way with DEFU, research associate of Danish Electric Supply Undertakings, to study the lifetime of wind turbines and components. The large experimental Gedser turbine, a different design from present U.S. experimental machines, has withstood 20 years of exposure to the elements; 10 years in operation and 10 years idle and without maintenance. In addition to refurbishing the wind turbine, the project provides for extensive instrumentation to gather information concerning rotor performance and blade loads on this "older-generation" wind turbine.

Wind turbine systems R&D efforts are being supported by a number of technology development studies, including a study of wake flow characteristics downwind from machines to provide information on appropriate spacing between wind turbines to achieve maximum effectiveness in multi-machine configurations. Mathematical models of turbine wakes are being developed, and the models will be validated by measuring actual wake flow downwind from the MOD-OA turbine in Clayton, New Mexico.

Technology Development. The prototype MOD-O wind turbine continued to operate as a testbed during 1978. Components for the MOD-OA were tested and verified, and a new series of advanced machine concept tests began. These tests will enhance technology development by providing data on free-yaw and upwind rotor operation, as well as advanced low-cost blade designs and modifications to vary tower flexibility. The modifications allow evaluation of machine performance in varied configurations and operating modes, providing technical information for design of other advanced wind turbines.

Advanced Systems Research and Development. The Solar Energy Research Institute (SERI) is now responsible for technical management of Wind Energy Innovative Systems. Innovative concepts investigated during 1978 included wind augmentation devices, such as diffuser and vortex augmentors, as well as direct wind-driven systems which vary from the so-called "tornado-type" turbine to charged aerosol and electrofluid dynamic generators, as well as reinvestigation of the Magnus effect using a Madaras rotor. The concepts undergo analytical and experimental investigation and emphasis is placed on performance and cost-effectiveness. Concepts which demonstrate a high potential for successfully meeting these requirements may be supported in field test and demonstration projects.

Farm and Rural Use Applications and Small Systems

Rocky Flats Small Systems Test and Development Center. Testing of small-scale commercial wind systems was expanded during FY 1978. Ten systems underwent instrumented, controlled testing, and five more systems have been acquired for tests during 1979. An expansion program, which will increase test site capacity from 12 sites to 30, is now under way to allow collection of data from additional machines. Long-term goals are to assess performance of small systems, identify development needs, assist small machine manufacturers and provide information to potential small WECS users.

Development of Advanced 1-kW, 8-kW and 40-kW WECS. The first cycle of contract awards in the small wind system test and development program was completed with nine awards for development of systems in 1-, 8- and 40-kW size ranges. These competitively selected advanced designs include vertical-axis and horizontal-axis turbines intended for applications which range from remote communications stations through residential use, to large farms or small businesses and factories.

Four of the projects will develop 8-kW systems; three are for high-reliability, minimum-maintenance 1-kW systems; and two projects are developing 40-kW systems. Four of the nine contracts were awarded to small businesses. During FY 1979, other size classes will be added to this group of small machines, and the entire small systems effort will be broadened and intensified.

United States Department of Agriculture. The United States Department of Agriculture's (USDA) Agricultural Research Service is charged with the responsibility for managing projects to identify requirements for farm and agricultural wind systems, with extensive application studies currently in progress. Principal milestones in FY 1978 consisted of experiments in the areas of: 1) wind powered water-pumping for remote applications; 2) wind powered farm building heating; 3) dairy milk cooling; and 4) wind powered reverse osmosis desalination. In addition, alternative techniques for wind powered irrigation systems and the feasibility of wind-driven brooder heating systems and farrowing pens have been examined. In March 1978, a 40-kW irrigation pumping system was installed at Bushland, Texas, to obtain performance data for evaluation of operation on a pumping system powered by a drive system combining a Darrieus windmill and an internal combustion engine or electric motor.

Medium-Scale Wind Systems

MOD-OA. Significant advances in the medium scale wind turbine development program were recorded during FY 1978 with commencement of field testing at Clayton, New Mexico, and Culebra, Puerto Rico. The 200-kW MOD-OA installed at Clayton operated for 1768 hours routinely connected to the local utility network. It generated 161,620 kWh of electricity between January 1978 (when operations began) and August 1978.

The second MOD-OA began operation in Culebra in July 1978; the MOD-OA became operational at Block Island, Rhode Island, and will shortly begin operation at Oahu, Hawaii in 1980.

The MOD-OA is designed for operation in connection with electric utility networks, and the field tests provide important information about machine operation in a variety of environments and with large and small utilities. The MOD-O, MOD-OA and later generation wind turbines are compared in Table 10-2.

Advanced Intermediate Size Machine Development. Refinement and expansion of medium size machine technology is continuing with plans for an advanced, second-generation, low-cost 100–200-kW turbine. A competitive solicitation for design, fabrication and testing of the new system was issued in FY 1979. The system is intended for small farm, utility, irrigation and remote applications.

Megawatt-Scale Systems

MOD-1 Fabrication and Field Testing. Site preparation for installation of the MOD-1, 2-MW wind turbine has been completed near Boone, North Carolina. Installation of the first MW-scale experimental turbine, which has a rotor diameter of 208 ft, began in September with placement of the tower. After final machine checkout and completion of blade fabrication, the machine began field tests in the spring of 1979.

MOD-2 Development. The 300-ft rotor diameter, 2.5–3-MW system is now in the design phase. Conceptual design, including extensive trade-off studies, has been completed. Fabrication is scheduled for 1980. This advanced technology machine will incorporate experience gained during previous field testing efforts to assess the practicality and cost-effectiveness of extremely large wind turbines (see Figure 10-14).

Advanced Large Size Machine Development. A new advanced, multi-MW scale system is planned for industrial and utility applications. A competitive solicitation for design, fabrication and testing was issued in FY 1979.

Field Testing of Experimental Wind Turbines. Five sites have now been selected for field tests of large experimental machines from the 17 candidate electric utility locations (see Figure 10-15). MOD-OA turbines have begun operation at Clayton, New Mexico, Culebra, Puerto Rico and on Block Island, Rhode Island, installation is planned for Oahu, Hawaii. Boone, North Carolina, will be the field test site for the MW-scale MOD-1.

Approximately 20 additional sites for future

Table 10-2. Characteristics of the MOD-O and later generation large wind turbines.

TYPE	ROTOR DIAMETER	CAPACITY/RATED WINDSPEED (at 30 ft.)	SITE WINDSPEED	LOCATION	YEAR OF FIRST RUN
MOD-O	125 ft.	100 kW/18 mph	Moderate	Plum Brook, OH	FY 76
MOD-OA	125 ft.	200 kW/18 mph	Moderate	Clayton, NM	FY 78
				Culebra, PR	FY 78
				Block Island, RI	FY 79
				Oahu, HA	FY 80
MOD-1	200 ft.	2 MW/26 mph	High	Boone, NC	FY 79
MOD-2	300 ft.	2.5 MW/20 mph	Moderate	TBD	FY 80
*	Approx. 200 ft.	TBD**	Low	TBD**	FY 82
*	Approx. 125 ft.	TBD**	Low	TBD**	FY 83

*Advanced Systems
**To be determined

Figure 10-14. Artists concept of the MOD-2 2.5-MW wind turbine being developed by Boeing under the Federal Wind Energy Program. *(Courtesy of DOE.)*

field tests will be selected through a competitive solicitation to be issued in 1980. The solicitation is designed to provide a broader set of locations for future field test installations. Detailed meteorological data will be taken at each candidate site to identify high-potential locations for field tests in a variety of environments.

Large Multi-Unit Systems. Multi-unit systems were studied in two research projects carried out in 1978. One study, completed in FY 1979 by General Electric, investigated the electrical stability requirements of wind turbines to provide machine designers and utility planners with information which will enable the use of large numbers of wind turbines in power systems. Another study by Westinghouse has examined the technical and economic feasibility of operating wind systems in various offshore environments and transmitting power (or hydrogen produced by elec-

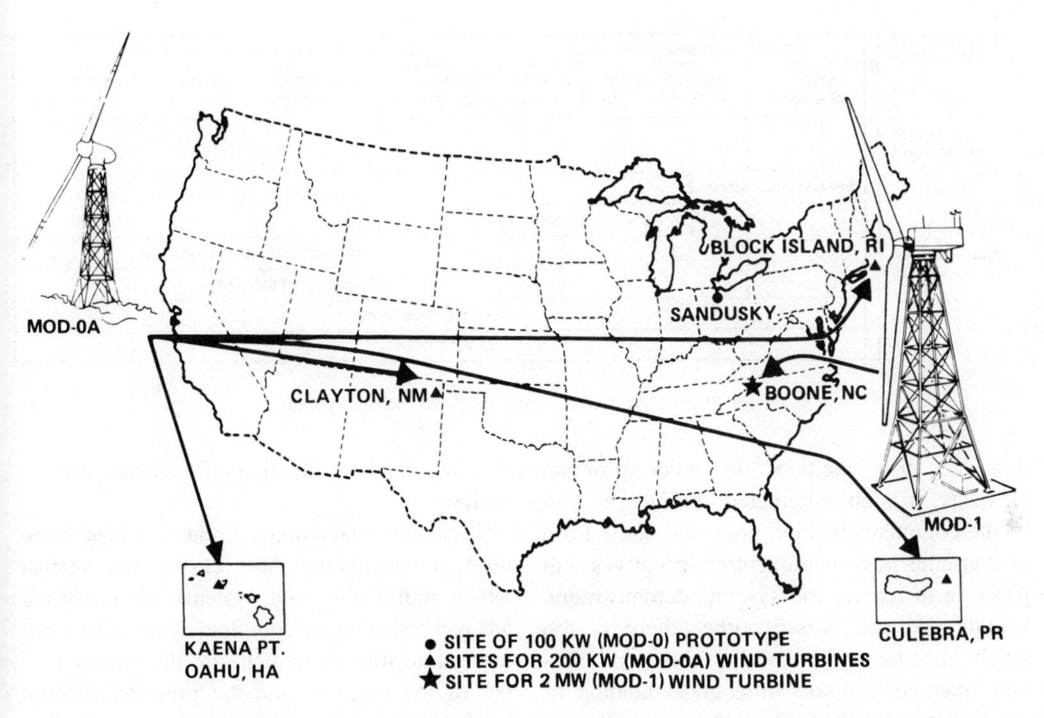

Figure 10-15. DOE wind turbine generator installation sites. *(Courtesy of DOE.)*

trolysis) to land-based users. This study was reported in FY 1979. These studies will provide a basis for future planning and analysis of possible multi-unit wind system applications.

Program Characteristics

Figure 10-16 provides an illustration of the research, development and commercialization cycle for the Federal Wind Energy Program. A budgetary breakdown of the Wind Energy Program into the six development phases is provided in Table 10-3. As shown, no funds are being devoted to basic research. The applied research phase contains: 1) the investigations into advanced and innovative concepts; 2) the wind characteristics program that is developing wind survey techniques culminating in national and regional wind resource assessments; 3) cost-effective siting tools/handbooks; and 4) improved wind loading requirements.

The technology development phase includes conceptual and detailed design of new systems developments, specialized design studies, component development and testing, fabrication studies, market analyses, application studies and legal/social/environmental investigations. At the completion of the detailed design for each new advanced system, the project is shifted from technology development to engineering development. The activities included in the engineering development phase are the fabrication, installation, testing and evaluation of each new system. The decrease in technology development funds after FY 1982 indicates the planned decrease in new systems design efforts. Correspondingly, the engineering development funds are generally increasing, which reflects the increasing number of new systems entering the fabrication and testing phase of development.

The final phase includes both large and small systems demonstrations. This also includes some planning activities, site qualification and component testing (especially blade fatigue tests) to ensure that systems to be demonstrated are of high quality.

Presently, no funds are specifically assigned to the commercialization phase in Table 10-3. The planning and special studies as well as user involvement and information dissemination necessary to support full commercializa-

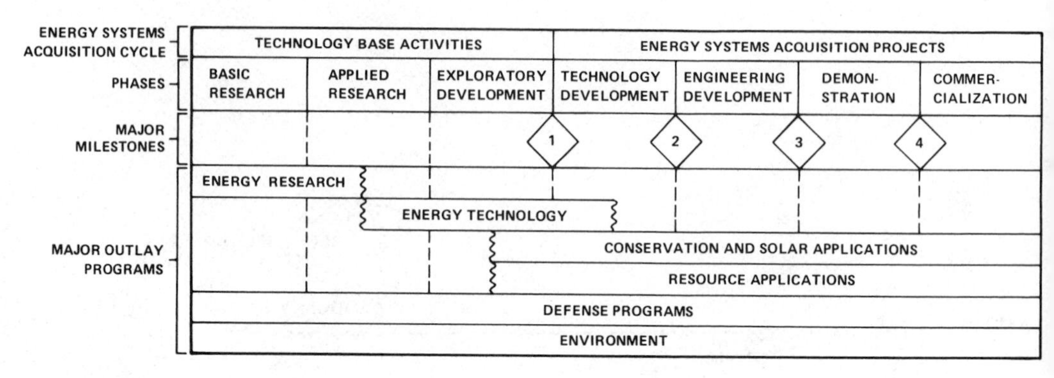

Figure 10-16. DOE RD&C cycle for wind energy systems.

tion are now included in other program elements. It is an anticipated and integral part of the commercialization plan that some form of consumer and manufacturer incentives will phase in to replace the systems demonstration activity. At the present time, there is too much uncertainty concerning the wind systems markets and too little understanding of the leverage available through various incentive options to recommend a specific incentive package. As a consequence, no funding requirements are shown.

Evaluation Criteria/Impact Assessment

Evaluation Criteria. As described in this chapter, the overriding measure of program effectiveness is the cost of generating electric power from wind energy versus more conventional competing methods. Thus, all technical, economic and social projects are ultimately considered and evaluated in terms of their contributions to cost reduction.

Figures 10-17 and 10-18 show program cost goals for small and large wind systems, respectively. The areas labeled "Quantity Production Range" are defined by two reasonable estimation procedures: a 95 percent learning curve for the 100*th* large unit or the 1000*th* small unit; and the mature product cost for equipment of smaller complexity. The quantity production range costs would not be achieved in the same time frame as the first production units, but would occur in the years following initial production, depending on the rate at which production efficiencies can be realized.

Technical evaluation criteria that have been used in formulating these curves and against which individual wind systems are measured are somewhat more complex. Key factors entering into this assessment are the energy output of the machine and the projected mature product or asymptotic cost of the system. The historical asymptotic costs for a number of mature products of similar complexity to WECS (such as power shovels and tractors) and of certain components that comprise part of the wind turbine system (such as towers and generators) are illustrated in Figure 10-19 in terms of unit price/unit weight ($/lb) versus unit weight. Note that mature product costs depend on steady, not necessarily mass, production quantities. Based on these data and on an analysis of the projected mature product costs of large wind turbine system components, it is estimated that the complete system can be built and installed for an asymptotic cost of $2.50–3.00/lb. The range in cost-per-pound goals reflects the different types of materials that may be used and the complexity of the design. In the design phase of systems development, it is possible to trade off the cost-per-pound and the total system weight; however, the stated goals appear to be reasonable averages. In general, achieving this goal will depend on simpler, more advanced designs; understanding system dynamics to permit lighter weight system components (especially the blades and tower); reduction of

Table 10-3. Wind Program budget by technology development phase.

PHASE	FISCAL YEAR 1979	1980	1981	1982	1983	1984
BASIC RESEARCH	--	--	--	--	--	--
APPLIED RESEARCH	3.7	4.7	5.3	5.3	5.8	4.2
TECHNOLOGY DEVELOPMENT	43.5	60.6	73.7	90.9	57.5	9.5
ENGINEERING DEVELOPMENT	7.9	13.7	23.0	22.7	18.9	38.3
DEMONSTRATION	4.2	16.0	25.0	30.0	40.0	60.0
COMMERCIALIZATION	--	--	--	--	--	--
TOTAL	59.3	95.0	127.0	148.9	122.2	112.0

(1) Definitions as applied to specific wind program elements are described in the text.

input loads through adequate dynamic and operational data; and the development of lower-cost fabrication techniques, especially for blades.

In concert with the system cost criterion of \$2.50–3.00/lb, an energy-production performance index has been established in terms of the annual energy output per pound of system weight. For an average wind speed of 12 mph, the performance index goal has been established at 25–50 kWh/lb. This goal and the estimated performance index values for various classes of machines are shown in Figure 10-20.

Taken together, the system cost criterion (\$/lb) and the performance index (kWh/lb) permit the computation of capital cost for any desired load. For example, a hypothetical load of 700,000 kWh annually (which corresponds to the MOD-OA rated at 200 kW with a capacity factor of 0.4) would weigh 28,000 lb at the goal of 25 kWh/lb. Based on this system weight and the higher system cost goal of \$3.00/lb, the system would have a capital cost of \$84,000 or \$420/kW. These numbers, combined with the appropriate amortization period and discount factors, can be used to calculate the energy cost, which falls in the 2¢/kWh goal range.

These cost and performance goals relate to all sizes of wind systems; however, there is some question as to whether they can be reached for small systems, especially those used in utility-connected situations. Conversely, if demand charges are not applied to wind systems as a result of public policy, then a higher cost target is viable. The cost and performance of small wind systems are now the subject of detailed analysis and testing to establish information similar to that presented for the larger systems.

With regard to the probability of meeting program goals, the key variable is market size. Technical feasibility is clearly established and technical progress is more a conventional engineering challenge than one of technical breakthroughs. Thus, in the technical areas, probability of success is considered very high. Likewise, other potential barriers to success, such as environmental and regulatory matters, do not appear to be significant deterrents to commercialization for large wind systems in rural areas. For small systems, a major issue is the possible demand charges by

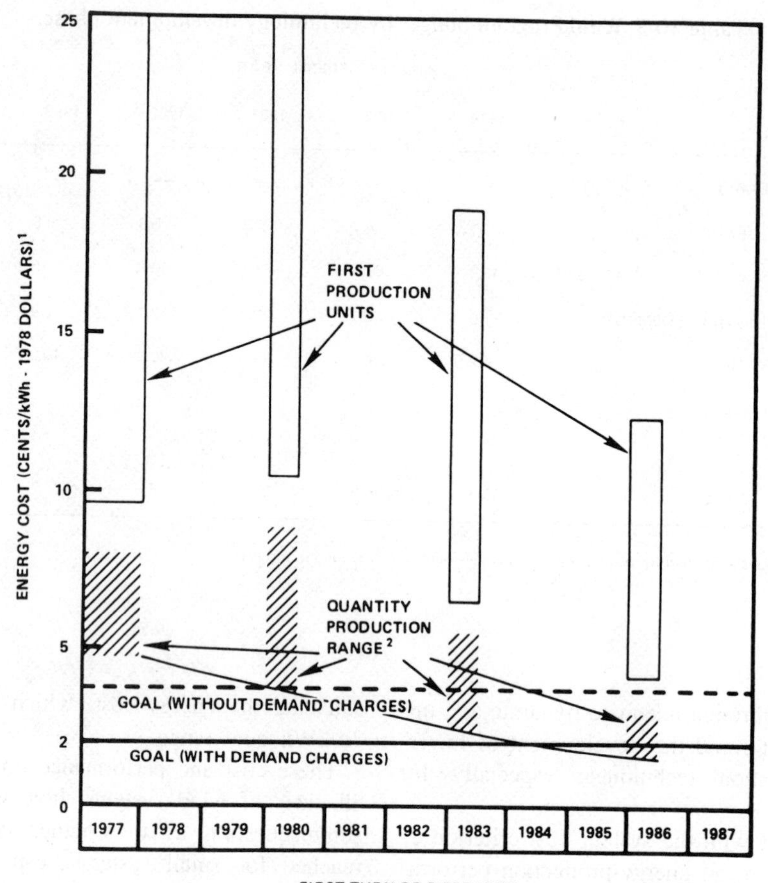

Figure 10-17. Cost goals for small wind systems.

utilities that would dramatically effect the cost goal, as shown in Figure 10-18, with a corresponding impact on the probability of meeting that goal. This leaves the more nebulous area of the commercial market and probability of success, which becomes mainly a function of the share or size of market attainable.

Potential wind energy markets are highly diverse, and considerable uncertainty over market size versus costs currently exists. The goal of 2¢/kWh has been established as a requirement for entering major energy markets. Extensive wind resource and market studies, now under way and planned, are directed toward firmly establishing the market size versus cost in the FY 1980 or FY 1981 time period.

An additional consideration is that the key variable of market size can be affected by government incentives such as direct subsidies or tax credits. Many options are available and will also be affected by the price and market success of other competing energy sources.

The probability of program success is tied most critically to the size of the market, which can in turn be significantly influenced by government action. The appropriate degree of government stimulation cannot be determined at this time; it can only be concluded as wind systems development evolves.

Expected Impact. The principal impact of wind energy is expected to come from large systems providing electricity to the utilities.

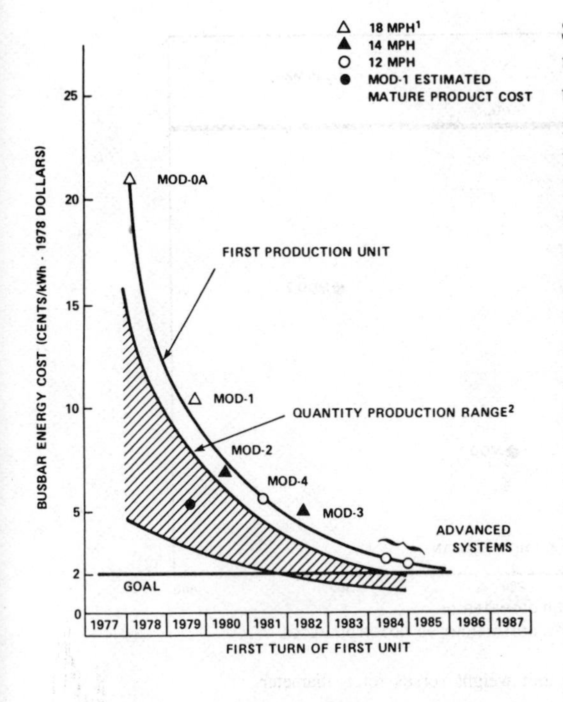

Figure 10-18. Cost goals for large wind systems. *(Courtesy of DOE.)*

Small systems are expected to provide electricity to farms and rural electricity in competition with utilities. The electric utility industry is large and diverse, with many different types of generating equipment, fuels, load patterns, operating policies and regulatory environments. To date, only a limited assessment of this market is available. There is also considerable uncertainty with respect to the small system's markets. Consequently, quantitative estimates of the impact of wind energy systems at any future date must be considered preliminary, at best. However, estimates from the Wind Program Commercialization Plan and the Domestic Policy Review (DPR) assessments are summarized in Table 10-4. These estimates are in fossil equivalent quads per year. In fact, wind energy provided directly to utilities or through reduced demand by farms and rural residences will be displacing a mix of all fuel types. Recognizing this restriction, the Domestic Policy Review Committee estimates for the year 2000 are the equivalent of 160–300 million barrels of oil per year, while the Commercialization Plan

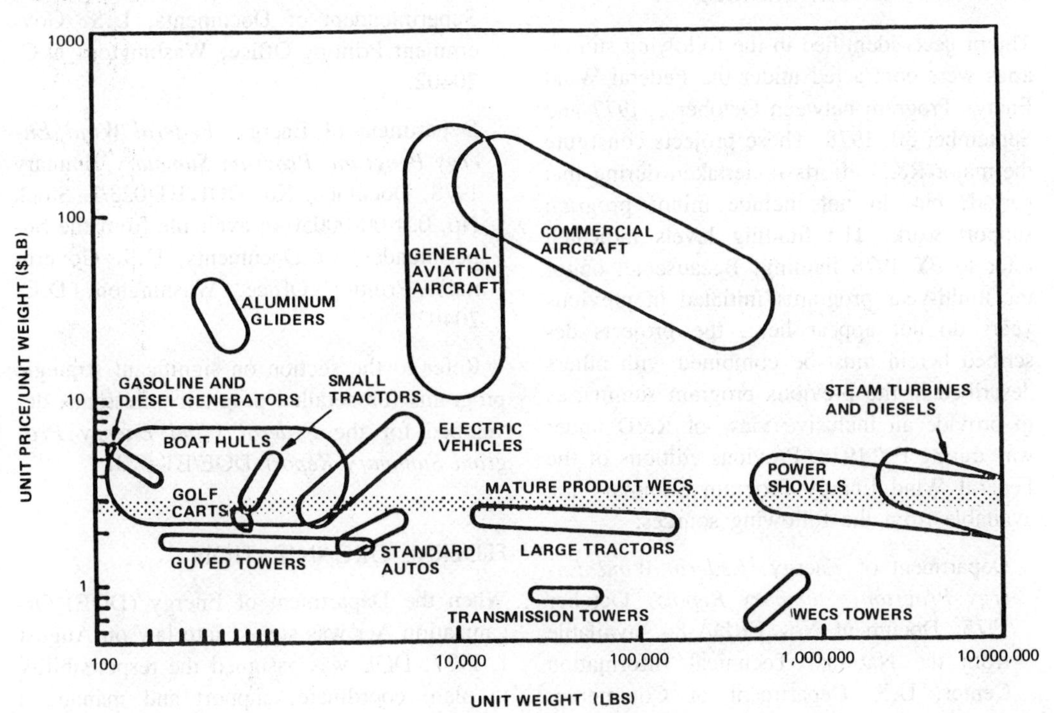

Figure 10-19. Mature product cost comparisons.

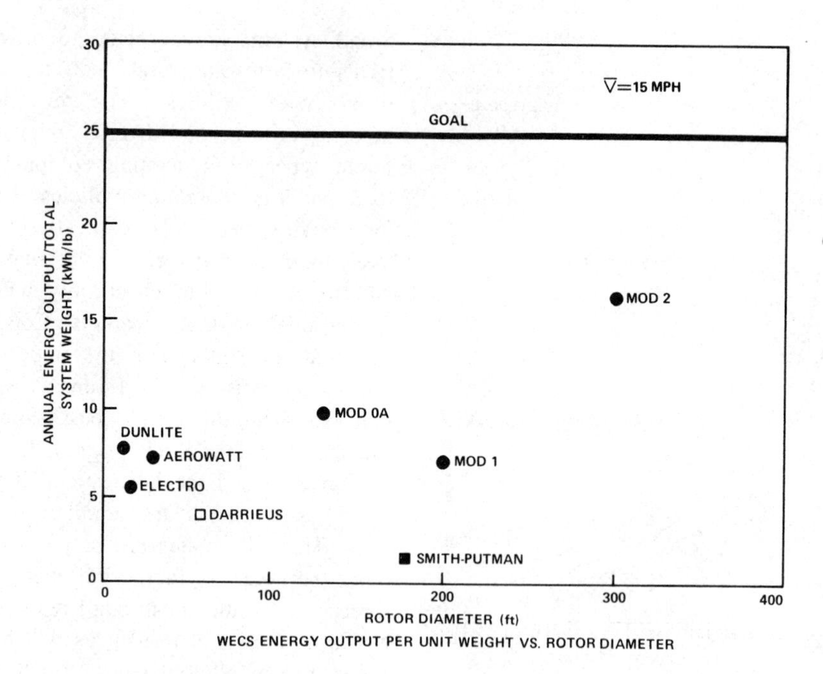

Figure 10-20. Energy output per unit weight versus rotor diameter.

estimates correspond to about 430 million barrels of oil per year.

DOE PROJECT SUMMARIES

The projects identified in the following summaries were contracted under the Federal Wind Energy Program between October 1, 1977 and September 30, 1978. These projects constitute the major R&D efforts undertaken during that period, but do not include minor program support work. The funding levels indicated refer to FY 1978 funding. Because all ongoing multi-year programs initiated in previous years do not appear here, the projects described herein must be combined with others described in the previous program summaries to provide an inclusive view of R&D under way during FY 1978. Previous editions of the Federal Wind Energy Program Summary are available from the following sources:

Department of Energy. *Federal Wind Energy Program, Summary Report,* October 1975, Document No. ERDA-84; available from the National Technical Information Center, U.S. Department of Commerce, 5285 Port Royal Road, Springfield, Virginia 22161.

Department of Energy. *Federal Wind Energy Program, Summary Report,* January 1977, Document No. ERDA-77-32, Stock No. 060-000-00048-4; available from the Superintendent of Documents, U.S. Government Printing Office, Washington, D.C. 20402.

Department of Energy. *Federal Wind Energy Program, Program Summary,* January 1978, Document No. DOE/ET-023/1, Stock No. 061-00-00050-0; available from the Superintendent of Documents, U.S. Government Printing Office, Washington, D.C. 20402.

Refer to the section on significant strategic programs for detailed program milestones developed for the *Federal Wind Energy Program Summary Report* DOE/ET-023.

FEDERAL ORGANIZATION

When the Department of Energy (DOE) Organization Act was signed into law on August 1, 1977, DOE was assigned the responsibility to plan, coordinate, support and manage a balanced and comprehensive energy research, development and demonstration (RD&D) pro-

Table 10-4. Estimated impacts of the Wind Energy Program.

	1985	1990	2000
Commercialization Plan			
Market penetration			
Large systems (quads)*	0.04	0.19	2.10
Small Systems (quads)*	0.03	0.32	0.94
(end use quads)	(0.01)	(0.11)	(0.32)
Total (quads)*	0.07	0.51	3.04
Approximate number of machines			
Large systems	600	3,000	35,000
Small systems	50,000	500,000	1,600,000
Domestic Policy Review (Quads)*			
Base case **			0.9
Maximum practical case			1.7

*Primary fossil fuel equivalents in quads/year.
**Assumes $0.32 per barrel of oil (1978 dollars).

gram for wind energy. In order to fulfill this responsibility, DOE has implemented a management structure which coordinates the program management activities of the various headquarters offices with project management in the field.

Program management is carried out via the organizational structure shown in Figure 10-21. Structure and functioning are primarily based on the DOE decentralization policy and the Wind Energy Program policy of maximizing industry and user participation.

The Wind Systems Branch is staffed by a branch chief and three program managers, assisted as necessary by field office and Federal laboratory personnel and by planning and management support contractors. Assuming that the program progresses as planned, the DOE manpower requirements for the next several years are summarized in Table 10-5. These requirements call for an increase of both headquarters and field personnel. The combined staff peaks in FY 1982 when the major demonstration procurements are scheduled to occur. Thereafter, the headquarters staff begins to decrease while the field staff expands slightly to support the planned incentives and information dissemination activities.

Headquarters

The headquarters operation carries out the significant management functions (planning, policy-making, budgeting, policy coordination, program monitoring), delegating a maximum amount of policy execution authority to the field organizations. Field inputs to the headquarters management process are encouraged by an "iterative planning process" that increases field inputs. Direct inputs from the wind industry and user groups are also encouraged through workshops, formal and informal meetings, and document reviews, as well as in direct interfacing in various projects.

Management control is exerted by:

- Basic policy from headquarters and annual operating plans developed jointly with the field organizations.
- A field reporting system (monthly summaries).
- Monthly (or quarterly or semi-annually, depending on the nature of the project) field review of ongoing projects.
- Quarterly program conferences.
- Informal visits by field and headquarters personnel, as appropriate.
- A Management Information System that monitors all projects at the program level and triggers management actions.

Field Operations

In accordance with the decentralized management policy of the Department of Energy, the Wind Energy Branch has—and will continue

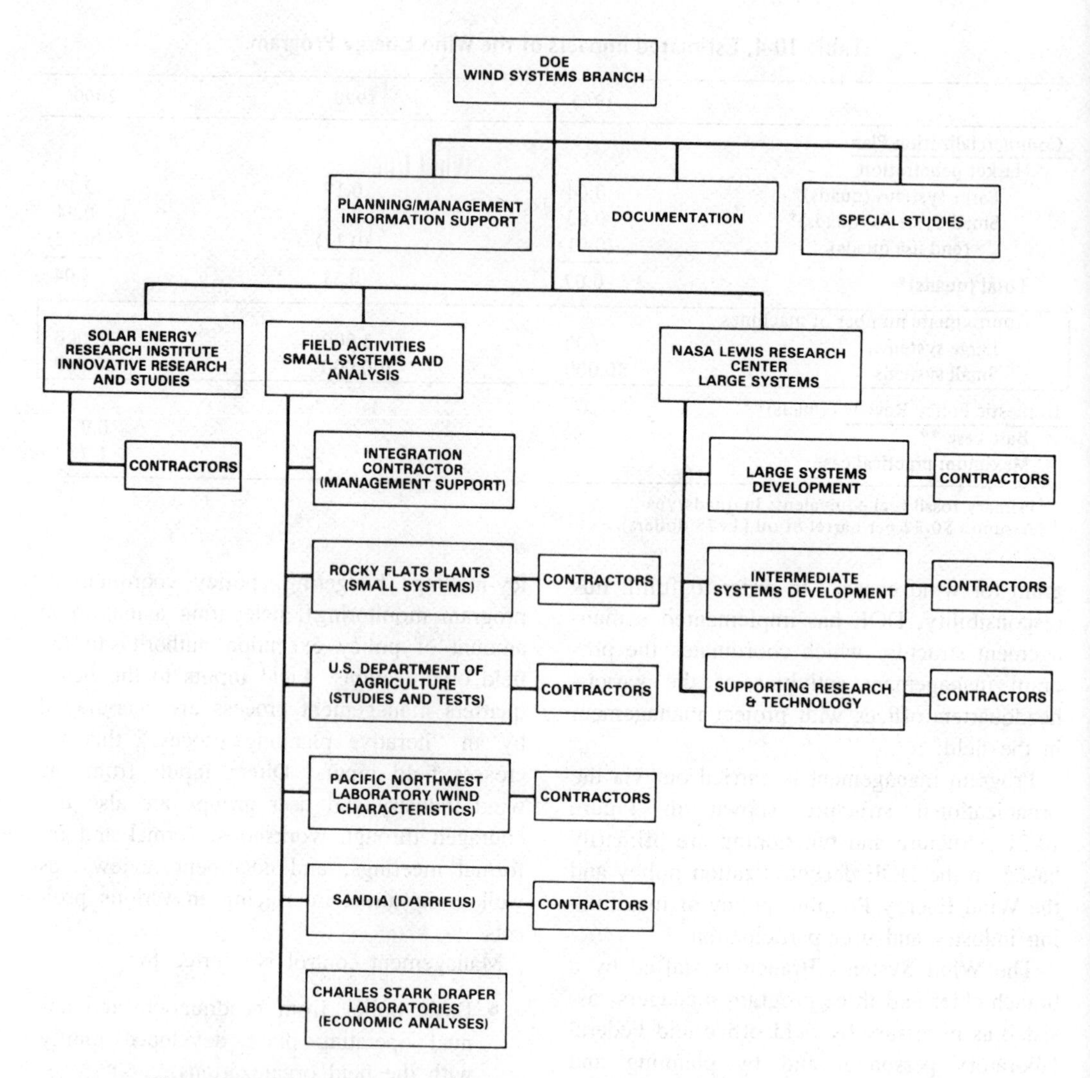

Figure 10-21. Wind energy systems branch management organization. (*Courtesy of DOE.*)

to—transfer the responsibility for key projects to the field. This has occurred, for example, in the transfer of responsibility for small wind systems development and demonstration programs to the Denver Project Office. In general, the DOE field offices are responsible for the administrative management of all programs, including the control of management plans for all fabrication and construction projects.

In many instances, national laboratories have been given additional management responsibilities as DOE field operation centers. For example, the DOE divisions which include the Federal Wind Energy Program have assigned responsibilities in the areas of re-

Table 10-5. Wind Energy Program manpower requirements.

FISCAL YEAR	HEADQUARTERS	FIELD OFFICES
1979	8	4
1980	12	7
1981	12	12
1982	8	14
1983	6	15
1984	6	15

search, analysis, assessment and information to the Solar Energy Research Institute (SERI) at Golden, Colorado. SERI is managed and operated by the Midwest Research Institute under a contract to DOE. DOE laboratories perform significant research and development programs in support of the Federal Wind Energy Program. These laboratories include Lawrence Livermore Laboratory, Sandia Laboratory, Battelle Memorial Institute, Pacific Northwest Laboratory, NASA, and the U.S. Department of Agriculture Laboratories.

The decentralization of DOE programs is moving forward. Personnel problems have been the greatest difficulty yet encountered. Technical management has been transferred out of headquarters (Figure 10-22), but there is still need for procurement, legal, administrative and other support personnel.

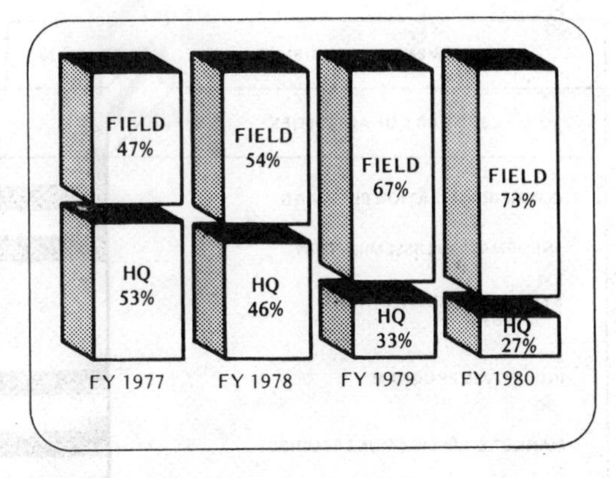

Figure 10-22. Percent of funds managed at DOE headquarters and field offices.

Future Management Responsibilities

As large and small wind systems come closer to commercialization readiness, overall program emphasis will switch from technology development to commercialization actions such as demonstrations, information dissemination, and incentives implementation. In keeping with DOE commercialization policy, the management responsibility for these commercialization activities will be transferred to Conservation and Solar Applications and Resource Applications, as appropriate. The man-

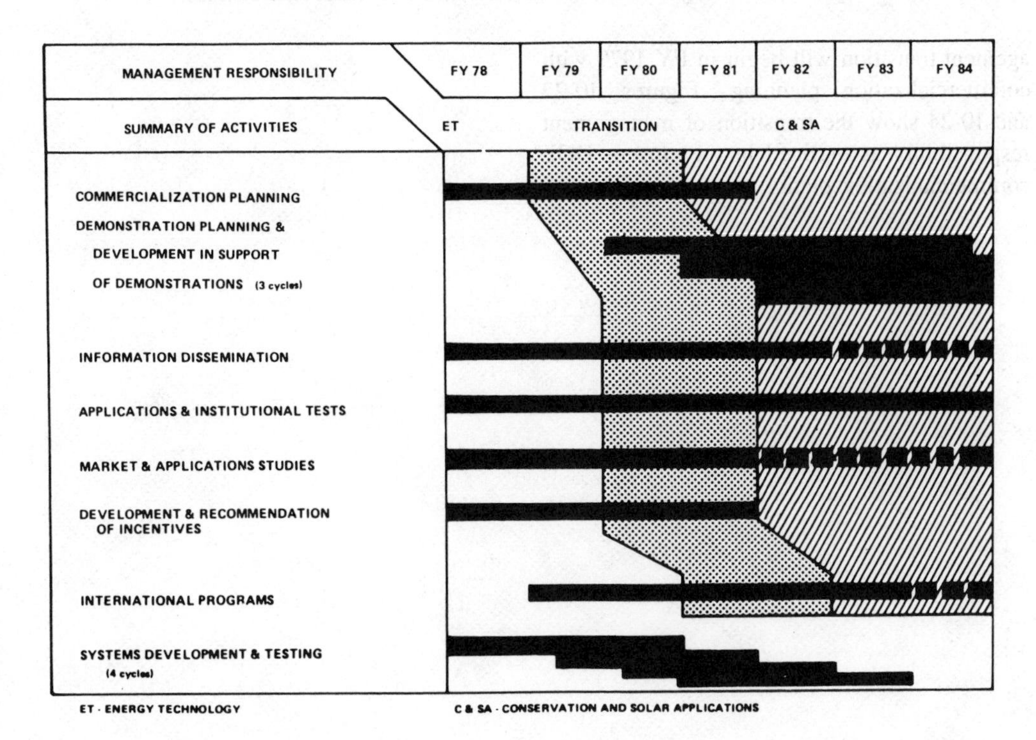

Figure 10-23. Management responsibility transition chart (small-scale systems).

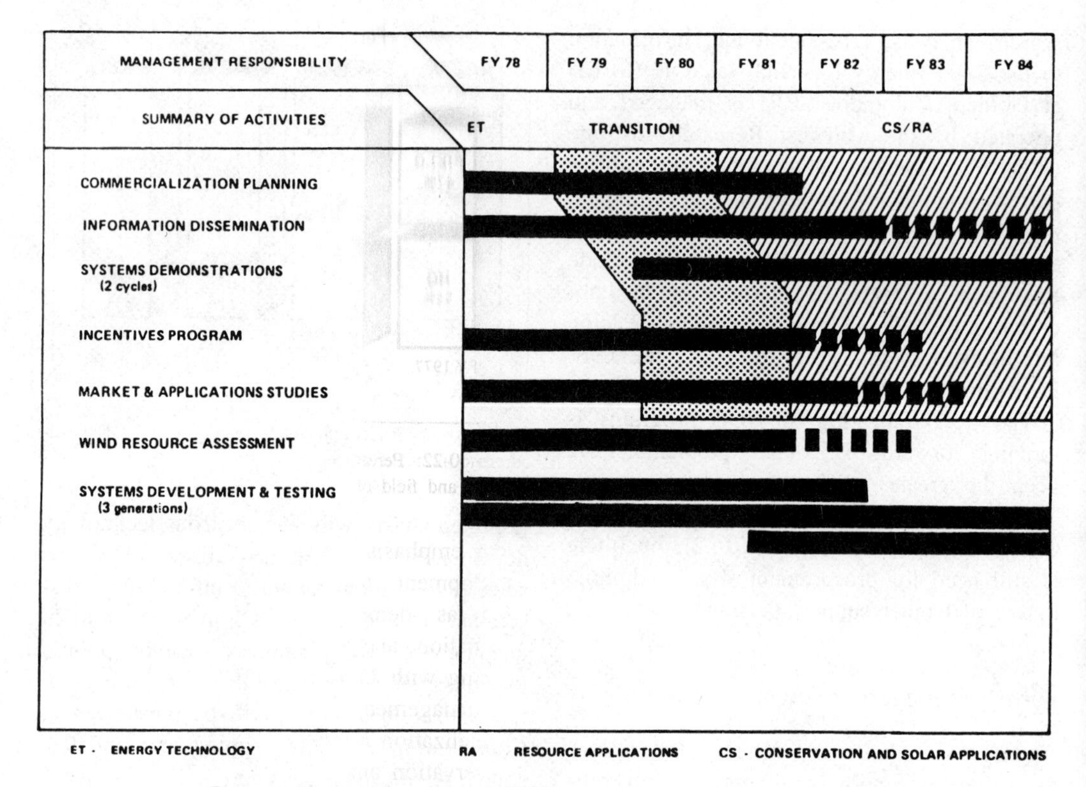

Figure 10-24. Management responsibility transition chart (large-scale systems).

agement transition will begin in FY 1979 with commercialization planning. Figures 10-23 and 10-24 show the transition of management responsibility as outlined in the current DOE commercialization plans.

Program Development And Technology

MISSION ANALYSIS

CONTRACTOR	TITLE
Charles Stark Draper Laboratory, Inc. 555 Technology Square Cambridge, MA 02139	Technical Management Support for Systems Analysis Elements of the Federal Wind Energy Program
	CONTRACT NO. EG-77-C-01-4125
PRINCIPAL INVESTIGATOR Arthur C. Parthe, Jr.	**PERIOD OF PERFORMANCE** September 1977 to September 1979
WORK LOCATION Cambridge, MA	**FISCAL YEAR 1978 FUNDING** $245,431
CONTRACTING OFFICE DOE Headquarters Washington, DC	**CUMULATIVE FUNDING** $490,729

BACKGROUND

The Federal Wind Energy Program continues to receive added attention and DOE support in pursuit of its goals to develop alternative sources of energy. These ever broadening responsibilities require the support of outside contractors to extend DOE's own management capabilities and resources.

OBJECTIVES

The objective of this project is to ensure the necessary systems capability to manage DOE contractors engaged in systems analysis studies, including: (1) mission analysis; (2) applications of wind energy; and (3) legal, social, and environmental issues.

APPROACH

Draper Laboratory is planning, providing work specifications for, coordinating and monitoring the activities of DOE contractors working on mission analysis, regional applications, and legal, social, and environmental studies. Draper is to ensure the quality of information produced by these contractors; prepare reports and reviews relevant to specific technical issues and economic questions; and ensure that work in interrelated study areas is motivated by a systems approach to technical and nontechnical requirements, problems, and issues.

OUTPUT

The project provides technical reviews and reports pertinent to the systems analysis program elements. In addition, the project develops planning documents, work plans and statements.

MISSION ANALYSIS

CONTRACTOR	TITLE
JBF Scientific Corporation 1925 N. Lynn Street Suite 308 Arlington, VA 22209	Wind Energy Systems Evaluation and Planning Studies **CONTRACT NO.** EX-76-C-01-2521
PRINCIPAL INVESTIGATOR	**PERIOD OF PERFORMANCE**
Theodore R. Kornreich	October 1976 to September 1979
WORK LOCATION	**FISCAL YEAR 1978 FUNDING**
Arlington, VA	$349,193
CONTRACTING OFFICE	**CUMULATIVE FUNDING**
DOE Headquarters Washington, DC	$590,387

BACKGROUND

The Federal Wind Energy Program is conducting a broad range of research and development activities to facilitate the commercialization of wind turbine systems. These activities have to be continually monitored, evaluated, and updated to implement a cost-effective research and development program which focuses on addressing the key unresolved programmatic issues.

OBJECTIVES

The studies performed under this contract are designed to provide the planning support information necessary for management decisions on program activities, emphasis, schedule, and application of resources.

APPROACH

Specific tasks are assigned in three areas: (1) an overview of the wind energy conversion system's state-of-technology to integrate and compare related information from different sources within and outside the Federal program and determine the reasons for variances in this information; (2) the development and assessment of alternative program approaches; and (3) the development of techniques for assuring that pertinent wind systems information developed inside and outside the Federal program is readily accessible for use by technical specialists and the general public.

OUTPUT

A major contribution of these studies is the preparation of a series of comprehensive reports and planning documents. These will be provided on a rapid response basis to satisfy specific program management needs and schedules.

APPLICATIONS/OF WIND ENERGY

CONTRACTOR	TITLE
JBF Scientific Corporation 2 Jewel Drive Wilmington, MA 01887	Wind Energy Systems Application to Regional Utilities (New England Region) **CONTRACT NO.** EX-76-C-01-2438
PRINCIPAL INVESTIGATOR	**PERIOD OF PERFORMANCE**
Edward E. Johanson	June 1976 to December 1978
WORK LOCATION	**FISCAL YEAR 1978 FUNDING**
Wilmington, MA	$24,877
CONTRACTING OFFICE	**CUMULATIVE FUNDING**
DOE Headquarters Washington, DC	$334,229

BACKGROUND

The Federal Wind Energy Program is conducting a broad range of research and development activities to facilitate the commercialization of wind turbine systems. One of the major areas to be addressed is an evaluation of the technical and economic feasibility of incorporating large-scale wind energy conversion systems (WECS) in utility grids with good wind resource availability. The New England region is one area which requires a detailed investigation.

OBJECTIVES

The objective of this study is to examine the New England region in general, and the Massachusetts coastal region in particular, to assess the feasibility of introducing wind energy systems on a regional basis.

APPROACH

With the participation of the Georgia Institute of Technology and the New England Gas and Electrical Association Service Organization, JBF is addressing the following areas: (1) assessment of regional wind potential, regional sites, and the ability of WECS to meet the utilities' generating needs; (2) analysis of WECS alternatives; (3) development of cost-benefit and cost-sensitivity analyses; (4) asssessment of future government actions which may affect the competitiveness of WECS; and (5) identification of policy implications and options for DOE.

OUTPUT

During the project period, sufficient information will be developed to allow the utilities under study to make a decision on WECS potential in the region. The final report will provide general planning principles which describe the actual step-by-step process necessary to bring WECS to on-line power production.

LEGAL/SOCIAL/ENVIRONMENTAL ISSUES

CONTRACTOR	TITLE
University of Michigan Radiation Laboratory Department of Electrical and Computer Engineering Ann Arbor, MI 48109	Electromagnetic Interference by Wind Turbines **CONTRACT NO.** EY-76-S-02-2846
PRINCIPAL INVESTIGATOR	**PERIOD OF PERFORMANCE**
Thomas B.A. Senior	January 1976 to April 1979
WORK LOCATION	**FISCAL YEAR 1978 FUNDING**
Ann Arbor, MI	$95,000
CONTRACTING OFFICE	**CUMULATIVE FUNDING**
DOE Chicago Operations Office Argonne, IL	$317,186

BACKGROUND

Prior contract work has demonstrated that wind turbines can cause TV reception interference in the area immediately adjacent to a wind energy conversion system (WECS) installation. Interference was more pronounced for the higher frequency UHF TV channels and is apparently confined to the amplitude modulated or video portion of the signal.

OBJECTIVES

This effort will continue to analyze and quantify the effect of wind turbines by means of theoretical analyses, laboratory simulations, and field tests. Emphasis will be on: (1) updating siting guides to include near-zone effects on TV reception; (2) determining wind turbine generator (WTG) effect on circularly polarized TV broadcasts; (3) determining criteria for assessing practical severity of interference at a given WTG site; (4) determining WTG effect on navigation aids such as LORAN C, OMEGA, and ILS; and (5) extending the analysis to include vertical-axis wind turbines.

APPROACH

Laboratory experimentation, field measurements, and analysis will be continued to extend the results of previous analyses of TV interference by a horizontal-axis WTG. The model for determining wind turbine-induced TV interference is being further developed to provide simpler expressions and greater accuracy. Extended laboratory simulations and field tests of this interference are performed as an input to the model.

OUTPUT

The WECS TV Siting Handbook developed in the previous contract will be updated and extended as required. A final report will be prepared that documents the results of the studies set forth in the project objectives. The results will be used to determine if additional work is needed to assess, quantify, and ameliorate potential legal, social, and environmental barriers to wind systems commercialization in this area.

WIND CHARACTERISTICS

CONTRACTOR	TITLE
Sandia Laboratories Environmental Research Albuquerque, NM 87115	Meteorological Studies for Wind Power **CONTRACT NO.** SOL 65-S189-76-32
PRINCIPAL INVESTIGATOR J.W. Reed	**PERIOD OF PERFORMANCE** July 1975 to September 1978
WORK LOCATION Albuquerque, NM	**FISCAL YEAR 1978 FUNDING** $64,000
CONTRACTING OFFICE DOE Albuquerque Operations Office Albuquerque, NM	**CUMULATIVE FUNDING** $168,000

BACKGROUND

This program began with a national assessment of wind power statistics, which was completed in 1973. The program has evolved into developing computer programs for normalizing wind power data to a uniform height for different observed heights and for assessing the long-term variability of wind power availability for various locations.

OBJECTIVES

The objective of this study is to develop data processing techniques that use standard National Weather Service data observations for defining various aspects of wind power availability, including time-dependent statistics, wind turbine speed limit effects, and run-duration statistics.

APPROACH

Data processing techniques have been developed and applied to records from 15 selected National Weather Service Stations around the United States. Wind speed distributions have been adjusted for changes in anemometer exposures. Time-dependent statistics of wind power availability have been calculated for months, seasons, years, and 10-year periods. Corrections to available wind power for local pressure altitude and nonstandard anemometer exposures, and estimates of power at 10 m, 20 m, and 50 m have been made for all stations in the U.S. used in the original "Wind Power Climatology of the United States."

OUTPUT

A series of reports is being submitted which documents the calculation procedures and the results of the data processing on the selected National Weather Service stations. A revision to "Wind Power Climatology of the United States" will be published.

WIND CHARACTERISTICS

CONTRACTOR	TITLE
Lawrence Livermore Laboratory P.O. Box 808 Livermore, CA 94550	Wind Power Studies **CONTRACT NO.** W-7405-Eng-48
PRINCIPAL INVESTIGATOR Christine A. Sherman	**PERIOD OF PERFORMANCE** July 1975 to September 1978
WORK LOCATION Livermore, CA	**FISCAL YEAR 1978 FUNDING** $240,000
CONTRACTING OFFICE DOE Headquarters Washington, DC	**CUMULATIVE FUNDING** $696,000

BACKGROUND

The island of Oahu, Hawaii, was selected as an initial study area to develop and demonstrate a complete wind energy assessment methodology. Oahu provides very rough mountainous terrain, a persistent trade wind environment, and an island geography helpful in the isolation and study of strong terrain effects. The basic concept is to use both field studies and numerical modeling in a balanced program that would be less costly than resource documentation by either measurements or theoretical analysis alone.

OBJECTIVES

The objective of this applied research project is to develop and verify a method of regional wind energy resource assessment and quantification suitable for application throughout the nation to regions with linear dimensions of about 100 km. The analysis of potential wind energy in mountainous areas is strongly emphasized.

APPROACH

Routinely collected meteorological data will be acquired from a region of potential wind energy. Numerical meteorological models will be applied to this multiple-station data set to derive the primary spatial wind patterns, their frequencies of occurrence and distribution during the year. The identified patterns will be numerically processed to define the subregion wind maximum and spatial extent. Test calculations for Oahu, Hawaii will be compared to measurements.

OUTPUT

The output of this project will be a demonstrated and verified method to assess regional wind energy potential in numerous mountainous areas of the nation using routinely available meteorological data.

WIND CHARACTERISTICS

CONTRACTOR	TITLE
Pacific Northwest Laboratory Subcontractor: Colorado State University Fort Collins, CO 80523	Sites for Wind-Power Installations: Physical Modeling and Validation **CONTRACT NO.** EY-76-S-06-2438
PRINCIPAL INVESTIGATOR Robert N. Meroney and Virgil A. Sandborn	**PERIOD OF PERFORMANCE** June 1976 to May 1979
WORK LOCATION Fort Collins, CO	**FISCAL YEAR 1978 FUNDING** $62,694
CONTRACTING OFFICE Pacific Northwest Laboratory Richland, WA	**CUMULATIVE FUNDING** $221,881

BACKGROUND

During the initial program, wind tunnel model studies were performed to study the influence of topography profile, surface roughness, and stratification on the suitability of various combinations of these variables for wind-power sites. For the range of cases examined (large turbulence integral scales with respect to surface feature scales) it was found that the flow is dominated by inviscid dynamics. Hence the influence of hill shape, surface roughness, and mild stratification can be reliably estimated by simple prediction procedures.

OBJECTIVES

It is now appropriate to critique the potential of physical modeling via a simultaneous comparison of field, physical model, and numerical model data. Results should provide a basis for immediate implementation of physical modeling as a siting tool, reevaluation of identified errors or trends via an extended program, or termination of the concept.

APPROACH

Many length and time scales of wind characteristics with respect to wind-power siting are not amenable to analytical or numerical methods. Physical modeling can thus provide guidance for siting handbooks, numerical model construction, and site-specific information at a reasonable cost in a short time period. This program is constructed to exploit the advantages of wind tunnel simulation of the atmospheric shear layer as well as illuminate its limitations.

OUTPUT

Data provided by tests over a physical model of Kahuku Point, Oahu, Hawaii, will be compared with field measurements and numerical model programs. Results will be used to determine credibility of physical modeling and to select potential WECS sites in the Kahuku area.

WIND CHARACTERISTICS

CONTRACTOR	TITLE
Battelle Memorial Institute Pacific Northwest Laboratory Battelle Boulevard Richland, WA 99352	Management and Technical Support for the Wind Characteristics Program Element **CONTRACT NO.** EY-76-C-06-1830
PRINCIPAL INVESTIGATOR	**PERIOD OF PERFORMANCE**
C.E. Elderkin	February 1976 - continuing
WORK LOCATION	**FISCAL YEAR 1978 FUNDING**
Richland, WA	$2,000,000
CONTRACTING OFFICE	**CUMULATIVE FUNDING**
DOE Richland Operations Office Richland, WA	$5,008,000

BACKGROUND

Pacific Northwest Laboratory (PNL) has provided technical and management support for the Wind Characteristics Program Element since February 1976. This effort encompasses five technical areas: (1) Wind Characteristics for Design and Performance Evaluation; (2) Mesoscale Wind Characteristics; (3) Siting Methodologies; (4) Wind Characteristics for Wind Energy Conversion System Operations; and (5) Special Studies.

OBJECTIVES

The primary objective of the Wind Characteristics Program Element is the development and dissemination of information on wind characteristics relating to the siting, design, and utilization of wind systems.

APPROACH

A significant portion of the PNL effort is in the area of coordinating and contract monitoring. In addition, PNL integrates technical information needed for other elements of the wind program, for the public, and for private industry. Close coordination is maintained with the Wind Systems Branch and its contractors.

OUTPUT

Information developed by the program will aid mission analysis planners, wind systems designers, and organizations and individuals interested in wind systems siting and operations. Documents are designed for ready assimilation into the ongoing efforts of other segments of the program as well as the wind energy community at large.

WIND CHARACTERISTICS

CONTRACTOR	TITLE
Pacific Northwest Laboratory Subcontractor: Oregon State University Corvallis, OR 97331	Vegetation as an Indicator of High Wind Velocities **CONTRACT NO.** EY-76-S-06-2227
PRINCIPAL INVESTIGATOR	**PERIOD OF PERFORMANCE**
E.W. Hewson	June 1976 to June 1979
WORK LOCATION	**FISCAL YEAR 1978 FUNDING**
Corvallis, OR	$115,525
CONTRACTING OFFICE	**CUMULATIVE FUNDING**
Pacific Northwest Laboratory Richland, WA	$214,357

BACKGROUND

This program was initiated to examine the feasibility of using the growth and appearance of vegetation as an aid to locating favorable wind energy areas.

OBJECTIVES

The objective of this work is to calibrate in terms of mean wind velocity the type and degree to which vegetation has been deformed by the wind.

APPROACH

Five indices of the effect of wind on conifers were identified in the first year of the study. At a number of sites in the region of the Columbia Gorge, wind data has been gathered over the course of the past 2 years. Two of the five indices have been successfully calibrated against annual mean wind speed. The techniques that have proved successful for conifers are being extended to deciduous trees. A study is also underway to determine if wind-flagged trees can be located by aerial photography.

OUTPUT

Published reports of the calibration results should prove valuable in providing an initial screening tool to those who are doing preliminary site survey work with similar vegetation.

WIND CHARACTERISTICS

CONTRACTOR	TITLE
Pacific Northwest Laboratory Subcontractor: Georgia Institute of Technology Atlanta, GA 30332	National Wind Energy Statistics for Large Arrays of Aerogenerators **CONTRACT NO.** EY-76-S-06-2439
PRINCIPAL INVESTIGATOR	**PERIOD OF PERFORMANCE**
C.G. Justus	May 1976 to April 1979
WORK LOCATION	**FISCAL YEAR 1978 FUNDING**
Atlanta, GA	$72,724
CONTRACTING OFFICE	**CUMULATIVE FUNDING**
Pacific Northwest Laboratory Richland, WA	$238,337

BACKGROUND

This work began in May 1976 to examine the power production characteristics of simulated large-scale arrays of wind turbines in various geographic regions of the United States and to develop and verify a simplified array simulation model.

OBJECTIVES

The objectives of this project are to continue to study the benefits of wind diversity for multiple turbine units and turbine arrays and to develop methodologies for simple modeling of array wind statistics and effects on energy production. In addition, this study is to evaluate probabilities and time and spatial correlations relating to year-to-year and month-to-month variations of mean wind speed, and to test the feasibility of adjustment of short-term "candidate site" data with long-term nearby data.

APPROACH

Wind performance statistics for arrays having maximum diversity have been evaluated. Development of simplified methods for simulating array performance from single site statistics has continued. The annual variability of wind power at 40 National Weather Service Stations with mean wind speed greater than 5 m/s has been studied. Simple models of wind gusts and shear on the power output are being investigated, along with parallel studies with a Grumman Windstream 25.

OUTPUT

A report will be submitted detailing the results of the array analyses, including a description of the techniques and the methods employed.

WIND CHARACTERISTICS

CONTRACTOR	TITLE
Pacific Northwest Laboratory Subcontractor: University of Virginia Charlottesville, VA 22903	Coastal Zone Wind Energy **CONTRACT NO.** EY-76-S-06-2344
PRINCIPAL INVESTIGATOR	**PERIOD OF PERFORMANCE**
Michael Garstang	September 1976 to September 1978
WORK LOCATION	**FISCAL YEAR 1978 FUNDING**
Charlottesville, VA	$34,246
CONTRACTING OFFICE	**CUMULATIVE FUNDING**
Pacific Northwest Laboratory Richland, WA	$158,527

BACKGROUND

This program began in September 1976, to study the wind power potential of the U.S. East and Gulf Coastal regions through extensive data analysis and application of a numerical sea-breeze circulation model.

OBJECTIVES

The objectives of this study are to establish quantitatively the space and time classifications of the East and Gulf Coastal zones for wind energy analyses, to investigate storm and inter-storm contributions, and to investigate the storm climatology using wind data and numerical modeling.

APPROACH

Regional and temporal classifications of the Coastal zone have been developed using statistical analysis. The numerical model is being applied to the coastal zone to identify "speed-up" regions and the vertical distribution of winds. Wind data is being analyzed to characterize storm and interstorm contributions. From these analyses, power output for representative wind energy conversion systems (WECS) in coastal regions is being predicted.

OUTPUT

A report will be submitted, which will detail the results of the data analyses and model applications in the East and Gulf Coastal zones. A supplement will be prepared for inclusion in the large and small WECS siting handbooks.

WIND CHARACTERISTICS

CONTRACTOR	TITLE
Pacific Northwest Laboratory Subcontractor: University of Alaska Fairbanks, AK 99701	Study of Alaskan Wind Power and Possible Applications
	CONTRACT NO. EY-76-S-06-2229
PRINCIPAL INVESTIGATOR	**PERIOD OF PERFORMANCE**
T. Wentink, Jr.	June 1976 to August 1978
WORK LOCATION	**FISCAL YEAR 1978 FUNDING**
Fairbanks, AK	$19,772
CONTRACTING OFFICE	**CUMULATIVE FUNDING**
Pacific Northwest Laboratory Richland, WA	$198,365

BACKGROUND

This project was initiated in 1976. Earlier studies (prior to 1976) have concentrated on wind analyses for selected locations, field measurements, and operation of a 6-kW windmill, applications of wind energy in Alaska, and economic studies.

OBJECTIVES

The primary objective of this study is to determine, by analysis of existing and new data sources, the potential of Alaskan wind power and its applications.

APPROACH

Wind data from 72 locations in Alaska were analyzed, and power output for representative wind energy conversion systems have been predicted. The data have been fit to several types of velocity distribution curves. Particular attention has been concentrated on the variation of the Weibull "k" parameter throughout Alaska, and the importance of this parameter in wind power calculations. Variations in anemometer height and exposure have also been evaluated.

OUTPUT

A final report on the evaluation of these 72 stations will be submitted. The report will establish a basis for an estimate of the wind power potential throughout Alaska. The report will include information that can be used to supplement small wind machine siting handbooks.

WIND CHARACTERISTICS

CONTRACTOR	TITLE
Pacific Northwest Laboratory Subcontractor: FWG Associates, Inc. R.R. 3 Box 331 Tullahoma, TN 37388	Technology Development for Assessment of Small-Scale Terrain Effects on Available Wind Energy **CONTRACT NO.** EY-76-C-06-2443
PRINCIPAL INVESTIGATOR Walter Frost	**PERIOD OF PERFORMANCE** August 1976 to September 1978
WORK LOCATION Tullahoma, TN	**FISCAL YEAR 1978 FUNDING** $161,315
CONTRACTING OFFICE Pacific Northwest Laboratory Richland, WA	**CUMULATIVE FUNDING** $254,672

BACKGROUND

Reliable information regarding the effect of small-scale terrain features on the wind near the surface is needed to evaluate the suitability of specific sites for wind energy conversion system (WECS) installations.

OBJECTIVES

The project objective is to characterize and catalog the effect of microscale terrain features on the near-surface wind.

APPROACH

Existing knowledge in the disciplines of fluid mechanics and meteorology of flow over two- and three-dimensional obstacles is being surveyed from the standpoint of its usefulness to WECS siting. This information is then to be compiled as guidelines and rules-of-thumb that would enable a user to determine the probable effect of local terrain on the wind. An experimental field program is being designed to verify the accuracy of these guidelines. The instrumentation required for these experiments is also being identified.

OUTPUT

A handbook will be produced to provide methods for engineers to use in selecting an optimum site (from the standpoint of wind power) for a WECS within a small area.

WIND CHARACTERISTICS

CONTRACTOR	TITLE
Pacific Northwest Laboratory Subcontractor: University of Wyoming Laramie, WY 82071	Locating Areas of High Wind Energy Potential by Remote Observations of Aeolian Features **CONTRACT NO.** EY-76-S-06-2343
PRINCIPAL INVESTIGATOR	**PERIOD OF PERFORMANCE**
R.W. Marrs	September 1976 to September 1978
WORK LOCATION	**FISCAL YEAR 1978 FUNDING**
Laramie, WY	$96,112
CONTRACTING OFFICE	**CUMULATIVE FUNDING**
Pacific Northwest Laboratory Richland, WA	$184,297

BACKGROUND

This study was initiated in September 1976 to develop an efficient way to identify high wind energy sites by inferring wind characteristics from aeolian geomorphologic features mapped from LANDSAT imagery.

OBJECTIVES

The objective of this study is to identify characteristics of aeolian features which can be interpreted as indicators of wind characteristics. Methods for rapid assessment of the wind energy potential of large regions are to be defined by interpretation of wind characteristics from wind-formed surface features observable from satellite and aircraft imagery.

APPROACH

Extensive field observations, including measurements from an instrumented aircraft and study of dune and playa lake characteristics, have been made in the "wind corridor" of southern Wyoming. Based on data collected in this study area, techniques for interpreting wind characteristics are being tested and refined. An assessment of the regional applicability of the methodology has been made. The techniques are being demonstrated in arid regions of the Pacific Northwest.

OUTPUT

Periodic special reports and a final report are being submitted which describe pertinent analyses, techniques, and results during the study. In addition, a preliminary handbook is being prepared which describes procedures for interpreting wind characteristics from aeolian features. A report assessing the wind energy potential in the Northwest determined from LANDSAT imagery is also being prepared.

WIND CHARACTERISTICS

CONTRACTOR	TITLE
Pacific Northwest Laboratory Subcontractor: Pennsylvania State University University Park, PA 16802	Wind Fluctuations in Complex Terrain **CONTRACT NO.** ET-78-S-06-1110
PRINCIPAL INVESTIGATOR H.A. Panofsky	**PERIOD OF PERFORMANCE** July 1978 to July 1979
WORK LOCATION University Park, PA	**FISCAL YEAR 1978 FUNDING** $58,836
CONTRACTING OFFICE Pacific Northwest Laboratory Richland, WA	**CUMULATIVE FUNDING** $58,836

BACKGROUND

While a fair amount of information concerning turbulence over smooth terrain is available, much less is known about the behavior of the wind over rougher terrain. Since many potentially attractive wind energy conversion system (WECS) sites are found in such areas, studies of wind characteristics, especially turbulence, are needed for proper and efficient WECS design.

OBJECTIVES

The objective of this study is to develop a model relating the fluctuation variance of turbulent intensity to parameters such as wind speed, surface roughness, stability, and mixing depth. A model will also be formulated to specify the dependence of spectra density shape on such parameters.

APPROACH

Data from a number of sources, in various terrain configurations, will be analyzed. The results will be compared to models and reported in the literature. The results of these analyses and comparisons will be used to develop the models.

OUTPUT

The output will consist of models specifying the various relationships outlined in the program objective. These should form the basis for a final report describing the methods used, the interpretation of data, and the model formula.

WIND CHARACTERISTICS

CONTRACTOR Pacific Northwest Laboratory Subcontractor: Science Applications, Inc. P.O. Box 2351 La Jolla, CA 92038	**TITLE** A New Wind Energy Site Selection Technology **CONTRACT NO.** EY-76-C-06-2440
PRINCIPAL INVESTIGATOR R.M. Traci	**PERIOD OF PERFORMANCE** May 1976 to June 1979
WORK LOCATION La Jolla, CA	**FISCAL YEAR 1978 FUNDING** $79,247
CONTRACTING OFFICE Pacific Northwest Laboratory Richland, WA	**CUMULATIVE FUNDING** $401,948

BACKGROUND

A three-year program to develop, test, and perform prototype applications of a wind energy conversion system site-selection methodology has been completed. Work is being extended to verify the computer codes developed in the program.

OBJECTIVES

The primary objective is to provide an improved siting methodology which makes use of mathematical wind-field modeling to extrapolate data from measurement locations to other, potentially windier sites throughout a mesoscale area, and to verify the wind-field models by comparing their predictions with actual observations.

APPROACH

The siting methodology is based on the use of a pseudo-potential flow objective analysis scheme, and a three-dimensional primitive equation boundary layer model. The objective analysis scheme is used to initiate the boundary layer model and to generate wind statistics by calibrating a number of runs corresponding to differing but typical boundary conditions. The boundary layer model is used to give a limited number of detailed "snapshots" of the area in question. These snapshots would correspond to the dominant climatological conditions observed in the area. The verification program is being conducted in cooperation with the Pacific Northwest Laboratory. Numerous simulations of the wind flow over various types of topography will be statistically compared with observations.

OUTPUT

All computer codes forming a part of this methodology have been documented so they can be applied by any competent user. Additionally, the siting methodology has been documented and demonstrated.

WIND CHARACTERISTICS

CONTRACTOR	TITLE
Pacific Northwest Laboratory Subcontractor: Northwestern University Evanston, IL 60201	Stochastic Modeling of Site Wind Characteristics **CONTRACT NO.** EY-76-S-06-2342
PRINCIPAL INVESTIGATOR	**PERIOD OF PERFORMANCE**
R.B. Corotis	September 1976 to October 1978
WORK LOCATION	**FISCAL YEAR 1978 FUNDING**
Evanston, IL	$73,986
CONTRACTING OFFICE	**CUMULATIVE FUNDING**
Pacific Northwest Laboratory Richland, WA	$172,631

BACKGROUND

This work began in October 1976, to develop stochastic and probabilistic methods for evaluating site wind characteristics.

OBJECTIVES

The objective of this study is to develop and apply complete procedures for site evaluation using probabilistic models and statistical methods and to establish the reliability of the characteristics.

APPROACH

New data sources have been tapped to test the models. These sources include data from the U.S. Forest Service fire weather network, data from candidate sites for testing large wind turbines, special hourly data, and small scale data. The reliability of the various models is being investigated using this data. For locations where high cross-correlation exists among data stations, a statistical relationship will be developed to enhance short-term data.

OUTPUT

A report will be submitted documenting all pertinent analyses, techniques, and calculated results. The report will discuss the reliability of the models and show results of a study to enhance short-term data from nearby long-term records.

WIND CHARACTERISTICS

CONTRACTOR	TITLE
Pacific Northwest Laboratory Subcontractor: North American Weather Consultants 600 Norman Firestone Road Goleta, CA 93017	Innovative Techniques for Identifying and Screening Potential Wind Energy Conversion Sites **CONTRACT NO.** B-50365-A-E
PRINCIPAL INVESTIGATOR M.W. Edelstein	**PERIOD OF PERFORMANCE** February 1978 to October 1978
WORK LOCATION Goleta, CA	**FISCAL YEAR 1978 FUNDING** $45,717
CONTRACTING OFFICE Pacific Northwest Laboratory Richland, WA	**CUMULATIVE FUNDING** $45,717

BACKGROUND

Innovative techniques are required to further our understanding of the wind energy potential in regions where little or no surface wind data exists. This contract began in February 1978 to develop a technique for utilizing standard upper-air observations obtained by the National Weather Service to identify regions where surface wind energy potential is high.

OBJECTIVES

The objective of this study is to obtain quantitative estimates of the long-term mean wind speeds and of the frequency distribution of wind speed and direction at a height appropriate to wind energy conversion systems (WECS) in data-sparse areas.

APPROACH

Two test sites have been selected where some surface data is available for verification purposes. A climatology of three-hourly geostrophic winds is obtained over each of these sites using standard upper-air rawinsonde observations interpolated to the sites. By incorporating the atmospheric stability and surface roughness characteristics over the site into standard boundary layer relationships, a correlation between surface and geostrophic winds will be obtained, and a long-term surface wind climatology will be generated.

OUTPUT

A final report will be submitted summarizing the technique, including how the technique can be applied to other areas and the usefulness of the technique for screening high wind areas for WECS implementation.

WIND CHARACTERISTICS

CONTRACTOR	TITLE
Pacific Northwest Laboratory Subcontractor: Marlatt and Associates 3611 Richmond Drive Ft. Collins, CO 80521	Applicability of the National Fire Weather Library to the Northwest Regional Wind Energy Study **CONTRACT NO.** B-50369-A-L
PRINCIPAL INVESTIGATOR	**PERIOD OF PERFORMANCE**
W.E. Marlatt	March 1978 to December 1978
WORK LOCATION	**FISCAL YEAR 1978 FUNDING**
Ft. Collins, CO	$17,000
CONTRACTING OFFICE	**CUMULATIVE FUNDING**
Pacific Northwest Laboratory Richland, WA	$17,000

BACKGROUND

The U.S. Forest Service fire weather data base has been identified as a potentially vast source of information for inclusion in large area analyses of wind energy potential. This consulting agreement was issued in January 1978 to assess the applicability of this data library.

OBJECTIVES

The objectives of this study are to review the U.S. Forest Service wind data library to identify stations with usable records so that an estimate of the seasonal and geographical wind energy in forest service regions can be developed.

APPROACH

The data have been obtained in magnetic tape format from the Forest Service's Rocky Mountain Forest and Range Experiment Station in Ft. Collins, Colorado. Statistical summaries have been developed from this tape for selected individual stations. The summaries have been screened for common windy areas. Frequency spectra of wind periods above certain threshold values have been developed. The analysis techniques have been applied to the Northwest region.

OUTPUT

A final report has been submitted that discusses the analysis procedures and demonstrates the application of the analysis to the Northwest. The report provides guidelines on how the procedure can be applied to other areas of the United States where fire weather data exists.

WIND CHARACTERISTICS

CONTRACTOR	TITLE
Pacific Northwest Laboratory Subcontractor: University of Texas Austin, TX 78712	Wind Power Potential in a Coastal Environment **CONTRACT NO.** B-36379-A-E
PRINCIPAL INVESTIGATOR	**PERIOD OF PERFORMANCE**
N.K. Wagner	February 1978 to February 1979
WORK LOCATION	**FISCAL YEAR 1978 FUNDING**
Austin, TX	$49,984
CONTRACTING OFFICE	**CUMULATIVE FUNDING**
Pacific Northwest Laboratory Richland, WA	$49,984

BACKGROUND

Innovative techniques for identifying high wind energy areas are important when little existing data is available in a region. This study began in February 1978, to test one promising technique--interviewing residents in a local area where high wind energy potential may exist and to apply the technique to a promising wind energy region, the western Gulf Coast of the United States.

OBJECTIVES

The objective of this study is to test oral interview techniques on fishing boat captains and pleasure craft operators in the western Gulf Coast and to compare the results of these interviews with available data and special measurements to further define the wind energy potential of this region.

APPROACH

Interview procedures for three specific coastal zones in the Port Aransas to Corpus Christi areas have been developed. These procedures cover the bay areas, the inshore areas, and the offshore areas. Existing meteorological data has been acquired and is being analyzed. Special short-term measurements of wind are being obtained to verify the results of the interviews.

OUTPUT

A report will be submitted which discusses the procedures for developing the interviews, the effectiveness of the interviews in producing additional information on wind energy potential in areas where little or no data exists, and results of their application in furthering knowledge of wind energy potential in the western Gulf Coast region.

TECHNOLOGY DEVELOPMENT

CONTRACTOR	TITLE
Massachusetts Institute of Technology Department of Aeronautics and Astronautics Aeroelastic and Structures Research Laboratory Cambridge, MA 02139	Wind Turbine Design Trade-off Analyses **CONTRACT NO.** EY-76-S-02-4131
PRINCIPAL INVESTIGATOR	**PERIOD OF PERFORMANCE**
Rene H. Miller	September 1976 to July 1978
WORK LOCATION	**FISCAL YEAR 1978 FUNDING**
Cambridge, MA	$40,000
CONTRACTING OFFICE	**CUMULATIVE FUNDING**
DOE Chicago Operations Office Argonne, IL	$186,780

BACKGROUND

A need has developed for a simplified handbook for use by wind turbine designers to select blade characteristics and to avoid undesirable dynamic characteristics and instabilities.

OBJECTIVES

The objective of this study is the performance of trade-off analyses of various wind turbine designs, and the provision of a design manual to furnish turbine designers with instructions on the application of design research performed by Massachusetts Institute of Technonolgy in resolving practical wind turbine design problems.

APPROACH

Tasks will include technical analyses of the trade-offs between upwind and downwind rotors; technical analyses of trade-offs between one-, two-, three-, and multi-bladed machines; an analyses of the trade-offs in locating the generator at the top versus the base of tower; and the development of an aeroelastic model for wind machines.

OUTPUT

This project will generate a simplified handbook on wind turbine dynamics analyses and background reports describing detailed theory, equation derivation, and related experimentation on aerodynamics, rotor blade structural dynamics, and drive system dynamics.

TECHNOLOGY DEVELOPMENT

CONTRACTOR	TITLE
Oregon State University Department of Mechanical Engineering Corvallis, OR 97331	Applied Aerodynamics of Wind Turbines - Aeromechanical
	CONTRACT NO. EY-76-S-06-2227
PRINCIPAL INVESTIGATOR Robert E. Wilson	**PERIOD OF PERFORMANCE** August 1977 to July 1979
WORK LOCATION Corvallis, OR	**FISCAL YEAR 1978 FUNDING** $90,000
CONTRACTING OFFICE DOE Headquarters Washington, DC	**CUMULATIVE FUNDING** $169,425

BACKGROUND

The current design techniques for determining performance and structural loading have their origins in high technology industries associated with commercial and military aircraft. These techniques are relatively complex and require large capacity, high-speed computers to perform the computations. The general trend in the design of large wind turbines is to make these computational routines more and more sophisticated. This trend reinforces the high technology approaches to design and design analysis of wind turbines.

OBJECTIVES

The objective of the project is to develop simplified, yet significant engineering analysis techniques for estimating aerodynamic performance, structural loads, and response of wind turbines.

APPROACH

Simple calculation schemes for hand-held calculators are being developed following an analytic review of existing computer models. This review is to consider the aerodynamics and loads of all configurations to be considered as well as the structural dynamics of horizontal and vertical-axis rotors and lift-translator arrays. Information to be provided will include: (1) the aerodynamic performance and loads associated with rotor operation and yaw, gusting winds, wind shear, and tower shadow; (2) criteria for selection of airfoil section, taper and twist, and the trade-offs among aerodynamic configuration, loading, and performance; (3) treatment of rotor/tower/nacelle dynamics to allow straightforward physical interpretation; and (4) design analysis of aerodynamic and structural aspects of the Darrieus, Savonius, and lift-translator concepts.

OUTPUT

The computational techniques will be presented in handbook form with associated curves and graphs, including examples illustrating the application of the techniques to situations representative of those encountered by a wind machine designer. The major emphasis of the handbook will be a clear presentation of physical basis for the major aerodynamic and structural loads and how they affect wind turbine design.

TECHNOLOGY DEVELOPMENT

CONTRACTOR	TITLE
National Aeronautics and Space Administration Lewis Research Center Subcontractor: Kaman Aerospace Corporation Old Windsor Road Bloomfield, CT 06002	150-foot Wind Turbine Blade Project **CONTRACT NO.** E(49-26)-1028
PRINCIPAL INVESTIGATOR	**PERIOD OF PERFORMANCE**
Herbert Gewehr	February 1977 to January 1979
WORK LOCATION	**FISCAL YEAR 1978 FUNDING**
Bloomfield, CT	$317,000
CONTRACTING OFFICE	**CUMULATIVE FUNDING**
NASA Lewis Research Center Cleveland, OH	$2,350,000

BACKGROUND

Wind turbine system studies indicated that large wind turbines with rotors of approximately 300-foot diameter would produce the lowest cost electricity. The blades were recognized as the component requiring most development to reduce cost and to qualify for use. A competitive procurement was initiated to develop a 150-foot prototype blade for evaluation and test.

OBJECTIVES

The objective of this project is to design, fabricate, test, and evaluate a potentially low-cost blade for a 300-foot diameter rotor used on a baseline wind turbine. The objective includes providing a technological base for blades of this size and identifying fundamental characteristics of the blade and design criteria that could be altered to improve both the low-cost potential and technical performance.

APPROACH

To achieve this objective, a contract for the design, development, fabrication, and test of a representative 150-foot long blade was awarded to Kaman Aerospace Company. The blade spar was fabricated by Structural Composite Industries, Inc., Azusa, California, under a Kaman subcontract. The spar was fabricated of fiberglass reinforced plastic using a transverse filament tape method. It was shipped by rail to the Kaman plant for final assembly and testing.

OUTPUT

A detailed assessment will be made of the design technology developed under this contract. Test results and fabrication cost analyses are to be developed for various production quantities of the blade. The blade fabrication was completed in August 1978, and blade static tests were furnished in January 1979.

TECHNOLOGY DEVELOPMENT

CONTRACTOR	TITLE
National Aeronautics and Space Administration Lewis Research Center 21000 Brookpark Road Cleveland, OH 44135	Supporting Research and Technology Program **CONTRACT NO.** E(49-26)-1028
PRINCIPAL INVESTIGATOR W.H. Robbins	**PERIOD OF PERFORMANCE** June 1975 - continuing
WORK LOCATION Cleveland, OH	**FISCAL YEAR 1978 FUNDING** $2,404,000
CONTRACTING OFFICE DOE Headquarters Washington, DC	**CUMULATIVE FUNDING** $8,545,000

BACKGROUND

The Supporting Research and Technology Program is part of the DOE High Power Horizontal-Axis Wind Turbine Program. The program started in FY 1975 with: (1) the design and construction of the MOD-O; (2) in-house and contract efforts to develop the computer codes to analyze wind turbine (WT) performance and structural dynamics; and (3) contract efforts to develop new WT blade designs. MOD-O has been used to qualify components, systems, and operating procedures for MOD-OA and to provide engineering data to verify computer codes for use in designing MOD-1 and MOD-2. The 150-foot blade design concept is being considered for back-up blades for the MOD-1 and MOD-2 wind turbines.

OBJECTIVES

The primary objectives of this project are: (1) to develop the technology to reduce WT capital and maintenance costs and to improve performance, reliability and service life; (2) to support WT projects, namely MOD-OA, MOD-1, and MOD-2; and (3) to provide a means for transferring the technology to industry.

APPROACH

This program will analyze the feasibility of new concepts to reduce cost and improve performance, reliability, and service life; test promising concepts on MOD-O; develop computer codes to analyze stiff and flexible structures and WT/utility stability and verify by analyzing MOD-O experimental results; run tests on MOD-O to increase understanding of WT operation and identify problems; perform studies to reduce the cost of components starting with blades, which are the most expensive; and study ways to reduce WT loads, weight, and number of components.

OUTPUT

The program will generate designs and fabricate prototypes of blades; prepare and produce reports of MOD-O tests in various configurations, e.g., upwind and downwind, stiff and passive yaw, stiff and soft towers, manual and microprocessor control, operations on load bank, and synched to diesels and the utility; and provide verified computer codes. This information will be used to support MOD-OA, MOD-1, MOD-2 and future WT's and to compare with Gedser 3-bladed, upwind, fixed pitch WT performance.

TECHNOLOGY DEVELOPMENT

CONTRACTOR	TITLE
National Aeronautics and Space Administration Lewis Research Center Subcontractor: Lockheed Missiles and Space Co., Inc. P.O. Box 1103 West Station Huntsville, AL 35807	Wake Flow Characteristics of Large Wind Turbines **CONTRACT NO.** E(49-26)-1028
PRINCIPAL INVESTIGATOR	**PERIOD OF PERFORMANCE**
Mel Brashears	March 1978 to January1979
WORK LOCATION	**FISCAL YEAR 1978 FUNDING**
Huntsville, AL	$135,000
CONTRACTING OFFICE	**CUMULATIVE FUNDING**
NASA Lewis Research Center Cleveland, OH	$135,000

BACKGROUND

To generate significant amounts of electricity with wind turbines it will be necessary to install large wind turbines in groups on a wind turbine farm. The spacing will be determined by several factors. One factor is the distance required for the wake of each turbine to be re-energized to near undisturbed freestream conditions.

OBJECTIVES

The objective is to develop and validate a turbulent mixing model that can be used to calculate wind speed profiles in the wake of a wind turbine, which can then be used to calculate the spacing between wind turbines.

APPROACH

The turbulent mixing model will be developed by adapting mixing models that are used to predict the wake characteristics downstream of solid objects. The wind speed in the wake of the MOD-OA wind turbine at Clayton, NM will be measured using laser doppler techniques that have been used to measure wind speeds in other atmospheric disturbance studies.

OUTPUT

The output will be a report detailing wind speed characteristics upstream and in the wake downstream of the wind turbine, the distance required to re-energize the wind stream to near free stream conditions and an experimentally validated computer model that predicts wind speed distributions in the wake.

TECHNOLOGY DEVELOPMENT

CONTRACTOR	TITLE
National Aeronautics and Space Administration Lewis Research Center Subcontractor: Paragon Pacific, Inc. 1601 E. El Segunda Blvd. El Segunda, CA 90245	Wind Turbine Systems Simulator **CONTRACT NO.** E(49-26)-1028
PRINCIPAL INVESTIGATOR	**PERIOD OF PERFORMANCE**
John A. Hoffman	March 1978 to April 1979
WORK LOCATION	**FISCAL YEAR 1978 FUNDING**
El Segunda, CA	$135,000
CONTRACTING OFFICE	**CUMULATIVE FUNDING**
NASA Lewis Research Center Cleveland, OH	$135,000

BACKGROUND

The U.S. Army and Navy are developing hybrid analog/digital simulators for helicopter rotor and system analysis. Advances in electronics have reduced the cost and size and improved the stability of analog components. This technology has the potential to increase the speed of wind turbine dynamics calculation by several orders of magnitude.

OBJECTIVES

The objective of this project is to develop a wind turbine hybrid simulator which will enable wind turbine dynamics calculations to be done in real time and at significantly lower cost than digital calculations.

APPROACH

The contractor will adapt the helicopter simulator technology being developed for Army and Navy and modify those developments for horizontal-axis wind turbine systems.

OUTPUT

The output will be one real-time hybrid simulator of a wind turbine rotor and one real-time hybrid simulator of a complete horizontal-axis wind turbine system including rotor, drive train, alternator, tower, and foundation.

TECHNOLOGY DEVELOPMENT

CONTRACTOR	TITLE
International Energy Agency Organization for Economic Cooperation and Development 2 Rue Andre Pascal Paris 16, France	Implementing Agreement for Cooperation in the Development of Large-Scale Wind Energy Conversion Systems **CONTRACT NO.**
PRINCIPAL INVESTIGATOR	**PERIOD OF PERFORMANCE**
WORK LOCATION	**FISCAL YEAR 1978 FUNDING**
Denmark, Sweden, United Kingdom, United States, West Germany	No exchange of funds
CONTRACTING OFFICE	**CUMULATIVE FUNDING**
DOE Headquarters Washington, DC	

BACKGROUND

Several nations are engaged in megawatt-scale machine development research, and have mutual interests in other similar undertakings. This agreement provides a vehicle which facilitates information-sharing activities.

OBJECTIVES

Participating nations seek to further the development of large-scale wind energy conversion systems by means of cooperative action within the research and development efforts of the International Energy Agency. By cooperation and coordination in the planning and execution of national programs, participants hope to generate research, development, and demonstration activities which are mutually supportive, while meeting national needs.

APPROACH

Participants exchange technical information and test data from their large wind system projects (excluding proprietary information) as well as economic and technical assessments, environmental assessments, and information on national research and development plans.

OUTPUT

Each party will provide reports on work performed on each activity undertaken and may participate in the exchange of personnel. Workshops and joint expert meetings are held to further facilitate information exchanges.

TECHNOLOGY DEVELOPMENT

CONTRACTOR	TITLE
International Energy Agency Organization for Economic Cooperation and Development 2 Rue Andre Pascal Paris 16, France	Implementing Agreement for a Program of Research and Development on Wind Energy Conversion Systems **CONTRACT NO.** ET-78-C-01-3285
PRINCIPAL INVESTIGATOR	**PERIOD OF PERFORMANCE** Tasks 1 through 4 1978 to 1980
WORK LOCATION 1.Stockholm, Sweden 3.Julich, W. Germany 2. Richland, WA 4.Julich, W. Germany	**FISCAL YEAR 1978 FUNDING** $157,400
CONTRACTING OFFICE DOE Headquarters Washington, D.C.	**CUMULATIVE FUNDING** $157,400

BACKGROUND

Much information on wind energy conversion systems is being developed on a unilateral basis by Japan, several European nations and the United States. To facilitate cooperation on an international basis, an implementing agreement was developed and signed by those nations wishing to participate.

OBJECTIVES

Objectives include cooperative research, development, demonstrations, and exchanges of information. To date, the agreement has produced four projects expected to yield substantial data over the next two years:

1. Environmental and Meteorological Aspects of Wind Energy Conversion Systems (WECS) -Sweden.

2. Evaluation of Wind Modeling - United States.

3. Integration of WECS in Utility Systems - West Germany.

4. Rotor Stressing and Smoothness of Operations in Large WECS - West Germany.

APPROACH

A single nation serves as operating agent for performance of specific tasks and other member nations interested in participating contribute to the funding of the task based on a set formula. All members participate in the development of tasks to be performed.

OUTPUT

The data, designs, specifications, etc., generated by the tasks are expected to contribute to the U.S. effort to achieve early implementation and commercialization of wind energy.

TECHNOLOGY DEVELOPMENT

CONTRACTOR	TITLE
University of Tennessee Department of Electrical Engineering Knoxville, Tennessee 37916	Power Systems Aspects of Multi-Unit Wind Systems **CONTRACT NO.** EY-76-S-05-5266
PRINCIPAL INVESTIGATOR	**PERIOD OF PERFORMANCE**
Thomas W. Reddoch	September 1976 to June 1979
WORK LOCATION	**FISCAL YEAR 1978 FUNDING**
Washington, DC and Knoxville, Tennessee	$28,000
CONTRACTING OFFICE	**CUMULATIVE FUNDING**
DOE Headquarters Washington, DC	$60,000

BACKGROUND

The development of the interface between the wind energy conversion system (WECS) and the conventional electric utility system is one vital part of the DOE wind program since the intermittent nature of the source represents a significant departure from classical electric power producing apparatus. Specific requirements of the interface as well as operational and planning methodologies for electric utilities utilizing WECS demand research, development, and demonstration.

OBJECTIVES

The primary objective is the review of technical information generated by DOE contractors investigating the systems dynamics of wind turbines and their interconnection with utility systems. From the assessment of the potential impact of WECS on electric utility practices, planning strategies are provided to the Wind Systems Branch for the WECS/utility interface requirements.

APPROACH

Programmatic needs are supplied through written reports, contractor quarterly review meetings, and regular meetings with the staff of DOE Wind Systems Branch. These meetings will define major issues pertaining to the interconnection of WECS to the electric utility system.

OUTPUT

Technical correspondence, topical reports, and planning details are provided on the requirements of the WECS/utility interface as well as operational and planning functions for utilities employing WECS.

ADVANCED AND INNOVATIVE CONCEPTS

CONTRACTOR	TITLE
Solar Energy Research Institute 1536 Cole Boulevard Golden, CO 80401	Program Management and Support **CONTRACT NO.** EG-77-C-01-4042
PRINCIPAL INVESTIGATOR Irwin E. Vas	**PERIOD OF PERFORMANCE** June 1978 - continuing
WORK LOCATION Golden, CO	**FISCAL YEAR 1978 FUNDING** $50,000
CONTRACTING OFFICE DOE Headquarters Washington, DC	**CUMULATIVE FUNDING** $50,000

BACKGROUND

During 1978, management responsibility for most projects in the program element Advanced and Innovative Systems (1.6) was transferred to the Solar Energy Research Institute (SERI).

OBJECTIVES

A primary function of the Wind Energy Innovative Systems Program is to solicit, investigate, evaluate, and develop innovative wind energy systems which are more cost-effective in comparison to existing systems.

APPROACH

This coordinating effort will: (1) effect the transition of detailed management responsibility from the Wind Systems Branch to SERI; (2) assess ongoing projects and plan objectives; and (3) support and technically direct projects investigating innovative wind concepts.

OUTPUT

This program will produce programmatic studies of background information for program decisions and for evaluation and response to unsolicited proposals. Further, SERI will conduct contractor reviews and will hold an innovative systems workshop.

ADVANCED AND INNOVATIVE CONCEPTS

CONTRACTOR	TITLE
Solar Energy Research Institute Subcontractor: West Virginia University Morgantown, WV 26506	Innovative Wind Turbines **CONTRACT NO.** EY-76-C-05-5135
PRINCIPAL INVESTIGATOR	**PERIOD OF PERFORMANCE**
Richard E. Walters	March 1976 to August 1978
WORK LOCATION	**FISCAL YEAR 1978 FUNDING**
Morgantown, WV	$99,888
CONTRACTING OFFICE	**CUMULATIVE FUNDING**
Solar Energy Research Institute Golden, Co	$412,641

BACKGROUND

Innovative wind turbine designs are being investigated as possible alternatives to conventional horizonal-axis propeller-type turbines. For vertical-axis turbines, blades with high lift airfoils are being considered to improve performance. One such airfoil is the circulation-controlled airfoil. Previous work on this airfoil at West Virginia University resulted in the development of a successful STOL aircraft which utilized a circulation-controlled flap design.

OBJECTIVES

The project objective is to investigate the potential of using circulation-controlled blades on a straight blade vertical-axis wind turbine. This includes turbine configuration, structural, and aerodynamic analyses, as well as a system component cost study.

APPROACH

Both theoretical and experimental methods will be used to study the performance. The major efforts will be to modify the existing vertical-axis wind turbine test model to allow direct measurements of blade aerodynamic parameters of lift, drag, and moment coefficient. Indoor tests will be performed with both conventional and circulation-controlled blades installed on the turbine.

OUTPUT

Test data will be analyzed to determine turbine performance gains which result from blades of circulation-controlled airfoil sections. Turbine power coefficients will be estimated from the blade aerodynamic data. Preliminary estimates of the cost benefits of the circulation-controlled turbine will be made. If results appear encouraging, future outdoor free-wind tests may be performed.

ADVANCED AND INNOVATIVE CONCEPTS

CONTRACTOR Solar Energy Research Institute Subcontractor: Polytechnic Institute of New York Route 110 Farmingdale, NY 11735	**TITLE** Vortex Augmentors for Wind Energy Conversion **CONTRACT NO.** ET-77-C-01-2358
PRINCIPAL INVESTIGATOR Pasquale M. Sforza	**PERIOD OF PERFORMANCE** May 1976 to February 1979
WORK LOCATION Farmingdale, NY	**FISCAL YEAR 1978 FUNDING** $43,924
CONTRACTING OFFICE Solar Energy Research Institute Golden, CO	**CUMULATIVE FUNDING** $379,927

BACKGROUND

Appropriate interaction of properly designed aerodynamic surfaces with natural wind of low power density can generate discrete vortical flow of relatively high power density. Suitable turbines may then be used to extract the energy from this compacted vortical field. This idea for energy concentration in natural flows is termed the vortex augmentor concept (VAC).

OBJECTIVES

The objective of this project is to determine the technical feasibility, performance, and economic potential of the delta wing VAC for wind energy conversion.

APPROACH

The VAC field test prototype instrumentation will be refined to provide detailed information on torque, speed, and power output under actual field conditions. Stability, control, and safety aspects of the prototype VAC system will be determined under power generation. Performance maps, power control flaps, and additional laboratory testing, as well as economic studies, will be generated.

OUTPUT

Results of field tests and economic studies will be utilized to determine the potential of the vortex augmentor concept for implementation in the Federal Wind Energy Program.

ADVANCED AND INNOVATIVE CONCEPTS

CONTRACTOR	TITLE
Solar Energy Research Institute Subcontractor: Grumman Aerospace Corporation South Oyster Bay Road Bethpage, NY 11714	Further Investigations of Diffuser Augmented Wind Turbines **CONTRACT NO.** EY-76-C-02-2616
PRINCIPAL INVESTIGATOR	**PERIOD OF PERFORMANCE**
K.M. Foreman	June 1975 to December 1978
WORK LOCATION	**FISCAL YEAR 1978 FUNDING**
Bethpage, NY	$201,964
CONTRACTING OFFICE	**CUMULATIVE FUNDING**
Solar Energy Research Institute Golden, CO	$467,931

BACKGROUND

A diffuser creates a low subatmospheric pressure behind a turbine rotor. A consequence of this suction is the capture of significantly more wind through a diffuser augmented wind turbine (DAWT) than a conventional wind turbine. The resulting increased mass flow increases the output power and reduces busbar cost. This allows smaller, cheaper turbine blades to be used for the DAWT than for a conventional wind energy conversion system of equal rating.

OBJECTIVES

This project will refine the performance and engineering design of a compact diffuser in order to improve confidence in scaled up designs and the cost/benefit ratio of the DAWT concept.

APPROACH

This continuing program employs wind tunnel testing, engineering design, and producibility analyses to increase the relative power coefficient of the concept and estimates costs of field demonstration and commercialized models that can become economically viable.

OUTPUT

Wind tunnel model tests will provide performance data that will be extrapolated to field site conditions. Manufacture cost estimates will be obtained for a prototype engineering design of a candidate diffuser configuration. Busbar energy costs will be determined for power ratings and production quantities that are considered commercially practical. A meaningful field demonstration model will be sized and its cost determined.

ADVANCED AND INNOVATIVE CONCEPTS

CONTRACTOR	TITLE
Solar Energy Research Institute Subcontractor: Marks Polarized Corporation 15 3-16 Tenth Avenue Whitestone, NY 11357	Tests and Devices for Wind/Electric Power Charged Aerosol Generator **CONTRACT NO.** EG-77-C-01-2774
PRINCIPAL INVESTIGATOR	**PERIOD OF PERFORMANCE**
Alvin M. Marks	July 1976 to September 1978
WORK LOCATION	**FISCAL YEAR 1978 FUNDING**
Whitestone, NY	$99,448
CONTRACTING OFFICE	**CUMULATIVE FUNDING**
Solar Energy Research Institute Golden, CO	$199,248

BACKGROUND

Under previous Energy Research and Development Administration and National Science Foundation grants, a wind tunnel test facility was developed and preliminary tests conducted on several electrofluid dynamic (EFD) aerosol charging devices. Resulting data demonstrated feasibility of the concept. The key technical problem is to develop a method of efficiently charging the aerosol.

OBJECTIVES

Four methods are to be investigated in order to derive the basic equations, obtain experimental data for the EFD machine, and to compare the performance of charged droplet methods: (1) waterjet/metal contact charging; (2) steam/metal contact charging; (3) condensation ion charging; and (4) induction charging/waterjet. The experimental results are to be compared with existing theories.

APPROACH

The basic equations of the aforementioned charging methods were derived and experimental devices were designed, fabricated, and set up in the wind tunnel. The test data on these devices included heat, air temperature, and relative humidity. The results of the methods will be compared with the predicted values.

OUTPUT

Experimental data was obtained on the charging devices. The basic equations were derived, and critical parameters and constraints determined. The induction charging/waterjet technique showed the most potential.

ADVANCED AND INNOVATIVE CONCEPTS

CONTRACTOR	TITLE
Solar Energy Research Institute Subcontractor: South Dakota School of Mines and Technology Rapid City, SD 57701	Energy from Humid Air **CONTRACT NO.** E(49-18)2553
PRINCIPAL INVESTIGATOR	**PERIOD OF PERFORMANCE**
Thomas K. Oliver	October 1976 to September 1979
WORK LOCATION	**FISCAL YEAR 1978 FUNDING**
Rapid City, SD	$99,547
CONTRACTING OFFICE	**CUMULATIVE FUNDING**
Solar Energy Research Institute Golden, CO	$99,547

BACKGROUND

A vast amount of energy is contained in the latent heat of vaporization of water vapor in humid air. This latent heat is the energy source for thunder storms and wind tropical storms (hurricanes and wind typhoons). Humid air at appropriate locations could possibly have the potential to provide useful energy.

OBJECTIVES

The objective of this project is to find a cost-effective way to convert the latent heat energy in humid air into mechanical work. This would then be used to drive an electrical generator or alternator. The objective includes the investigation and assessment of two possible techniques: the vertical natural draft and expansion-compression techniques.

APPROACH

Studies have been conducted by computer modeling. For humid air, which is made up of dry air plus water vapor, the dry air component is treated as an ideal gas. Properties of the water vapor component have been taken from a computer subroutine for the International Steam Tables. The Natural Draft Tower and Expansion-Compression techniques have been modeled and extensive parametric studies conducted.

OUTPUT

Taking into account present costs of construction, the Natural Draft Tower is not a cost-effective way of converting energy from humid air. A one-machine mechanization of an expansion-compression cycle making use of vortex flow may be a cost-effective method based on some preliminary cost studies that have been conducted.

ADVANCED AND INNOVATIVE CONCEPTS

CONTRACTOR	TITLE
Solar Energy Research Institute Subcontractor: Grumman Aerospace Corporation South Oyster Bay Road Bethpage, NY 11714	Tornado-Type Wind Energy System (Phase II) **CONTRACT NO.** E(49-18)2555
PRINCIPAL INVESTIGATOR	**PERIOD OF PERFORMANCE**
James T. Yen	March 1976 to June 1979
WORK LOCATION	**FISCAL YEAR 1978 FUNDING**
Bethpage, NY	$228,972
CONTRACTING OFFICE	**CUMULATIVE FUNDING**
Solar Energy Research Institute Golden, CO	$427,567

BACKGROUND

This concept incorporates a tall cylindrical tower with an open top, slotted side openings, and guide vanes to create a swirling, tornado-like vortex flow. Outside air rushes into the base of the tower and is drawn upward through the low pressure core, causing the rotor blades to spin and drive the generator.

OBJECTIVES

The major objective is to determine the cost-effectiveness and practicality of the tornado-type wind energy system (TTWES) in its ability to demonstrate multi-megawatt unit capacity, supply energy uniformly despite wind fluctuations, and be energy cost-competitive with a lifetime of 30 years.

APPROACH

Both theoretical and experimental investigations are to be carried out. Models of up to 10 feet if height and of both spiral and multi-vane tower cross-sections will be made and tested in wind tunnels. Turbines of up to one foot in diameter will be designed, manufactured, and installed in the models. Theoretical analyses will take into account the complex effects of turbulence on the complicated interactions among the vortex, the turbine flow, and the boundary layers. Scaling and cost estimates will be improved.

OUTPUT

A better understanding and design capability for the TTWES will be established. Details of passive or active inlet-vane design will be obtained; detailed tests of the turbine and total system will be conducted; improved scaling and cost estimates will be acquired; and experimental comparisons between spiral and multi-vane towers will be documented.

ADVANCED AND INNOVATIVE CONCEPTS

CONTRACTOR	TITLE
Sandia Laboratories Division 4715 Albuquerque, NM 87185	Vertical-Axis Wind Turbine **CONTRACT NO.** AT(29-1)-789
PRINCIPAL INVESTIGATOR	**PERIOD OF PERFORMANCE**
Richard H. Braasch	April 1975 to September 1978
WORK LOCATION	**FISCAL YEAR 1978 FUNDING**
Albuquerque, NM	$1,388,000
CONTRACTING OFFICE	**CUMULATIVE FUNDING**
DOE Albuquerque Operations Office Albuquerque, NM	$4,475,000

BACKGROUND

Supported by DOE, Sandia Laboratories has been engaged in the development of the Darrieus vertical-axis wind turbine (VAWT) since 1975. During the initial period, effort has been concentrated on developing a technical base with experimental machines. During 1978, the technology has matured to the point where more emphasis could be placed on commercialization.

OBJECTIVES

This program will: (1) generate cost and performance data associated with vertical-axis turbine operation; (2) demonstrate the feasibility of VAWT application in a utility grid; and (3) improve the design.

APPROACH

Models for economic, aerodynamic, and structural performance will be developed and verified. The 2-meter, 5-meter, and 17-meter Darrieus systems will be operated for performance data and testing of improved design features.

OUTPUT

A design manual, and summary of cost and performance data will be produced to disseminate VAWT technology. The machines produced serve as baseline designs for future low-cost VAWT machines.

ADVANCED AND INNOVATIVE CONCEPTS

CONTRACTOR	TITLE
Solar Energy Research Institute Subcontractor: University of Dayton Research Institute Dayton, OH 45469	Electrofluid Dynamic Wind Generator Program **CONTRACT NO.** EY-76-S-02-4130
PRINCIPAL INVESTIGATOR	**PERIOD OF PERFORMANCE**
John E. Minardi	September 1976 to November 1978
WORK LOCATION	**FISCAL YEAR 1978 FUNDING**
Dayton, OH	$102,264
CONTRACTING OFFICE	**CUMULATIVE FUNDING**
Solar Energy Research Institute Golden, OH	$197,295

BACKGROUND

Research conducted by the United States Air Force in the early 1970's demonstrated the electrofluid dynamic (EFD) generator concept and had provided the scaling laws. Research at the University of Dayton has provided EFD wind generator theory and experimentally confirmed the basic principle of EFD power generation.

OBJECTIVES

The primary objectives are: (1) to provide a sufficient density of charged water droplets of low mobility so that EFD generator geometries can be experimentally evaluated; and (2) to provide at low energy costs, satisfactory levels of density of charged water droplets of low mobility for wind generator applications.

APPROACH

Charged droplet production methods and performance of generator designs are being investigated in two Eiffel type wind tunnels. Theoretical investigations extend EFD theory and guide the experimental effort.

OUTPUT

Output current and voltage are measured and compared with theoretical predictions to establish research progress and direction of new efforts to lead to the development of practical EFD wind generators.

ADVANCED AND INNOVATIVE CONCEPTS

CONTRACTOR Sandia Laboratories Subcontractor: Aluminum Company of America Alcoa Laboratories Alcoa Center, PA 15069	**TITLE** Cost Study of Vertical-Axis Wind Turbine **CONTRACT NO.** SAND07-6942 (Phase I) SAND07-7283 (Phase II)
PRINCIPAL INVESTIGATOR P.N. Vosburgh	**PERIOD OF PERFORMANCE** September 1977 to September 1978
WORK LOCATION Alcoa Center, PA	**FISCAL YEAR 1978 FUNDING** $60,000
CONTRACTING OFFICE Sandia Laboratories Albuquerque, NM	**CUMULATIVE FUNDING** $74,979

BACKGROUND

Research and development on the Darrieus turbine has been pursued at Sandia Laboratories since 1975. The technical feasibility of the designs resulting from this work is now being demonstrated. There is a need to determine the economic performance of these designs to assess low-cost design and commercialization potential for Darrieus turbines.

OBJECTIVES

A detailed cost estimate of the Sandia Darrieus turbine is desired in a format consistent with other studies. Each element of design was separately examined to allow comparison with other design alternatives. The effect of turbine size and production quantity was studied and backup documentation established estimate sources.

APPROACH

In Phase I, Sandia Laboratories and the contractor jointly configured a detailed series of Darrieus turbine designs in a scenario of present day sales and manufacture. In Phase II, the contractor independently determined and documented the cost of each turbine element. Six turbine sizes and six production rates were considered.

OUTPUT

Phase I and Phase II reports will be combined by Sandia with other cost analyses to form the parametric cost optimization study.

ADVANCED AND INNOVATIVE CONCEPTS

CONTRACTOR Sandia Laboratories Subcontractor: A.T. Kearney, Inc. Two Embarcadero Center San Francisco, CA 94111	**TITLE** Cost Study of Vertical-Axis Wind Turbine **CONTRACT NO.** SAND07-7161
PRINCIPAL INVESTIGATOR John Ashby	**PERIOD OF PERFORMANCE** February 1978 to July 1978
WORK LOCATION San Francisco, CA	**FISCAL YEAR 1978 FUNDING** $58,500
CONTRACTING OFFICE Sandia Laboratories Albuquerque, NM	**CUMULATIVE FUNDING** $58,500

BACKGROUND

Research and development on the Darrieus turbine has been pursued at Sandia Laboratories since 1975. The technical feasibility of the designs resulting from this work is now being demonstrated. There is a need to determine the economic performance of these designs to assess low-cost design and commercialization potential for Darrieus turbines.

OBJECTIVES

A detailed cost estimate of the Sandia Darrieus turbine is desired in a format consistent with other studies. Each element of design was separately examined to allow comparison with other design alternatives. The effect of turbine size and production quantity was studied and backup documentation established estimate sources.

APPROACH

In Phase I, Sandia Laboratories and the contractor jointly configured a detailed series of Darrieus turbine designs in a scenario of present day sales and manufacture. In Phase II, the contractor independently determined and documented the cost of each turbine element. Four turbine sizes and six production rates were considered.

OUTPUT

Phase I and Phase II reports will be combined by Sandia with other cost analyses to form the parametric cost optimization study.

ADVANCED AND INNOVATIVE CONCEPTS

CONTRACTOR	TITLE
Sandia Laboratories Subcontractor: Texas Technological University Department of Mechanical Engineering Lubbock, TX 79409	Aerodynamic Model **CONTRACT NO.** SAND06-4178
PRINCIPAL INVESTIGATOR	**PERIOD OF PERFORMANCE**
J.H. Strickland	May 1977 to January 1979
WORK LOCATION	**FISCAL YEAR 1978 FUNDING**
Lubbock, TX	$18,862
CONTRACTING OFFICE	**CUMULATIVE FUNDING**
Sandia Laboratories Albuquerque, NM	$26,982

BACKGROUND

There is a need to improve upon the capabilities of multiple streamtube models which are currently used to predict aerodynamic performance and loads generated by Darrieus turbines. These models perform calculations which apply in a spatially average sense rather than locally. Knowledge of local contributions to total loads are required in order to optimize aerodynamic and structural design.

OBJECTIVES

The objective of this project is to develop a theoretical aerodynamic model which predicts experimental results more accurately than present multiple streamtube models. This new model would: (1) include blade mutual interference effects; (2) be programmable to run at reasonable computing costs; and (3) be verifiable through laboratory experimentation.

APPROACH

A viable approach is simulation of turbine blades and wakes by time-dependent vortices. This would allow modeling of blade interaction. Computing time could be shortened by using a two-dimensional version of the model where applicable. Necessary laboratory experiments would include strain-gaged blades and quantitative flow visualization techniques for small-scale turbines operating in a water tow tank.

OUTPUT

The final report will include computer codes written in ANSI FORTRAN with detailed instructions for their use. These codes will provide a capability for better estimating the vertical-axis wind turbine performance characteristics and aerodynamic loads.

ADVANCED AND INNOVATIVE CONCEPTS

CONTRACTOR	**TITLE**
Sandia Laboratories Subcontractor: Aluminum Company of America Alcoa Laboratories Alcoa Center, PA 15069	Blade Study **CONTRACT NO.** SAND07-7173
PRINCIPAL INVESTIGATOR	**PERIOD OF PERFORMANCE**
Daniel K. Ai	March 1978 to September 1978
WORK LOCATION	**FISCAL YEAR 1978 FUNDING**
Alcoa Center, PA	$94,712
CONTRACTING OFFICE	**CUMULATIVE FUNDING**
Sandia Laboratories Albuquerque, NM	$94,712

BACKGROUND

The current 17-meter vertical-axis turbine utilizes strutted composite blades. While satisfying aerodynamic and structural constraints, this type of blade construction is labor-intensive and costly to manufacture. The struts also represent additional cost and complexity while adding nothing to the performance of the turbine.

OBJECTIVES

The objective of this project is to design and construct a low-cost blade that is amenable to mass production techniques. This design is to be unstrutted, with no points other than those at the tower, to further reduce costs. The blade must still meet applicable aerodynamic and strength criteria.

APPROACH

Blades for the 17-meter turbine are constructed from continuous, one-piece aluminum (6063-T6) extrusions. The blades are shipped in straight lengths, then bent to the proper shape at Sandia Laboratories.

OUTPUT

The project will produce three formed blades that will be subsequently tested on the 17-meter turbine, two blades left in the original straight shape, and cost figures associated with the blades.

ADVANCED AND INNOVATIVE CONCEPTS

CONTRACTOR	TITLE
Sandia Laboratories Subcontractor: University of New Mexico Civil Engineering Research Facility P.O. Box 188 Albuquerque, NM 87131	Guy Cable Anchor and Foundation Design **CONTRACT NO.** SAND07-7271
PRINCIPAL INVESTIGATOR	**PERIOD OF PERFORMANCE**
Harry Auld	March 1978 to October 1978
WORK LOCATION	**FISCAL YEAR 1978 FUNDING**
Albuquerque, NM	$18,101
CONTRACTING OFFICE	**CUMULATIVE FUNDING**
Sandia Laboratories Albuquerque, NM	$18,101

BACKGROUND

Vertical-axis wind turbine (VAWT) foundation and guy cable anchor design is a significant part of the total cost of the VAWT. This design should be matched to the design of the VAWT and the earth at the building site, similar to the building of any structure, with the addition of the dynamic response of the VAWT and the discipline of civil engineering to the design.

OBJECTIVES

The objectives are to minimize the cost of the VAWT foundation and guy anchor designs and to achieve a design compatible with the static and dynamic requirements of the VAWT and the earth at the selected site.

APPROACH

The approach is to investigate the load capability and cost of five types of foundations and guy anchors for five sizes of VAWT's and six types of site soils. This gives the proper foundation and guy anchor design for a given size of VAWT on a given soil.

OUTPUT

The final report will outline the preferred anchor designs, the recommended sizes for average soil and poor soil, and the costs of these anchors as a function of cable tension.

ADVANCED AND INNOVATIVE CONCEPTS

CONTRACTOR	TITLE
Sandia Laboratories Subcontractor: Power Technologies, Inc. P.O. Box 1058 Schenectady, NY 12301	Induction and Synchronous Machines for Vertical-Axis Wind Turbine **CONTRACT NO.** SAND07-7230
PRINCIPAL INVESTIGATOR	**PERIOD OF PERFORMANCE**
E.N. Hinrichsen	April 1978 to September 1978
WORK LOCATION	**FISCAL YEAR 1978 FUNDING**
Schenectady, NY	$35,200
CONTRACTING OFFICE	**CUMULATIVE FUNDING**
Sandia Laboratories Albuquerque, NM	$35,200

BACKGROUND

Induction and/or synchronous machines installed between a wind turbine and a power grid can perform as both starting motors and synchronous power generators. The dynamics of the rotating mechanical hardware, the electrical generators, and the grid can impose undesirable voltage fluctuations if not properly matched. The aerodynamics of the wind turbine and the variability of the wind act as the forcing function.

OBJECTIVES

The objective is to study the performance of induction and synchronous machines as they interact with the vertical-axis wind turbine system and grid system to determine what effects, if any, are likely to cause problems in installed systems. Another objective is to make recommendations for the control or elimination of any deleterious effects.

APPROACH

The approach used was to model the system, verify the model, and study the effects of variations in the generation equipment. These calculated effects were then compared with actual data from the 17-meter research turbine, and these models were used for determination of characteristics in a larger system.

OUTPUT

Documentation will include the model used and will show the correlation of the model results and test results. The report will address the significance of voltage and power fluctuations and make pertinent recommendations.

Farm and Rural Use (Small) Systems

CONTRACTOR	TITLE
U.S. Department of Agriculture Science and Education Administration Subcontractor: Virginia Polytechnic Institute Dept. of Aerospace and Ocean Engineering Blacksburg, VA 24061	Apple Storage Cooling Proof-of-Concept Experiment **CONTRACT NO.** EX-76-A-29-1026
PRINCIPAL INVESTIGATOR	**PERIOD OF PERFORMANCE**
Joseph Schetz	November 1976 to October 1979
WORK LOCATION	**FISCAL YEAR 1978 FUNDING**
Blacksburg, VA	$40,000
CONTRACTING OFFICE	**CUMULATIVE FUNDING**
U.S. Department of Agriculture Beltsville, MD	$282,000

BACKGROUND

Refrigeration of agricultural products is a possible use of wind power because of the generally high annual hours of load, the economy of intermediate term energy storage in ice builders, and the possibility of system fabrication using commercially available components.

OBJECTIVES

The objective of this project is to obtain actual operating data on the performance of a windmill used for refrigeration of an apple storage warehouse.

APPROACH

A 6-kW wind turbine was installed and coupled, through a small battery bank and direct current motor, to the compressor of a refrigeration system for an apple storage warehouse. The wind velocity, ambient temperatures, cooling supplied to the building, and the cooling load of the building are measured. The measurements will be analyzed to provide empirical performance characteristics of such a system as a function of environmental conditions and wind velocities.

OUTPUT

The performance of the system will be analyzed to assess the technical feasibility of this use of wind power and to estimate its economic value to farmers.

FARM AND RURAL USE (SMALL) SYSTEMS

CONTRACTOR	TITLE
U.S. Department of Agriculture Science and Education Administration Agricultural Research Northeastern Region Beltsville, MD 20705	Program Management, Rural and Remote Areas of Wind Energy Research **CONTRACT NO.** E(49-26)-1026
PRINCIPAL INVESTIGATOR L.A. Liljedahl	**PERIOD OF PERFORMANCE** August 1975 to September 1979
WORK LOCATION Beltsville, MD	**FISCAL YEAR 1978 FUNDING** $80,000
CONTRACTING OFFICE DOE Headquarters Washington, DC	**CUMULATIVE FUNDING** $284,000

PROJECT SUMMARY

BACKGROUND

Agricultural power use is geographically dispersed, as is wind power availability, and historically agriculture has made considerable use of wind power, yielding heuristic arguments that agriculture might beneficially use wind power in the future. However, agricultural practices and power requirements have changed drastically since the time when the use of windmills was widespread.

OBJECTIVES

The objective of this project is to determine agricultural applications that indicate high potential for wind power use, define agricultural markets for wind energy conversion systems and their economic thresholds, and specify research and development needs which will meet these objectives.

APPROACH

Research literature on agricultural power and energy use and wind power technology will be reviewed, from which preliminary definitions of wind power applications will be formulated. Plans for needed research and development to assess the value of these applications will be submitted periodically to DOE, and research and development projects assigned to the U.S. Department of Agriculture by DOE will be managed, reviewed, and reported to DOE.

OUTPUT

Management of the program will yield annual program development plans, requests for proposals for externally executed work, and reports of completed work suitable for distribution by DOE.

FARM AND RURAL USE (SMALL) SYSTEMS

CONTRACTOR	TITLE
U.S. Department of Agriculture Science and Education Administration Agricultural Research Subcontractor: To be determined	Alternative Techniques for Agricultural Management of Wind-Powered Irrigation Systems **CONTRACT NO.** EX-76-A-29-1026
PRINCIPAL INVESTIGATOR	**PERIOD OF PERFORMANCE**
To be determined	January 1979 to December 1979
WORK LOCATION	**FISCAL YEAR 1978 FUNDING**
To be determined	$80,000
CONTRACTING OFFICE	**CUMULATIVE FUNDING**
U.S. Department of Agriculture New Orleans, LA	$80,000

BACKGROUND

Considerably less wind power is available during the summer than earlier in the spring. Consequently, windmill operation during the late summer requires a considerably larger rotor diameter to extract the same amount of power as one operating in the spring, which will increase the cost. This suggests that a greater economic return per unit of investment might be achieved with other crop combinations which have a less sharply peaked water demand than corn alone, which is the crop most commonly irrigated now.

OBJECTIVE

The objective of this project is to formulate alternative agricultural management systems to be used with wind-powered irrigation and identify their benefits, disadvantages, and costs.

APPROACH

A variety of alternative cropping systems will be formulated which might have advantages when used with wind-powered irrigation systems. The net return to the farmer for each alternative system will be estimated as a function of the cost and capacity factor of the wind-powered system and compared with the cost of conventionally powered irrigation systems, with the cost of energy as a variable.

OUTPUT

The results of the study will provide an estimate of the improvement in operating factor which could be achieved with optimized cropping patterns, and thus an improved estimate of the economic value of wind power for irrigation.

FARM AND RURAL USE (SMALL) SYSTEMS

CONTRACTOR	TITLE
U.S. Department of Agriculture Science and Education Administration Agricultural Research Subcontractor: To be determined	Preliminary Evaluation of Pumps for Wind-Driven Irrigation **CONTRACT NO.** EX-76-A-29-1026
PRINCIPAL INVESTIGATOR	**PERIOD OF PERFORMANCE**
To be determined	January 1979 to December 1979
WORK LOCATION	**FISCAL YEAR 1978 FUNDING**
To be determined	$50,000
CONTRACTING OFFICE	**CUMULATIVE FUNDING**
U.S. Department of Agriculture New Orleans, LA	$50,000

BACKGROUND

Pumps now used for irrigation are largely suitable for operation at constant speed and constant power. Consequently, they are not generally suited for variable power produced by a wind turbine.

OBJECTIVES

Deep-well and shallow-well pumps will be assessed for characteristics of value for wind-powered irrigation pumping.

APPROACH

The operating characteristics of all types of pumps customarily used for pumping water will be reviewed, using information from textbooks, engineering handbooks, engineering journals, and manufacturer's performance data. From analysis, and if necessary, computer simulation, the performance of such pumps when operated at varying speeds over a wide range of speeds will be computed, and from this the advantages and disadvantages of each type of pump will be assessed for its value in a wind-powered irrigation system. Modifications in the design and construction of these pumps which will make them more suitable for use with wind power will be investigated and analyzed, and recommendations will be made for future development (e.g., use of adjustable stators and/or impellers to maintain constant pressure at variable speeds or high capacity positive displacement pumps).

OUTPUT

The results of the analysis will be documented in reports and published for the guidance of windmill and pump manufacturers. If necessary, the results may also be used to plan additional work on pump development.

FARM AND RURAL USE (SMALL) SYSTEMS

CONTRACTOR	TITLE
U.S. Department of Agriculture Science and Education Agricultural Research Wind Erosion Laboratory Manhattan, KS 66502	Low-Lift Irrigation Pumping Proof-of-Concept Experiment **CONTRACT NO.** EX-76-A-29-1026
PRINCIPAL INVESTIGATOR	**PERIOD OF PERFORMANCE**
Lawrence J. Hagen	July 1975 to September 1979
WORK LOCATION	**FISCAL YEAR 1978 FUNDING**
Manhattan, KS	$40,000
CONTRACTING OFFICE	**CUMULATIVE FUNDING**
DOE Headquarters Washington, DC	$124,000

BACKGROUND

When wind power is used, irrigation pumping operations may be divided into two different types, distinguished from one another based upon the type of pump used and supply characteristics of the water source from which the water is drawn. Where irrigation water is pumped from surface sources (tail-water pits, rivers, reservoirs, etc.) the maximum water yield available, in gallons per minute, is so high that it need not be considered in the design of the irrigation system, and in particular, in the selection of the pump size. Usually simple centrifugal or propeller-type pumps are used. It appears that wind-powered irrigation systems of this type without a need for back-up power can be assembled from commercially available equipment. This needs experimental confirmation and performance testing.

OBJECTIVES

Actual operating data will be obtained on the performance of a windmill driving a low-lift irrigation pump, and operating problems will be identified.

APPROACH

A 20-foot diameter, 30-foot high Darrieus windmill has been installed, and directly coupled through an appropriate transmission system, to a low-lift irrigation pump. The pumping rate and wind velocities are recorded to determine performance as a function of wind velocities. The system will be operated through one or more growing seasons to accumulate performance records for use in predicting technical and economic feasibility.

OUTPUT

Performance data from the tests will be used to confirm or correct analytical studies of performance. The results will be used to predict the cost of wind power for this type of irrigation and to develop design criteria and procedures for future systems.

FARM AND RURAL USE (SMALL) SYSTEMS

CONTRACTOR	TITLE
U.S. Department of Agriculture Service and Education Administration Agricultural Research Subcontractor: To be determined	Agricultural Wind Energy Applications Analysis **CONTRACT NO.** EX-76-A-29-1026
PRINCIPAL INVESTIGATOR	**PERIOD OF PERFORMANCE**
To be determined	September 1976 to March 1979
WORK LOCATION	**FISCAL YEAR 1978 FUNDING**
To be determined	$40,000
CONTRACTING OFFICE	**CUMULATIVE FUNDING**
U.S. Department of Agriculture Beltsville, MD	$145,000

BACKGROUND

Potential manufacturers and government planners need estimates of the market for wind turbines or power sources on farms. The heterogeneity of farm enterprises, the geographical and temporal non-uniformity of wind power, and the dearth of records on farm power use make this estimation a complex task.

OBJECTIVES

Estimates will be made of the extent to which wind energy can be exploited in agriculture, feasible applications, and the sizes of windmills most likely to be used in various enterprises.

APPROACH

Energy and power-using operations in agriculture will be reviewed for major types of agricultural enterprises to estimate the cost of substituting wind power and wind energy for the power and energy sources now used. The general procedure will be to select a discrete wind turbine size which maximizes energy saving for each enterprise, given different levels of conventional energy cost.

OUTPUT

The expected number of wind turbines which will provide competitive energy for farming will be estimated by size (power) and rated by wind speed for various cost levels of conventional energy.

FARM AND RURAL USE (SMALL) SYSTEMS

CONTRACTOR	TITLE
U.S. Department of Agriculture Science and Education Administration Information Division 5133 South Agriculture Building Washington, DC 20520	Information Dissemination and Technology Transfer **CONTRACT NO.** EX-76-A-29-1026
PRINCIPAL INVESTIGATOR To be determined	**PERIOD OF PERFORMANCE** September 1978 to August 1979
WORK LOCATION Hyattsville, MD	**FISCAL YEAR 1978 FUNDING** $40,000
CONTRACTING OFFICE U.S. Department of Agriculture Beltsville, MD	**CUMULATIVE FUNDING** $50,000

BACKGROUND

Potential users, manufacturers, and marketers of windmills and wind-powered systems for farm use need an accelerated system for learning about progress and potential uses of research on wind power.

OBJECTIVES

The objective of this project is to communicate the results of this program to the potential equipment manufacturers, the agricultural industry, and the general public.

APPROACH

The project will prepare: (1) a pamphlet describing the Federally funded research program in general; (2) a bulletin describing all current Federally funded projects in detail; (3) a 15-minute film describing the current Federally funded projects; (4) biannual workshops on agricultural use of wind power, reviewing the status of all current major applications; and (5) bulletins on each major defined wind power application after completion of the related proof-of-concept and economic study.

OUTPUT

The film, publications, and workshop will be used to inform farm equipment manufacturers, agricultural extension specialists, and the agricultural industry of the goals and status of the Federal research and development program.

FARM AND RURAL USE (SMALL) SYSTEMS

CONTRACTOR	TITLE
U.S. Department of Agriculture Science and Education Administration Agricultural Research Southwestern Great Plains Research Center Bushland, TX 79012	Wind-Powered/Engine-Powered Hybrid-Drive Irrigation Pumping Proof-of-Concept Exepriment **CONTRACT NO.** EX-76-A-29-1026
PRINCIPAL INVESTIGATOR	**PERIOD OF PERFORMANCE**
R. Nolan Clark	September 1976 to March 1979
WORK LOCATION	**FISCAL YEAR 1978 FUNDING**
Bushland, TX	$80,000
CONTRACTING OFFICE	**CUMULATIVE FUNDING**
DOE Headquarters Washington, DC	$385,000

BACKGROUND

Where irrigation water is pumped from deep aquifers, it is necessary to maximize the area irrigated by each well by designing the irrigation system to pump the well at its maximum permissible yield. The lift from the aquifer free-water surface is a high proportion of the pressure generated by the pump. As the lift remains nearly constant, the pump must operate at constant pressure, which requires constant speed and torque. This can be done with a wind turbine using existing equipment only by sharing the load with another constant speed drive unit such as an electric motor or a governed internal combustion engine.

OBJECTIVE

The objective of this project is to obtain actual performance data on the operation of an irrigation pumping system driven by a drive system, combining a windmill and an internal combustion engine or electric motor, and to identify operating problems.

APPROACH

A 50-foot diameter windmill will be installed and connected to a dual drive pump to permit hybrid operation with an internal combustion engine or electric motor and operated in this manner for two irrigation seasons. The pumping rate, wind velocity, torque, and fuel savings can be computed as a function of the wind velocity distribution.

OUTPUT

Performance data from the tests will be used to confirm or correct analytical studies of performance. The results will be used to predict the cost of wind power for this type of irrigation and to develop design criteria and procedures for future systems.

2.0 FARM AND RURAL USE (SMALL) SYSTEMS

CONTRACTOR Rockwell International Energy Systems Group DOE Rocky Flats Plant Wind Systems Program P.O. Box 464 Golden, CO 80401	**TITLE** Technical and Management Support for the Development of Wind Systems for Farm, Remote and Rural Use **CONTRACT NO.** E(20-2)-3533
PRINCIPAL INVESTIGATOR Terry Healy	**PERIOD OF PERFORMANCE** April 1976 - continuing
WORK LOCATION Golden, CO	**FISCAL YEAR 1978 FUNDING** $6,350,000
CONTRACTING OFFICE DOE Albuquerque Operations Office Albuquerque, NM	**CUMULATIVE FUNDING** $8,888,000

BACKGROUND

While small (less than 100 kW) wind systems are commercially available, they are not yet cost-competitive with conventional power sources in most applications and locations. To reduce these costs and alleviate other barriers to the commercialization of wind systems, a Small Wind Systems Test and Development Center has been established at the DOE Rocky Flats Plant near Golden, Colorado. Most of the funding for this operation will be passed on to private industry through the purchase of test and prototype machines and research and development services.

OBJECTIVES

The overall objectives are to reduce the cost of energy generated by small wind systems and advance their commercialization and use. Specific objectives during FY 1978 were to: (1) reduce the uncertainties of performance and energy costs through instrumented testing of small wind energy conversion systems (WECS); (2) determine and achieve technical and cost goals through the development of advanced small WECS; (3) increase consumer knowledge and manufacturer capability; (4) assist private industry in the development of standards; and (5) plan future dispersed test activities.

APPROACH

To meet its objectives during FY 1978, the Rocky Flats Wind Systems Program was organized into seven technically related task areas, including: (1) test center establishment; (2) test operations; (3) systems engineering; (4) technology development; (5) standards development; (6) information dissemination; and (7) planning. In addition, an eighth element was added to initiate planning for future dispersed small WECS test and institutional interface programs.

OUTPUT

The output of the Rocky Flats effort includes test data on commercially available and advanced prototype wind machines together with a series of advanced designs created to meet specific performance and cost goals. Technical reports will contribute to consumer awareness of wind power potential and provide information to manufacturers on small WECS design, performance, applications and cost. Subcontracted system development efforts will help stimulate the development of the small WECS industry.

FARM AND RURAL USE (SMALL) SYSTEMS

CONTRACTOR	TITLE
U.S. Department of Agriculture SEA-Agricultural Research Agricultural Engineering Research Unit Iowa State University 213 Davidson Hall Ames, IA 50011	Wind-Powered Farmhouse Heating Proof-of-Concept Experiment **CONTRACT NO.** EX-76-A-29-1026
PRINCIPAL INVESTIGATOR Leo H. Solderholm	**PERIOD OF PERFORMANCE** August 1975 to September 1979
WORK LOCATION Ames, IA	**FISCAL YEAR 1978 FUNDING** $68,000
CONTRACTING OFFICE DOE Headquarters Washington, DC	**CUMULATIVE FUNDING** $217,000

BACKGROUND

Heating of buildings is a major use of energy in farming. The use of wind power has been proposed for this because of the correlation of wind speeds to heating loads and the simplicity of retrofit. There is a need for actual performance measurements upon which to base realistic assessments of technical and economic feasibility.

OBJECTIVES

The equipment and techniques that can be used for heating rural structures with wind energy at maximum efficiency and minimum cost will be determined.

APPROACH

A wind-powered alternator with a voltage and frequency which varies with wind velocity is being used to drive an electric resistance heating system using a water storage tank as an energy storage medium. Heated water will be circulated through a heat exchanger in a controlled flow to supply structure heat. An air-to-air heat pump modified to heat the water storage medium will provide supplementary heat to this complete heating system. The wind velocity, energy output of the wind system, thermal characteristics of the heating system, the utility power requirements, heat output of the heat pump, and ambient temperatures are being recorded and analyzed to characterize the performance of the system.

OUTPUT

The performance measurements recorded will be analyzed to estimate the economic value of wind power for heating farm structures and to assess other advantages and disadvantages of this use of wind power.

FARM AND RURAL USE (SMALL) SYSTEMS

CONTRACTOR	TITLE
DOE Rocky Flats Plant Subcontractor: Aerospace Systems, Inc. 1 Vinebrook Park Burlington, MA 01803	Development of a 1-kW High Reliability Wind Machine **CONTRACT NO.** E(20-2)-3533
PRINCIPAL INVESTIGATOR Rocky Flats: Warren Bollmeier Subcontractor: Jon Zvara	**PERIOD OF PERFORMANCE** January 1978 to May 1980
WORK LOCATION Burlington, MA and Rocky Flats, CO	**FISCAL YEAR 1978 FUNDING** $280,000
CONTRACTING OFFICE DOE Rocky Flats Plant Golden, CO	**CUMULATIVE FUNDING** $280,000

BACKGROUND

The high reliability machine being developed under this subcontract is a vertical-axis, 3-bladed Cycloturbine. The design will feature cyclic and collective pitch control to permit self-startup, rotor rpm control during operation, and shutdown in high winds. The rotor will be 8 feet high and 15 feet in diameter. The system is rated at 1 kW at a wind speed of 9 m/s (20 mph).

OBJECTIVES

The objective of this project is to accelerate the development of durable, high reliability wind turbine generators in the 1-2 kW size range and stimulate their use in rural and remote applications such as repeater and seismic monitoring stations, offshore navigation aids, and remote cabins and houses. The cost goal for this system is $1500 per kW (1977 dollars), excluding tower and storage equipment.

APPROACH

The project will be accomplished in two phases. During Phase I, the contractor will perform design and analysis to finalize a design. During Phase II, three prototype units will be fabricated, checked out, and shipped to Rocky Flats for testing. The contractor will provide technical support and monitoring during these tests.

OUTPUT

The project will provide documentation and detail drawings of an advanced 1-kW system and three prototype units ready for testing. The machine will help demonstrate how wind systems of this size range can provide reliable power for remote site applications and will advance private industry's technical knowledge and capability.

FARM AND RURAL USE (SMALL) SYSTEMS

CONTRACTOR	TITLE
DOE Rocky Flats Plant Subcontractor: University of Massachusetts Department of Mechanical Engineering Amherst, MA 01003	Investigation of the Feasibility of Using Wind Power for Space Heating in Colder Climates **CONTRACT NO.** E(20-2)-3533
PRINCIPAL INVESTIGATOR Duane E. Cromack	**PERIOD OF PERFORMANCE** July 1977 to July 1979
WORK LOCATION Amherst, MA	**FISCAL YEAR 1978 FUNDING** $130,748
CONTRACTING OFFICE DOE Rocky Flats Plant Golden, CO	**CUMULATIVE FUNDING** $577,279

BACKGROUND

The use of wind power for space heating minimizes the energy storage and interface problems associated with wind power used for other applications due to the inherent energy storage characteristics of the heating medium used (water). This project is the fourth and final year of an effort to investigate this application at the University of Massachusetts.

OBJECTIVES

The purpose of this project is to determine the feasibility of using wind power for space heating in colder climates, and provide performance and economic information which can be used to design heating systems for specific sites and buildings using any appropriate commercially available wind turbine.

APPROACH

Solar Habitat I at the University of Massachusetts was equipped with a 32-foot diameter wind turbine which generates power for direct resistance heating. Experimental and operational data are being obtained and evaluated for such factors as automatic operation, system dynamics and power output. A wind field analysis is being performed. The applicability of a previously developed economic model is being improved.

OUTPUT

The project will provide a methodology for determining the feasibility of wind-powered space heating at any site. Analytic performance and economic models will be simplified for use on hand-held computers. These models will enable the determination of the suitability of a site for wind-powered heating and the design of a system to meet its requirements.

FARM AND RURAL USE (SMALL) SYSTEMS

CONTRACTOR	TITLE
DOE Rocky Flats Plant Subcontractor: Enertech P.O. Box 420 Norwich, VT 05055	Development of a 2-kW High Reliability Wind Machine **CONTRACT NO.** E(20-2)-3533
PRINCIPAL INVESTIGATOR	**PERIOD OF PERFORMANCE**
Rocky Flats: Warren Bollmeier Subcontractor: Bill Drake	January 1978 to May 1980
WORK LOCATION	**FISCAL YEAR 1978 FUNDING**
Norwich, VT and Rocky Flats, CO	$150,000
CONTRACTING OFFICE	**CUMULATIVE FUNDING**
DOE Rocky Flats Plant Golden, CO	$150,000

BACKGROUND

Enertech's high reliability machine design features a 16.4-foot diameter, 2-bladed downwind horizontal-axis rotor. A new generator (utilizing a proven design) and an innovative hub design which will pitch the rotor blades to stall in the control and shutdown mode are specified for the system. The system is rated at 2 kW at a wind speed of 9 m/s (20 mph).

OBJECTIVES

The objective of this project is to accelerate the development of durable, high reliability wind turbine generators in the 1-2 kW size range and stimulate their use in rural and remote applications such as repeater and seismic monitoring stations, offshore navigation aids, and remote cabins and houses. The cost goal for this system is $1500 per kW (1977 dollars), excluding tower and storage equipment.

APPROACH

The project will be accomplished in two phases. During Phase I, the contractor will perform design and analysis to finalize a design. During Phase II, three prototype units will be fabricated, checked out, and shipped to Rocky Flats for testing. The contractor will provide technical support and monitoring during these tests.

OUTPUT

The project will provide documentation and detail drawings of an advanced 2-kW system and three prototype units ready for testing. The machine will help demonstrate how wind systems of this size range can provide reliable power for remote site applications and will advance private industry's technical knowledge and capability.

FARM AND RURAL USE (SMALL) SYSTEMS

CONTRACTOR	TITLE
DOE Rocky Flats Plant Subcontractor: North Wind Power Company Box 315 Warren, VT 05674	Development of a 2-kW High Reliability Wind Turbine Generator **CONTRACT NO.** E(20-2)-3533
PRINCIPAL INVESTIGATOR Rocky Flats: Warren Bollmeier Subcontractor: Don Mayer	**PERIOD OF PERFORMANCE** January 1978 to May 1980
WORK LOCATION Warren, VT and Rocky Flats, CO	**FISCAL YEAR 1978 FUNDING** $260,000
CONTRACTING OFFICE DOE Rocky Flats Plant Golden, CO	**CUMULATIVE FUNDING** $260,000

BACKGROUND

The horizontal-axis North Wind design features a 3-bladed, upwind rotor 16.9 feet in diameter. It is rated at 2 kW at a wind speed of 9 m/s (20 mph). An automatic tilt-back mechanism is used to rotate the complete rotor assembly to a horizontal orientation for protection in high winds.

OBJECTIVES

The objective of this project is to accelerate the development of durable, high reliability wind turbine generators in the 1-2 kW size range and stimulate their use in rural and remote applications such as repeater and seismic monitoring stations, offshore navigation aids, and remote cabins and houses. The cost goal for this system is $1500 per kW (1977 dollars), excluding tower and storage equipment.

APPROACH

The project will be accomplished in two phases. During Phase I, the contractor will perform design and analysis to finalize a design. During Phase II, three prototype units will be fabricated, checked out, and shipped to Rocky Flats for testing. The contractor will provide technical support and monitoring during these tests.

OUTPUT

The project will provide documentation and detail drawings of an advanced 2-kW system and three prototype units ready for testing. The machine will help demonstrate how wind systems of this size range can provide reliable power for remote site applications and will advance private industry's technical knowledge and capability.

FARM AND RURAL USE (SMALL) SYSTEMS

CONTRACTOR	**TITLE**
DOE Rocky Flats Plant Subcontractor: United Technologies Research Center Silver Lane East Hartford, CT 06108	8-kW Wind Turbine Generator Development **CONTRACT NO.** E(20-2)-3533
PRINCIPAL INVESTIGATOR Rocky Flats: Joe Boland Subcontractor: M.C. Cheney, Jr.	**PERIOD OF PERFORMANCE** October 1977 to April 1980
WORK LOCATION East Hartford, CT and Rocky Flats, CO	**FISCAL YEAR 1978 FUNDING** $410,000
CONTRACTING OFFICE DOE Rocky Flats Plant Golden, CO	**CUMULATIVE FUNDING** $410,000

BACKGROUND

The 31-foot diameter, horizontal-axis, downwind rotor of the United Technologies Research Center design incorporates flex beams which enable blade-pitch control without the use of bearings. The basic concept for this design was refined during projects funded by the Federal Wind Energy Program in 1976 and 1977. The system is rated at 9 kW in a 9 m/s (20 mph) wind.

OBJECTIVES

The objective of this project is to accelerate the development of cost-competitive wind turbine generators in the 8-kW size range and stimulate their use in providing power for homes and farm buildings. The cost goal for this system is $750 per kW (1977 dollars), excluding batteries, inverter, and other secondary components.

APPROACH

The project will be accomplished in two phases. During Phase I, the contractor will perform design and analysis to finalize a design. During Phase II, a prototype system will be fabricated, checked out, and shipped to Rocky Flats for testing. The contractor will provide technical support and monitoring during these tests.

OUTPUT

The project will provide documentation and detail drawings of an advanced 8-kW system and a prototype system ready for testing. The machine will help demonstrate how wind systems of this size range can provide reliable power for residential and rural building applications and will advance private industry's technical knowledge and capability.

FARM AND RURAL USE (SMALL) SYSTEMS

CONTRACTOR	TITLE
DOE Rocky Flats Plant Subcontractor: Windworks, Inc. Box 329, Rt 3 Mukwonago, WI 53149	8-kW Wind Turbine Generator Development **CONTRACT NO.** E(20-2)-3533
PRINCIPAL INVESTIGATOR	**PERIOD OF PERFORMANCE**
Rocky Flats: Joe Boland Subcontractor: Hans Meyer	October 1977 to April 1980
WORK LOCATION	**FISCAL YEAR 1978 FUNDING**
Mukwonago, WI and Rocky Flats, CO	$295,000
CONTRACTING OFFICE	**CUMULATIVE FUNDING**
DOE Rocky Flats Plant Golden, CO	$295,000

BACKGROUND

The Windworks design incorporates a direct-drive permanent magnet alternator, a hydraulic blade-pitch control system, and free-flapping blades fabricated of aluminum and fiberglass. The 3-bladed, horizontal-axis, downwind rotor is 33 feet in diameter. The system is rated at 8 kW in a 9 m/s (20 mph) wind when connected to a utility network through a synchronous inverter.

OBJECTIVES

The objective of this project is to accelerate the development of cost-competitive wind turbine generators in the 8-kW size range and stimulate their use in providing power for homes and farm buildings. The cost goal for this system is $750 per kW (1977 dollars), excluding batteries, inverter, and other secondary components.

APPROACH

The project will be accomplished in two phases. During Phase I, the contractor will perform design and analysis to finalize a design. During Phase II, a prototype system will be fabricated, checked out, and shipped to Rocky Flats for testing. The contractor will provide technical support and monitoring during these tests.

OUTPUT

The project will provide documentation and detail drawings of an advanced 8-kW system and a prototype system ready for testing. The machine will help demonstrate how wind systems of this size range can provide reliable power for residential and rural building applications and will advance private industry's technical knowledge and capability.

FARM AND RURAL USE (SMALL) SYSTEMS

CONTRACTOR	TITLE
DOE Rocky Flats Plant Subcontractor: Grumman Corporation Energy Systems Corporation 4175 Veterans Memorial Hwy. Ronkonkoma, NY 11777	8-kW Wind Turbine Generator Development **CONTRACT NO.** E(20-2)-3533
PRINCIPAL INVESTIGATOR	**PERIOD OF PERFORMANCE**
Rocky Flats: Peter Tu Subcontractor: Frank Adler	January 1977 to September 1979
WORK LOCATION	**FISCAL YEAR 1978 FUNDING**
Ronkonkoma, NY and Rocky Flats, CO	$310,000
CONTRACTING OFFICE	**CUMULATIVE FUNDING**
DOE Rocky Flats Plant Golden, CO	$310,000

BACKGROUND

This system design has a 3-bladed, horizontal-axis, downwind rotor 33.25 feet in diameter. The system is rated at 10.5 kW in a wind of 9 m/s (20 mph) and up to 20 kW at higher wind speeds.

OBJECTIVES

The objective of this project is to accelerate the development of cost-competitive wind turbine generators in the 8-kW size range and stimulate their use in providing power for homes and farm buildings. The cost goal for this system is $750 per kW installed (1977 dollars), excluding batteries, inverter, and other secondary components.

APPROACH

The project will be accomplished in two phases. During Phase I, the contractor will perform design and analysis to finalize a design. During Phase II, a prototype system will be fabricated, checked out, and shipped to Rocky Flats for testing. The contractor will provide technical support and monitoring during these tests.

OUTPUT

The project will provide documentation and detail drawings of an advanced 8-kW system and a prototype system ready for testing. The machine will help demonstrate how wind systems of this size range can provide reliable power for residential and rural building applications and will advance private industry's technical knowledge and capability.

FARM AND RURAL USE (SMALL) SYSTEMS

CONTRACTOR	TITLE
DOE Rocky Flats Plant Subcontractor: Aluminum Company of America Alcoa Laboratories Alcoa Center, PA 15069	8-kW Wind Turbine Generator Development **CONTRACT NO.** E(20-2)-3533
PRINCIPAL INVESTIGATOR	**PERIOD OF PERFORMANCE**
Rocky Flats: Peter Tu Subcontractor: Thomas Stewart	November 1977 to December 1979
WORK LOCATION	**FISCAL YEAR 1978 FUNDING**
Alcoa Center, PA and Rocky Flats, CO	$305,000
CONTRACTING OFFICE	**CUMULATIVE FUNDING**
DOE Rocky Flats Plant Golden, CO	$305,000

BACKGROUND

The Alcoa design is for a 33 x 33.7 foot, 3-bladed Darrieus system, rated at 11 kW in a 9 m/s (20 mph) wind and up to 20 kW at higher wind speeds.

OBJECTIVES

The objective of this project is to accelerate the development of cost-competitive wind turbine generators in the 8-kW size range and stimulate their use in providing power for homes and farm buildings. The cost goal for this system is $750 per kW installed (1977 dollars), excluding batteries, inverter, and other secondary components.

APPROACH

The project will be accomplished in two phases. During Phase I, the contractor will perform design and analysis to finalize a design. During Phase II, a prototype system will be fabricated, checked out, and shipped to Rocky Flats for testing. The contractor will provide technical support and monitoring during these tests.

OUTPUT

The project will provide documentation and detail drawings of an advanced 8-kW system and a prototype system ready for testing. The machine will help demonstrate how wind systems of this size range can provide reliable power for residential and rural building applications and will advance private industry's technical knowledge and capability.

FARM AND RURAL USE (SMALL) SYSTEMS

CONTRACTOR	TITLE
DOE Rocky Flats Plant Subcontractor: McDonnell-Douglas Aircraft Corporation Box 516 St. Louis, MO 63166	Development of a 40-kW Wind Turbine Generator **CONTRACT NO.** E(20-2)-3533
PRINCIPAL INVESTIGATOR	**PERIOD OF PERFORMANCE**
Rocky Flats: E.E. Bange Subcontractor: John Anderson	September 1978 to June 1980
WORK LOCATION	**FISCAL YEAR 1978 FUNDING**
St. Louis, MO and Rocky Flats, CO	$30,000
CONTRACTING OFFICE	**CUMULATIVE FUNDING**
DOE Rocky Flats Plant Golden, CO	$30,000

BACKGROUND

The vertical-axis McDonnell-Douglas design incorporates the gyromill configuration refined under previous Federal Wind Energy Program contracts. The system is rated at 40-kW in a 9 m/s (20 mph) wind. Its rotor is 32.5 feet high by 65 feet in diameter and has a cyclic pitch-change mechanism to maintain optimum lift of the blades as the rotor revolves.

OBJECTIVES

The objective of this project is to accelerate the development of cost-competitive wind turbines in the 40-kW size range and stimulate their use in such applications as deep-well irrigation pumping and providing power to small isolated communities and small factories. The cost goal for this system is $500 per kW (1977 dollars), excluding batteries, inverter, and other secondary components.

APPROACH

This project will be accomplished in two phases. During Phase I, the contractor will perform design and analysis to finalize a design. During Phase II, a prototype system will be fabricated, checked out, and shipped to Rocky Flats for testing. The contractor will provide technical support and monitoring during these tests.

OUTPUT

The project will provide documentation and detail drawings of an advanced 40-kW system and a prototype system ready for testing. The machine will help demonstrate how wind systems of this size range can provide reliable power for irrigation, remote community and small factory applications, and will advance private industry's technical knowledge and capability.

FARM AND RURAL USE (SMALL) SYSTEMS

CONTRACTOR	TITLE
DOE Rocky Flats Plant Subcontractor: Kaman Aerospace Corporation Old Windsor Road Bloomfield, CN 06002	Development of a 40-kW Wind Turbine Generator **CONTRACT NO.** E(20-2)-3533
PRINCIPAL INVESTIGATOR Rocky Flats: E.E. Bange Subcontractor: H. Howes	**PERIOD OF PERFORMANCE** July 1978 to February 1980
WORK LOCATION Bloomfield, CN and Rocky Flats, CO	**FISCAL YEAR 1978 FUNDING** $370,000
CONTRACTING OFFICE DOE Rocky Flats Plant Golden, CO	**CUMULATIVE FUNDING** $370,000

BACKGROUND

Both electrical and mechanical output prototypes of the Kaman system will be designed. The basic system features a 2-bladed, downwind, horizontal-axis rotor 64 feet in diameter. It is rated at 40-kW in a 9 m/s (20 mph) wind.

OBJECTIVES

The objective of this project is to accelerate the development of cost-competitive wind turbines in the 40-kW size range and stimulate their use in such applications as deep-well irrigation pumping and providing power to small isolated communities and small factories. The cost goal for this system is $500 per kW (1977 dollars), excluding batteries, inverter, and other secondary components.

APPROACH

The project will be accomplished in two phases. During Phase I, the contractor will perform design and analysis to finalize a design. During Phase II, a prototype system will be fabricated, checked out, and shipped to Rocky Flats for testing. The contractor will provide technical support and monitoring during these tests.

OUTPUT

The project will provide documentation and detail drawings of an advanced 40-kW system and two prototype systems ready for testing. The machine will help demonstrate how wind systems of this size range can provide reliable power for irrigation, remote community and small factory applications, and will advance private industry's technical knowledge and capability.

FARM AND RURAL USE (SMALL) SYSTEMS

CONTRACTOR	TITLE
Aluminum Company of America Alcoa Laboratories Alcoa Center, PA 15069	Low Cost Vertical-Axis Wind Turbine Fabrication **CONTRACT NO.** EM-78-C-04-4272
PRINCIPAL INVESTIGATOR Daniel K. Ai	**PERIOD OF PERFORMANCE** April 1978 to mid-1980's
WORK LOCATION Alcoa Center, PA	**FISCAL YEAR 1978 FUNDING** $184,693
CONTRACTING OFFICE DOE Albuquerque Operations Office* Albuquerque, NM	**CUMULATIVE FUNDING** $184,693

BACKGROUND

The future commercial value of the Darrieus vertical-axis wind turbine used to produce electrical energy depends on the ability to fabricate such systems at low cost. Studies can be utilized to obtain approximate cost estimates and to determine feasibility. However, better estimates can be derived from an actual design and fabrication effort.

OBJECTIVES

A major objective of the project is to obtain realistic fabrication cost data based on current technology, for an intermediate size turbine system (approximately 100-kW rated output) designed with the goal of minimizing costs. Another objective is to provide a low-cost system design which is suitable for continued production and/or to serve as a baseline for further cost reduction efforts.

APPROACH

The best way to accomplish these objectives is to have private, fabrication-oriented industry design and fabricate the systems with DOE providing technical support in the unique aspects of structural and systems design associated with the vertical-axis wind turbine. One to four turbines are planned for production, the first of which will be installed and tested at the Rocky Flats test site.

OUTPUT

The project will generate complete design and cost estimates, low-cost turbines for testing, and actual design and cost data.

*Technical monitoring performed by Sandia Laboratories.

100-kilowatt-scale Systems

CONTRACTOR	TITLE
National Aeronautics and Space Administration Subcontractors: Westinghouse Electric Corporation Industry Service Division Pittsburg, PA 15220 Lockheed Aircraft Corporation P.O. Box 33, Ontario, CA 91716	Experimental 200-kW Wind Turbines (MOD-OA Project) **CONTRACT NO.** E(49-26)-1004
PRINCIPAL INVESTIGATOR	**PERIOD OF PERFORMANCE**
W.H. Robbins	June 1975 to April 1981
WORK LOCATION	**FISCAL YEAR 1978 FUNDING**
Clayton, NM; Culebra, Puerto Rico Block Island, RI; Oahu, Hawaii	$4,417,000
CONTRACTING OFFICE	**CUMULATIVE FUNDING**
NASA-Lewis Research Center Cleveland , OH	$11,148,000

BACKGROUND

One phase of the Federal Wind Energy Program is to develop the technology necessary for the successful design, fabrication, and operation of large, horizontal-axis wind turbine systems. The four 200-kilowatt wind turbines in this project comprise the first system under development.

OBJECTIVES

The overall objective is to obtain early operation and performance data while gaining initial experience in the operation of large wind turbines in typical user environments.

APPROACH

Utility companies were involved in the program not only to provide test sites, but to identify their requirements while gaining direct operational experience. The Clayton, Culebra, Block Island, and Oahu installation sites were chosen from among 17 utility company sites selected for further wind data collection and evaluation. MOD-OA wind turbines will be connected to the utility network and operated at each of these sites for an initial field test period of two years.

OUTPUT

The field tests will provide data and information to allow optimization of wind turbine design features for durable and economical operation. Utility operation of the machines will enable identification of electrical system stability and control requirements.

Megawatt-scale Systems

CONTRACTOR	TITLE
National Aeronautics and Space Administration Subcontractor: Boeing Engineering and Construction P.O. Box 3999 Seattle, WA 95124	Experimental 300-Foot Diameter Wind Turbine (MOD-2 Project) **CONTRACT NO.** EG-77-A-29-1059
PRINCIPAL INVESTIGATOR	**PERIOD OF PERFORMANCE**
W.H. Robbins	October 1977 to September 1978
WORK LOCATION	**FISCAL YEAR 1978 FUNDING**
Seattle, WA and Cleveland, OH	$8,841,000
CONTRACTING OFFICE	**CUMULATIVE FUNDING**
NASA-Lewis Research center Cleveland, OH	$9,904,000

BACKGROUND

Boeing Engineering and Construction has been selected to develop a 2.5-MW, horizontal-axis wind turbine generator with a rotor diameter of 300 feet. This system's features will be optimized to generate electricity at a cost that is competitive with conventional systems while operating at sites with a mean wind speed at 14 mph.

OBJECTIVES

The objective of this project is to establish the design and then determine the peformance, present cost, future production cost, operation and maintenance cost, and practicality of manufacturing a large wind turbine, with a minimum rotor diameter of 300 feet, installed at a site with a mean wind speed of 14 mph. It is also the objective of this project to demonstrate the feasibility of such wind turbines operating in a utility network.

APPROACH

NASA is managing this project which will establish a baseline design for the MOD-2, determine the cost sensitivity of the design to various configuration alternatives, and up-date cost estimates at several NASA/DOE design review stages. Following approval of the design, Boeing will fabricate and install the wind turbine(s) on selected user site(s).

OUTPUT

One or more MOD-2 systems will be built and installed for testing. A detailed final design report will be published at completion of the design phase. Design and cost information and performance models for very large wind systems will be provided.

MEGAWATT-SCALE SYSTEMS

CONTRACTOR	TITLE
National Aeronautics and Space Administration Subcontractor: General Electric Company P.O. Box 8661 Philadelphia, PA 19101	Experimental 2-MW Wind Turbine (MOD-1 Project) **CONTRACT NO.** E(49-26)-1010
PRINCIPAL INVESTIGATOR	**PERIOD OF PERFORMANCE**
W.H. Robbins	July 1976 to March 1979
WORK LOCATION	**FISCAL YEAR 1978 FUNDING**
Boone, NC	$8,622,000
CONTRACTING OFFICE	**CUMULATIVE FUNDING**
NASA-Lewis Research Center Cleveland, OH	$21,402,000

BACKGROUND

General Electric is the primary subcontractor responsible for developing a 2-megawatt horizontal-axis, two blade, propeller-type experimental wind turbine generator with a rotor diameter of 200 feet. The system has been optimized for an 18 mph average site windspeed. The system will be installed at a competitively selected site near Boone, North Carolina.

OBJECTIVES

The primary objectives of the project are to: (1) determine the operating and economic characteristics of a utility-operated, megawatt-scale wind turbine; (2) involve industry in the design, fabrication, and installation of large wind systems; and (3) involve potential users of wind systems so that institutional, operational and technical interface requirements can be clearly defined.

APPROACH

A MOD-1 experimental wind turbine generator has been designed and fabricated and will be installed at Howard's Knob which is located in Boone, North Carolina. It will be interconnected with the Blue Ridge Electrical Membership Corporation's utility grid system to provide power to their customers. The utility will operate the system for a two-year field test period.

OUTPUT

Field testing of the turbine will provide engineering and performance data for use in refining the design features of future systems and will contribute valuable information to wind energy applications and multi-unit systems studies.

Multi-unit Systems

CONTRACTOR	TITLE
Westinghouse Electric Corporation Advanced Systems Technology Division 700 Braddock Avenue East Pittsburgh, PA 15112	Design Study and Economic Assessment of Multi-Unit Offshore Wind Energy Conversion Systems Application **CONTRACT NO.** E(49-18)-2330
PRINCIPAL INVESTIGATOR	**PERIOD OF PERFORMANCE**
Lance A. Kilar	August 1976 to July 1978
WORK LOCATION	**FISCAL YEAR 1978 FUNDING**
East Pittsburgh, PA	$49,000
CONTRACTING OFFICE	**CUMULATIVE FUNDING**
DOE Headquarters Washington, DC	$556,113

BACKGROUND

Throughout much of the U.S. offshore, strong surface winds prevail due to the absence of natural obstructions and the relatively smooth surface of the ocean which exists during normal weather conditions. The high winds, together with remoteness from population centers, suggest that offshore wind energy conversion systems (OWECS) may have merit. However, the cost penalty imposed by the submarine transmission cables and the massive support structures which are needed to carry the wind turbine plants above the waves, offsets and may even overshadow the economic gains from higher unit energy production.

OBJECTIVES

The objective of this study program is to assess the technical feasibility and costs of multi-unit wind energy conversion systems sited in the U.S. offshore.

APPROACH

A meteorological and oceanographic survey of the U.S. offshore was made to specify local environmental parameters influential in the design and cost of OWECS apparatus. Designs and cost were developed for support platforms, wind turbine generator plants, substations, and electrical and hydrogen transmission systems following a uniform economic scenario. Optimum systems were developed and analyzed parametrically.

OUTPUT

The final four-volume report consists of Apparatus Design and Costs, System Analysis, and Oceanographic and Meteorological Surveys. The report assesses the OWECS concept while functioning as a design and application guide.

MULTI-UNIT SYSTEMS

CONTRACTOR	TITLE
Power Technologies, Inc. P.O. Box 1058 Schenectady, N.Y. 12301	Study of Dynamics of Multi-Unit Wind Energy Conversion Plants Supplying Electric Utility Systems **CONTRACT NO.** ET-78-C-01-3247
PRINCIPAL INVESTIGATOR	**PERIOD OF PERFORMANCE**
F.P. De Mello	September 1978 to June 1979
WORK LOCATION	**FISCAL YEAR 1978 FUNDING**
Schenectady, N.Y.	$104,700
CONTRACTING OFFICE	**CUMULATIVE FUNDING**
DOE Headquarters Washington, DC	$104,700

BACKGROUND

Additional studies are needed to explore design requirements both for the wind energy conversion system's (WECS) machine and controls, and for the interface with electrical utility systems considering the application of a significant number of these devices into a power system. The success of the integration of wind energy into utility systems will depend not only on economics but principally on the reliability and trouble-free operation of these devices.

OBJECTIVES

Power Technologies, Inc. will study dynamics of wind generators in multi-unit power system applications with the objective of verifying that proposed schemes will perform satisfactorily in the power system environment. The study will cover the modification in equipment and control logic as may be indicated, and the establishment of guidelines for specifications of the WECS equipment and its interface utility.

APPROACH

The following major tasks are underway: (1) a review of available study and test results; (2) the definition of the physics of turbine and drive train with a mathematical block diagram; and (3) the definition of disturbance and performance criteria to be used in judging the adequacy of a given design.

OUTPUT

The principal output from the study will be design information for one or more types of WECS and their interface with electric utility power systems, which will allow a significant application of these energy sources in utility systems.

MULTI-UNIT SYSTEMS

CONTRACTOR	TITLE
National Aeronautics and Space Administration Subcontractor: Western Scientific Services, Inc. 328 Airpark Drive Fort Collins, CO	Candidate Wind Turbine Generator Site Meteorological Monitoring Program **CONTRACT NO.** E(49-26)-1004
PRINCIPAL INVESTIGATOR	**PERIOD OF PERFORMANCE**
James Wagner	December 1976 to September 1978
WORK LOCATION	**FISCAL YEAR 1978 FUNDING**
Fort Collins, CO	$81,000
CONTRACTING OFFICE	**CUMULATIVE FUNDING**
NASA-Lewis Research Center Cleveland, OH	$951,000

BACKGROUND

The technology necessary for the successful design, fabrication, and operation of large wind turbine systems can be developed most effectively by operation of several prototype systems in typical user environments. Thus, the need for utility company participation is established.

OBJECTIVES

The objective of this program is to obtain sufficient wind data to accurately assess the wind energy of various sites as an evaluation factor in the selection of sites for future wind turbine installations.

APPROACH

Selection of utilities as candidates for wind turbine installations must be based on the existence of sufficient wind energy at that site. A valid assessment of wind energy can only be made after continuous collection of wind data over a long period of time.

OUTPUT

Wind data analysis, including direction and speed at two elevations (30 feet and 150 feet) at each of the candidate sites will be tabulated and reported for a minimum 18-month period.

11
Commercialization

INTRODUCTION

A goal of the Federal Government's commercialization effort is to bring new technologies into general use with the cooperation of industry and the public.

Commercialization is the last stage of development when a technology is adopted into the marketplace by the private sector. For a technology to become commercial, it must be economical, and technological and engineering problems must be resolved. Additional conditions must be satisfied, including the following:

- The public must accept the technology as an effective way of meeting a need.
- The public must believe that action is being taken to guard against possible adverse effects.
- Industry must believe that the technology can be implemented without serious problems and that it can be operated profitably.

There is a difference of opinion over how far the Government legitimately should pursue commercialization and how much of the cost it should bear. It is generally agreed that some Federal incentives should be provided to move new energy technologies into the marketplace. After all, these new approaches have an influence on a wide range of social goals.

The question remains, however, of whether the Government should be relatively passive in providing incentives to the private sector through such devices as tax credits, or if it should intervene directly through strict regulation of private industry. Both approaches are being used, but there is disagreement over whether the balance should tip toward incentives or regulations.

There is additional disagreement over how much capital the Government should provide in bringing a technology to commercialization. Some people believe that private industry should bear the cost of research and development since it will be using the technology. Others say that only the Government would be willing to undertake such long-range, high-risk efforts. The Federal Wind Energy Program will take the middle ground on this issue by sharing the cost of demonstration plants with private companies.

The following material has been prepared as modular units, to provide easy reference for discrete phases of commercialization; therefore, repetition between sections on small-scale and large-scale WECS is expected.

The Process of Commercialization

The Technology Development Cycle. For the purposes of Federal energy programs, the technology development cycle is divided into seven phases:

- Phase 1: *Basic research*–the performance of fundamental studies to advance scientific knowledge in subjects related to energy.
- Phase 2: *Applied research*–using the knowledge gained through basic research, the systematic investigation of

problems related to specific applications. (These problems may be in the physical, biological, or social sciences.)

- Phase 3: *Exploratory development*–efforts comparable in content to applied research but directed toward a particular technology.
- Phase 4: *Technology development*–efforts to establish the technical feasibility of a particular energy technology. (This includes formulating alternative approaches and developing laboratory-scale models.)
- Phase 5: *Engineering development*–efforts directed toward the design and testing of pilot plants. (Emphasis is placed on engineering and systems analysis. Major facilities may be developed for testing and improving pilot plant components.)
- Phase 6: *Demonstration*–construction and operation of a first-of-a-kind facility to exhibit the acceptability of the technology in an operating environment.
- Phase 7: *Commercialization*–the adoption and introduction of a technology into the marketplace by the private sector. (This involves the participation of private industry, but allows the Government to provide a combination of incentives and information dissemination services that are considered most appropriate for a given technology. When the technology gains market acceptance, it is expected to merge into the mainstream of the general economy without further Government action.)

Requirements for Commercialization. Commercialization is not merely something that occurs at the end of the development cycle; requirements for successful commercialization influence efforts in earlier phases. Commercial introduction of a technology requires the following:

- All major technical and engineering problems associated with the technology must have been solved.
- The economics of the technology must be such that manufacturers can obtain a fair return on their investments at prices that offer users an attractive alternative to conventional technologies.
- Users must accept the technology as effective and reliable.
- The public must be satisfied that the socioeconomic and environmental effects of the new technology have been properly analyzed, and that actions are being taken to eliminate adverse effects.

When these conditions are satisfied and commercialization proceeds, an institutional infrastructure (producers, distributors, suppliers of raw materials and industry associations) will evolve. The development of this infrastructure can be aided by certain Government actions, such as the early establishment of industry standards for system performance, subsystems and components.

Major Decisions in the Development Cycle. Within the Department of Energy, wind energy technology commercialization is currently viewed as requiring four major decisions. The major decisions concern the readiness and suitability of a candidate technology to enter the phases of technology development, engineering development, demonstration and commercialization. The progress of a technology from research through engineering development is shown in Figure 11-1, and from engineering development through commercialization in Figure 11-2. These figures also depict situations for rejecting candidate technologies and situations where technologies could be placed in reserve. A decision to resume work on a technology placed in reserve could be made at a later stage, possibly on the basis of progress (or the lack of it) in other technologies.

These major transaction decisions will be made for the Federal Wind Energy Program by the Energy Systems Acquisition Advisory Board (ESAAB). This arrangement will allow high-level DOE policy-makers to make comparisons among technologies and classes of technologies, and to select the most suitable

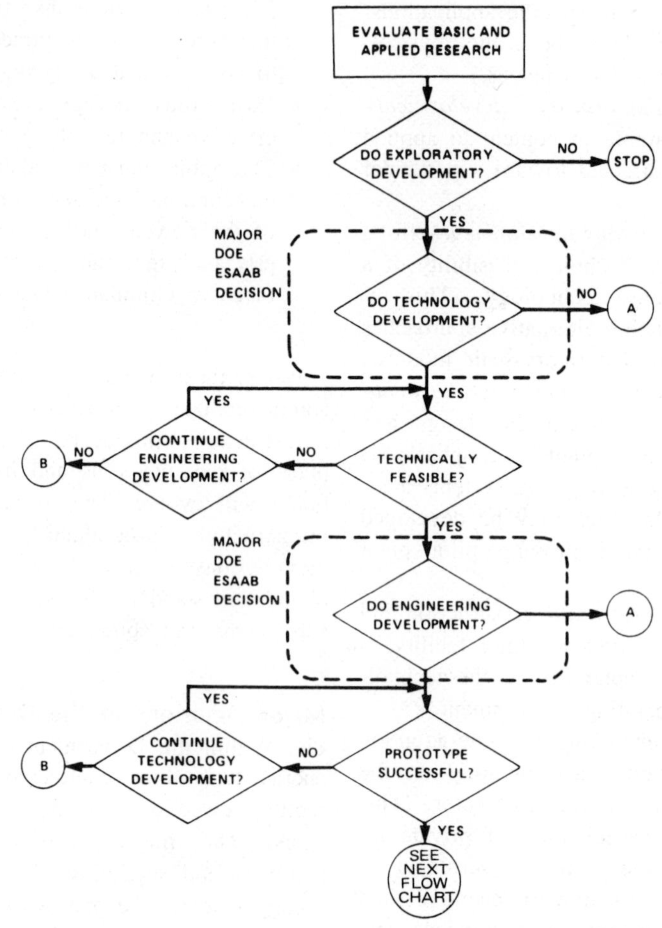

Figure 11-1. Progress of a technology from research through engineering development.

options for further development and financial support.

Commercialization of Wind Energy Conversion systems

Wind energy conversion systems can be grouped into three commercialization categories according to rated power: small-scale systems of less than 100 kW rated power, intermediate-scale systems of 100–200 kW rated power, and large-scale systems of 1–3 MW rated power and higher.

Small systems are well adapted to dispersed power markets, such as rural residences, agricultural applications and applications in remote locations. Systems of this type were initially used during 1850–1930, but they were displaced by inexpensive, centrally generated electric power. The Federal Wind Energy Program seeks to encourage the development of economical, small-scale wind energy systems and the emergence of a renewed wind system industry. Currently, there is a market for 75–125 electricity producing systems per year, with a range of 1–25 kW rated power.

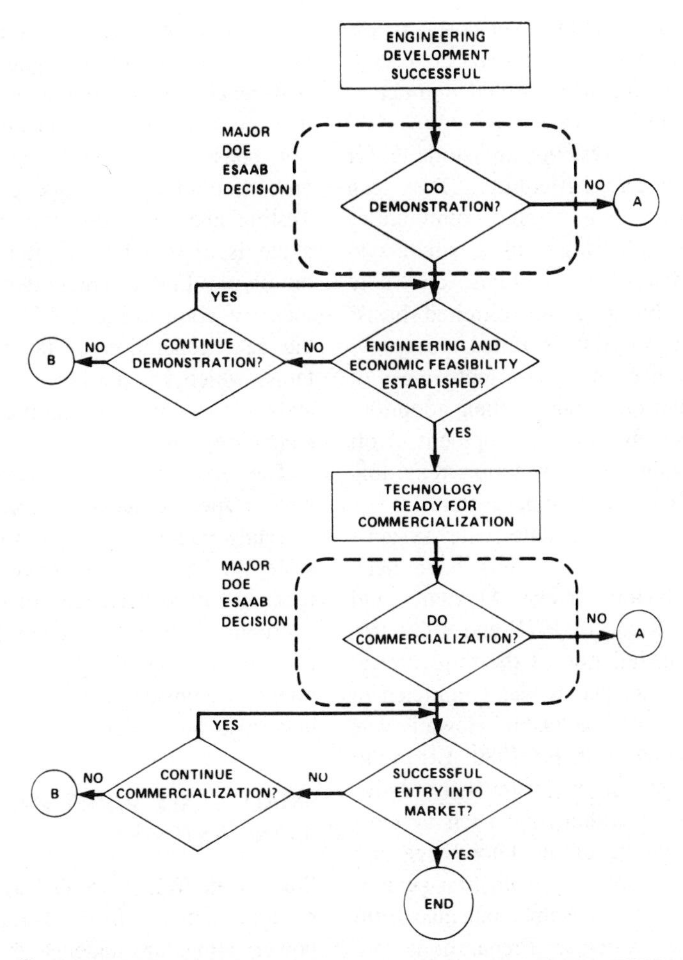

Figure 11-2. Progress of a technology from engineering development through commercialization.

Small-scale wind systems have the greatest economic advantage in remote locations, where the alternative of electric power from a utility requires the construction of expensive powerlines to the nearest point on the utility grid. For such applications, which constitute a very limited market, these systems are already more economical than conventional electric power. Other applications are expected to become cost-effective as system performance improves, system prices decrease and issues associated with utility grid backup and utility buy-back of excess power are resolved. It is possible to use on-site storage rather than grid backup to compensate for the intermittent nature of wind energy, but current costs for storage are so high that this option is, in most cases, uneconomical. (Efforts to improve storage technologies are discussed in Chapter 7.)

The first development cycle of small-scale WECS under the Federal Wind Energy Program includes units of 1 kW, 8 kW and 40 kW rated power. Prototypes of these units are being tested at the Rocky Flats Test Site. Work is also scheduled to begin on a second development cycle of small-scale WECS with improved performance and lower costs. In FY 1980, second-cycle systems are to be tested, and work is to begin on a limited number of third-cycle systems.

Intermediate-scale systems are suitable for use in irrigation and agriculture. They can also supply electricity to remote communities and small-scale industries. Unlike small systems, intermediate-scale systems are sold commercially only on a very limited basis. The Government program is aimed at improving the performance and cost-effectiveness of these systems and encouraging their adoption. As an added benefit, this development effort could provide valuable information regarding the development of large-scale systems.

First-generation intermediate-scale systems are currently being tested. Units have been installed at Clayton, New Mexico, and Culebra, Puerto Rico, in 1978 and 1979, respectively. The installation of the third unit at Block Island, Rhode Island, was completed in FY 1979, and another, at Oahu, Hawaii, was planned for completion in FY 1980. Efforts to develop second-generation systems are influenced by the consideration that such systems should take advantage of the knowledge and experience gained, not only from first-generation intermediate-scale systems, but also from early large-scale systems. Preparations are being made for an industry competition for second-generation intermediate-scale systems.

Large-scale systems offer the capability to supply electricity directly to existing power grids. Their primary application is in the electric utility industry. Units may be used singly or in groups. The use of statistical analysis and storage technologies may permit large-scale wind systems to be integrated into a utility power system so that fuel is saved and the utility is able to reduce the amount of new generating capacity required. These systems show promise for utility applications in regions that have favorable winds and now depend on oil-fired generation. These systems can also be used in conjunction with hydroelectric power generation.

A large-scale, 2.0-MW rated power WECS was installed at Boone, North Carolina in FY 1979. A more advanced system of 2.5 MW, based on the experience gained in developing first-generation intermediate-scale wind systems, is entering the final design and fabrication phase. A site was selected in FY 1979, and installation is to take place in FY 1980. Testing and evaluation will follow. This machine is expected to be the first commercial prototype. Efforts are under way to plan an industry competition for the development of the next generation of large-scale wind systems, which will be held if such systems are shown to represent significant advances in technology and cost.

The remainder of this chapter is designed to answer specific questions concerning the commercial readiness of small-scale and large-scale WECS. Barriers concerning commercial readiness have been identified which must be overcome before this technology is ready to be used commercially. Possible actions that might be considered to remove these barriers have also been identified.

SMALL-SCALE WIND ENERGY CONVERSION SYSTEMS

Small-scale WECS (SWECS) can have power ratings ranging from 1–100 kW. Typical power ratings are under 40 kW. Rotor diameters can be as large as 100 ft. Primary output is electricity; secondary output is direct heating and mechanical shaft power. SWECS are applicable to the dispersed power markets, particularly rural residences, agricultural applications and a wide variety of isolated power uses. Land and operational requirements usually dictate that single units be deployed where possible rather than groups in "wind-farms."

Current production activity is found primarily among SWECS having power ratings below 15 kW. These small wind turbines exhibit several significant design variations but may be best characterized as two- to three-bladed, horizontal-axis, variable rpm machines which produce dc or variable output ac and use inverters for 60-cycle ac applications. Larger units and a few of the smaller units maintain a constant rotation rate for power synchronization purposes. Power coefficients (i.e., a measure of aerodynamic efficiency) as high as 44 percent out of a theoretical maximum of 59 percent have been attained, and SWECS achieve an overall efficiency from wind to final output in the 30–40 percent range. Load factors (i.e., the ratio of average annual power to rated power) will vary between 20 and 40 percent for normal machines. Annual energy output is primarily a function of rotor diameter and site mean wind speed; the load factor is controlled by these paramenters and the rated power (i.e., generator size) of the wind turbine. SWECS performance-to-weight ratios and reliability should exhibit improvement through continued development, with significant economic improvements being realized as factory assembled wind turbines replace the largely hand-fabricated machines found on the market today. Future development efforts will be aimed at electricity-producing SWECS because this output form is the most versatile and represents the largest market. Mechanical output represents a relatively small market which can accept wind availability and output variations. Thermal output systems are under investigation and the economic capabilities of these systems are unresolved. Future increased emphasis on thermal output systems should not be entirely discounted.

Development History

Present Status. A significant wind turbine industry flourished in the U.S. from about 1850 until the 1930's but died at that time with the arrival of the REA and convenient, low-cost central power. The development of SWECS stopped until the early 1970's, when a number of small businesses (e.g., Enertech and Natural Power) entered the market—many, initially, as distributors of foreign machines. Additionally, several large industrial firms (e.g., Grumman) began showing interest in the development of new SWECS. The SWECS found on the market today reflect 1930's technology but lack the reliability and longevity (about 15 years later) of their predecessors. This is probably due to the limited capital and the limited engineering and market feedback available to small manufacturers at this stage. The Federal Wind Program, beginning in the mid-1970's, has brought some capital and technical support to the re-born SWECS industry. The first generation of government-financed and industry developed SWECS are under development (having 1 kW, 8 kW, and 40 kW power ratings). Compared to most solar technologies, current systems are economically viable in remote areas and require only moderate advances to begin becoming viable in the more general market.

Privately developed SWECS have achieved weak sales in the civilian environment by accessing the remote power market. Between 50 and 100 units are currently sold annually in the 1–15 kW range, mainly by foreign manufacturers (Dunlite, Lubing and Aerowatt). The U.S. SWECS industry consists of 10–15 small companies that are trying to develop and market their systems, and about 5 large industrial companies that are partially committed to production. At present, there is little SWECS industrial, distribution, maintenance, sales and installation infrastructure. A very large distribution and maintenance system is expected to develop, perhaps involving catalogue sales, as was common for SWECS in the 1930's. Although reasonably large production rates will be required, very large production facilities will not be required to meet demand. The large number of manufacturers found today will probably decrease to just a few manufacturers in the future due to cost and penetration competition.

It is difficult to forecast effective capitalization of SWECS production facilities in the face of cost, institutional and market uncertainties. However, an annual unit production rate of 1000 or even 10,000 is small compared to that found in many factory environments. Capital formation and construction of production facilities could present a significant problem in the late 1980's, when annual production conceivably could begin exceeding 100,000 units per year. However, this production level, divided among a number of companies, is not excessive. Civilian capital for developing increased production capability will be available only if a long-term market is demonstrated. Local installation, distribution and maintenance infrastructures will be difficult to build at low production rates, considering the dispersed nature of the market, but may suddenly be in great demand in the mid to late 1980's. It should be noted that several other industries have surmounted the "remote site" installation problem by developing factory-trained field teams which provide product transport, installation and on-site maintenance training. Some current SWECS marketing firms (e.g., Enertech) are currently providing this type of service. As installation is not technically complicated, catalogue-sales/farm-equipment outlets may also provide installation service. A tower erection infrastructure already exists in most urban areas, thus providing a potential labor pool.

Technical Risks/Open Questions. SWECS cost goal attainment remains an open question at this time, and considerable uncertainty exists at this early stage in trying to assess the economic capability which will be achieved after several cycles of advanced system development. Fortunately, the sensitivity of system economics to the geographically varied wind velocity and the present viability at remote sites means that this is primarily a market size uncertainty more than a technical risk. Although some engineering problems are not yet fully resolved, including utility electrical interfacing, wind load data for design, design for turn-key operation, low maintenance, technical simplicity and high reliability, there are no major technical risks which may act as fundamental feasibility questions.

Environment. SWECS do not appear to have any signficant environmental impacts, although some environmental questions are still under study. It should be noted that zoning ordinances will restrict SWECS to non-urban areas; rural and agricultural areas are primarily being considered. Consequently, SWECS environmental impacts will generally be experienced only by the SWECS user. These impacts include the following:

- *Noise*. Smaller SWECS can produce limited noise problems, but as carefully designed systems produce little noise, this is unlikely to significantly deter the owner or extend beyond the owner's property.
- *Aesthetics*. Careful site selection and design aids can reduce impacts when they occur.
- *Television interference*. TVI problems are functions of machine size. In most installations, the small systems should not encounter this problem. Mitigation could usually be achieved by revised siting or directional antennas.
- *Bird strikes, microclimate modification and small animal habitat change*. These and the other remaining environmental impacts are generally innocuous.

Technical Readiness Assessment

Operational Status. The commercial wind turbines of the 1930's could be characterized as efficient and reliable. The wind turbines available today have some engineering and reliability problems but may be considered to be fully "operational." Wind turbines with power ratings between 1 and 15 kW are currently undergoing testing at Rocky Flats to determine reliability and performance characteristics. Several brands of SWECS are in use nationwide today.

Capital Operating Cost Experience. Although costs are uncertain for the largest of the SWECS, capital costs for the smaller machines are well known. Historically, long-term operation and maintenance costs for small WECS have averaged about 1 percent of the capital cost per year.

Figure 11-3 indicates the median SWECS energy cost for currently available machines (largely hand-built machines using 1930's technology) and the sensitivity of this energy

Table 11-1. Small wind energy conversion system energy costs for existing hand-assembled machines.

	Base (V = 12 mph)	Hi-wind (V = 14 mph)	Low-loan +B (5%)	Tax credit +C
Retail conventional energy	4-6¢/kWh	4-6¢/kWh	4-6¢/kWh	4-6¢/kWh
SWECS energy cost without demand charges	15¢/kWh	11¢/kWh	8¢/kWh	7¢/kWh
Effective cost of SWECS energy with today's highest utility demand charge	26¢/kWh	19¢/kWh	14¢/kWh	12¢/kWh

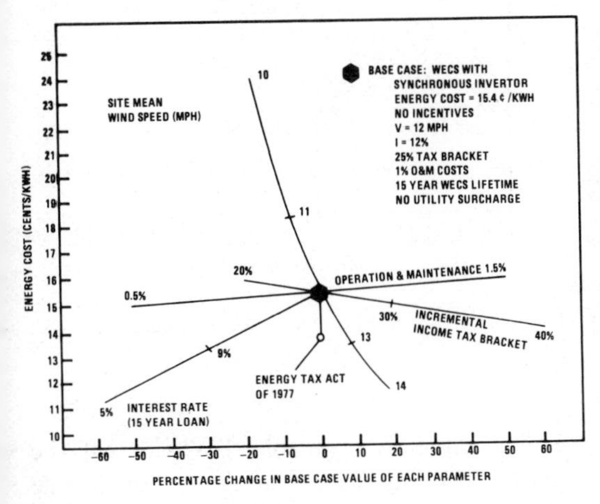

Figure 11-3. Sensitivity of wind energy conversion system energy cost.

cost to variations in major economic parameters. The "base case" energy cost, 15.4¢/kWh, is the median energy cost of the currently marketed small wind turbines under the base case conditions specified in Figure 11-3. Figure 11-3 indicates that SWECS energy costs are most sensitive to site mean wind speed and that interest rates are the second most significant economic factor. Table 11-1 compares maximum and minimum surcharge impacts on the base case energy cost as energy costs are driven toward the equivalent break-even energy cost by variations from the base case economic conditions. Typical retail energy costs for the mass market are about 4–6¢/kWh. Installed system capital costs for today's hand-built systems average about $33,000 for a 20,000 kWh/year system and about $4,500 for a 6000 kWh/year system, as described in the following paragraph.

As a point of reference, two example SWECS systems are described. These are not "typical" or even "representative" systems because efficiency, cost and capacity vary widely among present wind systems due to the technical, manufacturing and marketing immaturity of the industry. The examples provide an approximate "typical" case for current SWECS costs and are sized by rotor diameter rather than rated power (which often has little relation to output).

Diameter	*Item*	*Cost**
25–30-ft diameter (about 20,000 kWh/year @ 12 mph; about 7 kW @ 20 mph)	Turbine and controls	$20,000–35,000 fob
	Tower (60 ft)	$ 2000 fob
	Installation (site-dependent)	$ 1500–5000
	Synchronous inverter	Not utilized
	Installed system cost	$27,000–38,500 (total)
12–15-ft diameter (about 6000 kWh/year @ 12 mph; about 2 kW @ 25 mph)	Turbine and controls	$ 1600–3400 fob
	Tower (40 ft)	$ 500–900 fob
	Installation (site-dependent)	$ 300–500
	Synchronous inverter	$ 900
	Installed system cost	$ 3300–5700 (total)

The planned series of Federal wind energy development projects should decrease the base

*The current cost estimates for SWECS that are presented in other sections of this book are based on the full range of machines available today; the examples above are presented for the convenience of the reader.

case energy cost significantly. Figure 11-4 illustrates SWECS cost goals and the estimated time required to achieve the ultimate energy cost goal, set at 1–2¢/kWh, which is the actual cost goal required if full demand charges are applied. This goal will be difficult to achieve, but would allow SWECS to be an effective competitor in major markets. If demand charges are not applied, the energy cost goal can be 4–6¢/kWh, which has much less risk in its probably achievement, but represents, in effect, a subsidy for SWECS (or any other intermittent private source) by spreading its cost to the system over all subscribers.

The large band of cost uncertainty is due to the rapid startup of this program; DOE's emphasis on reliability, safety and performance engineering over that of prior commercial machines; and the fledgling state of the U.S. small manufacturers as compared to foreign manufacturers. Figure 11-4 indicates that each development cycle requires one to two years to completion, excluding time for extensive field testing. The base case conditions indicated in Figure 11-3 also generally apply to Figure 11-4.

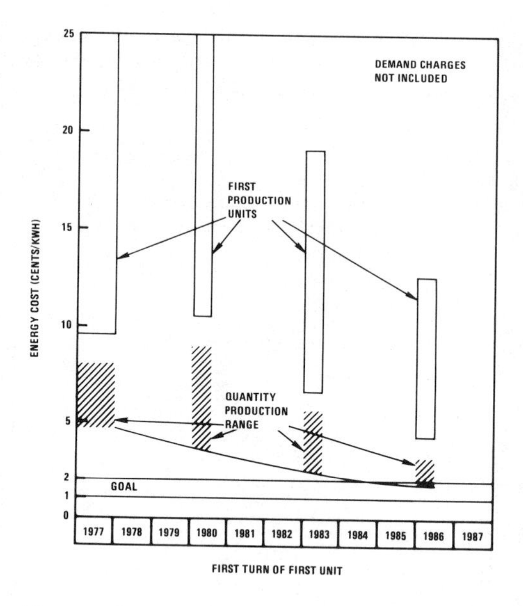

Figure 11-4. Small wind energy conversion system cost goals.

Current Developments, Tests and Demonstrations. The Rocky Flats Wind Energy Test and Development Center is the SWECS test facility. A wide range of currently marketed wind turbines are undergoing intensive testing at the Rocky Flats facility. An important result of this testing is the indication that design and fabrication maturity have not yet been fully achieved in the existing industry. An additional eight units are in test in farm applications through the USDA. A major SWECS development program is under way for the 1-kW, 8-kW and 40-kW rated wind machines. This program is expected to improve the cost, reliability, lifetime and commercial readiness of SWECS industry products through competitive, parallel, within-the-industry wind turbine design and development contracts. An institutional testing program is being developed to force clarification of important institutional questions affecting the economic viability of SWECS. As part of this program, a few currently available SWECS will be placed in utility service areas to provide rate structure data. Research and development efforts addressing key barriers will be an essential part of this program and are also a part of "applications testing" efforts.

Market/Economic Readiness

Market Description. In the U.S., SWECS are most suited to the dispersed power markets: farm, rural residential, small commercial and remote power applications. Currently, most SWECS sales activity is found among sub-kW "battery chargers" and multi-blade "water-pumpers" which have combined sales in excess of 3500 units/year (about 70 percent are water-pumpers), including imports, but which contribute little total electrical energy.

The primary SWECS market, however, requires larger wind turbines which can provide energy in a form which may be substituted for or supplement utility-supplied energy. Between 50 and 100 of these larger (1–15-kW) units are sold each year, principally to serve remote sites having good wind resources and high conventional energy costs. In the remote site application, the SWECS is usually competing with small diesel generators as a fuel saver. Diesel generated electricity, delivered, for example, to a remote site in Alaska, may cost as much as 20¢/kWh. Early sales to the remote power market (which includes irrigation) are expected to play an important part in SWECS commercialization by supporting production until costs drop to the level required for penetration of the remaining dispersed market applications (which usually use utility power).

In general, SWECS are economical in the remote power markets at this time, and, as succeeding generations of SWECS become available at lower cost, a larger market will be accessed. Each SWECS cycle will depend on the sales and service infrastructure built by its predecessor cycles until the SWECS industry is capable of supporting mass markets. Penetration of mass markets will require SWECS energy costs to drop to the 4–6¢/kWh range (without demand charges) and 1–2¢/kWh (with demand charges).

The first effective market penetration is expected to begin in the early 1980's, with significant market penetration occuring in the late 1980's. The several mission analysis estimates of the total potential SWECS market vary with a range of 22–120 gigawatts (saturation of good-wind, rural, non-urban areas). This market is regional in nature and more likely to be cost-effective in high-wind areas, including nearly all islands, portions of Alaska, the Pacific Northwest, the Great Lakes and portions of New England and (the major area) the Eastern Rockies and the bulk of the Great Plains from Texas north to Montana and east to near the Mississippi River.

Energy costs within the U.S. are below the world average, and, as a result, SWECS may be more competitive outside the U.S. The remote power market in developing countries is particularly attractive as an aid to SWECS commercialization and as a long-term export market, although design changes and some specialized development would be required to export SWECS. The World Bank estimates that remote power costs in developing countries average between 9¢/kWh and 21¢/kWh and that wide-scale electrification will not occur in many areas because energy supplied to remote sites from a grid may cost between 18¢/kWh and 40¢/kWh if new, lengthy power lines are required. The cost, in this case, is controlled by village demand and location. The largest-selling small wind turbine, the Australian Dunlite, exports a large percentage of its sales to developing countries as well as to "outback" areas of the British Commonwealth.

Potential Competing New Technologies. SWECS will be in direct competition with many new dispersed power technologies, including photovoltaic and solar thermal technologies. For many areas of the U.S., solar radiation and wind resources are roughly equivalent on a W/m^2 basis. High wind and high insolation regions are not so disjointed as to preclude competition between SWECS and other new technologies, although regional preferences may well be expected. Any system associated with utility power is an indirect competitor because it affects retail electricity rates.

Life-cycle Cost Comparison. Small-scale WECS compete with small amount of delivered energy or, in rare instances, with conventional on-site generation of similar capacity (small diesel generators). SWECS life cycle costs or energy costs must therefore be compared with, typically, the full mix of utility-owned generation equipment and part of the distribution system (i.e., the cost of energy as reflected in the utility rate structure).

Median SWECS life-cycle energy costs (not necessarily the criteria on which consumer purchase decisions are based) and the equivalent competing energy costs vary as a function of application. The base case employs a synchronous inverter or an induction generator which is interconnected with a utility grid. In this application, which presents by far the largest market, the median energy cost for commercially available, largely hand-built machines, is 15.4¢/kWh. In the base case, the SWECS life-cycle energy cost must be compared with utility energy, which almost universally costs less than 7¢/kWh even at low demand levels. The average utility energy cost in the U.S. is about 4¢/kWh in the residential sector and about 5¢/kWh in the commercial sector at average demand levels. The currently available SWECS—again, largely hand-built machines—are three times more costly than is presently acceptable in the major dispersed markets.

In applications which require short-term battery storage, an autonomous system requiring a low availability level (for example), the median energy cost increases to 22¢/kWh. In remote markets, SWECS energy costs may be compared to the expected median "life cycle" cost of diesel fuel. The expected average "farm" cost of diesel fuel over the next 15 years is about 8–13¢/kWh (assuming 6 percent escalation and 25–50 percent load factor).

The small market of applications allowing the use of intermittent energy has the lowest median life-cycle energy cost, about 12¢/kWh. In non-storage SWECS applications, the median SWECS life-cycle energy cost is beginning to break even with the cost of diesel fuel. A completely autonomous SWECS, with the high costs of long-term energy storage, is not competitive with diesel generation even when the generator itself is displaced. Exceptions occur, of course, when energy must be provided to very remote sites found in parts of Alaska, developing countries, and some portions of North America outside of the U.S. or for very small equipment power supplies.

Figure 11-3 illustrates the impact of the NEA (more specifically, the Energy Tax Act of 1977) on today's base case SWECS energy cost. For today's base case wind turbine, this act will reduce energy costs by 11 percent. The energy cost reduction generated by the Energy Tax Act of 1977 will increase to a maximum of 21.5 percent (for a $10,000 machine) and then decrease on a dollar basis as SWECS capital costs drop. This is due to the ceiling placed on the dollar value of the incentive. At today's SWECS prices, this will have little effect, but will improve market penetration somewhat as more cost-effective machines become closer to being viable in the future.

The NEA also has an impact on the institutional issue of utility surcharges by prescribing that "equitable" rate structures be developed by each utility within two years of passage of the NEA. The ultimate impact of the utility surcharge problem is potentially much more significant than the Energy Tax Act of 1977. Incentives which affect the cost of borrowing capital, as illustrated by Figure 11-3, also significantly affect SWECS energy costs.

Market Penetration. Although estimates of overall potential market size are available, SWECS market penetration studies are incomplete, and sufficient data are not yet available to allow penetration estimates without considerable uncertainty in the assumptions and outputs. Consequently, the estimates presented are based on "best judgment" and the only related "real world" energy market precedent available. Table 11-2 indicates various cost factors and market penetration as a function of time. The market penetration scenario used to develop the estimates is highly speculative and assumes that, if the cost goals presented in Figure 11-4 are met, SWECS market penetration will reach 1000 units/year late in 1982 (compared to 100 units/year) and that SWECS penetration will occur at the same rate as experienced by flat plate solar collectors. Flat plate solar collectors were selected as a model due to the similarity of the payback periods, technical risks, scales of investment and energy applications. The scenario further assumes that production growth will level off in 1987, and the cumulative total will reach 1.6 million units at the turn of the century in accordance with mission analysis projections. Production is initially expected to grow exponentially in this manner, and then level off (at about 100,000 units/year) as the easier high-wind markets are sat-

Table 11-2. Small wind energy conversion system market overview.

MARKET	1985	1990	2000
CAPITAL COST			
o TOTAL INVESTMENT[1] (1978 $ MILLIONS)	642	4529	9426
o CAPITAL PRODUCTIVITY (1978 $/MMBTU/YEAR)	64	41	29
COST TO END USER[2]			
o (1978 $/MMBTU)	17.6	8.8	5.9
o $/KW	1200	750	500
o c/KWH	6	3	2
EXPECTED MARKET PENETRATION			
o TOTAL MARKET[3] (QUAD)	2.5	2.5	2.5
o PENETRATION (QUAD)	.01	.11	.32
o PERCENT OF MARKET PENETRATED	.4	4	13
o APPROXIMATE NUMBER OF MACHINES	50,000	500,000	1,600,000

[1]CUMULATIVE INVESTMENT IN MILLIONS OF DOLLARS.

[2]12 MPH, 25 YEAR LOAN @ 9%, 25% TAX BRACKET, FOR END USER 1 c/KWH = 2.93 $/MMBTU.

[3]TOTAL FARM AND RURAL MARKET FROM GENERAL ELECTRIC WIND ENERGY MISSION ANALYSIS. WE ESTIMATE, DUE TO THE NATURE OF THE RURAL/FARM MARKET, ZERO GROWTH TO 2000 AD, the G.E. MARKET ESTIMATE DATE. NOT OIL EQUIVALENT.

urated and SWECS are forced to compete under more difficult conditions.

Market Barriers

The principal barriers to market penetration are as follows:

- The cost of currently marketed SWECS.
- Uncertain lifetime and reliability among existing SWECS.
- Absence of a marketing installation and service structure.
- Absence of marketing data.
- Lack of public familiarity with wind systems, their applications and siting needs.
- Technical risk for the consumer.
- Uncertainty in standards regarding both the protection for the consumer, and as a cost and deterrent to the manufacturer.
- Uncertainty of utility surcharges.
- Limited information on wind resource variations.

Cost reduction efforts are a major SWECS industry/DOE activity. Reliability and lifetime problems should be overcome by the ongoing SWECS development program. The absence of marketing, installation and service structures within the SWECS industry will present problems until production and sales are capable of supporting "local distributors." In the marketing area, specifically in farm hardware, a few large retail sales outlets are beginning to consider suitable SWECS product lines. Sears® and Montgomery Ward® sold SWECS in the 1930's and this same farm machinery marketing distribution system could be utilized today. Installation and service structures will be required initially in response to product unfamiliarity. It is anticipated that future generations of SWECS will be easier and less costly to install and will require infrequent service. Manufacturers and distributors are expected to be the main source of applications and siting needs; they are also a source for SWECS installation. The impact of utility surcharges in any one area should be identified within the next few years by the institutional testing program. Capital cost and utility surcharge issues currently appear to be the most significant barriers.

Environmental Readiness

Environmental Compliance. There appear to be no environmental standards which are applicable to SWECS with the possible exception of noise regulations, although properly designed systems are not particularly noisy. In a rural environment, SWECS-produced noise is not likely to cause problems beyond the owner's property because SWECS have a modest noise output. SWECS are likely to be excluded from high population density areas for institutional reasons (zoning, clearance zones, etc.). Television interference (TVI) problems can occur with the largest SWECS, thus limited siting opportunities near residential areas.

Potential Environmental Barriers. SWECS have no significant environmental impacts. However, standards for TVI could potentially have impacts on SWECS siting practices. The area affected by TVI may extend as far as a third of a mile for a 100-kW-scale wind turbine, although the smaller SWECS are not reported to have a TVI problem. Aesthetic "standards" issues, including occupational safety, may also present an "environmental" barrier.

Institutional Readiness

Institutional Barriers. Institutional barriers to SWECS commercialization are found in the insurance industry, the electric utility industry and regulatory needs (certification and standards). Due to the newness of the SWECS manufacturing industry, product liability presents a potential barrier to near-term commercialization. The ultimate impact of product liability problems appears as a possibly severe cost increase. Means for removing this barrier include government liability support (under study) and the general development of the SWECS industry and its product insurance industry confidence. DOE is sponsoring development of SWECS which will incorporate the safety and reliability features required of commercial products. Liability insurance is a significant short-term barrier, but not a long-term barrier.

Utility surcharges are employed by utilities to recoup the cost of providing standby service and the effect of an intermittent source on their system. At the present time, surcharges vary greatly in different areas and act as deterrents to the use of solar energy in some utility service areas. They are a state and local issue as well as a Federal policy question. Although the NEA contains provisions for development of "equitable rates," it is unclear whether this will lead to exact accounting of charges to each customer or dividing the effect of solar systems among all customers. Utility surcharges are likely to have a significant impact on costs of energy from solar equipment, potentially causing cost goals to be affected by a factor of two or more where utility backup is needed, which is the major market. An institutional testing program is being developed to address this problem, but the full impact will not be known until 1980. Utility surcharges are not an adversary issue. Significant surcharges result from the fact that only a portion (roughly one-third) of the cost of electricity is due to the cost of fuel, and that the SWECS user can only displace fuel, not the cost of generation and distribution equipment and overhead. "Capacity credit," an economic credit which results from the statistical probability that the SWECS will produce power during a period when the utility could not otherwise meet demand, is not expected to contribute significantly to SWECS energy cost reductions and is controlled as much by weather and system design as by utility policy. "Peak shaving," another means to gain economic credit by reducing utility capacity requirements, is not attractive due to wind resource variations.

Certification and standards setting occur in many industries and have a significant role in improving the public acceptance of products. Standards setting requires a careful blend of technical requirements and rules which do not cause heavy cost penalties to industry and consumers. DOE, the American Wind Energy Association, manufacturers and existing standards setting bodies are beginning to examine the certification and standards role. Progress in the develoment of performance, test and product information standards, while having a difficult schedule, should not cause more than modest delays. Product standards, however, may be a critical issue. One does not want poor products entering the market on an excessive scale. Conversely, premature setting of excessive product standards (e.g., fatigue load requirements) and test requirements for them could easily drive the price (and many manufacturers) out of competition. Given the time and difficulty expected to obtain data to establish reasonable product standards, it appears that only performance disclosure standards are practical in the near future. Quality must be determined by the consumer and warranties used to drive poor quality systems out of the market. Any attempts to impose product-type standards in the next few years could be a very major deterrent.

Manufacturer Status. Small manufacturers are currently marketing SWECS and are making an effort to expand the market. Large manufacturers have built prototypes and want to market SWECS if risks can be reduced. There is no evidence to suggest that the SWECS industry has a long-term planning horizon in anticipation of long production runs; however, as with any new market, there is substantial market uncertainty and considerable economic risk. Component suppliers are similarly cautious about developing specialized SWECS components for the OEM market. Early market activity in response to phased demonstrations, and the ongoing market studies supported by the Federal Wind Energy Program, should help refine the long-term market picture. A refined picture is important if manufacturers are to do the preproduction planning and tooling which will lead to sharp initial reductions in production costs.

Infrastructure Needs. Significant infrastructure problems could develop if growth is rapid and a long-term market has not been developed. As noted earlier, it is difficult to forecast effective capitalization of SWECS production facilities in the face of cost, institutional and market uncertainties. However, a unit production rate of 1000/year or even 10,000/year is quite small compared to that found in many factory environments. Capital formation and construction of production facilities could present a significant problem in the late 1980's, when annual production conceivably could begin exceeding 100,000 units/year. Divided among a number of companies, however, the number is not excessive. Civilian capital for developing increased production capability will be available only if a long-term market is demonstrated. Local installation, distribution and maintenance infrastructure will be difficult to build at low production rates, considering the dispersed nature of the market, but may suddenly be in great demand in the mid- to late 1980's. It should be noted that several other industries have surmounted the "remote site" installation problem by developing factory-trained field teams which provide product transport, installation and on-site maintenance training. Some current SWECS marketing firms are providing this type of service. Because installation is not technically complicated, catalogue-sales/farm-equipment outlets may also be capable of providing installation service. A tower erection infrastructure already exists in most urban areas, thus providing a potential labor pool.

It appears that Federal support could be needed in the infrastructure development area in the early to mid-1980's. Because infrastructure development will require longer time periods than that required to introduce improved SWECS, infrastructure problems will occur until "easy" markets are saturated or until large retail chains adopt and support SWECS products.

Benefits Analysis

Energy Impacts. Table 11-2 described total market penetration and total energy generated, in quads, for SWECS as a function of time. The primary form of SWECS output is expected to be electricity, which will displace utility fuels. SWECS will displace a statistical mix of peaking, intermediate and base load fuels, and the ratio of fuel types will depend on the particular region. In general (except in mountainous areas where the diurnal wind variations lead to some high night winds), mostly intermediate and peaking fuels will be saved. For some agricultural operations, natural gas, LPG and oil will be the primary fuels displaced. There will be a reduction in the utilities plant requirements due to the statistical availability of dispersed wind sources (on the order of 20–30 percent of the wind system's rated power), but it is unlikely that the utility planning economics can effectively consider this variable. It is noteworthy that the energy payback for SWECS is very rapid, about one year.

Recipients of Benefits. Due to the dispersed application of wind systems, the rural farmer and rural homeowner are expected to be the primary beneficiaries in terms of dollar savings, if not fuel savings.

In terms of sector fuel savings, farm applications are expected to receive the greatest benefit from SWECS. Self-generation benefits can be substantial in industries (including farms) which do not have priority access to shortage fuels in emergency situations, and can be more significant than fuel savings where crops or sales can be lost due to shutdowns.

In the utility-connected environment, potential fuel savings can be increased if utilities are willing to pay for surplus energy production, which occurs when SWECS are sized to provide a large portion of the user's power needs. (Note that, in a utility-connected system, the SWECS can place excess electricity on the grid, using the resulting utility fuel savings as a storage medium.)

Cost Impacts. First-year costs are high for SWECS. Funds for the initial purchase will undoubtedly come from lending institutions, and, in the case of demonstrations, some portion of the funding will be provided by the Federal Government. Incentives, financed by general revenue, may also play an important part in early SWECS financing. The impact of SWECS costs can be significant; first-year energy costs, even with a long-term, low-interest loans, will be considerably higher than the conventional cost of energy. The high cost of SWECS will require that users adopt life-cycle costing for energy purchase decisions.

Other National Benefits. SWECS commercialization will have a positive impact on the balance of trade through a reduction in oil import requirements as well as through potential export of SWECS equipment and technology to the energy-short developing countries. Additional benefits include improvement of job opportunities in the semi-skilled labor area, reduction in the use of polluting fossil fuels, reduction in the impact of temporary fuel shortages or curtailments, and rapid energy payback.

Readiness Assessment

The evaluation of the timeliness for the development of SWECS is delineated below.

- SWECS are ready for commercialization and are already competing in some markets without Federal support.
- SWECS present a low-risk, low-cost commercialization opportunity and allow considerable commercialization flexibility.
- Commercialization planning has begun in anticipation of the next SWECS generation readiness.
- Additional R&D cycles are needed to reach mass markets, but a significant remote market penetration can occur in the next few years.
- Early mini-demonstrations, requiring close monitoring, can help the industry prepare for major commercialization by resolving institutional issues; improving management expertise within the industry; providing growth capital; and allowing development of a preparatory sales and service infrastructure.
- SWECS have high public visibility and may be demonstrated in easily managed cycles at relatively low cost. Demonstrations will also provide an early and relatively large "guaranteed market" (considering the size of the existing industry).

Small-scale wind systems are close to being viable in many isolated, remote areas and await primarily the production of tested, U.S.-made units. More advanced systems can become cost-effective in the broad Midwest markets and similar rural and agricultural areas, if excessive utility demand charges are not applied.

Technical risks are low, but market size is uncertain. Individual system costs are relatively low and the time to develop more advanced systems is also low. The time for industry and the diverse market to develop, and for local issues to be resolved, can be longer than the development cycles.

Therefore, small-scale wind systems are

being considered for early commercialization, and a program is being developed which provides for, and is committed to, sequenced demonstrations and incentives, but is contingent upon defined cost goals being achieved. In this way, enough market visibility can be supplied to encourage the efforts of manufacturers and users, while allowing for major expansions or contractions in future years, as warranted by the achievement of goals.

Federal Commercialization Strategy

In the case of small wind systems, the mix and sequence of commercialization actions is strongly influenced by the characteristics of the current industry and the way in which it interacts with the market.

The Federal commercialization approach incorporates a blend of three basic actions: development, demonstration and incentives. Technology development must take the lead in small WECS commercialization to overcome the high cost and reliability unknown in the existing commercial machines. As safe, reliable, long-life, low-cost machines become available (the first development cycle is under way), demonstrations will be introduced. Demonstrations are a key element of the program, but will be introduced primarily as a means to provide user information and acceptance and to support the industry during the market formation period, not as a long-term commercialization lever. The most effective long-term commercialization lever appears to be the phased introduction of incentives. Incentives, "promised" early in the commercialization process, will be a powerful means to resolve market uncertainty problems and allow capital formation with the industry. Once visible, however, incentives could also destroy the limited market which is now supporting the industry, since consumers would delay purchase decisions in anticipation of incentives programs. Demonstrations will support the industry and provide data for user acceptance during this period and will be phased out as incentives begin to accelerate sales. Incentives will therefore become the most visible and long-lasting commercialization tool. Incentives programs, like the preceding technology development and demonstration program, will later be phased out to ensure a gradual transition to a self-supporting small wind system industry.

The most effective incentives program appears to be one in which both manufacturers and potential users can directly benefit. The American Wind Energy Association was particularly active in the construction of NEA tax credit provisions directed at potential users of small wind systems. Manufacturers are already being aided by development funds, and R&D is the most effective means to achieve initial cost reductions and is essential to improved product marketability. Market studies, a less direct aid (but one which is being pursued as a near-term benefit), will help manufacturers by defining market potential and risks which must be understood to achieve capital formation. Still, these aids and incentives are not sufficient to overcome "up-front costs," which is a problem to manufacturers, who must obtain tooling and marketing capital, and to users, who must still purchase a largely unknown product with lender financing. At this time, and subject to future review, long-term, low-interest loans appear to be the most attractive long-term commercialization lever. This option, in particular, helps the initial user's cash flow in the economic comparison between wind and conventional energy sources. Other options, (e.g., large demonstrations) may have undesirable impacts.

The demonstration action, including institutional testing, must be carefully sized relative to the production capability of the industry and allow an orderly transition to incentives. Phased demonstrations, and the introduction of incentives, should provide an economic buffer which will allow maximum effective growth whille isolating the industry from the economic shocks which can result from irregular Federal procurements. A closely monitored, moderate (roughly 500-unit) scale demonstration can provide needed growth and experience to what is presently a relatively small industry. A smaller-scale demonstration might not prepare the industry for the demand generated by incentives (which would make second-generation SWECS attractive in windy, high-energy-cost areas). A larger or premature demonstration could encourage expansion which could not be sustained by the market. Very large demonstrations could also drive the smaller manufacturers out of the market and commit the engineering, financial and product support capability of the industry to premature machines, which cannot penetrate major markets. Demonstration timing may be more critical than size. The commercialization strategy provides several review points to allow for adoption of more attractive commercialization options, should they appear.

In the long term, major Federal spending for small wind systems commercialization could be most appropriately targeted at incentive programs to promote normal industry growth. It is noteworthy that public interest, as evidenced by the "inquiry rate," is very strong, even though few demonstrations have taken place. Consumer interest wanes with realization of wind system costs, but it would be encouraged by consumer incentives which affect system economics.

Commercialization Profile. The commercialization strategy incorporates a full set of Federal actions, driving non-federal government and private sector actions, which together deal with the key barriers that inhibit a rapid development of the market.

It should be noted that the barriers are not wholly independent; the primary link is system cost. Therefore, the actions which affect the cost-production cycle will impact several barriers. Each barrier and action combination is presented in the commercialization profile matrix provided in Table 11-3. The effectiveness of each action in overcoming each specific barrier is indicated by a 1–5 score. A score of 1 indicates no effect and a score of 5 indicates that the action virtually eliminates the barrier. It should be noted that where individual barriers and actions appear to be similar between the large wind system and small wind system commercialization profiles, the individual effectiveness scores may vary. This variation is due largely to differences in the markets (i.e., small wind systems will serve primarily dispersed, technically unsophisticated markets, while large wind systems will be primarily used by utility companies).

Federal Strategy. The Federal commercialization strategy incorporates three steps—development, demonstration and the introduction of incentives—whose mix and emphasis vary with time. Each step may involve one or more cycles, as needed, and may overlap with adjacent steps. Due to the relatively low cost of SWECS development, the relatively modest cost of an individual machine and the associated short development and procurement times, the suggested commercialization approach should allow considerable flexibility and responsiveness within the commercialization process. Key features of the three steps are outlined below. It should be noted that the first cycle of the development program is under way. Each step has several built-in decision points which allow for overall program flexibility and low program risk. The first steps are currently under way; these are the development steps noted below.

Development

- Development is performed directly by industry to preclude technology transfer problems. DOE also provides technical support through development of information to support small manufacturers (e.g., simplified computer models, wind loading statistics and testing support) and develops and supports distribution of information to potential users.
- Parallel development is necessary within each turbine size range to provide competition (and examine alternate technical approaches).

Table 11-3. Commercialization profile for small-scale wind energy conversion systems (all markets).

BARRIERS		TECHNICAL		ECONOMIC			INITIAL DEPLOYMENT					ENVIRONMENT	INSTITUTION			
		UNCERTAIN RELIABILITY & LIFETIME	INDUSTRY TECHNICAL IMMATURITY	SITE SPECIFIC COSTING	ALTERED FINANCIAL PLANNING	HIGH COST	AWARENESS & ACCEP-TANCE	FRAGMENTED/DISPERSED MARKET	INTRASTRUCTURE	PRODUCT LIABILITY	MARKET UNCERTAINTY	TVI, NOISE, AESTHE-TICS, SAFETY	UTILITY SURCHARGE ISSUES	CODES: TESTING & LABELING	"PRODUCT"	ZONING & BUILDING
BARRIER IMPORTANCE		3	4	2	2	4	2	2	2	3	3	1	5	3	2	2
ACTIONS	SECTOR															
INFORMATION																
TECHNOLOGY DEVELOPMENT	F	5	5	2	1	5	2	2	2	4	3	4	3	5	4	3
DEMONSTRATIONS	F	4	4	3	4	4	5	4	4	3	4	4	3	4	3	4
MASS MEDIA/INFORMATION CENTERS	S	2	2	4	4	2	5	4	3	3	4	5	1	4	4	5
SITING AND ECONOMIC TOOLS	FS	1	4	5	4	4	4	5	3	1	3	3	1	4	1	3
MARKET STUDIES	F	1	2	4	2	4	3	5	3	1	5	3	1	1	1	3
APPLICATION TESTING	F	4	4	4	4	4	4	4	4	3	5	4	3	4	4	4
REGULATION																
STANDARDS (CONSENSUS)	P	3	3	2	2	2	3	2	3	4	3	4	2	5	5	5
RATE RECONSTRUCTING	FS	1	1	3	4	5	3	3	2	1	4	1	5	2	1	1
WARRANTIES	P	4	2	3	3	2	4	3	3	3	3	3	1	3	5	3
IMPORT QUOTAS	F	1	1	1	1	1	1	1	3	1	2	1	1	1	1	1
FINANCIAL INCENTIVES																
LOAN GUARANTEES (USER/MFG)	F	1	2	2	3	3	3	2	4	2	2	1	1	3	2	2
LONG-TERM LOW INTEREST LOANS	F	1	3	1	3	4	3	2	3	2	2	1	2	2	1	1
GRADUATED PAYMENTS LOANS	F	1	2	1	3	3	3	2	2	1	2	1	2	1	1	1
ACCELERATION DEPRECIATION	FS	1	2	1	3	4	2	2	3	1	2	1	2	1	1	1
FEDERAL INSURANCE	F	1	2	1	2	3	3	2	3	5	2	1	1	2	4	3
GOVERNMENT PROCUREMENTS	F	3	3	2	1	3	2	2	3	3	2	3	2	4	4	2
TAXES																
TARIFFS	F	1	1	1	1	1	1	1	1	1	2	1	1	1	1	1
FUEL SURCHARGES	FS	1	2	1	2	4	3	3	3	1	3	3	4	1	1	1
TAX CREDITS	FS	1	2	1	2	2	2	2	2	1	2	1	2	1	1	1

KEY: P Private
L Local
S State
F Federal

- Several development cycles should be initiated to improve systems, meet goals and access larger markets.
- Additional DOE support is needed, in terms of information for development of standards, market studies and data, applications testing, site wind prospecting techniques and public information. Providing testing services to the small manufacturers (similar to the role of wind tunnels in aviation) is included.
- Support the removal or resolution of institutional barriers such as uncertain product liability insurance and utility demand charges.

Demonstration

- Demonstrations should be open to any systems meeting minimum criteria and not limited to Federally-sponsored development.
- Very large demonstrations, employing existing machines, are specifically avoided by the commercialization strategy because existing machines do not meet safety requirements or have the materials and fabrication sophistication required to achieve the "in production" cost goals.
- Provide some early firming of the market to allow manufacturers to raise capital

and support the early lack of marketing and distribution structure.

- Provide visibility to develop user interest and acceptance and allow utilities to confront the wind technology and prepare a mean to incorporate SWECS into rate structures.
- DOE turbine selection should be made in early procurements to ensure that several high-quality brands survive to provide long-term competition, and to preclude user-aimed DOE program opportunity notices from overwhelming the small industry with user information requests.
- User selection should be made in later cycles to force competition and development of industry marketing and maintenance infrastructure.
- Demonstrations should be employed to support initial high-risk, high-growth expansion and to provide a sufficiently firm "guaranteed" market to allow private raising of capital.

Incentives

- The primary immediate need is for some visible, if future, commitment to provide industry with some assurance that a market will exist and to allow private industry planning and capital formation.
- Studies under way at this time are evaluating the relative leverage and impact of different forms of incentives for different market segments.
- Low-interest loans to the consumer are expected to have the highest leverage per unit cost. They also have the advantage of being limited, in that the extent of Federal cost can be defined, and increased or decreased at specific decision points as the economic and market developments warrant. They can also be used to relieve industry cashflow problems—but they could, if used as a major incentive, inhibit competition and infrastructure development.
- Tax credits, as in the present NEA, will provide additional incentives as the cost of wind systems begins to approach competitiveness.

Federal Goals. Progress of the commercialization program is effectively marked by "mature product" energy costs. The mature product energy cost of today's commercial machines, if produced in very large quantities, is estimated to be, at best, 5¢/kWh based on average industrial parametric cost experience. A goal of 3–4¢/kWh should be achieved (perhaps by 1980), and, neglecting utility surcharge issues, might be sufficiently low to promise effective commercialization. A mature product energy cost goal of 1–2¢/kWh, a much more difficult goal to reach, has been set to compensate for the high cost of utility backup which is needed for low wind periods if (and where) demand charges are applied and for a large moderate wind market. This energy cost goal assumes a 20-year system life and the "base case" conditions listed in Figure 11-3.

Small wind systems are not expected to make a significant contribution to national energy independence before 1985 (contributing about 0.01 quads). The 2¢/kWh goal should be reached in the later 1980 time-frame, as increasing production drops the energy costs of small wind systems. At this point, incentives will become less important and the industry should begin to be self-sustaining, reaching a penetration of 0.3 quads (13 percent of the small systems market) by 2000. The 2¢/kWh goal is designed to make SWECS "break even," and, when achieved, will not give SWECS a substantial cost advantage over conventional fuels unless some form of incentive is available to consumers.

Federal Actions. The set of Federal actions which is believed to be the most effective means to achieve effective commercialization of small wind systems is presented below with appropriate major task headings. Many of the listed activities are already in progress. The development, demonstration and incentives tasks make up the "critical path"; that is, they control the commercialization schedule. The support tasks provide essential activities which contribute to the basic development-demonstration-incentives strategy. The

Federal activities which evolve from the suggested commercialization strategy include the following:

Primary Tasks

- Research, development and applications testing
  - Four development cycles
  - Applications testing
- Demonstrations
  - Three demonstration cycles
- Incentives
  - Initial incentives study
  - Incentives policy analysis
  - Recommend incentives package
  - Implement incentives to manufacturers
  - Implement incentives to consumers
  - Implement rate structure-related incentives
  - Implement international marketing incentives

Support Tasks

- Information planning
  - Initial venture analysis
  - Detailed venture analysis
  - Liability insurance: policy and actuarial analysis
  - Preliminary information dissemination system in support of industry, consumers and utilities
  - Develop information dissemination system involving Federal and state agencies
  - Wide-scale user awareness and dissemination program
  - Development of siting and economic analysis tools
- Standards
  - Performance/test/label/nomenclature standards development
  - Planning and analysis activities
  - Warranties and guarantees
  - Safety and product testing standards
- Market studies
  - Initial remote market study
  - Initial agricultural applications study
  - Interconnected market study
  - REA market study
  - Regional and sector (including Federal) market studies/strategy development (multiple)
- Rate structures
  - Institutional testing program (multistate)
  - Economic and planning impacts of dispersed wind systems on utility systems
- International program
  - Initial international market requirements/criteria study
  - International wind and load data support
  - Preliminary support of industry export needs (studies, visibility, trade shows, testing)
  - Analysis of possible Federal actions in support of international commercialization
  - Expanded international market studies
  - Develop export-related machine modifications
  - Decision to implement international market commercialization

LARGE-SCALE WIND ENERGY CONVERSION SYSTEM

Large wind turbines, in the range of 100 kW (100–125 ft in diameter) to 3 MW (200–300 ft in diameter), are being developed to feed electricity directly into existing power grids. The primary application is in the electric utility industry, with the intermediate-sized units applicable to small communities, agriculture, and some remote uses. Multiple units, either grouped as "farms" or dispersed across a utility network, would be used to obtain higher power levels. Based on the statistical distributions of wind energy and electrical demand, wind energy systems may be used both to save fuel and to provide some capacity credit in utility operations. High-potential applications include high-wind, high-cost, oil-using regions and hydroelectric systems to generate additional power or to save fuel and water. Large wind turbines are generally considered to be reliable and safe; partly because of the extensive engineering effort devoted to these issues in the design process.

The conventional technology employs propeller-type blades with a horizontal axis, although alternate system concepts, including the Darrieus "egg-beater," are being investigated. Technical feasibility has been shown for individual horizontal-axis machines up to 125 ft in diameter. Remote, unattended and automatic operation has been demonstrated. Economy of scale, land costs and transmission costs are expected to drive "farms" toward use of the largest available units.

Development History

Present Status. There has been an almost complete lack of activity for 30 years in the U.S. Exceptions include the Federal Wind Energy Program began under the auspices of the National Science Foundation in 1972, some work on small-scale systems and two recent private ventures at the intermediate-size level. To provide early hands-on experience, the DOE MOD-0 prototype machine (125-ft diameter, 100 kW) began operation in October 1975 and is currently undergoing long-term tests. The first unit of the improved MOD-OA (125-ft diameter, 200 kW) has demonstrated operability and provided test data at the Clayton, New Mexico municipal utility. Second and third units of MOD-OA were also installed at other sites. The MOD-1 (200-ft diameter, 2 MW) was fabricated for testing in FY 1979; the MOD-2 (300-ft diameter, 2.5–3 MW), the first technological "second-generation" system, is in the advanced design phase. Design projects for additional second-generation systems at the 100-ft (MOD-4) and 200-ft (MOD-3) range were planned to start in FY 1979.

Several large organizations (GE, Boeing, Westinghouse, Grumman, etc.) and several small concerns are involved in the present development program and have indicated serious consideration of a possible product line. These organizations see a potential market in the private sector in addition to current government participation. Electric utilities have begun to show a serious interest, particularly in high fuel cost/high-wind areas. In the last competition for four test machines, 65 utilities proposed sites. According to the Electric Power Research Institute (EPRI), with the exception of heating and cooling, more utilities are involved in projects related to wind than with any other solar technology. Southern California Edison has purchased a Schachle privately designed machine for test purposes.

Technical Risks/Open Questions. No major feasibility questions are involved. Some technical issues related to fatigue loads and lifetime, electrical stability, operational interface and control require further resolution and verification. At this early stage, development problems will undoubtedly occur. Operations and maintenance costs are not well understood, although they should be low after experience is built up, due to the moderate complexity of the systems. Safety and reliability issues are being treated in the design process, and the nature of wind systems are such that these issues should not constitute major technical risks.

The primary challenge is obtaining high performance at low cost, which requires advanced development and cost reductions over the first-generation systems. Cost is expected to decrease as a result of the following:

- Development of advanced systems (e.g., reduced input loads; high aerodynamic performance; simpler components and subsystems; design for improved fabrication techniques; higher system optimization; and configuration tradeoffs) as well as more rapid and inexpensive wind "prospecting" techniques.
- Increased blade size (estimates predict economics of scale up to diameters of 300–400 ft).
- Competition among manufacturers in design, fabrication and sales.
- Quantity production efficiencies as compared to present "hand-built" systems (e.g., on the order of four to eight ma-

chines per month per company for steady-state production).

These factors form the rationale for the expected sequence of Federal wind program activities.

Environment. Environmental issues arise in two areas: public acceptance and TV interference. Initial reaction to the project at Clayton, New Mexico has been highly favorable, even at close ranges. At distances of a few miles, the system seems to blend into the landscape, resulting in minimal visual impact. Wind systems are, however, unusual and highly visible. Public reaction will depend on the aesthetics of the locale and may limit the number in some locations.

Wind machines with metal blades, in certain instances, can affect TV reception at distances of one to two miles. Fiberglass and other materials will help to reduce, but not completely eliminate, this problem. When coupled with the sensitivity to wind velocity, these environmental considerations would seem to stress applications in more rural areas. The problems of bird strikes and noise appear to be negligible. Proper design with appropriate margins and redundancy should reduce any significant safety problems. A modest easement zone around the wind turbine will probably be required but will be usable for other purposes (such as farming). To the extent that fossil fuel is displaced, the overall effect should be environmentally beneficial.

Technical Readiness Assessment

Operational Status. The initial wind program was directed toward the development of MOD-0, a basic test-bed machine with a 125-ft diameter rotor and a 100-kW power rating. This machine became operational in October 1975 at Plumbrook, Ohio, and has been used as a test bed, operating with diesels simulating remote sites as well as interconnection with utility lines. The first improved MOD-OA, with a 125-ft diameter and increased power rating of 200 kW, was dedicated in January 1978 and is operating in a remote, unattended mode with the Clayton, New Mexico municipal electric utility. It has demonstrated fully automatic synchronization and operation. Operational experience with this machine has indicated a power coefficient of 0.4, compared to a theoretical maximum of 0.6. A capacity factor of 0.3–0.4 should be typical in most installations. Although technical feasibility has been well demonstrated, a number of issues regarding fatigue, lifetime, loads, electrical control and stability, and operating and maintenance costs remain to be improved or verified. Costs for operation and maintenance are estimated to total about 1–2 percent of the capital cost per year. A 30-year design life seems achievable with existing technology.

Capital and Operating Cost Experience. The capital cost of the initial MOD-OA (200 kW, 125 ft) is approximately $1.25 million, and, using standard utility accounting practices, the energy cost is 20–21¢/kWh, as shown in Figure 11-5. Operating experience is very limited; however, the MOD-1 (2 MW, 200 ft) system is in fabrication providing a firm engineering basis for the estimated $3.7 million capital cost and the 10–11¢/kWh energy cost.

As a point of reference, Table 11-4 lists second-unit total installed costs of the MOD-OA, MOD-1 and MOD-2 wind turbines. These costs exclude transmission, land costs and the cost of a management contractor who could possibly be engaged by a utility to oversee the construction, delivery and installation of the system. Land and transmission costs are not shown and are extremely variable, depending on location, local land markets, the type of site (mountainous, coastal or lowland) and, most importantly, turbine spacing, safety zone requirements and options for shared land uses (i.e., what would be practical in farming areas). For single-unit installations, a management contractor could add 30 percent to the system's cost; this cost will drop considerably for larger installations, per-

Table 11-4. Large-scale wind energy conversion systems cost breakdown (second unit costs – prior to production).

	THOUSANDS OF DOLLARS
MOD-OA (200 kW)	
Total installed cost	$1738K
Rotor (blades, hub, PCM)	814
Mechanical equipment	337
Electrical equipment	70
Tower assembly	150
Control equipment	91
Shipping	18
Installation, startup	258
MOD-1 (2 MW)	
Total installed cost	$5770K
Rotor (blades, hub, PCM)	2120
Nacelle structure, drive train	780
Power generating equipment (generator, accessories)	360
Controls	230
Yaw drive system	330
Tower assembly	420
Assembly, testing	750
Site prep, installation, checkout	780
MOD-2 (2.5-3 MW)	
Total installed cost	$3540K
Rotor (blades, hub, PCM)	1040
Drive train (gearbox generator, shafts and rotor brake)	630
Nacelle (structure, shroud, hydraulics, yaw system)	190
Elec/electronic assembly	200
Tower assembly	200
Factory checkout, testing, Q.C., reporting	490
Site prep, installation, check-out, acceptance	790

haps increasing costs by only a few percent. Again, these costs are for the second unit rather than for a production unit and drop considerably for an assembly-line product. For example, the "100*th* unit" cost for the MOD-2 is about $1.4 million per copy, assuming a modest production facility.

The costs shown in Figure 11-5 reflect the judgment of experts, while the costs for advanced systems represent program goals. The cross-hatched area shown in Figure 11-5 as the "Quantity Production Range" is defined by a 95 percent learning curve for 100 units and the historical engineering parameter of cost-per-pound for mature products of the comparable nature and complexity. These two estimation procedures define a probable cost range for systems produced in reasonable quantities in a normal market environment. The goal of 1–2¢/kWh is based on "asymptotic cost" ($/lb for mature products) and appears achievable assuming that R&D can mitigate complexity, weight and other cost driving parameters. It should be noted that, with the exception of the blades, production of wind machines uses established manufacturing techniques. Thus, attainment of the asymptotic cost should not require massive production quantities. Based on comparisons with equipment of similar size and complexity, it appears that a steady-state production of about 100 units per year would be sufficient to achieve mature product costs.

Current Developments, Tests and Demonstrations. The MOD-0 and the first MOD-OA will continue in testing through 1980. Second and third units of MOD-OA have been installed this year, followed by two-year testing activities. The MOD-1, the largest wind system ever built, with a 200-ft diameter and a 2-MW power rating, was fabricated for installation at Boone, North Carolina, and for testing in FY 1979-1981. An even larger system, MOD-2, which represents second-generation technology, with a 300-ft diameter and a 2.5–3-MW power rating, is in the design phase. These new development activities are part of the near-term program strategy to improve performance and reduce cost. If produced in limited quantities, these systems are expected to be close to being competitive in early, specialized markets and to be used to begin the commercialization process, while advanced systems with further cost reduction are developed for penetration of the larger energy savings markets. This latter step remains the major technical challenge to the large-scale use of WECS.

Market/Economic Readiness

Market Description. The U.S. markets for large wind systems are directed toward the electrical utility and industrial markets. Be-

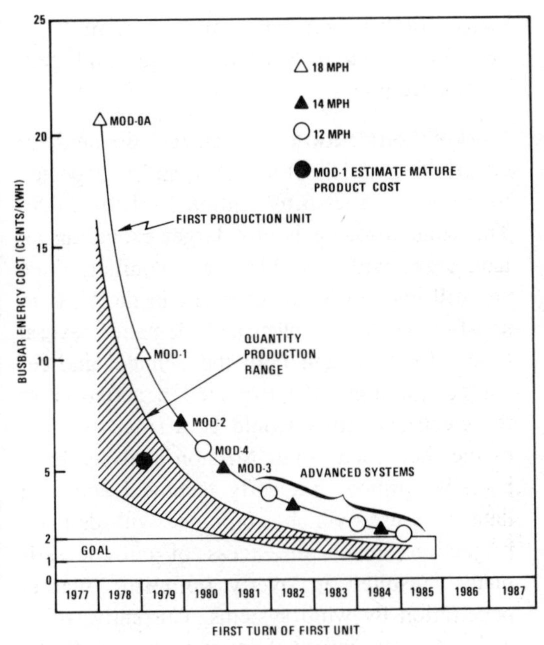

Figure 11-5. Large-scale wind energy conversion system cost goals.

cause of siting restrictions, the industrial applications are limited mainly to rural locations, and, therefore, do not represent a large market. Thus, the principal market for large wind systems, in terms of a major contribution to the nation's energy needs, will be the electric utilities.

The electric utility industry is a large and diverse market. In 1976, approximately 25 quads of energy were used to generate electricity and the amount is expected to more than double by 2000. The ability of wind systems to penetrate this market is influenced by many factors, including the following:

- The level of cost reduction that can be achieved.
- The different ownerships (public, private and governmental) and different methods of accounting for costs.
- The different mixes of generating equipment, types and costs of fuel and interties with other utilities.
- The different criteria used in selecting new generating equipment.
- The regional availability of wind resources.
- The daily and seasonal variability of demand loads.

These factors make the determination of market size as a function of wind system costs a multi-dimensional problem for which firm data are not yet available.

It should be noted that at penetration levels of, for example, 5–30 percent of a particular utility, WECS would act as a negative load. It would both save a statistical mix of intermediate, peaking and some baseload fuel and provide a reduction in conventional power plants represented by a decrease in base and intermediate requirements.

The potential for international marketing of large wind systems appears to be significant in the long term. In the near term, high-wind/high-cost U.S. utilities should be able to support wind industry growth, and the additional complexities of overseas marketing will tend to limit international marketing opportunities. Indeed, several European countries have competing wind system development programs. However, there appear to be potential markets in countries such as New Zealand, Australia and other moderate-sized countries.

Potential Competing Technologies. Large wind systems will compete initially against the standard utility electrical generating technologies, including nuclear, coal, oil and (to a lesser extent) natural gas-fired systems. Ultimately, wind systems would compete against a wide array of new technologies under development for the utility market. Some of the technologies relate to the fuel supply and not the generating equipment; e.g., coal liquids, various gasification products and oil shale. Other technologies represent significant new capital equipment; e.g., solar thermal and photovoltaics. It is not expected that all of the new technologies will achieve significant market shares; the ultimate competition is, therefore, unknown at this time.

Life-cycle Cost Comparison. Large wind systems compete primarily with a constantly varying mix of utility fuels, and, to a limited extent, with a mix of utility plant facilities composed primarily of intermediate and base-load capacity. A fraction of the wind system's rating is realized as added capacity; however, due to variations in the wind resource, this changes the overall economic target for the wind system by only a modest amount. Thus, the primary economic value of wind systems is found in fuel savings—fuel savings which are determined by the joint occurrence of load variations and wind variations. Coal, oil and nuclear fuels are saved. The following life-cycle cost comparisons are therefore based on the cost of displaced fuel, rather than the cost of any one type of generating plant and its specific fuels. The life cycle cost-estimating technique for the initial MOD-OA yields an energy cost of 20–21¢/kWh. This contrasts with the following estimates for present systems:

Energy Source	*¢/kWh*
Diesel generator	9–21
Utility Grid (fuel cost)	0.5–4

Thus, wind is approaching competitiveness with diesel generators, typical of remote and small communities. By 1982, life-cycle cost projections indicate a range of 2–5¢/kWh available from second-generation systems. If produced in quantity (in following years), these values should permit competition in some high-potential utility applications. Because of the mix of conventional power plant types in the typical utility, and applying constant loss-of-load and reserve requirements, a cost goal of 1–2¢/kWh must be reached to obtain the widespread market. As shown in Figure 11-5, this requires an advance in capability.

As a further point of comparison, a wind system competing against an oil-fired, intermediate-load utility plant as a fuel saver would have to produce electricity at an average cost of 5.8¢/kWh to be competitive (1978 dollars over a 30-year system life). Such a competition is most unlikely, since the most economical use of wind systems is whenever the resource is available, as contrasted to the predictable schedule for intermediate-load plants. Thus, large wind systems will logically compete against a mix of plant types and fuel types in various utilities and geographic regions.

Market Penetration. The current demand for electricity is 2.16 trillion kWh and is expected to grow to about 5.69 trillion kWh by 2000. The wind resource is also large; excluding all land areas with possible usage conflict, there are still enough wind resources in the U.S. to satisfy the entire electrical demand several times. Even though both the demand and resource are high, the expected penetration of the electrical utility would be a small fraction of the theoretical total. Mission analyses have been performed, but early technical and cost data were used which, combined with demand projections much in excess of current estimates, resulted in overly optimistic market penetration by wind systems. Currently, there is insufficient information and experience on which to base reasonably firm penetration estimates. A reliable market penetration estimate is one to two years away, when data will be available from market studies and economic analyses. The availability of site-specific meteorological data (one to three years of data) should not slow penetration in the near future, because some utilities are already gathering data. As further market penetration begins, additional favorable wind sites will be required. This need, and the lead time required for data collection, are recognized by the Federal Wind Energy Program, which is already pursuing the development of siting tools and strategies.

The best available market penetration for large wind systems is summarized in Table 11-5. Although tentative, these values generally agree with those produced by the studies summarized in Table 11-6. It should be noted that the referenced studies are based on different assumptions and methods, and must be considered in context, as must the values shown in Table 11-5.

Market Barriers. The major barriers to commercialization of large wind systems, together with potential means of addressing the barriers, are as follows:

Table 11-5. Large-scale wind energy conversion system market overview.

MARKET	1985	1990	2000
CAPITAL COST			
• TOTAL INVESTMENT (1978 $)BILLIONS	0.98	3.24	22.87
• CAPITAL PRODUCTIVITY (1978 $/MMBTU/YEAR)	24.50	17.05	10.89
COST TO END USER			
• (1978 $/MMBTU)	2.50	1.50	1.12
EXPECTED MARKET PENETRATION			
• TOTAL MARKET (QUAD)	30.7	37.7	56.9
• PENETRATION (QUAD)	0.04*	0.19	2.1
• PERCENT OF MARKET PENETRATED	0.13	0.50	3.69
• APPROXIMATE NUMBER OF MACHINES	600	3,000	35,000

*0.01 QUADS SUPPLIED BY THE DOE PROGRAM (INCLUDING INITIATIVE PACKAGE) BY 1983.

- Conservatism of utility industry-verification of cost, performance, lifetime and reliability through an adequate development demonstration and testing program; in particular, the effect of a stochastic energy source on reserve margin, loss of load probability and dispatching strategies must be verified.
- Market uncertainty—conduct applications testing and market studies to evaluate impact and provide sufficient market firming in early years to motivate industry decisions.
- Industrial infrastructure—foster through development and demonstration cycles.
- Fuel adjustment clauses—removal by state legislators.

Environmental Readiness

Environmental Compliance. Although siting considerations may require environmental assessments and, in some cases, environmental impact statements, wind systems do not violate any current environmental standards.

Potential Environmental Barriers. Wind systems are large and highly visible at close range, but they blend into the horizon at a range of a few miles. The highly subjective nature of aesthetics makes public acceptance very difficult to formulate and interpret. To date, public acceptance of wind systems has been favorable, a factor that would mitigate against any new standards in this area. If the wind systems were reasonable quantities in rural areas, public acceptance should not be a major barrier. Large numbers near urban or scenic areas would probably be unacceptable and public acceptance could eventually put an upper limit on market size, rather than resource limits.

Recent assessments have indicated that concerns about bird strikes, audible noise, microclimate and habitat disruption are minimal and new standards are unlikely. New safety standards may be developed; however, modern design techniques and the appropriate use of redundancy can minimize any potential safety problems.

Wind systems can cause interference with television in the immediate vicinity of the wind turbine (200 ft to 2 miles, depending on the site). Currently, there are standards concerning broadcast interference but none dealing with reflection of signals. New standards may be established that would impact the siting of wind systems. It is a site-specific phenomenon, and, therefore, not expected to

Table 11-6. Summary of market penetration estimates for wind energy conversion systems (quads – primary energy).

MODEL	1985	2000
MITRE/SYSTEM FOR PROJECTING THE UTILIZATION OF RENEWABLE RESOURCES (SPURR)		
NATIONAL ENERGY PLAN	0.0	1.7
RECENT TRENDS SCENARIO	0.0	2.2
SOLAR WORKING GROUP		
WITH BEHAVIORAL LAG	0.0	2.0
WITHOUT BEHAVIORAL LAG	0.1	2.3
ERDA/MARKET ORIENTED PROGRAM PLANNING STUDY (MOPPS)	0.03	0.1
ERDA/INEXHAUSTIBLE ENERGY RESOURCES PLANNING STUDY (IERPS)	..	0.31
NSF/NASA SOLAR ENERGY PANEL	0.0	0.76
COMMITTEE ON NUCLEAR AND ALTERNATIVE ENERGY SYSTEMS (CONAES)		
HIGH SOLAR	0.1	1.4
LOW SOLAR	0.0	0.0
PROJECT INDEPENDENCE		
BUSINESS AS USUAL	0.4	4.0
ACCELERATED SOLAR	0.5	5.0
ERDA-49, NATIONAL SOLAR ENERGY RD&D DEFINITION REPORT	0.054	1.25
COUNCIL ON ENVIRONMENTAL QUALITY (CEQ), (MID-POINT)	..	6.0

affect total penetration significantly. Fortunately, navigation and radio signals are not affected except in extreme cases.

Institutional Readiness

Institutional Barriers. Institutional barriers that may impede the adoption and use of wind energy and actions that may mitigate these barriers are as follows:

- Conservatism of the utility industry—demonstrate the cost, performance and reliability of wind systems.
- Wind data and prospecting—lack of information on high-wind sites, the time for the user to assess it, his unfamiliarity and the lack of meteorological consulting firms performing this function; develop more rapid techniques for wind prospecting and support site measurement programs.
- Wind rights—develop model laws and zoning codes that protect the rights to free-flowing wind that may be blocked by buildings or other wind systems.
- Liability insurance—develop and disseminate detailed safety and reliability data through testing; some form of governmental liability guarantees may be required during early market penetration.

None of these barriers are considered as serious deterrents to commercialization.

Manufacturer Status

Several large organizations have been participating in the program by designing and fabricating prototype wind systems. Other potential manufacturers are closely following the progress of the program and investigating the market potential.

If a commercially viable design were available, the manufacturing capability exists in the U.S. today. The tooling required is not new or exotic and the facilities are not large. The number of manufacturers that may actually enter the market will depend largely on the success of the development efforts to reduce costs and on the market acceptance.

Currently, all wind systems are essentially hand-built and all supply, installation and maintenance services are provided by the manufacturers. To gain additional cost reductions, the efficiencies of quantity production and a relatively predictable early production rate will be required. To develop a market that will sustain a steady-state industry, the entire support infrastructure must be developed. In the areas of installation and maintenance, no new skills are required; however, training may be necessary. Experience with the Clayton, New Mexico wind machine shows that a small utility can easily develop the required skills with, perhaps, some training in electronics.

Benefit Analysis

Energy Impacts. Since the principal market for large wind systems will be the utilities, the quantities and types of fuel saved will depend on the mix used by the utilities. Generally, each kW of rated wind power is the equivalent of 5 barrels of oil as primary energy per year, or 150 barrels in its lifetime. If the market penetration shown in Table 11-5 is realized, it would result in saving the equivalent of 7.2 million barrels of oil per year in 1985 and the equivalent of 378 million barrels of oil per year in 2000. The actual fuel saved will be some mixture of coal, oil, and nuclear fuels, or in water saved for other purposes in hydroelectric systems. Wind systems have short energy payback times (six to nine months) and thus should have less impact on U.S. energy needs during a rapid acceleration period.

Recipients of Benefits. The benefits from large wind systems would be derived from reduced fuel consumption by the utilities with the potential for fewer base-load conventional plants. Public, commercial and industrial consumers of electricity would benefit from reduced electric rates if cost goals could be exceeded. Users in remote communities and islands could benefit from lower electricity costs. In addition, the energy produced by wind will release fuels for other uses and reduce pollutants generated in the production of electricity from fuels.

Cost Impacts. Based on the market penetration shown in Table 11-5, the total capital investment required by the year 2000 has been estimated at $22–23 billion. The utilities would pass these costs on to their customers—the residential, commercial and industrial users of electricity. It should be noted, however, that for wind to penetrate the market, these costs will not be higher than the cost of alternative generating equipment, and might be lower.

Other National Benefits. The deployment of wind systems will produce additional national benefits by reducing the balance of payments problem through reduced demand for imported oil and by the export potential of wind systems. Also, reducing the worldwide competition of energy resources will promote international stability and economic growth, which are considered beneficial to our national interests.

Readiness Assessment. The key steps required to determine readiness for commercialization are noted below.

- Low technical risks are involved; some areas need additional engineering development and validation; advanced development is required; and moderate uncertainty exists in achieving large market energy cost goals of 1–2¢/kWh.
- Several generations of advanced machines are required to achieve the cost-effectiveness for significant market penetration; however, earlier systems could be utilized in high-wind/high-fuel-cost areas.
- Projections for systems in design show significant improvement over first-generation systems.
- Market size depends largely on cost reduction, for which both R&D and quantity production are required.
- Relative low cost of each development cycle (about $20 million) and low unit costs ($2–4 million) permit easy scaling of commercialization activities.

Although relatively early in their development, wind systems show a good probability of becoming cost-effective in high-wind areas and a reasonable probability in a very large market. Their relatively short development cycles mean that final determination of the eventual costs that can be achieved with advanced designs and the related market penetration can be determined in the next few years. The short fabrication times and modular nature mean that a flexible program can be developed with incremental decisions by users and the government.

The time required to build user acceptance and an industry infrastructure is of the same order as the time to develop the succeeding systems. The development of a commercialization program would have a significant impact on manufacturer and user alike. Because of the system's modular nature, the cost risk of a flexible commercialization program should not be high.

Early systems will begin to become economical in the more remote high-wind, high-fuel-cost areas. This initial use can form an early market to build up the industry infrastructure and user acceptance, while the development of more advanced systems and increasing production quantities continue to decrease costs to allow expansion into the broader general market.

It is, therefore, recommended that large wind systems be considered for early commercialization and that a program be established which provides for, and is committed to, sequenced demonstrations and incentives, but which is contingent on the achievement of defined cost goals. In this way, enough market visibility can be supplied to encourage the efforts of manufacturers and users, while allowing for major expansions or contractions in future years, as warranted by the achievement of goals.

Federal Commercialization Strategy

The key actions of the proposed strategy are continued technology development phasing into demonstrations, to be followed by financial and/or market firming incentives. The principal role of the technology development activity is to reduce the cost of energy to the point where moderate quantity production can bring the cost into the competitive range. While the technology development continues to bring the cost down to competitive levels of the broader markets, demonstrations are initiated to provide an early market (in high-wind, high-potential applications), to support the development of manufacturing and infrastructure capabilities and, most importantly, to support user acceptance. As the additional cost reductions are realized through development and production efficiencies, the demonstration activities will be replaced by financial incentives to utilities and, to a more limited degree, producers.

An important aspect of this strategy is the control available at each phase. At any stage, if the cost goals are not met, the level of program activity can be adjusted accordingly due to the relatively modular nature of the system. If an additional development cycle is required, it can be accomplished for the relatively low cost of approximately $20 million. Conversely, if the cost goals are met, the demonstration activity can be accelerated based on the comparatively low cost of $2–4 million per machine. These features of the wind energy program permit scaling of the commercialization activities at defined decision points.

The three key actions of the commercialization strategy discussed above (technology development, demonstration and incentives) would be augmented by a number of other activities, including market studies, venture analyses, utility application studies, wind prospecting, international marketing studies and requirements analysis, information dissemmination and the development of siting and economic planning tools.

The three major actions, in turn, form the thrust of the commercialization program. No two of these actions, by themselves, can promise effective commercialization. The evolution of program actions is closely keyed with the needs and capabilities of the industry.

Commercialization Profile. The commercialization profile for large wind systems is presented in Table 11-7. This table identifies the barriers to commercialization, and an array of possible Federal actions which could mitigate those barriers. The effectiveness of each possible action against each barrier is scored on a scale of 1 to 5, with 1 having essentially no effect and 5 eliminating the barrier.

Based on the current and projected characteristics of the energy market, the most significant commercialization barriers are those relating to technical and economic status and user acceptance, and those affecting initial market penetration. Barriers associated with environmental and institutional issues are not considered a primary deterrent to market penetration. Generally, the possible Federal actions relating to information and financial incentives appear to be more effective against the barriers than those relating to taxes and penalties or regulation. These observations form the basis on which the commercialization strategy has been developed.

As shown in Table 11-7, most of the possible Federal actions have an effect on more than one barrier and, conversely, most barriers can be addressed by more than one action. The element that is not visible in this table is the level of action required to affect the barrier as scored. For example, Federally sponsored technology development could reduce the present high cost of wind energy to competitive levels. However, to achieve all of the required cost reductions through technology development alone would be very expensive and time-consuming. Conversely, extremely large incentives or market firming could cause R&D to occur without Federal development support; and again, risk and cost would appear excessive. Thus, a plan can be selected encompassing a number of programmatic activities which, collectively, will mitigate the barriers to commercialization of wind systems without excessive reliance on a single course of action.

Federal Strategy. Because of its modular nature and short fabrication time, WECS is a logical candidate for the utility industry, where uncertainties regarding loads and pat-

Table 11-7. Commercialization profile for large-scale wind energy conversion systems.

BARRIER	TECHNICAL/ECONOMIC			INITIAL DEPLOYMENT			ENVIRONMENTAL			INSTITUTIONAL	
IMPORTANCE	4	2	3	3	3	3	2	2	2	1	2
COMMERCIALIZATION LARGE WIND SYSTEMS (ALL MARKETS) ACTION	HIGH COST	RELIABILITY/ SYSTEM LIFE-TIME	GRID STABIL-ITY & CONTROL	UTILITY CON-SERVATISM/ FAMILIARITY	MARKET UNCERTAINTY	INDUSTRIAL INFRASTRUC-TURE	TV INTER-FERENCE	AESTHETICS	SAFETY	INFRASOUND	PRODUCT LIABILITY (P)
INFORMATION											
TECHNOLOGY DEVELOPMENT (F)	5	5	5	4	3	3	4	2	5	5	4
MARKET INFORMATION/STRATEGIES (F)	3	2	1	2	5	3	3	4	2	2	1
DEMONSTRATION (F)	4	3	4	5	4	5	3	4	3	3	4
TECHNICAL INFORMATION/ASSISTANCE (FS)	3	3	4	4	4	2	3	2	3	1	2
WIND PROSPECTING (PS)	4	1	2	4	5	3	2	4	2	1	2
SITING/ECONOMIC TOOLS (F)	4	2	1	4	4	3	5	4	3	2	1
FINANCIAL INCENTIVES											
PLANNING GRANTS (F)	2	1	2	4	4	4	5	4	3	2	2
PRICE GUARANTEES (F)	2	1	1	4	4	3	2	1	1	1	2
LOAN GUARANTEES (PRODUCER) (F)	3	1	1	2	3	3	2	1	1	1	2
LOAN GUARANTEES (CONSUMER)	3	1	1	3	2	2	2	1	1	1	2
LOW INTEREST LOANS (F) (PRODUCER)	4	1	1	2	3	3	2	1	1	1	2
LOW INTEREST LOANS (F) (CONSUMER)	3	1	1	4	2	2	2	1	1	1	2
CONSTRUCTION/TOOLING EXPENSES (F)	4	3	1	1	4	4	3	2	2	2	2
INVESTMENT TAX CREDITS (FS)	3	2	1	4	2	2	2	1	1	1	2
ACCELERATED DEPRECIATION	3	2	1	4	2	2	1	1	1	1	2
FEDERAL POWER DEMONSTRATIONS (F)	4	3	4	5	3	4	3	4	3	3	4
GUARANTEED FEDERAL USE MARKET	4	3	4	5	4	4	3	4	3	3	4
FEDERAL LAND USAGE (F)	3	1	1	2	1	1	3	4	2	3	2
TAXES AND PENALTIES											
PRODUCT WARANTEES (FP)	2	3	2	4	2	3	2	1	3	2	3
FUEL SURCHARGE (F)	4	1	1	4	3	2	1	2	1	1	1
FUEL IMPORT QUOTAS (F)	4	1	1	4	3	2	1	2	1	1	1
REGULATION											
MODEL ZONING/BUILDING CODES (FSL)	2	3	1	2	3	2	3	3	3	3	3
RATE STRUCTURE GUIDELINES (FS)	2	2	1	4	3	2	1	1	1	1	1
FEDERAL INSURANCE (F)	2	1	1	3	3	2	1	1	4	1	5

KEY: F - Federal
L - Local
S - State
P - Private

terns of demand abound. WECS are expected to be "catalogue" items, built in factories to a standard design similar to diesels and gas turbines. This capability should prove to be an advantage to utility purchasers.

Figure 11-5, which illustrates cost goals, shows that cost reduction by R&D is the logical initial step. Cost reduction and market development, however, can be phased in within the next few years through production, demonstration and incentive programs, if cost and performance goals are met. The figure shows that they are relatively close to cost-effective as compared to many solar and other advanced or alternate energy technologies. Several key steps in the commercialization strategy are as follows:

- Competition among manufacturers which can have a major effect on cost reduction.
- System development performed directly by the companies expected to be the commercial manufacturers to preclude technology transfer factors.
- Sufficient machine system developments in parallel and in sequence to provide continuity and expanded capability in the industrial design teams and competition among companies.
- A phasing from individual experiments to a demonstration program to stimulate the initial market. The demonstration program would provide cost-sharing support directly to the user who would select the brand desired. This approach stresses competition and forces the development of the marketing and distribu-

tion networks. In addition, phasing the cost-sharing ration in multiple unit systems so that the user's share is low initially, but increases with added units if they meet goals, allows a matching of risk and investment levels.

- In addition to public and private utilities, the power administrations and other agencies (e.g., the Bureau of Reclamation and DOD) would be utilized both as potential future users and to provide special test and visibility advantages. They constitute potentially firm, core markets for the early commercialization of large wind turbines and would represent a visible commitment by the Federal Government.
- Capability would be provided to the users (e.g., siting, economic and wind forecasting methodologies) to assist in assessing wind power.
- The demonstration effect would be phased out in favor of appropriate incentives as the results of present incentive studies and as firmer prices versus estimates become available.

Wind systems and most other new technologies face an initial startup problem in achieving market penetration. That is, over and above the results of the technical development program, some portion of the cost reduction necessary for penetration must come from the efficiencies of quantity production. Conversely, there is little market without the necessary cost reduction. Fortunately, the major market is not monolithic. Small but important sub-markets exist in remote communities, islands and other high-fuel-cost areas, which, when coupled with high wind velocities in many of these regions, allow wind systems to become competitive earlier than in the majority of the utility market. For example, some residential customers in Alaska pay as much as 20¢/kWh compared to the national average of 3.4¢/kWh. In Puerto Rico, the average residential customer pays 5.4¢/kWh, which is 54 percent above the national average. Similar high costs exist in many areas and in the few remaining remote communities not connected to other utilities. It is anticipated that, within the next few years, wind systems capable of competing in these small, high-cost sub-markets can be made available. Even though these early markets are small, they are large enough to use the machines produced by a fledgling industry, and, with appropriate support, to cause industry to enter that market. The "appropriate support" can be indirect or direct, depending on the cost of machines, market size and capital resources. Additional incentives to enter the market may be required in the form of financing of tooling costs. The first production unit cost is often inversely related to the effort and resources devoted to pre-production engineering or assembly-line design. A substantial investment may be required by manufacturers to do this engineering.

Federal Goals. The goals of the large wind systems effort have been previously presented in Figure 11-5 and Table 11-5. In summary, the ultimate goal is an energy cost of 1–2¢/kWh to compete in the broad utility market. Currently, the first prototype of MOD-OA is providing electricity at 20–21¢/kWh; the experimental prototype MOD-1, now in operation, has an energy cost of 10–11¢/kWh. By 1981, the goal is set at 4–6¢/kWh for a prototype, which, with some quantity production, will permit penetration into the higher-cost, high-wind areas. The goal of 3–4¢/kWh for the prototype should be achieved in the 1983–1985 time period, with 1–2¢/kWh achieved with production in the late 1980's. If these goals are achieved, the best available market penetration estimates indicate an energy contribution of 0.04 quads in 1985, growing to 2.1 quads by 2000.

Federal Actions. The Federal commercialization strategy has been structured from this profile and is a balance between time and cost to the Government while stimulating the development of a visible, competitive wind industry. A number of decision points or "gates" can be incorporated into the strategy, at which the pace of the commercialization effort can be adjusted to reflect the success of the program at that time. The strategy is discussed in terms of the three key elements identified above and in terms of the information-generating activities.

The immediate effect of the technology development activities is to improve the design and fabrication techniques of wind systems to provide lower energy costs while supporting private industry in gaining experience and establishing competent design teams. Whenever possible, parallel development contractors will be used to maximize competition within the industry. A flexible patent policy by the Government also may stimulate industrial involvement and assist commercialization. In addition, with each generation of new systems, a few machines will usually be fabricated and placed in an actual application with a utility. Thus, these development cycles help to lower costs, acquaint users with the technology, provide performance and cost data, aid infrastructure development and generally reduce uncertainty.

System demonstration could begin in the FY 1980–82 time period and could be phased in cycles as the technology proceeded. Utilities wishing to participate would be selected through open competition, with each winner selecting its own "brand" to foster industrial competition and keep prices low. The option includes privately developed as well as Government sponsored machines, subject to some minimum criteria. Of the total machines, it is anticipated that some would be installed in conjunction with one or more Federal installations, including hydroelectric systems, and others would form a cluster or wind farm in a typical utility to demonstrate the capability of large-scale power production. The remainder would be distributed geographically as individual machines or small clusters to provide regional visibility. Cost-sharing with the private sector would be an integral part of the demonstration activity, with the Government's share decreasing in the later stages. In addition to their obvious benefits in familiarizing the utilities and stimulating the industry, the demonstration will offer an opportunity to resolve issues relating to grid stability and control and capacity credit.

The selection of a financial and market firming incentives package for large wind systems is still speculative at this time. An initial incentives study has been completed, and a more detailed study and policy analysis is scheduled before the final incentives selection is made. Based on current data, it appears that the most effective incentives for the utility sector are investment tax credits and accelerated depreciation (i.e., effective in that they produce the greatest market penetration per unit of government costs). It appears that low-interest, long-term loans, particularly for tooling and setup, would be the most effective incentive for the manufacturers.

Because uncertain user acceptance represents a major deterrent to early commercialization, the use of wind power, if cost-effective, by the Federal Power Administration and Bureau of Reclamation, in conjunction with their hydroelectric systems, represents a powerful potential incentive. This is not offered as a subsidy program because it could be made contingent upon achieving cost-effectiveness. Thus, the cost to the Government should be essentially the same as for meeting its energy needs through some other source. Rather, it is inteded to provide a known market segment at known cost to the potential manufacturers during the early commercialization period in the mid-1980's. This would provide the incentive to industry to commit to a product line once they visualize that cost-effectiveness can be achieved. It also provides an opportunity for private planning, capital formation and marketing while the private user segment is being developed. The size of the contingent commitment could range from 100 MW to over 1000 MW. The critical aspect is that a specific purchase commitment be made, contingent on cost-effectiveness, in FY 1981, to provide time for private industry to respond.

In support of three key program activities, the Federal Wind Energy Program will include a number of information-related activities designed to facilitate the entry of wind systems into the market. Specifically, mission analyses have been performed and venture analysis has been compiled to determine the conditions under which private industry would seek to enter the market and to quantify the effects of different strategies. Market studies have been conducted of various general application areas to provide initial estimates of wind market size versus system cost. The studies are being followed by a series of more detailed regional and sector market studies to provide firmer market data and possible marketing strategies. These activities are intended to reduce market uncertainties and to aid manufacturers in capital formation and related decisions.

For the user, primarily the electric utility industry, application studies have been conducted to analyze various planning methodologies. In parallel with this activity is a wind characteristics program which is developing siting and economic analysis tools suitable for use by utilities. This program also included a regional wind energy assessment to identify the general wind resources as a first-order siting input. (The wind resource varies so greatly from site to site that any attempt to identify all of the good wind sites would be extremely expensive. It would also require many years to complete and measure winds in areas which, because of other factors, are not necessarily good locations for wind systems.) These activities will culminate in a planning grant program to aid utilities in economics, integration and siting evaluation of wind systems. Such a program, when coupled with sufficient data from test and demonstration systems at other locations, will assist utilities in becoming familiar with wind technology and in overcoming their traditional conservatism.

As more detailed user-oriented data become available from the current studies and tests, an expanded user awareness and information dissemination program will be developed and implemented in conjunction with EPRI related activities.

The purpose and benefits of the remaining activities are reasonably straightforward. Major activities include:

- *Technology development (design, development and testing)*
  Intermediate-scale machines
  Large-scale machines
- *System demonstrations*
  Site qualification
  Planning
  Cycle 1
  Cycle 2
- *Incentives*
  Initial incentives study
  Incentives policy analysis
  Select recommended incentives package
  Implement utilities' incentives package
  Implement manufacturers' incentives package

These activities define the key elements of the recommended Federal commercialization strategy. They would be supported by a number of other planning and information generating activities, summarized as:

- *Information planning*
  Perform detailed venture analysis
  Liability insurance: policy and actuarial analysis
  Preliminary information dissemination systems
  Develop information dissemination system involving Federal and state agencies
  Wide-scale user awareness and dissemination program
  Development of siting and economic analysis
  Regional wind energy analysis
- *Market studies*
  Initiate remote market study
  Interconnected market study
  REA market study
  Regional and sector (including Federal) market studies/strategy development
- *Utility applications*
  Application analyses
  Analyses methodolgies
  Testing of methodologies
  Utility planning grants
- *Standards for information*
  Performance/test/label/nomenclature standards development
- *Standards*
  Planning and analysis activities

Warranties and guarantees
Product and safety testing standards

- *International program*

Initial international market requirements/criteria study
International wind and load data support
Preliminary support of industry export needs (studies, visibility, trade shows, testing)
Analysis of possible Federal actions in support of international commercialization
Expanded international market studies
Develop export related machine modifications
Decision to implement international market commercialization

There is another Federal action, not explicitly shown in the commercialization profile, that would have far-reaching benefits in stimulating the commercialization of wind systems. That action would be a commitment to a commercialization plan. If a specific demonstration and incentives packages were planned with a specific schedule, it could stimulate both the utilities and manufacturers. This is particularly true for wind energy because of its unusual nature and the significant concern of potential manufacturers over user acceptance. The adoption of such a plan, with its built-in assessment and decision points, would significantly reduce the market and user uncertainty, which is one of the major barriers to commercialization.

12
Environmental, Institutional and Legal Barriers

This chapter briefly describes the present status, program goals and potential environmental and institutional issues regarding WECS and identifies potential areas of concern relevant to their use. These concerns include occupational or public exposure to health and safety hazards, environmental effects, socioeconomic, and legal impacts. Identification and screening of possible impacts result in delineation of those of the most serious, irreversible and cumulative nature, those having near-term importance to the program and those for which current knowledge of effects and control technologies is inadequate.

There appear to be no environmental standards that are applicable to small-scale WECS with the exception of, perhaps, noise regulations, although properly designed systems are not particularly noisy. In a rural environment, SWECS-produced noise is not likely to cause problems beyond the owner's property, since SWECS have a modest noise output. SWECS are likely to be excluded from high population density areas for institutional reasons (zoning, clearance zones, etc.). Television interference problems can occur with the largest SWECS, thus limiting siting opportunities near residential areas.

Total program funding specifically identified with environmental issues currently totals about $2.6 million through FY 1984. The design and development activities stress safety and environmental issues integral to the design process; thus, the actual cost of environmental research could be higher than the direct funding indicated by program element. The rapid pace of the commercialization program requires that safety and other environmental concerns be addressed directly by the design and development programs.

Although siting considerations for large-scale WECS will require environmental assessments and, in some cases, environmental impact statements, large-scale wind systems do not violate any current environmental standards. Recent assessments have indicated that the impacts of bird strikes, audible noise, microclimate and habitat disruption are minimal. Concerns have been raised with respect to aesthetics, safety and television interference. New safety standards may be developed; however, modern design techniques and the appropriate use of redundancy can minimize any potential safety problems. Extensive fatigue testing is planned prior to any major demonstrations or commercialization efforts.

Large-scale wind systems are highly visible at close range, but even the largest systems should blend into the horizon at a range of a few miles. The highly subjective nature of aesthetics makes public acceptance very difficult to formulate and interpret. To date, public reaction to wind systems has been favorable, a factor that would mitigate against any new standards in this area. In reasonable quantities in rural areas, public acceptance should not be a major barrier. Large numbers near urban or scenic areas would probably be unacceptable, and public acceptance, not resource limits, eventually, could put an upper limit on market size.

Large-scale wind systems can cause interference with television in the immediate vicinity of the wind turbine (200 ft to 2 miles, depending on the site, TV channel, distance from transmitter, etc.). Currently, there are standards concerning broadcast interference, but none dealing with reflection of signals. New standards that would impact the siting of wind systems may be established. It is a site-specific phenomenon, and, therefore, is not expected to significantly affect total market penetration for WECS. Fortunately, navigation and radio signals are not affected except in extreme cases.

Potential physical impacts of WECS are therefore limited; they produce no air or water pollutants or solid waste products as a result of their operation. Fabrication is comparable to routine machine construction and poses no significant demands on scarce resources. Principal physical/environmental concerns are structural safety, electromagnetic radiation interference and bird collisions with WECS structures. Early planned research will survey existing technical research for safety-related data such as mechanical stress or failure potential analyses, and prescribe further research needed to define safety hazards, preventive design and operation measures and regulatory standards. DOE-sponsored technical and institutional studies of electromagnetic radiation interference are complete, requiring only final application to siting and design criteria and regulatory standards. Recently completed bird collision studies have revealed no significant hazard. However, bird collisions will be monitored at field test sites.

Socioeconomic concerns include public objection to WECS installations on aesthetic grounds of large WECS arrays and institutional impacts of integrating WECS with existing electric utilities, whether large utility-operated arrays, small, dispersed systems or offshore WECS emplacements. Planned research in these areas is geared toward ensuring WECS program compliance and integration with existing laws and regulations before widespread public use of the technology.

This chapter also presents user-oriented information which deals with the possible legal problems caused by purchasing, installing and operating a WECS.

ENVIRONMENTAL, HEALTH AND SAFETY BARRIERS

Identification and Screening

To ensure that environmental plans reflect the full range of environmental health and safety issues associated with a new technology, all possible impacts must be identified and screened. This can be accomplished by developing an issues identification and categorization matrix as shown in Table 12-1. The generic impact areas of air and water quality, land use, ecology, health and safety, esthetics, resource requirements, and social/institutional concerns were examined for each key phase of wind energy technology development (e.g., materials production, manufacturing, construction/installation, operation, and decommissioning) in order to identify the issues listed on the vertical axis.

The environmental, health and safety issues can be identified through a review of existing research data and projections of probable concerns associated with wind energy conversion technology. Because the technology is relatively well developed, many of its potential impacts have been determined largely on the basis of actual experience, as well as testing and research efforts.

Once issues are identified, a preliminary screening can isolate key issues. The essential screening criteria are as follows:

- Nature, magnitude and severity of environmental impact.
- Extent (scope) of impact.
- Current knowledge.
- Concern of other groups and Federal agencies.
- Irreversibility and cumulative nature of impact.

None of the above criteria are used alone to screen for key environmental issues. Taken together, however, they form the basis for a comprehensive screening.

Key Concerns

The above criteria for determining key issues have been systematically applied to the issues delineated in Table 12-1 by completion of a series of research and review efforts.

DOE has prepared an Environmental Factors Report which characterizes the nature and extent of potential WECS impacts based on cumulative experience and research results. Potential wind energy-related environmental research needs also have been addressed in DOE's Balanced Program Plan for biomedical and environmental research in the solar energy area. DOE also has conducted a detailed literature search, baseline environmental survey and ongoing monitoring program to identify and quantify environmental impacts at the DOE/NASA 100-kW Experimental Wind Turbine site near Sandusky, Ohio, and has prepared environmental impact assessments for 17 candidate sites for 200-kW – 1.5-MW experimental wind turbine generators. These studies also identify potential impacts and provide further experience and information for identifying and evaluating key concerns. Integrated evaluations, such as the above-mentioned reports, and research to date have been compared to yearly Wind Energy Systems Program Approval Documents to assess near-term importance of program elements and respective impacts. Finally, consultation with other Federal agencies on identified key issues has occurred via circulation of the Issues, Requirements and Projects (IRP's) List.

Application of this DOE screening criteria to the Wind Energy Program has resulted in the identification of seven key issues. Specifically, emphasis is placed on electromagnetic radiation interference, safety considerations and utility interface issues because of their near-term importance to the program and because research experience suggests that potential problems may exist. Potential bird collisions and operating noise are included as key issues because they produce measurable effects which, although minor to date, may have the potential to become cumulatively significant or affect specific site selection and design decisions when wind energy systems attain widespread use. Finally, aesthetic objections to load-based wind turbines as a barrier to public acceptance of the technology must be addressed as a key concern relative to long-term, large-scale wind program development. At this time, offshore installations are not a major concern because the economics of such systems are substantially inferior to those of onshore systems. However, institutional issues relative to offshore WECS emplacements have been identified as being potentially significant.

Although the following discussions focus on key environmental issues of the Federal Wind Energy Program, it is important to recognize that DOE has researched all issues identified in relation to the program. Thus, those issues identified in Table 12-1, but not discussed in the following section, are secondary issues.

Health and Safety Issues

Safety hazards may be associated with structural failure, electric transmission equipment, construction and/or maintenance of WECS towers and tower-mounted equipment. Potential causes of structural failures include mechanical stress that might result from rotational forces, wind shear and cataclysmic weather events such as severe storms or tornadoes. The hazardous zone, as shown in Figure 12-1, for tower failure would be a circular area with a maximum radius roughly equal to the height of the tower plus the radius of the disc formed by the blades. Blade throw would affect a much larger area. A blade thrown from the Smith-Putnam wind machine on Grandpa's Knob during the 1940's traveled 750 ft (230 m); the potential blade throw distance for DOE's MOD-OA 200-kW experimental machines has been estimated at up to 500 ft (167.75 m) under worst conditions.

Physical and Environmental Issues

Electromagnetic Interference. Electromagnetic radiation interference may occur when signals reflected from moving rotor blades in-

Table 12-1. Wind energy conversion systems – environmental issues and respective impact areas.

ISSUE	AIR QUALITY	WATER QUALITY	LAND USE/ SOLID WASTE	ECOLOGICAL IMPACTS	HEALTH & SAFETY	ESTHETICS	SOCIAL/ INSTITUTIONAL
ELECTROMAGNETIC RADIATION INTERFERENCE			X		X		X
SAFETY			X		X		X
BIRD COLLISIONS				X			
MICROCLIMATE EFFECTS			X	X			
OPERATIONAL NOISE			X	X		X	X
VISUAL DISRUPTION OF LANDSCAPES			X			X	X
OFFSHORE EMPLACEMENT				X	X	X	X
UTILITY INTERFACE							X
ZONING & LAND USE CONTROLS			X			X	X
UTILITY CERTIFICATION PROCEDURES							X
INSURING UNOBSTRUCTED WIND FLOW		X	X	X		X	X
AIR TRAFFIC INTERFERENCE					X		X
BUILDING AND SAFETY CODES					X		X
EQUIPMENT STANDARDS AND INDUSTRY REGULATIONS					X		X
DAMAGE BY FIREARMS					X		

Figure 12-1. WECS hazardous zone.

teract with the original signals, causing fluctuations in signal frequency and amplitude which degrade reception quality. Types of signals which may be affected are in the higher frequencies such as television, navigational aid (navaid) and microwave at points where geometries favorable for interference occur among the wind turbine, transmitter and receiver. Other factors affecting the magnitude and severity of this impact include blade area and speed, direct signal strength and reflected signal strength relative to the direct signal. Preliminary DOE studies indicate that reception difficulties may occur as far away as one-quarter mile (400 m) for low frequency (VHF) TV signals and 3 miles (4800 m) for higher frequency (UHF) signals. DOE's environmental assessment for its Block Island, Rhode Island experimental wind turbine site indicated that interference with area residents' television reception could occur and that mitigating measures would be required.

Hazard to Birds. Bird collisions with towers and moving or stationary rotors can be hazardous to local or migratory bird populations. Large kills of nocturnal bird migrants have occurred due to a collision with large (500-ft or 175-m) television towers (or their guy wires) under conditions of darkness or poor visibility. Night-flying birds also may be disoriented by, or attracted to, lights on tall structures, thereby increasing the possibility of fatal collisions. The number of birds which may be lost through such collisions is a function of the number of birds flying on a given night, the altitude of migration, weather conditions and the probability that an individual bird entering the area swept by the rotor will collide with the blades (this last item being a function of blade size and velocity and the angle at which the bird enters the plane of rotation). The largest planned government wind turbine is about 350 ft (107 m) in height. The probability of a bird colliding with a blade within the area swept by the rotor is much less than for a solid object (such as a building) of equal size. Birds have been observed to take evasive action when flying through the rotor disc of the MOD-O wind turbine.

Operational Noise. Operational noise from WECS may include both audible and infrasonic sound. Potential adverse impacts might include annoyance or interference with human activities and wildlife disturbance, and thus might limit siting or worker exposure. Noise levels from operating WECS and their impacts on the environment have not been adequately quantified. Low frequency vibrations have occurred at the DOE site in Boone, North Carolina.

Appearance. Physical appearance of wind structures, particularly large WECS arrays located in scenic areas, could become a barrier to public acceptance of WECS if a large segment of the public finds them aesthetically objectionable. Local zoning or subdivision codes whose goals may include maintaining desired aesthetic standards may preclude location of WECS in populated areas. In a preliminary study of wind energy devices in simulated landscapes, respondents generally preferred open landscapes to the same landscapes with various wind machines installed; groups of wind machines in a line were viewed as similar to lines of electric power transmission towers in their aesthetic impact. Among the various tower designs, conventional horizontal-axis wind machines mounted on an "Old Dutch" or closed columnar tower were preferred. Generally, the aesthetic impact will depend on the size and character of the turbine(s), the scenic character of the site and the degree of visibility from areas of human activity.

Social and Institutional Issues

Utility Interface. Interface with electric utility systems will require resolution of a number of social and institutional issues. These issues are most numerous for relatively small wind energy applications, particularly where interconnection with a conventional electric utility grid is needed. Utility service contracts are not designed for wind interconnections, and modifications in existing rate structures may be necessary. As a new technology, the safety and economic aspects of wind power must be examined in light of the usual requirements for power sources in utility electrical systems. The impact of WECS on these systems is expected to be similar to that of auxiliary generators and other small conventional systems.

Offshore WECS. Offshore emplacement of wind energy devices (as shown in Figure 12-2) will raise some social and institutional issues over and above those associated with land installations. These consist primarily of Federal-state jurisdictional and regulatory considerations, which may be unclear. Resolution of regulatory jurisdiction and development or adaptation of permitting and environmental control regulations will be needed before offshore WECS can be established. Legal/institutional complications also could arise concerning onshore support facilities and utility interconnections.

ENVIRONMENTAL, HEALTH AND SAFETY RESEARCH PLANNING

Setting Research Priorities

The economical and technical aspects of many technologies are strongly dependent on the degree of environmental control required; consequently, technology program goals must be considered carefully in defining environmental research priorities and schedules. The screening criteria discussed earlier not only identify environmental concerns but also provide pri-

Figure 12-2. Environmental issues regarding multi-unit offshore wind energy conversion systems have not been fully delineated. *(Courtesy of DOE/Westinghouse.)*

ority rankings. In addition to these criteria, three additional factors are crucial in defining environmental research priorities and scheduling needs. Each is discussed briefly below.

Regulatory Impetus. When Federal, state or local regulations or standards exist explicitly for new technologies, or implicitly through general regulation of some of the potential effects, an implicit ranking of possible impacts already has been performed by the regulating agency. Research directed toward determining the potential for compliance with existing or imminent regulations must be placed high on the priority list. Generally, when adequate regulations and standards exist, research is directed more toward producing cost-effective compliance options (control technology), rather than determining impacts of alternative levels of control. Federal regulations concerning matters applicable to wind energy technology development include the following:

- National Environmental Policy Act
- Non-nuclear Energy Research and Development Act
- Occupational Safety and Health Act
- Federal Power Commission regulations governing utility service and rates
- Federal Communications Commission regulations governing sources of electromagnetic radiation interference
- Federal Aviation Administration regulations governing tall structure siting and marking
- Endangered Species Act
- Fish and Wildlife Conservation Act
- Coastal Zone Management Act
- Coast Guard regulations governing offshore structures and vessels.

Environmental Resolution Time versus Time of Public Utilization. A primary consideration in determining research needs is the schedule for public utilization of the technology. An objective of environmental planning is to ensure that environmental issues are resolved before a technology is made available to the public. Therefore, whenever it is estimated that the time required to resolve an environmental issue (environmental resolution time) through RD&D is greater than or equal to the time before the technology is available to the public, then that issue should be a priority concern.

Cost/Effectiveness. Another approach to determining priorities is to look at the commonality of problem overlap among diverse technologies. Researching the problems of a particular technology may resolve similar problems involving other technologies. Such an approach could be viewed as cost-effective, and ranked accordingly.

Significant Research Issues

This sub-section discusses the key issues identified in the first section of this chapter, in terms of research areas and priorities needed to resolve them. A detailed issue-by-issue discussion is provided. Table 12-2 summarizes the required research to fulfill the identified needs.

Safety Hazards. Safety hazards include structural failure of the tower, blade shaft, hub and blade throw, and risks to workers or the public from tall structures and electrical generation and transmission equipment. Research needs include design analysis and testing to determine mechanical stress and failure potentials for a range of potential designs, materials and weather conditions. Hazardous zones for potential failures or accidents must be identified. Finally, existing safety regulations applicable to WECS must be identified and, if found to be inadequate or inappropriate to WECS operations, necessary modifications should be suggested.

Design analyses and testing will be performed as wind turbines are developed to determine mechanical stress and failure potentials for a range of operating and environmental conditions. These analyses will parallel the overall design trade-off analyses for each machine, and will have a major impact on machine design. Controlled area guidelines and safe operating procedures will be defined and specified for implementation at WECS sites as machines become available for testing and use. WECS will be designed to comply fully with existing safety regulations and requirements.

The procedures outlined have been followed as an integral part of DOE-NASA's WECS development activities. An overall safety plan was required as part of initial development and design studies for the MOD-O and MOD-OA machines, and each phase of design, fabrication, installation, testing and

Table 12-2. Issues and required research.

ITEM	ISSUES	REQUIRED RESEARCH
a	SAFETY	• ASSESS SAFETY ANALYSIS AND FAILURE MODE EVALUATION AND ANALYSIS (FMEA) ASSOCIATED WITH EXISTING DESIGNS FOR COMPLETENESS AND NEED FOR FURTHER EFFORT • DETERMINE PROBABILITY AND INCIDENCE OF TOWER AND BLADE FAILURE FOR VARIOUS DESIGNS • EMPLOY SAFETY ANALYSIS TO DETERMINE DESIGN OPTIONS TO MINIMIZE STRUCTURAL FAILURE HAZARDS • EMPLOY SAFETY ANALYSIS AND FMEA TO DETERMINE POTENTIALLY HAZARDOUS OPERATING CONDITIONS, ESTABLISH HAZARDOUS ZONES FOR WECS FIELD TEST SITES • ESTABLISH SAFETY GUIDELINES FOR WECS SITES • REVIEW SAFETY REPORTS AFTER SEVERAL MACHINES OF GOVERNMENT DESIGN ARE IN OPERATION TO ASSURE CONSISTENCY OF MACHINE SAFETY SPECIFICATIONS AND APPLICABILITY TO WIDESPREAD COMMERCIAL WECS USE
b	ELECTROMAGNETIC (E-M) RADIATION INTERFERENCE	• PERFORM DESIGN ANALYSIS, MODELING, AND MONITORING TO DETERMINE POTENTIAL FOR E-M INTERFERENCE • VERIFY THE ANALYSIS DURING OPERATION OF TEST MACHINES AT A VARIETY OF SITES • DETERMINE DESIGN AND SITING OPTIONS TO MINIMIZE E-M RADIATION INTERFERENCE • DEVELOP WECS SITING GUIDELINES TO MINIMIZE E-M INTERFERENCE
c	BIRD COLLISIONS	• PERFORM BASELINE MONITORING AND FIELD SURVEYS AT EXPERIMENTAL WECS SITES • PERFORM ARCHIVAL STUDIES TO EVALUATE RESEARCH LITERATURE CONCERNING BIRD KILLS AT TALL, LIGHTED STRUCTURES FOR APPLICATION TO WECS • DETERMINE REGIONAL, SEASON, TOPOGRAPHICAL, AND SPATIAL CHARACTERISTICS OF BIRD FLIGHT FOR INPUT TO SITING CRITERIA AND DECISIONS • MONITOR EXPERIMENTAL WECS FOR SIGNIFICANT BIRD COLLISIONS • ASSESS SITING GUIDELINES FOR WECS EXCLUSION REGIONS
d	OPERATIONAL NOISE	• IDENTIFY EXISTING NOISE CRITERIA AND EVALUATE THEIR APPLICABILITY TO WECS OPERATIONS • DESIGN AND IMPLEMENT NOISE MONITORING PROGRAMS AT OPERATING WECS SITES

Table 12-2, Issues and required research. (continued)

		• EVALUATE/PREDICT ACOUSTIC BEHAVIOR OF RESULTANT ENVIRONMENTAL IMPACT FOR SPECIFIC WECS APPLICATIONS • RECOMMEND MEASURES TO MITIGATE ANY UNDESIRABLE IMPACTS FOUND
e	UTILITY INTERFACE	• MONITOR INSTITUTIONAL PROGRESS OF UTILITY-LINKED EXPERIMENTAL WECS • MONITOR INSTITUTIONAL PROGRESS OF SMALL OR NON-UTILITY WIND ENERGY PRODUCERS • IDENTIFY REGULATORY CONFLICTS AND BARRIERS
f	PHYSICAL APPEARANCE OF WECS	• EVALUATE POTENTIAL VISUAL IMPACTS OF AND PUBLIC REACTION TO WECS APPLICATIONS • DEVELOP DESIGN AND SITING GUIDELINES TO MINIMIZE NEGATIVE VISUAL IMPACTS
g	OFFSHORE EMPLACEMENT	• PERFORM IN-DEPTH STUDIES TO IDENTIFY ISSUES SPECIFIC TO OCEAN EMPLACEMENT OF WECS, IF DEVELOPED, AND RECOMMEND NEEDED ENVIRONMENTAL, HEALTH, AND SAFETY RESEARCH, IF ANY

operation has been subjected to a comprehensive safety review and permitting system. Similar procedures are, or will be, employed in the MOD-1, MOD-2, and MOD-3 programs. No phase of program implementation may proceed without the required safety review.

The research plan should call for a review of ongoing safety-related documentation to define further research needs, if any, and to identify safety issues which should be addressed in order to ensure WECS safety and regulatory compliance in widespread applications. As noted, safety issues for specific DOE-developed systems are scheduled to be resolved before each system becomes operational; final issues resolution and application to widespread WECS use will be completed by the time a multiple MOD-1 central power site is selected at the end of 1981. For the dispersed applications, review of safety issues was completed in 1978 for incorporation into the 10-kW demonstration program and the 100-kW testing program. Safety-related research for offshore WECS will be scheduled in 1980, when a decision whether or not to proceed with an offshore development program is made.

Electromagnetic Radiation Interference. Electromagnetic (E-M) Radiation Interference research requires further technical evaluation and analysis. A comprehensive technical study of E-M interference potential from wind machines has been completed by the University of Michigan. A completed survey of legal/institutional barriers to WECS development defines applicable regulations and suggests WECS compliance strategies. Consequently, the only remaining environmental research needs involve studies that apply the existing information to siting and design decisions and the development of specific WECS guidelines where needed. These are scheduled for completion by mid-1980.

Bird Collisions. Bird collisions with WECS structures might affect local and migratory fauna adversely and also might impede public acceptance of WECS installations. A comprehensive study of existing literature has been made, and bird impacts are being monitored for future analysis at NASA's Plum Brook test facility and other government installations. Additional research is needed to characterize bird flight patterns as input to WECS siting and design decisions and the development of siting guidance to minimize impacts on birds. Continuous (day and night) monitoring studies of bird/turbine collisions are needed at WECS. These studies are scheduled for completion by late 1980 so as to provide input for the MOD-3 and multiple MOD-1 (10 10-MW wind units) system site selection decisions and testing programs.

Operational Noise. Operational noise from WECS operations has received little or no study. Planned noise research includes a survey of existing noise criteria to determine their applicability to WECS, monitoring and modeling of operational WECS noise to evaluate its environmental impact and development of measures to mitigate undesirable impacts which have been discovered at Boone, North Carolina.

Utility Interface. Conflicts may arise for applications involving production or sale of wind energy by non-utility entities. Although the greatest impact of these issues may be on the economic feasibility of WECS applications, environmental planning should be concerned primarily with social/institutional aspects, as they affect public acceptance and program fulfillment of institutional requirements.

Several comprehensive studies of legal/institutional impacts of WECS development have included utility interface issues in their scope. Synchronous operation of test systems with local utility grids (already initiated at the Plum Brook, Ohio MOD-O facility, and planned for the other central power test programs) is provided or will provide case-by-case data on utility interface progress and problems. Finally, several current regional application studies have addressed legal and institutional requirements for WECS.

Priority in future research is given to a survey of existing data collection programs or plans to ensure that utility interface is receiving adequate study and to establish monitoring programs where they are needed. This phase of the research effort should be completed by 1980, when the first multiple-model test site may be selected, providing time to incorporate utility interface case studies into project design and testing. Based on the information provided in these studies and the existing analyses, conflicts and barriers will be identified in time to facilitate widespread public utilization of WECS for power generation, anticipated to begin in the mid-1980's. The 10-kW dispersed applications demonstration program also will be monitored for utility interface problems.

Offshore Emplacement. A decision whether or not to develop offshore WECS technology will not be made until 1980 or later. General feasibility studies are under way; no specifically environmental studies are being conducted. Consequently, the only studies scheduled in the current environmental planning efforts provide for issue-identification research to guide future environmental research planning in the event of a decision to proceed with offshore WECS development.

Physical Appearance of WECS Structures.

Research is needed to further define public aesthetic reactions to wind energy systems and to identify unacceptable configurations and locations. A preliminary study of public reactions to a variety of WECS designs and locations has been completed. Monitoring of public response to the appearance of actual WECS arrays is scheduled to occur during the testing of the system. Siting and design recommendations based on this research will be developed by the end of 1983, in time for aesthetic considerations to be incorporated into the general public utilization phase of WECS use. Monitoring of public aesthetic reactions also will be performed during the dispersed applications demonstration program.

Scheduling for Research Effort

Figure 12-3 presents a schedule for environmental R&D efforts and compliance activities coordinated with major wind energy conversion technology developments and applications. The upper section of the figure esta-

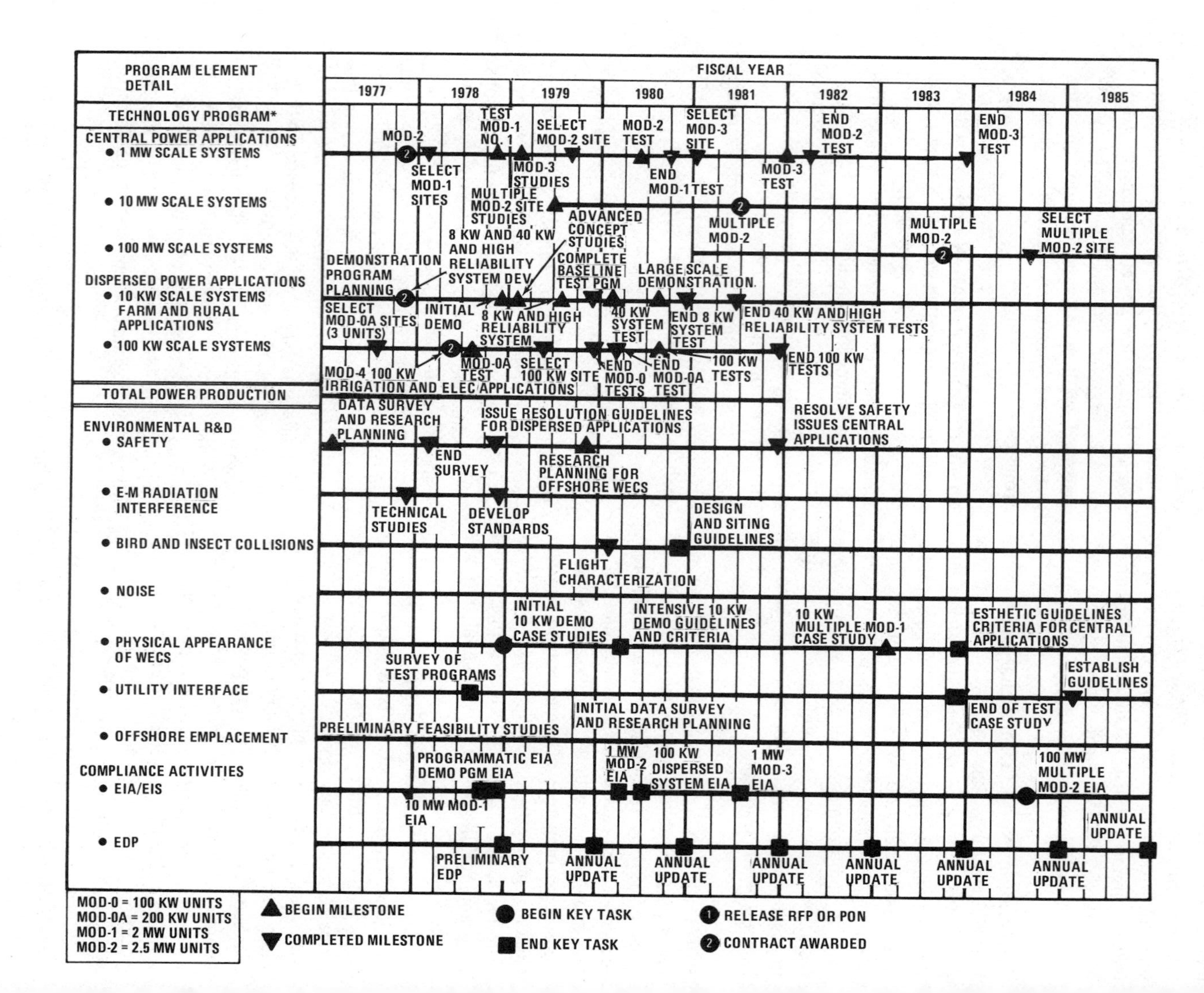
PROGRAM ELEMENT DETAIL
FISCAL YEAR
1977
1978
1979
1980
1981
1982
1983
1984
1985
TECHNOLOGY PROGRAM*
CENTRAL POWER APPLICATIONS
• 1 MW SCALE SYSTEMS
• 10 MW SCALE SYSTEMS
• 100 MW SCALE SYSTEMS
DISPERSED POWER APPLICATIONS
• 10 KW SCALE SYSTEMS FARM AND RURAL APPLICATIONS
• 100 KW SCALE SYSTEMS
TOTAL POWER PRODUCTION
ENVIRONMENTAL R&D
• SAFETY
• E-M RADIATION INTERFERENCE
• BIRD AND INSECT COLLISIONS
• NOISE
• PHYSICAL APPEARANCE OF WECS
• UTILITY INTERFACE
• OFFSHORE EMPLACEMENT
COMPLIANCE ACTIVITIES
• EIA/EIS
• EDP
MOD-2
SELECT MOD-1 SITES
TEST MOD-1 NO. 1
MOD-3 STUDIES
SELECT MOD-2 SITE
MOD-2 TEST
END MOD-1 TEST
SELECT MOD-3 SITE
MOD-3 TEST
END MOD-2 TEST
END MOD-3 TEST
MULTIPLE MOD-2 SITE STUDIES
ADVANCED CONCEPT STUDIES
MULTIPLE MOD-2
MULTIPLE MOD-2
SELECT MULTIPLE MOD-2 SITE
DEMONSTRATION PROGRAM PLANNING
8 KW AND 40 KW AND HIGH RELIABILITY SYSTEM DEV
COMPLETE BASELINE TEST PGM
LARGE SCALE DEMONSTRATION
SELECT MOD-0A SITES (3 UNITS)
INITIAL DEMO
8 KW AND HIGH RELIABILITY SYSTEM
40 KW SYSTEM TEST
END 8 KW SYSTEM TEST
END 40 KW AND HIGH RELIABILITY SYSTEM TESTS
MOD-4 100 KW IRRIGATION AND ELEC APPLICATIONS
MOD-0A TEST
SELECT 100 KW SITE
END MOD-0 TESTS
END MOD-0A TEST
100 KW TESTS
END 100 KW TESTS
DATA SURVEY AND RESEARCH PLANNING
END SURVEY
ISSUE RESOLUTION GUIDELINES FOR DISPERSED APPLICATIONS
RESOLVE SAFETY ISSUES CENTRAL APPLICATIONS
RESEARCH PLANNING FOR OFFSHORE WECS
TECHNICAL STUDIES
DEVELOP STANDARDS
DESIGN AND SITING GUIDELINES
FLIGHT CHARACTERIZATION
INITIAL 10 KW DEMO CASE STUDIES
INTENSIVE 10 KW DEMO GUIDELINES AND CRITERIA
10 KW MULTIPLE MOD-1 CASE STUDY
ESTHETIC GUIDELINES CRITERIA FOR CENTRAL APPLICATIONS
SURVEY OF TEST PROGRAMS
ESTABLISH GUIDELINES
END OF TEST CASE STUDY
PRELIMINARY FEASIBILITY STUDIES
INITIAL DATA SURVEY AND RESEARCH PLANNING
PROGRAMMATIC EIA DEMO PGM EIA
1 MW MOD-2 EIA
100 KW DISPERSED SYSTEM EIA
1 MW MOD-3 EIA
100 MW MULTIPLE MOD-2 EIA
10 MW MOD-1 EIA
PRELIMINARY EDP
ANNUAL UPDATE
ANNUAL UPDATE
ANNUAL UPDATE
ANNUAL UPDATE
ANNUAL UPDATE
ANNUAL UPDATE
ANNUAL UPDATE
ANNUAL UPDATE
MOD-0 = 100 KW UNITS
MOD-0A = 200 KW UNITS
MOD-1 = 2 MW UNITS
MOD-2 = 2.5 MW UNITS
BEGIN MILESTONE
COMPLETED MILESTONE
BEGIN KEY TASK
END KEY TASK
1 RELEASE RFP OR PON
2 CONTRACT AWARDED

blishes a time frame for the implementation of experimental applications and projected annual power production. The environmental schedule presented in the lower section of the figure establishes coordination between technology development and environmental resolution.

Environmental Development Plan (EDP)

The Environmental Development Plan (EDP) for wind energy conversion identifies health, environmental, safety, social and economic issues to be addressed as wind energy conversion technologies advance towards public utilization. It states the research needed to resolve these issues and provides a plan for the orderly development of required research. The EDP should support the overall wind energy technology development by ensuring that health, environmental and safety problems are specified in such a way that mitigation and control of potential problems are achieved in a timely manner. In addition, the plan should provide a concise reference to these issues for public interest groups outside the agency.

The current Wind Energy Conversion Environmental Development Plan (EDP) is published in DOE/EDP-0007 dated March 1978. This section includes the significant information provided in the EDP.

UTILIZATION OF RESEARCH INFORMATION

Environmental Impact Assessment and Statement Process

The National Environmental Policy Act (NEPA) of 1969 and the August 1, 1973, guidelines of the Council on Environmental Quality require that all agencies of the Federal Government prepare detailed environmental statements on major Federal actions significantly affecting the quality of human environment. The objective of NEPA is to build into the Federal agency decision-making process, at the earliest possible point, appropriate and careful consideration of all environmental aspects of a proposed action in order that adverse environmental effects may be avoided or minimized. The foundation of DOE's NEPA compliance program is the preparation of Environmental Impact Assessments (EIA's) and, if determined necessary, Environmental Impact Statements (EIS's) for significant energy RD&D and related activities.

Significant results or findings of the environmental research activities defined can serve as a data base for EIA/EIS preparation and update. In addition, the EIA/EIS process, incorporating environmental research results, serves as a means for DOE to provide "full disclosure" of the environmental implications of developing energy technologies to Congress and the public.

Environmental Performance Goals

Environmental Performance Goals (EPG's) are the technical design specifications for a new technology which will permit the hardware or system to function in an environmentally sound fashion. The specified EPG's must be compatible with existing environmental, health and safety regulations but will, in general, go beyond current rules, forming the bases for promulgation of new guidelines and regulations. In addition, because the EPG's generally impose constraints on the design of hardware or systems, they provide important input to the economic analysis of widespread use of the technology.

As EPG's determine the environmental acceptability of a technology's performance, they can, in some cases, represent the primary limiting constraint to technology utilization. Therefore, it is necessary that EPG's be developed to ensure maximum protection of public health and safety and environmental quality, while minimizing unnecessary impediments to technology development.

Dissemination of Information

Results of environmental research related to WECS should be disseminated on a wide scale to concerned Federal, state and local government agencies, involved private industry and the general public. This information will enable governments and industries to ensure that wind energy systems are acceptable from the environmental, health and safety standpoints and will educate consumers about the potential risks of, and necessary safeguards for, these systems.

Environmental research results can be disseminated in a variety of ways. Reports and assessments are available through various Government information services. In addition, pamphlets and/or newsletters summarizing results of environmental research can be prepared and distributed to concerned organizations and the general public. Finally, periodic conferences and seminars can be held to provide a forum for the public to comment on relevant environmental, health, safety and socioeconomic issues.

Standards

WECS must be designed and operated in compliance with standards developed by the Occupational Safety and Health Administration, the Federal Power Commission, the Federal Communications Commission, the Federal Aviation Administration and the U.S. Coast Guard. State and local governments may impose land use regulations as well as structural safety codes. Timely and accurate technical information may be provided by DOE, the National Institute of Occupational Safety and Health, the National Institute of Environmental Health Sciences and the National Bureau of Standards, as well as Underwriters Laboratories and the American National Standards Institute.

In addition to these standards, the wind energy conversion industry should develop and utilize quality workmanship and performance standards to assure the public that these new energy resources are reliable and safe.

LEGAL BARRIERS

This section deals with possible legal problems caused by purchasing, installing and operating a WECS. If you install a wind turbine within a few hundred feet of your property line, and your neighbor plants a row of fast-growing trees along that line, your wind energy could, in a few years, be greatly reduced. Protecting oneself from this occurrence is described in this section under "Wind Rights." A height limitation on structures in your local zoning ordinance is one of a number of possible problems that may affect your wind turbine construction plans. The section on building permits explains your courses of action. The optimized wind energy system may mean interconnection with the local utility company or sharing with your neighbors. Either can have economic or legal implications. Finally, before something goes wrong, warranty, liability and insurance matters deserve your consideration.

Nearly all the material in this section has been taken from a comprehensive report compiled by George Washington University.* The author does not represent himself as an attorney. You and your attorney should refer to that report for more detailed information. The material presented here is for general information only. Contact appropriate local and other governmental authorities if special permit problems are anticipated, or your attorney if easement or liability problems arise.

Wind Rights and the "Negative Easement"

Sometimes agreements have been made by the present or prior property owners of a potential wind turbine site not to conduct certain activities or erect buildings greater than certain heights on the property. These private agreements are commonly known as restrictions or restrictive covenants, some of which are said to "run with the land." Owners succeeding the person who entered into such an agreement are bound to comply with the restrictions. A title search of the deed should reveal any such agreement. However, these agreements must fulfill various legal requirements before a person can be bound by them. There-

*_Legal-Institutional Implications of Wind Energy Conversion Systems_, George Washington University, Washington, D.C., September 1977. (Report No. NSF/RA-770203.)

fore, their existence does not necessarily mean that the WECS owner is legally bound to follow them.

Also, there are no laws that describe your right to the wind that blows across your land. If wind turbines come into widespread use and conflicts arise, such laws might be enacted. The problem will be very real to you if your wind turbine is within a few hundred feet of an upwind neighbor who plants trees or builds a tall structure so that the smooth flow of your wind is interrupted, causing a reduction in the output of your wind turbine. Worse yet, if his structure is upwind during your strongest gale winds, the added turbulence might just be enough to destroy your blades.

The document that can provide the WECS user with the greatest protection is a "negative easement." Easements are interests in another person's property that give the easement holder limited right to use the property for a specific purpose—for example, a right-of-way. A negative easement gives the easement owner the power to prevent certain acts of an upwind landowner. Short of buying the property, this is the best way of protecting your wind source.

Consider trading partial negative wind easements with your neighbor. These easements would allow each of you to block each other's wind only by your own wind turbines, possibly specifying their diameter and minimum distance to the common property line. The natural growth of existing trees should probably be excluded. Obviously, an attorney should be consulted.

Community Building Permits and Controls

After determining that you have a satisfactory site for a wind turbine and deciding that you want to erect one, you should investigate any possible laws or local ordinances that may affect the erection of your tower and wind machine. A call to your local building inspector may be all that is required.

The categories of controls that affect the wind system owner are 1) local zoning ordinances, 2) Federal, state and local laws and 3) building codes. Each of these types of controls are discussed below.

Zoning. Zoning regulations are based on the state's jurisdictional powers, under which the state may regulate private activity for the purpose of enhancing or protecting the "public health, safety and welfare." Zoning laws, usually called ordinances, are, with a few exceptions, enacted and enforced by municipal and county governments. Where zoning is in force, the WECS owner must show that his proposed activity and structure conform to the restrictions applied to the site by zoning ordinances before obtaining a building permit.

The majority of municipalities and counties use the same basic process to enforce zoning ordinances. The prospective wind turbine owner, or his contractor, starts the process with an application for a building permit, filed with the planning department, zoning enforcement office, building inspector, etc., who will issue such a permit if the provisions of the applicable zoning ordinance are met. Construction may be checked periodically to ensure that the materials and workmanship meet the building codes.

The typical lattice-type windmill tower is generally termed an accessory building, since it is a separate structure and cannot be lived in. If it is part of a residence or other building, the zoning restrictions are applied to the whole structure.

Zoning ordinances typically regulate uses or activities that may occur on the land; population density; and such building requirements as height, number of stories, size of building, percentage of the lot that may be occupied and setback. Aesthetic considerations, not typically treated as a separate concern, are, however, a factor in writing and administering zoning ordinances. So-called architectural review is not standard practice, but some municipalities and "new towns" have enacted legislation designed specifically to regulate building appearance and compatibility with neighboring structures.

Conceivably, the proposed WECS may violate restrictions, particularly in a residential area. Resolution of this problem may depend on the wording of the ordinance and its interpretation. Also, the WECS may be permitted as an accessory use, one related to a permitted use of the land. Under this theory, for instance, ham radio towers have been permitted on residential property.

There are various options available to the WECS owner who is restricted by a zoning ordinance. He can appeal the interpretation and application of the ordinance to the board of zoning appeals or board of zoning adjustment, a local body that exists to oversee the zoning process. He may be able to utilize the so-called special exception or conditional use; a permitted use is explicitly mentioned in the ordinance, but its application to a particular area is allowed only after approval by the board of zoning adjustment. He may also be able to get a variance, a permitted variation from the ordinance. The granting of variances is governed by broad considerations of the purposes to be served by the zoning scheme, and the board of zoning appeals often has this power. Next, the WECS owner might attempt an amendment to the ordinance prohibiting his operations. In most states, the procedure for amending ordinances is the same as that for enacting them in the first place. This usually involves action by the city council or board of supervisors of the municipality. Also, the WECS owner might attempt to get a change ordered by the courts. This, of course, involves a court action, which might be undertaken with a variety of special procedures. Such an action is likely to be expensive and time-consuming and may succeed only in extreme circumstances.

Other Laws and Regulations Relevant to Land Use. Federal, state and local regulations and laws, or statutes other than the zoning ordinance, conceivably may affect the WECS owner. These include statutes that regulate the selection of sites for electric generating plants; laws designed to protect the environment; and legislation regulating the use of particular geographic areas, such as coastal lands, wildlife reserves, historical sites or navigable waters. It is very difficult to generalize about the impact of these provisions, but their overall effect on small WECS should be minimal. The laws that affect the WECS owner will depend on the wind system's location and size, and possibly on who wants to erect it: an individual, a cooperative or a company. If a permit from the Federal, state or local government is required before construction can begin, other laws may become involved in the permit process. For example, if the tower were located within the high-water mark of a river and a permit from the Corps of Engineers was required, Federal laws require the Corps to consult other governmental agencies before issuing a permit.

In addition, legislation designed to protect the environment exists at both the Federal and state levels. Most of these acts become relevant to the WECS owner only when a government permit granting agency is involved. Generally, compliance with the law is the direct responsibility of that organization, not of the developer. This involves the satisfaction of various paperwork requirements and the submission of such reports to a variety of interested agencies. Overall, because of the likely low environmental impact of small WECS, such procedures will probably be time-consuming at worst. Further, the small-scale WECS owner is not likely to be affected by a state's power plant site selection statute, since these typically apply to a minimum rated capacity of about 50 MW, or to those utilizing a certain fuel source.

Federal Aviation Administration (FAA) regulations require that the owner of any structure higher than 200 ft give notice to the FAA on forms provided for that purpose. While few small WECS towers are that tall, lower height limitations apply within the vicinity of an airport. For example, a wind turbine up to 100 ft high (to the top of the blade path) might be allowed at 5000 ft from a runway. As soon as notice (if necessary) is received, the FAA applies different height standards to determine whether the tower is an obstruction. These standards are generally less stringent than those governing notice. If the WECS were found to be an obstruction, the most likely requirement would be the placing of warning lights on it. Applications for building permits for structures in the vicinity of airports are usually forwarded to the FAA by municipalities.

Building Codes. Like zoning matters, the state's jurisdictional power is the basic authority under which building codes are enacted. Some state legislatures enact statewide building codes, while others delegate the authority to the local governments. One (or a combination) of the four model building codes has been adopted by most states or municipalities. These codes (and the geographic area) dominated by their association/author are: 1) the Uniform Building Code, written by the International Conference of Building Officials (adopted primarily in the West); 2) the Basic Building Code, compiled by the Building Officials and Code Administrators International, Inc. (found in the Northeast and North Central areas); 3) the Southern Standard Building Code, enacted by the Southern Building Code Conference (adopted in the South); and 4) the National Building Code, developed by the National Board of Fire Underwriters. Local variations exist despite the model codes. Some municipalities have adopted selected provisions rather than the entire code. Interpretations of the same code differ from city to city.

Unlike zoning ordinances, most building codes apply retroactively. Three types of information are provided in most codes: definition of terms, licensing requirements and standards. Taken together, the definitions and licensing requirements have the effect of prescribing who is authorized to conduct particular sorts of construction activity. For example, unless you are doing the work on your own system, the International Association of Plumbing and Mechanical Officials Code states that only licensed plumbers may do work defined as plumbing. Many codes require that structural design plans be prepared by a state certified engineer.

Two types of code standards exist: technical specifications and performance standards. Codes prescribing technical specifications set out how, and with what materials, a building is to be constructed. Performance standards represent a more progressive and technically more flexible approach. Codes based on these standards state product requirements that do not prescribe designs and materials. For example, "the structural frame of all buildings, signs, tanks and other exposed structures shall be designed to resist the horizontal pressures due to wind in any direction. . ." Typical construction components specified in codes are structural and foundation loads and stresses, construction material, fireproofing, building height (this represents a common duplication of the zoning ordinance) and electrical installation. The WECS developer is likely to be required to comply with the standards for structural and foundation loads and stresses, as well as the electrical installation code. The structural design standards set out the minimum force measure in $lb/in.^2$ that the WECS must bear under certain circumstances, e.g., wind or snow. The electrical code regulates the use of a generator and the electrical wiring when voltage levels are above 36 V.

Administration of the building codes is delegated to a board of review in some states, and to the building official in others. No building may be erected, constructed, altered, repaired, moved, converted or demolished without a building permit, and this can be obtained only after the building official is satisfied that the plans satisfy all applicable building codes. A trend is developing towards combining the administration of building codes and zoning ordinances in one municipal department.

Dissatisfaction with the building inspector's denial of a permit may result in an appeal before the local board of building appeals. The common bases of appeal provided by the codes are as follows: an incorrect interpretation of the code by the building official; the availability of an equally good or better form of construction not specified in the code; and the existence of practical difficulties in carrying out the requirements of the code. The local board members are usually appointed experts in the field of construction. The local board may uphold, modify or reverse the building official's decision. Further appeals to the state board of building appeals or to the courts are also available.

Sharing, Buying and Selling Power

You may be considering a wind system where you share excess power or sell it to neighbors, buy makeup power or sell excess power to a utility. Unfortunately, the state utility regulatory structure may cause you some problems.

The first of these possible problems has to do with the regulated monopoly structure of the public utility industry. This structure operates by the assignment (based on what is often called the certificate of public convenience and necessity) of a geographical area to a particular electrical supplier. In order to operate within an area occupied by an existing utility, all entities defined as public utilities typically must obtain this certificate, and to do this they usually must demonstrate a compelling need, such as the inadequacy of existing service. The result of this is often to prevent new electrical suppliers from operating within such a protected domain. Their status as a public utility is the crucial point here.

Selling Power to Neighbors. The small wind turbine owner who generates power for only personal use or shares it with neighbors at no charge is not defined as a public utility and thus will not be hindered by the regulated monopoly structure. Sales of electricity to others, however, may cause problems. Some state statutes limit the exact number of people (i.e., 10 or 25) to whom sales can be made before public utility status exists. More commonly, the statements in the law will contain language making it appear that *any* sale to *any* part of the public will result in public utility status and the need for a certificate, effectively prohibiting such sales. However, the courts of such states often interpret this language to require "dedication to the public use"; that is, an offering of service to the general public coupled with a willingness to serve those who apply. Under such standards, the small-scale WECS owner-operator selling to a few friends would probably escape. Sales of electricity by a landlord to his tenants may cause similar problems, and the states have taken a variety of approaches here. For instance, if each tenant is metered and billed apart from the rent charged, public utility status may be hard to avoid. Finally, it should be noted that if the WECS owner obtains supplemental power from the existing utility grid (see below), the service contract with the utility will almost certainly contain terms prohibiting such sales.

One possible way for small-scale WECS owners to avoid these problems is for them to start a cooperative. Basically, co-ops are nonprofit, membership corporations, the members being both the owners of the corporation and the consumers of the electricity produced by it. Most co-ops are fairly large, located in rural areas and funded by the Rural Electrification Administration. However, most states have special statutory schemes for the incorporation of co-ops, and these often allow incorporation by as few as three to five individuals. The various requirements, of course, must be complied with by the WECS owners.

The point of utilizing the cooperative form in this context is that cooperatives are not defined as public utilities in some states, although the number of states which make this exception is decreasing. In states which do make this exception, the WECS owners could generate power for themselves within the domain of a regulated utility without being checked by the certification requirement. However, in such a case, they would not be granted protection from competition with existing or future utilities.

Generally, this problem of restriction of WECS operations due to collisions with the existing regulated utility structure may be more hypothetical than real, at least for sales to a very few people. However, given the diversity of laws and practices in the 50 states, this may not always be true. Experts in the field, or perhaps the state Public Utility Commission, should be contacted before such sales are attempted. It should be remembered that the more substantial the sales to others, the greater the likelihood of problems.

Buying Power. At present, utilities generally do not object to user-owned power generation systems that provide the user's power needs part of the time while the utility provides the power when the system is turned off. However, increased use of solar and wind power generating systems may bring about a change in attitude of the utilities towards these systems, due to the fact that heavy demands could be placed on the utility during windless or sunless periods. The need for the utilities to maintain peak load capacity, even though it would generally be selling less electricity, would result in a loss of revenues. Conceivably, this objection could be overcome through the adoption of special rate structures—not unlike existing standby rates. In general, it can be said that utilities and state power commissions will need data on the power requirements of wind generator owners before adequate rate schedules can be set.

Selling Excess Power to a Utility. If a wind turbine generator owner requires the connection of his facilities with that of a utility so that he may produce some of his needed power while simultaneously buying the remainder of his needed power, or if he wishes to sell excess power, a proper interconnection between the two generating systems will have to be made. The utility will want to ensure that the connection is safe and will not jeopardize its own facilities or men working on the line. The state commission will oversee this process to its own satisfaction. The utility will require the right to inspect the connection at any subsequent time and make any necessary modifications to ensure safety and proper operation.

Power fed back into the utility's lines will have to be at the correct voltage and be synchronized in frequency and phase. A synchronous inverter is one device available to ensure these conditions. When such an interconnection is made, two meters will probably be used: one to measure the electricity bought from the utility and the other to measure the electricity fed back to the utility's lines. This would enable the utility to charge a particular price for the power it sells and buy the wind turbine generator owner's excess power at a wholesale rate (if reimbursed at all) to make up for its capital and distribution costs.

With this general description in mind, we will turn to the many legal aspects of this situation. First, it is likely that the WECS owner will have to bear (through the rate structure or otherwise) much of the cost of effecting the interconnection (e.g., the extra meter). Second, it is conceivable (though fairly unlikely) that the utility's general duty to serve all user's may not extend to this situation. Third, a "demand charge" might be applied here (which could involve a higher ¢/kWh rate and an additional charge based on the extra capacity required by the utility) and service might also be interruptible (i.e., capable of being shut off at the utility's option). Fourth, at least some of the utilities now prohibit a reverse flow of electricity back into the grid when they provide supplementary electricity to a self-supplying customer. Whether this will continue to be the case if the price of conventional fuel increases and wind/solar devices become more numerous is uncertain. Finally, there is a question as to the amount of the credit to be given the WECS owner's bill, assuming that the utility does permit such a sale to it. State utility commissions will probably be required to decide on all of these questions.

Warranty, Liability and Insurance

Liability of the Manufacturers and Installer. Usually, one who is injured (financially or physically) by a product can receive money to cover damages on the basis of negligence, warranty or strict liability. The injured person must show that the product was defective, that the defect caused the injury and that the defendant being sued is responsible for the defect. The term defect has come to mean anything initially wrong with the product that can occur during the process of manufacture and sale. To recover on a claim, an injured person must prove that the defendant should have taken reasonable care to take precautions against creating foreseeable and unreasonable risks of injury to another, and that his not doing so was the cause of the injury (either financial or physical).

In circumstances where there may be a dangerous nature to a product (i.e., blade and tower failure consequences), the seller may have an obligation to give adequate warning of unreasonable dangers of which the seller knows or should know. This obligation to warn the potential buyer extends to all advertising.

An express warranty is a claim, promise, description or sample made by the seller, which is made part of the bargain with the buyer. The injured person must have knowledge of this claim or promise, and only be injured as a result of reasonable reliance on it. Liability is established when the product is demonstrated to be not as good as was claimed. Potential sources of express warranties include: the name of the product; descriptions of the product found in advertising brochures, catalogues or packaging; drawings or other pictorial representations accompanying the product; and all representations made by the seller or his agent to the buyer. However, not all claims about a particular product are treated as express warranties giving rise to liability. Mere sales talk and opinion have been distinguished as representations which are not meant to be relied upon by the buyer.

An action for breach of warranty proceeds on a contract theory, as distinguished from the laws governing negligence. As such, it focuses upon the express or implied promises made by the defendant to the injured person and not on the defendant's fault. In such an action, a consumer need only prove that the product was defective when sold and did not conform to the defendant's representation about the product, and that he was injured as a result of that defect. The advantage of warranty action over a negligence claim is the absence of the need to prove that the seller failed to use reasonable care. The rules of warranty have been codified by the Uniform Commercial Code, which has been adopted by all states with the exception of Louisiana.

Whom to sue in a product liability case depends upon the parties connected with the particular product, as well as the plaintiff's evaluation of the economic worth of each potential defendant. In most cases, the defendant will be the manufacturer, distributor, wholesaler, retailer or other supplier in the direct chain of distribution. Anyone in the chain of distribution who represents a product as his own is subject to the same measure of liability as that of the manufacturer. Liability will depend on whether anyone in the chain of control has the duty to discover defects in the product.

Liabilities of the WECS Owner. Each owner/occupier of land enjoys the privilege of using land for his own benefit. A standard of reasonable care qualifies that privilege by imposing the duty to make a reasonable use of property, which causes no unreasonable harm to others in the vicinity. The duty of reasonable care is affected by the location, since the hazards to be anticipated in crowded, commercial areas are not the same as those involved in rural areas. Reasonable care, on the other hand, does not require such precautions as will absolutely prevent injury or render accidents impossible.

Owners of property are under no legal obligation to trespassers other than to do such persons no willful harm. The status of trespasser has been held to include those who enter upon the premises unintentionally, such as persons who wander too far from the highway. The standards applicable in the case of trespassing children, however, are not the same as those for adults. A number of states follow the "attractive nuisance" doctrine, which imposes liability for the creation of conditions that are so alluring to children (despite the danger apparent to those possessing greater discretion) that they are induced to approach and be exposed to the possibility of injury. Liability has been held not to exist in more isolated places where the owner had no reason to anticipate the presence of children.

Insurance. The standard homeowner's insurance package usually covers liability connected with an accessory building with the following conditions attached:

- The installation is not to be used for commercial purposes; and
- The structure is not highly susceptible to fire (for example, a woodwork shop or a storage area for flammable materials).

It is not certain if a WECS would be considered to be engaged in commercial activity if excess power is sold to a utility and credit obtained against the cost of power that is bought. Written clarification of this point in such circumstances would assure the owner of necessary coverage. Antennas and masts are not covered against damage from wind, rain, hail and snow. However, it is possible that a WECS will not be included in the category of antennas, and, therefore, will be covered fully against fire and acts of nature that are covered by the policy.

Some insurance underwriters will not want to accept the added risk of a wind turbine and may simply force you to find another insurance company. Those companies that have a sizable business in rural areas with a good wind potential will most likely be prepared to offer coverage for your WECS. The insurance company, of course, will require that the structure conform to all applicable state and Federal ordinances and regulations.

Typically, a homeowner's insurance policy provides coverage only for outsiders performing minor amounts of yard work and not jobs that would be covered by workmen's compensation. In this case, help used for the erection of a tower and installation of the wind turbine will not be covered by the homeowner's policy. A contractor will have his own insurance. If other help is used for these tricky and potentially very dangerous jobs, special attention must be paid to insurance coverage.

13
International Development

WORLD WIND ENERGY RESOURCES

Wind energy is not new. Man has used it for centuries for such tasks as driving mill stones and pumping water, and for other small-scale, intermittent uses. The latent power of the wind and its potential for doing useful work are great; of the total solar energy falling on the earth each year—5300 quads (10^{18} Btu)—about 88 quads or 25,800 kWh are converted into air in motion. Some fraction of this energy is available in the first 100 m or so above the ground surface where wind power plants might be installed. The World Meteorological Organization has estimated that a little less than 1 percent of the wind energy (that is, 0.6 quads or 10^{18} Btu or 175 × 10^{12} kWh) is available at selected sites throughout the world.* Wind, therefore, is an untapped renewable energy source that might, in the long term, make a contribution to the world's energy needs.

The distribution of wind is not uniform over the earth—wind velocities and frequencies are higher in polar and temperate zones than in tropical zones and are generally higher in coastal areas than inland. The map of wind energy availability, shown in Figure 13-1, gives data in kWh of output per year per kW of rated plant capacity. If winds blew constantly throughout the year, the maximum output would be 8,760 kWh of installed capacity.

As noted earlier, the output of wind power machines increases as the cube of the wind velocity. Because of friction losses, wind power machines usually do not operate at wind velocities much less than 6 mph; if winds are of gale force, the rotors of wind machines are usually feathered to prevent damage. Wind energy resources, therefore, do not include the entire range of wind velocities.

INTERNATIONAL INTEREST IN WIND ENERGY

During the past 100 years, the industrialized world has come to depend very heavily on fossil fuel systems as basic power resources. Historically, this has occurred because the industrialized countries discovered that a sequence of fossil fuels could be used abundantly and conveniently, and many of them discovered large indigenous fossil fuel resources that seemed adequate for centuries to come. These countries believed that short supplies in their own resources could be increased through purchases at low prices on the world market. This, in turn, spurred large capital investments for developing, commercializing and installing a wide range of power systems based on fossil fuels; large R&D investments were also incurred by the industrial nations to improve the performance and efficiency of conventional fossil fuel systems.

Developing countries, attempting to optimize their own economic development efforts, found it expedient to spend their limited capital on already developed fossil fuel power systems. This appeared to be an obvious approach because fossil fuels were thought to be widely available at a relatively low cost, even though most developing countries had minimal indigenous fossil fuel resources. Their economic development, as well as that of the industrialized countries, is now being threat-

*"Energy from the Wind," Tech. Note 4, World Meteorological Organization, Geneva, Switzerland, 1954.

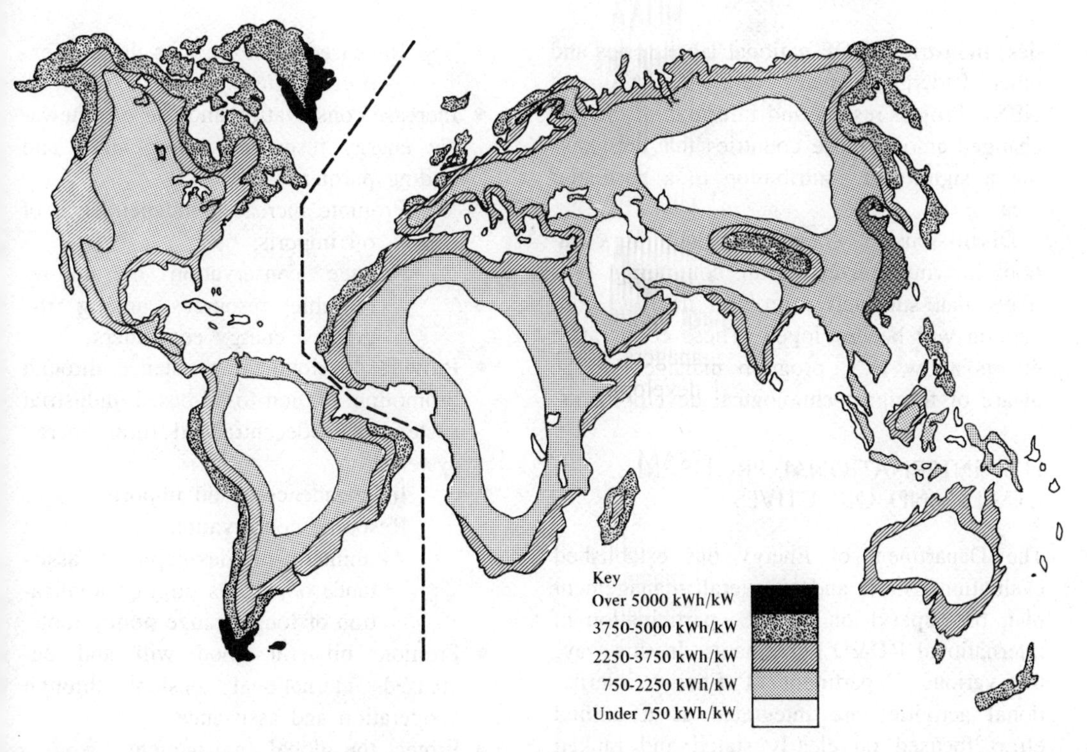

Figure 13-1. Availability of wind energy (annual specific output of windmills rated at 25 mph). *(Courtesy of WMO/NSF/DOE.)*

ened by the recent crisis in availability of one fossil resource; namely, oil. Moreover, their economic development is jeopardized by the negative patterns in the discovery and production of the most desirable fossil fuels relative to world energy needs and the resulting increases in the prices of all fossil fuels. Because of a lack of resources and embedded economics, the developing countries are less able to make the necessary adjustments to continue their programs for social and industrial development.

Industrial countries are making substantial investments in their quest for alternative energy resources, especially in solar/wind energy. The U.S. has taken an overall lead position in developing alternative energy systems. In particular, the U.S. has the foremost commitment to develop solar energy technologies, including direct low- and high-temperature thermal energy conversion systems for heating and cooling buildings, agricultural and industrial process heat and electric power generation. The U.S. also has established a lead position in photovoltaic systems, for direct conversion of sunlight to electricity, and in conversion systems, which provide electrical, thermal or mechanical power from wind, ocean and biomass-based systems (indirect forms of renewable solar energy).

INTERNATIONAL ENERGY AGENCY

The International Energy Agency (IEA) was established in 1974 with 19 member states. Among the purposes of the organization are long-term cooperative efforts in alternative energy sources. The IEA participating nations formed research and development working parties and designated lead countries for specific technologies. Cooperative projects are under way in heating, cooling and solar thermal technologies.

All U.S. funds for these projects are spent in the U.S. under normal program funding and management. The cooperation with other countries involves designating specific projects whose objectives and results are significant to a given IEA project area. U.S. participants in these projects include universities, industries, DOE national laboratories and other Federal agencies (e.g., NOAA and NBS). Project results and information are exchanged among those countries that are making a significant contribution to a particular area.

Discussions of cooperation with IEA nations in wind conversion are continuing. It is likely that substantive projects in wind conversion will be developed. These cooperative efforts allow U.S. program managers to be aware of foreign technological developments.

U.S. INTERNATIONAL PROGRAM GOALS AND OBJECTIVES

The Department of Energy has established evaluation criteria and an overall management plan for supervision of U.S. participation in international RD&D agreements. In this way, the various Department of Energy international activities are integrated as a unified effort focused on clearly stated and ranked mission goals. The principal goals addressed in the DOE management plan in order of increasing scope and the resulting benefits are listed below.

- Advance the objectives of the U.S. domestic program through cooperative agreement activities and projects:
  - Decreased RD&D costs through shared information and project cost;
  - Increased breadth of information and experience;
  - Reduced time required for some phases of domestic projects through the use of existing facilities overseas.
- Promote the balance of U.S. trade by increasing exports of U.S. goods and services, along with reducing unnecessary imports:
  - Re-allocation of resources from consumption to investment;
  - Increased opportunities for small and large U.S. businesses;
  - Expanded U.S. manufacturing capacity that parallels or prepares for the emerging U.S. market;
  - Increased domestic jobs through export expansion.
- Increase conservation and use of renewable energy resources among allies and trading partners:
  - Promote increased independence of oil imports;
  - Provide conservation of non-renewable resources among the greatest energy consumers.
- Provide development assistance through promotion of non-fossil-based industrial systems and decentralized rural energy systems:
  - Independence of oil imports;
  - Resource conservation;
  - Attainment of development assistance objectives through application of longer-range policy tools.
- Promote bilateral good will and decreased international tensions through cooperation and assistance.
- Protect the global environment.

These goals will be accomplished through DOE's support of those projects which show the greatest promise for international commercial application. DOE is currently designing guidelines for project evaluation. The results will assist in determining the extent of U.S. participation in cooperative agreements as well as further project funding. Evaluation guidelines are to include qualitative assessments of each agreement's impact on domestic programs, expected cost savings, balance of payments, conservation, environmental quality and possible impacts on international relations. These assessments will use a numeric weighting scale for each agreement.

CURRENT U.S. INTERNATIONAL AGREEMENTS

In general, developing countries around the world have substantial indigenous energy resources, including wind energy and other solar resources. If harnessed, these wind energy resources could provide a significant portion of their energy needs. As shown in Figure 13-1, substantial wind energy is available internationally.

The U.S., as one of the few industrialized countries with substantial solar/wind technical resources, has an opportunity to extend assistance to developing countries so that they can use their abundant solar/wind resources. This can be accomplished by cooperation through multilateral and bilateral agreements. The Department of Energy directs the implementation and review of international cooperative participation on behalf of the U.S. The following international commitments to global energy research, development and demonstration (RD&D) have recently been initiated:

- Solar Energy Domestic Policy Review, International Panel
- International Energy Development Program
- Foundation for International Technological Cooperation
- Department of Energy Reorganization Act (Public Law 95-91)
- President Carter's Export Promotion Program
- Section 119, International Development and Food Assistance Act of 1977 (Public Law 95-88).

These commitments are binding in both multilateral and bilateral arrangements. Bilateral agreements are made as dictated by the level of interest in particular subject areas and other negotiations. Multilateral programs are administered through the IEA, an autonomous body within the Organization for Economic Cooperation and Development (OECD). Headquartered in Paris, IEA was founded as part of the 1974 Agreement between principal industrialized countries for an International Energy Program. Its objectives are to provide international RD&D to ensure the development and application of innovative energy technologies which will provide significant contribiutions to near-term energy needs and will also reduce long-term dependence on oil. Each member country selects RD&D project areas in which it wants to participate; participants then provide project financial support so that only that activities of high mutual interest are undertaken. The two main IEA activity areas for the wind energy program are 1) large-scale wind demonstration, and 2) technology assessment.

As of July 1, 1977, IEA countries include Austria, Belgium, Canada, Denmark, the Federal Republic of Germany, Greece, Ireland, Italy, Japan, Luxembourg, the Netherlands, New Zealand, Norway, Spain, Sweden, Switzerland, Turkey, the United Kingdom and the U.S.

The Department of Energy is currently participating in five wind energy-related RD&D international agreements. Each of these activities is either ongoing or under negotiation. A summary listing of these programs is noted in Table 13-1.

U.S. Multilateral Agreements

Brief descriptions of the multilateral RD&D agreements entered into by the Department of Energy on behalf of the U.S. are provided in Table 13-2.

U.S. Bilateral Agreements

The U.S. currently participates in two international bilateral agreements for wind energy. These agreements are with Saudi Arabia and Spain and are summarized in Table 13-3.

INTERNATIONAL COMMERCIALIZATION POTENTIAL

The multilateral and bilateral technical exchange agreements in which the U.S. participates will undoubtedly assist U.S. marketing efforts in the participating countries as well as non-participating developing countries. DOE's responsibility to support promising RD&D indirectly influences international commercialization. This influence is an inherent characteristic in the links which join basic research, applied research, development, demonstration and commercialization. There is great potential for DOE's International Program RD&D funding to aid U.S. export interests.

Commercial opportunities appear to exist in developing countries for WECS which meet the need for dispersed electric uses and which can efficiently operate in the windspeeds of 6–10 mph common to tropical zones.

Table 13-1. International RD&D agreements.

TECHNOLOGY (SOLAR ELECTRIC APPLICATIONS)	SPONSORING ORGANIZATION OR COOPERATING COUNTRY	STATUS
Large-scale wind energy conversion	IEA	Ongoing
Wind energy conversion systems	IEA	Ongoing
Wind energy	Saudi Arabia	Ongoing
Wind energy	Spain	Ongoing
Large wind–200 kW	IEA	Under negotiation

The two IEA WECS agreements focus on the availability of data concerning large-scale wind systems appropriate for centralized power grid applications. This is a reflection of the domestic program emphasis on reducing the costs of intermediate-size and large-size wind energy systems. The U.S. is participating in a project with Denmark to obtain data on operating efficiencies of the Gedsar windmill. This windmill was built by Denmark in 1961 to evaluate a wind turbine design. The facility was not operated for more than a decade until the recent renewed interest in wind turbine development emerged in both the U.S. and Denmark. The wind turbine was renovated in 1977 and is now being operated to provide the U.S. and Denmark with performance data for an improved wind turbine design.

In addition to the development of intermediate-scale and large-scale WECS, small businesses in the U.S. and other countries have made major contributions toward commercialization of WECS.

Aerowatt, a French firm, produces high-quality, high-price WECS for commercial applications. The Aerowatt systems produce full output at low wind speeds compared to some of their competitors.

The Dunlite Company of Australia has manufactured and sold many simple and effective 2-kW WECS around the world, including some of the first effective electricity-producing systems used in the U.S.

Elektro of Switzerland has produced electricity-producing WECS for more than four decades. These systems provide more power (2–10 kW) than the Dunlite system and have

Table 13-2. U.S. Multilateral agreements.

SUBJECT AND BASIC TERM*	SIGNATORIES	DESCRIPTION
Large-scale wind energy conversion, October 1977-October 1979	Denmark, Federal Republic of Germany, Sweden, U.S.	Exchange of information and personnel, including periodic meetings of program directors to coordinate execution of national projects to design, construct and operate machines with a rated power of 1 MWe.
Wind energy conversion systems, October 1977-October 1980	Austria, Canada, Denmark, Federal Republic of Germany, Ireland, Japan, Netherlands, New Zealand, Sweden, U.S.	Exchange of information, joint funding of studies, joint development of models and workshops on environmental and meteorological aspects, evaluation siting models, integration into electricity supply systems and rotor-stressing and operation of large-scale systems.

*IEA agreements typically cover a specified period with the provision for automatic extension.

Table 13-3. U.S. bilateral agreements.

COUNTRY	SCOPE, TERM AND DESCRIPTION
Saudi Arabia	Wind energy conversion systems, October 1977-October 1982
Spain	Wind energy conversion systems, September 1976-September 1981. This effort is part of the U.S.-Spain Treaty of Friendship and Cooperation. Information exchange will be provided for wind energy conversion and meteorology technologies.

been used in many remote power installations.

The Wincharger is a small, 12-V, 200-W WECS used for charging batteries and other low-power applications.

Water-pumping windmills have been manufactured by Aeromotor, Dempster, Heller-Aller, Sparco and other firms here and abroad. These systems have not only served effectively in the U.S., but are also marketed worldwide.

The rising cost of electricity and the installation of power lines is causing consumers here and abroad to reconsider the use of WECS. (Refer to Chapter 8 for complete descriptions of the characteristics of small-, intermediate- and large-scale WECS currently available.)

SUMMARY OF INTERNATIONAL PROGRAMS

Some wind power units of considerable size are already operating. Sweden has a 63-kW research unit with an 18-m turbine diameter. Germany had a 100-kW unit operating for several years; the U.S. has one 100-kW research unit and two 200-kW demonstration units in service, both of similar design. A private Danish group has put a 2-MW, 54-m-diameter wind power unit into test operation, which is the largest working wind power unit in the world at this time.

Today, wind turbines in the 100-kW range are not cost competitive in the energy market, except perhaps on isolated islands or in other remote places. It is expected, however, that the costs per kWh of electricity will decrease considerably with increasing turbine sizes. Whereas smaller wind turbines may continue to be used in special situations, larger-scale units have a potential for making a major impact on energy requirements. Therefore, efforts are being directed at improving economics by designing larger machines with a greater electrical output (e.g., in the MW range), which could contribute to integrated electricity networks in countries with favorable wind conditions.

Horizontal-axis machines appear to offer the most promise for success in the MW range. In the wind energy programs of Denmark, Germany, Sweden and the U.S., large units of horizontal-axis machines are being designed and built, ranging from 600 kW to 4 MW. These are to be tested to solve problems concerning turbine aerodynamics and aeroelastics, total system dynamics, material problems and electrical network interface.

Besides the technical questions concerning the actual construction of larger machines, problems related to the optimum siting of machines; environmental considerations connected with large-scale use of wind energy; and safety problems must be faced. Considerable R&D must, therefore, be completed before wind energy may be incorporated into energy networks.

IEA Activities

In October 1977, two IEA agreements were signed for projects aimed at accelerating the development of large-scale WECS through the design, construction and operation of relatively large wind machines; and through joint R&D on the technical, environmental and economic aspects of siting and operating these machines.

Development of Large-Scale Wind Energy Conversion Systems

To provide information essential to the development of large machines, Denmark, Germany, Sweden and the U.S. are each designing, constructing and operating at least one horizontal-axis machine in the range of 1

MW or above. They will exchange the results of their respective technical approaches, economic evaluations and environmental assessments and keep one another informed about national plans for R&D on large machines. Technical personnel will also be exchanged.

Sweden will install at least two prototype units in the range of 2–4 MW with 70–90 m rotor diameters. Proposals are being invited and contracts will be let in 1979 for operations to begin late 1980 or early 1981.

Germany is concentrating on the design and construction of a 3-MW turbine. The design will be finalized in 1978.

Denmark is constructing three experimental turbines of about 0.6 MW. Construction of these machines began in 1977 and will continue through 1979/1980.

Technology of WECS

This project area involves four tasks dealing with the environmental and meteorological aspects of WECS, mathematical models for siting such systems, the integration of wind power conversion machines into national electric power networks and the design of advanced wind machines.

The first tasks of the project are to evaluate the safety of large rotors; study efficient land use, including the possibility of using wind farms for other energy-related activities (i.e., biomass production); and assess existing meteorological forecast methods and related computer programs. The possible effect of "wind farms" on local weather conditions, broadcasting, telecommunication and radar systems, as well as the "optical pollution" of the environment by large numbers of wind machines, are also being studied.

Participants in this task are Austria, Canada, Denmark, Germany, Ireland, Japan, the Netherlands, New Zealand, Sweden and the U.S.

Wind speed and direction are two fundamental variables in the optimum siting of wind machines. Selecting the most appropriate site for a wind turbine is a long and costly process. To reduce the number and complexity of the necessary measurements and calculations, the participants of this project, as a second task, are comparing various computer models based on known general meteorological conditions and differing terrain to determine suitability for siting selection. An international validation plan for these models will be established and the results of tests with the various models will be verified by testing the models with specific data sets of meteorological conditions. Recommendations will then be made as to which model can be used for different meteorological conditions and terrain, and a user's manual will be prepared for each model.

Participants in this tasks are Canada, Japan, Sweden and the U.S.

As a third task, the participants are analyzing the integration of wind power into existing national electric power networks. Comprehensive computer models are being developed for this purpose, taking into account such factors as supply and demand variables and energy-storage requirements in the various participating countries. The objective is to identify in each participating country the prospects for, and the practicability of, using wind power to supplement other energy sources in generating electricity.

Participants in this task are Germany, Japan, the Netherlands, Sweden and the U.S.

The fourth task involves the accumulation of theoretical design data for machines in the 3-MW range for wind systems using advanced aerodynamic theories. Emphasis is on developing lighter rotors to minimize stress and vibration between rotor and tower. Results will be verified by scale models using wind tunnel test methods.

Participants in this task are Denmark, Germany, Japan, the Netherlands, Sweden and the U.S.

The following sub-sections provide technical overviews of the wind energy programs in Canada, Denmark, the Federal Republic of Germany, the Netherlands, Saudi Arabia, Spain and Sweden.

The Programs

Canada. Only a few years ago, it was assumed that Canada's situation with regard to oil and gas supply was a healthy one, and that new explorations would discover new resources. Although $2 billion has now been spent in frontier exploration, only limited discoveries have been made, and in 1975, Canada passed from a net exporter to a net importer of oil. It has been estimated by Shell Canada that 1975 marked the peak production year for oil and gas, with a steady decline predicted thereafter. A recent assessment by one of the deputy ministers in the Ministry of Energy, Mines and Resources, predicts a growing gap, at least to 1990 and beyond, between total national energy demand and domestic supply. The gap exists even if a successful energy conservation program is assumed, and if it is also assumed that total installed nuclear plant capacity is built up at a rapid rate (with a doubling time of roughly five years).

There is, therefore, a growing realization that all potential sources of energy, including renewable sources such as tidal, solar and wind, should be exploited wherever this is economically feasible.

*The Potential for Wind Energy.** Canada is a country with a huge area and a small population. This makes it easy to show that the theoretical potential for wind energy development is large in relation to total electricity demand. One recent estimate suggests that an average power of at least 50,000 MW could be developed in those areas lying within 100 miles of existing transmission lines, with very conservative assumptions regarding wind plant spacing. However, probably less than one-fourth of this potential lies in regions where wind energy density is high enough for reasonable economic conversion.

Figure 13-2 shows the distribution of raw annual wind energy density over most of Canada. It is, of course, unevenly distributed, but one interesting region is along the east coast and in the Gulf of St. Lawrence. It is in these maritime provinces that the main interest in wind power potential has now developed, partly because a large fraction of their electricity is generated by oil-fired plants with relatively high fuel costs. In Prince Edward Island, near the southern edge of the Gulf, a provincial agency has begun a detailed program of wind measurement at selected sites.

One of the windiest locations in Canada is on the Magdalen Islands, an island chain about 40 miles long, connected together by sand dunes, and lying close to the geographic center of the Gulf of St. Lawrence. It is part of the province of Quebec, and the provincial power authority, Hydro-Quebec, supplies the 13,000 inhabitants with electricity from a 25-MW diesel plant at high cost. Since the charge to the users has been artificially maintained the same as that on the mainland, the company suffers a substantial operating loss on the islands, and has been interested for some years in wind energy conversion, and in the possibility of energy storage by means of compressed air in the salt caverns underlying the islands.

Industrial Activity. Two Canadian companies, Bristol Aerospace in Winnipeg, and Dominion Aluminum Fabricating Limited in Toronto, began development of commercial versions of the vertical-axis Darrieus-type wind turbine about four years ago. A small Bristol unit, about 9 ft in diameter, has been operating successfully over a couple of years as the power supply for a remote unmanned weather station in the Arctic Sea. The station transmits its data to a stationary satellite, and has had no maintenance since it was first put into operation. In fact, its precise location is not presently known, since it has drifted on the ice surface about 1000 miles and is now somewhere north of Siberia. The company is now developing models suitable for intended operation at remote microwave communications sites. The Bristol turbines are equipped with aerodynamic self-starters and with low-speed multi-pole alternators, developed in cooperation with the University of Manitoba.

Dominion Aluminum Fabricating (DAF) has concentrated its recent development work on constant-speed, vertical-axis turbines using induction alternators, both for starting and for normal operation. Commercial units rated at roughly 10 kW and 50 kW have been developed. The largest wind turbine in Canada, a

*Wind Energy Research in Canada was prepared by R.J. Templin of Canada's National Research Council in Ottawa, Canada.

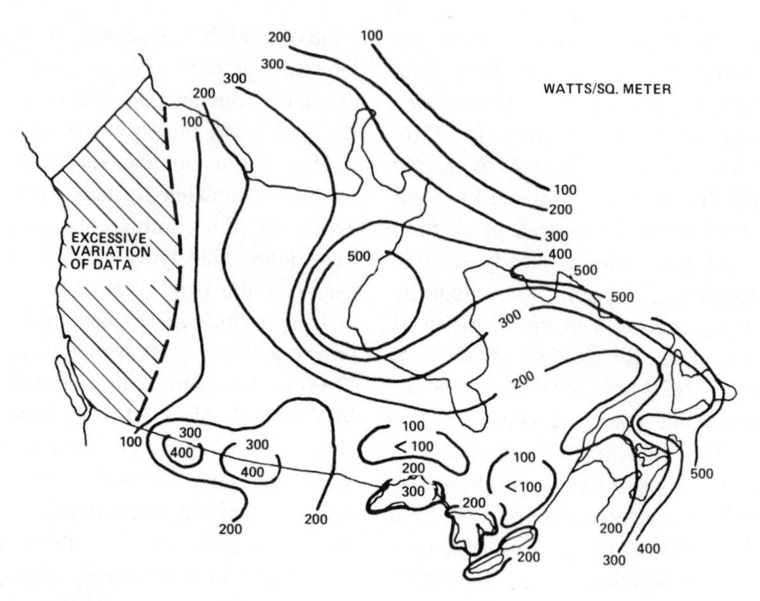

Figure 13-2. Annual average wind power density in Canada (measured at 30-m height). *(Courtesy National Research Council.)*

230-kW plant now installed on the Magdalen Islands, was designed and built by DAF under contract to the federal government.

The Magdalen Islands Wind Turbine. The Magdalen Islands turbine is the only large-scale demonstration wind plant in Canada. The project is being carried out cooperatively by the National Research Council and by Hydro-Quebec. The turbine was erected in May of 1978 by Hydro-Quebec using a 200-ft crane to lift the rotor after assembly at ground level. Installation of the control system and connection to the island transmission system took place in the next two months, and turbine operation began on July 14.

The turbine rotor has a diameter of 80 ft and a height of 120 ft. The lower pedestal, housing the generator and mechanical transmission, is 30-ft high. The generator is actually a commercial G.E. three-phase, 300-hp induction motor with a synchronous speed of 720 rpm. The mechanical transmission is in two stages: a gearbox with a speed-up ratio of about 17:1, and a multiple V-belt stage with variable speed ratio (by changing pulleys on the generator shaft). Access to the upper rotor bearing assembly is by safety ladder inside the central column, which is 5 ft in diameter.

The turbine is instrumented with strain gauges at a large number of locations, with transmission of data by means of a telemetry unit in the central column.

Normal operating speed is about 38 rpm, which is estimated to produce the rated 230-kW output at a wind speed of 30 mph. Most of the operation to date has been at positive power output, at mean wind speeds in the range of 15–45 mph. Strain gauge data have so far given no indication of structural resonances, and preparations are under way for operation at high speed. During initial operation at low speed, the guy wires have been de-tensioned to check for resonant conditions, and it has been found that, when their first-mode natural frequency is equal to two per revolution, small-amplitude oscillations will occur intermittently. Blade speeds have so far been too low for meaningful noise measurements to be made and, in fact, the rotor is silent. The site is very open and serves also as a test location for high-voltage transmission towers and lines. The turbine is located close to a test line of four 100-ft-high towers, and an interesting observation is that from a distance of more than a mile or two, it is almost invisible on the landscape in comparison with the towers.

Denmark. The wind power program of the Ministry of Commerce and the electric utilities in Denmark is a program for the development of larger electricity-producing wind power plants.* The program includes development work and tests on large-scale wind power plants, with special emphasis on concrete solutions of the technical problems which are feasible in practice. By means of measurements and service experience, the wind power program should further procure background material for new steps and advanced investigations concerning technology, economy and environment.

The first stage of the wind power program (1977–1979) comprises three projects:

1. Measurements on the 20-year-old Gedser windmill which has been refurbished with a view to limited test operation.
2. The construction of two larger, partly different wind power plants.
3. A survey of suitable sites for experimental windmills.

In the second stage of the wind power program (1978–1980), the work is progressing based on the experience from the first stage. The second stage is comprised of the following:

1. A measuring program for the Nibe windmills.
2. Investigation of the potentialities of wind power within the Danish electricity supply system.
3. Theoretical investigations concerning aerodynamics.

The wind power program is jointly financed by the Danish Ministry of Commerce and the Danish electric utilities, the Ministry having granted 28 million Dkr. and the power utilities 7 million Dkr. Further, the U.S. Department of Energy contributes toward the Gedser project with almost one million Dkr.

The program is administered and coordinated by DEFU (research Association of the Danish Electricity Supply Undertakings). Engineers from ELSAM, ELKRAFT and DEFU are in charge of the management of the individual projects. In the implementation of the projects, a number of institutions are also participating, primarily the Technical University of Denmark and the Riso National Laboratory.

The Gedser Windmill. After 10 years of inactivity, the 20-year-old Gedser windmill was refurbished for limited test operation in November 1977. (See Figure 13-3 for a photo of refurbishment of the Gedser WECS.) The mill has been equipped with several measuring instruments, and a measuring program is being carried out with the aim of procuring information about the essential properties of the mill.

The Gedser measurements are carried out in cooperation with the U.S. Department of Energy. The cooperation comprises a mutual exchange of measuring results and analysis for the Gedser windmill and the new American wind power plants. DOE is particularly interested in a comparison between its own plants and the Gedser windmill because of the favorable experience from the operation of the latter mill, and because of the many fundamental differences between the two types of mills. For example, the DOE windmills have rotors with two blades and pitch change mechanisms, whereas the Gedser mill has a three-blade rotor with fixed rotor blade angle; further, the rotor of the DOE mills is placed to the lee or downward side of a lattice tower, whereas the rotor of the Gedser windmill is situated on the upwind side of a concrete tower.

The Gedser windmill was built in 1956–1957 by the Wind Power Committee of the Danish Association of Electricity Supply Undertakings. Following a test period, the mill was in ordinary operation from 1959 until 1967; in this period, it supplied a total of about 2,242,000 kWh to the electricity supply network. In 1964, the highest annual production of 367,140 kWh was attained. The basic system characteristics are as follows:

- Three-blade rotor placed downwind and with a diameter of 24 m.

**The Wind Power Program of the Ministry of Commerce and the Electric Utilities in Denmark,* February 1979.

Figure 13-3. Refurbishment of Danish Gedser turbine sponsored by DEFU and DOE. *(Courtesy of DEFU.)*

- Blade tip speed of 38 m/sec.
- Rotor revolution of 30 rpm.
- Double chain gear, ratio 1:25.
- Eight-pole induction generator, 200 kW.
- Pre-stressed concrete tower; height to rotor hub is 25 m.

The Gedser Test Program. The test program for the Gedser mill is carried out by the Gedser Test Group, a measuring group formed by the Riso National Laboratory, the Danish Ship Research Laboratory, and the Department of Structural Research at the Technical University of Denmark.

The measurements aim at a determination of the following conditions which are of essential importance for the construction of wind power plants.

A preliminary power curve, shown in Figure 13-4, has been established on the basis of simultaneous measurements of the windmill power and the wind velocity.

The measuring points (with the exception of a few, presumably erroneous measurements) are located in very close proximity to an almost straight line. The electricity production of the mill starts at a wind velocity of 5–6 m/sec, and the power output is 200 kW at about 16 m/sec. There are currently only a few measurements at very high wind velocities, at which level the power output is fairly constant at 200 kW.

The rotor, which is of special interest as a special windmill component, has proved to possess relatively low material stresses. This is due to the very rigid construction, with many stays; the stays placed in a sloping position from the blades over, whereas the load caused by the force of gravity is mainly taken up by the stays between the blades.

An example of the variations in the power output of this windmill is shown in Figure 13-5, together with the wind velocity measured at the hub height of 25 m behind the windmill.

In this figure variations in the electric power are seen caused by changes in the wind velocity level. Further, power fluctuations of up to 40 kW are seen to be caused by the effect of wind turbulence on the rotor construction.

The Gedser measurements have also provided valuable experience concerning the practical performance of a test program for windmills on which measurements are not simple due to the great number of rotating parts and to the sometimes rough ambient conditions. The experience gained will be utilized in the test program for the new Nibe windmills.

Noise and Radio/TV Disturbances in Connection with the Gedser Mill. Measurements performed by the Tele-Administration of radio/TV disturbances reveal that the windmill primarily disturbs the TV reception. The disturbances which are caused by reflections from the rotor blades, and by the passage of the signal through the blade plane, are strongly felt up to 100 m from the windmill and are still visible at a distance of 200–300 m. The Tele-Administration is of the opinion that, for

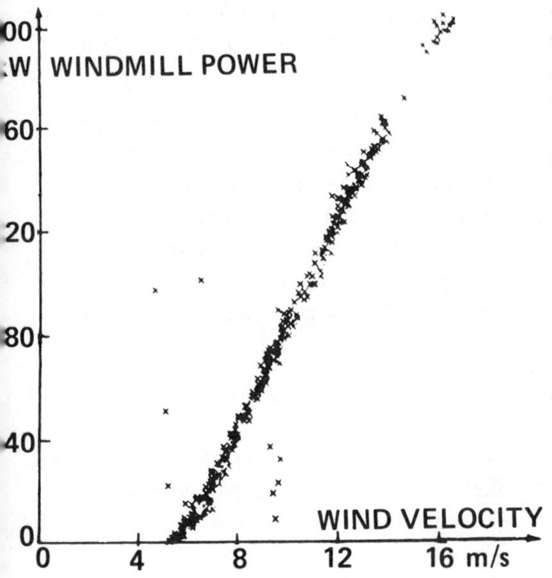

Figure 13-4. Gedser power curve. *(Courtesy of DEFU.)*

modern windmills, a maximum disturbance radius of 10 m/m windmill height will be a realistic figure.

The nearest neighbor, living a little less than 100 m from the windmill, saw the disturbances as a "wandering stripe" over the TV screen and heard them as a hissing in the TV and radio sound. After the erection of an aerial a few hundred meters from the residence, the inconveniences have been eliminated.

The noise level has been measured by Dansk Kedelforening (Danish Boiler Association). At a distance of 50 m from the mill, a level of 50–52 dB (A) was measured to wind velocities of 5–6 m/sec when the mill is just able to produce electricity; for increasing wind velocities up to 10–15 m/sec, noise levels of more than 62 dB (A) were measured. AT a distance of 100 m from the mill, the noise level is about 5 dB (A) lower. If the mill is stopped, the noise levels are reduced by 10–15 dB (A), which demonstrates the windmill's essential influence on the noise levels.

Thus, the noise level due to the mill increases with the wind velocity, and according to the measurements becomes increasingly directional (in the direction of the wind). Gears and pumps, etc., in the machinery cabin of the mill give an essential contribution to the noise, and at increasing wind velocities; also the whirr of the rotor blades contributes to the audible noise.

The Nibe Windmills. The most extensive project within the scope of the wind power program is the building of two big windmills which are being erected near the provincial town of Nibe. Front and side views of the two windmills are shown in Figure 13-6.

Design. The design of the windmills has been the subject of extensive consideration in order to ensure that the experience gained from these new plants to the greatest possible extent could form the basis of future decisions concerning wind power plants for electricity production on a large scale.

The most important choices made are summarized as follows:

- *Type:* Of the many different types of windmills, only a few are suitable for large plants. Among those are the traditional propeller mill and the Darrieus mill. The propeller mill-type has been chosen, partly due to the desired possibility of utilizing earlier experience (e.g., from the F.L. Smidth mills and from the Gedser mill), and partly because theoretical considerations point to a possibility of a very high utilization of the energy content of the wind.
- *Tower:* A conical, concrete tower has been chosen partly for aesthetic reasons, and partly with a view to the damping of possible oscillations. Steel towers may be a possibility in connection with future windmills.
- *Generator:* A simple and sturdy induction generator was chosen.
- *Rated power:* The generator size of 630 kW corresponds to a maximum windmill power output of 500 W/m^2 of swept area. Under the given wind conditions, it would hardly pay to include a larger generator; for one thing, the strongest winds are rather infrequent, and for another, the mill would operate with a

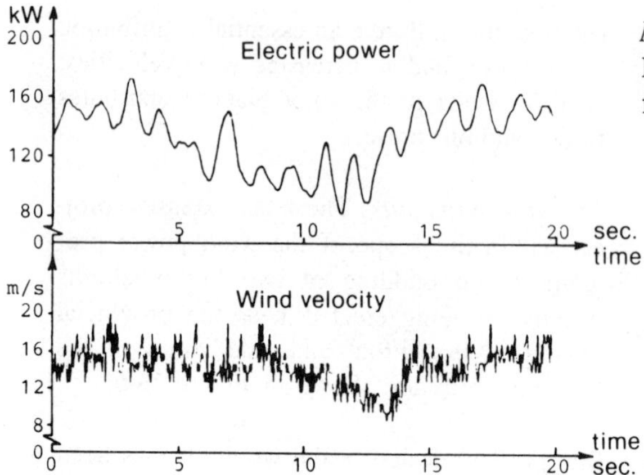

Figure 13-5. Variations in power output. *(Courtesy of DEFU.)*

poor utilization of the heavy machinery for a considerable part of the time.

- *Rotor locations:* The rotor is placed in front of the tower so that the blades will not have to pass through the "wake" of the tower and thus be subjected to heavy shock influences as in the case of a rotor location downwind.
- *Number of rotor blades:* A construction with three rotor blades has been chosen since this will give a more stable rotation than a construction with two blades.
- *Pitch control:* The reason for the building of two windmills which are partly different is the desire to test two different control methods and hence two different rotor constructions. Briefly explained, one mill (windmill A) is controlled by placing the rotor blades broadside to the wind (as in the Gedser mill), whereas the other (windmill B) is controlled by placing the rotor blades with the edges towards the wind (as in most foreign plants and in the Tvind mill).
- *Material of rotor blades:* The rotor blades are built of fiberglass on the 12-m end section of the blade, whereas the innermost 8 m of the blade is supplied with a bearing steel tube. Fiberglass was chosen because low-cost series production seems possible in this material.

Main Data. The primary characteristics of the Nibe windmill (shown in Figure 13-7) are as follows:

- Three-blade rotor, placed upwind, diameter of 40 m.
- Blade tip speed of 71 m/sec.
- Rotor revolution 34 rpm.
- Three-stage gear, ratio about 1:44.
- Four-pole, induction generator, 630 kW.
- Concrete tower, height to rotor hub of 45 m.

The electricity production of the mills will start at a wind velocity of 5–6 m/sec and reach the maximum effect at about 13 m/sec wind velocity. The mills are stopped at wind velocities above 25 m/sec.

Time Schedule. The preliminary design study started in the beginning of 1977. In the autumn of 1977, the main data for the two mills were established, and offers were then invited for the main components. The manufacture of components and the building activities were commenced in the spring and summer of 1978. The first mill (windmill A) was assembled and erected in May/June of 1979, and the installation of the other mill was completed in September.

Budget. The price of the two mills will be around 13 million Dkr. (plus VAT); planning and development costs will amount to about 5 million Dkr. The first amount includes, in addition to the cost of construction, a certain detailed planning on the part of the suppliers and tools such as slip form for towers and mandrel for the fiberglass beam. The 13 million Dkr. are distributed on the main parts of the plants, as shown in Figure 13-8, approximately as follows:

- Towers—18 percent
- Nacelles—45 percent
- Gears—5 percent
- Generators—2 percent
- Electrical systems and control systems—7 percent
- Rotor blades—15 percent

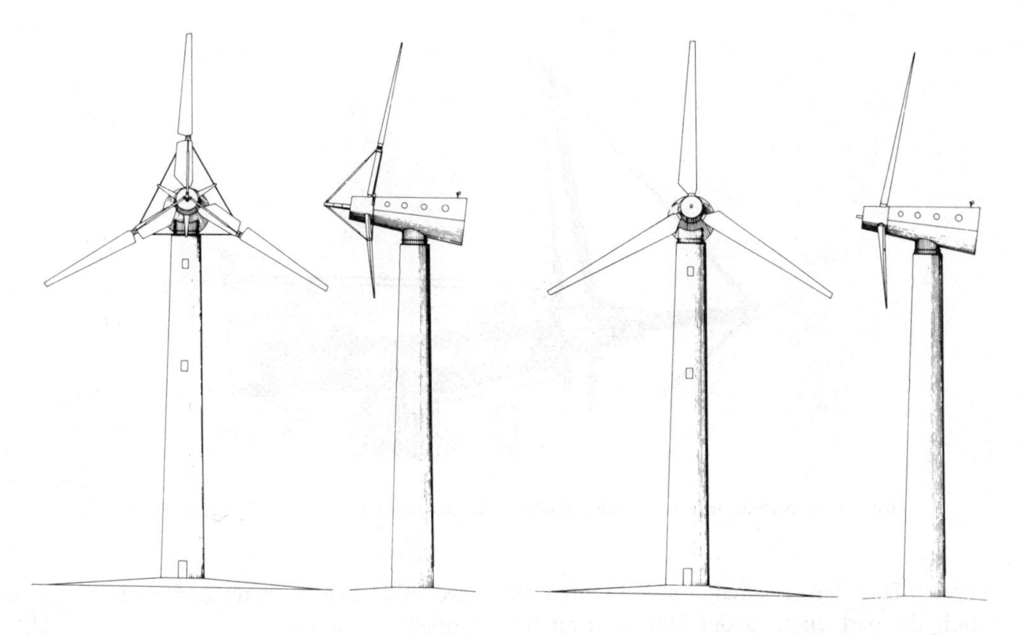

Windmill A *Windmill B*

Figure 13-6. Front and side views of the two Nibe windmills. *(Courtesy of DEFU.)*

- Access to plant and connection to mains—6 percent
- Miscellaneous—2 percent.

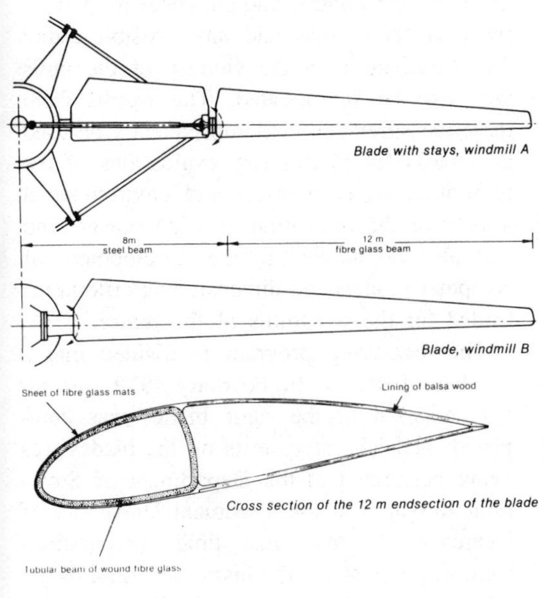

Figure 13-7. Rotor blade configurations. *(Courtesy of DEFU.)*

The site for the two experimental windmills, as shown in Figure 13-9, was chosen after an extensive investigation of the possibilities all over the country. It was difficult to find a site which would fulfill the requirements for wind conditions, safety conditions, environmental adjustment (according to the Nature Conservancy Board, the mills must not be placed within the conservancy interest zone I, which includes a major part of all coastal areas), foundation conditions, accessibility, etc.

In the first instance, the wind power program—based on assessments of experts within the field of wind techniques—tried to obtain the authorities' approval of a site near Velling Maersk at the eastern side of Ringkobing Fiord. With a view to the use of the nearby Stauning airport, the windmills, both of a total height of 65 m, had to be placed at least 5 km north of the airport. At the same time, the authorities demanded that the mills be placed not less than 2 km south of Velling church in order not to visually "dwarf" the

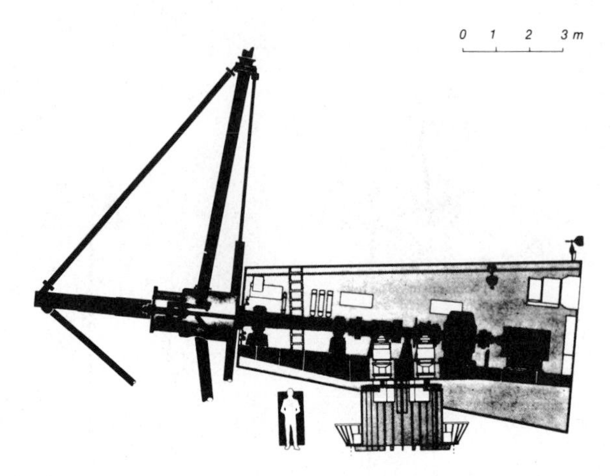

Figure 13-8. Nacelle with rotor hub (Windmill A) and tower top section. *(Courtesy of DEFU.)*

church. This left a small, narrow area of which the part up to about 700 m from the coast was classified as zone I and thus had to be left out. The area at a greater distance from the coast could not be recommended by wind experts, because (among other reasons) a number of windbreaker hedgerows and a road embankment would impede the interpretation of the measurements on the mills.

In the second place, the site at Nibe Bredning was chosen, which with regard to the wind conditions is as good as the site at Vel ling Maersk. After the proceedings had extended over eight months, the authorities' approval of the local plan for the area as well as the building license, etc., were obtained in July 1978. Building activities then began at the site.

As it appears in Figure 13-9, the Nibe site is situated south of Klitgard Fiskerleje (a fishing hamlet) on the coast of Nibe Bredning. To the west, from which most of the wind power originates, are open waters over a distance of at least 6 km; on the landward side, the ground is flat and the countryside rather open over a distance of 1–2 km.

The mills are placed on a north-south line at a distance of 220 m from each other. The close proximity results in savings on buildings activities and electricity lines, and at the same time nearly identical test conditions are obtained so that the measuring results from the two mills can be compared with the greatest possible accuracy.

Measuring Program for the Nibe Windmills. The two Nibe mills will operate fully automatically and produce electricity into the local high-voltage network whereby representative service experience and production data for the windmills will be obtained. Simultaneously, a measuring program for the mills will be put into operation in order to clarify their aerodynamic, mechanical and electrical properties; the noise conditions and any possible radio/TV disturbances in the vicinity of the mills will also be investigated. The results from these measurements will subsequently be used as a basis for preliminary evaluations of the technological, economical and environmental aspects of the utilization of wind power, and will also be applied to the development of computer models and dimensioning criteria intended for the windmills of the future.

The measuring program is divided into a number of stages. In February 1979, the instrumentation of the rotor blades was completed, and laboratory tests on the blades was being performed at the Department of Structural Research at the Technical University of Denmark. At the same time, preparations were in progress for the instrumentation of the other parts of the mills. At the Riso National Laboratory, an automatic system for data col-

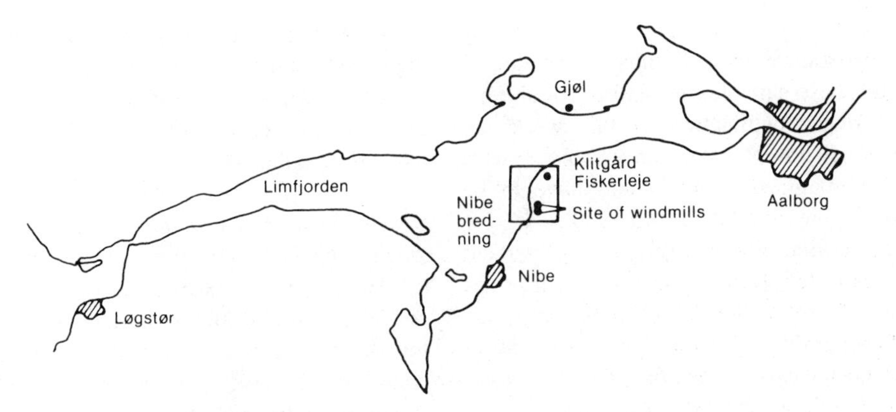

Figure 13-9. Site of Nibe WECS. *(Courtesy of DEFU.)*

lection which collects and processes measuring data for the Nibe windmills was constructed.

Investigation of the Wind Power Potentialities within the Danish Electricity Supply System. The preliminary site investigations made in connection with the final choice of the Nibe location have shown that the siting of any great number of windmills will involve a conflict between the consideration for the electricity production of the mills (wind conditions are best in coastal areas facing west or southwest) and the considerations for conservation and other nature interests. With a view to a more thorough evaluation of these and other considerations in connection with the siting, a general site investigation will be carried out to identify 1000–2000 possible sites. Many sites will be needed if, for instance, 10 percent of the Danish electricity consumption is going to be supplied by means of wind power.

As part of the investigation, the compilation of a wind atlas has been commenced describing the Danish wind resources and their geographical distribution. The wind atlas was prepared by the Riso National Laboratory and the Danish Institute of Meteorology. At the same time (February 1979), negotiations were carried out with the Danish Ministry of Environment Preservation (Miljoministeriet) about the performance of a nationwide, detailed analysis of the siting possibilities for larger, electricity-producing wind power plants. Finally, as a third sub-investigation, the problems and conditions in connection with the integration of the wind power into the existing electricity supply system will be looked into.

Theoretical Investigation Concerning Aerodynamics. A fairly accurate optimization of windmills with a view to technology and economy presupposes the availability of methods providing a recalculation of outputs and loads (and thus dimensions) considerably more accurate than is the case today. Denmark participates in the wind power agreements within IEA, including an improvement of computer models and dimensioning criteria for windmills, and, as a sequence to these agreements, activities within the wind power program are expanding Danish knowledge within this field.

Perspectives. The two stages of the wind power program now in operation are expected to be completed during 1980. On the basis of an evaluation of the results—Danish as well as international—obtained until then, and on the basis of the energy-political situation in Denmark at that time, a decision must be made in 1980 concerning the extent and form of a possible continuation of the wind power development activities.

A continuation might, for instance, consist of the erection of a number of wind power plants of the size and design which, according to the experience gained, must be evaluated as to the best suitable to Danish conditions. As a further continuation of this work, an actual wind power extension of the system could be imagined which would have to be decided in the middle of the 1980's. If, for instance, a wind power coverage of 10 percent of the expected Danish electricity consumption in 1995 of about 30 TWh is aimed at, and if windmills of 50-m height and 50-m rotor diameter are chosen, this would necessitate the building of 1500–2000 windmills. In the exemplified case, 150–200 mills must be erected in Denmark each year during the period 1985–1994.

Federal Republic of Germany. The Wind Energy Program of the Federal Republic of Germany consists of two major thrusts:

1. Adaption of small WECS, horizontal- or vertical-axis, to special uses (e.g., island operation, developing countries); and
2. Electricity generation by large WECS for the national grid: the GROWIAN Program.

These two thrusts are based on former theoretical investigations and experiences with WECS. The successful operation of a 100-kW horizontal-axis WECS from 1957–1968 by Professor Hütter demonstrated the possibility of generating electricity for the national grid. Furthermore, in 1976, the fundamental study, "Nutzung der Windkraft," recommended the utilization of wind energy in Germany by large-scale WECS for electricity production.

The activities of the first thrust on small WECS will be explained briefly, whereas the important second line on large WECS will be described in more detail.

In the summer of 1976, the Grosse Windenergie-Anlage (GROWIAN) Program was initiated as the main line of R&D. It was recognized that wind energy, in a highly industrialized society, can contribute usefully only by supplying megawatts of power. The general aims of the GROWIAN Program are to evaluate the technical feasibility of large-scale electricity generation by the wind with a 2–3-MW WECS prototype unit and to assess plant installation and power generation costs.

The wind program of the Federal Republic of Germany consists of ten major programs that are detailed below.*

Development Construction and Test Prototype, 265 kW. Objective: The objectives are to develop and construct a two-vane wind converter with an almost horizontal rotor axis for electrical current generation at exploitable wind velocities of 2–8 m/sec and optimum matching of the operating factors, wind availability and energy output, with the aid of an automatically operating plant which can work without maintenance for lengthy periods of time under all possible environmental conditions, and which can also be self-supporting in operation.

Status: With the axial-flow, high-speed wind turbine with two rotor blades, System Voith-Hutter, 14 variations of the arrangement of component groups, such as gearbox, drive shaft, generator and regulator, were investigated. For the location of the rotary carriage, a decision had to be made on whether the mast should be free or guyed:

- Guyed, rigid tubular steel mast, hydraulically folding in one piece, which is easily transportable yet which can be operated subcritically.
- Bevel gearbox $i = 4.1$, drive tube, helical gearbox $i = 9.8$, and synchro-generator in the foot of the mast; rotary carriage above guy attachments.
- Regulator, control and measuring equipment together with high-tension plant in modular transformer station at the foot of the mast.
- Rotor blades of GFP-shall structure, 7-m-long tips flange-mounted with loop flange (Voith-Hutter patent applications) to be driven subcritically, with laminar profiles.
- Rotor hub initially semi-cardanic, later

*Program for Energy Research and Technologies 1977–1980, Project Management for Energy Research, Federal Minister of Research and Technology.

fully cardanic Voith-Hutter patent applications suspension.

- Hydraulic pitch adjustment, electronic regulator with mechanical overspeed inhibition.
- Starting and operating safety equipment, wind, vibration and load plotter.
- 52-m rotor /and 8-m/sec wind velocity gives 316 kW at the rotor at 37 1/min; generator 350 kVA, 1500 1/min; terminal voltage 265 kW.
- Stability up to 200 km/hour, machine weight approximately 30,000 kg; foundation weight approximately 180,000 kg.
- Earliest test run end 1978.

Meteorological Measurements in Coastal Areas for the Selection of Locations for Wind Energy Plants. Objective: This objective is to prepare the first detailed meteorological basis-work for the use of wind energy by large wind energy plants, up to 150 m above the ground. The more exact wind profiles should be referred to in the choice of location for a large test plant in the power range 1–3 MW, as well as in the technical layout of the plant.

Status: The hardware has been obtained, and the sites for the towers have been determined with the approval of the landowners. The elaboration of the computer programs for analyzing the data is finished. To interpolate the wind profile between the height of the geostrophic wind (~1500 m) and the height of the wind measured (~50 m), a model was used enabling the wind energy supply between 30 m and 200 m to be estimated. Statistics of the geostrophic wind are being computed. It is planned to obtain quantitative information on the accuracy of the predicted geostrophic wind data at the five sites. The statistical analysis also includes a spectral analysis to obtain some information on the time-dependent fluctuations of the area mean of the wind energy supply.

Preparing Documentation for Construction of a Large Wind Energy Plant–"GROWIAN" Objective: The objective is to prepare documentation for the construction of a large wind energy plant which reaches its nominal rating of 2–3 MW at a wind velocity of 12 m/sec.

Status: The main system-related data and the basic concept for a plant exposed to winds having an average velocity of 6 m/sec at 10 m above the ground have been established.

Detailed definitions are being determined for the various sub-systems on the basis of the above data, using profiles specially developed for the plant.

All the main components have been conceptually determined. Work is being carried out on the detailed design of the components, having regard to reciprocal influences. Auxiliary and supplementary equipment is being defined.

Load assumptions for the first computation have been compiled.

Programs for determining strength and rigidity of the rotor blades have been written and are being tested.

To facilitate design of the control system and to determine the stresses on components, a compilation of all important operating conditions and probably operational behavior is being made.

Estimates of vibration behavior of the plant and of the reciprocal influences of rotor and tower have been made based on a computer model for plants with two-blade rotors.

Engineering and Economical Possibilities of Large-scale Wind Energy Systems Converting Wind Power into Electrical Power, With Special Attention to Reserve and Storage Systems. Objective: Based on original hourly data of wind velocity, wind direction, energy production and energy demand, the engineering and economic possibilities of large-scale wind energy systems converting wind power into electrical power will be analyzed. Special attention will be given to complementary power regulation capacity, particularly with respect to coastal areas of Northern Germany. At the same time, the integration of wind energy into national energy supply systems will be examined.

Status: The models of wind energy production and the models of conventional energy production for the North German coastal area have been finished. Original, hourly wind data of the German Weather Service of five coastal weather stations for the period of 1969–1976, as well as demand data of utility companies of North Germany for 1975, have been provided and processed for the computer.

Besides this, detailed availability data of conventional power plants have been submitted, so that the "value" of wind energy can be calculated. Based on these facts, the maximum admissible investment costs of wind power plants for an economic operation of the plants can be calculated by making use of the known investment costs and the fuel costs of conventional power plants. These comparisons of costs will be conducted on the basis of other known cost comparisons, such as the recently published comparisons between nuclear power and coal power.

Investigations for the Construction of Large Rotor Blades for GROWIAN and for the Dynamic Behavior of the Complete System GROWIAN. Objective: By means of a large-scale wind energy converter in the power range of 2–3 MW, methods of obtaining a machine with minimal cost and with a long operational life will be demonstrated. This shall be achieved by a special rotor-concept, which reduces all loads except the driving momentum as far as possible.

Status: Within the framework of the development of a model for a wind energy converter, the behavior during the running-in period and during stationary operation, as well as the absorption of gusts, are being investigated using calculations and experiments with a small wind-tunnel model. For rotor blades constructed of different materials and of different basic design, decision loading bases are being investigated using the FE-method; some results are already available. Furthermore, a new type of tower is currently being investigated. Studies of materials and constructions are being undertaken, in addition to preliminary investigations for the data collection and data transport necessary for the instrumentation of wind-tunnel models.

Construction and Investigation of the Operation of a Wind Energy Converter. Objective: A survey of the performance of one from a series of wind energy converters will be produced by the Allgaier Company. The converter is to be equipped with rotor blades of new profiles and construction technique and subsequently tested. A new wind energy converter in module technique is to be developed to achieve a quick, simple and low-priced interchangeability of all components considering its application in developing countries.

Status: The measurement and registration equipment for measuring the wind power plant has been installed. First measurements have been performed and evaluated. Two female moulds for the blade construction and subsequent production of two fiberglass reinforced plastic rotor blades (length 5.56 m, mass 26.6 kg) have been completed. A prototype for a modified rotor blade connection has been built, and load and dynamic rotor blade tests have been prepared.

Operation and wind loads have been calculated or defined. The project definition phase was completed, and the design for the module wind energy converter has been started. Two-thirds of the detail construction have been completed. Extensive feasibility studies concerning the rotor blade connections have been performed. The special problems of gust loads required additional evaluation of gust records.

Development of a 5.5-m-diameter, Vertical-axis wind energy converter. Objective: The objective is to develop a 5.5-m-diameter, vertical-axis wind energy converter, including structural design, construction and optimization of the aerodynamic layout by consideration of wind tunnel test results.

Status: The design specifications have been elaborated for the layout of the vertical-axis wind turbine for the following purposes:

- For measurement test runs in the wind tunnel and a continuous open field test (with consideration of the meteorological conditions of the proposed test site).
- Derivation of data for a 20-kW rotor by extrapolation.

- To obtain manufacturing procedures applicable in developing countries for most of the parts.

The design work is finished. It incorporates the balancing of aerodynamic and structural layout requirements. The cross-section of the blades corresponds to a NACA 0015 profile with a chord length of 320 mm. Each blade consists of an extruded aluminum-alloy bar, bent to the troposkiene shape. Two sections of Savonius rotors, adjustable with regard to their length, serve as an aerodynamic starting device. Three blades for each rotor are best suited for the three-bar grid rotating tower.

After the test assembly of the Darrieus rotors in a three- and four-blade version, and the Savonius rotors, the height of which can be varied due to the separated constructional features, the construction of the plant was completed. The wind tunnel tests were performed in the first weeks of 1978.

Evaluation of a 15-kW Wind Power Plant; Determination of Transferable Power Data and Evidence of the Profitability of the Power Generation. Objective: The objectives are as follows:

- To develop a simple method for the predetermination of annual output of a wind power plant on-site.
- To develop and construct a low-priced wind power plant also suitable for production in under-developed countries.
- To prepare a survey to decide whether a profitable power production is possible with this plant.

Status: The construction of the wind power plant has now been completed, and the column has been erected. The major portion of the construction works, for the individual parts of the plant have been finished (with the exception of the wings). The measuring chamber has been connected to the network and the cabling between the chamber, the tower for wind measurement and the column of the wind power plant have been installed.

The installation of all measuring devices has been completed. The measurement of wind velocity and wind direction has been taking place since July 1977.

The decision on the assessment of the profitability was made in 1978.

A preliminary evaluation of all data on the velocity and direction of wind measured during the second half of 1977 has been made.

Wind Conditions in the Federal Republic of Germany in View of the Use of the Wind Energy. Objective: The objective is to analyze the wind conditions in the Federal Republic of Germany (inland/offshore) as a meteorological basis for studies of the use of wind energy.

Status: As soon as the first analysis of wind data in the north German coastal area for the determination of appropriate areas for wind power plants, 50–150 m high, was finished, data processing began. It was conducted with the aim of gaining the yearly averages, yearly and daily ranges, frequency distributions, etc., and was carried out separately for inland and coastal areas. After concluding this work, the necessary tables, diagrams and maps were prepared; scientific elaborations were then made.

Standardized Meteorological Measurement for the Verification of the Specification of Wind Energy Converter Systems. Objective: The object is to work out recommendations which allow a correct measurement of wind energy plant specification data with regard to the meteorological conditions.

Netherlands. Traditionally, Holland is known as a windmill country. In the middle of the last century, about 9000 windmills produced mechanical energy for water-pumping, the sawing of wood and the grinding of grain. The introduction of other means to produce mechanical energy (i.e., the steam engine and electric motor) reduced interest in the application of wind energy. About 900 windmills are left, but very few are operational. The remaining windmills are protected as monuments from the past and a special foundation takes care of them. They are currently one of the tourist attractions of Holland. From the historical presence of windmills in the Netherlands, one may suggest that much knowledge should be available about the proper use of wind energy. Unfortunately, this is not true and the old windmills are of no use in terms of modern technology.

Shortly after the oil crisis, a revival in the interest in wind energy occurred. A specially established council for energy research advised the government that, in the scope of diversification of energy sources, the application of wind energy should be reconsidered.

A rough estimation of the potentiality of wind energy in the Netherlands indicates that about 20 percent of the present day electricity production could be obtained from the wind in areas with high wind energy density; that is, areas along the coastlines of the North Sea and the Zuyder Zee. The wind energy drops off fast, going land inwards. To produce 20 percent of the electricity, about 5000 windmills with a rotor diameter of 50 m and an installed capacity of 1 MW each are needed. Although 20 percent of the electricity production is a significant quantity, it means a saving in fossil energy sources of only a few percent. The possible contribution of wind energy to the total energy consumption will always be marginal.

In February 1976, the Dutch government approved the execution of a National Research Program on Wind Energy. The objective of the program is to study the technical and economical feasibility of large-scale use of wind energy in the Netherlands. The study will be restricted to the use of wind power for the generation of electricity to be fed into the existing utility grid.

The usable amount of (mechanical) energy in the windy regions of the Netherlands—the coast lines of North Sea and Ijsselmeer—roughly equals 0.7×10^{10} kWh/per year. This is nearly 15 percent of the amount of electricity produced in 1974.

The program should be accomplished during a five-year period and is carried out in phases. The first phase (March 1976–March 1977) was of a preparatory nature. During this period, a literature study was made and the individual projects for research and development defined more precisely. Two basic wind turbine designs were considered: the conventional wind turbine (horizontal-axis rotor) and the type based on the principal of the Darrieus rotor (vertical-axis rotor). The latter was patented in 1929, but little is known of its potentialities. Consequently, in order to gain more knowledge, a test facility with a vertical-axis rotor of 5-m in diameter was designed and built by Fokker Aircraft.

A computer model was developed of the aerodynamic behavior of the rigid rotor, followed by a dynamic and aero-elastic stability study. Studies on the gear train and the electric generator were also made.

An important aspect under investigation is the influence of one wind turbine on the other(s) with respect to the positioning. Wake measurements have been performed in a wind tunnel on a 20-cm model of a Darrieus-type wind turbine. the second phase of the wind research program covered the period March 1977–January 1979, when the siting, economical and technological aspects were investigated.

The main obstacle in using wind energy on a large scale in a populated country with a high energy consumption will be finding areas for the erection of WECS. Wind turbines cannot be placed in built-up areas or in areas used for industry and traffic. They cannot be located in uncultivated areas because most of them are either unsuitable (e.g., woods) or protected areas. The only possibility lies, therefore, in the cultivated areas. This seems to be a good solution because the combined use of land for agricultural purpose and the production of energy from wind is possible.

However, a complete change of the current regional planning policy would be necessary because only buildings and structures directly related to (in this case) agricultural uses of land are permitted. Although windmills of the old-Dutch type are accepted and even appreciated in the rural areas, modern wind energy turbines could meet with heavy opposition from the public and the authorities. In investigating the possibility of siting wind turbines on land, the following items will be examined: the location of sites qualifying for wind energy parks; the characteristics of each site; and the amount of power to be harvested from such a site. The effects on telecommunication-, navigation- and direction-finding systems as well as migrating birds, etc., will be studied. An alternative solution is the location of wind turbines at sea, but this will inevitably result in a tremendous impact on costs. Also many restrictions for navigation, fishing, offshore oil and natural gas prospecting and naval defense requirements must be met.

The use of wind energy will mainly depend on the cost of the energy produced. Therefore, economic studies, including the evaluation of capital and operational costs of WECS, will be made. The cost of the large rotor blades will contribute substantially to the total energy costs. Only rotors constructed according to cheap fabrication techniques can be expected to generate economically acceptable power. The electric equipment and cables needed to feed the electric power into the grid must also be considered in a cost analysis. The costs could be of the same magnitude as those of the wind turbine itself.

In studying the technological aspects, parallel approaches of software and hardware investigations are being followed. The second phase of the national program comprises the following activities:

- Initially, an extensive test program with the vertical-axis test facility of 5-m diameter will be executed. The main objectives are to obtain a thorough insight into the performance and dynamic behavior of a vertical-axis rotor so that the results can be reliably applied to far larger designs (up to rotor diameters of 50 m and more). The rotor blades are amply provided with instrumentation. The results of the experiments will be compared with theoretical results and computed data. Furthermore, the concept of a variable geometry will be tested. By varying the effective swept rotor area, the loss of aerodynamic efficiency may be compensated for by an increase of energy supplied at a constant rotor speed.
- Continuation of the fabrication research on all-plastic rotor blades for vertical- and horizontal-axis turbines, whereby use is made of the modern bonding techniques developed for aircraft structures.
- The design, fabrication and assembly of an experimental horizontal-axis wind turbine with a rotor diameter of 25 m and a rated power of 150 kW. (The machine began operation in the fall of 1978.)
- Investigation of the tip-vane concept. This project deals with a horizontal-axis wind turbine where relatively small vanes are attached to the tips of the rotor blades. The vanes deflect the air radially outwards and this diffuser effect results in an increased mass flow of air through the swept plane. This leads to a larger energy output per unit of swept area of the rotor blades.
- Development of an electric system for the conversion of mechanical to electrical energy in which a modified line-commutated dc-ac converter will be used.
- Extensive study of all the technological problems related to the erection, running and maintenance of wind turbines located offshore.
- Study of wake effects.
- Development of a double-helical epicyclic, hydraulically controlled system of gears for the conversion of a variable rotor speed to a constant generator speed.

The continuation of the national program on wind energy into a third phase will depend on the results obtained in the second phase. During this third phase, the design, fabrication, assembly and testing of a vertical-axis wind turbine of the same rotor diameter and rated energy output as the horizontal-axis wind turbine will be undertaken. Thus, sufficient information is available to determine the most suitable type.

Saudi Arabia

The Department of Energy, the Department of Treasury and the Kingdom of Saudi Arabia have undertaken a $100 million, joint, five-year solar energy program. SERI has been assigned the responsibility for the development of a program management plan. This solar energy R&D program, based in Saudi Arabia, will concentrate on mutually beneficial technologies such as wind energy. WECS studies will continue until October 1982.

Spain

The United States-Spain Treaty of Friendship, signed in 1976, provides for cooperation in solar energy technology development in Spain. A U.S. team from DOE is assisting Spanish scientists in establishing a wind energy program. Information exchange will be provided for wind energy conversion technology and meteorology characterization. The benefit to the U.S. program is involvement of U.S. industry in the Spanish demonstration of wind and other solar applications. Such efforts give the U.S. industry further incentives to develop an export market.

Sweden

Sweden is situated within the "belt of westerly winds" and thus has favorable wind conditions. Generally, the winds are strongest along the coasts and at the great lakes. The wind conditions in the northern parts of Sweden are unfavorable, with the exception of some small areas along the Gulf of Bothnia (Bottenviken). Due to the sparse number of meteorological stations, the knowledge about wind conditions in the mountainous area is incomplete.*

In the short run, the availability of wind energy is very fluctuating. Strong winds seldom prevail for such a long time that a wind power unit will deliver full power for more than 24 hours at a time. In a few hours, the wind may decrease from a high wind speed to a low value that cuts off power delivery. This behavior of the wind is a challenge to the possibilities of the power system to balance the production. It also demands the development of accurate forecasting methods for wind speed.

*Swedish status report developed from nämnden för energiproduktionsforskning report NE 1977:2, entitled "Vindenergi i Sverige."

Due to the sea breeze and other phenomena, the wind is stronger during daytime than in the night. This difference, however, decreases with the height above ground. When observed for longer periods, the wind shows significant seasonal variations. Thus, the availability of wind energy is greater in the winter than in the summer.

In open areas along the coasts, available wind energy at 100-m height is normally about three times greater than at 10-m height when measured over a certain cross-section of the wind. Inland, the difference may increase to five or six times. This phenomena favors large windpower units that reach the higher wind speeds aloft.

System Analysis of Large Wind Power Units. Today only the propeller-like, horizontal-axis wind turbine has reached such a state of development that it is worthwhile to perform detailed design work and cost calculations. As far as it is known today, it also gives the best possibilities to produce cost-effective designs due to the fact that the amount of construction material for a given energy production can be kept small because of small blade area and a comparatively high rpm.

In order to gain design and construction experience at an early stage, as well as a platform for testing purposes at a later stage, a wind power test unit was ordered in April 1976 from the Saab-Scania company. The two-bladed wind turbine measures 18 m in diameter. It is located at Kalkugnen by Älvkarleby, close to the Baltic. The testing started in April 1977 and has so far proceeded according to plans.

The wind turbine design is important for the competitiveness of a wind power unit. The analysis shows that cost-effective designs may be realized with several alternative designs on the general theme: horizontal-axis wind turbines with few blades. Welded steel, aluminum alloys or composite materials with carbon or glass fibers may be used as blade material. Due to differences in weight and other properties, the choice of material leads to fairly different designs. The choice of hub design is also of great importance.

Units equipped with 50-m and 100-m wind turbines were thoroughly investigated in order to (among other things) serve as a basis for cost calculations. The smaller unit is normally equipped with a 1-MW generator, and the bigger one with a 4-MW generator. In both sizes, alternative designs with heavy blades made of welded steel, and light blades in carbon reinforced plastics or aluminum alloy, were evaluated using either a hub of an uncomplicated or a more sophisticated design. Besides this analysis, which was performed by Saab-Scania, some complementary studies were performed by other parties in Sweden and abroad.

Siting Possibilities for Wind Power Units. The meteorological analysis shows that especially windy areas exist along the West Coast of Sweden, in many areas in the southern province of Skäne, on the islands of Öland and Gotland, east of lake Vättern and in northern Uppland. These areas, amounting to a total of 12,000 km^2, are characterized by winds blowing at least 7 m/sec during half of the year (median wind) on a height of 100 m above the ground. The possibilities to site wind power units in these areas were investigated by testing preliminary criteria for siting against maps and plans. The criteria allowed restricted siting in areas for recreational activities and in areas where historical or scenic landmarks are of great importance, but not at all in planned urban areas and nature conservancy areas. According to these criteria, siting is possible without restrictions within 55 percent, and, with restrictions, with 15 percent, of the area.

A detailed analysis investigated the possibilities to site groups of wind power units in Skäne, northern Uppland and on the island of Gotland. To avoid energy losses due to one unit "shadowing" another, the units on an average were positioned 13 turbine diameters from each other. Because of restrictions in the preliminary critieria (e.g., at least 200 m to single buildings and 1000 m to communities), only about 80 percent of the intended sites could be used.

The following estimate of possible wind power installations presupposes that siting should be allowed without limitations in areas that are devoid of restrictions, and in 50 percent of the areas that have some restrictions, according to the preliminary criteria. No consideration has been taken to the possibilities to compensate for the random behavior of the wind.

Areas with at least 7-m/sec median wind at 100-m height are estimated to be able to accommodate about 3300 turbines with 100-m diameters. The corresponding electricity production is about 32 TWh during a normal year. If 8m/sec is the economic limit, the number of units decreases to about 500 and energy production to 6 TWh/year. if 50-m-diameter wind turbines are chosen, the number of units roughly increases by a factor of four and the energy production decreases to two-thirds. Areas with 6–7 m/sec might very roughly accommodate 13,000 units of 100-m diameter that increase energy production to 120 TWh/year, although at a higher cost.

Interaction with the Power Grid. A small installation of wind power obviously will be accepted by the power grid without any extra provisions. For a large installation, special capacity for regulation and standby power has to be reserved or installed. In 1974, the Swedish State Power Board (Battenfall) estimated that 3000–5000 MW (corresponds to 7.5–12.5 TWh/year) of wind power could be accepted by the national power grid at a small cost for regulation and standby capacity almost entirely within the existing hydropower system. New computer runs based on more accurate data will show whether this estimate is realistic or not.

Impact on Safety and Environment. The installations of wind power units will affect the view of the landscape. It should be possible to keep the probability for severe hub or blade failures extremely low by the adaption of carefully made designs and periodic inspections. Ultimately, a broken turbine blade might touch ground several hundred meters from the unit. To avoid ice being thrown from the blades during icing conditions, it may be necessary to install de-icing systems; e.g., utilizing heat losses from the generator.

Annoying noise has not been generated from earlier wind power units. Theoretical calculations indicate that a large unit might be audible over the background wind noise, within a distance of 500 m in the direction of the wind. Infrasound is not expected to present any problems, according to the same calculations. These phenomena will be very carefully measured and analysed.

No disturbance of bird life is reported from earlier wind power units. During daylight, it is not expected that birds will collide with the tower or the turbine. In the night, the risk of collision is defined by the random probability for a hit. Most night-migrating birds, however, fly at altitudes higher than those affected by wind power units.

Theoretically, a wind power unit might influence the local climate, as energy is withdrawn from the atmosphere. It is, however, extremely difficult to prove such an impact.

Interference with telecommunications will be investigated at the wind power test unit at Kalkugnen.

Energy analysis shows that a wind power unit will pay off energywise in less than a year.

Future demolition of a concrete tower, which is the only component that might cause any substantial cost for demolition, is calculated at a few percent of the cost of the unit (today's prices and money value).

Arrangements made in order to compensate the random availability of the wind may affect the environment. This impact should be added to the total environmental impact of wind power.

Economical Considerations. When calculated according to the methods used by Swedish utilities, the cost for electricity is estimated at 0.14 Sw.Cr./kWh, when produced by 100-m-diameter wind power units localized in areas with at least 8-m/sec median wind at 100-m height. Areas with 7–8 m/sec median wind increase the cost to 0.18 Cr., and areas with 6–7 m/sec wind further increase the cost to 0.23 Cr. (1 1976/1977 U.S. $ = 4.40 Sw. Cr.) The cost is valid for the designs mainly analyzed. It seems to be possible to decrease cost somewhat (e.g., by increasing the size). Other Swedish and foreign evaluations indicate lower costs.

The cost for regulating and reserve capacity has to be added to the figures mentioned. Depending on the size of wind power installation, the cost is estimated at 0.00–0.05 Sw. Cr./kWh.

Today, not even a commercial size prototype unit exists. Thus, the cost figures mentioned cannot be entirely relied upon. However, they indicate good reasons to continue the wind energy program, initially towards development of commercial-size prototypes.

Collaboration within IEA. Collaboration in the wind energy field within the International Energy Agency (IEA) is currently negotiated mainly between the U.S., Canada, the Netherlands, the German Federal Republic, Great Britain, Ireland, Denmark and Sweden. Agreements have been made on projects concerning the integration of wind power units into different kinds of national grid systems, the development of methods for forecasting wind velocity and a technical/social assessment of wind energy.

Units for Local Use. Generally, small units are less competitive than larger ones due to less available wind energy at low heights.

Evidently, electrification of, e. g., lonely farms, cannot form a large market for wind power units in Sweden, as most of the 400 permanent dwellings that lack electricity today can be attached to the grid within reasonable costs. According to estimates, heating of permanent dwellings with individual wind power units seem too expensive. Supporting heating of recreation houses probably can form some market for middle-size wind power units, especially if the use of electricity from the grid for such purposes no longer is permitted. The use of wind turbines for heating of greenhouses is not competitive today. The demand concerning wind power units for radio sets, lighthouses, etc., seems small. There is probably a potential market amounting to several hundred units per year in conjunction with export of such things as telephone stations to developing countries or to remote areas.

The Future of Sweden's Wind Program. The Swedish Wind Energy Program, funded by NE, is presently ending its studies and experiments phase, and beginning the prototype phase. The first phase has shown that large-scale wind energy is technically feasible, although some technical problems will require a substantial development effort. The conclusions from the first phase have resulted in a specification for full-scale prototypes of wind power units. An invitation to tender for the design, construction and testing of full-scale prototypes was made in April 1978. Tender deadline is October 31, 1978. The prototypes are planned to enter operation and evaluation in early 1981.

The first phase has also shown that the use of small-scale wind power units is technically rather simple, but that existing units need development, and that the Swedish market is very small for autonomous systems, as all permanent dwellings except 400 are connected to the electric grid. The work in this part of the program is centered around specifying a very reliable and simplified unit of 8–10 kW size, equipped with an induction (asynchronous) generator for straightforward connection to a stable grid.

In parallel with the hardware-oriented parts of the program, a comprehensive measurement project to map the winds over Sweden at 40-, 100- and 150-m levels has started, together with development of wind forecasting methods. Longer-range technical development projects are also funded by NE.*

*Private communications between author and Sven Hugosson, Program Management Consultant, National Swedish Board for Energy Source Development (NE), Stockholm, Sweden.

14
The Future of Wind Power

INTRODUCTION

In creating this handbook on wind energy conversion systems in the quiet of my office I found that there was a need to give other individuals an opportunity to express their hopes, fears and dreams about the future of wind energy. A major effort was undertaken to identify and contact those who could provide insight on this topic. In addition to the organizations which have or are currently studying wind energy systems, I have been fortunate in receiving commentary from a diverse group of individuals. Their comments are valuable and interesting, and provide other views of the future of wind energy.

VIEWS ON THE FUTURE OF WIND POWER

As noted in the introduction, excerpts have been taken from books and reports and specific contributions by various individuals interested in wind energy. The entries, in alphabetical order, are from: Alternative Energy Institute, American Wind Energy Association, R. Nolan Clark, Council on Environmental Quality, De Wayne Coxon, Frank R. Eldridge, Department of Energy, J. Stuart Fordyce, Kenneth M. Foreman, Marcellus L. Jacobs, C. G. Justus, Herman Kahn, Richard Katzenberg, Amory B. Lovins, Donald J. Mayer, Richard A. Oman, People and Energy, Palmer Cosslett Putnam, Pasquale M. Sforza, and Volta Torrey.

Alternative Energy Institute

The Alternative Energy Institute (AEI) was created in the fall of 1978 by West Texas State University. AEI is the outgrowth of efforts, which began in 1970, by Dr. Vaughn Nelson and Dr. Earl Gilmore (Amarillo College) in the area of wind energy. Dr. Robert E. Barieu joined the staff in 1974. The purpose of AEI is to conduct research and development in the use of solar, wind and biomass. Within this context the primary emphasis is on energy systems for rural operations. The program to date has been directed almost entirely toward wind energy. Since we have been working together for some time, the only fair way to present views on the future of dispersed wind power would be joint authorship.

In the national programs to expand the energy base to include alternative energy resources, the philosophy is widespread that small-scale systems can make no important contribution. In particular, the potential benefits from large numbers of dispersed wind systems (less than 100 kW) have been overlooked or ignored in favor of large wind systems. However, the long-term contribution already made by the farm windmill should dispel this notion.

An order of magnitude calculation gives a value of approximately 30,000 windmills in the Southern High Plains. The power output of a single windmill is around 0.25 kW and collectively they provide an average of 5 MW. The prime function of the farm windmill is to pump water in isolated areas, and even today there is no economic replacement. Suppose these installations were to be converted to electric power. A reasonable estimate of the minimum conversion cost would be $500 million, and this does not include the energy fuel costs in the future (on the average, 10 MW thermal). This analysis shows what a valuable commodity wind power can be even for small systems.

In the future, small wind energy conversion systems (SWECS), will be as numerous as other power equipment now routinely purchased and operated in rural areas: tractors, combines, etc. SWECS will be operated in conjunction with present utility service and will be used in the wind assist mode to supplement existing power sources. The size of SWECS purchased will change from smaller (10–25 kW) to larger (50–100 kW) units as the farmer realizes the economic benefits of producing his own power. Also, the institutional problems will be solved such that the owner will be able to sell excess power produced by wind turbines back to the utility.

More than likely some mix of ownership will evolve. The farmer could purchase the unit; the unit could be leased (or lease-purchased) from local electric cooperatives; or the cooperatives could own and operate the wind units. In any case, we foresee that the manufacturing and service of SWECS will be similar to the farm implement industry.

This deployment of SWECS in rural areas will offer a number of advantages, as outlined below.

Cost of SWECS. Many people promote large wind units because of the economics of scale, but not enough wind units have been built to determine the optimum size. Presently the cost/kw for small units (≃ 25 kw) is cheaper than the large units. Since the capital cost is much less, a number of SWECS will be purchased by the consumer. The farmer will become a producer consumer and will farm the wind for energy, much as he now cultivates the land for food and fiber. In addition, the farmer will furnish the land, labor and most of the maintenance, which will make dispersed SWECS very cost-comparative.

Operation of SWECS. Much of the energy produced will be used on-site, and thus a more efficient system will result. The farmer customarily operates and maintains mechanical equipment and works within the variability of nature. He will probably have to correlate his electric use to the availability of the wind so as not to increase the peak demand.

Benefits of Society. The economic benefits are obvious in that fossil fuels are displaced and power is still available to the user. Dispersed SWECS means a decentralized power source which would be under local control.

Such a system is less vulnerable to catastrophic failure since the failure of a single unit is a small part of the entire dispersed system. Also, the power will be more dependable on a statistical basis since the variability of the wind will be averaged over a large area.

It is difficult to imagine how a program to utilize a significant fraction of the nation's wind power potential can succeed unless it includes a powerful thrust in the realm of the low-technology, small wind energy systems. Such a thrust would invite public participation even to the individual level and would then help to generate public support for the large machines which it will see as being paid for with tax money for the benefit of large utility companies.

American Wind Energy Association*

There is no question that wind energy can make a substantial contribution to the United States' energy supply in the future.

*Statement of the American Wind Energy Association before the Senate Committee on Energy and Natural Resources Subcommittee on Energy Research and Development, 2 April 1979.

Estimates range from 1 to 7 quads in the year 2000, saving from $34 to $110 billion dollars in fuel and power plant construction. However, both the timing and magnitude of the actual contribution depends on the scope and pace of the Federal Wind Program.

On August 8 and 9, 1978, the Department of Energy sponsored a meeting of individuals involved in various solar technologies including an eleven person panel on wind energy. Included were representatives of government, large and small business, researchers, and public interest groups. A key recommendation of this panel was that the "absolute minimum threshold" FY 1980 budget required to attain commercial viability of wind systems by the early 1980's is $100 million.

The Domestic Policy Review Response Memorandum, sent to the White House Domestic Council on December 6, 1978, suggests in its second option that the FY 1980 budget for wind energy should be $100 million. These figures represent a conservative case. An aggressive program would require a FY 1980 budget of $150 million according to the public review panel; $128 million according to the Domestic Policy Review; and something in excess of $120 million according to the Department's high budget request to OMB.

In light of this agreement among those familiar with wind energy technology, its potential, and the needs of the industry, what did the administration's request for $67 million, considered by this committee, mean?

Based on our knowledge of the Federal Wind Program, testimony of the assistant secretary for Solar Energy, and the justification offered in the FY 80 request, the AWEA felt the $67 million request signaled three things:

1. The emphasis is shifting within the program from building a strong, competitive industry to increased research and development to design the "ultimate" machine.
2. The application and use of wind energy is taking a back seat to trade-off studies, parametic analyses, and cost-optimization studies.
3. The Department is unable to make a policy commitment to getting cost-effective, renewable resources "on-line."

We estimated that if the Department's FY 1980 budget request and the program it represents is maintained, the contribution from wind energy will be approximately .003 quads in the near-term (1985). If the recommendations of the Public Meeting on Solar Technolgoy, the Domestic Policy Review, and the American Wind Energy Association are followed, that impact could be increased by more than an order of magnitude, the primary fuel equivalent of all the coal and gas used in the state of Idaho in 1976 (62 trillion BTU).

Up to this point, we have been discussing the Federal Wind Program in broad terms. We would now like to turn to specific programs to illustrate our concern.

Program Development and Technology. In the past, the focus of the DOE wind program has been on technology development. Based in part on the success of the early program, the technology is becoming increasingly cost-effective. As the manufacture and use of this technology becomes increasingly viable as a commercial venture, the limitations of continuing to focus on technology development are becoming not mere limitations, but actual barriers.

For example, a report by the Electric Power Research Institute (EPRI, the research arm of the utility industry), states: "An overriding conclusion of this study is that greatly improved understanding of the wind resource is required, including temporal, geographical, and topographical variations. Without this understanding, wind generation cannot responsibly be integrated into electric utility systems, irrespective of success in hardware development."

Given the length of time required to collect, reduce and interpret data, we find it most curious that the budget for this activity has been reduced by nearly 30%.

In another area, Legal/Social/Environmental, the budget request is unchanged from FY 1979. The principal activity identified in the budget justification is a further investigation of television interference.

In the commercialization assessment studies done for Deputy Under Secretary Jackson Gouraud, the possible imposition of "demand charges" for owners of distributed wind systems is identified as a "show stopper." That is, if the utilities and state-level regulatory bodies are not given information and assistance in assessing the impact of the use of wind energy by their customers on the cost of providing service, the option of user-owned wind machines, large and small, could be completely foreclosed.

These are only two areas within Program Development and Technology that reinforce our perception that the Department is failing to address the questions of the actual use of wind energy in favor of increasing emphasis on research and development.

Small-Scale Systems. Small-scale wind machines have been identified as a candidate technology for early commercialization. The number of private companies, large and small, entering the field is growing rapidly. The first round of DOE-sponsored technology development is nearing completion, showing machine costs in production that could deliver energy at a cost competitive in many parts of the country.

A program to address the institutional issues associated with "user-owned" power plants, known as "field evaluation" got started in FY 1979, although it was several months behind schedule because of internal DOE management delays.

In spite of the early success in technology development, and a great deal of public support, the program funds for small machines are being reduced. The timing of the reduction is puzzling in that the program is just beginning to meet its objective of ensuring the availability of cost-effective, reliable hardware. One reason may be that the gap, in terms of the cost of delivered energy, between the small and large machines is narrowing, and the Department has to assess its very basic "economies of scale" assumptions.

There are specific programs the Department could undertake to turn the promise of small wind systems into realized benefits. These include a second generation development of early prototypes; additional contracts for the current "design competitions"; development of machines in the upper end of the small spectrum (e.g., 60 kW); and grant support for private prototypes and applied concepts. These additional programs would cost a total of approximately $15.7 million, or about the cost of one megawatt-scale development program. We feel this is a relatively small amount of money to spend on a technology which, by DOE's own estimate could contribute the primary fuel equivalent of all electricity produced in Hawaii in 1976 (36 trillion BTU) by 1985.

Intermediate and Large-Scale Systems. While overall program development and technology and the small machine program are being hampered by the proposed FY 1980 wind program, the large machines, long the cornerstone of the Federal Program, are being devastated. The cycle began with the cancellation of the procurement action for advanced systems development known as the Mod-3 and Mod-4. The reason for the cancellation given publicly are that the proposals did not meet the cost goals specified in the Request for Proposals. However, on closer inspection, that reason simply does not hold water. First, the cost goals are too low. There will be significant markets open to wind machines long before they meet the stated goals (1–2¢/kWh for the Mod-4). In fact, the Department has recognized its error and is in the process of redefining its cost goals.

Second, such a goal hinders the development of a viable industry by suggesting, incorrectly, that buyers should wait until costs meet the DOE goal before buying.

Site-specific commercial potential exists now.

Finally, there is evidence of capital and kilowatt hour costs at or below the near-term DOE goals available from non-DOE-funded wind turbine manufacturers.

What is worse, in our opinion, is that the Department has refused to admit it made a mistake in the cancellation of the procurement for whatever reasons. Rather than regroup and reissue the request for proposals which ensured multiple contractors, the FY 1980 budget requested represents a program which emphasizes a "front-end loaded" development cycle, in both intermediate and megawatt scale, with single contractors performing parametric analyses, trade-off studies, etc.

Summary. In that machines, large and small are being designed and built today which by industry projections will produce energy for 3–6¢ per kilowatt hour in moderate wind regimes, the industry and the Nation is looking to Congress for help.

The Department of Energy needs direction from Congress to shift its attention from endless R&D to the questions of building a strong competitive industry reducing costs through competition; to the assessment of the wind resource in this country; to addressing the legal, social, and environmental barriers to the use of wind energy in both dispersed and centralized applications; and to a commitment that will ensure a contribution from this renewable nonpolluting source of energy.

R. Nolan Clark*

I would emphasize that the use of mechanical drive wind turbines is more efficient when wind power is used directly at the site. Water-pumping is one example of this direct use. There are many uses in agriculture for wind energy and I believe that agriculture will be the first major user of wind power. Most machines sold in this market will be 25–70 ft in diameter and have a rated power of 15–30 kW.

The greatest need today is for a reliable machine that can be easily erected and maintained. This really means quality construction of the wind turbine and its component drive system.

Council on Environmental Quality*

The Council on Environmental Quality (CEQ) stated in their report entitled *Solar Energy—Progress and Promise* that "a resurgence of interest is occurring in wind energy, an already well-developed technology." Both private and public funds are being invested to improve windmills, which in the near future promise to generate electricity at costs comparable to those of conventional power plants. Both small (kilowatt-sized) and large (megawatt-sized) machines have the potential for marked cost reductions through mass production.

The rate at which wind energy can be introduced into the economy depends heavily on the results of current R&D programs and on subsequent commercialization efforts. The wind potential of the nation, excluding offshore regions, has been estimated to lie between 1 and 2 trillion kWh/year. By comparison, total U.S. consumption of electricity in 1976 was about 2 trillion kWh. Several studies have concluded that, under conditions of rapid implementation, an electrical output of between 0.5 and 1.0 trillion kWh could be achieved by the turn of the century. This is equivalent to a savings of 5–10 quads of fuel. CEQ estimates a possible contribution of 4–8 quads by 2000 (0.4–0.8 trillion kWh) and 8–12 quads by 2020 (0.8–1.2 trillion kWh). (See Table 14-1).

A number of small machines for producing electricity already are on the market. Machines sized at 2 and 3 kWe are being sold as reconditioned units for $1000–1800-kWe. Somewhat larger machines (15–30

---

*R. Nolan Clark is the principal investigator for wind energy irrigation pumping studies being conducted by the USDA Southwestern Great Plains Research Center in Bushland, Texas.

*The Council on Environmental Quality is part of the Executive Office of the President.

Table 14-1. CEQ estimates of maximum solar contribution to U.S. energy supply under conditions of accelerated development. (Units: quads/year of displaced fuel.*)

	1977	2000**	2020**
Wind	Small	4-8	8-12
Heating and cooling (active and passive)	Small	2-4	5-10
Thermal electric	None	0-2	5-10
Intermediate temperature systems	None	2-5	5-15
Photovoltaic	Small	2-8	10-30
Biomass	1.3	3-5	5-10
Hydropower	3	4-6	4-6
Ocean thermal energy conversion	None	1-3	5-10

NOTE: Total U.S. energy demand in 1977 was 76 quads. Estimated total U.S. energy demand is 80-120 quads for the year 2000 and 70-140 quads for the year 2020.

*A quad is a quadrillion or 10^{15} Btu's. Electricity is converted to equivalent fuel that would have to be burned at a power plant to supply the same amount of power. The conversion rate used here is 10,000 Btu per kWh.

**The estimates in these columns are not strictly additive. The various solar electric technologies will be competing with one another, and their actual total contributions will be less than the sum of their individual contributions.

kWe) are becoming available at comparable unit costs. As a substantial market develops, the mass production of small-scale windmills should lead to significantly lower costs.

Large windmills are being developed by both the Department of Energy and private industry. A 100-kWe machine has been undergoing tests for more than several years near Sandusky, Ohio. In January 1978, a 200-kWe wind turbine began generating electric power for Clayton, New Mexico. During 1978, two similar machines were constructed, one in Puerto Rico and the other on Block Island, Rhode Island. The objective of the next phase of the large windmill program is the construction of a 2 MWe windmill with a rotor 200 ft in diameter. The largest of the series, with a rotor 300 ft in diameter, was built in late 1979 at a cost of $10 million. Private companies are also installing large turbines, reportedly at costs lower than those of government-sponsored machines.

DOE is studying windmill designs, different from the horizontal-axis type, which may have technical and economic advantages. One concept being developed is the vertical-axis wind turbine. A 55-ft-diameter Darrieus rotor, with blades shaped like an egg-beater, began operation in mid-1977. In a 22-mph wind, this machine produces 30 kW of electrical power. During FY 1978, studies examined how the design might be improved to reduce costs and permit mass production.

DOE has also begun testing six commercially available, small windmills, with contracts being issued for the development of 40-, 8-, and 1-kWe machines. A proposal also has been accepted by the Bureau of Reclamation which involves the demonstration of a significant number of wind energy systems integrated into Federal hydroelectric systems.

A recent report prepared for DOE recommends a substantially expanded program to speed the widespread use of wind energy.

DeWayne Coxon*

One does not need to read many books nor interview many experts in energy before realizing that there are several facets to a Wind Energy Conversion System (WECS). Marcellus Jacobs, manufacturer of the Jacobs wind machine (which grossed five million dollars in sales in a 25-year period ending in 1956) maintains that the market for decentralized wind generation is not competitive with utility company prices nor will it be due to the Rural Electrification Act. This act effectively eliminated production of most windmills in the United States by providing nationwide electrical lines to areas that did not have access to alternate current (AC). The early wind machine power plant capitalized on a market created by people living at a distance from the centralized power plant. The cost of poles

*De Wayne Coxon is Co-President of Jordan College.

and lines for the AC was too expensive so that the direct current (DC) from the Jacobs wind system was used when the wind wasn't blowing.

Now, buying from public utilities is far more economical for urban people. The most conservative kilowatt-hour purchase price figured over a 20-year payback for a WECS is about 25¢ per kilowatt as opposed to approximately 7¢ for company power. We can deduce then that people do not choose to build their own power plants for economical reasons. The reasons for building WECS's appear to be independence from energy conglomerates and environmental concerns. The desire for independence erupted after people experienced production problems resulting in brownouts, blackouts, and strikes. Finally, there is a growing concern that environmentalists and fossil fuel analysts are right in their prediction that the earth may be in the final stages of freely bestowing her reserves. If this is true, we will have to rely on the wind and sun.

No one has been able to estimate the amount of energy contained in the wind. Man has created windmills and windmachines that are able to operate in 7 to 70 mile per hour winds. But, we are currently unable to use the gale force winds, hurricanes, and tornados. Such energy, if controlled and stored, could satisfy most of man's electrical needs.

Wind studies completed by the Government and reported at the Helioscience Conference in Palm Springs, California in 1977, proved that both velocity and constancy increase with height. That seems logical when one realizes there are few obstacles to impede the wind progress in jet streams (30,000 ft) as they circle the earth at 150–250 miles per hour. (Recent Federal Government wind studies showed that, at fifty feet, most of the United States was unsuited for large WECS. At one hundred feet, nearly every state had some probability of wind success. But, at 200 feet there was usually a prevailing wind in some part of each state.

Most of the work that is done at the Jordan Energy Institute with people who wish to install WECS's results in bottom-line totals that scare many people. Machines such as the 4.5 kW on the campus of Jordan College produce about half of the power that a conservative small household uses per month. The system, including the tower and batteries, costs about $7,500. One cannot buy a refurbished wind machine for much less than $1.00 per watt. Add to that a $2,000 price for the tower and several hundred for batteries, depending on storage needs, and the total return is suspect.

Although the government, by helping with research and development funds to private companies, may be able to perfect a 10 kW WECS for about $9,000, including $3,000 for tower and batteries or syncronous inverter, there is still a question of bottom-line economics. As Americans watch energy prices continue to rise and inflation depress saving accounts, many may decide to become utility independent.

Both batteries and inverters have disadvantages. Machines with inverters built into the head do not operate without utility lines; thus, one cannot get power from the wind if the utility power is lost. Obviously, this does not provide security when there is a power failure.

The option to go from the power head into an inverter and either to the battery bank or to the house may be more satisfactory, but it creates double expense in providing batteries and inverter from DC to AC.

To solve these problems, we recently built two electrical systems into the living area. One system could be used as AC in the utility grid. This demands a minimum charge for service costs, unfortunately. But, if the wind system is small and the battery storage small, a consistent backup is available. The other electrical system is a DC from the mill to the battery to the house. The availability of DC appliances currently makes it possible to use wind power at the peak seasons and not suffer inconveniences during low wind seasons.

Wind is a usable source of energy. Americans are so innovative that WECS will probably soon be commonplace in our country. The rich may be challenged to seek independence and the wise will continue to look for their security in the sun and wind.

Frank R. Eldridge*

Let's begin with applications. I think that in addition to the use of wind machines for generating electricity for large utilities, there's a lot of other possibilities, many that we haven't begun to explore. There are numerous applications for a wide range of wind machine sizes, for everything from 1-kW rated output to as big as you can get them, maybe up to 5 MW. DOE has been concentrating on the problem of large machines for generating electricity for utilities, but I would like to see much more effort being given to the whole range of sizes. In particular, I think there might be a big market for smaller machines, though now there's a great deal of uncertainty as to what the ultimate price of the smaller machines will be. At Mitre we've been looking into that and gathering information from people who are building small systems like the Pinson cycloturbine and the Zephyr. Some of the estimates that have been made through the DOE Mission Analysis study, as far as I can see, look to be about an order of magnitude greater than what the builders themselves are projecting in terms of the ultimate price per kilowatt installed. . . .

DOE's thinking that large-scale machines will be more cost-effective may not be true after all. In my opinion there will be a cost differential between the large- and small-scale, but I don't think it's as large as some of the early predictions. When it comes to estimating what the ultimate price will be, there's a lot of factors to be considered. One of the most important of these is the amount of material that goes in and what's the ultimate price of that material. While the developmental costs start out higher for these machines, I think over the long run, as these machines get into production, they will become a very small part of the capital costs of the machine. And of course you also have in production work the learning curve, which will decide the rate at which prices will drop. I don't think they'll drop along smooth functions of ten; there will be plateau periods where the prices stabilize, then there will be breakthroughs in new designs that will give sharp decreases, then another plateau. Another factor that must be taken into account is the wind regime. This will greatly affect the cost per kilowatt; it's really one of the most important factors in maximizing the return on the investment. There are some interesting possibilities here: the variations of wind speed with altitude favor the large machines. Small machines have the advantage that they can be used in a dispersed mode and save on transmission costs, although these costs aren't necessarily overwhelming.

Then there's the whole question of storage. There are a number of applications for small wind machines that don't require storage, like a wide range of pumping applications for water or oil, say stripper wells. There is a large potential application for small machines for heating and cooling buildings. We've done some studies into solar heating and cooling, and there might be some cost savings in using wind systems that pump a hydraulic fluid rather than using a generator and a resistance coil to make heat. You can put the high-pressure fluid through a resistance orifice and heat a heat transfer cylinder. There you'd be saving the cost of the generator, wiring and so forth.

There's also the possibility of driving heat pumps directly. And air compression has a tremendous possibility. We've been looking at it as a storage medium, and one of the things that would help the wind industry at this point is to get the natural gas utilities involved and use their experience in storing compressed gases. They've been doing it for just a few dollars a kilowatt rather than

---

*Frank R. Eldrige is a pioneer regarding the evaluation and development of the Federal Wind Energy Program. He is a member of the Division Staff, the Mitre Corporation.

the few hundred dollars for other wind power storage systems. The gas people have been storing gas underground in naturally occurring cavities, aquifiers, depleted gas wells and so forth. The large machine might make use of this technology. They will also be effective in combined wind-hydroelectric modes. We call this the "water saver" mode where the windmill and the hydro-turbines operate in tandem, reducing the flow of water.

We need some kind of market at this point to bring down the cost to enable manufacturers to put in mass production facilities. This will, in turn, enable the price to drop quite rapidly to something close to the ultimate cost. One of the kinds of markets that could do this soon for the large-scale industry is the wind-hydro systems. If the Bureau of Reclamation, for example, puts in a large number of MOD-1 machines up in Wyoming, this will create a market very quickly, and it won't take production of very many to bring the price down to where the energy will be quite competitive with peak power costs. One of the ways that this could get started is to have the Federal Government build these types of systems for the Federal utilities. The Bureau of Reclamation has a number of large hyrdo-dams like Hoover, Boulder and Grand Coulee. There are about ten gigawatts in the Colorado River storage project, and they're planning to expand their capacity there by another 30 gigawatts by 1985. This could mean a large potential for operation of wind-hydro systems. . . .

There's a great need for commercialization of small systems as well as large systems. So far it's all been aimed at large systems with little effort along commercialization lines for small-scale. I think DOE had felt that the small systems are going to develop on their own, but this isn't necessarily so. As well as money for R&D of small scale systems, there should be money available for demonstration projects. The Federal Government should be thinking of ways of providing incentives for using small machines.

Mr. Eldridge has recently revised his book *Wind Machines* (Van Nostrand Reinhold, 1979), and the following paragraphs are from his chapter "Future Utilizations."

There has been considerable speculation concerning the total amount of wind energy that will be used to help supply energy needs of the future and how and when this wind energy will be utilized. There is general agreement that wind is a very clean, replenishable source of energy and, that though it is intermittent in nature and relatively dilute as compared to fossil and nuclear fuels, it constitutes a very large but practically untapped energy resource. How effectively and to what extent it will be used in the future depends on our energy needs and our ingenuity.

Dr. Marvin Gustavson of the University of California has estimated that about 4,000 quads per year of energy are potentially available, worldwide, in the near-surface winds of the earth, of which about 60 quads per year are available over the land areas of the 48 contiguous states of the United States. Lockheed estimates that by siting wind machines only on open-range lands of the conterminous United States, a total of over 150 quads per year of primary wind energy could be converted to forms useful to man. If the Betz coefficient and the conversion efficiencies of these machines are taken into account, as in the Gustavson estimate, this amounts to about 50 quads per year of usable energy. In the Lockheed definition of "open range land," populated areas, privately owned land, national or state parks, military reservations, scenic reserves, and transportation corridors and rights-of-way are all excluded. Lockheed further estimates that offshore, within 15 miles of the coastlines of the United States, about 70 quads of additional primary wind energy is available, which could be converted into about 20 quads of useful WECS output energy.

On the basis of these estimates, the total amount of useful WECS output energy potentially available to the United States is roughly equivalent to the present and foreseeable demands for all types of energy in the United States. A total of about 75 quads per year of primary energy resource is being used in the United States at present, to satisfy end-use energy demands of about 45 quads per year. By the year 2000, it is expected that end-use demands will rise to about 70 quads per year, requiring about 110 quads per year of primary energy resources.

Because of the relative simplicity of wind energy systems compared to energy systems of other types, and because of the worldwide availability of wind resources, it is expected that large markets for WECS will develop in other countries, as well—particularly in the underdeveloped countries of the world.

Even though wind resources are large, there are a number of problems that must be addressed if wind energy is to be used effectively and extensively in the future. New, efficient, reliable designs for wind machines must be developed and proven in a wide variety of potential applications. Manufacturing facilities must be provided and adequate marketing, distribution, installation, operations, and maintenance infrastructures must be developed. In addition, suitable sites must be chosen for the wind machines that are installed. All of this will require time and effort.

Taking these factors into account, a recent MITRE/Metrek study concludes that wind energy, under an accelerated development program (with federal initiatives of the type provided in the National Energy Plan of 1977) might, by the year 2000, be expected to displace about 2 quads per year of fossil fuel used by utilities for the generation of electricity—and about 6 quads by the year 2020. Another recent study, by the Council for Environmental Quality of the Executive Office of the President, indicates that, with an accelerated Federal Wind Energy Program, wind energy might be expected to displace 4 to 8 quads per year of fossil fuels by the year 2000, and 8 to 12 quads per year by the year 2020. Estimates by Lockheed, General Electric, Stanford Research Institute, and o... indicate that the fossil fuel displacement by wind energy by the year 2020 might be as high as 30 quads per year.

In summary, while wind machines are inherently among the simplest of energy conversion devices, the wind itself is a diffused and variable medium. The selection of optimal siting for wind machines and the choice of optimum designs for particular applications are often difficult and complex problems. Much research and development work on wind machines still remains to be done, but wind machines appear to have a high potential for making significant contributions to the goal of meeting our future energy needs through the use of clean and essentially inexhaustible sources of energy.

Department of Energy*

As the Federal Wind Energy Program advances toward the commercialization goal, increasing emphasis will be placed on studies and development efforts which directly support WECS production and use. Further marketing, economic, and applications studies will identify potential users and provide data to assist industry in planning and development. Studies are now underway to identify areas which demonstrate the greatest potential for wind energy use and to assist in developing strategies to fulfill that potential. Market size, characteristics (for example, power demand schedule), strategies, sales, and maintenance networks will be analyzed.

The small machine market will continue to receive strong attention. Since interface with electrical utilities appears to be the most promising method for generating electricity while assuring uninterrupted electrical service, future studies will concentrate on important interface questions, including costs, rate structure, and operational and safety factors. Analyses of these problems will provide important data for potential

*Department of Energy, Wind Energy Systems Program Summary, DOE/ET-0093, December 1978.

WECS users in planning applications and for manufacturers in venture analyses.

Economic incentives and demonstration projects for large and small machines are under considerations to stimulate WECS use. Demonstrations of turbine operation in a variety of applications and environments may be a valuable method for increasing public knowledge about and acceptance of wind energy. As reliable, cost-effective machines are developed, such projects will help to identify high potential applications and further refine the technology. Economic incentives, including tax benefits and direct subsidies, are being evaluated. Market data will assist program planners in assessing the potential impact of incentives on a variety of markets, including utilities and residential, agricultural and industrial users.

Through regional applications studies, detailed questions associated with integrating WECS into utility networks will be addressed. Analyses of technical and institutional factors relating to particular utility configurations, mixes of existing power generation equipment, and long-term plans for alternative mixes will provide a data base for long-term planning. Although the studies will be conducted using sample utilities on a regional basis, results are expected to provide information on generic classes of problems which can be generalized for application by utilities.

International cooperative efforts will be expanded in the future, and information sharing programs are expected to enhance program efforts by providing access to alternative technology and data, as well as opportunities to validate analytical techniques. The importance of these programs, and international interest in such efforts, is likely to increase as a result of the worldwide energy crisis.

Technology development programs will investigate components and systems and the relationship between the wind and the dynamics of the wind system to produce lighter, less expensive, and more reliable and efficient machines. This goal will require development of improved design tools and manufacturing techniques, as well as development and evaluation of new hardware designs and configurations. Promising concepts and ideas will be tested and evaluated on experimental wind turbines, and electrical interface problems will be explored in anticipation of integrating significant numbers of turbines with conventional power systems. The involvement of numerous commercial manufacturers will be encouraged in an effort to develop expertise and a technology base in the private sector. A technology workshop will be held to present significant new results of the program to the rapidly evolving wind turbine industry.

Large systems will be developed in a cyclical pattern to facilitate transfer of cost, design, and performance information from one generation of machines to the next and to provide a continuous flow of machine development. Second-generation machines are being developed, and third-generation machines are in the planning phase. The small machine development program will follow a similar cyclical pattern. In addition to the 1-, 8-, and 4-kW machines being developed by nine manufacturers and tested at DOE's Rocky Flats Plant, the program will initiate additional cycles of small machine development, taking advantage of design, performance, and cost data derived from the initial cycle and based on needs identified in market and application studies.

The wind characteristics program will emphasize development and dissemination of wind characteristics information [see Figure 14–1]. Training guides for siting small WECS will be produced, and a major project to analyze wind characteristics over large areas (including distribution, magnitude, and reliability) will identify areas of high wind energy potential for further detailed study. Additional major projects include development of wind characteristics information relevant to wind turbine design and performance evaluation and determination of wind forecast reliability for use in WECS operations.

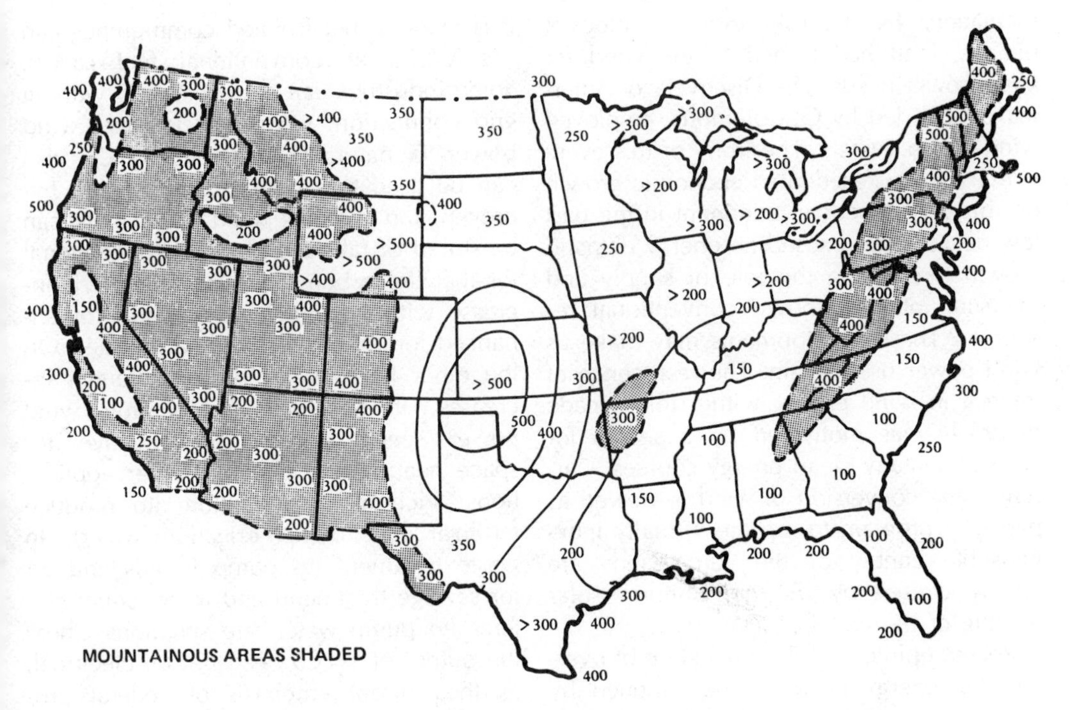

Figure 14-1. U.S. annual average wind power (W/m²) at 50 m. (Courtesy of DOE.)

The ultimate measures of the Federal Wind Energy Program's success will be a capable, successful wind energy industry and a sufficiently informed and active market to motivate industry to produce efficient machines for commercial implementation of wind energy. The final goal will be reached when wind energy is effectively harnessed to significantly contribute to fulfilling the nation's energy needs.

J. Stuart Fordyce*

Since the preferred use of wind power appears to be electrical generation, it is reasonable to assume that electrical energy storage is the main area of concern. Battery systems of various kinds, beginning with conventional rechargeable batteries and their advanced systems operating at elevated temperatures, to more unusual concepts like the Redox flow system or electrolysis-fuel cell concepts, are all technically interesting for the application. The economics, however, will be the dominant consideration, all other things being equal, and the newer technologies may ultimately come in at lower cost. . . .

It seems to me that any storage system applicable to the utility load-leveling application will satisfy the wind system requirements. A key consideration is, of course, whether or not the wind system is tied in to the utility grid.

For very near-term storage technology, the lead-acid rechargeable battery is simple, expensive and not maintenance-free.

Kenneth M. Foreman*

Centuries before the urgency imposed by the 1973 oil embargo, and more recent cost

*J. Stuart Fordyce is Chief, Electrochemistry Branch of the Solar and Electrochemistry Division at the NASA Lewis Research Center.

*Kenneth M. Foreman is Head, Advanced Concepts Branch of the Grumman Aerospace Corporation in Bethpage, New York.

escalations by the oil producing bloc of nations, man had turned to the wind for basic power needs. The Discovery of America project, led by C. Columbus, employed wind-driven vessels. The irony of this event is that the unprecedented economic growth on this very American continent in the past few decades has caused an energy demand growth threatening the rate of supply and the sum total of known conventional resources. These developments may make us wind power users again. The resurgence of interest in wind energy within this decade originally was motivated by a search for environmentally clean energy sources. Currently, the conversion of wind is viewed as part of a program to tap the virtually inexhaustible energy of the sun; winds are largely caused by the non-uniform solar heating of the earth.

Recent opinions about the extent of overall U.S. energy needs to be supplied by wind power conversion systems tend to be conservative. In most policy-making circles, the projections of one to ten percent of electrical grid capacity have been associated with wind power for the post-2000 period. Obviously, such modest assignments and goals can be unduly restrictive in that governmental programs are formulated to conform to preconceived objectives rather than to exploit the full potential indicated by, for example, general theoretical considerations. Also, these projected programs to provide wind energy resources two or more decades in the future are based on a technology that is essentially frozen as of the mid-1970's. Obviously, this static implementation model ignores the growth possibilities inherent in advanced and innovative wind energy conversion systems currently only in research or conceptual stages.

There is strong reason to suspect that the wind energy field is in a period of great transition. Government timidity, as well as the emotional, unbounded enthusiasm of wind energy activists, must be balanced by serious and objective workers in this field addressing the economics issue that has thwarted prior promotion in modern times.

It is evident that isolated communities can use WECS as conventional fuel savers. Some industries can employ WECS without grid connection, whenever sufficient wind power is naturally available. WECS also can be used to produce fuels such as hydrogen and methane, and these fuels can be stored or take the place of conventional fossil hydrocarbon fuels. Heating needs increase with wind velocity because of enhanced forced convective heat transfer. On the other hand, wind power potential increases with wind velocity. Therefore, wind power seems particularly attractive for space heating and cooling. Other applications, such as in agriculture (to produce fertilizer and pump irrigation water), in waste treatment (to pump liquids and air for sewage treatment) and in aqueduct systems (to pump water) are situations where the output of WECS need not be electrical, as the current emphasis of Federal programs appears directed, but can hold great economic value.

With a more flexible Federal support policy encouraging advanced and innovative approaches and the resolve and skill of the engineering profession, there is every indication that economically competitive WECS can be developed to supply a significant part of U.S. energy needs for the foreseeable future at moderate rates.

Marcellus L. Jacobs*

We appreciate the opportunity of outlining our views on the future of wind energy. We feel that a considerable portion of the public is ready to accept the general concept of using electric energy generated by wind-operated plants to supplement other sources. This new interest in wind energy has been greatly accelerated during the past years by the energy crisis.

So as to give you a better understanding of the vantage point from which our later comments about Wind Energy Electric Generating Systems are based, the first portion of our comments briefly reviews our more than 40 years' experience in the engineering, manufacture and

*Marcellus L. Jacobs is President of Jacobs Wind Electric Company, a firm which has participated in the research and engineering of WECS since 1922.

sale of tens of thousands of wind electric plants which we shipped all over the world.

My brother, J. H. Jacobs (now deceased) and I started designing wind electric systems in 1922 on our Montana ranch, testing out prototypes on neighboring ranches for several years. After proving the practicability of wind-operated individual electric systems for home use, we established the Jacobs Wind Electric Company, a Montana corporation, and moved to Minneapolis in 1930 to set up a manufacturing plant to produce our three-blade, propeller-driven, direct drive, flyball speed-governed, variable-pitch propeller system, with automatic voltage and charging control.

More than 30 patents, past, current and pending, have resulted from our engineering, development and manufacturing designs. For nearly 30 years we manufactured our 2- and 3-kW dc battery storage wind electric systems, selling them worldwide. Admiral Richard E. Byrd used a Jacobs Wind Electric Plant at his Little America base in the Antarctic, where it operated from 1933 until 1955, when the propeller blades were removed by Richard E. Byrd, Jr., and brought back as souvenirs.

The ice had built up over the years, almost to the propellers. The original tower was 70 ft high. This plant operated with no repairs or service of any kind and was still spinning away when it was stopped to remove the blades in 1955. The aircraft spruce propellers were still in good working shape after all those years.

A mission in Ethiopia that installed a Jacobs Wind Electric plant in 1938 purchased their first repairs, a set of generator brushes, in 1968. Thousands of farm and ranch homes in the U.S. and in many parts of the world, mining camps in Alaska, small hotels in Finland, a U.S. Weather Station at Eureka (in the Artic Circle) and many island homes all over the world have operated their lights, home freezers, refrigerators, water systems and appliances with the Jacobs Wind Electric System. Up to 500 kWh/month (more in windy areas) is produced by our 3-kW, completely automatic, attention-free electric generating system. The ten-year guaranteed storage batteries frequently lasted 12–15 years and the plants were designed for 25 years or more of dependable service.

Back in the mid-30's, I engineered, developed and manufactured a system of cathodic protection for pipe lines, using my wind electric plants installed at intervals to maintain a negative pipe to soil electric potential in the lines. The wind generated direct current applied to the steel pipe line effectively reduced underground corrosion. Many hundreds of miles of pipe lines were protected by this system in the U.S., Mexico, Venezuela and Arabia. The plants operated in isolated areas under all wind and storm conditions, completely unattended year after year. These pipeline plants were identical to those built for home use except for their lower voltage. At the time, I engineered the special plants for pipe line cathodic protection. I also helped to organize the National Association of Corrosion Engineers, became a Charter Member and served on its first board of directors for two years.

We designed, built and tested, more than 20 years ago, at considerable expense, a high-line grid system booster ac wind electric plant that was designed to be installed on grid towers to assist central station power generation. We sent a report to Congress suggesting that we build and install 1000 plants, one mile apart, from Minneapolis, Minnesota to Great Falls, Montana. Some of the plants would be producing all the time, generating a considerable output total to reduce central station fuel requirements (1000 plants of 10 kW each). This method eliminated the difficult problem of a single large plant that, when operating, would inject too much energy into a local area, upsetting local grid control systems every time wind spurts occurred. The large plant is still subject to no output during calm periods. We believe the equipment cost per kWh produced by the smaller, completely automatic and virtually attention-free unit system will be less than that of a large single plant with its complicated gear, control systems and operating personnel. No one was then concerned with an "energy crisis" and we did not even receive a reply to our outline. We felt at that time that the wind electric booster system would fit in with the growing use of electric home heating because, when the wind blows, home heating requirements are greatly accelerated and the wind energy output would help to counteract such increases. We still feel that our plan of wide distribution of smaller plants over a grid system is much easier to absorb and adjust to widely varying wind energy outputs and at less cost per kWh produced than a large system.

We do not feel that the present technical state-of-the-art (which has regressed since 1960 when we closed down our Minneapolis plant and moved to Florida to engage in other product engineering programs) has a suitable plant that will meet the electric requirements of the modern home and be acceptable to the interested public. Our former wind electric plants were designed to supply the electric requirements of that era and were installed as a "Complete Package System" with bottle gas for home heating, water heating and cooking.

The manufacturing of our plants was stopped in 1960 because the building of R.E.A. lines on a nationwide basis throughout rural America, financed by the U.S. Government, put an end to farm light plant sales. Although the law required three meter hook-ups per mile, before an R.E.A. line could be built, in order for it to be self-liquidating, many lines were constructed with miles between meter connection, thus requiring large annual subsidies to keep them in operation, and even today after 20 years of service some of the subsidies continue. . . .

The energy requirements of the modern home are considerably greater than was supplied by our "system" in former years and will call for a larger capacity wind electric plant with a larger storage system automatically coordinated with a solar heat system and coupled with a bottle gas auxiliary unit that will ensure a complete and adequate total energy source. The use of bottle gas for cooking and as a backup emergency energy source for the wind-solar system would complete the "Package System" for the modern home or other user. While natural gas is getting in short supply, the amount used would not be great, and the future, as now projected, for gasification of coal or oil shale indicates that bottle gas of some form will be available and it is about the cleanest fossil-based fuel there is. As hydrogen gas becomes available as an alternative source of bottle gas, this can of course be worked into the system. We believe the energy requirements of the modern home can be supplied by this "system" and that such a complete system, when designed and manufactured to be installed as a "Package Unit," could be operated on a cost basis competitive to any other energy supply method. A dc to ac converter for supplying ac current to the home would be part of the "package system."

Due to the present incomplete state-of-the-art, few manufacturers are willing to spend the amounts of venture capital necessary to develop a practical and dependable wind electric system suitable for commercial acceptance in an untried market. Since no one apparently is now working on perfecting a complete package wind-solar system and is instead developing only parts of such a system, a means is needed to tie together "under one roof" the entire project to supply the energy needs of the modern home. Such a system should appeal to the home buyer where access to the normal electric grid system is difficult or expensive.

With sufficient financing backed by experienced engineering efforts, a combined wind and solar energy system could be perfected for commercial acceptance in 3–5 years. An important aspect of this entire program is to develop methods for securing direct, non-polluting solar (wind and sun) energy to meet the requirements of modern man in his modern home and thus reduce the need for energy produced by fossil fuels and other limited supply resources. Solar energy, wind and sun, is unlimited and is a non-polluting power resource available almost everywhere and should be harnessed as fast as possible on a large scale to save dwindling fossil fuels for purposes where only such fuels can be used. We feel that considerable U.S. Government funding to experienced men and organizations can rapidly develop the kind of practical systems needed to harness in effective quantities these inexhaustible solar energies for man's use.

In 1972, when we set up a considerable fund (half a million dollars) to engineer and develop a new Jacobs Wind Energy System, we felt the investment was well worth our efforts, with the energy costs starting to raise. Our foresight at that time has been proven by the considerable increase in energy costs and the absolute certainty of further raises in its cost. Our new 8–10 kVA Wind Energy System fully justifies the investment in time and money to develop it. We feel this new plant represents the most advanced concept of design and performance yet produced in this field. With 50 years of experience backing this project and many thousands operating all over the world since 1930, it represents "experience" which is vitally important if a design is to be able to withstand the storms and high winds that may exceed 100 mph on top of towers.

Too many of the newly engineered designs are not "time and weather" proven. It takes several years and a number of installations to prove a design against the storms that "Mother Nature" will sooner or later throw at it. You do not develop a successful wind energy system on the engineer's drawing board; rather, you prove it up on the tower. Look at the millions in

Government R&D that have been wasted in the past 5 years on impractical wind plant designs, ideas that we older wind plant engineers have thrown away 40 years ago, yet these ideas are being rebuilt as if they were entirely new ideas. Why doesn't the Government require research of the past developments before giving large amounts to try what some "Johnny-come-lately" presents as a "new idea." Not one of the original experienced manufacturers of wind plants in the 30's and 40's has been approached by the Government for engineering experience in this field. They make large grants to "outfits" that never saw a wind plant, to develop a new one.

C. G. Justus*

Wind power is the alternate energy process which is closest to being commercially feasible for the production of electricity. Wind energy will initially be used for electric power production at remote sites, where it can compete economically with high (and escalating) fossil fuel alternatives. Other near-term applications will include large (MW-scale) electric power production by arrays of wind turbines on the utility service networks, and medium-to-small (1–100 kW) wind powered systems for farm and rural applications.

There appears to be no technological reason why wind power cannot produce a significant fraction (5 to 20 percent) of the electric power needs of the United States. When one estimates the number of MW-size machines this would require, since a typical single power plant today is 1000 MW, the task of implementing such a program becomes staggering. However, if one looks back at the development and implementation of the nation's highway and airway systems, all since the beginning of this century, then such an enormous effort does not seem so unrealistic.

California has already set a State goal of 10 percent electric production (30 billion kWh/yr) by wind power. This is to be accomplished by the installation and operation of several thousand MW-size wind turbines. The Federal DOE program is heavily involved in the development of both large and small wind energy systems for early commerical development. However, no specific national goals have been set for wind-electric power production. The current emphasis on national mobilization of energy resources should include continued and expanded efforts at developing this potential for meeting some of our energy demand, while developing exportable technology.

*C. G. Justus is a professor at the Georgia Institute of Technology, School of Geophysical Services.

Herman Kahn*

Long-term energy prospects—resting on sources that are inexhaustible—are very good; but the recent oil embargo showed how vulnerable the world may be in the short term. In these circumstances, the task ahead in energy is twofold: the development of alternate energy sources and the achievement of a degree of energy independence.

It may seem strange to go back to a power source which is thousands of years old. However, the new technology available for generating and storing power from windmills makes this an attractive and economical power source for regions where the wind blows rather steadily or at a higher than average speed, or both—for instance, the Texas Gulf coast, the Aleutian Islands, the Great Plains and the eastern seaboard of the U.S. A Federal Government program is under way to determine the optimum locations and to test various windmill systems. The current expectation is that the first commercial system will be installed during the early 1980's, and the cost will probably be less than that of most current conventional sources. [See Figure 14–2.]

*Herman Kahn is founder and Director of the Hudson Institute, a non-profit policy-research organization. *The Next 200 Years—A Scenario for America and the World* provides a long range view of Herman Kahn, William Brown and Leon Martel regarding inexhaustible sources of energy such as wind power.

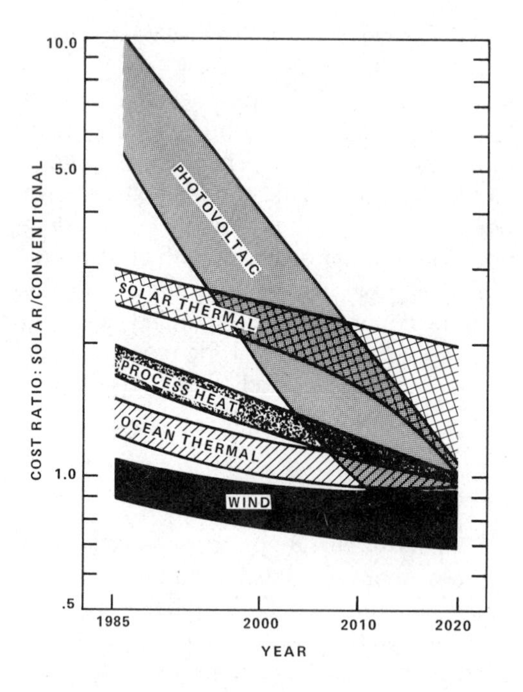

Figure 14.2. Projected cost ranges for several types of solar energy including wind energy. (Courtesy *The Next 200 Years* by Herman Kahn).

Richard Katzenberg*

Is the wind industry progressing? Are we any closer to seeing wind generators dot the horizon, providing renewable non-polluting energy for home and industries? The answer to both questions is yes!

The industry has taken great strides towards making wind energy a reality. Most of the successes have been excruciatingly slow and much work remains to be done. But in the near future, the public will hear a great deal more about wind energy and its potential.

States of the Art. A wind machinery is a very complex piece of machinery. Its many components must complement one another perfectly or problems will develop. The blades, the most visible components, are the solar collectors of the wind machine. They must be designed to handle very rapid changes of wind speeds due to gustiness, shifts of direction, and tower shadow. Their ability to accept these rapid changes is also dependent upon the speed with which interrelated components can react. The interrelationship factor is also important in the tower-rotor interface. Towers have natural frequencies that must be compatible with the natural frequencies generated by the machine. The problem of interfacing components to arrive at a "perfect" machine is one of the many problems that must be worked out through experience.

Machines must be tested all the time and in a variety of environmental conditions before they can be delivered to the public with confidence. The industry has learned a great deal about patience since the Arab oil embargo, but we are now reaching the point where both manufacturers and consumers can have confidence in commercially-available wind machines. At first, the marketing will be done carefully, and monitoring of each machine's performance will be prudent. Highly reliable cost-effective hardware is coming, and with it will come the new era of American wind machines.

Manufacturing Capabilities. It is not uncommon for today's wind machines to be produced largely by hand and in very small production runs. Some are produced one at a time. The American Wind Energy Association reports that the industry's best interest would be served by stimulating production, not demand—there are enough applications for the limited number of wind machines that the industry can now produce. Stimulating demand could aggravate the problem. A means must be found to provide capital for manufacturers, so they can expand and modernize their facilities for mass production.

Cost and Cost-effectiveness. A 2–4-kW wind machine can be purchased and installed today for $6000–10,000, depending upon component selection and site conditions. This dollar outlay will be reduced by over 20 percent by the solar tax incentive. The reimbursement will help, but there's

*Richard Katzenberg is President of Natural Power, Inc. His views on the future of wind energy were summarized in an article entitled "In the Wind" in the February 1978 issue of *Solar Age*.

still a long way to go before machines "make cents" for the public. I believe that a wind machine of that size should be available in the $2500–3000 range. After studying production costs carefully, I believe it is a distinct possibility within a few years. That is what it will take before we have a truly broad-based demand for wind power.

However, it should be noted that many applications for wind energy are cost-effective today. In a study prepared for DOE, it was reported that electricity from a wind machine can be produced for 15.4¢/kWh, using today's production techniques. The two greatest variables affecting cost per kWh are wind speed, assumed to be 12 mph and interest rates, also assumed to be 12 percent. There are many areas in the country where average wind speed is greater than 12 mph, and many possible government programs to assure interest rates far below 12 percent. These factors, coupled with mass manufacturing facilities, could bring the cost per kWh down sharply.

I must emphasize 15.4¢/kWh is not terribly out of line with today's electrical costs. We are not a "pie in the sky" technology. As fossil fuels cease to be subsidized, and the real cost of energy is taken into consideration, wind energy will be there as a viable source.

Government Programs. The broad range of Federal research programs has started to produce meaningful results. The large machine program spearheaded by NASA Lewis Research Center is full of vim, vigor and confidence. They embark on the next stage of the program by testing large systems in several sites: Puerto Rico; Long Island; Clayton, New Mexico; and Boone, North Carolina. The Rocky Flats small machine testing program has added a 40-, 8-, and 1-kW design program to its funding efforts. The winners of this competition have been announced, and the program should produce a depth of design engineering and testing that will assure marketable quality hardware. The test facility at Golden, Colorado has accepted machines that are tied together by microprocessors. The wind energy conversion systems' operating characteristics are analyzed in one of the most difficult wind regimens America has to offer. Through this program, manufacturers and consumers will be able to learn the characteristics of each brand of machine and all should benefit from this information flow. The wind characteristic program at Battelle Pacific Northwest Laboratories is starting to supply the industry with usable information. This will help manufacturers understand the forces they must use (and overcome) to build hardware for a 20–30-year life expectancy. Battelle's publications will teach consumers to evaluate and understand whether their site will be a cost-effective one for use with a wind energy system.

Strengths—What Does This Energy Source Have Going for It?

- Enormous industry enthusiasm.
- A broad range of technical expertise going into hardware development.
- Cooperation among many diverse views—government and industry, manufacturers and consumers, large- and small-scale wind system programs.
- A nation eager to promote renewable energy sources (including wind) for large-scale implementation.
- A willing Congress.

Weaknesses

- Remarkably little capital for product development and expansion.
- Little understanding of the markets and applications available to the industry; little fund-raising information reaching those who need it.
- Underfunded, understaffed, overworked federal offices (particularly DOE), resulting in a severe slowdown of information dissemination.
- DOE's underestimation and underfunding of wind power. In comparison

with most other energy development programs at DOE, we can produce energy far less expensively, have a greater potential for sharply reducing that energy cost, and can make a faster return on the investments. Yet the wind department at DOE has done little to prompt a larger funding effort for wind power—slowing the industry's progress.

- Lack of a broad-based demonstration project has severely hampered the industry's efforts at gaining visibility, and slowed the development of wind systems. Although the delay is being rectified now, it will take three years to catch up. Placing 10-kW wind machines (and smaller ones, too) around the country would be relatively inexpensive, simple and highly visible. In contrast, a singular large machine would be riskier and more expensive, and would provide little visibility.
- Small businesses did most of the initial research; thus, research and development took a long time, without the benefits of sophisticated engineering or test equipment. As additional funds are being channeled into the wind program, the problem is being corrected, but again there is a lag time.

In all, there is reason for great optimism—exciting years for the wind industry.

Amory B. Lovins*

The aerospace community envisages high-technology wind-electric machines of MW capacity costing of the order of $400–600/kW(e) peak and yielding 2–4¢/kWh electricity in typical windy-site wind regimes averaging 7 m/sec. (Some observers estimate slightly higher costs.) But large machines, even if better designed than the >$5000/kW(p) (kilowatt of peak output) DOE/NASA prototype, may have more severe materials problems than small machines, owing to greater blade stress and vibration. Though technical lead times are only 1–2 years, such big machines are also accessible mainly to utilities and can be mass-produced only in the same sense as large jet aircraft, so they may be slower to deploy than small Model-T-style machines in the spirit of those that used to dot the Great Plains (such as the Jacobs machines, many of which are now being resurrected and overhauled so that they can send out <3–6¢/kWh electricity for another 30 years). Perhaps a more promising design approach is illustrated by a 17-kW(p) (in 17 m/sec wind), 6-m-diameter Chalk-wheel machine that two people can bring home in a pickup truck and put up in 15 hours on a bolt-together tower, using only two crescent wrenches and a screwdriver. Such a machine has been homemade for $1500 and has an estimated mass-production cost (with tower) of about $5000 ($295/kW(p)), yielding ~2–3¢/kWh in a 6–7 m/sec average wind (O. J. M. Smith, unpublished data). Smith estimates there are over 200,-000 identified sites in rural California alone, and plausible deployment rates on the Great Plains might be of the order of several GW/year.

Unofficial wind programs show the same wind range of costs as solar collectors. Reported values include $350/kW(p), not counting much contributed labor (B. Sørenson, personal communication, 1978), for the Tvind (Denmark) 2-MW machine, the largest in the world, finished in 1978 as a school project; ~$760/kW(p) (price, 1978 $) for complete, recently installed 22.5-kW machines sold by Riisager El-Windmølle (Skaerbek, near Herning, Jutland—prices FOB factory are ~DKr 60,000 or 40 percent less); DKR 36,00 (factory price) or ~$310/kW(p), for two simpler 20-kW machines sold by Niels Borre (L. Albertsen, personal communication, 1978); $1050–1350/kW average (estimated installed cost) for a built-today version of the successful 200-kW Gedser (Denmark) machine (now

*Amory B. Lovins, author of *Soft Energy Paths—Toward A Durable Peace*, is an active proponent of renewable, distributed, low technology, and end-use-oriented energy supplies.

being revived); substantially under \$1000-kW(p) (installed 1978 price) for a 200-kW updated Gedser machine recently installed at WTG Energy Systems (Angola, New York) on Cuttyhunk Island, Massachusetts; ~\$250/kW(p) (estimated 1979 retail price including tower) for the 25-kW(e) hydraulic drive machines to be sold at cost by Charles Schachle (Moses Lake, Washington); \$205/kW(p) for the 1978 installed commercial price of 3-MW machines subsequent to the first unit sold by Schachle in spring 1978 to Southern California Edison for \$356/kW(p) (including first-unit design costs); \$500/kW(p) (factory cost) for a typical 4-kW system (114); ~ \$50/kW(p) (materials cost only) for a typical small machine homemade from old car parts; and about \$750/kW(p) for the run-on installed price corresponding to the 200-kW Darrieus machine of Dominion Aluminum Fabricating, Ltd., Missisauga, Ontario, which operates on the Magdalen Islands for Hydro-Quebec. The costs of electricity even toward the top of this range seem competitive with marginal and sometimes with historic values, especially if dispersion credits are taken into account. Substantial commercial ventures are therefore beginning. Wind would normally be run base-loaded and used as a fuel saver on the grid during the fossil fueled transition period, then peaked with hydro in a soft-technology economy.

Donald J. Mayer*

Wind energy is perhaps the most underrated, underutilized and underfunded energy source in the United States. This is true in spite of the following:

1. The use of wind as an energy source is technically feasible in the near-term.
2. The wind energy resource in the United States is significant—somewhere on the order of 10–20 quadrillion Btu's of energy.
3. An aggressive program of commercialization of wind energy with a goal of a contribution of 7 quads by the year 2000 would save \$130 billion in capital equipment and fuel expense.
4. Wind energy is a nonpolluting, non-depleting, safe solar energy source.
5. The disbursed nature of wind energy conversion systems means more jobs and a more versatile and secure national energy resource.

The reason for this discrepency in the utilization of wind energy is primarily because the potential contribution of wind energy is a political rather than a technical decision. An example may best illustrate my point.

The 1979 Federal Budget for Wind Energy Systems, large and small, was less than 100 million dollars, whereas the budget for photovoltaic arrays was 300 million dollars. This is in spite of the fact that the optimistic goal for the year 2000 for the costs of photovoltaics is 50¢ per peak watt, which is roughly equivalent to 15¢/kWh. Many wind systems can produce energy *today* for less than 15¢/kWh. Even more significant, with only moderate levels of production wind machines can be expected to produce energy at a cost of 3–6¢/kWh.

Virtually all of the photovoltaic manufacturers are owned in whole or part by large oil companies; wind energy manufacturers have until recently been primarily small businesses.

The realization of wind energy's potential contribution will come; the question is can we make an early enough commitment to utilize wind energy so as to have enough capacity available for the 1980's and beyond. Again the answer is political, not technical. A commitment to wind energy should include:

1. The commitment by the Federal Government to a multi-year program to develop and apply wind energy tech-

*Donald J. Mayer is Vice President of North Wind Power Co.

nology. This leadership role of the government will stimulate private resources.

2. A commitment to purchase large and small wind machines in correlation with this plan to ensure installed capacity in excess of 7 quads by the year 2000.
3. A continuation and expansion of incentives to utilities, consumers and manufacturers of wind energy conversion systems.
4. An active and aggressive public education program regarding the benefits of the use of wind energy as a renewable nonpolluting energy source.

Wind energy is economically competitive right now in many applications and areas of the country. The economics of wind machines will only improve as energy costs rise and production of WECS increases.

The energy problem facing us today is not only one of diminishing supply and increasing demand, but also a much more serious question of priorities. Abandoning our concern for protection of our ecological system for our gluttonous appetite for energy is only inviting disaster. Committing ourselves to large centralized power sources instead of disbursed local resources can only result in a massive disruption of our society.

An ambitious imaginative and clear commitment must be made immediately to the use of wind energy as well as other renewable energy sources if we are to adequately respond to energy needs of our children.

Richard A. Oman*

Expectations are that petroleum prices will continue to increase, making the use of wind energy economically competitive in an ever-widening segment of the market. My concerns are that the variety of policy alternatives and wind energy devices available will contribute to a waste of valuable time and a loss of public and private confidence in wind energy before major successes can be achieved. In particular, I am disturbed by the very difficult structural and mechanical problems associated with large horizontal-axis wind turbines, and doubt whether they can be solved within the tight life-cycle constraints of today's marketplace. Several of us strongly advocate the development of realistic wind energy alternatives on a scale such that their real promise can be determined. My own preferences for such alternatives include two aerodynamically-augmented wind turbine concepts: the Diffuser-Augmented Wind Turbine (DAWT), and the Tornado Wind Energy System (TWES).

The U.S. Department of Energy is quite properly proceeding with the development of conventional wind turbines in a wide range of sizes. It has also, relatively recently, escalated the development of vertical-axis wind turbines to the intermediate prototype scale. When given extra funds in FY 1978 and 1979, the extra funds were allocated to "increased competition in large, horizontal-axis wind turbines," while the total budget for non-conventional (i.e., "Advanced and Innovative Concepts") remained approximately 1 percent of the wind energy budget. In my opinion, this apportionment is extremely short-sighted in view of the serious cost and technological risks of large machine development, the qualitative advantages promised by several innovative concepts and the central fact that the real costs of innovative systems will be determined only by building some of them in respectable sizes. It would seem wiser to have a budget allocation of the order of 10 percent of the total for meaningful demonstration of at least two of the more promising innovative concepts.

The primary reason for concern about the future of large horizontal-axis machines is the complexity and intensity of the loading experienced by the blades and their mounting structures, particularly as sizes increase into the multi-megawatt range. Aero-

*Richard A. Oman is the Director of the Fluid Dynamics Research Department, Grumman Aerospace Corporation.

space experience is clear on these matters: All problems can be solved by sufficient engineering attention to detail; sufficient weight and cost provisions for design and manufacture of the parts; and continuous inspection and replacement of parts weakened by cracks throughout the life of the structure. It is my opinion, shared by many experienced colleagues, that the costs of properly performing these functions throughout the life of large horizontal-axis machines will prove to be several times greater than those currently being projected by their proponents.

We at Grumman have chosen to investigate two concepts in aerodynamically-augmented wind turbines because we believe they *might* prove to be cheaper in the long run. Unfortunately, that issue is not an easy one to decide. Aerodynamic augmentation means using large static surfaces (usually curved walls or ducts) to concentrate wind energy so that a given size of rotor can extract a great deal more energy from the same external wind. The success of such devices then depends on finding acceptable aerodynamic effectiveness with configurations that can be designed and built at low cost in spite of their relatively large physical size. The aerodynamic part of the problem has proven easier than the structural costing and the evaluation of secondary advantages, so we conclude that the only way to find out if innovative systems are really practical is to build a few at respectable sizes. Compared to the value of a successful outcome, the costs of this experiment would be insignificant.

I will refer the reader to any of the recent DOE or archival publications by Yen (TWES) or Foreman *et al.* (DAWT) for all the reasons why I believe in these concepts, and for a description of the concepts themselves. My theme here is that innovation must be given a realistic opportunity to prove its point, because the cost and reliability projections currently being associated with conventional machines are based on some courageous assumptions. It is primarily the simplicity of the structural mechanics and the improved predictability of life-cycle loading that leads me to have faith in aerodynamic augmentation. The TWES, in particular, offers the technological freedom to build units as large as described, and to combine synergistically with available sources of waste heat (e.g., gas turbine exhausts) to provide augmentation during periods of little or no wind. Advantages such as these need to be tried.

People and Energy*

Wind energy has pumped water for centuries, and it powered the world's fleets for several millenia. In the early 1900's, six million windmills generated electricity and pumped water all over the U.S., but by the 1950's cheap fossil fuel and rural electrification replaced most of them. So quickly have people forgotten this that now wind technology is refered to as exotic.

Wind power could supply 19 percent of the U.S.'s energy by 1990 (see "Wind Energy Mission Analysis," by Lockheed-California, available from National Technical Information Service, 5285 Port Royal Rd., Springfield, VA 22151; NTIS #SAN/1075-1/1, $15). William Heronemus, professor of engineering at the University of Massachusetts, has proposed a scheme whereby offshore windmills could supply 159 billion kWh, all of New England's energy needs in the year 1990, for a capital cost of $22.4 billion. By comparison, after 25 years and $100 billion worth of R&D, nuclear power provides 4 percent of the nation's energy. Wind turbines with 24-hour storage capacity could provide electricity more reliably than nuclear power plants, according to a study by Bent Sorensen, physics professor at the Niels Bohr Institute in Denmark. In the study, he compared the 200-kW Danish Gedser windmill to the Zion I nuclear plant, one of the more reliable U.S. reactors.

**People and Energy* is a liberal newsletter which stresses decentralized systems and curtailment of nuclear energy. This material is from an article entitled "Wind—Big Companies Blow It," which appeared in Volume III, No. 5 of the newsletter.

Wind energy will be a bargain. Already, small systems cost as little as $1500 per installed kW (see "An Analysis of the Economics of Current Small Wind Energy Systems," by JBF Scientific Corp., 1925 North Lynn St., Arlington, VA 22209), and costs may sink as low as $300 per installed kW, according to Frank Eldridge, who has directed studies of wind for the Mitre Corporation. By contrast, nuclear power plants now cost at least $1000 per installed kW.

How should wind power be commercialized? Small companies have always manufactured and sold windmills and they continue to do so. However, the Federal Wind Program (now in the Department of Energy) is dominated by large aerospace companies. A renaissance of small companies producing windmills for individual and community use would allow "a number of small businesses to compete in the market for energy generating equipment—a market which has been denied to small business for decades," according to Lee Johnson of RAIN (2270 N.W. Irving, Portland, OR 97210). Small businesses producing wind power equipment would generate much local employment. The Montana Energy Plan found that a small company approach would result in windmills which could be constructed in local shops using local resources and labor, and would generate about three times as many jobs for Montanans as the NASA-ERDA or the Lockheed WECS (wind energy conversion systems) being developed by the Federal Wind Program.

One of the most telling arguments for small-scale windmills is economics. Although almost every early study of wind economics found that electricity was cheaper from large systems, experience and recent studies point to other conclusions. Studies by the JBF Scientific Corporation point to costs of $1500 per installed kW for small systems compared to $1800 for large systems (see "Summary of Current Cost Estimates of Large Wind Systems," by JBF Scientific Corp.). Eventually, small systems (1–10 kW) will cost $300–1000 per installed kW, while large machines will cost $600–1000, according to Frank Eldridge.

Utilities themselves are uninterested in wind power, according to a survey by the Mitre Corporation ("Preliminary Wind Energy Commercialization Plan," by Frank Eldridge). The study found that they prefer to rely on their fossil fuel and nuclear power commitments. Indeed, many utilities are reluctant to welcome even small windmills into the grid. Windworks (Box 329, Route 3, Mukwonago, WI 53149), a small firm in Wisconsin, developed the Gemini Synchronous Inverter, which allows a windmill to feed electricity into the grid when it produces more than its user consumes. When Windworks first tried to hook an inverter-equipped windmill into the grid, Wisconsin electric refused to allow it. Eventually the utility permitted the windmill to feed excess electricity to the grid, but the utility refused to buy the surplus.

In New York City, when members of a tenant housing cooperative installed a 2-kW windmill on their roof, Con Edison threatened to disconnect service to the building. Richard N. Arcari, director of Con Ed's central commercial services, said a feedback of windmill power could create "adverse effects on Con Ed's transformers and computerized control equipment, and even pose a hazard to repair crews working elsewhere in the same grid." However, the New York Public Service Commission ordered Con Ed to buy surplus power from the wind generator, and Con Ed has remained unharmed by this.

"Wind energy is treated as a stepchild of the solar program; worse, small wind energy systems are treated as the offspring of the stepchild," said U.S. Senator Thomas J. McIntyre. Between July 1, 1975 and December 1, 1976, roughly two-thirds of the wind program's grant money went to firms with names like Lockheed-California, Boeing Vertol, Grumman Aerospace, McDonnell-Douglas and Rockwell International. During this period, only $2 million out of more than $20 million was earmarked to "Farm and Rural Use (small Systems)," and Rockwell International received more than half of this—$1.3 million.

The Wind Program has emphasized huge studies to carefully define the parameters of wind power use. For example, the Battelle Memorial Institute received $1.4 million for "developing information on wind descriptors for designers and manufacturers," among several tasks. The major thrust of the program was outlined in the "Mission Analyses" by General Electric and Lockheed. Their purposes included to "define functional, cost, and performance requirements of these systems; assess the benefits to be derived from and the barriers to large scale implementation; and examine the impact of possible future use of wind energy conversion systems on a large scale . . . " The findings of the Mission Analyses may have been influenced by the size of the companies performing the studies. Although Lockheed found that small windmills could produce 209 billion kwh in 1990, this is only one-sixth of the total wind power that Lockheed foresaw for that year.

The first of the big windmills to emerge from the Federal Wind Program is the 100 kW "Mod O," developed by the NASA-Lewis Research Center near Cleveland, Ohio. It operated only 57 hours in its first 8 months of existence. Severe forced oscillations, blade fatigue and other problems raised its cost to $2 million. Three 200-kW versions of Mod O will soon be feeding utility grids. One will go to a private utility on Block Island, Rhode Island; a second will go to a municipal utility at Clayton, New Mexico; and a third will go to a municipal at Culebra, Puerto Rico. The windmills, which cost ERDA $2 million, will be given to the utilities. The largest windmill in history will be designed and built for the Federal Wind Program by Boeing, for a total cost of $10 million. The 2.5-MW generator will sit atop a 200-ft-high tower and its blades will span 300 ft. The windmill will go to the Blue Ridge Electrical Membership Cooperative in South Carolina.

These large machines face severe stress. Airplane-wing-sized blades cost a fortune to engineer. The high technology of these systems has no doubt increased their cost and problems. Lee Johnson explained how problems are handled more simply by the Gedser than Wind Program mills: "Instead of hydraulically rotating the blades to change their angle of attack for greater efficiency in higher winds . . . the familiar stalling technique was put to work. With unmovable blades set to get the most power out of a narrow band of wind speeds, the windmill automatically slows in higher winds . . . the interaction of wind and airfoil do the work almost naturally."

Many local efforts are demonstrating that simple and small-scale approaches to wind power are more economical than what the aerospace companies are developing. Just off the coast of Massachusetts, the island of Cuttyhunk now receives approximately half of its electricity from a windmill. The 200-kW machine was manufactured by WTG Energy Systems, Inc. (Box 87, 1 LaSalle St., Angola, NY), a firm which employs only 10 people. The windmill provides the 50 islanders with over 400,000 kW per year. Projected cost of electricity is 5.3¢/kWh, based on test runs, which is competitive with electric rates in most parts of the country. The windmill cost $280,000, compared to its $2 million neighbor on Block Island. It has a simple, Gedser-type design.

Home owners are looking to wind for a power source. Congressman Henry Reuss of Wisconsin recently installed a 2-kW windmill from Windworks. The $5000 machine provides Reuss's summer home with 400 kWh per month; 70 percent of the home's electrical needs. Cost per kWh is 6.25¢—almost double the going rate in Milwaukee, but less than New York's Con Edison. Reuss's windmill was one of 745 sold for electricity generation (as opposed to pumping or mechanical work) in 1975. By 1985, a market will exist for 50,000–100,000 windmills, according to the Mitre study, Preliminary Wind Energy Commercialization Plan.

Small-scale efforts have produced large windmills. The largest windmill ever operated is still a 1.25-MW known as Grandpa's Knob, erected in 1941. It was designed by Palmer Putnam and the S. Morgan Smith Co., of York, Pennsylvania. From conception to completion, only two years elapsed. It generated power for a Vermont utility for two years, until a faulty bearing—irreplaceable during the war—put it out of commission. Later, it threw a blade. It was never repaired because, by that time, cheap oil had out-competed wind. However, it was a technical success, and engineers calculated that more "Grandpa's Knobs" could be built for $190/kW.

A windmill almost as big as the Wind Program's giant is nearing completion in Denmark. The 2-MW machine is a joint project by teachers, students, carpenters, engineers and others in the college community of Tvind. The group's aim was to reduce the college's $48,000 fuel bill. Since the wind at Tvind blows 280 days/year at ground level speeds greater than 9 ft/sec, a windmill seemed the logical choice. The Tvind mill will cost $300,000, compared to $10 million for the Boeing giant.

There are signs that Federal Wind Program administrators see the handwriting on the wall. For fiscal 1978, the small WECS program will receive $8 million out of $33 million in the Wind Program. Among the small wind programs is a design competition, in which 8 companies will receive grants to develop 1-, 8-, and 40-kW wind generators. Four grants have been set aside for small businesses. The first of these went to Windworks, whose half-dozen architects and engineers live and work cooperatively on a farm in Northern Wisconsin. Windworks received approximately $380,000 to develop an 8-kW windmill. The small business set-aside is only the first step in the right direction, and the irony of the small WECS program is that it is administered by Rockwell International. The wind program should stress grants to companies like Windworks and local efforts. The one generalization about wind development is that for windmills of every size, small companies have produced more reliable hardware more quickly and cheaply than large companies.

Palmer Cosslett Putnam*

1. Neither solar nor tidal-power will be harnessed on a large scale in the near future.
2. As long as water-power remains economically justified, special partnerships between wind and water will be justified.
3. The market for wind-power plants may fall into four groups, characterized by the size of the unit:
   A. The largest unit would be rated at 2000 or 3000 kilowatts, for addition to existing power systems, principally in conventional support of water and steam. This application is limited to those selected sites near heavy load centers which occur most frequently in windy regions between latitudes 30 and 60 degrees, North and South. The market may range in size from 1,000,000 kilowatts to 10,000,000 kilowatts.
   B. The medium unit would be rated at 100 to 500 kilowatts, for use in conjunction with small hydroelectric installations, or Diesel sets, in windy, isolated communities, such as the Shetlands, the Orkneys and some of the islands in the trade winds. The market may range in size from 250,000 kilowatts to 2,500,000 kilowatts.
   C. A small unit of about 10 kilowatts would have a special limited use in charging batteries for untended airway beacons, as in the Arctic, and perhaps some desert regions. The market may range from 1000 kilowatts to 10,000 kilowatts.
   D. The smallest unit of 1 kilowatt or less is for farm lighting. The market may range in size from 250,000 kilowatts to 2,500,000 kilowatts.
4. Grandpa's Knob has demonstrated that the technical problems of the 1250-kilowatt wind turbine are under-

*Palmer Cosslet Putnam was manager of the Smith-Putnam Wind Turbine Project at Grandpa's Knob, Vermont. More than 30 years ago, he summarized his conclusions regarding the future of wind power in his book, *Power from the Wind*. It appears that Mr. Putnam's insight was then as applicable as to today's energy situation. Some of his comments are given here.

stood. To solve the economic problems of putting this or a larger wind turbine into low-cost production probably requires Government aid.

Pasquale M. Sforza*

For hundreds of years wind power has been utilized because of its availability in certain locations and because of its capability of providing direct shaft power at acceptable efficiency. At the present, wind power is looked to as a replacement for other, more indirect, forms of energy transformation and is thereby constrained to provide power at much higher efficiencies and at reasonably low cost. Thus, any future of wind power is inextricably tied to high efficiency and low cost. In this light it seems impossible to proceed effectively without a commitment to an enhanced program of research and development aimed at a much better fundamental understanding of the interaction between atmospheric winds and the transformation device itself. The wind, except in certain restricted locations, is variable in time and direction and an energy transformer such as a rotor must interact properly with that wind in order to function efficiently. This problem is much more difficult than that involving the interaction of the wind with a static structure, such as a building or a bridge. It should be noted that only recently has the latter problem received the attention it deserves, and that is so because of the recognition of the high costs of wind-induced damages evident in modern times. Coupled to this aspect of wind energy and its future is the control of the wind turbine to achieve high performance. This will involve the application of advanced materials for passive turbine control as well as use of dedicated microprocessors for activating control systems. Of course, technological advancements in such items as transmissions and generating equipment to reflect the unique problems of wind energy conversion systems will have to be achieved, rather than reliance on off-the-shelf equipment designed for vastly different purposes and conditions.

Furthermore, since the use of wind turbine systems will very often take the form of arrays and cascades of individual units, it is important to address the problem of the spacing of wind turbine units. Such spacing must consider the interaction of the wind with a unit and the re-energization of the wake behind the unit prior to reaching the next unit in the array. This, in turn, places constraints upon land use and requires due consideration of terrain features at the site. In addition, the extraction of power from the wind by a given array will influence local conditions at neighboring parcels of land.

The problems of wind energy conversion are many and varied, which suggests that it is a fruitful field for both basic and applied investigations of man's interaction with his environment. There is no doubt that wind energy conversion will continue to assume greater importance in the overall energy picture of the world and that the way to the future will again be by way of the frontier—those regions where central generating systems are not entrenched, where transmission lines are too expensive to be strung, and where the winds blow freely.

*Dr. Pasquale M. Sforza is Professor of Mechanical and Aerospace Engineering at Polytechnic Institute of New York, Aerodynamics Laboratories, Farmingdale, N.Y. 11735. Dr. Sforza also is the Principal Investigator for DOE's Vortex Augmentors for the Wind Energy Conversion Program.

Volta Torrey*

From the Moon the Earth would look different every time you glanced up at it because the wind here never rests. Somewhere in our sky it rearranges the clouds every minute, like an artist forever trying to make a picture more beautiful. People have used wee bits of the wind's inexhaustible energy for thousands of years, some still do, and many more may before long.[a]

*Mr. Torrey is the author of *Wind Catchers,* an excellent historical, readable book on wind machines.

[a]Reprinted by permission of the Stephen Green Press from *Wind-Catchers: American Windmills of Yesterday and Tomorrow* by Volta Torrey © 1976 by Volta Torrey.

References

CHAPTER 1

Annual Review of Solar Energy. Solar Energy Research Institute, November 1978.

Coty, Ugo A. *Wind Energy Mission Analysis*. SAN/1075-76/1, Lockheed California Company, September 1976.

Elderkin, G. E. and Ramsdell, J. V. *Annual Report of Wind Characteristics Program Element for the Period April 1976 thru June 1977*. Battelle Pacific Northwest Laboratories, July 1977.

Eldridge, Frank R. *Wind Machines*. The Mitre Corporation for the National Science Foundation, October 1975.

Eldridge, Frank R. *Wind Machines*, 2nd edition. New York: Van Nostrand Reinhold Co., 1980.

Garate, John A. *Wind Energy Mission Analysis*. COO/2578-1/1, General Electric Co., February 1977.

Gustavson, M. R. *Wind Energy Resource Parameters*. Report No. M77-29, The Mitre Corporation, METREK Division, February 1977.

Halacy, D. S., Jr. *Earth, Water, Wind and Sun*. New York: Harper and Row, 1977.

Metz, William D. Wind Energy: Large and Small Systems Competing. *Science* 197: 971–73 (Sept. 2, 1977).

Proceedings: The Fourth Biennial Conference and Workshop on Wind Energy Conversion Systems. October 1979.

SERI Wind Energy Information Directory. October 1979.

Torrey, Volta. *Wind-Catchers, American Windmills of Yesterday and Tomorrow*. Brattleboro, Vermont: The Stephen Greene Press, 1976.

U.S. Department of Energy. *Wind Energy Systems Programs Summary*. DOE/ET-0093, December 1978.

CHAPTER 2

Barbour, Erwin Hinchley. *The Homemade Windmills of Nebraska, 1899*. Reprinted by Farallones Institute. 15290 Coleman Valley Road, Occidental, CA.

Blackwell, B. F. *The Vertical-Axis Wind Turbine—How it Works*. Sandia Laboratories, SLA-74-0160, April 1974.

Darrieus, G. J. M., U.S. Patent No. 1,835,018 (December 8, 1931).

Dupree, W. D. and West, J. A. *United States Energy Through the Year 2000*. U.S. Department of Interior, December 1972.

Eldridge, Frank R. *Wind Machines*, 2nd edition. New York: Van Nostrand Reinhold Co., 1980.

Federal Energy Administration. "Project Independence Blueprint Final Task Force Report." Supt. of Documents, Washington, D.C., 1974.

Gimpel, G. and Stodhard, A. H. *Windmills for Electricity Supply in Remote Areas*. Electrical Research Association, Technical Report C/T120, 1958.

Golding, E. W. *The Generation of Electricity by Wind Power*. New York: Philosophical Library, New York, 1956.

Klemin, A. The Savonius Wing Rotor. *Mechanical Engineering* 47: 11 (1925).

Lindsley, E. G. Wind Power. *Popular Science*. July 1974.

Newman, B. G. "Measurements on a Savonius Rotor with Variable Gap." Presented at Sherbrooke University Symposium on Wind Energy, May 1974.

Puthoff, R. L. and Sirocky, P. J. *Preliminary Design of a 100 kW Wind Turbine Generator*. NASA, NASA TMX-71585, August 1974.

Putnam, P. C. *Power From the Wind*. New York: Van Nostrand Company, Inc., 1948.

Reed, J. Wind Power Climatology. *Weatherwise* 27 (6): 237-242, December 1974.

Reed, J. W.; Maydew, R. C.; and Blackwell, B. F. *Wind Energy Potential in New Mexico*. Sandia Laboratories, SAND-74-0077, July 1974.

South, P. and Rangi, R. *The Performance and Economics of the Vertical-Axis Wind Turbine Developed at the National Research Council, Ottawa, Canada*. Presented at the 1973 Annual Meeting of the Pacific Northwest Region of the American Society of Agricultural Engineers, Calgary, Alberta, October 10–12, 1973.

South, P. and Rangi, R. S. *A Wind Tunnel Investigation of a 13-Ft. Diameter Vertical-Axis Windmill*. National Research Council of Canada, LTR-LA-105, September 1972.

A Synopsis of Energy Research 1960–1974. Engineering Energy Laboratory, Oklahoma State University, 1974.

Templin, R. J. Aerodynamic Performance Theory for the NRC Vertical-Axis Wind Turbine. National Research Council of Canada, LTR-LA-160, June 1974.

Torrey, Volta. *Wind-Catchers, American Windmills of Yesterday and Tomorrow*. Brattleboro, Vermont: The Stephen Greene Press, 1976.

CHAPTER 3

Bennett, Iven. *Glaze: Its Meteorology and Climatology, Geographical Distribution, and Economic Effects*. Technical Report EP-105, Quartermaster, Research and

Engineering Center, Natick, MA, March 1959.

Changery, M. J. *Index-Summarized Wind Data*. BNWL-2220 Wind-II, National Climatic Center, September 1977.

Cliff, W. D. *The Effect of Generalized Wind Characteristics on Annual Power Estimates from Wind Turbine Generators*. PNL-2436, Battelle, Pacific Northwest Laboratories, Richland, WA, October 1977.

Daniels, G. E. (ed.). *Terrestrial Environment (Climatic) Criteria Guidelines for Use in Space Vehicle Development*. NASA-TM-X-64589, Marshall Space Flight Center, AL, 1971 (revised).

Defant, F. "Local Winds," *Compendium of Meteorology*. T. F. Malone, ed., American Meteorology Society, Boston, MA, pp. 655–672, 1951.

Eimern, J. Van; Karschon, R.; Razumova, L. A.; and Robertson, G. W. *Windbreaks and Shelterbelts*. Technical Note 59, World Meteorological Organization, Geneva, Switzerland, 1964.

Eldridge, Frank R. *Wind Machines*, 2nd edition. New York: Van Nostrand Reinhold Co., 1980.

Elliott, D. L. *Synthesis of National Wind Energy Assessments*. BNWL-2220 WIND-5, Battelle, Pacific Northwest Laboratories, Richland, WA, 1977.

Frenkeil, J. Wind Profiles over Hills (in Relation to Wind-Power Utilization). *Quarterly Journal of the Royal Meteorological Society 88* (376): 156–169, April 1962.

Frost, W. and Nowak, D. *Handbook of Wind Turbine Generator Siting Techniques Relative to Two-Dimensional Terrain Features*. Prepared for Battelle, Pacific Northwest Laboratories by FWG Associates, Inc., Tullahoma, TN, November 1977.

Golding, E. W. *The Generation of Electricity by Wind Power*. New York: Philosophical Library, New York, 1956.

Hewson, E. W.; Wade, J. F.; and Baker, R. W. *Vegetation as an Indicator of High Wind Velocity*. Prepared for the Energy Research and Development Administration by Oregon State University, Corvallis, OR, June 1977.

Justus, C. G. *Winds and Wind System Performance*. Philadelphia, PA.: The Franklin Institute Press, 1978.

Meroney, R. N. "Wind in the Perturbed Environment: Its Influence on WECS." Presented at American Wind Energy Association Conference, Boulder, CO, May 11–14, 1977, Colorado State Univ., Fort Collins, CO, 1977.

The National Atlas of the United States of America. Compiled by the U.S. Dept. of Interior, Geological Survey, Washington, D.C., 1970.

Orgill, M. M. and Sehmel, G. A. Frequency and Diurnal Variation of Dust Storms in the Contiguous U.S.A. *Atmos. Environ. 10* (10): 813–825, 1976.

Park, Jack and Schwind, Dick. *Wind Power for Farms, Homes, and Small Industry*. DOE Report RFP-2841/1270/78/4, September 1978.

Pautz, M. E., *Severe Local Storm Occurrences, 1955–1967*. ESSA Technical Memorandum, WBTM FCST 12, reprinted by and available from NOAA, Silver Spring, MD, May 1974.

Personal Communications between Harry L. Wegley and author.

Putnam, P. C. *Power From the Wind*. New York: Van Nostrand Company, Inc., 1948.

Reed, J. Wind Power Climatology. *Weatherwise* 27 (6): 237-242, December 1974.

Sandborn, V. A. *Placement of Wind-Power Systems*. EY-76-S-06-2438, Colorado State University, Fort Collins, CO, 1977.

Stevenson, John D. "Application of Tornado Technology to Nuclear Industry." *Proceedings of the Symposium on Tornadoes, Assessment of Knowledge and Implications for Man,* June 22–24, 1976, Institute for Disaster Research, Texas Tech University.

Survey of Historical and Current Site Selection Techniques for the Placement of Small Wind Energy Conversion Systems. Report to Battelle, Pacific Northwest Laboratories by American Wind Energy Association, Bristol, IN, February 1977.

Thom, H. C. S. New Distributions of Extreme Winds in the United States. *Journal of the Structural Division. Proceedings of the Amer. Soc. of Civil Eng. 94* (ST7): 1787-1801, July 1978.

Verholek, M. G., *Summary of Wind Data from Nuclear Power Plant Sites*. BNWL-2220 WIND-4, Battelle, Pacific Northwest Laboratories, Richland, WA, 1977.

Wegley, Harry L.; Orgill, Montie M.; and Drake, Ron L. *A Siting Handbook for Small Wind Energy Conversion Systems*. Battelle Memorial Institute, DOE, PNL-2521, Richland, Washington: May 1978.

CHAPTER 4

Barbour, Erwin Hinchley. *The Homemade Windmills of Nebraska, 1899*. Reprinted by Farallones Institute. 15290 Coleman Valley Road, Occidental, CA.

Eldridge, Frank R. *Wind Machines*, 2nd edition. New York: Van Nostrand Reinhold Co., 1980.

Kovarik, Tom; Pipher, Charles; and Hurst, John. *Wind Energy*. Northbrook, Illinois: Quality Books Inc., Domus Books, 1979.

Park, Jack and Schwind, Dick. *Wind Power for Farms, Homes, and Small Industry*. DOE Report RFP-2841/1270/78/4, September 1978.

Planning a Wind Powered Generating System. Norwich, Vermont: Enertech Corporation, February 1977.

Selecting Water-Pumping Windmills. New Mexico Energy Institute Report NMEI 11-0-6M, January 1978.

CHAPTER 5

Betz, A. "Schraubenpropeller mit Geringstem Energieverlust," Nach. der Kgl. Gesellschaft der Wiss. zu Gottingen, Math.-Phys. Klasse, pp. 193–217; reprinted in Vier Abhandlungen zur Hydrodynamik und Aerodynamik by L. Prandtl and A. Betz, Gottingen, 1927 (reprint Ann Arbor: Edwards Bros., 1943), pp. 68–92.

Blackwell, B. F., and Reis, G. E. *Blade Shade for a Troposkien Type of Vertical-Axis Wind Turbine,* Sandia Laboratories Energy Report SLA-74-0154, April 1974, Albuquerque, New Mexico.

Froude, R. E. *Transactions, Institute of Naval Architects,* Vol. 30, p. 390, 1889.

Froude, W. *Transactions, Institute of Naval Architects,* Vol. 19, p. 47, 1878.

Glauert, H. *Aerodynamic Theory,* Vol. 6, Division L, p. 324. Julius Springer, Berlin, 1935.

Glauert, H. *The Analysis of Experimental Results in the Windmill Brake and Vortex Ring States of an Airscrew,* Br. R & M 1026, 1926.

Golding, E. W. *The Generation of Electricity by Wind Power.* New York: Philosophical Library, New York, 1956.

Goldstein, W. *On the Vortex Theory of Screw Propellors.* Roy. Soc. Proc. (A) 123, pp. 440, 1929.

Goorjian, P. M. *AIAA Journal,* Vol. 10, No. 4, April 1972, pp. 543-4.

Hackleman, Michael. *The Homebuilt, Wind-Generated Electricity Handbook.* Mariposa, California: Earthwind, 1977.

Joukowski, N. W. *Travanx du Bureau des Calculs et Essais Aeronautiques de l'Ecole Superieure Technique de Moscou.* 1918.

Kuchemann, D. and Weber, J. *Aerodynamics of Propulsion.* New York: McGraw-Hill, 1953.

Lerbs, H. W. "Moderately Loaded Propellers with a Finite Number of Blades and an Arbitrary Distribution of Circulation," Trans. Soc. Naval Architects and Marine Engrs., 60, 73–117.

Lilley, G. M. and Rainbird, W. J. "A Preliminary Report on the Design and Performance of Ducted Windmills." Cranfield, CoA Report No. 102, 1956.

Lock, C. N. H.; Bateman, H.; and Townsend, H. C. H. *An Extension of the Vortex Theory of Airscrews with Applications in Airscrews of Small Pitch, Including Experimental Results.* Br. A. R. C., R. & M 1014. 1925.

Monin, A. S. and Obukhov, A. M. *Basic Laws of Turbulent Mixing In the Ground Layer of the Atmosphere.* Translated from Akademiia Nauk SSSR, Leningrad, Geofizicheskii Institut, Trudy, Vol. 151, No. 24, 1954. pp. 163–187.

New York University. *Final Report on the Wind Turbine.* Washington, D.C.: Office of Production, Research and Development, War Production Board, PB25370, January 31, 1946.

Park, Jack and Schwind, Dick. *Wind Power for Farms, Homes, and Small Industry.* DOE Report RFP-2841/1270/78/4, September 1978.

Personal Communications between Philip W. Metcalfe, Unarco-Rohn and author on January 17, 1978.

Planning a Wind Powered Generating System. Norwich, Vermont: Enertech Corporation, February 1977.

Prandtl, L. Appendix to *Schraubenpropellor mit gerngstein Energieverlust* by A. Betz, Gottinger Nachr. pp. 193–217, 1919.

Putnam, P. C. *Power From the Wind.* New York: Van Nostrand Company, Inc., 1948.

Rankine, W. J. *Transactions, Institute of Naval Architects,* Vol. 6, p. 13, 1865.

Ribner, N. S. *Propellors in Yaw.* NACA Report 820, Washington, D. C., 1948.

Selecting Water-Pumping Windmills. New Mexico Energy Institute Report NMEI 11-0-6M, January 1978.

Shapiro, J. *Principles of Helicopter Engineering.* New York: McGraw-Hill, 1955.

Slade, D. H. *Journal of Applied Meteorology,* Vol. 8, April 1979, pp. 293–7.

South, P. and Rangi, R. *The Performance and Economics of the Vertical-Axis Wind Turbine Developed at the National Research Council, Ottawa, Canada.* Presented at the 1973 Annual Meeting of the Pacific Northwest Region of the American Society of Agricultural Engineers, Calgary, Alberta, October 10–12, 1973.

Theodorsen, Theodore. *Theory of Propellers.* New York: McGraw-Hill Book Co., Inc., 1948.

Weinig, F. Aerodynamics of the Propeller. Trans. from 1939 German book, *Aerodynamik der Luftschraube* and revised by author. Dayton, Ohio: Air Documents Div. Air Atl. Command.

Weinig, F. Die Stromung um die Schaufeln von Turbomaschinen. Leipzig: J. A. Barth.

Wilson, E. Robert and Lissaman, Peter. *Applied Aerodynamics of Wind Power Machines.* National Science Foundation Grant No. GI-41840.

Wolkovitch, J. Analytical Prediction of Vortex Ring Boundaries for Helicopters In Steep Descents. *J. Amer. Helicopter Soc.,* Vol. 17, No. 3, 1972.

CHAPTER 7

Annual Review of Solar Energy. Solar Energy Research Institute, November 1978.

Personal Communications between ESB Incorporated, Exide Power Systems Division, and author on October 1979.

Personal Communications between Hans Meyer and author in August 1979.

Planning a Wind Powered Generating System. Norwich, Vermont: Enertech Corporation, February 1977.

CHAPTER 8

Barchet, R. "Mod-1 Project." Paper submitted to Third Biennial Conference and Workshop on Wind Energy Conversion Systems, September 19–21, 1977, Washington, D.C.

Cahill, T. "Large Wind Turbine Supporting Research and Technology." Paper submitted to Third Biennial Conference and Workshop on Wind Energy Conversion Systems, September 19–21, 1977, Washington, D.C.

Couch, J. "Mod-2 Project." Paper submitted to Third Biennial Conference and Workshop on Wind Energy Conversion Systems, September 19–21, 1977, Washington, D.C.

Donham, R. E.; Schmidt, J.; and Linscott, B. S. "100 kW Hingeless Metal Wind Turbine Blade Design, Analysis and Fabrication." Presented at the 31st Annual National Forum of the American Helicopter Society, Washington, D.C., May 1975.

Gilbert, L. J. "A 100 kW Experimental Wind Turbine: Simulation of Starting, Overspeed, and Shutdown Characteristics." NASA TMX-71864.

Glasgow, J. C and Linscott, B. S. "Early Operation Experience on the ERDA/NASA 100 kW Wind Turbine." NASA TMX-71601, September 1976.

Hoffman, J. A. "Coupled Dynamics Analysis of Wind Energy Systems." Prepared for NASA by Paragon Pacific Inc., El Segundo, California, NASA CR-135152, Feb. 1977.

Hwang, H. H. and Gilbert, L. J. "Synchronization of the ERDA-NASA 100 KW Wind Turbine Generator with Large Utility Networks." NASA TMX-7613 March 1977.

Linscott, B. S.; Shapton, W. R.; and Brown, D. "Tower and Rotor Blade Vibration Test Results for a 100-Kilowatt Wind Turbine." NASA TMX-3426, Oct. 1976.

Personal Communications between contributors to Chapter and author in December 1979.

Personal Communications between manufacturers of SWECS and author.

Puthoff, R. L., "Fabrication and Assembly of the ERDA/NASA 100-Kilowatt Experimental Wind Turbine." NASA TMX-3390, April 1976.

Puthoff, R. L. and Sirocky, P. J. "Status Report of 100 kW Experimental Wind Turbine Generator Project." NASA TMX-71758, June 1975.

Savino, J. M. and Wagner, L. H. "Wind Tunnel Measurements of the Tower Shadow on Models of the ERDA/NASA 100 kW Wind Turbine Tower." NASA TMX-73548, November 1976.

Spera, D. A. "Rotor Dynamics Analysis Methods." Paper submitted to Third Biennial Conference and Workshop on Wind Energy Conversion Systems, September 19–21, 1977, Washington, D.C.

Spera, D. A.; Janetzke, D. C.; and Richards, T. R. "Dynamic Blade Loading in the ERDA-NASA 100 kW and 200 kW Wind Turbines." NASA TM-73711.

Thomas, R. L. "Large Experimental Wind Turbines—Where We Are Now." NASA TMX-71890, March 1976.

Thomas, R. L. and Richards, T. R. "ERDA/NASA 100 kW Mod-0 Wind Turbine Operations and Performance." Paper submitted to Third Biennial Conference and Workshop on Wind Energy Conversion Systems, September 19–21, 1977, Washington, D.C.

Wilson, D. J. et al. "Full-Scale Measurements for 100 kW Wind Turbine via Laser Velocimetry." Prepared for NASA by Lockheed Missiles and Space Company, Inc., Huntsville, Alabama, September 1976.

CHAPTER 9

Park, Jack and Schwind, Dick. *Wind Power for Farms, Homes, and Small Industry*. DOE Report RFP-2841/1270/78/4, September 1978.

Personal Communications between R. Nolan Clark and author in June 1979.

Planning a Wind Powered Generating System. Norwich, Vermont: Enertech Corporation, February 1977.

Selecting Water-Pumping Windmills. New Mexico Energy Institute Report NMEI 11-0-6M, January 1978.

CHAPTER 10

U.S. Department of Energy. *Federal Wind Energy Program, Summary Report*. DOE Report ERDA-84, October 1975.

U.S. Department of Energy. *Federal Wind Energy Program, Summary Report*. DOE Report ERDA-77-32, January 1977.

U.S. Department of Energy. *Federal Wind Energy Program, Program Summary*. DOE Report DOE/ET-023/1, January 1978.

U.S. Department of Energy. *Federal Wind Energy Program Summary*. December 1979.

U.S. Department of Energy. *Wind Energy Systems Programs Summary*. DOE/ET-0093, December 1978.

CHAPTER 11

U.S. Department of Energy. *Commercialization Strategy Reports for Small and Large Solar Wind Energy Conversion Systems*. Task Force Chairman Louis V. Civone, 1978 and 1979.

CHAPTER 12

Energy Research and Development Administration, Division of Biomedical and Environmental Research. *Balanced Program Plan: Analysis for Biomedical and Environmental Research*. ERDA 116, Volume 8, October, 1976.

Energy Research and Development Administration, Division of Solar Energy and Division of Technology Overview. "Draft Solar Energy Health, Environmental, and Safety Issues and Requirements." Washington, D.C., December 1976.

Energy Research and Development Administration, Division of Solar Energy, Wind Systems Branch. *Environmental Impact Assessments for Large Experimental Wind Turbine Generator Systems at Seventeen Candidate Sites*. 1976–1977.

Energy Research and Development Administration, Division of Solar Energy. *Federal Wind Energy Program Summary Report*. ERDA 77-32, Washington, D.C., 1977.

Energy Research and Development Administration. "Fiscal Year 1977 Program Approval Document: Solar Electric Applications." (Operating Draft.) Washington, D.C., November 22, 1976.

Energy Research and Development Administration, Division of Solar Energy Wind Systems Branch, Program Files. "MOD-0/OA Safety Documentation, 1974–1975," "MOD-0/OA Safety Documentation, 1976–1977," "MOD-0/OA Failure Mode Effects Analysis (October 14, 1977)," and procurement materials for MOD-1, MOD-2, MOD-3, MOD-4.

Energy Research and Development Administration. *Solar Program Assessment: Environmental Factors—Wind Energy Conversion*. ERDA 77-4716, March 1977.

Legal—Institutional Implications of WECS. National Science Foundation Report NSF/RA-770204, 1977.

Mayo, Louis. "Program of Policy Studies in Science and Technology for NSF/ERDA, Legal-Institutional Implications of Wind Energy Conversion Systems." March 31, 1977.

Park, Jack and Schwind, Dick. *Wind Power for Farms, Homes, and Small Industry*. DOE Report RFP-2841/1270/78/4, September 1978.

Public Reactions to Wind Energy Devices. National Science Foundation, 1979.

Rogers, S. E., et al. *The Potential Environment Effects of Wind Energy System Development*. Prepared by Battelle Columbus Laboratories for NSF/ERDA, August 1976.

Survey Research Laboratory, University of Illinois. "Public Reactions to Wind Energy Devices—Final Report (Preliminary Version)." Prepared for NSF/ERDA, March 1977.

Taubenfeld, R. and Taubenfeld, H. "Barriers to the Use of Wind Energy Machines: The Present Legal/Regulatory Regime and a Preliminary Assessment of Some Legal/Political/Societal Problems." Prepared for NSF/ERDA, March 11, 1976.

University of Michigan Radiation Laboratory. "Broadcast Interference by Windmills." Ongoing research under ERDA contract # E(11-1)-2846).

U.S. Department of Energy. Environmental Development Plan for Wind Energy Conversion. March 1977.

CHAPTER 13

Daniel, C. William. "The Role of Oil and Gas in the Canadian Energy Situation to the Year 2000." Paper delivered at the Third Canadian National Energy Forum, Halifax, April 4–5, 1977.

MacNabb, G. M. "The Canadian Energy Situation in 1990." Paper delivered at the Third Canadian National Energy Forum, Halifax, April 4–5, 1977.

Papers presented at the Second International Symposium on Wind Energy Systems, October 1978, Netherlands Energy Research Foundation.

Templin, R. J. "Wind Energy." Paper delivered at the Third Canadian National Energy Forum, Halifax, April 4–5, 1977.

Templin, R. J. and South, P. "Canadian Wind Energy Program." Proceedings of the Vertical-Axis Wind Turbine Technology Workshop. Sandia Laboratories, Albuquerque, N. Mex., May 1976.

CHAPTER 14

Eldridge, Frank R. *Wind Machines*, 2nd edition. New York: Van Nostrand Reinhold Co., 1980.

Kahn, Herman. *The Next 200 Years—A Scenario for America and the World*. New York: William Morrow and Company Inc., 1976.

Lovins, Amory B. *Soft Energy Paths - Toward a Durable Peace*. New York: Friends of the Earth, 1977.

Personal Communications between contributors to Chapter and author in December 1979.

Solar Energy—Progress and Promise. Washington, D.C.: Council on Environmental Quality, April 1978.

Bibliography

The entries in this bibliography are divided into categories which describe elements of the Federal Wind Energy Program and chapter headings. Many of these reports are available from the National Technical Information Service, 5285 Port Royal Road, Springfield, VA 22161, telephone: (703) 557-4650, and/or The Superintendent of Documents, U.S. Government Printing Office, Washington, D.C. 20402, telephone: (202) 783-3238. Report identification numbers are listed where available.

Mission Analysis

Department of Energy. *Federal Wind Energy Program, Program Summary,* January 1978, 71 pp. U.S. Government Printing Office (Stock No. 061-000-00050-0).

Department of Energy. *Third Wind Energy Workshop* (Washington, D.C., September 19–21, 1977). Coordinated by JBF Scientific Corporation, May 1978, 979 pp. U.S. Government Printing Office (Stock No. 061-000-00089-5, Document No. CONF-770921/1 and /2; 2-volume set).

Energy Research and Development Administration. *Federal Wind Energy Program, Summary Report,* January 1, 1977, 56 pp. U.S. Government Printing Office (Stock No. 060-000-00048-4).

Energy Research and Development Administration. *Federal Wind Energy Program, Summary Report.* Division of Solar Energy, October 1975, 78 pp. (ERDA-84).

General Electric, Space Division. *Wind Energy Mission Analysis,* February 1977. Contract No. E(11-1)-2578. (Executive Summary: COO/2578-1/1, 26 pp; Final Report: COO/2578-1/2, 219 pp. Appendices A–J: COO/2578-1/3, 480 pp.)

JBF Scientific Corporation. *Summary of Current Cost Estimates of Large Wind Energy Systems* (Special Technical Report), February 1977, 62 pp. Contract No. E(49-18)-2364; (DSE/2521-1).

Lockheed California Company. *Wind Energy Mission Analysis,* October 1976. Contract No. EY-76-C-03-1075. (Executive Summary: SAN/1075-1/3, 30 pp; Final Report: SAN/1075-1/1; Appendix: SAN/1075-1/2.)

Mitre Corporation. *Wind Energy Conversion Systems, Proceedings of the Second Workshop* (on June 9–11, 1975). F. R. Eldridge, June 1975, 536 pp. Contract No. NSF-AER-75-12937 (NSF-RA-N-75-050).

Mitre Corporation. *Wind Machines.* F. R. Eldridge, October 1975, 84 pp. (NSF-RA-N-75-051). U.S. Government Printing Office (Stock No. 038-000-00272-4).

NASA-Lewis Research Center. *Wind Energy Conversion Systems, Workshop Proceedings* (Washington, D.C., June 11–13). J. M. Savino, December 1973, 258 pp. Grant No. NSF-AG465 (NSF-RA-N-73-006) (PB 231 341).

NASA-Lewis Research Center. *Wind Energy Utilization, A Bibliography.* Technical Applications Center, University of New Mexico, for NASA-LeRC (TACW-75-700).

Planning, Management, and Analysis

Battelle Memorial Institute, Columbus Laboratories. *An Evaluation of the Potential Environmental Effects of Wind Energy Systems Development* (Final Report). S. Rogers, et. al., August 1976. Contract No. NSF-AER-75-07378. (ERDA/NSF/07378-75/1).

Department of Energy. *Environmental Development Plan Wind Energy Conversion.* July 1979. 34 pp., DOE/EDP-0030.

General Electric, Electric Utility System Engineering Dept. *Requirements Assessment of Wind Power Plants in Electric Utility Systems. Final Report.* January 1979. EPRI ER-978. Volume 1, Summary Report, 47 pp.; Volume 2, Requirements Assessment of Wind Power Plants in Electric Utility Systems, 339 pp.; Volume 3, Appendices, 94 pp., January 1979, EPRI ER-978.

George Washington University. *Legal-Institutional Implications of Wind Energy Conversion Systems, Final Report.* L. H. Mayo, et. al., September 1977, 33 pp. Contract No. APR 75-19137. (NSF/RA-77-204).

Hawaii University, Honolulu. *Wind and Solar Energy Applications Study.* August 1977. 133 pp. PB-287593.

Market Facts, Inc., Washington, D.C. *Small Wind Energy: Focus Group Results.* 1978. DOE Contract No. EV-78-C-01-6458. 43 pp., Order No. DOE/TIC-10018.

Michigan University of, Radiation Laboratory. *TV and FM Interference by Windmills* (Final Report). T. B. A. Senior, et. al., February 1977, 150 pp., Contract No. EY-76-S-02-2846. (COO/2846-76/1).

Michigan University of, Radiation Laboratory. *Wind Turbine Generator Siting and TV Reception Handbook,* Technical Report No. 1. T. B. A. Senior and D. L. Sengupta, January 1978, 36 pp., Contract No. EY-76-S-02-2846 (COO-2846-1).

Michigan State University. *Planning Manual for Utility Application of WECS.* June 1979. COO/4450-79/1.

Michigan State University, Division of Engineering Research. *Application Study of Wind Power Technology to the City of Hart, Michigan, 1977.* Jes Asmussen, et. al., January 1978. 244 pp. Contract No. EY-76-S-02-2992. COO-2992-78-1.

NASA Lewis Research Center. *Wind Turbines for Elec-*

tric Utilities: Development Status and Economics. J. R. Ramler and R. M. Donovon. Prepared for Terrestrial Energy Systems Conference, sponsored by American Institute of Aeronautics and Astronautics, Orlando, Florida, June 4–6, 1978. DOE/NASA/1028-79/23, NASA TM-79170.

NASA Lewis Research Center. *Safety Considerations in the Design and Operation of Large Wind Turbines*. Dwight H. Reilly, June 1979. DOE/NASA/20305-79/3, NASA TM-79193.

Societal Analytics Institute, Inc. *Barriers to the Use of Wind Energy Machines: The Present Legal/Regulatory Regime and a Preliminary Assessment of Some Legal/Political/Societal Problems*. R. F. and H. J. Taubenfeld, July 1976, 159 pp., Contract No. NSF-AER75-18362 (PB-263 567).

Solar Energy Research Institute. *Wind Energy: Legal Issues and Institutional Barriers*. June 1979. DOE Contract No. EG-77-C-01-4042. 32 pp., Order #SERI/TR-62-241.

Solar Energy Research Institute. *Conversion System Overview Assessment. Volume II. Solar-Wind Hybrid Systems*. SERI/TR-35-078, August 1979.

Westinghouse. *Design Study and Economic Assessment of Multi-Unit Offshore Wind Energy Conversion Systems Application*. Volume I—Executive Summary; Volume II—Apparatus Designs and Costs; Volume III—Systems Analysis; Volume IV—Meteorological and Oceanographic Surveys, ERDA Contract No. E(49-18)-2330, June 1979, WASH-2330-78/4.

Pacific Northwest Laboratory—DOE. *Annual Report of the Wind Characteristics Program Element for the Period July 1977 through July 1978*. Wendell, L. L., et. al., December 1978, 140 pp. PNL-2545.

Pacific Northwest Laboratories—DOE *Assessment of the Applicability of the National Fire Weather Data Library to Wind Energy Analyses. Final Report*. Marlat and Associates, Fort Collins, Colorado. May 1979. Contract No. EY-76-C-06-1830. 115 pp., PNL-2538.

Poseidon Research Institute. *Effect of Atmospheric Density Stratification on Wind Turbine Siting. Final Report*. Agopian, K. G.: Crow, S. C. January 1978. 103 pp., Contract No. EY-76-C-06-2444.

Sandia Laboratories. *Wind Power Climatology of the United States: Supplement*. Reed, J. W. April 1979. 85 pp., Contract No. AC-4-76DP00789. SAND-78-1620.

Sandia Laboratories. *Some Variability Statistics of Available Wind Power*. Reed, J. W. March 1979. 50 pp., Contract No. EY-76-C-04-0789. 50 pp., SAND-788-1735.

Applications of Wind Energy

The Aerospace Corporation. *Wind Machines for the California Aqueduct*, Volume 2 (Final Report). Charles A. Lindley, February 1977. Contract No. EY-76-03-1101-005, 192 pp. NTIS (Order No. SAN/1101-76/2).

Colorado State University. *Wind-Powered Aeration for Remote Locations* (Final Report). P. M. Schierholz *et al.*, October 1976, 130 pp. Contract No. NSF-G-AER-75-00833 (Order No. ERDA/NSF/00833-75/1).

Michigan State University, Division of Engineering Research. *Application Study of Wind Power Technology to the City of Hart, Michigan*. J. Asmussen, P. D. Fisher, G. L. Park and O. Krauss, December 1975, 103 pp. Contract No. E(11-1)-2603 (COO-2603-1).

NASA-Lewis Research Center. *Benefit-Cost Methodology Study with Example Application of the Use of Wind Generators*. R. P. Zimmer, C. G. Justus, R. N. Mason, S. L. Robinette, P. G. Sassone and W. A. Schaffer of Georgia Institute of Technology, July 1975, 411 pp. (NASA CR-134864).

Legal/Social/Environmental Issues

Battelle Memorial Institute, Columbus Laboratories. *An Evaluation of the Potential Environmental Effects of Wind Energy Systems Development* (Final Report). S. Rogers *et al.*, August 1976. Contract No. NSF-AER-75-07378 (ERDA/NSF/07378-75/1).

George Washington University. *Legal-Institutional Arrangements Facilitating Offshore Wind Energy Conversion Systems (WECS) Utilization* (Final Report). Louis H. Mayo, Septemer 1977. Contract No. APR75-19137, 93 pp. NTIS (DOE/NSF/19137-77/3).

George Washington University. *Legal-Institutional Implications of Wind Energy Conversion Systems* (Final Report). L. H. Mayo *et al.*, September 1977, 333 pp. Contract No. APR 75-19137 (NSF/RA-77-204).

University of Michigan, Radiation Laboratory. *TV and FM Interference by Windmills* (Final Report). T. B. A. Senior *et al.*, February 1977, 150 pp. Contract No. EY-76-S-02-2846 (COO/2846-76/1).

Societal Analytics Institute, Inc. *Barriers to the Use of Wind Energy Machines: The Present Legal/Regulatory Regime and a Preliminary Assessment of Some Legal/Political/Societal Problems*. R. F. and H. J. Taubenfeld, July 1976, 159 pp. Contract No. NSF-AER75-18362 (PB-263 567).

Wind Characteristics

University of Alaska, Geophysical Institute. *Study of Alaskan Wind Power and Its Possible Applications* (Final Report), May 1, 1974–January 30, 1976. T. Wentink, Jr., February 1976, 139 pp. Contract No. NSF-AER-74-00239 (NSF/RANN/SE/AER-74-000239) (PB 253 339).

American Wind Energy Association. *Survey of Historical and Current Site Selection Techniques for the Placement of Small Wind Energy Conversion Systems*, December 1977. Contract No. EY-76-C-06-1830, 150 pp. NTIS (BNWL-2220 WIND-9).

Battelle-Pacific Northwest Laboratories. *Annual Report of the Wind Characteristics Program Element for the Period April 1976–June 1977*. J. V. Ramsdell, June 1977. Contract No. EY-76-C-06-1830 (BNWL-2220-WIND-10).

Colorado State University. *Sites for Wind Power Installa-*

tions: Wind Tunnel Simulation of the Influence of Two-Dimensional Ridges on Wind Speed and Turbulence (Annual Report). R. N. Meroney *et al.*, July 1976, 80 pp. Contract No. NSF-RANN-GAER-75-00702 (ERDA/NSF/00702-75/1).

Georgia Institute of Technology. *Wind Energy Statistics for Large Arrays of Wind Turbines (New England and Central U.S. Regions)*. C. G. Justus, August 1976, 129 pp. Contract No. NSF-AER75-00547 (PB 260 679).

NOAA-National Climatic Center. *Initial Wind Energy Data Assessment Study*. M. J. Changery, May 1975, 132 pp. Contract No. NSF-AG-517 (NSF-RA-N-75-020) (PB 244 132).

Northwestern University, Department of Civil Engineering. *Stochastic Modeling of Site Wind Characteristics*. R. B. Corotis, September 1977, 150 pp. Contract No. EY-76-S-06-2342, NTIS (RLO/2342-77/2).

Sandia Laboratories. Contract No. S189-76-32: *Wind Energy Potential in New Mexico*. J. W. Reed, R. C. Maydew and B. F. Blackwell, July 1974, 40 pp. (SAND-74-0071).

Sandia Laboratories. *Wind Power Climatology*. J. W. Reed, December 1974 (SAND-74-0435).

Sandia Laboratories *Wind Power Climatology of the United States*. J. W. Reed, May 1975, 163 pp. (SAND-74-3078).

Colorado State University. *Wind Characteristics over Complex Terrain: Laboratory Simulation and Field Measurements at Rakaia George, New Zealand*. R. N. Meroney, et. al., May 1978, 220 pp. RLO/2438-77/2.

FWG Associates, Inc. *Summary of Guidelines for Siting Wind Turbine Generators Relative to Small-Scale, Two Dimensional Terrain Features*. March 1979. Walter Frost, Dieter K. Nowak. RLO/2443-77/1. DOE Contract No. EY-76-C-06-2443.

Georgia Institute of Technology. *Energy Statistics for Large Wind Turbine Arrays*. C. G. Justus, May 1978, 155 pp. Contract No. NSF-AER75-00547. RLO/-2439-78/3.

NASA. *Summary of Atmospheric Wind Design Criteria for Wind Energy Conversion System Development*. Frost, W.; Turner, R. E. (Tennessee Univ., Tullahoma (USA). Space Inst.; NASA, Huntsville, Alabama. George C. Marshall Space Flight Center). January 1979, 53 pp., NASA-TP-1389.

NASA. *The Use of Wind Data with an Operational Wind Turbine in a Research and Development*. Harold Neustadter. June 1979. DOE/NASA/1004-79/6. TM-73832.

National Climatic Center, Asheville, NC. *National Wind Data Index. Final Report*. December 1978. Changery, M. J. HCO/T1041-01.

Northwestern University, Department of Civil Engineering. *Stochastic Modeling of Site Wind Characteristics*. R. B. Corotis, September 1977. Contract No. EY-76-S-06-2342, 150 pp., (RLO/2342-77/2).

Oregon State University, Dept. of Atmospheric Sciences. *Vegetation as an Indicator of High Wind Velocity. Annual Progress Report, June 15, 1978—March 14, 1979*. March 1979. Contract No. EY-76-S-06-2227-024. 15 pp., RLO-2227-T24-79-2.

Oregon State University. *A Handbook on the Use of Trees as Indicators of Wind Power Potential*. Hewson, E. W., Wade, J. E., and Baker, R. W., May 1979. 22 pp. RLO/2227-T24-79/3.

Pacific Northwest Laboratory—DOE. *Wind Directionn Change Criteria for Wind Turbine Design*. W. C. Cliff. January 1979, 26 pp., PNL-2531.

Pacific Northwest Laboratory—DOE. *Accuracy of Wind Power Estimates*. J. C. Doran, et. al., October 1977, 22 pp. Contract No. EY-76-C-06-1830. Order No. PNL-2442.

Pacific Northwest Laboratory—DOE. *A Siting Handbook for Small Wind Energy Conversion Systems*. Harry L. Wegley, et. al., March 1980. Contract No. EY-76-C-06-1830. Order No. PNL-2521.

Pacific Northwest Laboratory—DOE. *Synthesis of National Wind Energy Assessments*. D. L. Elliott, July 1977. Contract No. EY-76-C-06-1830. 58 pages, Order No. BNWL2220/WIND-5.

Pacific Northwest Laboratory—DOE. *Simulation of Hourly Wind Speeds for Randomly Dispersed Sites*. Cliff, W. C., Justus, D. G., Elderkin, C. E., May 1978, 43 pp. PNL-2523.

Pacific Northwest Laboratory—DOE. *Wind Velocity—Change (Gust Rise) Criteria For Wind Turbine Design*. Cliff, W. C. and Ficht, G. H., July 1978. 25 pp. PNL-2526.

Pacific Northwest Laboratories. *Gust Rise Exceedance Statistics for Wind Turbine Design*. Huang, C. H.; Fichtl, G. H. July 1979. DOE Contract No. EY-76-C-06-1830. PNL-2530.

Engineering Development

General Electric, Co. *MOD-1 Wind Turbine Generator Analysis and Design Report*. Contract NAS 3-20058, Executive Summary DOE/NASA/0058-79/3, NASA CR-159497 March 1979; Volume I, DOE/NASA/0058-79/2, NASA CR 159495, May 1979.

General Electric Company. *MOD-1 Wind Turbine Generator Failure Modes and Effects Analysis*. February 1979. 91 pp. Contract No. EX-77-A-29-1010. DOE/NASA/0058-79-1. NASA CR-159494.

Kaman Aerospace Corp. *Design Study of Wind Turbines, 50 kW to 3000 kW for Electric Utility Applications*. Volume I—Executive Summary, 97 pp., NASA-CR-134936. Volume II—Analysis and Design, NASA CR-134937, Kaman Report No. R-1382. July 1977.

NASA Lewis Research Center. *Utility Operational Experience on the NASA/DOE Mod-OA 200 kW Wind Turbine* by J. C. Glasgow and W. H. Robbins. Technical paper presented at Sixth Energy Technology Conference, Washington, D.C., February 26–28, 1979, DOE/NASA/1004-79/1, NASA TM-79084.

NASA Lewis Research Center. *Safety Considerations in the Design and Operation of Large Wind Turbines*. June 1979. Contract No. EX-76-A-29-1007. 39 pp., (NASA-TM-79193).

NASA Lewis Research Center. DOE/NASA/1028-79-1. *200-kW Wind Turbine Generator Conceptual Design Study*. January 1979. Contract No. E(49-26)-1028. 104 pp. NASA TM-79032.

NASA Lewis Research Center. *Installation and Checkout of the DOE/NASA MOD-1 2000 kW Wind Turbine Generator*. Richard L. Puthoff, John L. Collins, and Robert A. Wolf. Prepared for Wind Energy Conference cosponsored by American Institute of Aeronautics and Astronautics and Solar Energy Research Institute, Boulder, Colorado, April 9–11, 1980. DOE/NASA/1010-80/6, NASA TM-81444.

Technology Development

General Electric, Space Division. *Design Study of Wind Turbines 50-kW to 3000-kW for Electric Utility Applications*. December 1976. [Volume I (Summary Report): NASA CR-134934; Volume II: NASA-CR134035; Volume III: NASA CR-134936.]

Lockheed-California Company. *10-kW Metal Wind Turbine Blade Basic Data, Loads, and Stress Analysis*. A. W. Cherritt and J. A. Caidelis, June 1975. NASA Contract No. NAS3-19325 (NASA CR-134956).

Lockheed-California Company. *100-kW Metal Wind Turbine Blade Dynamics Analysis, Weight/Balance and Structural Test Results*. W. D. Anderson, June 1975. NASA Contract No. NAS3-19235 (NASA CR-134957).

Martin Marietta Laboratories. *Segmented and Self-Adjusting Wind Turbine Rotors* (Final Report). P. F. Jordon and R. L. Goldman, April 1976, 113 pp. Contract No. EY-76-C-02-2613 (COO/2613-2).

Massachusetts Institute of Technology. *Research on Wind Energy Conversion Systems*. R. H. Miller, December 1976. Contract No. NSF-AER-75-00826.

NASA-Lewis Research Center. Interagency Agreement No. E(49-26)-1028: *Free Vibrations of the ERDA-NASA 100-kW Wind Turbine*. C. C. Chamis and T. L. Sullivan, February 1976 (NASA TM-X-71879).

NASA-Lewis Research Center. *Transient Analysis of Unbalanced Short Circuits of the ERDA-NASA 100-kW Wind Turbine Alternator*. H. H. Hwang and Leonard J. Gilbert, July 1976 (NASA TM-X-73459).

NASA-Lewis Research Center. *ERDA/NASA 100-Kilowatt MOD-O Wind Turbine Operations and Performance*. R. L. Thomas and T. R. Richards, September 1977. Contract No. E(49-26)-1028, NTIS (ERDA/NASA-1028-77/9).

Grumman Aerospace Corporation. *Investigation of Diffuser-Augmented Wind Turbines*. R. A. Oman, January 1977. Contract No. EY-76-C-02-2616; Executive Summary: (COO/2616-1); Technical Report: (COO/2616-2).

McDonnell Aircraft Company. *Feasibility Investigation of the Giromill for Generation of Electric Power* (Final Report), April 1975–April 1976; Volume I, Executive Summary. R. V. Brulle, December 1977. Contract No. EY-76-C-02-2617, 22 pp. NTIS (COO/2617/1/1).

Polytechnic Institute of New York. *Vortex Augmentors for Wind Energy Conversion* (Progress Report, May–November 1976). P. M. Sforza, December 1976. Contract No. EX-76-S-01-2358, 114 pp. NTIS (NSF/RANN/GI-41891/FR/75/4).

AAI Corporation, Baltimore, Maryland and Institute of Gas Technology, Chicago, Illinois. *Production of Methane Using Offshore Wind Energy, Final Report*. R. B. Yound, A. F. Tiedemann, L. G. Marianowski, E./H. Camera, November 1975. Contract No. NSF-C993. 131 pages, Order No. ERDA/NSF/993-75/TI.

California University, Los Angeles. *Nonlinear Equations of Equilibrium for Elastic Helicopter or Wind Turbine Blades Undergoing Moderate Deformation*. December 1978. 105 pp. Contract No. EX-76-A-29-1028. NASA-CR-1549478; UCLA-ENG-7718.

Dayton, University of. *An Analysis of the Madaras Rotor Power Plant–An Alternate Method for Extracting Large Amounts of Power From the Wind. Progress Report: October 1976-April 1977*. May 1977. D. H. Whitford; J. E. Minardi; F. L. Starner; B. S. West. HQS-2554-77/1,2.

Dayton, University of. *Electrofluid Dynamic (EFD) Wind Driven Generator. Final Report*. John E. Minardi, Maurice O. Lawson, Gregory Williams, October 1976. Contract No. EX-76-S-02-4130. Order No. COO/4130-77/1.

General Electric Company. *System Dynamics of Multi-Unit Wind Energy Conversion Systems Application*. Executive Summary. February 15, 1978. DSE-2332-T1.

Hamilton Standard. *Experimental and Analytical Research in the Aerodynamics of Wind Turbines*. (Mid-term technical report, June 1—December 31, 1975). C. Rohrbach, February 1976. Contract No. E(11-1-2615). 111 pages, Order No. COO-2615-76-T-1.

Kaman Aerospace Corporation. *Design, Fabrication, Test, and Evaluation of a Prototype 150-Foot Long Composite Wind Turbine Blade*. Herbert W. Gewehr, September 1979. DOE/NASA/0600-79/1, NASA CR-159775, R-1575.

Lawrence Livermore Lab. *Methods of Estimating the Reliability of Wind Energy Systems with Storage*. Glassey, C. R.; Moyer, G. F. 1978. 61 pp., Contract No. W-7405-ENG-48. UCRL-15005.

Massachusetts Institute of Technology. *Wind Energy Conversion*. (Progress Report, July 15, 1975—February 15, 1976) R. H. Miller, et. al., February 1976. Conntract No. NSF-G-AER-75-00826. 181 pages. Order No. ERDA/NSF/00826-75/2.

NASA Ames Research Center. *Nonlinear Dynamic Response of Wind Turbine Rotors*. Chopra, I. February 1977. 233 pp. Contract NSF AER-75-00826. N-79-12542.

NASA Lewis Research Center. *200-kW Wind Turbine Generator Conceptual Design Study*. January 1979. 109 pp. Contract EX-76-A-29-1028. NASA/1028-79/1. NASA-TM-79032.

NASA Lewis Research Center. *Transient Response to Three-Phase Faults on a Wind Turbine Generator*. Gilbert, L. J. June 1978. 146 pp. NASA N-78-26542.

NASA Lewis Research Center. *Design and Operating Experience on the U.S. Department of Energy Experimental MOD-0 100 kW Wind Turbine*. John C. Glasgow and Arthur G. Birchenough. DOE/NASA/1028-78/18. Technical paper presented at Thirteenth Intersociety Energy Conversion Engineering Conference, San Diego, CA, August 20–25, 1978. NASA TM-78915.

NASA Lewis Research Center. *Design, Fabrication, and Test of a Composite Material Wind Turbine Rotor Blade*. D. G. Griffee, Jr., R. E. Gustafson, and E. R. More. November 1977. DOE/NASA/9773-78/1, NASA CR-135389, HSER 7383.

NASA Lewis Research Center. *Engineering Handbook on the Atmospheric Environmental Guidelines for Use in Wind Turbine Generator Development*. Walter Frost, B. H. Long, and R. E. Turner. December 1978. NASA Technical Paper 1359.

NASA Lewis Research Center. *Wind Turbine Generator Rotor Blade Concepts with Low-Cost Potential*. T. L. Sullivan and T. P. Cahill, NASA; D. G. Griffee, Jr., United Technologies Corp.; and H. H. Geroehr, Kaman Aerospace Corp. Paper presented at the Twenty-Third National SAMPE Symposium, Anaheim, California, May 2–4, 1978. DOE/NASA/1028-77/13, NASA TM-73835.

NASA Lewis Research Center. *Wind Turbine Structural Dynamics*. Workshop held at Lewis Research, November 15–17, 1977. NASA CP 2034, DOE Publication CONF-771148.

NASA Lewis Research Center. *Wake Characteristics of a Tower for the DOE-NASA MOD-1 Wind Turbine*, Joseph M. Savino, Lee H. Wagner, and Mary Nash, NASA, April 1978. DOE/NASA/1028-78/17, NASA TM-78853.

NASA Lewis Research Center. *A 100-Kilowatt Experimental Wind Turbine: Simulation of Starting Overspeed and Startdown Characteristics*. L. Giblert, February 1976. Order No. DOE/NASA/1028-77/6.

NASA Lewis Research Center. *Evaluation of Urethane for Feasibility of Use in Wind Turbine Blade Design*. Lieblein, S.; Ross, R. S.; Fertis, D. G. April 1978, 158 pp. Contract No. EX-76-A-29-1028. NASA CR-159530.

NASA Lewis Research Center. *Design, Fabrication and Initial Test of a Fixture for Reducing the Natural Frequency of the MOD-0 Wind Turbine Tower*. July 1979. 21 pp. Contract No. EX-76-A-29-1028. DOE/NASA/1028-79/24.

Oregon State University. *Applied Aerodynamics of Wind Power Machines*. R. E. Wilson, P. B. S. Lissaman, July 1974, 116 pp. Contract No. NSF-AER-74-04014 A03 (PB 238 595).

Oregon State University. *Aerodynamic Performance of Wind Turbines*. R. E. Wilson, P. B. S. Lissaman, S. N. Walker, June 1976, 170 pp. Contract No. NSF-AER-74-04014 A03 (PB 259 089).

Princeton University. *Optimization and Characteristics of a Sailwing Windmill Rotor*. M. D. Maughmer, March 1976. Contract No. GI-41891 (NSF/RANN/GI-41891/FR/75/4).

Sandia Laboratories, Contract No. AT(29-1)-789: *Vertical-Axis Wind Turbine Technology Workshop* (held at Sandia Laboratories, Albuquerque, New Mexico, May 18–20, 1976). L. Wetherhold, July 1976, 439 pp. (SAND-76-5586).

Sandia Laboratories. *The Vertical-Axis Wind Turbine—How it Works*. B. F. Blackwell, April 1974, 8 pp. (SLA-74-0160).

Sandia Laboratories. *Blade Shape for a Troposkein Type of Vertical-Axis Wind Turbine*. B. F. Blackwell and G. E. Reis, April 1974, 24 pp. (SLA-74-0154).

Sandia Laboratories. *An Electrical System for Extracting Maximum Power from the Wind*. A. F. Veneruso, December 1974, 29 pp. (SAND-74-0105).

Sandia Laboratories. *Some Geometrical Aspects of Troposkeins as Applied to Vertical-Axis Wind Turbines*. B. F. Blackwell and G. E. Reis, March 1975 (SAND-74-0177).

Sandia Laboratories. *Practical Approximations of a Troposkein by Straight-Line and Circular-Arc Segments*. G. E. Reis and B. F. Blackwell, March 1975, 34 pp. (SAND-74-0100).

Sandia Laboratories. *Wind Energy—A Revitalized Pursuit*. B. F. Blackwell and L. V. Feltz, March 1975, 16 pp. (SAND-75-0166).

Sandia Laboratories. *An Investigation of Rotation-Induced Stresses of Straight and of Curved Vertical-Axis Wind Turbine Blades*. L. V. Feltz and B. F. Blackwell, March 1975, 20 pp. (SAND-74-0379).

Sandia Laboratories. *Application of the Darrieus Vertical-Axis Wind Turbine to Synchronous Electrical Power Generation*. J. F. Banas, E. G. Kadlec and W. N. Sullivan, March 1975, 14 pp. (SAND-75-0165).

Sandia Laboratories. *Nonlinear Stress Analysis of Vertical-Axis Wind Turbine Blades*. W. I. Weingarten and R. E. Nickell, April 1975, 20 pp. (SAND-74-0378).

Sandia Laboratories. *Methods of Performance Evaluation of Synchronous Power Systems Utilizing the Darrieus Vertical-Axis Wind Turbine*. J. F. Banas, E. G. Kadlec and W. N. Sullivan, April 1975, 22 pp. (SAND-75-0204).

Sandia Laboratories. *The Darrieus Turbine: A Performance Prediction Model Using Multiple Streamtubes*. J. H. Strickland, October 1975 (SAND-75-0431).

Sandia Laboratories. *Engineering of Wind Energy Systems*. J. F. Banas and W. N. Sullivan, January 1976. (SAND-75-0530).

Sandia Laboratories. *Wind Tunnel Performance Data for the Darrieus Wind Turbine with NACA-0012 Blades*. B. F. Blackwell, L. V. Feltz and R. E. Sheldahl, 1976 (SAND-76-0130).

Sandia Laboratories. *Synchronization of the DOE/NASA 100-Kilowatt Wind Turbine Generator with a Large Utility Network*. Leonard J. Gilbert, December 1977. Contract No. E(49-26)-1028, 16 pp. NTIS (ERDA/NASA-1004-77/3).

Sandia Laboratories. *Tower and Rotor Blade Vibration Test Results for a 100-Kilowatt Wind Turbine*. B. S. Linscott, W. R. Shapton and D. Brown, October 1976. (NASA TM X-3426).

Sandia Laboratories. *Wind Tunnel Measurements of the Tower Shadow on Models of the ERDA/NASA 100-kW Wind Turbine Tower*. J. M. Savino and L. H. Wagner, November 1976 (NASA TM X-73548).

Sandia Laboratories. *Vibration Characteristics of a Large Wind Turbine Tower on Non-Rigid Foundations*. S. T. Yee, T. Yung, P. Change *et al.*, May 1977 (ERDA/NASA 1004-77/1).

Sandia Laboratories. *Dynamic Blade Loading in the ERDA/NASA 100-kW and 200-kW Wind Turbines*. D. A. Spera, D. C. Janetzke and T. R. Richards, May 1977 (ERDA/NASA 1004-77/2).

Sandia Laboratories. *Drive Train Normal Modes Analysis*

for the ERDA/NASA 100-Kilowatt Wind Turbine Generator. T. L. Sullivan, D. R. Miller and D. A. Spera, July 1977 (ERDA/NASA/1028-77-1).

Sandia Laboratories. *Investigation of Excitation Control for Wind Turbine Generator Stability*. V. D. Gebben, August 1977 (ERDA/NASA 1028-77/3).

Sandia Laboratories. *Nastran Use for Cyclic Response and Fatigue Analysis of Wind Turbine Towers*. C. C. Chamis, P. Manos, J. H. Sinclair and J. R. Winemiller, October 1977, 20 pp. (ERDA/NASA 1004-77/3).

Sandia Laboratories. *Darrieus Wind Turbine Program at Sandia Laboratories*. Sandia Labs, Albuquerque, NM. 1979. Contract No. EY-76-C-04-0789. SAND-79-0997C.

Sandia Laboratories. *Induction and Synchronous Machines for Vertical Axis Wind Turbines*. Final Report. June 1979. Contract No. EY-76-C-04-0789. SAND-79-7017.

Sandia Laboratories. *A User's Manual for the Computer Code Parep*. April 1979, 56 pp., SAND-79-0431.

Sandia Laboratories. *FY79 Program Plan; Technical Management and Support for the Vertical Axis Wind Turbine Program*. Emil G. Kadlec. November 1979. SAND-79-1594.

Sandia Laboratories. *Economic Analysis of Darrieus Vertical Axis Wind Turbine Systems for the Generation of Utility Grid Electrical Power*. W. N. Sullivan. August 1979. SAND-78-0962. 4 Vols.

Sandia Laboratories. *Characteristics of Future Vertical Axis Wind Turbines*. Emil G. Kadlec. July 1978. SAND-79-1068.

Sandia Laboratories. *Aerodynamic Performance of the 17-Metre Diameter Darrieus Wind Turbine*. Mark H. Worstell, January 1979. Contract No. AT(29-1)-789. Order No. SAND-78-1737.

Sandia Laboratories. *Aeroelastic Analysis of the Troposkein Type Wind Turbine*. N. D. Ham, April 1977. Contract No. AT(29-1)-789. Order No. SAND-77-0026.

Sandia Laboratories. *Application of the Darrieus Vertical-Axis Wind Turbine to Sybnchronous Electrical Power Generation*. J. F. Banas, E. C. Kadlec, W. N. Sullivan, March 1975. 14 pages. Order No. SAND-75-0165.

Sandia Laboratories. *Engineering Development Status of the Darrieus Wind Turbine*. B. Blackwell, W. N. Sullivan, R. C. Reuter, J. F. Banas, March 1977. 68 pages. Order No. SAND-76-0650.

Sandia Laboratories. *Proceedings of the Workshop on Mechanical Storage of Wind Energy*. January 1979. Contract No. AT(29-1)-789. Order No. SAND-79-0001.

Solar Energy Research Institute. *Summary of Currently Used Wind Turbine Performance Prediction Computer Codes*. Perkins, F. May 1979. 29 pp. Contract No. EG-77-C-01-4042. SERI/TR-35-225.

Solar Energy Research Institute. *Giromill Overview*. 1979. Contract No. EG-77-C-01-4042. SERI/TP-35-263.

Oklahoma State University. *Development of an Electrical Generator and Electrolysis Cell for a Wind Energy Conversion System* (Final Report), July 1, 1973–July 1, 1975. W. Hughes, H. J. Allison and R. G. Ramarkumar, July 1975, 280 pp. Contract No. NSF-AER-75-00647 (NSF/RA/N-75-043) (PB 243 909).

Oregon State University. *Applied Aerodynamics of Wind Power Machines*. R. E. Wilson and P. B. S. Lissaman, July 1974, 116 pp. Contract No. NSF-AER-74-04014 A03 (PB 238 595).

Oregon State University. *Aerodynamic Performance of Wind Turbines*. R. E. Wilson, P. B. S. Lissaman and S. N. Walker, June 1976, 170 pp. Contract No. NSF-AER-74-04014 AO3 (PB 259 089).

Paragon Pacific, Inc. *Coupled Dynamics Analysis of Wind Energy Systems,* February 1977. NASA Contract No. NAS3-197707 (NASA CR-135152).

United Technologies Research Center. *Self-Regulating Composite Bearingless Wind Turbine* (Final Report), June 3, 1975–June 2, 1976. M. C. Cheney and P. A. M. Spierings, September 1976, 62 pp. Contract No. EY-76-C-02-2614 (COO/2614-76/1). Executive Summary, 13 pp. (COO/2614-76/2).

West Virginia University, Morgantown, West Virginia. *Design, Instrumentation, and Calibration of a Vertical-Axis Wind Turbine Rotor*. D. G. Elko, 1977. Contract No. EY-76-C-05-5135. 112 pages. Order No. TID-27754.

Advanced Systems

AAI Corporation and Institute of Gas Technology. *Production of Methane Using Offshore Wind Energy*. R. B. Young, A. F. Tiedman, Jr., T. G. Marianawski and E. H. Camara, November 1975. Contract No. NSF-C993. Final Report: PB 252 307, 131 pp. Executive Summary: PB 252 308, 29 pp.

University of Dayton, Research Institute. *Electrofluid Dynamic (EFD) Wind Driven Generator*. J. E. Minardi, M. O. Lawson and G. Williams, October 1976. Contract No. EY-76-S-02-4130.

University of Dayton, Research Institute. *Wind Tunnel Performance Data for Two- and Three-Cup Savonius Rotors*. B. F. Blackwell, L. V. Feltz and R. E. Sheldahl, July 1977, 108 pp. (SAND-76-0131).

University of Dayton, Research Institute. *Engineering Development Status of the Darrieus Wind Turbine*. B. F. Blackwell, W. N. Sullivan, R. C. Reuter and J. F. Banas, March 1977, 68 pp. (SAND-76-0650).

University of Dayton, Research Institute. *Darrieus Vertical-Axis Wind Turbine Program at Sandia Laboratories*. E. G. Kadlec, August 1976, 11 pp. (SAND-76-5712).

University of Dayton, Research Institute. *Status of the ERDA/Sandia 17-Meter Darrieus Turbine Design*. B. F. Blackwell, September 1976, 16 pp. (SAND-76-5683).

West Virginia University. *Innovative Wind Machines* (Executive Summary and Final Report). R. E. Walters *et al.,* June 1976. Contract No. EY-76-C-05-5135 (ERDA/NSF/00367-76/2).

Farm and Rural use (small) Systems

Colorado State University. *Wind-Powered Aeration for Remote Locations* (Final Report), March 15, 1975–August 31, 1976). P. M. Schierholz, October 1976, 130 pp. Contract No. NSF-G-AER-75-00833 (ERDA/NSF/00833-75/1).

Institute of Gas Technology. *Wind-Powered Hydrogen-Electric Systems for Farm and Rural Use*. J. B. Pangborn, April 1976, 158 pp. Contract No. NSF-AER-75-00772 (PB 259 318).

University of Massachusetts, Amherst. *Investigation of the Feasibility of Using Windpower for Space Heating in Colder Climates*. (Third quarterly progress report covering the final design and manufacturing phases of the project, September–December, 1975). W. E. Heronemus, December 1975, 165 pp. Contract No. NSF-AER-75-00603 (ERDA/NSF/00603-75/T1).

NASA-Lewis Research Center. *Installation and Initial Operation of a 4100 Watt Wind Turbine*. H. B. Tryon and T. Richards, December 1975 (ANSA TM-X-71831).

100-kW-Scale Systems

NASA-Lewis Research Center: *Preliminary Design of a 100-kW Turbine Generator*. R. L. Puthoff and P. J. Sirocky, 1975, 22 pp. (NASA TM X-71585; E-8037) (N-74-31527).

NASA-Lewis Research Center. *Structural Analysis of Wind Turbine Rotor for NSE-NASA MOD-O Wind-Power System*. D. A. Spera, March 1975, 39 pp. (NASA TM X-3198; E-8133) (N-75-17712).

NASA-Lewis Research Center. *Plans and Status of the NASA-Lewis Research Center Wind Energy Project*. R. Thomas, R. Puthoff, J. Savino and W. Johnson, 1975, 31 pp. (NASA TM X-71701; E-8309) (N-75-21795).

NASA-Lewis Research Center. *A 100-kW Experimental Wind Turbine: Simulation of Starting Overspeed and Startdown Characteristics*. L. Gilbert, February 1976 (NASA TM X-71864).

NASA-Lewis Research Center. *Large Experimental Wind Turbines—Where We Are Now*. R. L. Thomas, March 1976 (NASA TM X-71890).

NASA-Lewis Research Center. *Fabrication and Assembly of the ERDA/NASA 100-kW Experimental Wind Turbine*. R. L. Puthoff, April 1976 (NASA TM X-3390).

NASA-Lewis Research Center. *Design Study of Wind Turbines 50-kW to 3000-kW for Electric Utility Applications* (See Technology Department).

MW-Scale Systems

General Electric, Space Division. *Design Study of Wind Turbines 50-kW to 3000-kW for Electric Utility Applications* (see Technology Development).

NASA-Lewis Research Center. *Large Experimental Wind Turbines* (see 100-kW-Scale Systems).

Implementation and Market Development

The Aerospace Corporation. *Wind Machines for the California Aqueduct*, Volume 2 (Final Report). Charles A. Lindley, February 1977. Contract No. EY-76-03-1101-005, 192 pp., NTIS (Order No. SAN/1101-76/2).

Booz, Allen, and Hamilton. *Economic Incentives to Wind Systems Commercialization*. Final Report. August 1978. Michael Lotker, et. al., DOE/ET/4053-78/1. NTIS

Department of Agriculture. *Gusts of Power*. 16 mm color movie. 14 minutes. DOE Film Library. P.O. Box 62, Oak Ridge, TN 37830; Sales Order Department, National Audiovisual Center, Washington, D.C. 20409. Order No. A01302.

Department of Energy. *Commercialization Strategy Report for Small Wind Systems*. 1978. TID-28844 Draft.

Michigan State University. *Planning Manual for Utility Application of WECS*. Gerald L. Park, Otto Krauss, Jack Lawler, Jes Asmussen. June 1979. COD/4450-79/1.

NASA Lewis Research Center. *200-Kilowatt Wind Turbine Project*. January 1978. N-78-29583. NASA-TM-79757.

Nielsen Engineering and Research, Inc. *Wind Power for Farms, Homes, and Small Industry*. Jack Park and Dick Schwind, September 1978. Contract Nos. EY-76-C-03-1270 and EY-76-C-04-3533. Order No. RFP-2841/1270/78/4.

Oak Ridge National Lab. Tennessee. *Wind Turbines*, Yeoman, J. C. Jr., December 1978. DOE Contract No. W-31-109-ENG-38. 61 pp. Order No. ANL/CES/TE-78-9.

Rockwell International. *A Guide to Commercially Avaiilable Wind Machines*. Prepared with the assistance of the American Wind Energy Association. April 1978. Contract No. EY-76-C-04-3533. RFP-2836/3533/78/3.

Solar Energy Research Institute. *Overview Assessment of Potential Small Electric Utility Applications of Wind Energy*. November 1978. Contract No. EG-77-C-01-4042. SERI/TR-35-086.

Solar Energy Research Institute. *Status of Information for Consumers of Small Wind Energy Systems*. February 1979. Contract No. EG-77-C-01-4041. SERI/TP-51-158.

Solar Energy Research Institute. *Wind: An Energy Alternative*. 16 mm color movie. 12 minutes. DOE Film Library. P.O. Box 62, Oak Ridge, TN 37830; Sales Order Department, National Audiovisual Center, Washington, D.C. 20409. Order No. A02709.

Solar Energy Research Institute. *Wind Energy Information Directory*. October 1979. Superintendent of Documents, Washington, D.C., 20402. Stock No. 061-000-00350-9.

Overview Reports

Department of Energy. *Third Wind Energy Workshop* (Washington, D.C., September 19–21, 1977). Coordinated by JBF Scientific Corporation, May 1978, 979 pp. U.S. Government Printing Office [Stock No. 061-000-00089-5, Document No. CONF-770921/1 and /2 (2-volume set)].

Department of Energy. *Small Wind Turbine Systems 1979, A Workshop on R&D Requirements and Utility Interface/Institutional Issues*. Volume I R&D Requirements, Volume II Utility Interface/Institutional Issues. Coordinated by Rockwell International, Energy Systems Group, Rocky Flats Plant. Vol I, 271 pp., Vol II, 220 pp. Contract No. DE-AC04-76DP03533, Report No. RFP/3014/3533/79-8.

Department of Energy. *Wind Energy Innovative Systems Conference Proceedings* (Colorado Springs, Colorado, May 23–25, 1979). Coordinated by Solar Energy Research Institute December 1979, 361 pp. SERI/TP-49-184, DOE CONF 790501.

Department of Energy and American Meteorological Society. *Conference and Workshop on Wind Characteristics and Wind Energy Siting* (Portland, Oregon, June 19–21, 1979). Coordinated by DOE Pacific Northwest Laboratory, 471 pp. DOE CONF-790665. PNL-3214.

Department of Energy and NASA-Lewis Research Center. *Large Wind Turbine Design Characteristics and R&D Requirements*. (A workshop held at Lewis Research Center April 24–26, 1979) 464 pp. NASA Conference Publication 2106, DOE CONF-7904111.

Electric Power Research Institute. *Proceedings of the Workshop on Economic and Operational Requirements and Status of Large Scale Wind Systems* (Monterey, California, March 28–30, 1979), July 1979, 447 pp. EPRI ER-1110-SR, DOE CONF-790352.

General Electric, Space Division. *Wind Energy Mission Analysis,* February 1977. Contract No. E(11-1)-2578. (Executive Summary: COO/2578-1/1, 26 pp; Final Report COO/2578-1/2, 216 pp. Appendices A-J: COO/2578-1/3, 480 pp.).

Lockheed California Company. *Wind Energy Mission Analysis,* October 1976; Contract No. EY-76-C-03-1075. (Executive Summary: SAN/1075-1/3, 30 pp.; Final Report: SAN/1075-1/1; Appendix: SAN/1075-1/2).

Mitre Corporation. *Wind Machines*. F. R. Eldridge, October 1975, reprinted 1976, 77 pp. (NSF-RA-N-75-051). U.S. Government Printing Office (Stock No. 038-000-00272-4).

NASA-Lewis Research Center. *Wind Energy Utilization,* A Bibliography. Technical Applications Center, University of New Mexico, for NASA-LeRC (TACW-75-700).

Glossary of Key Terms*

Active solar system. An assembly of wind energy conversion systems which converts solar energy into electrical energy.

Airfoil. A curved surface designed to create lift as air flows over its surface.

Air mass. The length of the path through the earth's atmosphere traversed by the direct solar radiation, expressed as a multiple of the path length with the sun at zenith.

Amp-hours (A-hours). Amp-hours are calculated by multiplying current flow in amperes by the number of hours the current flows.

Amperes (amps, A). A measure of electric current flow.

Anemometer. An instrument for measuring wind speed.

Annual energy displacement. Quantity of energy replaced by WECS and/or energy storage discharge.

ASHRAE. Acronym denoting the American Society of Heating, Refrigeration and Air Conditioning Engineers, 345 E. 47th Street, New York, New York 10017. ASHRAE Handbooks are sources of basic data on heating and air conditioning.

Asynchronous. An electric generator designed to produce an alternating current that matches an existing power source (e.g., utility mains) so that two sources can be combined to power one load (e.g., your home). The generator does not have to turn at a precise rpm to remain at correct frequency or phase; see *synchronous generator*.

Average wind speed. The mean wind speed over a specific period of time.

Availability factor. The ratio of time a WECS is producing power (or is producing some specified amount of power) to the total time interval under consideration. This excludes reliability factors such as maintenance.

Azimuth. The angle between the south-north line at a given location and the projection of the earth-sun line in the horizontal plane.

Baseload plant (baseload electricity). An electrical generation facility designed to provide a constant portion of generated power output.

Battery energy storage test (facility). An experimental laboratory in Hillsborough Township, New Jersey, for the testing of electric storage batteries: jointly funded by the Electric Power Research Institute and the Department of Energy.

bbl. Barrels (of oil). One barrel equals 42 American gallons, 306 lb, 5.6 ft^3. The heat content is approximately 5.8×10^6 Btu/bbl.

*Courtesy of The Energy Institute, Glossary of Wind Energy Terminology, 1980.

Bedplate. A base-plate for supporting a system component or structure.

Black body. Term describing an ideal substance which would absorb all the radiation falling on it and reflect nothing. An alternative definition is a body which emits the maximum possible radiation; i.e., its emissivity is 1.0.

Booster mill. Second water-pumping windmill added in tandem to provide added capacity for transporting water some distance uphill or overland.

Break-even costs. The system costs at which the price of a system's product is equal to the price of the equivalent energy product of another type of system.

Btu. British thermal unit. The amount of heat required to raise the temperature of one pound of water by 1°F.

Bus. A major electrical interconnection or tie.

Busbar price. The price of electricity at a generating plant; does not include the price increment resulting from transmission and distribution costs.

Capacity credit. A credit earned for ability to replace a conventional generating unit.

Capacity factor. The actual amount of electricity generated by a power plant during a time interval, divided by the amount of electricity that would be generated by the plant during the same interval if it operated at maximum capacity.

Capital costs. Investment costs required to build a system or device.

Capital-intensive. Indicates that a relatively large percentage of the total costs of production, as for solar technologies, is associated with the initial costs rather than the operating costs. Also used in another sense to differentiate from technologies which are labor-intensive.

Carbon dioxide (CO_2). A relatively minor constituent of air that is used by plants to produce biomass. Also released in the combustion of fossil fuels in such quantities as to cause concern over possible future earth climatic changes.

Check valve. Mechanical valve that prevents water spilling back into a wind-driven well.

Chord. The distance from the leading to the trailing edge of an airfoil.

Coefficient of heat transmission. The rate of heat loss in Btu/hour through a square foot of a wall or other building surface when the difference between indoor and outdoor air temperatures is 1°F. Often called a U-value.

Coefficient of performance (COP). The ratio of the useful energy output from a device to the incoming energy. For heat pumps, the COP can be as large as 4 or 5;

1 unit of electricity can transfer 3 or 4 units of outside energy indoors while still converting the unit of electricity to usable thermal energy.

Concentrator. A device or structure that concentrates a windstream.

Conductance. A property of a slab of material equal to the quantity of heat in Btu/hour that flows through 1 ft^2 of the slab when a 1°F temperature difference is maintained between the two sides.

Conduction. Transmission of energy through a medium which does not involve movement of the medium itself.

Continuous power. Power available from a WECS facility on a continuous basis under the most adverse wind conditions contemplated (usually significant only if storage is an integral part of the WECS facility).

Convection. Transmission of energy or mass through a medium involving movement of the medium itself, or heat transfer owing to fluid.

Conversion efficiency. The actual net output provided by a conversion device divided by the gross input required to produce the output.

Conversion system. A device or process that converts a raw energy form into another more useful form of energy. Examples: conversion of wood into methanol or sunlight into electricity.

Converter. A class of devices for performing DC/AC power conversion or "inversion."

Counter-flow heat exchanger. Device in which two fluids flow in opposite directions, with one fluid transferring heat to the other.

Crosswind. Crosswise to the direction of the windstream.

Cube factor. The ratio of the cube root of the mean cube of the wind speed to the mean wind speed, the cube root of the energy pattern factor.

Current. Flow of electricity through wires.

Cut-in speed. The lowest wind speed below which no usable power is produced by a wind turbine. Care should be taken to define whether the cut-in speed is expressed at hub height or at some reference level; e.g., 10 m.

Cut-out speed. The highest wind speed above which a wind turbine produces no power (avoid the term "furling speed" unless actual furling of sails is involved). Care should be taken to define whether the cut-out speed is expressed at hub height or at some reference level; e.g., 10 m.

Cylinder. The bottom element in a water-pumping windmill. The cylinder contains a check valve and plunger and valve. The pump rod connects to the plunger and valve, and moves vertically in sync with the wind.

Darrieus machine. A vertical-axis wind machine that has long, thin blades in the shape of loops connected at the top and bottom of the axle; often called "egg-beater" windmill because of its appearance.

Data base. A set of numbers, variables and information that is used to provide the operational criteria for processing and decision-making.

dc. Direct electric current; does not alternate direction of electric flow as does ac.

Decentralized systems. In solar systems, refers to establishment of autonomous units, such as households or neighborhoods, to provide electricity or heat.

Dedicated storage. An energy storage system charged solely from WECS or any single energy source.

Demand. The amount of energy required to satisfy the energy needs of a stated sector of the economy. See also *end-use demand*.

Department of Agriculture. Responsible for several portions of the solar program including small wind machines and some biomass activities. Congress mandated a major solar demonstration program in 1977.

Department of Energy (DOE). Created by law; started operation in October 1977. Responsible for all U.S. Federal energy activities. It combined all solar activities previously existing in the Energy Research and Development Administration and the Federal Energy Administration, both of which it replaced.

Dependable capacity. The load-carrying ability of the WECS facility under adverse conditions for the time interval and period specified when related to the characteristics of the load to be supplied.

Design tip speed ratio. Tip speed ratio for which the power coefficient is a maximum.

Diffuser. A device or structure that diffuses a windstream.

Diode. An electric device which changes ac to dc. See also *rectify*.

Discharge/charge rate. The time rate for transferring energy to or from storage rated power.

Distributed system. See *decentralized systems*.

Diurnal. Active or occurring during the daytime rather than at night; daily.

Diversified load. A mix of different types of power-consuming devices, in residential use, various appliances, motors, etc., as opposed to space heating or water heating loads.

Downwind. On the opposite side of the direction from which the wind is blowing.

Drag. A force which "slows down" the motion of wind turbine blades, or actually causes motion and power to be produced by drag-type wind machines.

Drag-type devices. Devices, such as the Savonius-type rotor, that are actuated by aerodynamic drag in a windstream.

Drop pipe. A galvanized pipe with smooth interior walls of 2-in. diameter which provides the tube in which the pump rod moves. The drop pipe fits onto the lowest element of the water-pumping windmill, which is the cylinder.

Duty cycle. The duration and periodicity of operation of a device.

Economic Regulatory Administration. A successor to the Federal Energy Administration; has a mandate which includes intervention in rate and siting cases which can affect solar deployment. Part of DOE.

Effective carrying capacity. The power capacity that can be reliably furnished from storage.

Efficiency (e). A number arrived at by dividing the power output of a device by the power input to that device (usually the larger of the two numbers); usually expressed as a percentage value. See also *power coefficient*.

Electric Power Research Institute. Located in Palo Alto, California; the principal research arm of U.S. electric utilities. Has a sizable solar program.

Electrolysis. The use of an electric current to produce hydrogen and oxygen from water.

Electrolyte. A substance that conducts electricity by the transfer of ions.

End-use demand. The amount of energy used by final consumers, often given by type of end use. Sometimes measured in primary energy equivalents, which can include conversion losses. Electrical end-use demands provided by solar sources are sometimes converted by 1/0.65–1/0.75 to reflect typical boiler or furnace efficiencies. Care must be exercised in interpreting end-use demand data. See *energy supply* and *primary energy*.

Energy. A measure of the amount of work that can be, or has been, done; expressed in kilowatt-hours (kWh) or horsepower-hours (hp-hours).

Energy density. A ratio of energy per pound, usually used to compare different batteries.

Energy efficiency ratio. The amount of useful work or product divided by the fuel or energy input. Example: in electrical generation, it is the amount of electricity produced per unit of fuel consumed; for an air conditioner, it is the amount of cooling provided per unit of electricity used.

Energy pattern factor. The ratio of the mean cube of the wind speed to the cube of the mean wind speed, the cube of the cube factor.

Energy recovery factor. The ratio of the average load on a WECS for the period of time considered to the power which would have been produced by a WECS with a specified constant power coefficient.

Energy Research and Development Administration. Predecessor (before October 1977) to the Department of Energy. Succeeded the Atomic Energy Commission in January 1975.

Energy rose. A diagram which presents wind energy measurements from a site analysis in relation to the direction from which the wind occurs at the site. See also *wind rose*.

Energy supply. The total amount of primary energy resources used. For solar sources, energy supply data are often expressed in fossil equivalents. See end-use demand.

Escalation rate. A number which defines the annual increase in monetary value of a specified quantity.

Eutectic salts. A mixture of two salts with a melting point lower than that of any other combination of the same components. Attractive in solar applications at both low and high temperatures because of the large amount of heat which can be stored and recovered in the phase change in a relatively small volume.

Fantail. A propeller-type wind turbine mounted sideways on a larger wind machine (horizontal-axis-type) to keep that machine aimed into the wind.

Federal Energy Administration. Absorbed into the Department of Energy in October 1977. Previously responsible for solar commercialization.

Federal Energy Office. The predecessor of the Federal Energy Administration.

Federal Power Commission. A utility regulatory organization whose duties were assumed by the Federal Energy Regulatory Commission in October 1977.

Firm power. Power intended to have assured availability to the consumer to meet all, or any agreed-on portion, of his load requirements.

FOB. Free on board. An economic convention for specifying the cost of hardware at the factory, but not including transportation charges.

Forced outage rate. The annual amount of unscheduled out-of-service time for power generation units.

Furling speed. The wind speed at which the wind machine must be shut down to prevent high wind damage. Also see *cut-out speed*.

FY. Fiscal year of Federal Government (July through June before 1976; October through September after 1976). A three-month transition quarter (TQ) occurred in 1976.

Gear box. Converts mechanical rotary motion from a water-pumping wheel to vertical motion for pumping. The gear box can also contain a spring mechanism which allows the tail to fold parallel to the wheel in a high wind.

Gear ratio. A ratio of speeds (rpm) between the rotor power shaft and the pump, generator or other device power shaft; applies both to speed-increasing and speed-decreasing transmissions.

Gigawatt (GW). Power unit equal to one billion (10^9) W, one thousand MW or one million kW.

Gin pole. A pipe, board or tower used to improve leverage while raising a tower.

Gross National Product. A national economic measure often related to energy consumption with implications for solar policy.

Head. A measure of height a pump must lift water.

Head loss. A measure of friction loss caused from water flow through pipes.

Heat exchanger. A device used to transfer heat from a fluid flowing on one side of a barrier to a fluid, or fluids, flowing on the other side.

Heat pump. An engine that transfers heat from a relatively low-temperature reservoir to one at a higher temperature, the heat sink.

Heat rate. The amount of thermal input to a power generating unit necessary to produce 1 kWh of output (3413 Btu/kWh divided by heat rate = unit efficiency).

Heat sink. Any device by means of which heat is removed from a thermal system. A medium or container to which heat flows.

High head. Refers to generation of hydroelectric power using large dams.

Horsepower (hp). A measure of power capacity; 550 lb raised 1 ft/sec.

Horsepower-hours (hp-hours). A measure of energy. See also *energy*.

Hybrid system. A combination of a solar technology with a conventional technology to provide the controlled availability needed for everyday use.

Hydroelectricity. The conversion of the kinetic energy in moving water (generally first held behind a dam) to mechanical (rotary) energy and then to electricity by a generator. Sometimes considered a solar technology, the development of the potential of small dams is the responsibility of the DOE.

Hydrostorage. Technique to store power utilizing a dam by pumping water into a reservoir, to be drawn out when power needs to be generated.

Infiltration. Uncontrolled air leakage into or out of a building.

Isolation. The solar power density incident on a surface of stated orientation, usually measured in W/m^2 or Btu/ft^2/hr. This implies total isolation, which is the sum of diffuse and direct components.

Installed capacity. The total of the rated powers of the units in a WECS facility.

Institute of Electrical and Electronics Engineers (IEEE). A major professional organization for publishing literature and standards on a wide range of electricity generating devices.

Intermediate plant. An electrical plant which is used to meet daily or seasonal variations in electrical load. The annual average use of this type of plant is less than that of base-load plants, but more than peak-load plants.

International Solar Energy Society. The principal worldwide organization for exchanging technical information on all forms of solar energy. U.S. headquarters are located at Killeen, Texas. The international headquarters' mailing address is P.O. Box 52, Parkville, Victoria, Australia 3052.

Inverter. A device which converts dc to ac and generates its own frequency and voltage references. See also *synchronous inverter*.

Irradiance. Radiant flux density. The amount of radiant power per unit area that flows across or onto a surface.

Isotropic distribution. Approximation (when referring to diffuse radiant energy fluxes from the sky) in which the radiance of the sky is taken to be independent of position in the sky.

Joule. Unit of energy; work done when a force of one Newton is displaced through a distance of 1m.

Kilowatt (kW). A measure of power; one hp equals 776 W, or 0.776 kW.

Kilowatt-hours (kWh). A measure of electric energy, (1000 W-hours). See also *kilowatt, horsepower* and *horsepower-hours*.

Labor-intensive. Indicates that a relatively large percentage of costs are labor-related.

Langley. Unit of solar radiation intensity equivalent to 1.0 g-cal/cm^2.

Latitude. The angular distance north (+) or south (−) of the equator, measured in degrees.

Learning curve. The ability of manufacturers to reduce the unit cost of a given "item" as total production volume accumulates. A product is said to be on an 80 percent learning curve if the cost of the product drops 20 percent with a doubling of cumulative production.

Levelized annual cost. An annual sum, which, if expended each year over a specified time for equipment or services, would be equivalent to the summation of all actual costs during the same period, for fixed and variable charges, including burdens.

Lewis Research Center. Part of the National Aeronautics and Space Administration, located near Cleveland, Ohio. Responsible for work in wind and storage areas.

Life-cycle cost. A measure of what something will cost totally, not only to buy but also to operate over its lifespan. The accumulation generally includes a discounting of future costs to reflect the relative value of money over time.

Lift. The force which "pulls" a wind turbine blade along, as opposed to drag.

Life-type devices. Devices that use air-foils or other types of shapes that provide aerodynamic life in a windstream.

Load-leveling. Technique for smoothing a power profile of a customer or an entire utility over a period of time; this usually requires constraints on power usage (shift usage). Solar devices with storage have a load-leveling potential.

Load. The amount of electric power delivered at a specific location in the system (compare *demand*).

Load duration. The time during which the load (utility power demand) exceeds a given magnitude. Usually summed for time periods of particular interest.

Load factor. The average power output of an energy system divided by its rated power output.

Local apparent time (lat). System of astronomical time, in which the sun always crosses true north-south meridian at 12 noon. This system of time differs from local time according to longitude and time zone. The precise displacement also varies with the time of year.

Low head. Refers to generation of hydroelectric power with relatively small dams.

LWECS. Large-scale wind energy conversion system.

Market penetration. How much of a product will be sold on a yearly basis as it gains consumer acceptability over a specified time.

Maximum designed wind speed. The highest wind speed which a turbine is designed to withstand intact.

MBtu. One Million (10^6) Btu. The predominant unit of energy in the U.S. Generally, one million Btu are worth between $1 and $10. Frequently written MMBtu.

Mean wind speed. Arithmetic average wind speed over the specified time period and at specified height above ground level. The international (WMO) standard 10-m level may be implied; however, the majority of U.S. National Weather Service anemometers are at a height of 6.1 m (20 ft).

Median wind speed. The 50 percentile (i.e. 50 percent probable) wind speed value.

Megawatt (MW). Power unit equal to one million W or one thousand kW.

Meteorological station. Location where the weather is recorded.

MOPPS. Market Oriented Program Planning Study. A 1977 energy analysis developed for ERDA to simultaneously compare and project all ERDA energy technologies.

National Energy Plan. The Presidential energy message given to Congress on April 20, 1977. The enactment of the National Energy Plan by Congress is known as the National Energy Act.

National Science Foundation. The principal research organization of the Federal Government. Had responsibility for the national solar program until 1974, when the Energy Research and Development Administration was formed.

Office of Energy Research. A part of the Department of Energy responsible for basic research with energy applications; a portion of its activities is solar-related.

Office of Management and Budget. Reports to the President. Has final authority in recommendations of all budgets to Congress.

Office of Technology Assessment. An analysis organization reporting to Congress. Has prepared several solar energy reports and a major study on on-site solar electricity.

Office of the Assistant Secretary for Energy Technology. In the Department of Energy; was responsible for the development of numerous energy technologies, including all solar electric and biomass technologies.

Off-peak. Refers to utility load demand or power generation occurring at other than peak load hours of the day.

Organization for Economic Cooperation and Development. Members—the nations of Western Europe, the U.S. and Japan—support several solar programs.

Packerhead. A brass fitting which is installed over the top of the drop pipe to prevent overflow. It also is used to avoid well contamination.

Panemone. A name for drag-type vertical-axis wind machines; coming from pan (all directions) and anemone (wind), it could describe Darrieus-type machines also, but is not generally used except for drag machines. System reacts to wind from any direction.

Payback. A traditional measure of economic viability of investment projects. A payback period is defined in several ways, one of which is the number of years required to accumulate fuel savings which exactly equal the initial capital cost of the system. Payback often does not give an accurate representation of total life-cycle value.

Peaking units. Utility generating units assigned solely to respond to the periods of highest load demand.

Peak-sharing. The process of supplying power from an extraneous source to help meet the peak demand on a system.

Peaking capacity. The load-carrying ability of the WECS facility during peak load periods (usually significant only if storage is an integral part of the WECS facility).

Phase change. The change from one state—solid, liquid or gaseous—to another. Associated with such a change of state is a large energy input (or release), known as the heat of fusion or vaporization.

Polder. A tract of low land reclaimed from the sea by dikes, dams, etc.

Polished rod. Brass rod connecting the red rod, in a water-pumping windmill, to the sucker rod. The polished rod is inserted through the packerhead to provide the mechanical movement for pumping.

Power. The rate work is performed—mechanical power is force times velocity (see *horsepower*); electric power is volts times amps (V × A).

Pitch angle. Angle between the chord and plane of the cone of rotation at the 0.75 R station, where R is the maximum blade radius.

Plant factor. The ratio of the average load on a multi-unit WECS facility for the time period considered to be the aggregate rated power of all the WECS units in the facility.

Power coefficient. The ratio of power extracted by a wind turbine to the power in the reference area of the free windstream.

Power density. The amount of power per unit area of the free windstream (the energy flux in the free windstream).

Power duration curve. The cumulative probability curve of power output, expressed in probability units or in hours/year.

Primary energy. Fuels as they are extracted from their original sources; i.e., fuels not derived from other fuels (coal, oil and natural gas, for example).

Projection. An estimation of probable future events.

Quad. One quadrillion (10^{15}) Btu. Commonly used as measure of annual energy consumption, usually expressed as primary fuel equivalent. Present U.S. consumption is about 78 quads.

Rated output capacity. The output power of a wind machine operating at the constant speed and output power corresponding to the rated wind speed.

Rated power. The power output (W or hp) of a wind turbine can be its maximum power, or a power output at some wind speed less than the maximum speed before governing controls reduce the power.

Rated speed. Wind speed at which rated power occurs; can be speed at which a governor takes over, or can be a wind speed lower than this; an industry standard for rated speed does not exist at this time.

Rated wind speed. The lowest wind speed at which the rated output power of a wind machine is produced.

Reactive power. The quantity in ac circuits that is obtained by taking the square root of the difference between the square of the kilovolt-amperes (kVA) and the square of the kW. It is expressed as reactive volt-amperes or vars.

Rectify. Convert ac to dc; see also *diodes*.

Redox. Abbreviation for reduction-oxidation.

Redox battery. In this battery, the electrolytes in the anode and/or cathode compartments contain ions of metals which can exist in more than one valence state. At the anode, such a redox couple is oxidized, while another couple is reduced at the cathode.

Red rod. The uppermost part of the water-pumping assembly. It is designed to be the weakest link in the pump train and is usually constructed of wood to permit it to break in case of system malfunction.

Reference area. πR^2 where R is the maximum blade radius, not including effects of coning, flapping or tethering.

Reliability. The availability factor less power generation time lost to planned or unplanned down-time.

Renewable resources. Sources of energy that are regenerative or virtually inexhaustible, such as solar energy.

Reserve. Spinning reserve (or operating reserve); generating capacity connected to the grid and ready to take the load, including capacity availability.

Reserve capacity. The difference between the total system dependable capacity and the actual or anticipated total system peak load for a specified period.

Resistor. An electric device which "resists" electric current flow, used to control current (e.g., field-current in a generator).

Retrofit. To fit solar collectors to existing buildings, or, more generally, any addition of a new technology to an existing structure.

Return time. Time before the wind returns to a higher, specified value, such as the cut-in speed of a windmill.

Rotor. The power-producing blades and rotating members of a wind turbine (e.g., the blades).

Rotor efficiency. The efficiency of the rotor only; does not include transmissions, pumps, generators or line or head loss.

Rotor power coefficient. Same as *rotor efficiency*.

rpm. Revolutions per minute.

Run duration. Length of time after wind speed or power goes below (or above) a certain level (called the run level) until it returns above (or below) that level.

Run of the Wind. The distance the wind travels during a specific time period; this usually refers to the dial reading from a wind anemometer.

Sensible heat. Refers to storage of thermal energy through only a change in temperature of a substance (not a phase change). Examples of storage materials are water and rock.

SERI. Solar Energy Research Institute, a division of Midwest Research Institute, Kansas City, Missouri; funded by the Department of Energy, and located in Golden, Colorado.

SHACOB. Solar Heating and Cooling of Buildings. Both the process and the name of one of the major budget categories in the Department of Energy's solar program.

Shear. A relative motion parallel to the surface of contact.

Shelter belt. A tree row planted in windy country to shelter crops and soil.

Shroud. A structure used to concentrate or deflect a windstream.

Sine wave. The type of ac generated by utility companies, rotary inverters, sophisticated solid-state inverters and ac generators.

Solar constant. The intensity of solar radiation beyond the earth's atmosphere, at the average earth-sun distance, on a surface perpendicular to the sun's rays. The value for the solar constant is 1353 W/m^2, 1.940 cal/cm^2-min, or 429.2 Btu/ft^2/hr($\pm$1.6 percent).

Solar Energy Industries Association. Organization of manufacturers of solar devices, primarily for heating and cooling of buildings, headquartered in Washington, D.C.

Solar rights. Refers to the legal question regarding the right of a person who uses a solar energy device not to have his sunlight blocked by another structure.

Solidity. Ratio of rotor blade surface area to frontal (or swept) reference area of the entire windwheel.

SOLDAY. A meteorological data base containing daily values for meteorological and solar radiation values. Available from the National Climatic Center.

SOLMET. A meteorological data base containing hourly values for meteorological and solar radiation values. Available from the National Climatic Center.

Specific output. Total energy output in kWh over a specified time period, divided by the rated power of the WECS in kW.

Spectral energy distribution. A curve showing the variation of spectral irradiance with wavelength.

Spectal irradiance. The monochromatic irradiance of a surface per unit bandwidth at a particular wavelength. Units often W/m^2—nanometer bandwidth.

Square wave. Type of ac output from low-cost solid-state inverters; usable for many appliances, but may affect stereos and TV sets.

Stock. A bar used to support a windmill sail.

Storage system cost. A current estimated cost of a storage system or a projected future cost.

Sucker rod. In a water-pumping windmill, provides the lifting force connected to the pump cylinder.

SWECS. Small-scale wind energy conversion system.

Synchronous generator. An ac generator which operates together with an ac power source (similar to asynchronous generator) must turn at a precise rpm to hold the frequency and phase relationship to the ac source.

Synchronous inverter. Also called "line commutated inverter," inverts dc to ac (see *inverter*) but must have another ac source (e.g., utility mains or ac gas generator) for voltage and frequency reference; ac is created synchronously (that is, in phase and at the same frequency as outside the ac source).

System required reserve (or reserve margin). The system reserve capacity needed as standby to ensure an adequate standard of service.

System-wide storage. An energy storage system accessible to, and chargeable by, any generating source in the system having available and/or excess capacity.

Tail. Mechanical fin which orients wheel to the direction of the wind.

Technical fix. The use of technical solutions to problems which might be solved through institutional or behavioral changes. Sometimes used in a derogatory manner.

Technocracy. A government or social system managed by "technicians" or a technically trained elite.

Technology. The application of knowledge for practical purposes; for example, engineering designs to convert solar energy into more useful forms of energy such as electricity or space heating.

Tennessee Valley Authority. A major potential Federal purchaser of solar systems, already active in experiments.

Thermal capacitance. The ratio of the amount of heat added to the resulting rise in temperature in a unit mass of material, often called specific heat: Btu/lb/°F.

Thermal mass (also thermal inertia). The tendency of a building with large quantities of heavy materials to remain at the same temperature or to fluctuate only very slowly; also, the overall heat storage capacity of a building.

Thermochemical dissociation. A process for generating storable chemicals by using the sun's energy in a (sometimes complex) chemical reaction. A possible reaction could produce hydrogen and oxygen from water, using only a high-temperature, solar-driven chemical reaction.

Thrust. Force in the streamwise direction.

Tip speed to wind speed. Ratio of the speed of the tip of a propeller blade to the speed of a windstream in which it is located: $R\Omega/V$.

Tons of cooling. One standard commercial ton of refrigeration is defined as 288,000 Btu absorbed at a uniform rate during 24 hours.

Top assembly. The components of a water-pumping windmill, which include the wheel, gear box and tail.

Torque. A measure of force from windwheel, causing rotary motion of power shaft.

Tower. Mechanical structure which supports wind energy conversion system.

Translational motion. In a straight line.

Turbulence. Rapid wind speed fluctuations; gusts are maximum values of wind turbulence; randomness in the winds.

Turnaround efficiency. The resulting efficiency when energy is converted from one form or state to another form or state, and then reconverted to the original form or state.

Upwind. On the same side as the direction from which the wind is blowing.

Voltage (V). The electrical pressure which causes current flow (amps).

Watt (W). Unit of electric power; see also *horsepower*.

Watt-hours (W-hours). Unit of electric energy; see also *kilowatt-hours*.

Watt per square meter (W/m²). A measure of the energy in the wind passing through a square meter of area.

WECS. Wind energy conversion system. A single or multi-unit wind-powered system including all system components (storage, controllers, etc.) which are integral parts of the system).

Well casing. The well driller usually installs a plastic or galvanized pipe in the well hole to support the well wall.

Well seal. Mechanical plate inserted into the well casing. The drop pipe is inserted through well seal.

Wheel. The aerodynamic blade of a water-pumping multi-vane WECS. The diameter of the wheel is a major factor in the capacity of the wind machine.

Wind energy conversion system. A wind system which produces electrical or mechanical power. See *WECS*.

Windmill. Archaic term for wind system; still used to refer to high-solidity rotor water-pumpers and older mechanical output machines.

Windpower. Power in the wind, part of which can be extracted by a wind turbine; see also *power*.

Windpower profile. How the wind power changes with height above the surface of the ground or water; the wind power profile is proportional to the cube of the wind speed profile. See *wind speed profile*.

Wind rose. A plot showing the average (usually a monthly or yearly average) wind speed from each direction (usually 16 directions are used) and percentage time the wind blows from each direction.

Wind speed duration curve. The cumulative probability curve for wind speed, expressed in probability units or in hours/year.

Wind speed frequency curve. The probability density function curve for wind speed, expressed in probability units per speed interval or in hours/year/speed interval.

Wind speed profile. How the wind speed changes with height above the surface of the ground or water.

Wind turbine, wind system or wind machine. Accepted modern terms for devices which extract power from the wind; can refer to devices which produce mechanical or electrical power output.

Wind turbine generator. A wind system which produces electrical power; sometimes abbreviated WTG. Preferred use is wind energy conversion system.

Windwheel. Same as rotor.

Work. Force lined up with the direction of movement times the distance moved; for example, by lifting a 1-lb weight up 1 ft, 1 ft-lb of work is performed. A 2-lb weight lifted up 3 ft requires 2 × 3 = 6 ft-lb of work.

Zero differential escalation. A condition where the general inflation rate and the escalation of a specific commodity (such as fuel) are identical.

Abbreviations and Acronyms

A Ampere
ac Alternating current
A&E Architect and engineer
AGA American Gas Association
AID Agency for International Development
AL Ames Laboratory
ANL Argonne National Laboratory
API American Petroleum Institute
ASME American Society of Mechanical Engineers
atm Atmosphere (of pressure)
AWEA American Wind Energy Association

bbl Barrel
bbl/d Barrels per day
Bcf Billion cubic feet
BETC Bartlesville Energy Technology Center
BEST Battery Energy Storage Test Facility
BNL Brookhaven National Laboratory
BOM Bureau of Mines
BOP Balance of Plant
Btu British Thermal Unit

CBO Congressional Budget Office
CEQ Council on Environmental Quality
CFPD Cubic feet per day
COP Coefficient of performance
CSA Department of Energy Office of Conservation and Solar Applications
CY Calendar Year

dc Direct current
DOA Department of Agriculture
DOC Department of Commerce
DOD Department of Defense
DOE Department of Energy
DOI Department of Interior

EA Office of Environmental Activities
ECP Engineering Change Proposal
ECS Environmental Control Systems
EDP Environmental Development Plan
EEI Edison Electric Institute
EH&S Environmental Health and Safety
EIA Environmental Impact Assessment
EIS Environmental Impact Statement
ENCON Environmental Constraints
EPA Environmental Protection Agency
EPRI Electric Power Research Institute
ER Energy Research
ERA Economic Regulatory Administration
ERDA Energy Research and Development Administration
ET Energy technology

FPC Federal Power Commission
FY Fiscal year

GNP Gross National Product

IEA International Energy Agency
IEEE Institute of Electrical and Electronics Engineers
ISES International Solar Energy Society

JPL Jet Propulsion Laboratory

kWh Kilowatt-hours
kVA One thousand volt-amperes, a measure of power capacity

LASL Los Alamos Scientific Laboratory
lb/hr Pound per hour
LBL Lawrence Berkeley Laboratory
LDC Less Developed Country
LETC Laramie Energy Technology Center
LLL Lawrence Livermore Laboratory
LOLP Loss of load probability
LRC Lewis Research Center
LWECS Large wind energy conversion system

M Thousand
mA Milliampere
Mcf Thousand cubic feet
MIT Massachusetts Institute of Technology
MIUS Modular Integrated Utility Systems
MM Million
MMbbl/d Millions of barrels per day
MMscf Millions of standard cubic feet
MOPPS Market Oriented Program Planning Study
Mph Miles per hour
msec Millisecond
MSR Management Status Reviews
MW Megawatt, a measure of power, equal to 10^6 watts
MWh Megawatt-hours
MWe Megawatts electrical
MY Man-year

NASA National Aeronautics and Space Administration
NBS National Bureau of Standards

NEA	National Energy Act
NEP	National Energy Plan
NEPA	National Environmental Policy Act
NIOSH	National Institute of Occupational Safety and Health
NL	National Laboratory
NOAA	National Oceanic and Atmospheric Administration
NSF	National Science Foundation
NTIS	National Technical Information Service
OCS	Outer Continental Shelf
O & M	Operations and Maintenance Cost
OMB	Office of Management and Budget, Executive Office of the President
OPEC	Oil Producing and Exporting Countries
ORNL	Oak Ridge National Laboratory
OS/IES	On-site Integrated Energy Systems
OSHA	Occupational Safety and Health Administration
OTA	Office of Technology Assessment
PDU	Process Development Unit
PL	Public Law
PNL	Pacific Northwest Laboratory
PON	Program Opportunity Notice
PPS	Pure pumped storage
PSH	Pumped storage—hydro
psi	Pounds per square inch
psig	Pounds per square inch, gauges
PUC	Public Utility Commission
PV	Photovoltaics
Q_e	Quads of electrical energy
Q_f	Quads of fossil fuel consumption
Q_r	Quads of heat rejection
Quad	10^{15} (quadrillion) Btu's
RA	Resource Assessment
REA	Rural Electrification Administration
R&D	Research and Development
RD&D	Research, Development and Demonstration
RFP	Request for proposal
RPM	Revolutions per minute
SA	Solar array
scf	Standard cubic feet
scf/d	Standard cubic feet per day
scf/h	Standard cubic feet per hour
SEO	State Energy Office
SEIA	Solar Energy Industries Association
SERI	Solar Energy Research Institute
SIC	Standard Industrial Code
SL	Sandia Laboratory
SWEC	Small wind energy conversion system
Tcf	Trillion cubic feet
T/D	Tons per day
T and D	Transmission and distribution
TEC	Total estimated cost
T/H	Tons per hour
U.K.	United Kingdom
UTC	United Technologies Corporation
V	Volts
WECS	Wind energy conversion system
WRC	Water Resources Council
WTG	Wind turbine generator

Points of Contact

Aermoter Division
Valley Industries
Box 1364
Conway, AR 72032
(501) 329-9811

Aeroelectric
13517 Winter Lane
Cresaptown, MD 21502
(609) 547-3488

Aero Power Systems Inc.
2398 4th Street
Berkeley, CA 94710
Mr. Mario Agnello
(415) 848-2710

Aero-Power
432 Natoma Street
San Francisco, CA 94103

Aerospace Systems, Inc.
1 Vinebrook Park
Burlington, MA 01803
Mr. John Zvara

Aerowatt S. A.
37 Rue Chan
75-Paris 11^{e}, France

Aerowatt %
Pennwalt Corporation
P. O. Box 18738
Houston, Texas 77023
(713) 228-5208

Dr. Daniel K. Ai
Aluminum Company of America
Alcoa Laboratories
Alcoa Center, PA 15069
(412) 339-6651, extension 2303

Mr. Robert B. Allen
Dynergy Corporation
1269 Union Avenue
Laconia, NH 03246
(603) 524-8313

Alternative Energy Institute
West Texas State University
Box 248
Canyon, TX 79016
(806) 656-3904

Alternative Sources of Energy
Rt. 2, Box 90A
Milaca, MN 56353

ALTOS
P.O. Box 905
Boulder, CO 80302
Mr. Michael Blakely
(303) 442-0855

Aluminum Company of America
Alcoa Technical Center,
Alcoa Center, PA 15069
Mr. Paul N. Vosburgh
(412) 339-6651

Amerenalt Corporation
Box 905
Boulder, Colorado 80302
(303) 442-0820

American Wind Energy Association
1621 Connecticut Ave., N.W.
Washington, D.C. 20009

American Wind Turbine
P.O. Box 466
St. Cloud, Florida 32769

American Wind Turbine Co., Inc.
1016 E. Airport Road
Stillwater, OK 74074
Ms. Nancy Thedfor
(405) 377-5333

American Energy Alternatives
P.O. Box 905
Boulder, Colorado 80302
% Mr. John Saylor

Astral Wilcon, Inc.
Millbury, MA
Mr. Bill Stern
(617) 865-9412

Automatic Power Inc.
P.O. Box 18738
Houston, TX 77023
(713) 228-5208

Mr. R. Barchet
Space Division
General Electric Company
P.O. Box 8661, Room 7310
Philadelphia, PA 19101
(215) 962-5547/8

Mr. William R. Batesole
Kaman Aerospace Corp.
Old Windsor Road
Bloomfield, CT 06002
(203) 242-4461, extension 1735

Mr. Ross L. Bisplinghoff
Pinson Energy Corporation
P.O. Box 7
Marstons Mills, MA 02648
(617) 477-2913

Mr. B. F. Blackwell
Division 1333
Sandia Laboratories
Albuquerque, NM 87115
(505) 264-4394

Mr. Steve Blake
Sunflower Power
Route 1, Box 93-A
Oskaloosa, KS 66066
(913) 597-5603

Mr. Kenneth W. Boras
BDM Corporation
7513 Jones Branch Road
McLean, VA 22101
(703) 821-5304

Boston Wind
2 Mason Court
Charlestown, MA 02129

Mr. R. H. Braasch
Division 5715
Sandia Laboratories
Albuquerque, NM 87115

Brace Research Institute
MacDonald College
McGill University
Ste. Anne de Bellevue 800
Montreal, Quebec (CANADA)

Mr. Gary Bregg
Jordan College
Box Y
Cedar Springs, MI 49319
(616) 696-1180

Mr. Robert V. Brulle
McDonnell Aircraft Co.
Box 516, Dept. 241, Bldg. 32
St. Louis, MO 63166
(314) 232-3575

Budgen & Associates
72 Broadview Avenue
Pointe Claire 710
Quebec, Canada

Mr. Robert C. Bundgaard
Kaman Aerospace Corp.
50 Old Windsor Road
Bloomfield, CT 06002
(203) 242-4461

Mr. R. Nolan Clark
Agricultural Research Service
Bushland, Texas 79012
(806) 376-2534/2524

Mr. Henry M. Clews
Enertech Corporation
P.O. Box 420
Norwich, VT 05055
(802) 649-1145

Mr. James P. Couch, Manager
MOD-2 Project
NASA-Lewis Research Center
Cleveland, Ohio 44135

DAF INDAL, Ltd.
3570 Hawkeistone Road
Mississauga, Ontario
L5C2V8, Canada
Mr. C. F. Wood
(416) 275-5300

Dakota Wind & Sun Ltd.
P.O. Box 1781
Aberdeen, SD 57401
Mr. R. L. Kirchner
(303) 632-0590

Dr. Edgar A. DeMeo
Electric Power Research Inst.
P.O. Box 10412
Palo Alto, CA 94303
(415) 493-4800, extension 254

Dempster Industries
P.O. Box 848
Beatrice, NB 68310
(402) 223-4026

Mr. Louis Divone
Div. of Distributed Solar Technology
Department of Energy
600 E Street, N.W.
Washington, D.C. 20545
(202) 376-4878/4460

DOE/NASA-Lewis Research Ctr
Cleveland, OH 44135
(216) 433-4000

DOE/Sandia Laboratories
Albuquerque, NM 87185
(505) 264-3850

Domenico Sperandio & Ager
Via Cimarosa 13-21
58022 Folloncia (GR) Italy

Mr. Ronald L. Drake
Batelle, Pacific Northwest Laboratories
P.O. Box 999
Richland, WA 99352
(509) 942-2861

Mr. William E. Drake
Enertech Corporation
Box 420, River Road
Norwich, VT 05055
(802) 649-1145

Mr. Herman M. Drees
Pinson Energy Corporation
P.O. Box 7
Marstons Mills, MA 02648
(617) 477-2913

Dunlite Electrical Products
21 Frame St.
Adelaide 5000, Australia
% Enertech

Dyna Technology, Inc.
P.O. Box 3263
Sioux City, Iowa 51102

Dynergy
P.O. Box 428
1269 Union Ave.
Laconia, NH 03246
(603) 524-8313

Earthmind
5249 Boyer Road
Mariposa, CA 95338

Mr. Frank R. Eldridge
The Mitre Corporation
Westgate Research Park
McLean, VA 22101
(703) 790-6283

Elektro G.m.b.h.
St. Gallerstrasse 27,
Winterthur, Switzerland
(See U.S. Distributors)

ELTEECO Ltd.
Aldborough Manor
Boroughbridge, North Yorkshire,
England, Y05 9EP
Mr. J. R. Thompson
(090-12) 3223 and 2716

ENAG S.A.
Rue de Pont-L'Abbe
Quimper (Finistere) France

Energy Alternatives, Inc.
P.O. Box 233
Leverett, MA 01054

Energy Development Co.
179 E. R.D. 2
Hamburg, PA 19526
(215) 562-8856

Enertech, Inc.
P.O. Box 420
Norwich, VT 05055
(802) 649-1145

Environmental Energies, Inc.
P.O. Box 73, Front Street
Copemish, MI 49625

Mr. Michael J. Evans
Wind Power Digest
54468 CR 31
Bristol, IN 46507
(219) 848-4360

Mr. K. M. Foreman
Grumman Aerospace Corp.
Bldg. 35, Mail Stop A-8
Bethpage, NY 11714
(516) 575-2221

Mr. Allan L. Frank
Solar Energy Intelligence Rpt.
Box 1067, Blair Station
Silver Spring, MD 20910
(301) 587-6300, extension 31

Mr. Glenn Gazley
North Wind Power Co., Inc.
P.O. Box 315
Warren, VT 05674

General Electric Company
P.O. Box 8661
Philadelphia, PA 19101

Mr. Willard D. Gillette
Zephyr Wind Dynamo Co.
21 Stanwood Street
Brunswick, ME 04011
(207) 725-6534

Mr. L. Michael Glick
Aeroelectric Company
126 N. Smallwood Street
Cumberland, MD 21502

Grumman Energy Systems
4175 Veterans Memorial Hwy.
Ronkonkoma, NY 11779
Mr. Paul Henton
(516) 575-7261

Mr. Terry J. Healy
Rockwell International
P.O. Box 464
Golden, CO 80401
(303) 497-4470

Dr. E. Wendell Hewson
Dept. of Atmospheric Sciences
Oregon State University
Corvallis, OR 97331
(503) 754-4557

Mr. William L. Hughes
Room 216, Engineering South
Oklahoma State University
Stillwater, OK 74074
(405) 624-5157

Helion
Box 4301
Sylmar, CA 91342

Heller Aller Company
Perry & Oakwood Streets
Napoleon, Ohio 43545
(419) 592-1856

Mr. V. Daniel Hunt
The Energy Institute
5716 Jonathan Mitchell Road
Fairfax Station, Virginia 22039
(703) 250-5136

Independent Energy Co.
11 Independence Court
Concord, MA 01742

Independent Energy Systems
6043 Sterrettania Road
Fairview, PA 16415
(814) 833-3567

Independent Power Developer
Box 1467
Noxon, MT 59853

The International Molinological Society
The Gristmill at Lobachsville
R.D. 2
Oley, PA 19542

Jacobs Wind Electric
Marcellus Jacobs
Route 11, Box 722
Fort Myers, FL 33901

Loren L. Johnson
RAIN: Journal of Appropriate Technology
2270 N. W. Irving
Portland, Oregon
(503) 227-5110

C. G. Justus
School of AE
Georgia Tech
Atlanta, GA 30332
(404) 894-3014

Kaman Aerospace Corporation
Old Windsor Road, Bloomfield
Connecticut 06002

Richard Katzenberg
Natural Power, Inc.
Francestown Turnpike
New Boston, NH 03070
(603) 487-2456

KEDCO, Inc.
9016 Aviation Blvd.
Inglewood, CA 90301
(213) 776-6636

Mr. Herschel H. Klueter
USDA ARS
Bldg. 001, Room 128, BARC-W
Beltsville, MD 20811
(301) 344-3504

KMP Parish Windmills
Box 441
Earth, TX 79031

Dr. James Lerner
California Energy Commission
111 Howe Avenue
Sacramento, CA 95825
(916) 322-6316

Dr. Louis A. Liljedahl
Agr. Wind Energy Project
Bldg. 001, Room 124, BARC-W
Beltsville, MD 20705
(301) 344-3504

Mr. Charles A. Lindley
Wind & Energy Applic. Office
The Aerospace Corporation
P.O. Box 92957
Los Angeles, CA 90009
(213) 648-6745

Mr. Peter B. S. Lissman
AeroVironment, Inc.
145 N. Vista Avenue
Pasadena, CA 91107
(213) 449-4392

Lubing Maschinenfabrik
Ludwig Berring, 2847
Bransdorf, P.O. Box 171, Germany
(See U.S. distributor)

Mr. Alvin N. Marks
Marks Polarized Corp.
153-16 10th Avenue
Whitestone, NY 11357
(212) 767-9600

Mr. Donald J. Mayer
Northwind Power Co., Inc.
P.O. Box 315
Warren, VT 05674
(802) 496-2955

Professor Louis H. Mayo
George Washington University
22nd & H Streets, N.W.
Washington, D.C. 20052
(202) 676-7382

McDonnell Douglas Aircraft Corp.
Box 516
St. Louis, MO 63166
Mr. John Anderson

Mr. Philip W. Metcalfe
UNARCO-ROHN
P.O. Box 2000
Peoria, IL 61656
(309) 697-4400

Mr. Hans Meyer
Windworks, Inc.
Route, 3, Box 329
Mukwonago, WI 53149
(414) 363-4408

Millville Windmills & Solar
Equipment Co.
10335 Old 44 Drive (Box 32)
Millville, CA 96062
Mr. Devon Tassen
(916) 547-4302

National Aeronautics and Space
Administration, Lewis Research
Center
Mail Stop 500-201
21000 Brookpark Road
Cleveland, Ohio 44135
(216) 433-4000 extensions: 6833,
6832, 6134

National Climatic Center
Federal Office Building
Asheville, NC 28801

National Technical Information
Service (NTIS)
Springfield, VA 22161
(703) 321-8500

Natural Power, Inc.
Francestown Turnpike
New Boston, NH 03070
(603) 487-2426

Dr. Vaughn Nelson
West Texas State University
Box 248
Canyon, TX 79016
(806) 656-3904

North Wind Power Co.
Box 315
Warren, VT 05674
Mr. Don Mayer
(802) 496-2995

Oakridge Windpower
Route 1
Underwood, MN 56586
(218) 826-6446

O'Brock Windmill Sales
North Benton
Ohio 44449

Dr. Richard A. Oman
Grumman Aerospace Corp.
M/S A-08-35
South Oyster Bay Road
Bethpage, NY 11714
(516) 575-7253

Pinson Energy Corporation
P.O. Box 7
Marstons Mill, MA 02648
(617) 428-8535

Prairie Sun & Wind Company
4408 62nd Street
Lubbock, Texas 79414

Real Gas & Electric Co.
Mr. Solomon Kagin
Box A
Guerneville, CA 95446

REDE Corporation
P.O. Box 212
Providence, RI 02901

Mr. Jack W. Reed
Division 5443
Sandia Laboratories
P.O. Box 5800
Albuquerque, NM 87115
(505) 264-3042

Mr. William H. Robbins
NASA-Lewis Research Center
21000 Brookpark Road
Cleveland, Ohio 44135
(216) 433-4000, extension 6134

Rockwell International, Rocky
Flats Plant
P.O. Box 464
Golden, CO 80401
(303) 497-4943

Sandia Laboratories
Albuquerque, NM 87185
(595) 264-3850

Sencenbaugh Wind Electric
P.O. Box 1174
Palo Alto, CA 94306
Mr. Jin Sencenbaugh
(415) 964-1593

Dr. Pasquale M. Sforza
Polytech. Inst. of N.Y.
Route 110
Farmingdale, NY 11735
(516) 694-5500

Mr. Robert W. Sherwin
Enertech Corporation
Box 420, River Road
Norwich, VT 05055
(802) 649-1145

Society for the Preservation of Old
Mills
232 Roslyn Ave.
Glenside, PA 19038

Mr. L. H. Soderholm
USDA-ARS
Room 213, Davidson
Iowa State University
Ames, Iowa 50011
(515) 294-5723

Sparco
Box 420
Norwich, VT 05055
(802) 649-1145

Mr. Allen P. Spaulding
WTG Energy Systems
Angola, NY 14006
(716) 549-5544

Mr. Forrest S. Stoddard
University of Massachusetts
299 Amity Street
Amhert, MA 01002
(413) 549-5070

Mr. David P. Suey
Valley Industries, Inc.
927 Hanley Ind. Court
St. Louis, MO 63144
(314) 968-5222

Mr. William N. Sullivan
Sandia Laoratories
Division 5715
Albuquerque, NM 87112
(505) 264-6434

Technology Application Ctr.
University of New Mexico
Albuquerque, NM 87131

Mr. R. J. Templin
Nat'l Research Council
M-2 Montreal Road
Ottawa, Ontario K1A OR6
(613) 993-2423

Tetrahelix
P.O. Box 241
Brunswick, ME 04011
(207) 725-6534

Mr. Ronald L. Thomas
NASA-Lewis Research Center
21000 Brookpark Road
Cleveland, Ohio 44135
(216) 433-4000, extension 6134

Topanga Power
P.O. Box 712
Topango, CA 90290

TWR Enterprises
72 W. Meadow Lane
Sandy, Utah 84070

United Technologies Research Center
Silver Lane
East Hartford, CT 06108
(203) 727-7536

U.S. Department of Energy
Wind Energy Program
600 E. Street N.W.
Washington, D.C. 20545

Wadler Manufacturing Co.
Route 2, Box 76
Galena, KS 66739
(316) 783-1355

Dr. Tunis Wentink, Jr.
Geophysical Institute
University of Alaska
Fairbanks, AK 99701
(907) 479-7607

Mr. Robert E. Wilson
Prof. of Mechanical Eng.
Oregon State University
Corvallis, OR 97331
(503) 754-2218

WINCO-Div. of Dyna Technology
E. 7th & Division St.
Sioux City, Iowa 51102
(712) 252-1821

Wind Power Digest
Mr. Mike Evans
R.R. #2, Box 489
Bristol, IN 46507

Wind Power Products Co., Inc.
213 Boeing Field Terminal
Seattle, WA 98108
(206) 762-1491

Wind Power Systems, Inc.
Box 17323
San Diego, CA 92117
(714) 452-7040

Windlite Alaska
Box 43
Anchorage, Alaska 99510

Windworks, Inc.
Box 44A, Route 3
Mukwonago, WI 53149
Hans Meyer
(414) 363-4088

Mr. Ben Wolfe
Windworks, Inc.
Route 3, Box 329
Mukwonago, WI 53149
(414) 363-4408

Mr. C. F. Wood
DAF INDAL Ltd.
3570 Hawkestone Road
Mississauga, Ontario L5C 2V8
(416) 275-5300

Mr. C. A. Wright
Unarco Rohn
P.O. Box 2000
Peoria, IL 61601
(309) 697-4400

W. T. G. Energy Systems Inc.
251 Elm Street
Buffalo, NY 14203
Mr. Allen Spaulding
(716) 856-1620

Dr. J. Yen
Grumman Aerospace Corporation
Bethpage, NY 11714
(516) 575-3858

Zephyr Wind Dynamo Co.
P.O. Box 241
Brunswick, Maine 04011
Mr. Willard Gillette
(207) 725-6534

Electrical WECS Manufacturers

Aeroelectric
13517 Winter Lane
Cresaptown, MD 21502
Contact: Kevin Moran
Telephone: (609) 547-3488

Aero Power
2398 Fourth Street
Berkeley, CA 94710
Contact: John Harold or Tom Cummins
Telephone: (415) 848-2710

Aerowatt, S.A.
% Automatic Power, Inc.
P.O. Box 18738
Houston, TX 77023
Contact: Robert Dodge or Ernest Tindle
Telephone: (713) 228-5208

Altos: The Alternate Current
P.O. Box 905
Boulder, CO 80302
Contact: Edward Gitlin
Telephone: (303) 442-0885

American Wind Turbine, Inc.
1016 East Airport Road
Stillwater, OK 74074
Contact: Nancy Thedford, Office Manager
Telephone: (405) 377-5333

Bertoia Studio
644 Main Street
Bally, Pennsylvania 19503

Jay Carter Enterprises
P.O. Box 684
Burkburnett, Texas 76354

Dakota Wind & Sun, Ltd.
P.O. Box 1781
811 First Avenue, NW
Aberdeen, SD 57401
Contact: Paul Biorn or Orv Lynner
Telephone: (605) 229-0815

Dominion Aluminum Fabricators
3570 Hawkestone Road
Mississauga, Ontario
Canada L5C 2V8
Contact: Chuck Wood, Program Manager
Telephone: (416) 270-5300

Dunlite Electrical Products Co.
Enertech Corporation
P.O. Box 420
Norwich, VT 05055
Contact: Ed Coffin or % Robert Sherwin
Telephone: (802) 649-1145

Dynergy Corporation
P.O. Box 428
1269 Union Avenue
Laconia, NH 03246
Contact: Robert Allen
Telephone: (603) 524-8313

Enertech Corporation
P.O. Box 420
Norwich, VT 05055
Contact: Ed Coffin or Robert Sherwin
Telephone: (802) 649-1145

Grumman Energy Systems
4175 Veterans Memorial Highway
Ronkonkoma, NY 11779
Contact: Ed Diamond or Ken Speiser
Telephone: (516) 575-6205

Hinton Research
417 Kensington
Salt Lake City, Utah 84115

Kedco, Inc.
9016 Aviation Boulevard
Inglewood, CA 90301
Contact: Wind Program Manager
Telephone: (213) 776-6636

Megatech Corporation
29 Cook Street
Billerica, Massachusetts 01866

Millville Windmills & Solar Equipment Company
P.O. Box 32
10335 Old Drive
Millville, CA 96062
Contact: Devon Tassen
Telephone: (916) 547-4302

North Wind Power Company
P.O. Box 315
Warren, VT 05674
Contact: Don Mayer
Telephone: (802) 496-2955

Pinson Energy Corporation
P.O. Box 7
Marstons Mills, MA 02648
Contact: Herman Drees
Telephone: (617) 477-2913

Product Development Institute
508 South Byrne Road
Toledo, OH 43609
Contact: Tom Nichols
Telephone: (419) 382-0282

Sencenbaugh Wind Electric
P.O. Box 11174
Palo Alto, CA 94306
Contact: Jim Sencenbaugh
Telephone: (415) 964-1593

Tumac Industries, Inc.
650 Fort Street
Colorado Springs, Colorado 80915

Winco
Div. of Dyna Technology
7850 Metro Parkway
Minneapolis, MN 55420
Contact: Len Attema
Telephone: (612) 853-8400

Wind Power Systems, Inc.
P.O. Box 17323
San Diego, CA 92117
Contact: Ed Salter
Telephone: (714) 452-7040

WINFLO Power Ltd
90 Esna Drive, Unit 15
Markham, Ontario
Canada L3R 2R7

W.T.G. Energy Systems, Inc.
P.O. Box 87
1 LaSalle Street
Angola, NY 14006
Contact: Al Wellikoff
Telephone: (716) 549-5544

Zephyr Wind Dynamo Company
P.O. Box 241
21 Stamwood Street
Brunswick, ME 04011
Contact: Bill Gillette
Telephone: (207) 725-6534

Mechanical WECS Manufacturers

Aermoter Division
Valley Industries
P.O. Box 1364
Conway, Arkansas 72032
Contact: Stan Anderson
Telephone: (501) 329-9811

American Wind Turbine, Inc.
1016 East Airport Road
Stillwater, OK 74074
Contact: Nancy Thedford, Office Manager
Telephone: (405) 223-4026

Bowjon
2829 Burton Avenue
Burbank, California 91504

Dempster Industries, Inc.
P.O. Box 848
Beatrice, NB 68310
Contact: Roy Smith
Telephone: (402) 223-4026

Dynergy Corporation
P.O. Box 428
1268 Union Avenue
Laconia, NH 03246
Contact: Bob Allen
Telephone: (603) 524-8313

Heller-Aller Company
Perry & Oakwood Street
Napoleon, OH 43545
Contact: James Bradner or Charles Buehrer
Telephone: (419) 592-1856
Mfg. of Baker Windmills

Sparco (Denmark) % Enertech, Inc.
P.O. Box 420
Norwich, VT 05055
Contact: Edmund Coffin at Enertech
Telephone: (802) 649-1145

Wadler Manufacturing Co., Inc.
Route 2, Box 76
Galena, KS 66739
Contact: Jerry Wade
Telephone: (316) 783-1355

Wind Machine Dealers/Distributors

COMPANY	*PRODUCT NAME*
Aermotor 1243 Majesty Drive Dallas, TX 75247	Aermotor Contact: Mr. William E. Barney Telephone: (214) 634-1950
Aermotor 2385 South Cherry Fresno, CA 93706	Aermotor Contact: Telephone: (209) 486-7200
Aermotor 900 Nabco Avenue P.O. Box 1321 Conway, Arkansas 72032	Aermotor Contact: Miles Patten Telephone: (501) 329-2969
Aermotor 6448 Warren Drive Norcross, GA 30093	Aermotor Contact: Telephone: (404) 449-1840
Aermotor 518-M North Douglas Avenue Altamonte Springs, FL 32701	Aermotor Contact: Telephone: (305) 862-0171
Aermotor 2421 West Main Ft. Wayne, IN 46808	Aermotor Contact: Telephone: (219) 432-2595
Aermotor 4655 Colt Road Rockford, IL 61109	Aermotor Contact: Telephone: (815) 874-9502
Aermotor 801 Howard Street Omaha, NB 68102	Aermotor Contact: Marvin Vesenik Telephone: (402) 341-1716
Aermotor 8105 Lewis Road Minneapolis, MN 55427	Aermotor Contact: Telephone: (612) 544-4106
Aermotor 2803 South Longview Drive Middletown, PA 17057	Aermotor Contact: Telephone: (717) 939-9311
Aermotor 1575 Avon Street Extended Charlottesville, VA 22901	Aermotor Contact: Telephone: (804) 977-0445
Alternate Energy Systems 150 Sandwich Street Plymouth, MA 02360	Elektro, Dunlite, Winco Contact: Telephone: (617) 747-0771
Automatic Power, Inc. P.O. Box 18738 Houston, TX 77023	Aerowatt Contact: Robert Dodge Telephone: (713) 228-5208
Dean Bennet Supply Company 4725 Lipan Street Denver, CO 80211	Aermotor, Dunlite, Winco Contact: Deana Bennet Telephone: (303) 433-8291
Chinook Wind Power, Inc. 1084 W. 101st Place Denver, Colorado 80221	Aero Power, Sencenbaugh, Dakota Wind and Sun, Enertech Contact: Alice Unger Telephone: (303) 451-0660
Clean Energy Products 3534 Bagley, N. Seattle, WA 98103	Jacobs, Kedco, Sencenbaugh, Wincharger Contact: Ed Kennell Telephone: (206) 633-5505

COMPANY	*PRODUCT NAME*
Cojo Wind Company Hollister Ranch #6 Gaviota, California 93017	Contact: David A. Kay Telephone: (805) 963-5248
Coulson Wind Electric RFD 1, Box 225 Polk City, IA 50226	Re-conditioned Jacobs, Winco, Winpower Contact: R. Coulson Telephone: (515) 547-3488
Crowdis Conservers RR 3, MacMillan Mt. Cape Breton, Nova Scotia Canada BOE 1B0	North Wind Contact: Daniel Atkins
Edmond Scientific Company 380 EDS Corp. Bldg. 101 East Gloucester Pike Barrington, NJ 08007	Aeroelectric (Wind Wizard) Contact: Robert F. McKelvery Telephone: (609) 547-3488
Empire Energy Development Corp. 3371 West Hampden Avenue Englewood, CA 80110	Altos, Winco Contact: David L. Flook Telephone: (303) 789-1363
Environmental Energies, Inc. P.O. Box 73 Front Street Copemish, MI 49625	Elektro, Dunlite, Re-conditioned Jacobs Contact: Timothy J. Horning Telephone: (616) 378-2000
Energy Alternatives 52 French King Highway Greenfield, MA 01301	Elektro, Dunlite, Winco, Sencenbaugh Contact: Frank Kaminsky or Klaus Kroner Telephone: (413) 733-5175
Energy Development Company 179E Road #2 Hamburg, PA 19526	Winco, Homebuilt Contact: Terrance or Helena Mehrkam or Karen Votyas Telephone: (215) 562-8856
Energy–2000 Route 800, RFD #3 Winstead, CT 06098	Re-conditioned Jacobs (North Wind) Contact: Robert Herbert Telephone: (203) 379-5185 or (413) 528-3440
Enertech Corporation P.O. Box 420 Norwich, VT 05055	Dunlight, Winco, Sencenbaugh, Sparco Contact: E. Coffin Telephone: (802) 649-1145
Environmental Resource Group Box 3A, RD 2 Williston, VT 05495	North Wind, Sparco Contact: Perry Kleine Telephone: (802) 879-0511 or (802) 878-4000
Fenton's Feeders Route 1, Box 124 Arcadia, FL 33821	Aermotor (Water Pumping) Contact: Catherine Fenton, Bill Autry, or Nell Gammage Telephone: (817) 494-2727
Independent Energy Company 314 Howard Avenue Ewarthmore, PA 19081	Elektro, Dunlite, Winco Contact: Kendall B. Hampton Telephone: (617) 368-6992
Independent Energy Company 6043 Sterrettania Road Fairview, PA 16415	Re-conditioned Jacobs, Dakota Wind Electric Contact: John D'Angelo Telephone: (814) 833-0829
Kramco P.O. Box 1536 Allentown, PA 18150	Re-conditioned Jacobs, Winco Contact: Telephone: (215) 437-6758

COMPANY	*PRODUCT NAME*
Laholms Motor & Bilelektriska A/B Export Office & Information Wind Kraft A.E.S.C. Box #104 S-312 01 Laholm, Sweden	Wind Kraft A.E.S.C. Contact: Eric Alkstad Int. Phone: 00946-43020371 Cable: WERKOMP
Makia Ocean Engineering Box 1194 Kailua, Oahu, Hawaii 96734	Dunlite Contact: Henry Horn Telephone: (808) 259-5904 or (808) 259-5722
Mountain Valley Energy, Inc. Sugarloaf Star Route Boulder, Colorado 80302	Enertech Contact: Robert B. Keller Telephone: (303) 442-6161
Natural Power Systems, Inc. 3316 Augusta Avenue Omaha, NB 68144	Dakota Wind Electric, Dunlite, Sencenbaugh, Sparco Contact: John Traudt Telephone: (402) 334-5881
O'Brock Windmill Sales Route 1, 12th Street North Benton, OH 44449	Baker Contact: Ken O'Brock Telephone: (216) 584-4681
Pacific Energy Systems 615 Romero Canyon Road Santa Barbara, CA 93018	North Wind Contact: Fred Carr Telephone: (805) 969-6603
Prairie Sun & Wind Company 448 –2nd Street Lubbock, TX 79409	Re-conditioned Jacobs, Winco, Aeropower, Dakota Wind Electric Aermotor Contact: Ken Ketner Telephone: (806) 795-1412
Real Gas & Electric P.O. Box 193 Shingletown, CA 96088	Elektro, Dunlite Contact: Solomon Kagin Telephone: (916) 474-3852
Rede Corporation P.O. Box 212 Providence, RI 02901	Dominion Aluminum Fabricating (DAF) Contact: Ronald Beckman Telephone: (401) 751-7333
Schupbach, Ralph 321–13th Street Alva, OK 73717	Winco Contact: Ralph Schupback Telephone: (405) 327-1685
Shingletown Electric P.O. Box 237 Shingletown, CA 96008	Elektro, Dunlite Contact: Robert E. Eckert Telephone: (916) 474-3852
Sunflower Power Company Route 1, Box 93-A Oskaloosa, KS 66066	Re-conditioned Jacobs, North Wind Contact: Steve Blake Telephone: (913) 597-5603
Wind Engineering Corporation Box 5936 Lubbock, TX 79417	Re-conditioned Jacobs, DAF, Dynergy, Dakota Wind Electric Contact: Telephone: (806) 763-3182
Windependence Electric P.O. Box M1188 Ann Arbor, MI 48106	Re-conditioned Jacobs Contact: Craig Toepfer Telephone: (313) 769-8469

WECS Towers

All wind machines must be placed on a support structure, generally a tower. A variety of factors influence the final choice of a tower for a particular wind machine and application. Two tower types, guyed and free-standing, are applicable for most wind energy installations.

Wind system distributors can help select the best tower for a particular site, but several factors should be considered. Wind machines should be at least 30 feet above the nearest obstruction. The tower must also support the weight of the wind machine and withstand loads imposed by the wind. These loads are a function of both the wind velocity and wind machine rotor diameter. Aesthetics, building costs, and zoning are also considerations in some areas.

There are four American Manufacturers of towers designed specifically for WECS. A number of foreign manufacturers offer their towers through WTG distributors. In addition, there are a number of tower manufacturers who have not been active in the WECS field but may have suitable equipment.

American Tower Company
Shelby, Ohio 44875

Astro Research Corporation
P.O. Box 4128
Santa Barbara, California 93103

Bayshore Concrete
Bayonne, NJ

Natural Power Inc.
Francestown Turnpike
New Boston, NH 03070

North Wind Power Company
Box 315
Warren, VT 05674

Solargy Corporation
17914 E. Warren Avenue
Detroit, Michigan 48224

Unarco-Rohn
6718 West Plank Road
P.O. Box 2000
Peoria, IL 61601

Valmont Industries, Inc.
Valley, Nebraska 68064

Batteries

Because the wind is an intermittent energy source, it is often necessary to find a means of storing its energy. Although many energy storage systems such as heat, compressed air and flywheels are now being investigated, the state of the art for electricity storage is the lead acid battery.

There are several types of storage batteries now commercially available. Each type has characteristics that make it best suited for a particular application. The ultimate choice of batteries depends on the total wind system characteristics including wind at the site, the wind turbine generator, and the load.

Batteries Manufacturing Company
14694 Dequindu
Detroit, MI 48212

Bright Star
602 Getty Avenue
Clifton, NJ 07015

Burgress Div. of Clevite Corp., Gould
Box 3140
St. Paul, MN 55101

C & D Batteries Eltuce Corp.
Washington & Chewy Street
Conshohocken, PA 19428

Century Storage Battery Company, LTC.
Birmingham Street
Alexandria, Australia

Delatron Systems Corporation
553 Lively Boulevard
Elk Grove Village, IL 60007

Delco-Remy Division of GM
Box 2439
Anderson, IN 46011

Eggle-Pichen Industries
Box 47
Joplin, MO 64801

ESB Incorporated–Willard
Box 6949
Cleveland, OH 44101

Exide
5 Pen Center Plaza
Philadelphia, PA 19103

Ever Ready – Union Carbide Corp.
270 Park Avenue
New York, NY 10017

Globe-Union
5757 N. Greenbay Avenue
Milwaukee, WI 53201

Gould Incorporated
485 Calhoun Street
Trenton, NJ 08618

Gulton
212 T Dorham Avenue
Metuchen, NJ 08840

Keystone Battery Company
8301 Imperial Drive
Waco, TX 76710

Mule Battery Company
325-T Valley Street
Providence, RI 02908

RCA
415 South 5th Street
Harrison, NJ 07029

Surrette Storage Battery Co., Inc.
Box 711
Salem, MS 01970

Trojan Batteries
1125 Mariposa Street
San Francisco, California 94107

Inverters

Inverters are devices that convert direct current power (dc) to the alternative current (ac) more commonly used in this country.

There are a number of considerations in selecting an inverter for a WECS including the amount and quality of the power required, overload capabilities, and cost.

The majority of these systems are designed to operate with battery bank storage system and are known as "stand alone" inverters. There is also available an inverter known as a synchronous inverter or "line commutated" inverter. These inverters are designed to feed the power produced to an ac line and requires the ac line for a voltage signal.

It should be noted that there are quite a few inverter manufacturers and most wind turbine generator distributors also sell inverters.

ATR Electronics, Inc.
300 East 4th Street
St. Paul, Minnesota 55101

275 to 1100 watt; 12 volt input

Allied Electronics
2400 W. Washington BLVD.
Chicago, IL 60612
(312) 421-4200

350 to 1000 watt; 12 volt input

Carter Motor Company
2711 West George Street
Chicago, Illinois 60618

210 to 750 watt; variable voltage input

Dynamote Corporation
1130 N.W. 85th
Seattle, Washington 98117

Automatic DC to 120VAC 60Hz static inverters

Eico Electronic Instrument Co.
283 Malta Street
Brooklyn, NY 11207

110 to 220 watt; 12 volt input

Electro Sales Co., Inc.
100 Fellsway West
Somerville, MA 02145
(617) 666-0500

20 to 2000 watt, 12 to 200 volt input

Elgar Corporation
8225 Mercury Court
San Diego, California 92111

600 to 1000 watt; 24 and 48VDC input

Heath Company
Benton Harbor, MI 49002
(616) 983-3961

175 watt; 12 volt input

LeMarche Mfg. Company
106 Bradrock Drive
Des Plaines, IL 60018
(312) 279-0831

100 to 10,000 watt; 24 to 120 volt input

Newark Electronics
500 N. Pulaski Road
Chicago, IL
(312) 638-4411

100 to 250 watt; 12 volt input

Nova Electric
263 Hillside Avenue
Nutley, NJ 07110
(201) 661-3432

30 to 120 watt; 12 to 110 volt input

Ratelco, Inc.
610 Pontius Avenue, N.
Seattle, WA 98109
(206) 624-7770

250 to 2500 watt; 24 to 100 volt input

Real Gas & Electric, Inc.
P.O. Box F
Santa Rosa, California 95402

5 KVA maximum AC output; 75 to 200 VDC input

Soleq Corporation
5969 Elston Avenue
Chicago, IL 60646
(312) 792-3811

1500 to 6000 watt; 12 to 112 volt input

Topaz Electronics
3855 Ruffil Road
San Diego, CA 92123
(714) 279-0831

200 to 3000 watts; 12 to 125 volt input

Willmore Electronics
Box 2973
Durham, NC 27705
(919) 489-3318

45 to 1500 watt; 12 to 120 volt input

LINE COMMUTATED INVERTERS

Gemini Synchronous Inverters
Windworks
Box 329, Route 3
Mukwonago, WI 53149
(414) 363-4408

4 to 1000 kw; variable voltage input

Anemometers and Recorders

Anemometers are devices which measure the wind. The manner in which they accomplish this task ranges from an odometer which records the number of miles of wind that pass the recorder, to three-axis anemometers that record wind velocity on three separate planes.

As with all wind system subcomponents, anemometers vary widely in terms of accuracy, the type of information obtained, and cost. The value of a certain anemometer or anemometer system will depend largely upon the specific application for which it is used. Again, wind system distributors can help select the proper anemometer or anemometer system.

The companies in this list market anemometers and recording equipment. Many operate on commercial or battery power. Prices and ordering information should be obtained from the manufacturers.

Aeolian Kinetic
P.O. Box 100
Providence, RI 02901
(401) 274-3690

Aircraft Components
North Shore Drive
Benton Harbor, MI 49022

Bendix Environmental Science Div.
1400 Taylor Avenue
Baltimore, MD 21204

Climet, Inc.
1620 W. Colton Ave.
Redlands, CA 92373
(714) 793-2788

Danforth
Div. of the Eastern Company
Portland, ME 04103

Davis Instrument MFG. Co., Inc.
513 E. 36th Street
Baltimore, MD 21218

Dwyer Instruments, Inc.
P.O. Box 373
Michigan City, IN 464360
(219) 872-9141
(609) 448-9200 (Hightstown, NJ)
(404) 427-9406 (Marietta, GA)
(714) 991-6720 (Anaheim, CA)

Kahl Scientific Instruments Corp.
P.O. Box 1166, El Cajon
San Diego, CA 92022
(714) 444-2158 and 444-5944

Kenyon Marine
P.O. Box 308
Guilford, CT 06437
(203) 453-4374

2730B S. Main Street
Santa Ana, CA 92702
(714) 546-1101

Maximum, Inc.
42 S. Avenue
Natick, MA 01760
(617) 785-0113

Meteorology Research, Inc.
P.O. Box 637
Altadena, CA 91001
(213) 791-1901

Natural Power, Inc.
New Boston, NH 03070
(603) 487-5512

Sencenbaugh Wind Electric
P.O. Box 11174
Palo Alto, CA 94306

Sign X Laboratories, Inc.
Essex, CT 06426
(203) 767-1700

R. A. Simerl Instrument Div.
238 West Street
Annapolis, MD 21401
(301) 849-8667

M. C. Stewart Co.
Ashburnham, MA 01430

Taylor Instruments
Arden, NC 28704

Texas Electronics, Inc.
5529 Redfield Street
Box 7151 Inwood Station
Dallas, TX 75209

Thermo-Systems, Inc.
500 Cardigan Road
P.O. Box 3394
St. Paul, MN 55165
(612) 483-0900

Westberg Manufacturing Co.
3400 Westach Way
Scnoma, CA 95476
(707) 938-2121

Robert E. White Instruments, Inc.
33 Commerical Wharf
Boston, MA 02110

Windflower Wind-Computer
Lund Enterprises, Inc.
1180 Industrial Ave.
Escondido, CA 92025
(714) 746-1211

Wind Power Systems, Inc.
P.O. Box 17323
San Diego, CA 92117
(714) 452-7040

Reference Information

This section contains reference information which the author has found of value in regard to wind energy and related conversion systems. Included is the following information:

- Monthly kilowatt-hour usage chart
- Units and conversion factors
- Conversion factor cross-reference
- BTU equivalents of fuels
- Conversion factors for oil
- Electrical units
- Metric terminology
- Symbols
- Monthly average windpower in the U.S. and Southern Canada.

MONTHLY KILOWATT-HOUR USAGE CHART. (Courtesy of DOE.)

COMPOSITE OF KILOWATTHOUR RATINGS FOR VARIOUS APPLIANCES

Home

Name	Typical Watts	Hrs./Mo.	KWHRS/Mo.
Air conditioner, central			620*
Air conditioner, window	1566	74	116*
Battery charger			1*
Blanket	190	80	15†
Blanket	50-200		15†
Blender	350	3	1†+
Bottle sterilizer	500		15
Bottle warmer	500	6	3†
Broiler	1436	6	8.5
Clock	1-10		1-4†
Clothes drier	4600	20	92†
Clothes drier, electric heat	4856	18	86†
Clothes drier, gas heat	325	18	6†
Clothes washer			8.5*
Clothes washer, auto.	250	12	3†
Clothes washer, conventional	200	12	2†
Clothes washer, auto.	512	17.3	9†
Clothes washer, ringer	275	15	4†
Clippers	40-60		1/2
Coffee maker	800	15	12†
Coffee maker, twice a day			8†
Coffee percolator	300-600		3-10†
Coffee pot	894	10	9†
Cooling, attic fan	1/6-3/4 hp		60-90†+
Cooling, refrigeration	3/4-1½ ton		200-500†
Corn popper	460-650		1†
Curling iron	10-20		1/2†—
Dehumidifier	300-500		50*
Dishwasher	1200	30	36†
Dishwasher	1200	25	30†
Disposal	375	2	1*
Disposal	445	6	3*
Drill, elecric, 1/4"	250	2	5—
Elec. baseboard heat	10,000	160	1600†
Electrocuter, insect	5-250		1*
Electronic oven	3000-7000		100*
Mixer	125	6	1+•
Mixer, food	50-200		1+•
Movie projector	300-1000		
Oil burner	500	100	50*
Oil burner			50*
Oil burner, 1/8 HP	250	64	16*
Pasteurizer, 1/2 gal.	1500		10-40†
Polisher	350	6	2+
Post light, dusk to dawn			35—
Power tools			3—
Projector	500	4	2†
Pump, water	450	44	20†
Pump, well			20†
Radio			8*
Radio, console	100-300		5-15*
Radio, table	40-100		5-10*°
Range	8500-1600		100-150†
Range, 4 person family			100†
Record player	75-100		1-5*
Record player, transistor	60	50	3*
Record player, tube	150	50	7.5*
Recorder, tape	100	10	1*°†
Refrigerator	200-300		25-30*†•
Refrigerator, conventional			83*†•
Refrigerator-freezer	200	150	30*†•
Refrigerator-freezer 14 cu. ft.	326	290	95*†•
Refrigerator-freezer, frost free	360	500	180†
Roaster	1320	30	40*
Rotisserie	1400	30	42*
Sauce pan	300-1400		2-10†
Sewing machine	30-100		1/2-2—
Sewing machine	100	10	1—
Shaver	12		1/10—°
Skillet	1000-1350		5-20†
Skil Saw	1000	6	6—
Sunlamp	400	10	4—
Sunlamp	279	5.4	1.5—
Television	200-315		15-30†°
TV, BW	200	120	24†°
TV, BW	237	110	25†°
TV, color	350	120	42†°

Name	Typical Watts	KWHRS/Mo.
Ventilating fan	fractional	10-25 per mo.†
Water heater	1000-5000	1 per 4 gal.†
FOR POULTRY		
Automatic feeder	1/4-1/2 HP	10-30 KWHR/mo.†
Brooder	200-1000 watts	1/2-1-1/2 per chick† per season
Burgler alarm	10-60 watts	2 per mo.*°
Debeaker	200-500 watts	1 per 3 hrs.*
Egg cleaning or washing	fractional HP	1 per 2000 eggs†
Egg cooling	1/6-1 HP	1¼ per case†
Night lighting	40-60 watts	10 per mo.—per 100 birds
Ventilating fan	50-300 watts	1-1-1/2 per day† per 1000 birds
Water warming	50-700 watts	varies widely†
FOR HOGS		
Brooding	100-300 watts	35 per brooding† period/litter
Ventilating fan	50-300 watts	1/4-1-1/2 per day†
Water warming	50-1000 watts	30 per brooding†• period/litter
FARM SHOP		
Air compressor	1/4-1/2 HP	1 per 3 hr†•
Arc welding	37½ amp	100 per year†
Battery charger	600-750 watts	2 per battery charge*
Concrete mixing	1/4-2 HP	1 per cu yd†
Drill press	1/6-1 HP	1/2 per hr.†
Fan, 10"	35-55 watts	1 per 20 hr.†
Grinding, energy wheel	1/4-1/3 HP	1 per 3 hr†
Heater, portable	1000-3000 watts	10 per mo.†•

Fan, attic	370	65	24†
Fan, kitchen	250	30	8†
Fan, 8"-16"	35-210		4-10†
Food blender	200-300		1/2†
Food warming tray	350	20	7•
Footwarmer	50-100		1†•
Floor polisher	200-400		1•
Freezer, food 5-30 c ft	300-800		30-125†
Freezer, ice cream	50-300		1/2
Freezer	350	90	32*
Freezer, 15 cu ft	440	330	145*
Freezer, 14 cu ft			140*
Freezer, frost free	440	180	57*
Fryer, cooker	1000-1500		5†
Fryer, deep fat	1500	4	6†
Frying pan	1196	12	15†
Furnace, elec. control	10-30		10*
Furnace, oil burner	100-300		25-40*
Furnace, blower	500-700		25-100†
Furnace, stoker	250-600		3-60†
Furnace, fan			32†
Garbage disposal equipment	1/4-1/3 HP		1/2*
Griddle	450-1000		5†
Grill	650-1300		5†
Hair drier	200-1200		1/2-6†
Hair drier	400	5	2†
Heat lamp	125-250		2†
Heater, aux	1320	30	40†•
Heater, portable	660-2000		15-30†•
Heating pad	25-150		1†
Heating pad	65	10	1†
Heat lamp	250	10	3†
Hi Fi Stereo			9†
Hot plate	500-1650		7-30†•
House heating	8000-15,000		1000-2500†
Humidifier	500		5-15*
Iron	1100	12	13†•
Iron			12†•
Iron, 16 hrs. month			13†•
Ironer	1500	12	18†•
Knife sharpener	125		1/4*
Lawnmower	1000	8	8†
Lighting	5-300		10-40—
Lights, 6 room house in winter			60—
Light bulb, 75	75	120	9—
Light bulb, 40	40	120	4.8—
TV, color			100†°
Toaster	1150	4	5†•
Typewriter	30	15	.5*
Vacuum cleaner	600	10	6—
Vacuum cleaner, 1 hr/wk			4—
Vaporizer	200-500		2-5†
Waffle iron	550-1300		1-2†
Washing machine 12 hrs/mo			9†
Washer, automatic	300-700		3-8†
Washer, conventional	100-400		2-4†
Water heater	4474	89	400†
Water heater	1200-7000		200-300†
Water pump (shallow)	1/2 HP		5-20†
Water pump (deep)	1/3-1 HP		10-60†

AT THE BARN

Name	Capacity HP or watts	Est. KWHR
Barn cleaner	2-5 HP	120/yr.†
Clipping	fractional	1/10 per hr.*
Corn, ear crushing	1-5 HP	5 per ton*
Corn, ear shelling	1/4-2	1 per ton†
Electric fence	7-10 watts	7 per mo.*†+
Ensilage blowing	3-5	1/2 per ton†
Feed grinding	1-7½	½-1½ per 100 lbs.†
Feed mixing	1/2-1	1 per ton†
Grain cleaning	1/4-1/2	1 per too but†
Grain drying	1-7½	5-7 per ton†
Grain elevating	1/4-5	4 per 1000 but†
Hay-curing	3-7½	60 per ton†
Hay hoisting	1/2-1	1/3 per ton†
Milking, portable	1/4-1/2	1½ per cow/mo.†
Milking, pipeline	1/2-3	2½ per cow/mo.†
Sheep shearing	fractional	1½ per sheep—
Silo unloader	2-5 HP	4-8 per ton†
Silage conveyer	1-3 HP	1-4 per ton†
Stock tan heater	200-1500 watts	varies widely†
Yard lights	100-500 watts	10 per mo†—
Ventilation	1/6-1/3 HP	2-6 per day† per 20 cows

IN THE MILKHOUSE

Milk cooling	1/2-5 HP	1 per 1000 lbs. milk†
Space heater	1000-3000	800 per year†•

Heater, engine	100-300 watts	1 per 5 hr.†•
Lighting	50-250 watts	4 per mo.—
Lathe, metal	1/4-1 HP	1 per 3 hr.†•
Lathe, wood	1/4-1 HP	1 per 3 hr.†•
Sawing, circular 8" - 10"	1/3-1/2 HP	1/2 per hr.†
Sawing, jig	1/4-1/3 HP	1 per 3 hr.†
Soldering, iron	60-500 watts	1 per 5 hr.†•

MISCELLANEOUS

Farm chore motors 1/2-5		1 per HP per hr.†
Insect trap	25-40 watt	1/3 per night*
Irrigating	1 HP up	1 per HP per hr.†
Snow melting, sidewalk and steps, heating—cable imbedded in concrete	25 watts per sq. ft.	2.5 per 100† sq. ft. per hr.
Soil heating, hotbed	400 watts	1 per day per season†
Wood sawing	1-5 HP	2 per cord†

* AC power required (can be inverted from battery bank with DC to AC Inverter)
† Normally AC - usually convertible to DC
— Normally AC-DC, sometimes newer mfg. items are not
\+ Normally AC but these items can be purchased new to operate of 115V DC
• Older ones were usually AC-DC
° Can be bought battery operated

UNITS AND CONVERSION FACTORS.

Weight

1 kilogram (kg) = 1000 grams (g) = 1,000,000 milligrams (mg) = 1,000,000,000 micrograms (μg)
1 kg = 2.205 pounds (lb); 1 g = 0.035 ounce (oz)
1 lb = 453.6 g = 0.4536 kg
1 metric ton = 2205 lb = 1000 kg
1 short ton = 2000 lb = 907.2 kg
1 megaton = 1,000,000 tons

Length

0.001 kilometer (km) = 1 meter (m) = 100 centimeters (cm) = 1,000 millimeters (mm) = 1,000,000 microns (μ)
1 km = 0.6214 statute mile; 1 m = 39.37 inches (in.) = 3.281 feet (ft); 1 cm = 0.3937 in.
1 mile = 1.609 km; 1 ft = 0.3048 m; 1 in. = 2.54 cm

Area

1 hectare = 10,000 square meters (m^2)
1 hectare = 2.47 acres = 0.003861 square mile
1 square mile = 640 acres = 259 hectares

Volume and Cubic Measure

1 cubic meter (m^3) = 1,000,000 cubic centimeters (cm^3)
1 m^3 = 35.31 cubic feet (ft^3); 1 cm^3 = 0.061 cubic inch ($in.^3$)
1 liter (l) = 1000 cm^3
1 l = 61.02 $in.^3$ = 0.2642 gallon (gal)
1 ft^3 = 0.02832 m^3 = 28.32 l; 1 gal = 231 in^3 = 3.785 l

Energy and Work

1 British thermal unit (Btu) = 252 calories (cal) = 0.0002931 kilowatt-hour = 1055 joules (kWh)
1 joule = 1 watt-sec = 10^7 erg = 0.73 ft lb
1 gm-cal = 4.184 joules
1 kWh (kilowatt-hour) = 3.6 $\times$ 10^6 joules = 3412 Btu
1 watt-yr = 3.15 $\times$ 10^7 joules = 3 $\times$ 10^4 Btu
1 hp = 550 ft lb/sec = 746 watt

Power

1 megawatt (MW) = 1000 kilowatts (kW) = 1,000,000 watts (W) = 3,413,000 Btu/hour (Btu/hr) = 1341 horsepower (hp)

Pressure

1 atmosphere (atm) = 76 cm mercury = 14.70 lb/$in.^2$ = 1,013 millibars (mb)

Temperature scales

	Absolute Zero	Ice Point (water)	Steam Point (water)
Degrees Fahrenheit (°F)	-459.7	32	212
Degrees Celsius or Centigrade (°C)	-273.15	0	100
Degrees Kelvin (°K)	0	273.15	373.15
Degrees Réaumur (°R)		0	80

CONVERSION FACTOR CROSS-REFERENCE

Conversion Factors

To Convert	Into	Multiply by
acres	ft^2	43,560
acres	m^2	4,047
acres	hectares	0.4047
British thermal unit (Btu)	cal	252
Btu	joules (J)	1,055
Btu	kilowatt-hours (kWh)	2.93×10^{-4}
Btu	megawatt-years (MW yr)	3.34×10^{-11}
Btu/ft^2	langleys (cal/cm^2)	0.271
calories (cal)	Btu.	3.97×10^{-3}
cal	ft lb	3.09
cal	J	4.18
cal/minute (cal/min)	watts (W)	0.0698
centimeters (cm)	inches (in.)	0.394
feet (ft)	meters	0.305
foot-pounds (ft lb)	cal	0.324
ft lb	J	1.36
ft lb	kilogram-meters (kg m)	0.138
ft lb	kWh	3.77×10^{-7}
gallons (gal)	liters	3.79
gal	lb	0.00220
gigawatts (GW)	W	1×10^9
hectares	acres	2.47
horsepower (hp)	kW	0.745
J	Btu	9.48×10^{-4}
J	cal	0.239
J	ft lb	0.738
J	W hr	2.78×10^{-4}
1000 cal (Kcal)	Btu	3.97
Kcal/min	kW	0.0698
kg m	ft lb	7.23
kg	lb	2.20
kg	tons	0.00110
kilowatts (kW)	hp	1.34
kWh	Btu	3,413
kWh	ft lb	2.66×10^6
kWh	W hr	1×10^3
kW	Kcal/min	14.3
langleys	Btu/ft^2	3.69
langleys/min	W/sc cm	0.0698
megawatt-hours	Btu	3.41×10^6
MW yr	Btu	2.99×10^{10}
miles	meters	1,609
cm^2	ft^2	0.00108
cm^2	$in.^2$	0.155
ft^2	$in.^2$	144
ft^2	m^2	0.0929
$in.^2$	cm^2	6.45
m^2	ft^2	10.8
m^2	$mile^2$	3.86×10^{-7}
$mile^2$	acres	640
$mile^2$	ft^2	2.79×10^7
$mile^2$	m^2	2.59×10^6
tons (short)	kg	907
tons (short)	lb	2000
tons (metric)	tons (short)	1.1025
tons (short)	tons (metric)	0.907
W hr	J	3600
°F	°C	subtract 32 and multiply by 0.555
°C	°F	multiply by 1.8 and add 32
cal/cm^2-sec-°C	Btu/ft^2-hr-°F	7380
Btu/ft^2-hr-°F	cal/cm^2-sec-°C	1.35×10^{-4}

BTU EQUIVALENTS OF FUELS

Btu Equivalents of Common Fuels[1]

Fuel	Common Measure	Btus
Crude oil	barrel (bbl.)–42 gallons	5,800,000
Natural gas	1000 cubic feet	1,035,000
Natural gas	therm	100,000
Coal (bituminous)	ton (short ton -907.2 kg)	26,000,000
Coal (lignite)	ton	14,000,000
Gasoline	gallon	124,000
Electricity	kilowatt-hour (kWh)	3,412[2]

[1]A Btu is the amount of heat required to raise the temperature of 1 lb of water 1°F.

[2]Because of conversion losses in the generation of electric power from heat, about 10,000 Btu are required to produce one kilowatt-hour.

Energy Consumption Factors in Btus

8000 Btu = energy/day needed to sustain 1 man = 2,000,000 calories
50×10^3 Btu = daily per capita consumption by primitive agricultural man
1.0×10^6 Btu = daily per capita consumption by 1973 American
70×10^6 Btu = annual consumption by American family car
80×10^6 Btu = annual consumption for heating typical American house
75×10^{15} Btu = United States gross consumption in 1973 = 2.5×10^{12} watt-yr.
250×10^{15} Btu = world annual gross consumption (1973)

Equivalent Factors for Energy Consumption in Quad Units (10^5 Btu)

1 quad = 10^{15} Btu = 10^{-3} Q-units
= 33×10^9 watt-yr.
= 1.0×10^{18} joules
= 250×10^{15} calories
= 3.0×10^{12} cubic feet of hydrogen
= 1.0×10^{12} cubic feet of natural gas = 1 TCF (trillion cubic feet)
= 300×10^9 kWh–(thermal) = 100×10^9 kWh (electrical)
= 8.0×10^9 gallons of gasoline
= 7.0×10^9 gallons of crude oil
= 170×10^6 barrels of crude oil[a]
= 50×10^6 short tons subbituminous coal
= 40×10^6 short tons bituminous coal
= 4.0×10^6 pounds of uranium (light-water reactor LWR)
= 50×10^3 pounds of uranium (in a nuclear breeder with 60% efficiency)

[a]One million bbl per day corresponds to 2.12 quad per year.

HEAT CONTENT OF FUELS IN BTUS

	Btu	Per Unit
Coal		
Anthracite (Pa.)	25,400,000	ton
Bituminous	26,200,000	ton
Blast furnace gas	100	ft^3
Briquettes and package fuels	28,000,000	ton
Coke	24,800,000	ton
Coke-breeze	20,000,000	ton
Coke-oven gas	550	ft^3
Coal tar	150,000	gal.
Coke-oven and manufactured gas products, light oils	5,460,000	bbl.
Natural gas (dry)	1,035	ft^3
Natural gas liquids (average)	4,011,000	bbl.
Butane	4,284,000	bbl.
Propane	3,843,000	bbl.
Petroleum:		
Asphalt	6,640,000	bbl.
Coke	6,024,000	bbl.
Crude oil	5,800,000	bbl.
Diesel	5,806,000	bbl.
Distillate fuel oil	5,825,000	bbl.
Gasoline, aviation	5,048,000	bbl.
Gasoline, motor fuel	5,253,000	bbl.
Jet fuel:		
Commercial	5,670,000	bbl.
Military	5,355,000	bbl.
Kerosene	5,670,000	bbl.
Lubricants	6,060,000	bbl.
Miscellaneous oils	5,588,000	bbl.
Refinery still gas	5,600,000	bbl.
Heavy fuel oil	6,287,000	bbl.
Road oils	6,640,000	bbl.
Wax	5,570,000	bbl.
Shale oil	5,800,000	bbl.
Uranium:		
Total contained energy	60,000,000,000,000	short ton U_3O_8
Energy available with present technology	500,000,000,000	short ton U_3O_8

Sources: The U.S. Energy Problem, Volume II (NTIS), 1972, p. A-82.
The U.S. Energy Dilemma (Illinois State Geological Survey), 1973, p. 28.

ENERGY UNIT CONVERSION CHART*

CUBIC FEET NATURAL GAS** (CF)	BARRELS OIL (bbl)	SHORT TONS BITUMINOUS COAL (T)	BRITISH THERMAL UNITS (Btu)	KILOWATT HOURS ELECTRICITY (kWhr)
–	–	–	1	0.000293
1	0.00018	0.00004	1000	0.293
3.41	0.00061	0.00014	3413	1
1000 (1 MCF)	0.18	0.04	1 MILLION	293
3413	0.61	0.14	3.41 MILLION	1000 (1 MWhr)
5600	1	0.22	5.6 MILLION	1640
25,000	4.46	1	25 MILLION	7325
1 MILLION (1 MMCF)	180	40	1 BILLION	293,000
3.41 MILLION	610	140	3.41 BILLION	1 MILLION (1 GWhr)
1 BILLION (1 BCF)	180,000	40,000	1 TRILLION	293 MILLION
1 TRILLION (1 TCF)	180 MILLION	40 MILLION	1 QUADRILLION (QUAD) (Q)	293 BILLION

* Based on the following nominal fuel heating values:
1 Cubic Foot Natural Gas - 1000 Btu
1 Barrel Crude Oil - 5.6 Million Btu
1 Pound Bituminous Coal - 12,500 Btu

** Substitute Natural Gas (SNG) and Liquefied Natural Gas (LNG) will have approximately the same heating value.

Institute of Gas Technology, 3424 South State Street, IIT Center Chicago, Ill. 60616.

CONVERSION FACTORS FOR OIL

Energy Equivalents of Oil

One 42 gallon barrel of oil	5.8×10^6 Btu
	5.6×10^3 cubic feet of natural gas
	1.70 thermal megawatt-hours
	0.58 electrical megawatt-hours[1]
	0.232 tons bituminous coal
	0.42 tons lignite
	6.119×10^9 joules
One million barrels of oil per day	2.1×10^{15} Btu/year
	2.1 quad/year
	0.365 billion barrels/year
	232,000 tons bituminous coal/day
	420,000 tons lignite/day
	5.6 trillion cubic feet (TCF) natural gas/day
	580 million kilowatt hours of electricity/day[1]

[1] At a heat rate of 10,000 Btu per kWh of electricity

Approximate Calorific Equivalents

One million tonnes of oil equals approximately:		Heat units and Other fuels expressed in terms of million tonnes of oil.		
Heat Units				million tonnes of oil
39	million million Btu	10	million million Btu approximates to	0 26
395	million therms	100	million therms approximates to	0 25
10 000	Teracalories	10 000	Teracalories approximates to	1 00
Solid Fuels				
1·5	million tonnes of coal	1	million tonnes of coal approximates to	0·67
4·9	million tonnes of lignite	1	million tonnes of lignite approximates to	0·20
3·3	million tonnes of peat	1	million tonnes of peat approximates to	0·30
Natural Gas	(1 cub. ft = 1 000 Btu) (1 cub. metre = 9 000 kcal)			
1·167	thousand million cub. metres	1	thousand million cub. metres approximates to	0·86
41·2	thousand million cub. ft.	10	thousand million cub. ft. approximates to	0·24
113	million cub. ft./day for a year	100	million cub. ft./day for a year approximates to	0·88
Town Gas	(1 cub. ft. = 470 Btu) (1 cub. metre = 4 200 kcal)			
2·5	thousand million cub. metres	1	thousand million cub. metres approximates to	0·40
88 3	thousand million cub. ft.	10	thousand million cub. ft. approximates to	0·11
242	million cub. ft./day for a year	100	million cub. ft./day for a year approximates to	0·41
Electricity	(1 kWh = 3 412 Btu) (1 kWh = 860 kcal)			
12	thousand million kWh	10	thousand million kWh approximates to	0·82

One million tonnes of oil produces about 4,000 million units (kWh) of electricity in a modern power station.

NOTE

In previous years, the BP Statistical Review has employed a conversion for other energy sources into oil equivalent of 10 500 kcal/kg. This year a value of 10 000 kcal/kg has been employed which downrates the heating value of oil, thus slightly increasing the oil equivalent value of other fuels.

Source: The British Petroleum Company Limited, Britannic House, Moor Lane, London EC2Y 9BU.

ELECTRICAL UNITS

Power. The basic unit of power is the watt. For an electrical appliance the power rating is found by multiplying the voltage by the current (in amperes). Thus, a 125 volt appliance drawing 10 amperes has a power rating of 1250 watts. The kilowatt is simply 1000 watts and a megawatt is a million watts (10^6 watts). The horsepower is equivalent to 746 watts.

We can summarize these facts as follows:

$$\text{number of watts} = \text{voltage} \times \text{current}$$
$$1 \text{ kilowatt} = 1000 \text{ watts} = 10^3 \text{ watts}$$
$$1 \text{ megawatt} = 1{,}000{,}000 \text{ watts} = 10^6 \text{ watts}$$
$$1 \text{ horsepower} = 1 \text{ hp} = 746 \text{ watts}$$

Energy. The basic unit for energy is the kilowatt-hour (kWh), which is the energy used when a device rated at 1000 watts operates for an hour (or a 100 watt appliance operates for 10 hours). The table below gives various equivalents.

Unit or Process	Number of kWh
Calorie (diet)	0.0012 (1.2×10^{-3})
Calorie (ordinary)	1.2×10^{-6}
Btu	2.9×10^{-4}
One hour manual labor	0.06
Combustion of 1 gallon of gasoline	38.3
Heat one gallon of water 1°F	0.0024

GROWTH RATE UNITS

Multiplication Factors for Energy Consumption at Various Growth Rates (Reference Year 1974)

Growth Rate per Year (%)	1974	1979	1984	1999
0.5	1.00	1.025	1.051	1.133
1.0	1.00	1.051	1.105	1.284
1.5	1.00	1.078	1.162	1.455
2.0	1.00	1.105	1.221	1.650
2.5	1.00	1.133	1.284	1.868
3.0	1.00	1.162	1.350	2.117
4.0	1.00	1.217	1.480	2.666
5.0	1.00	1.276	1.629	3.386

For example: if energy consumption is 75 quad in 1974, then it will be (75×2.117) in 1999 if the growth rate is 3% per year.

METRIC TERMINOLOGY

The following terms are frequency used in employing the metric system of units.

1 billion	1000 million = 10^9
1 trillion	1000 billion = 10^{12}
1 kcal	1 kilocalorie = 3.968 Btu (British thermal units)
1 therm	10^5 Btu = 25,200 kcal
1 thermie	1000 kcal
1 metric ton	1000 kilograms = 0.985 ton = 1.1023 short ton
1 toe	1 ton of oil equivalent = 10^7 kcal
1 Mtoe	1 million metric tons of oil equivalent = 10^{13} kcal
1 tce	1 ton of coal equivalent = 0.7 toe
1 Mtce	1 million tons of coal equivalent = 0.7 Mtoe
1 cubic foot	0.0283 cubic meter
1 Mcf	1 thousand cubic feet = 28.3 cubic meters
1 TCF	1 trillion cubic feet = 28.3 X 10^9 cubic meters
$1 per Mcf	is approximately equivalent to $40 per toe
f.o.b.	free on board
c.i.f.	cost including insurance and freight
GNP	Gross National Product
GDP	Gross Domestic Product

Source: Energy Prospects to 1985, Organization for Economic Co-operation and Development, Paris, France, 1974.

SYMBOLS

A Projected rotor disc area
a Axial interference at the rotor $a \equiv u/V_\infty$
a' Tangential interference factor at the rotor $a\ \omega/2\Omega$
B Number of blades
b Axial interference factor in the wake $b \equiv u_1/V_\infty$
c Blade chord
C_L Sectional lift coefficient
C_D Sectional drag coefficient
C_x Sectional force coefficient in the direction of rotation
C_y Sectional force coefficient normal to the plane of rotation
C_p Power coefficient, $P/(1/2)\rho A V_\infty^3$ *or* $P/(1/2)\rho S V_\infty^3$
D' Sectional drag force per unit length
E Lift to drag ratio, L/D
L' Sectional lift force per unit length
m Mass flow rate
P Power extracted from the air
p Pressure
R Rotor radius
S Rotor or translator projected surface area
r Local rotor radius
u Axial flow velocity at the rotor
u_1 Axial flow velocity in the wake
V_∞ Freestream wind velocity
v Translator velocity
W Resultant velocity relative to the rotor element
X Tip speed ratio, $R\Omega/V_\infty$
x Local speed ratio, $r\Omega/V_\infty$

Greek

α Angle of attack
ρ Fluid density
Ω Rotor angular velocity
ω Fluid angular velocity downwind of the rotor
Φ Angle between the plane of rotation and the relative velocity
β Angle between the wind and the normal to the translation velocity
(Chapter 2); blade pitch angle (Chapters 5, 6)
θ Blade pitch angle (Chapter 3, 4); blade rotation angle (Chapters 5, 6)
η Wind-height relation exponent

Monthly Average Windpower in the United States and Southern Canada

The following table* lists the wind power at 750 locations in the U.S. and Southern Canada. The data in the table have been extracted from Sandia Laboratories Report SAND-74-0348, *Wind Power Climatology in the United States.*

The wind data locations are listed by region within each state. The states are listed first, alphabetically, then the Southern Canadian provinces. Listed on each line are the following:

1. State—U.S. Postal abbreviation (obvious abbreviations of Canadian provinces).
2. Location—the most common abbreviations are:
   APT (airport)
   AFB (Air Force Base)
   AFS (Air Field Station)
   IAP (International Airport)
   IS (island)
   NAF (Naval Air Field)
   PT (point)
   WBO (Weather Bureau Office).
3. International station number.
4. Latitude in degrees and minutes (3439 = 34° 39′ N).
5. Longitude in degrees and minutes (8646 = 86° 46′ W).
6. Average wind speed in knots (multiply by 1.15 to convert to mph.
7. Twelve average monthly wind power values in W/m² (multiply by 0.0929 to convert to W/ft²).
8. Average of the previous twelve monthly power values.

NOTE: These data have not been corrected for varying heights of the wind anemometer. Also, possible distortions in the wind pattern by natural terrain features, trees and buildings are not accounted for. Because of this, no particular set of these data can be blindly accepted as representative of a particular region.

**Courtesy of DOE.*

MONTHLY AVERAGE WIND POWER IN THE UNITED STATES AND SOUTHERN CANADA

State	Location	Sta. No.	Lat	Long	Ave. Speed knots	J	F	M	A	M	J	J	A	S	O	N	D	Ave.
						Wind Power, Watts per Square Meter												
AL	Huntsville	3856	3439	8646	6.6	80	109	118	87	48	37	29	31	56	50	78	87	66
AL	Foley	93826	3358	8605	8.0	133	182	153	174	142	106	73	66	109	96	116	115	122
AL	Gadsden	75258	3358	8605	5.8	84	104	114	99	43	34	25	21	40	45	63	53	61
AL	Birmingham APT	13876	3334	8645	7.3	127	157	156	137	80	64	49	44	68	68	108	106	97
AL	Tuscaloosa, Vn D Graf APT	93806	3314	8737	5.1	79	79	93	69	33	21	15	21	28	36	55	69	49
AL	Selma, Craig AFB	13850	3221	8659	5.7	74	86	91	68	42	34	29	26	37	31	48	54	51
AL	Montgomery	13895	3218	8624	6.1	74	90	85	68	39	35	34	26	39	36	51	62	53
AL	Montgomery, Maxwell AFB	13821	3223	8621	4.8	58	67	69	51	28	25	20	19	27	26	39	45	39
AL	Ft. Rucker, Cairns AAF	3850	3116	8543	4.7	40	50	55	41	23	18	12	12	19	19	29	35	30
AL	Evergreen	13885	3125	8702	5.3	66	69	78	57	29	18	17	17	21	24	38	51	40
AL	Mobile, Brookley AFB	13838	3038	8804	7.3	105	104	128	119	94	58	43	41	69	51	72	89	80
AK	Annette IS	25308	5502	13134	9.5	320	264	216	199	110	97	71	77	128	297	324	319	199
AK	Ketchikan	952	5521	13139	5.8	59	53	42	52	47	37	38	44	43	67	75	75	57
AK	Craig	25317	5529	13309	7.9	185	159	167	132	82	95	71	55	113	186	174	165	128
AK	Petersburg	960	5649	13257	3.7	26	40	37	41	32	23	22	22	24	29	22	21	33
AK	Sitka	961	5703	13520	3.5	109	26	34	42	27	23	22	14	34	44	46	93	37
AK	Juneau APT	25309	5822	13435	7.5	119	134	123	127	95	70	60	67	108	170	157	159	115
AK	Haines	955	5914	13526	8.0	218	202	203	148	74	61	94	54	72	160	238	159	146

MONTHLY AVERAGE WIND POWER IN THE UNITED STATES AND SOUTHERN CANADA (Continued)

State	Location	Sta. No.	Lat	Long	Ave. Speed knots	Wind Power, Watts per Square Meter J	F	M	A	M	J	J	A	S	O	N	D	Ave.
AK	Yakutat APT	25339	5931	13940	7.0	177	144	114	100	90	71	56	64	96	181	183	169	114
AK	Middleton IS AFS	25403	5927	14619	11.9	625	597	468	355	238	141	96	134	243	519	582	608	376
AK	Cordova, Mile 13 APT	26410	6430	14530	4.4	46	48	42	41	37	23	18	17	32	53	47	48	36
AK	Valdez	26442	6107	14616	4.3	72	28	75	41	36	16	13	7	7	45	100	72	53
AK	Anchorage IAP	26451	6110	15001	5.9	61	95	48	61	108	76	61	52	46	38	38	50	61
AK	Anchorage, Merrill Fld	26409	6113	14950	4.9	57	66	29	27	41	40	23	22	30	30	59	23	37
AK	Anchorage, Elmendorff AFB	26401	6115	14948	4.4	46	60	50	41	40	34	24	22	26	30	46	33	36
AK	Kenai APT	26523	6034	15115	6.6	96	109	94	66	61	63	56	54	53	83	85	80	74
AK	Northway APT	26412	6257	14156	3.9	16	21	30	44	40	42	33	32	27	22	18	16	28
AK	Gulkana	26425	6209	14527	5.8	45	88	85	105	111	98	83	100	95	76	48	40	81
AK	Big Delta	26415	6400	14544	8.2	447	322	239	147	148	85	68	102	163	209	300	333	215
AK	Fairbanks IAP	26411	6449	14752	4.3	10	16	25	37	50	44	33	29	28	22	13	10	27
AK	Fairbanks, Ladd AFB	26403	6451	14735	3.5	10	17	24	28	38	35	23	29	23	24	12	9	23
AK	Ft. Yukon APT	26413	6634	14516	6.7	30	41	64	81	91	84	86	81	74	52	31	31	64
AK	Nenana APT	26435	6433	14905	5.1	68	44	48	45	46	34	27	26	33	42	45	44	42
AK	Manley Hot Springs	567	6500	15039	4.8	76	54	34	109	93	89	53	42	63	104	62	52	62
AK	Tanana	976	6510	15206	6.6	92	85	89	83	56	53	47	29	50	64	56	80	73
AK	Ruby	508	6444	15526	6.5	58	133	119	84	40	51	46	42	64	76	119	54	78
AK	Galena APT	26501	6444	15656	5.4	56	69	66	77	51	53	48	61	61	59	59	49	59
AK	Kaltag	458	6420	15845	4.7	46	103	26	81	31	33	30	28	37	56	40	51	56

MONTHLY AVERAGE WIND POWER IN THE UNITED STATES AND SOUTHERN CANADA (Continued)

State	Location	Sta. No.	Lat	Long	Ave. Speed knots	Wind Power, Watts per Square Meter J	F	M	A	M	J	J	A	S	O	N	D	Ave
AK	Unalakleet APT	26627	6353	16048	10.5	520	502	336	191	112	96	116	146	175	234	395	376	265
AK	Moses Point APT	26620	6412	16203	10.6	329	363	275	279	149	129	181	233	217	222	246	263	24[illegible]
AK	Golovin	502	6433	16302	9.6	188	229	246	250	142	117	178	264	271	258	269	256	236
AK	Nome APT	26617	6430	16526	9.7	328	308	228	225	153	119	117	162	189	230	263	238	21[illegible]
AK	Northeast Cape AFS	26632	6319	16858	11.0	468	263	246	347	239	137	218	240	288	462	632	387	328
AK	Tin City AFS	26634	6534	16755	15.0	763	919	811	658	427	271	260	334	352	522	722	728	549
AK	Kotzebue	26616	6652	16238	11.2	455	418	310	294	161	187	212	234	228	270	397	366	291
AK	Cape Lisburne AFS	26631	6853	16608	10.5	432	268	335	266	227	210	303	216	266	432	444	333	314
AK	Indian Mountain AFS	26535	6600	15342	5.4	115	113	88	58	57	37	30	36	52	86	95	104	70
AK	Bettles APT	26533	6655	15131	6.3	28	44	57	62	66	62	44	38	43	43	43	47	48
AK	Wiseman	979	6726	15013	3.1	28	26	16	15	24	15	23	14	12	12	26	16	2[illegible]
AK	Umiat	26537	6922	15208	6.0	113	121	43	77	80	93	62	53	57	51	121	78	76
AK	Point Barrow	27502	7118	15647	10.5	215	194	162	167	169	143	145	208	211	258	286	183	19[illegible]
AK	Barier IS	26401	7008	14338	11.3	512	468	379	279	216	145	123	208	287	470	486	425	341
AK	Sparrevohn AFS	26534	6106	15534	4.7	69	73	108	76	47	35	36	41	54	63	74	92	63
AK	McGrath	26510	6258	15537	4.2	13	27	29	37	39	35	34	35	32	24	15	12	27
AK	Tataline AFS	26536	6253	15557	4.4	25	37	36	37	38	27	27	29	33	35	24	21	31
AK	Flat	16	6229	15805	8.1	206	266	205	150	116	100	81	108	143	168	185	184	17[illegible]
AK	Aniak	26516	6135	15932	5.6	51	59	63	59	49	37	27	34	41	47	47	42	40
AK	Bethel APT	26615	6047	16148	9.8	229	258	224	166	125	108	110	137	140	158	185	211	17

MONTHLY AVERAGE WIND POWER IN THE UNITED STATES AND SOUTHERN CANADA (Continued)

State	Location	Sta. No.	Lat	Long	Ave. Speed knots	J	F	M	A	M	J	J	A	S	O	N	D	Ave.
						Wind Power, Watts Per Square Meter												
AK	Cape Romanzof AFS	26633	6147	16602	11.7	692	699	493	476	246	124	110	154	234	305	520	654	380
AK	Cape Newenham AFS	25623	5839	16204	9.8	400	371	330	288	168	119	101	142	165	212	300	315	241
AK	Kodiak FWC	25501	5744	15231	8.8	328	271	258	198	124	87	52	77	120	210	294	329	189
AK	King Salmon APT	25503	5841	15639	9.2	250	260	235	180	182	138	92	139	156	180	230	206	191
AK	Port Heiden APT	25508	5657	15837	12.9	576	564	493	361	289	273	225	381	466	451	439	565	429
AK	Port Mollor	25625	5600	16031	8.8	158	168	171	195	135	81	108	144	164	222	260	219	172
AK	Cold Bay APT	25624	5512	16243	14.6	736	731	699	580	506	465	428	507	462	606	652	631	573
AK	Dutch Harbor NS	25611	5353	16632	9.6	355	376	295	223	135	125	69	105	169	390	419	266	233
AK	Driftwood Bay	25515	5358	16651	8.0	204	203	154	148	115	72	88	77	71	120	161	182	131
AK	Umnak IS, Cape AFB	25602	5323	16754	13.5	651	688	577	514	454	251	163	249	466	603	606	723	497
AK	Nikolski	25626	5255	16847	14.0	538	560	532	566	437	321	239	283	361	634	732	662	482
AK	Adak	25704	5153	17638	12.2	426	467	528	453	366	223	218	258	331	502	481	525	404
AK	Amchitka IS	45702	5123	17915	18.0	1764	1517	1418	1062	653	448	405	457	740	1053	1165	1569	1025
AK	Attu IS	45709	5250	17311	11.2	553	582	508	403	235	162	135	129	360	366	414	554	368
AK	Shemya APT	45715	5243	17406	15.7	887	932	878	641	483	266	235	285	432	301	977	870	633
AK	St. Paul IS	25713	5707	17016	15.0	758	867	684	518	355	207	175	282	399	693	691	791	547
AZ	Grand Canyon	378	3557	11209	6.2	38	43	49	71	66	55	35	31	57	58	44	28	49
AZ	Winslow APT	23194	3501	11044	7.3	104	104	232	169	161	141	93	77	63	73	63	78	111
AZ	Flagstaff, Pulliam APT	3103	3508	11140	6.4	71	70	96	95	93	86	40	33	52	56	69	69	69
AZ	Maine	178	3509	11157	8.9	132	186	218	253	240	224	111	68	116	178	139	158	151

MONTHLY AVERAGE WIND POWER IN THE UNITED STATES AND SOUTHERN CANADA (Continued)

State	Location	Sta. No.	Lat	Long	Ave. Speed knots	Wind Power, Watts Per Square Meter J	F	M	A	M	J	J	A	S	O	N	D	Ave.
AZ	Ashfork	171	3514	11233	7.5	111	116	154	201	142	126	82	68	86	100	103	82	114
AZ	Kingman	381	3516	11357	8.9	126	156	172	203	153	166	126	99	102	124	115	107	138
AZ	Prescott	23184	3439	11226	7.5	54	95	117	144	138	124	75	56	67	57	59	44	85
AZ	Yuma APT	23195	3240	11436	6.8	55	62	68	77	71	69	93	77	45	40	55	51	62
AZ	Phoenix	23183	3326	11201	4.8	16	28	34	39	37	35	43	31	28	24	22	17	29
AZ	Phoenix, Luke AFB	23111	3332	11223	4.6	21	31	41	52	49	43	49	39	25	21	20	18	34
AZ	Chandler, Williams AFB	23104	3318	11140	4.1	17	21	28	35	34	33	41	33	28	22	18	16	26
AZ	Tucson APT	23160	3207	11056	7.3	71	59	69	90	87	73	75	54	62	78	82	72	74
AZ	Tucson	23160	3207	11056	7.1	71	59	69	90	87	73	75	54	62	78	82	72	74
AZ	Tucson, Davis-Monthan AFB	23109	3210	11053	5.7	48	48	57	63	56	60	51	35	42	40	43	45	49
AZ	Ft. Hauchuca	3124	3134	11020	5.7	49	58	84	96	80	66	39	27	30	30	34	41	53
AZ	Douglas	93026	3128	10937	6.4	84	94	166	143	128	86	61	46	47	63	62	75	87
AR	Walnut Ridge APT	93991	3608	9056	6.0	93	81	103	104	58	44	27	22	28	43	64	75	62
AR	Blytheville AFB	13814	3558	8957	6.4	85	106	108	111	66	41	26	25	36	39	67	71	65
AR	Ft. Smith APT	13964	3520	9422	7.4	76	86	116	104	81	62	51	45	50	55	66	75	73
AR	Little Rock	13963	3444	9214	7.6	82	91	105	96	70	58	46	46	48	50	73	71	70
AR	Jacksonville, Ltl. Rk. AFB	3930	3455	9209	5.8	61	67	85	70	47	34	28	24	28	29	45	49	48
AR	Pine Bluff, Grider Fld	93988	3410	9156	6.5	102	89	102	87	51	39	31	29	35	45	71	81	64
AR	Texarkana, Webb Fld	13977	3327	9400	7.7	92	108	128	115	77	69	48	49	62	61	74	87	80
CA	Needles APT	23179	3446	11437	6.7	108	125	128	112	108	98	67	67	58	78	113	124	97

MONTHLY AVERAGE WIND POWER IN THE UNITED STATES AND SOUTHERN CANADA (Continued)

State	Location	Sta. No.	Lat	Long	Ave. Speed knots	J	F	M	A	M	J	J	A	S	O	N	D	Ave.
						Wind Power, Watts Per Square Meter												
CA	El Centro NAAS	23199	3249	11541	7.7	98	126	171	208	225	189	80	73	79	86	98	76	127
CA	Thermal	3104	3338	11610	9.1	66	79	103	149	191	153	125	114	119	92	76	63	111
CA	Imperial Bch., Ream Fld	93115	3234	11707	5.9	48	51	54	54	52	45	35	32	30	31	43	42	43
CA	San Diego, North IS	93112	3243	11712	5.3	32	41	56	56	49	41	33	32	35	31	30	32	39
CA	San Diego	23188	3244	11710	5.4	30	33	40	47	47	40	30	29	27	25	22	21	31
CA	Miramar NAS	93107	3252	11707	4.4	23	24	28	30	26	19	16	17	18	19	20	24	22
CA	San Clemente IS NAS	93117	3301	11835	6.3	53	72	89	97	67	48	33	32	33	32	54	69	55
CA	San Nicholas IS	93116	3315	11948	9.9	152	199	295	306	348	244	161	166	164	140	180	159	209
CA	Camp Pendleton	3154	3313	11724	5.2	30	35	45	61	53	43	43	44	36	24	28	29	38
CA	Oceanside	189	3318	11721	8.0	129	122	108	82	67	59	49	49	64	66	96	116	87
CA	Laguna Beach	195	3332	11747	5.0	35	38	44	37	30	30	27	26	25	25	22	32	34
CA	El Toro MCAS	93101	3340	11744	4.8	45	38	33	30	26	22	19	19	19	23	36	43	28
CA	Santa Ana MCAF	93114	3342	11750	4.6	43	43	47	46	37	31	30	26	25	26	36	43	37
CA	Los Alimitos NAS	93106	3348	11807	4.8	36	39	47	44	41	32	28	25	22	23	36	37	34
CA	Long Beach APT	23129	3349	11809	4.9	27	40	45	48	43	35	34	32	31	27	29	26	35
CA	Los Angeles IAP	23174	3356	11824	5.9	40	57	69	70	63	49	43	44	39	36	38	35	48
CA	Ontario	93180	3404	11737	7.7	36	117	109	118	148	124	135	135	92	71	46	127	103
CA	Riverside, March AFB	23119	3353	11715	4.4	35	43	40	44	49	51	52	49	37	28	29	32	41
CA	San Bernardino, Norton AFB	23122	3406	11715	3.5	43	43	33	27	25	22	22	20	19	17	28	29	28
CA	Victorville, George AFB	23131	3435	11723	7.7	99	134	170	183	163	135	87	85	74	70	87	90	118

MONTHLY AVERAGE WIND POWER IN THE UNITED STATES AND SOUTHERN CANADA (Continued)

State	Location	Sta. No.	Lat	Long	Ave. Speed knots	Wind Power, Watts Per Square Meter J	F	M	A	M	J	J	A	S	O	N	D	Ave.
CA	Daggett	23161	3452	11647	9.6	94	173	315	290	355	236	177	159	145	121	107	74	187
CA	China Lake, Inyokern NAF	93104	3541	11741	7.1	121	156	238	249	225	186	124	126	113	124	103	93	155
CA	Muroc, Edwards AFB	23114	3455	11754	7.9	90	118	187	206	236	230	155	131	99	87	82	83	141
CA	Palmdale	81	3438	11806	10.2	163	205	226	267	315	328	254	200	165	158	130	109	225
CA	Palmdale APT	23182	3438	11805	8.8	121	146	233	234	234	229	173	141	107	104	113	132	163
CA	Saugus	83	3423	11832	6.3	105	128	88	96	96	108	101	88	67	76	105	90	89
CA	Van Nuys	23130	3413	11830	4.6	105	82	66	50	43	21	22	19	18	22	90	69	49
CA	Oxnard AFB	23136	3413	11905	4.4	63	56	49	46	43	26	20	19	19	31	50	76	41
CA	Point Mugu NAS	93111	3407	11907	5.6	100	79	71	78	51	33	28	26	28	35	76	82	55
CA	Santa Maria	23273	3454	12027	6.5	75	80	114	94	93	93	63	57	56	66	79	91	82
CA	Vandenberg, Cooke AFB	93214	3444	12034	6.1	62	67	99	97	115	67	34	33	41	51	58	58	65
CA	Pt. Arguello	93215	3440	12035	7.2	72	105	138	135	133	79	58	54	51	74	76	66	85
CA	San Louis Obispo	93206	3514	12039	6.9	60	69	134	127	146	173	105	120	131	129	89	73	115
CA	Estero	395	3526	12052	4.3	83	66	69	76	60	50	22	31	42	47	44	77	53
CA	Paso Robles, Sn Ls Obispo	93209	3540	12038	5.5	34	39	57	76	105	127	106	83	59	42	32	30	64
CA	Jolon	93218	3600	12114	2.8	9	6	10	6	11	11	8	8	6	4	6	6	7
CA	Monterey NAF	23245	3635	12152	5.0	30	33	45	48	51	45	35	32	23	21	20	30	35
CA	Ft. Ord, Fritzsche AAF	93217	3641	12146	5.7	30	31	46	61	67	63	66	59	41	34	25	24	47
CA	Taft, Gardner Fld.	23126	3507	11918	4.4	20	18	20	29	45	46	31	22	18	16	18	17	26
CA	Bakersfield, Meadows Fld.	23155	3525	11903	5.4	27	33	46	55	69	65	47	43	34	25	24	28	41

MONTHLY AVERAGE WIND POWER IN THE UNITED STATES AND SOUTHERN CANADA (Continued)

State	Location	Sta. No.	Lat	Long	Ave. Speed knots	Wind Power, Watts Per Square Meter J	F	M	A	M	J	J	A	S	O	N	D	Ave.
CA	Bakersfield, Minter Fld.	23102	3530	11911	5.0	26	31	38	49	61	73	38	25	22	21	19	25	34
CA	Lemoore NAS	23110	3620	11957	4.8	21	30	38	40	45	47	35	29	25	27	18	19	30
CA	Fresno, Hammer Fld.	93193	3646	11943	5.5	24	28	42	48	60	62	42	33	25	23	17	20	35
CA	Bishop APT	23157	3722	11822	7.5	74	106	161	145	129	100	80	81	85	101	88	80	103
CA	Merced, Castle AFB	23203	3722	12034	6.0	56	66	72	74	69	78	59	52	44	44	34	42	59
CA	Livermore	196	3742	12147	7.9	109	108	115	124	158	180	173	143	107	85	64	71	122
CA	San Jose APT	23293	3722	12155	6.4	51	47	61	61	86	84	52	43	46	34	45	47	54
CA	Sunnyvale, Moffett Fld.	23244	3725	12204	5.4	47	50	54	59	65	73	62	54	41	35	32	56	51
CA	San Francisco IAP	23234	3737	12223	9.5	96	129	183	228	268	280	236	211	171	141	80	91	176
CA	Farallon IS	495	3740	12300	9.6	61	406	287	193	188	208	100	91	83	106	204	275	212
CA	Alameda FWC	23239	3748	12210	7.4	92	94	122	125	129	124	99	87	65	65	69	81	93
CA	Oakland	23230	3744	12212	6.8	52	75	77	92	101	98	74	69	57	50	40	51	71
CA	San Rafael, Hamilton AFB	23211	3804	12231	4.8	51	52	54	52	50	50	39	39	30	34	32	51	45
CA	Fairfield, Travis AFB	23202	3816	12156	10.7	114	153	176	232	347	488	577	481	332	182	106	91	270
CA	Point Arena	499	3855	12342	13.0	401	398	361	488	500	614	388	513	321	368	320	467	421
CA	Sacramento	23232	3831	12130	7.8	145	145	126	118	116	128	92	83	64	70	61	123	95
CA	Sacramento, Mather AFB	23206	3834	12110	6.0	117	108	89	69	63	69	56	45	38	46	59	88	72
CA	Sacramento, McClellan AFB	23208	3840	12124	6.5	107	102	98	79	83	90	62	56	49	67	75	84	79
CA	Auburn	190	3857	12104	8.4	106	148	109	76	77	64	65	65	64	57	69	67	83
CA	Blue Canyon APT	23225	3917	12042	8.4	237	212	168	106	92	75	57	64	65	110	130	188	122
CA	Donner Summit	23226	3920	12022	12.1	1100	619	729	269	266	226	173	168	154	439	579	645	463

MONTHLY AVERAGE WIND POWER IN THE UNITED STATES AND SOUTHERN CANADA (Continued)

State	Location	Sta. No.	Lat	Long	Ave. Speed knots	Wind Power, Watts Per Square Meter J	F	M	A	M	J	J	A	S	O	N	D	Ave.
CA	Beale AFB	93216	3908	12126	5.1	75	59	64	56	49	52	31	29	34	39	43	62	50
CA	Williams	498	3906	12209	8.2	163	172	179	112	126	120	78	64	78	105	112	116	111
CA	Ft. Bragg	590	3927	12349	5.9	75	88	82	96	46	44	25	26	25	33	50	52	51
CA	Eureka, Arkata APT	24283	4059	12406	6.0	93	93	109	102	115	87	56	42	39	50	61	75	75
CA	Mt. Shasta	595	4116	12216	11.9	456	535	349	309	343	297	177	163	182	214	295	262	309
CA	Redding	592	4034	12224	7.9	71	86	94	81	88	89	69	62	68	68	72	70	74
CA	Montague	197	4144	12231	5.8	65	130	120	122	130	131	123	100	76	75	71	57	98
CA	Montague, Siskiyou Co APT	24259	4146	12228	5.3	106	108	123	115	78	63	59	50	45	64	82	89	80
CO	La Junta	23067	3803	10331	8.3	115	136	222	204	168	164	94	84	85	78	139	115	134
CO	Alamosa APT	23061	3727	10552	7.4	92	110	195	254	214	167	84	70	85	91	77	74	127
CO	Pueblo, Memorial APT	93058	3817	10431	7.7	101	122	180	231	166	129	105	84	82	81	93	104	121
CO	Colo Springs, Peterson Fld	93037	3849	10443	9.0	142	163	217	212	189	163	99	86	105	105	138	128	142
CO	Ft. Carson, Butts AAF	94015	3841	10446	7.3	85	93	145	218	127	131	63	71	68	112	74	87	107
CO	Denver	23062	3945	10452	8.8	117	139	182	183	132	126	94	83	85	88	118	136	126
CO	Denver, Lowry AFB	23012	3943	10454	8.1	115	94	131	163	112	100	95	87	102	88	126	121	109
CO	Aurura Co, Buckley Fld.	23036	3942	10445	6.7	60	60	79	121	79	67	54	52	51	51	57	59	66
CO	Akron, Washington Co APT	24015	4010	10313	11.7	216	313	383	359	276	239	226	184	243	212	252	280	242
CO	Rifle Co, Garfield Co. APT	23069	3932	10744	4.1	17	32	37	69	51	39	26	23	31	25	23	15	31
CO	Craig	24046	4031	10733	7.7	57	63	70	97	80	58	50	52	54	61	56	51	62
CT	Hartford, Bradley Fld.	14740	4156	7241	7.7	115	127	142	129	96	75	54	53	61	74	93	100	93

MONTHLY AVERAGE WIND POWER IN THE UNITED STATES AND SOUTHERN CANADA (Continued)

State	Location	Sta. No.	Lat	Long	Ave. Speed knots	Wind Power, Watts Per Square Meter J	F	M	A	M	J	J	A	S	O	N	D	Ave.
CT	New Haven, Tweed APT	14758	4116	7253	8.7	117	122	142	120	83	65	52	60	78	89	114	106	98
CT	Bridgeport APT	94702	4110	7308	10.4	244	274	256	219	158	114	96	101	139	192	214	251	186
DE	Dover AFB	13707	3908	7528	7.7	135	152	148	125	85	69	49	49	73	78	99	104	96
DE	Delaware Breakwater	404	3848	7506	12.7	449	570	477	430	283	196	163	190	270	403	391	410	343
DE	Wilmington, New Castle APT	13781	3940	7536	8.1	127	149	175	147	105	85	66	59	61	84	118	126	109
DC	Washington, Andrews AFB	13705	3848	7653	7.2	130	156	161	126	77	51	39	36	45	62	101	109	90
DC	Washinton, Bolling AFB	13710	3850	7701	7.5	125	173	171	140	84	58	45	40	51	72	119	112	101
DC	Washington National	13743	3851	7702	8.6	142	151	163	134	95	82	62	44	67	85	103	107	105
DC	Washington, Dulles IAP	93738	3857	7727	6.7	104	115	118	111	66	42	37	41	40	42	66	78	68
FL	Key West NAS	12850	2435	8147	9.5	158	172	172	176	122	98	78	71	133	133	139	147	131
FL	Homestead AFB	12826	2529	8023	6.4	61	74	90	89	72	51	31	35	66	59	60	56	60
FL	Miami	12839	2548	8016	7.8	87	98	111	116	80	59	58	54	90	88	78	79	80
FL	Boca Raton	12803	2622	8006	8.2	80	108	125	135	109	72	51	55	109	140	108	106	99
FL	West Palm Beach	12865	2643	9003	8.3	123	129	151	145	106	79	70	67	80	105	126	102	108
FL	Ft. Myers	12835	2635	8152	7.0	93	111	153	156	104	79	58	70	99	96	90	101	101
FL	Ft. Myers, Hendricks Fld.	12802	2638	8142	7.1	59	68	98	91	74	51	38	47	76	85	58	60	69
FL	Tampa	12842	2758	8232	7.6	85	100	100	101	76	67	40	38	61	65	73	80	68
FL	Tamps, Macdill AFB	12810	2751	8230	6.9	73	95	98	83	59	51	35	40	67	73	62	67	67
FL	Avon Park Range AAF	12804	2738	8120	5.4	50	51	55	64	45	30	18	24	61	73	43	48	45
PL	Orlando, Herndon APT	12841	2833	8120	8.2	86	110	131	120	99	83	69	85	91	107	89	99	97

MONTHLY AVERAGE WIND POWER IN THE UNITED STATES AND SOUTHERN CANADA (Continued)

State	Location	Sta. No.	Lat	Long	Ave. Speed knots	Wind Power, Watts Per Square Meter J	F	M	A	M	J	J	A	S	O	N	D	Ave
FL	Orlando, MCCoy AFB	12815	2827	8118	5.9	61	76	71	67	46	41	29	24	43	48	46	51	49
FL	Titusville	109	2831	8047	6.7	57	72	72	58	44	42	43	32	47	56	50	56	49
FL	Cocoa Beach, Patrick AFB	12867	2814	8036	8.8	127	149	144	134	115	80	52	62	130	191	143	127	119
FL	Cape Kennedy AFS	12868	2829	8033	7.4	82	107	103	90	71	55	41	37	82	94	75	76	73
FL	Daytona Beach APT	12834	2911	8103	8.9	112	141	146	142	125	94	91	95	113	161	108	116	120
FL	Jacksonville, Cecil FLD NAS	93832	3013	8157	5.2	43	65	56	30	35	31	21	19	39	39	37	39	39
FL	Jacksonville NAS	93837	3014	8141	6.9	61	80	81	70	53	60	40	38	76	77	62	64	63
FL	Mayport NAAS	3853	3023	8125	7.2	82	105	92	90	67	67	40	39	110	90	74	68	76
FL	Tallahassee	93805	3023	8422	5.8	51	59	76	66	41	28	24	28	39	43	51	51	45
FL	Marianna	13851	3050	8511	6.9	92	104	115	86	65	48	43	36	55	61	71	84	7[illegible]
FL	Panama City, Tynoall AFB	13846	3004	8535	6.7	79	101	120	97	62	47	42	37	65	55	64	75	71
FL	Crestview	13884	3047	8631	5.6	68	77	85	57	31	22	16	16	35	38	60	65	4[illegible]
FL	Valparaiso, Eglin AFB	13858	3029	8631	6.2	66	74	78	71	56	48	40	37	55	46	55	59	56
FL	Valparàiso, Duke Fld	3844	3039	8632	7.0	104	123	105	115	78	46	33	38	40	48	84	88	7[illegible]
FL	Valparaiso, Hurlburt Fld	3852	3025	8641	5.5	55	62	55	51	36	31	23	21	34	33	39	45	4[illegible]
FL	Milton, Whiting Fld NAAS	93841	3042	8701	7.1	107	114	125	93	62	44	36	32	65	57	84	92	7[illegible]
FL	Pensacola, Saufley Fld NAS	3815	3026	8711	6.8	98	109	110	94	57	42	37	35	79	63	81	99	7[illegible]
FL	Pensacola, Ellyson Fld	3840	3032	8712	7.8	87	104	116	112	86	62	48	44	65	57	74	81	7[illegible]
FL	Pensacola, Forest Sherman Fd	3855	3021	8719	8.0	110	119	113	106	79	75	56	57	73	71	86	99	8[illegible]
GA	Valdosta, Moody AFB	13857	3058	8312	4.8	40	51	54	43	29	28	21	19	33	32	29	35	3[illegible]

MONTHLY AVERAGE WIND POWER IN THE UNITED STATES AND SOUTHERN CANADA (Continued)

State	Location	Sta. No.	Lat	Long	Ave. Speed knots	Wind Power, Watts Per Square Meter												
						J	F	M	A	M	J	J	A	S	O	N	D	Ave.
GA	Moultrie	13835	3108	8342	6.6	73	89	84	79	45	32	34	30	47	60	59	75	58
GA	Albany, Turner AFB	13815	3135	8407	5.3	50	68	73	55	33	27	23	19	33	27	36	41	41
GA	Brunswick, Glynco NAS	93836	3115	8128	5.5	39	54	53	52	40	36	28	25	38	39	35	37	40
GA	Ft. Stewart, Wright AAF	3875	3153	8134	3.7	20	30	32	23	22	14	12	10	14	16	15	23	20
GA	Savannah	3822	3208	8112	7.5	88	108	98	87	56	49	46	45	61	62	63	72	69
GA	Savannah, Hunter AFB	13824	3201	8108	5.8	59	76	86	70	44	40	34	31	38	43	48	47	51
GA	Macon	13836	3242	8339	8.0	92	112	103	117	69	59	56	44	61	56	68	73	75
GA	Warner Robbins AFB	13860	3238	8336	4.9	54	76	75	59	35	26	22	18	27	31	42	44	41
GA	Ft. Benning	13829	3221	8500	3.9	45	64	71	51	28	21	13	13	21	22	32	36	33
GA	Winder	12	3400	8342	7.6	99	113	93	91	57	50	51	44	43	79	92	92	78
GA	Adairsville	110	3455	8456	6.2	87	95	109	74	56	42	36	33	33	49	100	71	64
GA	Augusta, Bush Fld	3820	3322	8158	5.9	68	83	87	83	43	41	36	32	43	39	45	49	53
GA	Atlanta	13874	3339	8426	8.5	170	169	165	151	84	67	56	46	73	80	109	127	106
GA	Marietta, Dobbins AFB	13864	3355	8432	5.8	89	99	105	96	52	38	34	30	40	50	66	72	66
HI	Honolulu IAP	22521	2120	15055	9.8	118	131	164	163	155	172	189	194	141	128	133	144	153
HI	Barbers Point NAS	22514	2119	15804	8.3	106	99	104	102	93	97	100	102	77	76	95	104	95
HI	Wahiawa, Wheeler AFB	22508	2129	15802	5.9	48	49	61	59	60	70	72	65	43	40	39	49	54
HI	Waialua, Mokoleia Fld	22507	2135	15812	7.7	59	52	97	141	115	136	151	158	113	84	89	108	109
HI	Kaneohe Bay MCAS	22519	2127	15747	10.0	131	144	157	156	140	137	143	143	116	113	135	168	141
HI	Barking Sands AAF	22501	2203	15947	5.6	112	62	42	40	33	24	20	22	21	38	43	69	44

MONTHLY AVERAGE WIND POWER IN THE UNITED STATES AND SOUTHERN CANADA (Continued)

State	Location	Sta. No.	Lat	Long	Ave. Speed knots	Wind Power, Watts Per Square Meter J	F	M	A	M	J	J	A	S	O	N	D	Ave.
HI	Molokai, Homestead Fld	22502	2109	15706	12.3	110	195	249	291	250	312	361	342	266	268	233	238	266
HI	Kahului NAS	22516	2054	15626	11.1	203	204	240	276	335	366	375	377	283	219	247	200	276
HI	Hilo	21504	1943	15504	7.7	82	86	77	71	65	67	63	67	59	56	52	74	67
HI	Hilo, Lyman Fld	21504	1943	15504	7.8	82	86	77	71	65	67	63	67	59	56	52	74	67
ID	Strevell	179	4201	11313	9.7	275	255	189	175	161	148	127	120	127	128	188	209	168
ID	Pocatello	24156	4255	11236	8.6	211	224	230	209	163	159	113	87	102	103	148	176	160
ID	Idaho Falls	671	4331	11204	9.7	226	185	321	295	241	214	132	139	166	184	178	172	200
ID	Burley APT	24133	4232	11346	8.0	185	162	246	199	156	116	73	58	72	89	114	150	133
ID	Twin Falls	15634	4228	11429	8.7	168	181	232	237	155	139	85	74	86	114	131	172	147
ID	King Hill	185	4259	11513	8.8	220	222	357	363	330	216	169	147	212	158	165	185	221
ID	Mountain Home AFB	24106	4303	11552	7.3	94	136	154	172	144	121	92	76	80	105	89	81	110
ID	Boise APT	24131	4334	11613	7.8	103	112	127	119	95	79	64	56	60	76	84	95	91
IL	Chicago Midway	14819	4147	8745	9.0	129	145	151	144	115	70	53	52	74	95	149	134	112
IL	Glenview NAS	14855	4205	8750	8.4	164	164	203	206	137	83	56	52	72	105	143	137	128
IL	Chicago, Ohare	14810	4159	8754	9.7	220	242	268	272	197	140	99	89	140	162	258	213	193
IL	Chicago, Ohare IAP	94846	4159	8754	9.5	189	199	227	229	174	118	83	71	113	129	227	176	162
IL	Waterman	139	4146	8845	9.1	236	269	222	269	134	100	50	60	77	99	210	175	166
IL	Rockford	94822	4212	8906	8.8	112	107	135	164	126	85	61	70	82	92	126	121	107
IL	Moline	14923	4127	9031	8.9	121	151	215	200	155	93	63	54	91	113	185	141	130
IL	Bradford	146	4113	8937	10.2	210	271	284	290	203	129	68	88	96	123	237	183	196

MONTHLY AVERAGE WIND POWER IN THE UNITED STATES AND SOUTHERN CANADA (Continued)

State	Location	Sta. No.	Lat	Long	Ave. Speed knots	Wind Power, Watts Per Square Meter												
						J	F	M	A	M	J	J	A	S	O	N	D	Ave.
IL	Rantoul, Chanute AFB	14806	4018	8809	8.5	158	164	193	210	143	91	50	47	66	87	145	127	121
IL	Effingham	436	3909	8832	9.3	170	210	251	217	116	95	73	68	89	95	217	144	136
IL	Springfield, Capitol APT	93822	3950	8940	10.6	215	253	308	295	212	131	92	82	119	152	263	242	198
IL	Quincy, Baldwin Fld	93989	3956	9112	9.9	209	229	275	220	137	98	71	61	95	136	211	194	161
IL	Belleville, Scott AFB	13802	3833	8951	7.2	129	140	162	143	85	61	36	33	46	6	109	96	90
IL	Marion, Williamson Co APT	3865	3745	8901	7.6	159	186	230	243	136	88	59	45	88	93	187	28	139
IN	Evansville	93817	3808	8732	8.1	129	139	165	154	98	69	46	38	60	71	118	117	100
IN	Terre Haute, Holman Fld	3868	3927	8717	8.2	160	151	203	182	105	74	43	33	60	79	130	134	115
IN	Indianapolis	93819	3944	8617	7.1	174	198	247	205	147	96	68	59	81	108	176	161	143
IN	Columbus, Bakalar AFB	13803	3916	8554	7.0	97	106	128	117	71	50	36	32	44	58	91	88	74
IN	Milroy	125	3928	8522	9.5	243	270	230	209	116	115	73	67	94	101	189	163	148
IN	Centerville	130	3949	8458	9.0	196	237	209	182	101	87	64	57	79	93	176	136	134
IN	Marion APT	94852	4029	8541	8.4	211	254	279	255	160	116	64	50	79	95	253	186	170
IN	Peru, Grissom AFB	94833	4039	8609	7.7	123	137	158	165	109	65	40	36	53	69	131	133	100
IN	Lafeyette	530	4025	8656	10.3	290	317	290	316	175	142	91	98	112	126	296	222	215
IN	Fort Wayne	14827	4100	8512	9.5	149	167	230	205	154	101	73	66	96	116	214	171	146
IN	Helmer	535	4133	8512	9.6	256	243	263	242	141	100	79	78	133	153	242	215	161
IN	Goshen	132	4132	8548	8.9	229	209	208	221	124	104	75	73	89	103	182	148	146
IN	South Bend	138	4142	8619	9.8	243	243	283	256	155	128	92	92	107	127	229	156	188
IN	MCCool	136	4133	8710	10.7	284	297	311	290	183	149	82	96	130	157	311	231	223

MONTHLY AVERAGE WIND POWER IN THE UNITED STATES AND SOUTHERN CANADA (Continued)

State	Location	Sta. No.	Lat	Long	Ave. Speed Knots	Wind Power, Watts Per Square Meter												
						J	F	M	A	M	J	J	A	S	O	N	D	Ave
IA	Dubuque APT	94908	4224	9042	9.4	208	209	239	317	241	150	112	135	170	198	312	231	21[illegible]
IA	Burlington	14931	4046	9107	9.5	147	160	257	164	94	85	52	44	72	97	200	144	12[illegible]
IA	Iowa City APT	14937	4138	9133	8.6	175	195	231	229	118	82	71	61	81	100	205	159	14[illegible]
IA	Cedar Rapids	14990	4153	9142	9.2	160	171	23	249	157	97	53	49	59	102	138	131	13[illegible]
IA	Ottumwa	14948	4106	9226	9.1	209	243	257	239	169	140	118	112	156	169	174	168	17[illegible]
IA	Montezuma	145	4135	9228	11.0	270	330	330	390	256	203	103	117	150	158	271	223	237
IA	Des Moines	14933	4132	9339	9.9	192	193	251	289	180	126	81	81	109	142	219	178	168
IA	Ft. Dodge APT	94933	4233	9411	10.3	253	260	331	334	258	140	77	74	104	181	185	188	199
IA	Atlantic	140	4122	9503	11.3	296	350	363	457	295	256	136	123	155	190	264	270	256
IA	Sioux City	14943	4224	9623	9.7	180	172	247	283	206	143	89	82	114	155	212	170	169
KS	Ft. Leavenworth	13921	3922	9455	6.3	73	84	116	111	73	57	31	33	50	52	78	66	69
KS	Olathe NAS	93909	3850	9453	9.2	143	157	211	187	139	117	69	69	91	102	152	136	135
KS	Topeka	13996	3904	9538	9.8	138	147	237	229	170	159	107	112	136	137	159	163	157
KS	Topeka, Forbes AFB	13920	3857	9540	8.6	117	134	186	185	132	115	69	79	88	95	125	104	120
KS	Ft. Riley	13947	3903	9646	8.0	112	122	224	233	171	130	86	102	139	138	125	106	139
KS	Cassoday	152	3802	9638	13.0	370	436	550	550	350	310	231	257	283	284	371	311	377
KS	Wichita	3928	3739	9725	12.0	243	273	344	337	262	276	168	177	203	221	249	237	253
KS	Wichita, McConnell AFB	3923	3737	9716	10.9	222	234	336	317	252	237	151	136	176	188	200	207	222
KS	Hutchinson	93905	3756	9754	10.7	287	335	372	375	330	351	215	195	309	280	308	269	305
KS	Salina, Schilling AFB	13922	3848	9738	9.1	134	168	230	221	176	150	100	112	148	135	147	111	155

MONTHLY AVERAGE WIND POWER IN THE UNITED STATES AND SOUTHERN CANADA (Continued)

State	Location	Sta. No.	Lat	Long	Ave. Speed knots	Wind Power, Watts Per Square Meter J	F	M	A	M	J	J	A	S	O	N	D	Ave
KS	Hill City APT	93990	3923	9950	9.7	122	199	337	262	210	226	152	125	153	140	153	131	184
KS	Dodge City APT	13985	3746	9958	13.5	281	360	441	458	360	368	259	245	296	296	334	318	336
KS	Garden City APT	23064	3756	10043	12.4	227	326	451	450	415	456	277	272	309	259	216	204	295
KY	Corbin	3814	3658	8408	4.4	71	54	65	58	26	15	16	12	16	18	43	44	36
KY	Lexington	93820	3802	8436	8.9	161	158	156	169	103	75	61	47	72	73	148	146	11[illegible]
KY	Warsaw	23	3846	8454	6.8	123	132	137	123	65	57	49	39	40	54	106	101	85
KY	Louisville, Standiford Fld	93821	3811	8544	6.5	75	84	104	96	53	32	26	27	29	36	58	66	5[illegible]
KY	Ft. Knox	13807	3754	8558	6.6	108	121	126	111	64	46	30	25	39	47	98	96	7[illegible]
KY	Bowling Green, City Co APT	93808	3658	8626	6.6	131	115	136	113	65	39	38	32	46	57	89	93	7[illegible]
KY	Ft. Campbell	13806	3640	8730	5.8	78	88	107	89	51	32	27	25	29	39	60	69	5[illegible]
KY	Paducah	3816	3704	8846	6.7	109	106	122	106	59	44	35	33	40	48	90	93	7[illegible]
LA	New Orleans	12916	2959	9015	8.0	129	137	144	114	76	52	44	43	81	91	128	109	9[illegible]
LA	New Orleans, Callender NAS	12958	2949	9001	4.6	47	55	50	35	26	14	10	10	26	24	31	40	3[illegible]
LA	Baton Rouge	13970	3032	9109	7.4	105	106	102	95	72	52	40	36	50	53	79	92	7[illegible]
LA	Lake Charles, Chenault AFB	13941	3013	9310	8.3	184	156	204	176	125	91	58	57	67	67	133	140	12[illegible]
LA	Polk AAF	3931	3103	9311	5.7	51	69	78	68	47	37	23	15	21	29	55	51	4[illegible]
LA	Alexandria, England AFB	13934	3119	9233	4.6	45	57	64	52	37	20	15	13	17	21	39	41	3[illegible]
LA	Monroe, Selman Fld	13942	3231	9203	7.0	88	104	108	90	61	46	36	36	46	51	73	79	6[illegible]
LA	Shreveport	13957	3228	9349	8.4	128	138	145	131	92	71	57	55	58	69	105	111	9[illegible]
LA	Shreveport, Barksdale AFB	13944	3230	9340	6.0	69	74	83	72	48	36	27	27	35	34	53	59	[illegible]

MONTHLY AVERAGE WIND POWER IN THE UNITED STATES AND SOUTHERN CANADA (Continued)

State	Location	Sta. No.	Lat	Long	Ave. Speed knots	Wind Power, Watts Per Square Meter J	F	M	A	M	J	J	A	S	O	N	D	Ave.
ME	Portland	14764	4339	7019	8.4	127	145	158	140	103	80	197	61	83	101	112	120	107
ME	Brunswick, NAS	14611	4353	6956	6.8	109	116	106	107	87	64	55	48	58	69	80	97	82
ME	Bangor, Dow AFB	14601	4448	6841	7.1	132	138	136	113	83	70	54	59	63	82	100	110	93
ME	Presque Isle AFB	14604	4641	6803	7.8	151	167	151	161	123	88	77	69	97	115	110	134	120
ME	Limestone, Loring AFB	14623	4657	6753	6.9	97	107	110	88	69	55	48	45	60	68	73	78	74
MD	Patuxent River NAS	13721	3817	7625	8.1	159	177	186	148	97	76	59	59	83	102	138	139	119
MD	Baltimore, Martin Fld	93744	3920	7625	6.9	107	111	119	95	53	44	37	39	34	46	57	64	61
MD	Baltimore, Friendship APT	93721	3911	7640	9.6	206	253	265	209	152	117	96	79	110	117	179	188	164
MD	Ft. Mead, Tipton AAF	93733	3905	7646	4.4	57	58	69	65	37	19	14	14	14	23	42	41	38
MD	Aberdeen, Phillips AAF	13701	3928	7610	7.9	126	170	173	157	95	66	52	55	69	95	121	118	109
MD	Camp Detrick, Fredrick	13749	3926	7727	5.4	101	122	144	110	51	33	25	22	30	47	96	75	72
MD	Ft. Ritchie	93745	3944	7724	4.6	38	34	33	37	21	16	14	23	18	27	27	52	28
MA	Chicopee Falls, Westover AAF	14703	4212	7232	7.1	122	143	131	133	96	70	52	48	60	81	104	114	96
MA	Ft. Devons AAF	4779	4234	7136	5.4	45	40	66	84	44	31	29	32	33	39	48	53	45
MA	Bedford, Hanscom Fld	14702	4228	7117	6.1	109	120	117	94	70	48	39	36	44	65	80	92	76
MA	Boston, Logan IAP	14739	4222	7102	11.8	314	321	314	268	195	150	128	108	131	131	230	277	227
MA	Boston	14739	4222	7102	11.7	314	321	314	268	195	150	128	108	131	131	230	277	227
MA	South Weymouth NAS	14790	4209	7056	7.6	125	125	146	136	84	58	43	56	53	71	92	102	90
MA	Falmouth, Otis AFB	14704	4139	7031	9.2	188	199	198	193	147	110	87	90	112	139	148	185	149
MA	Nantucket	14756	4116	7003	11.6	304	346	298	277	190	140	104	113	169	214	261	298	223
MA	Nantucket Shoals	14658	4101	6930	16.7	1024	1025	977	838	632	551	592	544	482	769	856	927	757

MONTHLY AVERAGE WIND POWER IN THE UNITED STATES AND SOUTHERN CANADA (Continued)

State	Location	Sta. No.	Lat	Long	Ave. Speed knots	J	F	M	A	M	J	J	A	S	O	N	D	Ave.
						Wind Power, Watts Per Square Meter												
MA	Georges Shoals	14657	4141	6747	17.1	1168	1175	1058	891	619	575	519	378	473	739	891	1156	783
MI	Mt. Clemens, Selfridge AFB	14804	4236	8249	8.2	157	151	160	145	96	71	56	53	71	84	156	144	115
MI	Ypsilanti, Willow Run	14853	4214	8332	9.5	169	169	244	194	139	101	85	77	104	113	188	173	147
MI	Jackson	133	4216	8428	8.8	196	149	182	215	106	92	57	69	77	99	175	147	127
MI	Battle Creek, Kelogg APT	14815	4218	8514	8.9	161	189	205	179	124	99	76	63	106	99	152	180	137
MI	Grand Rapids	94860	4253	8531	8.7	120	134	180	158	112	77	62	54	83	87	162	135	113
MI	Lansing	14836	4247	8436	10.8	273	298	356	287	178	112	69	74	123	146	251	269	203
MI	Flint, Bishop APT	14826	4258	8344	9.6	233	206	246	195	140	109	85	71	129	140	210	223	167
MI	Saginaw, Tri City APT	14845	4326	8352	9.7	218	196	223	196	152	111	92	74	121	128	199	189	158
MI	Muskegon Co APT	14840	4310	8614	9.4	156	164	140	171	121	96	68	70	80	155	177	166	129
MI	Gladwin	14828	4359	8429	5.9	67	71	92	76	63	40	31	24	34	39	61	53	53
MI	Cadillac APT	14817	4415	8528	9.4	210	204	239	193	172	151	104	89	139	161	208	203	171
MI	Traverse City	14850	4444	8535	9.5	229	206	249	207	147	132	100	91	160	187	250	225	178
MI	Oscoda, Wurtsmith AFB	14808	4427	8322	7.6	117	123	121	116	91	76	55	58	71	94	109	108	94
MI	Alpena, Collins Fld	94849	4504	8334	7.3	76	76	92	108	88	59	49	46	52	62	67	58	70
MI	Pellston, Emmett Co APT	14841	4534	8448	8.9	183	159	192	165	154	115	105	81	115	144	175	185	147
MI	Sault Ste Marie	14847	4628	8422	8.3	114	105	119	125	113	77	62	57	79	93	115	108	98
MI	Kinross, Kincheloe AFB	94824	4615	8428	7.6	88	105	106	120	106	69	53	56	68	81	109	93	89
MI	Escanaba APT	94853	4544	8705	7.8	126	164	148	186	191	150	116	93	135	163	232	143	154
MI	Gwinn, Sawyer AFB	94836	4621	8723	7.5	94	116	109	116	100	72	52	57	65	92	102	105	90

MONTHLY AVERAGE WIND POWER IN THE UNITED STATES AND SOUTHERN CANADA (Continued)

State	Location	Sta. No.	Lat	Long	Ave. Speed knots	Wind Power, Watts Per Square Meter J	F	M	A	M	J	J	A	S	O	N	D	Av
MI	Marquette	14838	4634	8724	7.6	78	84	118	125	117	96	73	70	89	85	88	66	
MI	Calumet	14858	4710	8830	8.5	116	126	136	139	106	90	78	66	97	108	116	112	1
MI	Houghton Co APT	14858	4710	8830	8.5	116	126	136	139	106	90	78	66	97	108	116	112	1
MI	Ironwood, Gogebic Co APT	94926	4632	9008	8.5	164	198	167	290	280	174	130	139	200	213	270	203	2
MN	Minneapolis, St. Paul IAP	14922	4453	9313	9.4	127	142	152	211	186	133	88	86	112	130	167	123	1
MN	St. Cloud, Whitney APT	14926	4535	9411	6.9	70	70	102	129	99	71	44	38	55	66	84	58	
MN	Alexandria	751	4553	9524	10.7	221	215	262	290	249	202	128	161	182	263	263	196	2
MN	Brainerd	94938	4624	9408	6.9	90	92	111	167	134	98	62	58	107	87	123	89	1
MN	Duluth IAP	14913	4650	9211	10.7	219	229	249	299	233	147	122	111	154	196	254	206	1
MN	Bemidji APT	14958	4730	9456	7.2	104	120	111	244	201	155	117	120	133	141	162	122	1
MN	International Falls IAP	14918	4834	9323	8.4	95	103	110	175	155	103	82	88	119	122	163	117	1
MN	Roseau	14955	4851	9545	6.5	33	31	50	58	51	35	16	20	27	34	49	41	
MN	Thief River Falls	8243	4803	9611	8.7	215	207	194	305	279	299	135	157	189	221	309	219	2
MS	Biloxi; Keesler AFB	13820	3024	8855	6.8	82	79	83	81	64	49	38	35	58	55	66	68	
MS	Jackson	13927	3220	9014	6.2	85	92	88	78	46	31	26	25	31	39	63	77	
MS	Greenville APT	13939	3329	9059	6.6	89	100	104	90	66	48	32	34	49	50	65	77	
MS	Meridian NAAS	3866	3323	8833	3.5	29	40	37	25	12	8	9	5	8	10	17	21	
MS	Columbus AFB	13825	3338	8827	4.7	52	60	61	48	25	17	15	13	23	21	31	40	
MO	Malden	13848	3636	8959	8.3	151	117	162	152	109	75	57	53	62	74	124	118	1
MO	St. Louis, Lambert Fld	13994	3845	9023	7.9	95	116	143	138	94	60	41	36	55	60	92	96	

MONTHLY AVERAGE WIND POWER IN THE UNITED STATES AND SOUTHERN CANADA (Continued)

State	Location	Sta. No.	Lat	Long	Ave. Speed knots	Wind Power, Watts Per Square Meter												
						J	F	M	A	M	J	J	A	S	O	N	D	Ave.
MO	New Florence	147	3853	9126	10.1	198	231	238	231	138	105	85	85	111	112	199	165	158
MO	Kirksville	540	4006	9232	10.5	250	271	297	310	191	137	111	103	118	158	231	191	184
MO	Vichy, Rolla APT	13997	3808	9146	8.6	153	158	211	170	92	72	54	45	68	76	142	156	117
MO	Ft. Leonard Wood, Forney AF	3938	3743	9208	6.0	67	65	81	88	50	37	22	21	26	50	62	67	52
MO	Springfield	440	3714	9315	9.7	183	243	230	263	123	96	70	77	97	-10	183	170	150
MO	Butler	93995	3818	9420	9.3	212	208	266	226	123	124	74	65	81	131	156	160	152
MO	Knobnoster, Whiteman AFB	13930	3844	9334	7.4	96	109	146	151	94	65	40	45	61	70	94	81	87
MO	Marshall	144	3906	9312	9.5	203	223	263	250	115	109	82	90	89	95	156	136	143
MO	Grandview, Rchds-Gebaur AFB	3929	3851	9435	8.1	105	101	156	173	113	76	55	59	74	91	110	110	100
MO	Kansas City APT	13988	3907	9436	9.4	115	126	165	182	145	129	105	99	113	112	143	122	132
MO	Knoxville	142	3925	9400	10.1	210	211	284	278	144	124	84	91	98	125	171	151	158
MO	Tarkio	14945	4027	9522	8.2	121	137	258	225	192	150	87	71	85	113	135	94	135
MT	Glendive	24087	4708	10448	7.7	131	141	145	222	217	146	125	137	139	144	111	122	149
MT	Miles City APT	24037	4626	10552	8.6	123	137	120	161	134	102	87	98	104	109	93	116	115
MT	Wolf Point	94017	4806	10535	8.1	117	112	143	326	248	139	126	171	245	223	183	124	179
MT	Glasgow AFB	94010	4824	10631	8.6	133	131	125	178	198	130	102	101	138	126	119	125	133
MT	Billings, Logan Fld	24033	4548	10832	10.0	230	210	185	202	165	137	110	99	128	152	218	237	173
MT	Livingston	678	4540	11032	13.5	778	819	574	415	327	239	233	253	321	500	713	1058	500
MT	Lewiston APT	24036	4703	10927	8.6	198	185	141	163	135	108	82	95	114	125	189	153	140
MT	Havre	777	4834	10940	8.7	148	106	155	141	123	115	75	74	86	120	132	127	114

MONTHLY AVERAGE WIND POWER IN THE UNITED STATES AND SOUTHERN CANADA (Continued)

State	Location	Sta. No.	Lat	Long	Ave. Speed knots	Wind Power, Watts Per Square Meter J	F	M	A	M	J	J	A	S	O	N	D	Ave.
MT	Great Falls IAP	24143	4729	11122	11.6	444	439	300	281	194	193	136	143	197	300	456	509	304
MT	Great Falls, Malmstrom AFB	24112	4731	11110	8.9	253	240	181	176	120	112	80	80	115	178	215	263	169
MT	Helena APT	24144	4636	11200	7.3	145	95	142	134	63	113	85	44	111	34	61	65	90
MT	Whitehall	24161	4552	11158	11.4	710	543	352	274	221	245	193	167	174	260	410	602	344
MT	Butte, Silver Bow Co APT	24135	4557	11230	6.9	98	101	116	158	141	120	86	85	93	93	86	76	104
MT	Missoula	24153	4655	11405	5.0	49	36	64	75	72	70	64	41	57	23	19	26	50
NB	Omaha	14942	4118	9554	10.0	191	186	264	280	186	148	104	104	122	158	217	191	177
NB	Omaha, Offutt AFB	14949	4107	9555	7.6	118	123	189	198	138	98	67	58	68	93	112	112	115
NB	Grand Island APT	14935	4058	9819	11.1	177	195	270	312	251	217	161	158	179	130	232	200	211
NB	Overton	154	4044	9927	10.5	209	195	323	389	276	243	155	149	162	202	299	182	222
NB	North Platte	562	4108	11042	10.5	193	233	374	435	321	208	153	153	206	246	234	166	254
NB	Lincoln AFB	14904	4051	9646	9.4	163	173	258	251	193	143	97	102	102	119	173	146	162
NB	Columbus	73084	4126	9720	10.0	184	192	316	301	246	172	113	111	120	184	150	143	186
NB	Norfolk, Stefan APT	14941	4159	9726	9.7	236	235	308	387	275	215	142	173	204	281	361	255	256
NB	Big Springs	161	4105	10207	11.7	270	284	430	450	349	270	210	210	217	264	297	251	290
NB	Sidney	563	4108	10302	10.5	275	267	369	395	294	227	193	160	680	227	241	188	248
NB	Scottsbluff APT	24028	4152	10336	9.8	147	225	271	254	189	180	119	124	122	166	254	199	165
NB	Alliance	24044	4203	10248	10.6	203	209	274	358	289	233	189	204	233	228	238	210	238
NB	Valentine, Miller Fld	24032	4252	10033	10.0	181	237	267	323	286	242	199	226	230	271	338	242	253
NV	Boulder City	382	3558	11450	7.6	109	162	185	230	247	293	186	197	137	95	155	81	173

MONTHLY AVERAGE WIND POWER IN THE UNITED STATES AND SOUTHERN CANADA (Continued)

State	Location	Sta. No.	Lat	Long	Ave. Speed knots	Wind Power, Watts Per Square Meter J	F	M	A	M	J	J	A	S	O	N	D	Ave.
NV	Las Vegas	23169	3605	11510	8.7	105	142	186	229	225	209	166	141	111	122	67	92	150
NV	Las Vegas, Nellis AFB	23112	3615	11502	5.7	73	89	137	138	125	123	77	75	61	63	66	57	88
NV	Indian Springs AFB	23141	3635	11541	5.3	38	75	141	196	154	109	64	46	54	28	63	54	80
NV	Tonopah APT	23153	3804	11708	8.7	99	150	196	196	174	142	98	94	104	113	109	101	133
NV	Fallon NAAS	93102	3925	11843	4.7	48	50	74	72	60	49	29	23	25	30	27	36	44
NV	Reno	23185	3930	11947	5.2	77	108	123	99	93	81	56	54	52	52	45	44	74
NV	Reno, Stead AFB	23118	3940	11952	5.9	79	105	125	132	110	91	72	72	56	64	51	69	85
NV	Humboldt	580	4005	11809	6.7	61	76	140	102	103	118	98	83	66	57	47	48	79
NV	Lovelock	24172	4004	11833	6.4	109	91	121	99	98	113	80	72	56	67	43	47	83
NV	Winnemucca APT	24128	4054	11748	7.2	79	91	117	115	105	97	89	81	73	80	55	60	86
NV	Buffalo Valley	24181	4020	11721	6.4	66	97	94	94	102	97	77	62	57	59	53	52	76
NV	Battle Mountain	24119	4037	11652	7.3	148	80	147	113	132	113	85	70	61	102	64	66	98
NV	Beowawe	181	4036	11631	6.2	57	98	112	106	99	86	79	68	63	62	45	51	76
NV	Elko	582	4050	11548	6.2	68	76	100	92	99	98	96	76	75	73	52	60	76
NV	Ventosa	70	4052	11448	6.8	139	160	178	179	166	112	104	97	96	88	93	99	109
NH	Portsmouth, Pease AFB	4743	4305	7049	6.6	90	112	99	82	73	50	39	37	42	54	63	90	68
NH	Manchester, Grenier Fld	14710	4256	7126	6.7	105	150	134	137	83	68	49	37	51	72	95	127	86
NH	Keene	94721	4254	7216	4.8	63	86	69	74	67	48	29	31	34	42	43	50	53
NJ	Atlantic City	93730	3927	7435	9.1	185	207	207	166	109	81	62	61	81	107	144	164	129
NJ	Camden	103	3955	7504	8.0	132	131	167	160	86	73	64	55	62	82	118	111	104

MONTHLY AVERAGE WIND POWER IN THE UNITED STATES AND SOUTHERN CANADA (Continued)

State	Location	Sta. No.	Lat	Long	Ave. Speed knots	J	F	M	A	M	J	J	A	S	O	N	D	Ave.
						Wind Power, Watts Per Square Meter												
NJ	Wrightstown, McGuire AFB	14706	4000	7436	6.6	100	114	114	101	48	42	30	28	40	52	73	89	69
NJ	Lakehurst NAS	14780	4002	7420	7.4	133	158	166	131	94	62	49	41	48	60	99	109	93
NJ	Belmar	4739	4011	7404	6.1	80	82	83	57	37	31	23	23	30	50	56	55	50
NJ	Trenton	501	4017	7450	8.1	125	146	146	194	85	71	47	55	70	92	140	120	105
NJ	Newark	14734	4042	7410	8.7	145	145	157	126	107	80	73	68	71	96	100	110	109
NM	Clayton	23051	3627	10309	13.0	447	397	519	483	427	360	230	207	255	279	350	395	354
NM	Tucumcari	364	3511	10336	10.6	273	321	359	365	293	227	167	154	166	207	206	205	260
NM	Anton Chico	160	3508	10505	8.9	204	257	306	227	130	132	78	67	74	101	143	140	155
NM	Clovis, Cannon AFB	23008	3423	10319	9.9	171	215	320	279	228	204	126	93	111	122	160	180	186
NM	Hobbs, Lea Co APT	93034	3241	10312	10.4	195	234	353	276	250	215	138	109	118	109	166	188	190
NM	Roswell APT	23043	3324	10432	8.5	148	191	273	260	216	172	101	82	84	102	126	172	163
NM	Roswell, Walker AFB	23009	3318	10432	7.3	79	102	145	145	129	135	93	71	65	72	82	86	98
NM	Rodeo	272	3156	10859	9.4	195	216	250	325	259	191	166	131	129	155	210	173	203
NM	Las Cruces, White Sands	23039	3222	10629	6.1	100	106	169	149	123	88	50	43	41	42	82	99	89
NM	Alamogordo, Holloman AFB	23002	3251	10605	5.6	45	59	92	101	85	72	54	43	38	35	41	40	57
NM	Albuquerque, Kirtland AFB	23050	3503	10637	7.6	83	115	154	190	160	134	101	72	88	97	80	74	112
NM	Otto	166	3505	10600	9.6	248	311	491	372	271	264	116	102	94	176	234	228	243
NM	Santa Fe APT	23049	3537	10605	10.3	218	200	308	308	248	217	138	113	135	154	184	194	201
NM	Farmington APT	23090	3645	10814	7.1	53	74	136	151	106	101	78	55	50	71	81	42	83
NM	Gallup	23081	3531	10847	6.2	92	133	237	293	248	217	92	82	75	114	84	54	143

MONTHLY AVERAGE WIND POWER IN THE UNITED STATES AND SOUTHERN CANADA (Continued)

State	Location	Sta. No.	Lat	Long	Ave. Speed knots	Wind Power, Watts Per Square Meter J	F	M	A	M	J	J	A	S	O	N	D	Ave.
NM	Zuni	93044	3506	10848	8.4	127	109	234	220	183	138	58	54	80	104	97	126	127
NM	El Morro	373	3501	10826	7.4	76	113	229	210	185	136	95	66	66	92	91	83	107
NM	Acomita	170	3503	10743	9.6	150	169	283	223	156	143	96	82	75	109	143	136	169
NY	Westhampton, Suffolk Co AFB	14719	4051	7238	8.1	146	145	154	133	100	82	69	67	87	110	120	118	110
NY	Hempstead, Mitchell AFB	14708	4044	7336	9.2	194	221	211	189	134	115	99	88	100	129	185	194	155
NY	New York, Kennedy IAP	94789	4039	7347	10.3	242	259	260	204	151	139	122	106	120	140	173	180	168
NY	New York, La Guardia	14732	4046	7354	10.9	300	282	283	211	160	123	105	110	135	174	217	278	197
NY	New York, Central Park	94728	4047	7358	8.1	114	108	117	99	57	43	38	37	61	63	83	95	76
NY	New York WBO	94706	4043	7400	11.6	436	428	384	259	211	173	146	107	143	209	329	336	261
NY	Bear Mountain	100	4114	7400	12.5	476	463	550	444	311	183	171	163	271	289	396	523	350
NY	Newburgh, Stewart AFB	14714	4130	7406	7.8	164	206	193	176	108	76	61	52	64	101	136	163	124
NY	New Hackensaok	106	4138	7353	6.0	83	91	93	87	50	42	34	32	40	63	93	84	63
NY	Poughkeepsie, Duchess Co APT	14757	4138	7353	6.1	68	90	106	90	52	43	33	28	36	46	66	74	60
NY	Columbiaville	115	4220	7345	8.7	185	220	226	173	131	104	69	69	97	138	164	172	138
NY	Albany Co APT	14735	4245	7348	7.9	148	163	173	138	95	80	68	63	81	96	103	111	108
NY	Schnectady	4782	4251	7357	7.4	156	116	160	155	123	81	82	67	84	68	112	114	112
NY	Plattsburg AFB	4742	4439	7327	6.0	63	78	76	82	70	52	42	36	41	54	66	60	60
NY	Massena, Richards APT	94725	4456	7451	9.5	176	192	217	193	150	129	108	101	111	154	170	193	158
NY	Watertown APT	94790	4400	7601	10.0	409	312	373	298	153	134	119	99	171	197	278	350	236
NY	Rome, Griffiss AFB	14717	4314	7525	5.7	91	108	109	94	66	44	30	26	36	50	71	82	65

MONTHLY AVERAGE WIND POWER IN THE UNITED STATES AND SOUTHERN CANADA (Continued)

State	Location	Sta. No.	Lat	Long	Ave. Speed knots	J	F	M	A	M	J	J	A	S	O	N	D	Ave.
						Wind Power, Watts Per Square Meter												
NY	Utica, Cneida Co APT	94794	4309	7523	8.2	117	130	113	100	74	81	43	49	59	67	106	111	87
NY	Syracuse, Hancock APT	14771	4307	7607	8.4	158	174	166	162	108	80	66	61	76	91	129	138	115
NY	Binghampton, Bloome Co APT	4725	4213	7559	9.0	157	183	194	191	138	77	73	70	77	104	155	160	132
NY	Elmira, Chemung Co APT	14748	4210	7654	5.6	73	78	91	80	50	45	29	25	34	57	79	69	59
NY	Rochester	14768	4307	7740	9.8	205	229	240	201	138	123	98	82	102	123	197	194	163
NY	Buffalo	528	4256	7843	11.5	468	430	417	411	422	281	278	383	377	334	354	306	382
NY	Buffalo	14733	4256	7849	10.9	254	258	322	239	160	151	132	118	145	160	227	251	205
NY	Niagara Falls	4724	4306	7857	8.3	193	175	147	130	106	82	72	67	82	105	134	176	126
NY	Dunkirk	127	4230	7916	11.2	488	348	368	302	173	154	121	127	174	269	396	361	281
NC	Wilmington	13748	3416	7755	8.1	118	151	163	169	102	87	77	80	98	97	101	97	108
NC	Jacksonville, New Rvr. MCAF	93727	3443	7726	6.0	59	72	84	79	53	44	32	31	43	40	49	46	51
NC	Cherry Point NAS	13754	3454	7653	7.0	92	105	124	125	84	68	57	57	84	66	67	74	83
NC	Cape Hatteras	93729	3516	7533	10.6	195	229	209	202	144	138	117	135	180	160	166	168	169
NC	Goldsboro, Symr-Jhnsn AFB	13713	3520	7758	5.4	55	71	80	72	45	32	30	23	30	29	42	46	45
NC	Ft. Bragg, Simmons AAF	93737	3508	7856	5.8	63	82	76	70	46	32	27	24	27	33	53	48	46
NC	Fayetteville, Pope AFB	13714	3512	7901	4.3	43	54	60	55	34	25	24	21	21	23	29	30	33
NC	Charlotte, Douglas APT	13881	3513	8056	7.4	101	101	120	118	69	57	50	53	70	76	78	82	82
NC	Asheville	13872	3536	8232	5.5	77	77	110	90	41	24	17	16	18	33	76	74	54
NC	Hickory APT	3810	3545	8123	7.2	69	69	89	79	57	50	50	49	49	54	63	61	62
NC	Winston Salem	93807	3608	8014	8.1	141	166	149	169	88	68	66	54	97	106	98	117	111

MONTHLY AVERAGE WIND POWER IN THE UNITED STATES AND SOUTHERN CANADA (Continued)

State	Location	Sta. No.	Lat	Long	Ave. Speed knots	Wind Power, Watts Per Square Meter J	F	M	A	M	J	J	A	S	O	N	D	Ave.
NC	Greensboro	13723	3605	7957	6.7	67	90	94	94	47	37	34	29	35	43	69	57	58
NC	Raleigh	13722	3552	7847	6.7	89	81	106	113	53	51	49	37	45	41	64	61	64
NC	Rocky Mount APT	13746	3558	7748	4.2	72	74	97	86	49	62	51	43	45	50	57	60	52
NC	Elizabeth City	13786	3616	7611	7.4	81	89	95	98	76	65	50	58	67	71	63	63	74
ND	Fargo, Hector APT	14914	4654	9648	11.7	280	264	293	389	286	215	144	160	225	280	337	270	263
ND	Grand Forks AFB	94925	4758	9724	8.9	167	182	183	197	166	103	71	88	123	147	147	172	146
ND	Pembina	758	4857	9715	11.7	308	381	321	341	335	261	187	241	261	329	403	409	308
ND	Bismarck APT	24011	4646	10045	9.5	147	140	186	250	217	174	118	119	157	167	186	143	170
ND	Minot AFB	94011	4825	10121	9.1	191	192	166	199	185	117	95	98	127	164	163	181	157
ND	Williston, Sloulin Fld	94014	4811	10338	8.2	80	86	109	143	141	104	76	83	101	98	88	78	98
ND	Dickinson	24012	4647	10248	13.0	402	365	462	486	401	402	246	208	300	332	426	334	362
OH	Youngstown APT	14852	4116	8040	9.2	187	177	218	178	115	84	66	57	81	95	180	188	133
OH	Warren	21	4117	8048	9.3	196	183	197	197	116	95	67	59	82	122	164	149	136
OH	Akron	14895	4055	8126	9.1	163	184	192	151	101	75	55	55	70	86	156	147	118
OH	Perry	128	4141	8107	10.7	296	296	290	277	136	115	82	90	136	108	311	270	223
OH	Cleveland	14820	4124	8151	10.1	189	237	244	211	147	111	80	72	104	122	230	202	152
OH	Vickery	20	4125	8255	10.7	284	310	303	284	150	136	89	88	130	157	284	217	217
OH	Toledo	94830	4136	8348	7.7	109	114	138	108	76	51	39	37	49	60	89	93	80
OH	Archbold	120	4134	8419	8.8	182	176	182	189	99	86	57	62	84	98	182	135	127
OH	Columbus	14821	4000	8253	7.2	109	118	136	116	74	52	37	35	44	55	100	91	82

MONTHLY AVERAGE WIND POWER IN THE UNITED STATES AND SOUTHERN CANADA (Continued)

State	Location	Sta. No.	Lat	Long	Ave. Speed knots	Wind Power, Watts Per Square Meter J	F	M	A	M	J	J	A	S	O	N	D	Ave.
OH	Columbus, Lockbourne AFB	13812	3949	8256	6.8	109	127	135	120	70	53	36	33	43	58	93	94	80
OH	Hayesville	124	4047	8218	10.0	257	257	236	204	123	124	76	76	104	151	258	204	163
OH	Cambridge	122	4004	8135	6.2	110	103	110	94	54	50	40	32	40	59	30	73	71
OH	Zanesville, Cambridge	93824	3957	8154	7.7	160	142	188	158	87	67	46	32	56	63	136	130	105
OH	Wilmington, Clinton Co AFB	13841	3926	8348	7.8	133	148	166	157	94	60	44	37	48	67	117	119	93
OH	Cincinnati	93814	3904	8440	8.4	133	135	150	144	90	63	51	42	61	77	126	112	99
OH	Dayton	93815	3954	8413	9.0	179	192	207	173	108	73	59	46	69	84	170	160	125
OH	Dayton, Wright AFB	13813	3947	8406	7.6	161	188	226	182	122	83	58	50	71	86	162	157	128
OH	Dayton, Patterson Fld	13840	3949	8403	7.4	171	186	202	176	108	73	47	43	61	79	160	145	120
OK	Muskogee	13916	3540	9522	8.5	149	189	230	210	132	94	53	71	80	108	114	99	130
OK	Tulsa IAP	13968	3612	9554	9.5	157	178	196	185	155	116	94	85	110	118	145	149	141
OK	Oklahoma City	13967	3524	9736	12.2	306	333	376	386	274	250	165	153	182	216	248	265	263
OK	Oklahoma City, Tinker AFB	13919	3525	9723	11.4	263	277	370	412	319	318	176	153	197	230	243	248	264
OK	Ardmore AFB, Autrey Fld	13903	3418	9701	8.7	140	165	204	203	123	113	71	72	86	98	132	114	127
OK	Ft. Sill	13945	3439	9824	9.2	174	215	272	247	193	182	104	91	125	140	164	165	173
OK	Altus AFB	13902	3439	9916	8.0	99	134	198	180	140	126	71	64	79	92	91	92	113
OK	Clinton-Sherman AFB	3932	3520	9912	9.8	184	202	283	262	224	158	84	77	108	113	139	161	166
OK	Enid, Vance AFB	13909	3620	9754	9.0	162	173	229	188	138	138	88	81	100	107	136	142	139
OK	Waynoka	358	3638	9850	12.4	295	416	562	556	389	310	290	250	275	309	308	261	356
OK	Gage	13975	3618	9946	10.5	203	207	281	323	257	321	168	132	173	161	167	188	221
OR	Ontario	983	4401	11701	6.2	52	70	96	143	139	107	115	122	75	69	49	41	8[illegible]

MONTHLY AVERAGE WIND POWER IN THE UNITED STATES AND SOUTHERN CANADA (Continued)

						Wind Power, Watts Per Square Meter												
State	Location	Sta. No.	Lat	Long	Ave. Speed knots	J	F	M	A	M	J	J	A	S	O	N	D	Ave.
OR	Baker	685	4450	11749	7.0	44	52	47	54	47	40	40	40	39	45	38	43	46
OR	La Grande	24148	4517	11801	8.1	328	261	173	131	93	66	55	54	60	82	201	312	152
OR	Pendleton Fld	24155	4541	11851	8.7	96	164	196	188	174	180	134	124	134	98	127	134	145
OR	Burns	24134	4335	11903	5.9	46	46	68	69	57	60	49	44	46	50	43	40	51
OR	Klamath Falls, Kingsley Fld	94236	4209	12144	4.8	84	77	99	86	62	44	33	30	36	53	66	76	60
OR	Redmond, Roberts Fld	24230	4416	12109	5.6	56	76	63	61	46	36	29	30	38	36	58	45	47
OR	Cascade Locks	192	4539	12150	13.1	651	718	330	331	365	351	387	353	344	451	645	750	465
OR	Crown Point	194	4533	12214	9.6	746	765	209	148	107	50	36	50	113	304	712	650	308
OR	Portland IAP	24229	4536	12236	6.8	139	104	91	61	44	39	45	38	38	51	91	131	75
OR	Eugene, Mahlon Sweet Fld	24221	4407	12313	7.6	83	86	110	94	79	74	89	74	79	58	73	77	81
OR	North Bend	691	4325	12413	8.4	76	128	108	107	88	185	192	149	88	52	80	94	113
OR	Roseburg	690	4314	12321	4.2	22	23	32	26	27	28	29	26	22	16	18	19	23
OR	Astoria, Clatsop Co APT	94224	4609	12353	7.2	125	109	95	82	69	66	71	61	54	70	105	111	84
OR	Salem, McNary Fld	24232	4455	12301	7.1	160	122	100	72	56	47	52	44	47	59	104	137	85
OR	Newport	695	4438	12404	8.5	110	109	107	95	127	145	151	111	69	72	95	129	113
OR	Wolf Creek	87	4241	12323	2.5	10	11	15	16	19	19	22	18	11	9	8	8	14
OR	Sexton Summit	90	4236	12322	11.6	323	283	243	196	236	243	255	269	248	223	316	310	276
OR	Brookings	598	4203	12418	6.4	91	131	96	68	58	55	35	26	37	43	72	92	63
OR	Medford	597	4221	12251	4.9	33	44	53	50	57	51	50	49	39	28	24	40	46
OR	Siskiyou Summit	91	4205	12234	8.8	96	115	102	82	129	150	170	123	102	68	89	82	109

MONTHLY AVERAGE WIND POWER IN THE UNITED STATES AND SOUTHERN CANADA (Continued)

State	Location	Sta. No.	Lat	Long	Ave. Speed knots	Wind Power, Watts Per Square Meter J	F	M	A	M	J	J	A	S	O	N	D	Ave.
PA	Philadelphia	13739	3953	7515	8.5	131	139	170	133	93	78	62	52	61	84	95	109	103
PA	Willow Grove NAS	14793	4012	7508	6.8	111	136	148	114	74	46	35	30	43	54	88	90	81
PA	Allentown	14737	4039	7526	7.3	157	141	227	124	77	73	38	35	55	61	108	151	104
PA	Scranton	14777	4120	7544	7.7	87	107	94	95	80	62	47	37	50	64	85	84	74
PA	Middletown, Olmstead AFB	14711	4012	7646	5.5	97	124	113	96	53	37	31	28	28	41	75	82	66
PA	Harrisburg	14751	4013	7651	6.4	96	120	125	88	53	41	27	24	31	42	69	68	66
PA	Parkplace	10	4051	7606	12.9	464	477	451	411	251	198	132	137	211	318	411	424	345
PA	Sunbury, Selinsgrove	14770	4053	7646	5.4	72	85	115	73	38	29	19	18	22	32	56	58	50
PA	Woodward	13	4055	7719	13.4	630	564	596	550	330	257	157	163	257	403	551	590	417
PA	Bellefonte	113	4053	7743	6.8	133	139	139	169	83	68	44	46	60	96	128	126	103
PA	Buckstown	114	4004	7850	9.4	261	308	321	241	131	83	73	62	83	179	235	247	192
PA	McConnellsburg	119	3950	7801	7.2	168	148	183	150	92	72	49	42	63	97	174	120	106
PA	Altoona, Blair Co APT	14736	4018	7819	7.9	155	180	241	164	95	79	53	46	56	82	124	143	118
PA	Kylertown	512	4100	7811	9.8	268	241	302	262	147	98	85	77	98	146	202	247	200
PA	Dubois	4787	4111	7854	7.5	118	89	127	118	79	34	36	35	41	49	77	109	78
PA	Bradford	4751	4148	7838	6.1	81	69	76	70	49	30	21	20	25	34	54	68	49
PA	Erie IAP	14860	4205	8011	9.1	234	191	208	147	92	77	67	62	94	111	176	216	139
PA	Mercer	525	4118	8012	9.0	189	169	190	176	109	80	65	58	80	107	170	163	128
PA	Brookville	121	4109	7906	7.4	146	119	119	132	75	60	42	45	46	75	112	104	89
PA	Pittsburg APT	94823	4630	8013	8.5	166	170	187	162	105	75	59	50	67	82	146	150	120

MONTHLY AVERAGE WIND POWER IN THE UNITED STATES AND SOUTHERN CANADA (Continued)

State	Location	Sta. No.	Lat	Long	Ave. Speed knots	Wind Power, Watts Per Square Meter J	F	M	A	M	J	J	A	S	O	N	D	Ave.
PA	Greensburg	4718	4016	7933	9.1	239	226	207	160	94	75	72	61	80	117	186	174	141
RI	Quonset Point NAS	14788	4135	7125	8.4	164	164	163	156	121	84	61	70	83	115	135	141	123
RI	Providence	14765	4144	7126	9.5	173	189	129	180	140	117	98	88	100	177	150	163	144
RI	Providence, Green APT	14765	4144	7126	9.6	173	189	192	180	140	117	98	88	100	117	150	163	144
SC	Beaufort MCAAS	93831	3229	8044	5.7	44	70	62	61	43	35	28	23	35	35	42	43	43
SC	Charleston	13880	3254	8002	7.5	93	124	130	117	69	64	54	52	63	59	71	81	81
SC	Myrtle Beach AFB	13717	3341	7856	6.1	49	65	70	78	54	51	49	44	44	40	39	40	52
SC	Florence	300	3411	7943	7.6	95	99	114	93	74	67	52	43	50	74	73	78	72
SC	Sumter, Shaw AFB	13849	3358	8029	5.4	49	57	64	62	40	31	26	24	34	36	38	41	41
SC	Eastover, McIntire ANG	3858	3355	8048	4.9	37	68	54	52	33	24	26	14	32	27	31	28	34
SC	Columbia	13883	3357	8107	6.2	65	74	90	96	52	42	42	34	41	38	44	50	55
SC	Anderson	112	3430	8243	7.7	99	107	99	113	80	60	51	51	52	73	85	98	79
SC	Greenville, Donaldson AFB	13822	3446	8223	6.3	67	70	80	79	44	38	32	27	36	37	39	50	49
SC	Spartanburg	313	3455	8157	8.2	121	122	156	129	95	66	65	52	60	81	108	100	94
SD	Sioux Falls, Foss Fld	14944	4334	9644	9.5	158	156	215	266	190	129	94	91	121	144	212	147	161
SD	Watertown	14946	4455	9709	10.1	189	224	284	332	251	233	129	123	185	210	240	151	212
SD	Aberdeen APT	14929	4527	9826	11.2	215	234	341	413	290	244	173	177	240	249	295	202	258
SD	Huron	14936	4423	9813	10.2	164	169	229	285	214	165	131	131	165	197	239	174	187
SD	Pierre APT	24025	4423	10017	9.8	216	205	255	294	202	141	124	129	149	168	230	207	191
SD	Rapid City	24090	4403	10304	9.6	176	173	239	234	178	144	123	139	168	193	285	205	191

MONTHLY AVERAGE WIND POWER IN THE UNITED STATES AND SOUTHERN CANADA (Continued)

State	Location	Stn. No.	Lat	Long	Ave. Speed knots	Wind Power, Watts Per Square Meter J	F	M	A	M	J	J	A	S	O	N	D	Ave
SD	Rapid City, Ellsworth AFB	24006	4409	10306	9.9	306	256	382	354	245	191	164	172	196	233	320	294	2[illegible]
SD	Hot Springs	94013	4322	10323	8.2	90	117	155	297	219	163	94	128	126	136	179	125	1[illegible]
TN	Bristol	318	3630	8221	5.9	76	105	93	85	53	40	30	30	22	37	51	59	6
TN	Knoxville APT	13891	3549	8359	7.0	133	135	156	161	85	66	53	42	47	54	99	100	9
TN	Chattanooga	13882	3502	8512	5.6	64	70	76	83	42	31	26	20	26	31	49	52	4
TN	Chattanooga	324	3503	8512	5.4	70	86	106	83	46	44	37	29	35	42	67	60	6
TN	Monteagle	126	3515	8550	5.4	77	84	84	63	32	19	18	16	19	33	60	70	4
TN	Smyrna, Sewart AFB	13827	3600	8632	5.2	76	83	87	82	42	29	22	20	22	31	58	62	5
TN	Nashville, Berry Fld	13897	3607	8641	7.4	116	114	132	117	70	67	41	33	44	57	87	80	8
TN	Memphis NAS	93839	3521	8952	6.2	84	85	93	84	54	37	25	24	29	36	67	73	5
TN	Memphis IAP	13893	3503	8959	7.9	130	137	148	125	89	57	45	42	54	61	102	110	8
TX	Brownsville, Rio Grande IAP	12919	2554	9726	10.7	231	229	281	292	277	211	185	141	102	104	150	186	19
TX	Harlington AFB	12904	2614	9740	8.8	124	175	212	207	172	160	132	129	82	75	101	109	13
TX	Kingsville NAAS	12928	2731	9749	8.5	111	130	162	183	171	157	142	115	107	77	104	98	12
TX	Corpus Christi	12924	2746	9730	10.4	189	229	260	252	199	177	158	150	107	114	157	154	17
TX	Corpus Christi NAS	12926	2742	9716	11.3	209	232	272	286	263	225	189	153	150	144	206	172	21
TX	Laredo AFB	12907	2732	9928	10.0	90	122	153	185	206	223	216	174	122	101	93	82	14
TX	Beeville NAAS	12925	2823	9740	7.3	81	101	124	129	111	85	71	61	60	52	76	72	9
TX	Victoria, Foster AFB	12912	2851	9655	7.9	134	173	198	138	112	97	67	71	53	58	103	124	10
TX	Houston	12918	2939	9517	10.1	191	216	240	258	186	139	85	74	101	117	185	159	1[illegible]

MONTHLY AVERAGE WIND POWER IN THE UNITED STATES AND SOUTHERN CANADA (Continued)

State	Location	Sta. No.	Lat	Long	Ave. Speed knots	J	F	M	A	M	J	J	A	S	O	N	D	Ave.
						Wind Power, Watts Per Square Meter												
TX	Houston, Ellington AFB	12906	2937	9510	6.8	86	96	114	104	81	56	38	41	56	54	82	75	72
TX	Galveston AAF	12905	2916	9451	11.0	262	261	298	257	227	200	149	144	147	148	239	217	210
TX	Pt. Arthur, Jefferson Co APT	12917	2957	9401	9.3	153	186	184	190	156	95	64	54	115	85	121	134	128
TX	Lufkin, Angelina Co APT	93987	3114	9445	6.1	65	72	76	70	45	30	27	22	26	38	56	62	49
TX	Saltillo	249	3312	9519	8.7	131	172	171	185	100	86	72	58	71	84	102	104	112
TX	San Antonio	12921	2932	9028	8.1	99	113	114	123	104	104	83	62	65	71	94	87	98
TX	San Antonio, Randolph AFB	12911	2932	9817	7.3	94	105	115	109	94	82	64	58	61	56	90	82	84
TX	San Antonio, Kelly AFB	12909	2923	9835	6.9	84	85	105	109	96	83	63	53	52	53	72	63	77
TX	San Antonio, Brooks AFB	12931	2921	9827	8.9	141	144	185	189	187	166	136	108	95	107	135	103	142
TX	Hondo AAF	12903	2920	9910	6.7	59	82	88	94	99	96	51	45	42	32	50	49	64
TX	Kerrville	12961	2959	9905	7.1	86	95	134	129	117	92	97	48	52	63	74	54	86
TX	San Marcos	12910	2953	9752	7.2	115	121	142	115	127	102	64	65	56	78	108	105	98
TX	Austin, Bergstrom AFB	13904	3012	9740	7.8	145	140	167	145	118	119	88	73	58	74	117	116	115
TX	Bryan	13905	3038	9628	7.0	85	104	108	103	89	74	49	48	38	47	73	89	76
TX	Killeen, Fort Hood AAF	3933	3108	9743	8.1	123	138	151	155	130	109	81	59	60	75	96	114	106
TX	Ft Hood, Gray AAF	3902	3104	9750	9.2	163	179	198	208	162	153	121	87	72	102	147	158	146
TX	Waco, Connally AFB	13928	3138	9704	7.7	117	111	135	128	101	90	72	61	56	68	104	101	95
TX	Dallas NAS	93901	3244	9658	9.1	155	166	210	199	154	144	97	82	85	99	135	131	137
TX	Ft. Worth, Carswell AFB	13911	3246	9725	8.2	138	154	216	194	141	133	71	60	72	89	129	121	124
TX	Mineral Wells APT	93985	3247	9804	9.2	120	146	203	201	162	154	103	79	78	88	111	110	[illegible]

MONTHLY AVERAGE WIND POWER IN THE UNITED STATES AND SOUTHERN CANADA (Continued)

State	Location	Sta. No.	Lat	Long	Ave. Speed Knots	Wind Power, Watts Per Square Meter J	F	M	A	M	J	J	A	S	O	N	D	Ave
TX	Mineral Wells, Ft Walters AAF	3943	3250	9803	8.7	117	135	192	184	149	136	92	71	71	92	99	103	11
TX	Santo	155	3237	9814	7.1	92	136	157	163	87	83	55	55	60	55	112	77	9
TX	Sherman, Perrin AFB	13923	3343	9640	9.1	172	164	217	213	142	121	79	73	81	106	153	153	13
TX	Gainsville	153	3340	9708	11.1	244	303	338	357	226	184	141	134	155	163	222	188	22
TX	Wichita Falls	13966	3358	9829	9.9	160	178	244	222	176	158	110	95	107	114	168	154	15
TX	Abilene, Dyers AFB	13910	3226	9951	7.7	91	104	145	147	124	102	61	51	58	66	87	87	9
TX	San Angelo, Mathis Fld	23034	3122	10030	8.9	117	154	195	184	170	147	96	86	90	90	113	108	12
TX	San Angelo, Goodfellow AFB	23017	3124	10024	8.7	108	152	191	179	169	157	85	81	91	88	114	102	12
TX	Del Rio, Laughlin AFB	22001	2922	10045	7.6	71	107	114	120	118	117	91	68	59	57	56	60	8
TX	Canadian	162	3500	10022	13.0	390	416	570	596	430	344	263	231	310	337	370	290	37
TX	Dalhart APT	93042	3601	10233	12.9	370	371	476	477	477	559	334	278	280	228	266	305	35
TX	Amarillo, English Fld	23047	3514	10142	11.7	229	279	359	329	290	240	166	135	180	200	221	222	24
TX	Childress	23007	3426	10017	10.2	152	194	272	269	221	204	118	93	117	126	128	147	17
TX	Lubbock, Reese AFB	23021	3336	10203	9.4	155	211	291	268	204	188	89	67	88	99	140	169	16
TX	Big Spring, Webb AFB	23005	3213	10131	10.1	155	197	264	266	226	216	128	101	112	124	135	139	17
TX	Midland	23023	3156	10212	8.9	91	143	146	155	133	123	94	76	85	84	89	102	10
TX	Wink, Winkler Co APT	23040	3147	10312	8.5	95	148	204	181	181	193	119	78	72	75	81	128	11
TX	Marfa APT	23022	3016	10401	7.9	128	182	165	192	146	117	84	65	85	84	88	119	12
TX	Guadalupe Pass	163	3150	10448	15.8	887	892	999	932	868	603	422	342	401	555	760	827	71
TX	El Paso	23044	3148	10629	9.8	176	257	296	299	222	173	131	111	102	128	153	163	18

MONTHLY AVERAGE WIND POWER IN THE UNITED STATES AND SOUTHERN CANADA (Continued)

State	Location	Sta. No.	Lat	Long	Ave. Speed knots	Wind Power, Watts Per Square Meter J	F	M	A	M	J	J	A	S	O	N	D	Ave.
TX	El Paso, Biggs AFB	23019	3150	10624	5.8	68	99	144	144	96	74	47	38	32	33	49	59	72
UT	St George	93198	3703	11331	5.1	26	39	73	67	72	74	53	62	40	25	34	23	49
UT	Milford	475	3826	11300	10.2	179	241	228	275	302	241	220	175	167	173	172	152	214
UT	Bryce Canyon APT	23159	3742	11209	6.4	69	65	93	79	94	90	40	47	53	48	53	49	66
UT	Hanksville	23170	3822	11043	4.6	43	41	115	84	102	108	35	38	43	40	36	25	57
UT	Tooele, Dugway PG	24103	4011	11256	4.8	38	48	69	81	71	69	52	57	46	41	31	28	53
UT	Darby, Wendover AFB	24111	4043	11402	5.3	59	62	90	90	71	82	62	61	49	50	58	36	62
UT	Wendover	24193	4044	11402	5.4	48	55	91	103	84	80	57	57	43	43	47	40	62
UT	Locomotive Springs	187	4143	11255	9.3	113	129	205	193	221	215	195	202	167	125	115	93	185
UT	Ogden, Hill AFB	24101	4107	11158	8.0	102	118	127	126	130	124	124	130	122	123	99	92	119
UT	Salt Lake City	24127	4046	11158	7.7	77	84	100	100	96	96	82	106	73	68	66	71	85
UT	Coalville	174	4054	11125	3.9	26	35	39	32	30	28	17	18	37	24	17	24	28
VT	Montpelier, Barre APT	94705	4412	7234	7.2	162	167	152	118	99	97	69	60	94	103	102	125	111
VT	Burlington, Ethan Allen AB	14742	4428	7309	7.7	114	111	103	100	85	70	55	52	70	82	100	114	90
VA	Norfolk NAS	13750	3656	7618	8.8	154	182	171	139	103	84	75	80	111	127	130	128	121
VA	Oceana NAS	13769	3650	7601	7.6	135	136	150	126	84	62	52	52	83	93	97	107	98
VA	Hampton, Langley AFB	13702	3705	7622	8.5	156	187	193	166	122	86	73	81	117	137	135	146	136
VA	Ft. Eustis, Felker AAF	93735	3708	7636	6.5	79	90	89	74	51	42	32	32	43	46	61	64	58
VA	South Boston	108	3641	7855	5.0	49	49	65	65	34	33	32	24	28	36	40	38	40
VA	Danville APT	13728	3634	7920	6.1	69	63	84	78	43	36	32	31	33	35	41	39	48

MONTHLY AVERAGE WIND POWER IN THE UNITED STATES AND SOUTHERN CANADA (Continued)

State	Location	Sta. No.	Lat	Long	Ave. Speed knots	J	F	M	A	M	J	J	A	S	O	N	D	Ave
						Wind Power, Watts Per Square Meter												
VA	Roanoke	13741	3719	7958	7.1	152	207	171	144	75	55	51	46	46	53	98	132	10
VA	Richmond	13740	3730	7720	6.7	59	68	79	74	49	40	34	32	39	40	47	47	5
VA	Quantico MCAS	13773	3830	7719	6.0	55	63	75	67	45	35	29	29	35	35	46	45	4
VA	Ft.Belvoir, Davison AAF	93728	3843	7711	3.8	42	60	60	42	24	15	12	12	13	19	34	44	3
WA	Spokane IAP	24157	4738	11732	7.2	92	111	106	102	73	68	55	53	58	63	80	95	7
WA	Spokane, Fairchild AFB	24114	4738	11739	7.2	118	138	134	121	92	87	67	62	74	84	92	123	10
WA	Moses Lake, Larson AFB	24110	4711	11919	6.2	68	62	98	101	79	80	56	45	61	55	53	55	6
WA	Walla Walla	24160	4606	11817	6.7	87	100	114	93	68	65	56	55	51	47	86	89	7
WA	Pasco, Tri City APT	24163	4616	11907	6.8	342	331	346	372	243	212	190	224	229	186	247	164	25
WA	North Dalles	188	4537	12109	8.0	65	72	171	189	264	279	334	284	166	93	63	64	15
WA	Yakima	24243	4634	12032	6.4	60	53	89	115	78	71	54	48	52	48	41	40	6
WA	Chehalis	792	4640	12205	6.4	109	86	82	59	53	44	45	44	43	57	80	110	6
WA	Kelso, Castle Rock	24223	4608	12254	6.9	128	107	87	65	65	46	46	39	52	66	119	133	8
WA	North Head	791	4616	12404	13.0	547	521	495	369	430	357	330	275	215	366	460	773	42
WA	Hoquium, Bowerman APT	94225	4658	12356	8.2	141	112	108	88	86	66	59	56	53	91	94	103	8
WA	Moclips	794	4715	12412	7.6	82	82	75	81	68	38	37	36	41	65	69	102	6
WA	Tatoosh IS	798	4823	12444	12.3	753	603	443	269	276	117	130	109	195	422	569	735	38
WA	Tacoma, McChord AFB	24207	4709	12229	4.6	50	48	53	49	38	30	25	23	25	31	41	41	3
WA	Ft. Lewis, Gray AAF	24201	4705	12235	3.9	36	28	30	30	23	19	17	18	17	21	23	27	2
WA	Seattle Tacoma	24233	4727	12218	9.5	194	210	211	173	130	120	94	84	104	135	147	195	14
WA	Seattle FWC	24244	4741	12216	5.6	69	61	58	48	32	28	25	24	28	44	53	69	

MONTHLY AVERAGE WIND POWER IN THE UNITED STATES AND SOUTHERN CANADA (Continued)

State	Location	Sta. No.	Lat	Long	Ave. Speed knots	Wind Power, Watts Per Square Meter												
						J	F	M	A	M	J	J	A	S	O	N	D	Ave.
WA	Everett, Paine AFB	24203	4755	12217	6.3	70	67	66	57	44	40	38	34	39	43	60	65	51
WA	Whidbey IS NAS	24255	4821	12240	7.1	185	158	148	124	75	54	43	33	50	104	160	188	108
WA	Bellingham APT	24217	4848	12232	6.3	131	132	92	66	42	43	43	35	26	56	89	119	71
WV	Charleston	13866	3822	8136	5.6	46	61	62	53	37	28	24	15	22	21	45	45	38
WV	Elkins, Randolph Co APT	13729	3853	7951	5.8	80	90	101	92	57	32	24	22	25	39	71	68	58
WV	Morgantown APT	13736	3939	7955	5.8	73	73	86	67	37	26	18	16	22	34	64	81	48
WI	Green Bay	14898	4429	8808	5.6	174	150	212	197	173	129	89	72	122	131	192	151	149
WI	Green Bay, Straubel APT	14898	4429	8808	9.3	174	150	212	197	173	129	89	72	122	131	192	151	149
WI	Milwaukee, Mitchell Fld	14839	4257	8754	10.2	198	212	246	238	195	120	95	91	129	159	229	200	175
WI	Madison, Traux Fld	14837	4308	8920	8.8	139	148	199	197	153	97	72	62	93	112	170	136	130
WI	Janesville, Rock Co APT	94854	4237	8902	7.6	158	170	203	271	234	138	108	90	112	137	212	170	167
WI	Lone Rock	143	4312	9011	7.8	117	118	125	166	116	82	60	60	75	96	133	97	103
WI	Camp Douglas, Volk Fld	94930	4356	9016	6.3	62	76	79	77	64	33	28	26	34	64	76	56	54
WI	La Crosse APT	14920	4352	9115	8.8	120	116	145	201	171	100	70	69	100	128	179	134	127
WI	Eau Claire	14991	4452	9129	8.3	96	113	102	168	155	90	83	85	100	109	134	106	112
WI	Hager City	141	4436	9232	8.4	151	132	179	221	125	98	57	77	83	125	126	119	119
WY	Cheyenne APT	24018	4109	10449	11.9	433	453	434	399	242	176	125	132	157	220	402	463	302
WY	Laramie	164	4118	10540	11.5	498	506	520	339	313	300	152	173	212	259	338	379	312
WY	Medicine Bow	165	4153	10611	12.7	773	758	825	518	343	296	250	223	328	423	544	726	490
WY	Cherokee	173	4143	10740	14.0	662	703	610	463	350	337	257	277	283	364	510	577	430

MONTHLY AVERAGE WIND POWER IN THE UNITED STATES AND SOUTHERN CANADA (Continued)

State	Location	Sta. No.	Lat	Long	Ave. Speed knots	Wind Power, Watts Per Square Meter J	F	M	A	M	J	J	A	S	O	N	D	Ave
WY	Bitter Creek	172	4140	10833	12.7	477	550	623	397	297	230	150	223	210	284	370	477	3[illegible]
WY	Rock Springs	574	4138	10915	10.6	503	476	551	357	311	264	216	189	236	250	320	394	3[illegible]
WY	Granger	177	4136	10958	9.6	309	283	410	296	244	257	157	191	170	176	254	252	2[illegible]
WY	Knight	573	4124	11050	10.5	278	385	375	301	261	241	175	187	276	208	233	265	2[illegible]
WY	Casper AAF	24005	4255	10627	11.4	445	417	359	266	201	222	145	150	219	208	362	473	2[illegible]
WY	Riverton APT	24061	4303	10827	5.4	54	52	79	75	54	49	37	31	47	38	30	40	[illegible]
WY	Sheridan	24029	4446	10658	6.6	71	71	80	94	88	73	60	56	62	65	76	64	[illegible]
WY	Cody APT	24045	4431	10901	9.3	292	303	274	342	238	189	155	176	183	218	288	303	2[illegible]
NS	Yarmouth	14647	4350	6605	9.0	230	173	190	153	109	84	63	67	84	123	162	196	1[illegible]
NS	Greenwood	14636	4459	6455	8.8	278	240	276	207	148	120	87	98	102	167	206	243	1[illegible]
NB	Frederickton	14648	4552	6632	7.7	153	124	148	116	99	86	69	64	70	94	98	127	1[illegible]
QU	Mont Jolt	14639	4836	6812	11.2	356	358	310	220	197	157	134	152	188	245	297	381	2[illegible]
QU	Bagotville	94795	4820	7100	9.3	206	180	202	166	164	141	95	97	128	147	181	148	1[illegible]
QU	St. Hubert	4712	4531	7325	9.3	224	213	163	153	145	138	100	83	110	136	203	177	1[illegible]
ON	Ottowa	4706	4519	7540	8.2	123	120	125	111	99	79	58	53	70	88	117	100	
ON	Trenton	4715	4407	7732	8.9	203	176	166	148	125	103	94	80	100	122	181	151	1[illegible]
ON	Muskoka	4704	4458	7918	7.0	52	63	68	75	61	44	41	37	44	54	69	55	
ON	Toronto	94791	4341	7938	8.6	198	177	157	150	110	87	69	68	85	105	177	152	1[illegible]
ON	London	94805	4302	8109	9.2	239	229	233	214	139	85	66	65	82	107	173	176	[illegible]
ON	Whar[illegible]	94809	4445	8106	9.5	234	173	164	165	127	92	72	85	113	148	222	209	1[illegible]

MONTHLY AVERAGE WIND POWER IN THE UNITED STATES AND SOUTHERN CANADA (Continued)

State	Location	Sta. No.	Lat	Long	Ave. Speed knots	Wind Power, Watts Per Square Meter J	F	M	A	M	J	J	A	S	O	N	D	Ave.
ON	North Bay	4705	4622	7925	8.5	108	121	122	121	98	85	68	67	84	90	126	104	100
ON	Sudbury	94828	4637	8048	12.2	318	392	330	324	328	290	218	195	252	294	355	312	301
ON	White River	94808	4836	8517	4.3	19	26	28	32	37	33	24	21	24	28	30	24	27
ON	Lakehead	94804	4822	8919	7.5	117	97	104	135	132	77	65	59	80	104	155	119	104
ON	Kenora	14999	4948	9422	8.6	91	95	91	111	104	74	63	70	88	96	112	86	90
MN	Winnipeg	14996	4954	9714	10.7	206	218	227	293	266	177	116	138	174	209	239	210	206
MN	Portage La Prairie	94912	4954	9816	9.5	164	162	183	217	207	122	91	107	135	169	157	157	156
MN	Rivers	25014	5001	10019	10.5	216	171	187	266	276	194	137	149	200	235	218	200	204
SA	Regina	25005	5026	10440	11.9	327	286	315	350	368	243	162	186	286	232	279	300	278
SA	Moose Jaw	25011	5023	10534	12.3	399	343	314	344	390	299	197	214	340	319	340	367	322
AL	Medicine Hat	25118	5001	11043	8.9	164	159	146	215	176	138	96	110	159	168	187	177	158
AL	Lethbridge	94108	4938	11248	12.5	625	563	356	450	370	319	202	246	279	510	546	567	419
BC	Penticton	94116	4928	11936	7.5	265	187	141	104	76	64	52	49	64	132	233	274	137
BC	Abbotsford	24288	4901	12222	5.8	135	108	89	72	46	38	32	25	28	57	86	93	67
BC	Vancouver	24287	4911	12310	6.5	72	75	85	83	53	48	52	39	48	62	80	75	64
BC	Victoria	24297	4839	12326	6.5	84	80	74	75	50	51	34	36	36	47	68	79	60

Index

(Please note the use of the following abbreviations in the index: WECS=Wind Energy Conversion System; DOE=Department of Energy)

TK1541 .H86
SUNY GENESEO
Windpower :handbook of wind energy conve
YGMM
3 0260 00275010 1

TK
1541
H86

TO RENEW BOOKS
CALL (716) 245-5594
ON OR BEFORE DUE DATE
DATE DUE

Milne Circ.
FEB 16 2000

Milne Circ.
JAN 5 2001

NOV 19 2005
Milne Circ.

DEMCO, INC. 38-2931

GENESEO, N. Y. 14454

BORROWER HAS SOLE RESPONSIBILITY
FOR RETURN BY DATE DUE SHOWN ABOVE.